# Options That SAVE Your Students Money

| Book Rentals | Online Learning | Study and Practice Materials |
|---|---|---|
| **UP TO 60% OFF** | **FREE WHEN BUNDLED** | **UP TO 60% OFF** |

## CENGAGE brain.com

**Students have the CHOICE to purchase the eBook or rent the text at CengageBrain.com**

## OR

an eTextbook in PDF format is also available for instant access for your students at **www.coursesmart.com**.

## CourseSmart

---

## CENGAGE LEARNING'S GLOBAL GEOSCIENCE WATCH
### GLOBAL GEOSCIENCE RESOURCE CENTER

**Help your students see the relevancy of what they learn in your course!**

Updated several times a day, the Global Geoscience Watch is a focused portal into GREENR - our Global Reference on the Environment, Energy, and Natural Resources - an ideal one-stop site for classroom discussion and research projects for all things geoscience! You can easily get to the most relevant content available for your course.

**Features Include:**
- Articles
- Case Studies
- Podcasts
- Interactive World Map
- Videos
- Citation Tools
- Sharing Options (email, Twitter, Facebook, etc.)

---

## CourseMate

**Give your students an interactive learning experience!**

Interested in a simple way to complement your text and course content with study and practice material?

Cengage Learning's CourseMate brings course concepts to life with interactive learning, study and exam preparation tools that support the printed textbook. Watch student comprehension soar as your class works with the printed textbook and the textbook-specific website. CourseMate goes beyond the book to deliver what you need!

**Learning Resources Include:**
- Interactive eBook
- Video Exercises
- Practice Quizzes
- Animations
- Flashcards
- Google Earth Activities

---

Students can rent Petersen/Sack/Gabler's *Physical Geography*, 10th Edition for up to 60% off list price.

Add this ISBN to your SSO account to get access today!

**Global Geoscience Watch**
ISBN: 1-111-42905-7

**Global Geoscience Watch Bundle**
ISBN: 1-133-39722-0

To view a demo, visit
www.cengage.com/coursemate

**CourseMate Bundle**
ISBN: 1-111-97915-4

---

## *Custom Solutions to Fit Every Need*

Contact your Cengage Learning representative to learn more about what custom solutions are available to meet your course needs.

- Adapt existing Cengage Learning content by adding or removing chapters
- Incorporate your own materials
- Add Earth Science Review for your less prepared students

# PHYSICAL GEOGRAPHY

EDITION
10

# PHYSICAL GEOGRAPHY

**JAMES F. PETERSEN**
Texas State University, San Marcos

**DOROTHY SACK**
Ohio University, Athens

**ROBERT E. GABLER**
Western Illinois University, Macomb

BROOKS/COLE
CENGAGE Learning·

Australia • Brazil • Japan • Korea • Mexico • Singapore • Spain • United Kingdom • United States

BROOKS/COLE
CENGAGE Learning™

*Physical Geography,* **Tenth Edition**
**James F. Petersen, Dorothy Sack,
Robert E. Gabler**

Earth Science Editor: Aileen Berg

Developmental Editor: Amy Collins

Assistant Editor: Kristina Chiapella

Editorial Assistant: Shannon Elderon

Media Editor: Alexandria Brady

Marketing Manager: Jack Cooney

Marketing Coordinator: Julie Stefani

Marketing Communications Manager:
Darlene Macanan

Content Project Manager: Hal Humphrey

Design Director: Rob Hugel

Art Directors: John Walker, Pam Galbreath

Print Buyer: Karen Hunt

Rights Acquisitions Specialist: Tom McDonough

Production Service: Kate Mannix,
Graphic World Inc.

Text Designer: Diane Beasley

Photo Researcher: Jeremy Glover,
Bill Smith Group

Text Researcher: Pablo D'Stair

Copy Editor: Graphic World Inc.

Illustrators: Accurate Art, Precision Graphics,
Rolin Graphics, Pre-Press PMG,
Graphic World Inc.

Cover Designer: William Stanton

Cover Image: Sam Striker / shutterstock.com

Compositor: Graphic World Inc.

For product information and technology assistance, contact us at
**Cengage Learning Customer & Sales Support, 1-800-354-9706.**

For permission to use material from this text or product,
submit all requests online at **www.cengage.com/permissions.**
Further permissions questions can be e-mailed to
**permissionrequest@cengage.com.**

Library of Congress Control Number: 2011930759

ISBN-13: 978-1-111-42750-4

ISBN-10: 1-111-42750-X

**Brooks/Cole**
20 Davis Drive
Belmont, CA 94002-3098
USA

Cengage Learning is a leading provider of customized learning solutions with office locations around the globe, including Singapore, the United Kingdom, Australia, Mexico, Brazil, and Japan. Locate your local office at **www.cengage.com/global.**

Cengage Learning products are represented in Canada by Nelson Education, Ltd.

To learn more about Brooks/Cole visit **www.cengage.com/brookscole.**

Purchase any of our products at your local college store or at our preferred online store **www.CengageBrain.com.**

Printed in Canada
1 2 3 4 5 6 7 15 14 13 12 11

# Preface

Earth, our planetary home, is a wondrous life-support system that is complex and ever-changing. Our planet's natural environments are robust enough to adapt to many changes, but if certain limits are approached our habitat may be threatened or damaged. Today, the modern technologies that we have come to rely on tend to insulate us from fully experiencing our environment; we can become lulled into forgetting about our dependence on Earth's natural systems and resources. It is also easy to forget that on Earth—our planetary oasis—we are continuously traveling through the vastness of space, dependent on our sun for life-supporting energy. Understanding our planet, the nature of its environments, and how they operate is as critical now as it has ever been for humankind.

For as long as people have existed, the resources provided by their physical environments have been the key to survival. Preindustrial societies, such as those dependent on hunting and gathering or small-scale agriculture, tended to have small populations that exerted relatively little impact on their natural surroundings. In contrast, present industrialized societies have large populations, consume huge quantities of natural resources, and can influence or cause environmental change, not only on a local scale, but also on a global one. A great concern today is how human-induced changes in global climates will impact other aspects of our environments. As the world's population has increased, so have the scales, degrees, and cumulative effects of human impacts on the environment. We have polluted the air and water. We have used up tremendous amounts of nonrenewable resources and have altered many natural landscapes without fully assessing the potential consequences. Too often, we have failed to respect the power of Earth's natural forces when constructing our homes and cities or while pursuing our economic activities. It is now evident that if we continually fail to comprehend Earth's potential and respect its limitations as a human habitat, we may be putting ourselves and future generations at risk. The ways we use and affect our physical environment provide the keys to our survival.

Acceptance of this important message continues to grow. We understand that Earth does not offer limitless natural resources. The news media have expanded their coverage of environmental characteristics and issues, including human impacts. People continue to work to enact legislation that will address environmental problems. Scientists and governmental leaders from around the world meet to discuss environmental issues that increasingly cross international boundaries. Humanitarian organizations, with funding from governments as well as private citizens, struggle to alleviate the suffering that results from natural disasters or from environmental degradation. The more we know about our world and its environments, the more effective we can be in working toward stewardship and preservation.

Geography is a highly regarded subject in most nations of the world, and in recent decades it has undergone a renaissance in the United States. National education standards that include physical geography support the offering of high-quality geography curricula in U.S. elementary, secondary, and postsecondary schools. Employers are increasingly recognizing the value and importance of geographic knowledge, skills, and techniques in the workplace. Physical geography as an applied field makes use of the latest digital technologies, such as geographic information systems (GIS), computer-assisted cartography (mapmaking), the global positioning system (GPS), and satellite image acquisition and interpretation. At the college level, physical geography offers an introduction to the concerns, ideas, knowledge, and tools that are necessary for further study of our planet. More than ever before, physical geography is being recognized as an ideal science course for general-education students—students who will not only make decisions that consider human needs and desires, but also environmental limits and possibilities. It is for these students that *Physical Geography* has been written.

## Features

**Comprehensive View of the Earth System** *Physical Geography* introduces all major aspects of the Earth system, identifying physical phenomena and natural processes and stressing their characteristics, relationships, interactions, and distributions. The text covers a wide range of topics, including the atmosphere, the solid Earth, oceans and other water bodies, and the living environments of our planet.

**Engaging Graphics** Because the study of geography is greatly enhanced with visual aids, the text includes an array of illustrations and photographs that make the concepts come alive. Clear and simple diagrams illuminate challenging concepts. Stunning photographs bring the physical world to the student. New to this edition, selected photographs are accompanied by locator maps to provide spatial context and help students identify the feature's geographic position on Earth.

**Clear Explanation** The text uses an easily understandable, narrative style to explain the origins, development, significance, and distribution of processes, physical features, and events that occur within, on, or above Earth's surface. The writing style is targeted toward rapid comprehension and making the study of physical geography meaningful and enjoyable.

**Introduction to the Geographer's Tools** Digital technologies have revolutionized the ways that we can study our planet, its features, its environmental aspects, and its natural processes. A full chapter is devoted to maps and other forms of spatial imagery and data used by geographers. Illustrations throughout the book include images gathered from space, accompanied by interpretations of the environmental aspects that the scenes illustrate. Also included are introductory discussions of techniques geographers use to analyze or display location and environmental aspects of Earth, including remote sensing, geographic information systems, computer-assisted cartography, and the global positioning system.

**Focus on Student Interaction** The text continually encourages students to think, conceptualize, hypothesize, and interact with the subject matter of physical geography. Activities at the end of each chapter can be completed individually or as a group, and were designed to engage students and promote active learning. Review questions reinforce concepts and prepare students for exams, and practical application assignments require active solutions, such as sketching a diagram, performing a calculation, or exploring geographic features using Google Earth®. Questions following most figure captions prompt students to think beyond or to use the map, graph, diagram, or image, and to give further consideration to the topic. Detailed learning objectives at the beginning of the chapters provide a means to measure comprehension of the material. Latitude/longitude coordinates for selected photographs invite students to further explore physical geography using satellite imagery provided by Google Earth.

**Three Unique Perspectives** Physical geography is a field that seeks to develop an understanding and appreciation of our Earth and its environmental diversity. In approaching this goal, this textbook employs article boxes that illustrate the three major perspectives of physical geography. Through a **spatial perspective,** physical geography focuses on understanding and explaining the locations, distribution, and spatial interactions of natural phenomena. Physical geography can also be approached from a **physical science perspective,** which applies the knowledge and methods of the natural and physical sciences, for example, by using the scientific method and systems analysis techniques. Through an **environmental perspective,** physical geographers consider impacts, influences, and interactions among human and natural components of the environment; in other words, how the environment influences human life and how humans affect the environment.

**Map Interpretation Series** Learning map interpretation skills is a priority in a physical geography course. To meet the needs of students who do not have access to a laboratory setting, this text includes map activities with accompanying explanations, full-color maps printed at their original map scale, satellite images, and interpretation questions. These maps give students an opportunity to develop valuable map-reading skills. In courses that have a lab section, the map interpretation features offer a supplement to lab activities and a link between class lectures, the text, and lab work.

# Objectives

Since the first edition, the authors have sought to accomplish four major objectives:

**To Meet the Academic Needs of the Student** Instructors familiar with the style and content of *Physical Geography* know that this textbook is written specifically for the student, and it is designed to satisfy the major purposes of a liberal education. Students are provided with the knowledge and understanding they need to make informed decisions involving the environments that they will interact with throughout their lives. The text assumes little or no prior background in physical geography or other Earth sciences. Numerous examples from throughout the world are included to illustrate important concepts and help nonscience majors bridge the gap between scientific theory and practical application.

**To Strongly Integrate the Illustrations with the Written Text** Numerous photographs, maps, satellite images, scientific visualizations, block diagrams, graphs, and line drawings have been carefully chosen to clearly illustrate important concepts in physical geography. Figures are called out in the conceptual discussions so students can easily make the connection between the illustration and the text that describes it. Some examples of topics that are clearly explained through the integration of visuals and text include map and image interpretation (Chapter 2), the seasons (Chapter 3), Earth's energy budget (Chapter 4), surface wind systems (Chapter 5), storms (Chapter 7), soils (Chapter 12), plate tectonics (Chapter 13), rivers (Chapter 17), glaciers (Chapter 19), and coastal processes (Chapter 20).

**To Communicate the Nature of Geography** The nature of geography and three major perspectives of physical geography (spatial, physical science, and environmental) are discussed in Chapter 1. In subsequent chapters, important topics of geography involving all three perspectives are discussed. For example, location is a dominant topic in Chapter 2 and remains an important theme throughout the text. Spatial distributions are stressed as the elements of climate are discussed in Chapters 4 through 6. The changing Earth system is a central focus in Chapter 8. Characteristics of climate regions and environments constitute Chapters 9 and 10. Spatial interactions are demonstrated in discussions of weather systems (Chapter 7), soils (Chapter 12), and volcanic and tectonic activity (Chapters 13 and 14). Feature boxes in every chapter present interesting and important examples of each geographic perspective.

**To Fulfill the Major Requirements of Introductory Physical Science College Courses** *Physical Geography* offers a full chapter on the tools and methodologies of physical geography. The Earth as a system and the physical processes that are responsible for the location, distribution, and spatial relationships of physical phenomena beneath, at, and above Earth's surface are examined in detail. Scientific method, hypothesis, theory, and explanation are continually stressed. In addition, end-of-chapter questions that involve understanding and interpreting graphs of environmental data (or graphing data for analysis), quantitative transformation or calculation of environmental variables, and/or hands-on map analysis directly support science learning. Models and systems are frequently cited in the discussion of important concepts, and scientific classification is presented in several chapters—some of these topics include air masses, tornadoes, and hurricanes (Chapter 7), climates (Chapters 8, 9, and 10), biogeography (Chapter 11), soils (Chapter 12), rivers (Chapter 17), and coasts (Chapter 20).

# Tenth Edition Revision

This edition has been revised so that the latest and most important information is presented to those who are studying physical geography. Not only is our planet ever-changing, but so are the many ways that we study, observe, measure, and analyze Earth's characteristics, environments, and processes. New scientific findings and new ways of communicating those findings are continually being developed. The tenth edition introduces or expands discussions of newer technology, such as LIDAR imaging and satellite weather maps. Students have greater access to geographic tools than ever before; therefore, latitude/longitude coordinates are provided for selected photographs and maps so students may further explore an area or phenomenon on their own, using an interactive mapping tool such as Google Earth.

As authors we continually seek to include coverage on physical geography topics that will spark student interest. In the tenth edition, sustainable development is discussed in greater depth, as is the global movement to better manage modern industry's environmental impact on Earth. Attention has been given to the most recent environmental concerns, findings, and natural hazards—we explain the events, the conditions that led to those events, and how they are related to physical geography. This edition has been updated to include discussion of the most recent natural disasters, including the deadly earthquake and tsunami in Japan, the earthquakes in Haiti and Chile, mudslides and wildfires in Southern California, flooding in many areas, and recent volcanic eruptions, hurricanes, and tornado outbreaks. Recent disasters continue to be discussed in terms of human impact and new efforts toward avoidance of such tragedies in the future. These events and others are addressed as examples of Earth processes and human–environment interactions.

**New Lead Author** Formerly helmed by Robert Gabler, the *Physical Geography* authoring team is now under the leadership of James Petersen.

**New and Revised Text** Revising *Physical Geography* for a tenth edition involved thoughtful consideration of the input from many reviewers with varied opinions. Many topics have been rewritten for greater clarity, discussions have been expanded where more explanation was required, and several new feature boxes have been conceived. Great concern has been given to unusual weather conditions and the potential impacts of global warming. Earth systems approaches are reinforced with additional content, illustrations, and examples.

Many other sections of the book contain new material, line art, photographs, and feature boxes. For example, the following discussions and features have been included or improved in this edition: the importance of sustainable development (Chapter 1), thermal images (Chapter 2), recent information about $CO_2$ levels in the atmosphere, as well as change in the ozone hole (Chapter 4), the world's limited water supply (Chapter 6), satellite weather maps and Doppler radar (Chapter 7), invasive exotic species (Chapter 11), measuring earthquakes (Chapter 13), rockfalls and avalanches (Ch. 15), and the nature of springs (Chapter 16). Additionally, there are three new map interpretation features (Chapters 2, 7, and 14).

**Enhanced Program of Illustrations** The illustration program has undergone substantial revision. The new and expanded topics required many new figures and updates to others, including numerous photographs, satellite images, and maps. One hundred ninety-seven figures have been replaced by new photographs and there are 91 new or revised line drawings. Many photographs now have the added value of being accompanied by a locator map, which provides a spatial reference for the feature or place pictured.

# About Locate and Explore Activities

Throughout this textbook you will find Locate and Explore activities at the end of many chapters, which require you to use Google Earth. Google Earth is a virtual globe browser that allows you to interactively display and investigate geographic data from anywhere in the world. To perform these exercises, you should have the latest version of Google Earth installed on your computer. Some of the exercises require you to use some data layers that are included with Google Earth, as well as some additional data layers that you must download. For detailed instructions about using Google Earth, and to download the necessary data, visit the book's Geography CourseMate at www.cengagebrain.com.

# Ancillaries

Instructors and students alike will greatly benefit from the comprehensive ancillary package that accompanies this text.

## For the Instructor

### Class Preparation and Assessment Support

**Instructor's Manual with Test Bank** The online Instructor's Manual contains Chapter Outlines and summaries, Lecture Suggestions, Key Terms and Concepts, and answers to end-of-chapter review questions. The online Test Bank contains a variety of Multiple Choice, True/False, Fill-in-the-Blank, and Essay questions to meet your testing needs. It is available exclusively for download from our password-protected instructor website. A Lab Manual Answer Key is also available for download.

**ExamView® Computerized Testing** Create, deliver, and customize tests and study guides (both print and online) in minutes with this easy-to-use assessment and tutorial system. Preloaded with the *Physical Geography* test bank, *ExamView* offers both a *Quick Test Wizard* and an *Online Test Wizard*. You can build tests of up to 250 questions using as many as 12 question types. *ExamView's* complete word-processing capabilities also allow you to enter an unlimited number of new questions or edit existing questions.

## Dynamic Lecture Support

**PowerLecture with JoinIn™** A complete all-in-one reference for instructors, the PowerLecture DVD contains PowerPoint® slides with lecture outlines, images from the text, stepped art from the text, zoomable art figures from the text, and active figures that interactively demonstrate concepts. Besides providing you with fantastic course presentation material, the PowerLecture DVD contains electronic files of the Test Bank and Instructor's Manual, as well as JoinIn, the easiest Audience Response System to use, featuring instant classroom assessment and learning.

**Active Earth CD** The Active Earth Collection allows you to pick and choose from more than 120 earth science animations and active figures, ABC® natural hazard video clips, and in-depth Google Earth lecture activities, and includes a link to the Earth Science Newsroom. Grab your students' attention by creating your lectures using these dynamic tools.

## Laboratory and GIS Support

**GIS Investigations** Michelle K. Hall-Wallace, C. Scott Walker, Larry P. Kendall, Christian J. Schaller, and Robert F. Butler of the University of Arizona, Tucson.

The perfect accompaniment to any physical geography course, these four groundbreaking guides tap the power of *ArcGIS®* to explore, manipulate, and analyze large data sets. The guides emphasize the visualization, analysis, and multimedia integration capabilities inherent to GIS and enable students to "learn by doing" with a full complement of GIS capabilities. The guides, designed for students with access to ArcGIS software, contain all the data sets needed to complete the exercises. ISBN for use with ArcGIS 9.x site license: 0-495-11506-1

## For the Student

**Exploring the Dynamic Earth:**
**GIS Investigations for the Earth Sciences**
ISBN with ArcView GIS CD: 0-534-39138-9
ISBN for use with ArcGIS 9.x site license: 0-495-11509-6

**Exploring Tropical Cyclones:**
**GIS Investigations for the Earth Sciences**
ISBN with ArcView GIS CD: 0-534-39147-8
ISBN for use with ArcGIS 9.x site license: 0-495-11543-6

**Exploring Water Resources:**
**GIS Investigations for the Earth Sciences**
ISBN with ArcView GIS CD: 0-534-39156-7
ISBN for use with ArcGIS 9.x site license: 0-495-11512-6

**Exploring the Ocean Environment:**
**GIS Investigations for the Earth Sciences**
ISBN with ArcView GIS CD: 0-534-42350-7
ISBN for use with ArcGIS 9.x site license: 0-495-11506-1

**Lab Manual**
The lessons contained in the Lab Manual are designed to build and heighten understanding of the text chapters. Use these lessons to see how the textbook content can be applied to the everyday problems in the world around you. Lab Manual lessons help build valuable skills such as map reading, map and graph interpretation, three-dimensional thinking, problem solving, and predictive modeling. ISBN: 1-111-57226-7

**Geography CourseMate** *Physical Geography* 10e includes Geography CourseMate, which helps you make the grade.

Geography CourseMate includes:

- an interactive eBook, with highlighting, note taking and search capabilities
- interactive learning tools including:
  - Quizzes
  - Flashcards
  - Videos
  - Animations
  - Google Earth activities
  - and more!

Go to login.cengagebrain.com to access these resources.

# Acknowledgments

This edition of *Physical Geography* would not have been possible without the encouragement and assistance of editors, friends, and colleagues from throughout the country. Great appreciation is extended to Martha, Emily, and Hannah Petersen, Greg Nadon, and Sarah Gabler and for their patience, support, and understanding.

Special thanks go to the splendid freelancers and staff members of Brooks/Cole Cengage Learning. These include Yolanda Cossio, Earth Sciences Publisher; Amy Collins, Development Editor; Kristina Chiapella, Assistant Editor; Alexandria Brady, Media Editor; Hal Humphrey, Content Project Manager; Diane Beasley, Designer; Jeremy Glover of Bill Smith Group, Photo

Researcher; illustrators Graphic World Inc., Pre-PressPMG, Accurate Art, Precision Graphics, and Rolin Graphics; Kate Mannix of Graphic World Inc., Production Editor; Shannon Elderon, Editorial Assistant; and Dr. Chris Houser, creator of our wonderful *Locate and Explore* activities.

Colleagues who reviewed the plans and manuscript for this and previous editions include: Peter Blanken, University of Colorado; Brock Brown, Texas State University; Richard A. Crooker, Kutztown State University; Joanna Curran, Texas State University; J. Michael Daniels, University of Wyoming; Ben Dattilo, University of Nevada, Las Vegas; Leland R. Dexter, Northern Arizona University; James Doerner, University of Northern Colorado; Percy "Doc" Dougherty, Kutztown State University; Daniel Dugas, New Mexico State University; Richard Earl, Texas State University; Tom Feldman, Joliet Junior College; Mark Fonstad, Texas State University; Roberto Garza, San Antonio College; Greg Gaston, University of North Alabama; Beth L. Hall, Towson University; Perry Hardin, Brigham Young University; David Helgren, San Jose State University; Chris Houser, University of West Florida; Kenneth Hundreiser, Roosevelt University; Karen Johnson, North Hennepin Community College; Fritz C. Kessler, Frostburg State University; Elizabeth Lawrence, Miles Community College; Jeffrey Lee, Texas Tech University; Michael E. Lewis, University of North Carolina, Greensboro; Elena Lioubimtseva, Grand Valley State University; Kangshu Lu, Towson University; John Lyman, Bakersfield College; Charles Martin, Kansas State University; Christopher F. Meindl, University of South Florida; Debra Morimoto, Merced College; Andrew Oliphant, San Francisco State University; James R. Powers, Pasadena City College; Joyce Quinn, California State University, Fresno; Dorothy Sack, Ohio University; George A. Schnell, State University of New York, New Paltz; Peter Siska, Austin Peay State University; Richard W. Smith, Harford Community College; Ray Sumner, Long Beach City College; Michael Talbot, Pima Community College; Colin Thorn, University of Illinois at Urbana-Champaign; David L. Weide, University of Nevada, Las Vegas; Thomas Wikle, Oklahoma State University, Stillwater; Glynis Jean Wray, Ocean County College; Amy Wyman, University of Nevada, Las Vegas; and Craig ZumBrunnen, University of Washington, Redmond.

Photos courtesy of: Bill Case and Chris Wilkerson, Utah Geological Survey; Center for Cave and Karst Studies, Western Kentucky University; Hari Eswaran, USDA Natural Resources Conservation Service; Richard Hackney, Western Kentucky University; David Hansen, University of Minnesota; L. Elliot Jones, U.S. Geological Survey; Susan Jones, Nashville, Tennessee; Bob Jorstad, Eastern Illinois University; Carter Keairns, Texas State University; Parris Lyew-Ayee, Oxford University, UK; Greg Nadon, Ohio University; Anthony G. Taranto Jr., Palisades Interstate Park–New Jersey Section; Justin Wilkinson, Earth Sciences, NASA Johnson Space Center.

The detailed comments and suggestions of all of the above individuals have been instrumental in bringing about the many changes and improvements incorporated in this latest revision of the text. Countless others, both known and unknown, deserve heartfelt thanks for their interest and support over the years.

Despite the painstaking efforts of all reviewers, there will always be questions of content, approach, and opinion associated with the text. The authors wish to make it clear that they accept full responsibility for all that is included in the tenth edition of *Physical Geography*.

*James F. Petersen*
*Dorothy Sack*
*Robert E. Gabler*

# Foreword to the Student

## Why Study Geography?

In this global age, the study of geography is absolutely essential to an educated citizenry of a nation whose influence extends throughout the world. Geography deals with location, and a good sense of where places or features are, especially in relation to other places or features in the world, is an invaluable asset whether you are traveling, conducting international business, or browsing the Internet.

Geography examines the characteristics of places and areas on Earth, their roles as part of the Earth system, how they interact with other locations, and the changes and processes involved in these interactions. Geography also gives strong consideration to the relationships between humans and their environments. Today, everyone shares the responsibility of learning more about our physical environment so that we can preserve and protect it for future generations.

Geography provides essential information about the distribution of features on Earth's surface and the interconnections of places. The distribution pattern of volcanoes, for example, provides an excellent indication of where Earth's great crustal plates come in contact with one another; and the violent thunderstorms that plague Illinois on a given day may be directly associated with the low pressure system spawned in Texas two days before. Geography, through a study of regions, provides a focus and a level of generalization that allows people to examine and understand the immensely varied environmental characteristics of Earth.

As you will note when reading Chapter 1, there are many approaches to the study of geography. Some courses are regional in nature; they may include an examination of one or all of the world's political, cultural, economic, or physical regions. Some courses are topical or systematic in nature, dealing with human geography, physical geography, or one of the major subfields of the two.

A great advantage of taking a general course in physical geography is the permanence of the knowledge learned. Although change is constant and often sudden in human aspects of geography, alterations of the physical environment on a global scale are generally slow unless they are influenced by human intervention. Theories and explanations may differ, but the broad patterns of atmospheric and oceanic circulation and of world climates, landforms, soils, natural vegetation, and physical landscapes will be the same tomorrow as they are today.

## Keys to Successful Study

Good study habits are essential if you are to master science courses such as physical geography, where the topics, explanations, and terminology are often complex and unfamiliar. To help you succeed in the course in which you are currently enrolled, we offer the following suggestions.

## Reading Assignments

- Read the assignments before the material is covered in class by the instructor.
- Compare what you have read with the instructor's presentation in class. Pay particular attention if the instructor introduces new examples or course content not included in the reading assignment.
- Ask questions in class and seek to understand any material that was not clear from your first reading of the assignment.
- Reread the assignment as soon after class as possible, concentrating on areas that were emphasized in class. Highlight only those items or phrases that you now consider to be important, and skim those sections already mastered.
- Record in your class notes important terms, your own comments, and summarized information from each reading assignment.

## Understanding Vocabulary

- Mastery of basic vocabulary often becomes a critical issue in the success or failure of students in a beginning science course.
- Focus on terms that appear in boldface type in your reading assignments. Do not overlook additional terms that the instructor may introduce in class.
- Develop your own definition of each term (representing a feature, process, or characteristic), or phrase and associate it with related terms in physical geography.
- Identify any physical processes associated with the term. Often, knowing the process will help to define the term.
- Whenever possible, associate terms with location.
- Consider the significance to humans of terms you are defining. Recognizing the significance of terms and phrases can make them relevant and easier to recall.

## Learning Earth Locations

- A good knowledge of place names and of the relative locations of physical and cultural phenomena on Earth is fundamental to the study of geography.
- Take personal responsibility for learning locations on Earth. Your instructor may identify important physical features and place names, but you must learn their locations for yourself.
- Thoroughly understand latitude, longitude, and the Earth grid. They are fundamental to finding and describing locations on maps and globes. Practice locating features by their

latitude and longitude until you are entirely comfortable using the system.

■ Develop a general knowledge of the world political map. The most common way of expressing the location of physical features is by identifying the political unit (state, country, or region) in which it can be found.

■ Make liberal use of outline maps. They are a key to learning the names of states and countries and the locations of specific physical features. Personally placing features correctly on an outline map is often the best way to learn location.

■ Cultivate the habit of using an atlas. The atlas does for the individual who encounters place names or the features they represent what the dictionary does for the individual who encounters a new vocabulary word.

## Utilizing Textbook Illustrations

The secret to making good use of maps, diagrams, and photographs lies in understanding why an illustration has been included in the textbook or incorporated as part of your instructor's presentation.

■ Concentrate on the instructor's discussion. Taking notes about photographs, maps, graphs, and other illustrations will allow you to follow the same line of thought at a later date.

■ Study all of the textbook illustrations. Be sure to note which of them received considerable attention in the lecture. Do not quit examining an illustration until it makes sense to you, until you can read the map or graph, or until you can recognize what a diagram or photograph has been selected to explain.

■ Hand-copy important diagrams and graphs. Few of us are graphic artists, but you might be surprised at how much better you understand a graph or line drawing after you reproduce it yourself.

■ Read the captions of photographs and illustrations thoroughly and thoughtfully. If the information is included, be certain to note where a photograph was taken and in what way it is representative. What does it tell you about the feature, process, region, or site being illustrated?

■ Attempt to place the concept that is being illustrated in new situations. Seek other opportunities to test your skills at interpreting similar maps, graphs, and photographs and think of other examples that support the text.

■ Remember that all illustrations are reference tools, particularly tables, graphs, and diagrams. Refer to them often.

## Taking Class Notes

A good set of class notes is based on selectivity. You cannot, and should not, try to write down every word uttered by your classroom instructor.

■ Learn to paraphrase. With the exception of specific quotations or definitions, put the instructor's ideas, explanations, and comments into your own words. You will understand them better when you read them over at a later time.

■ Be succinct. Never use a sentence when a short phrase will do, and never use a phrase when a word will do. Start your recall process with your note-taking by forcing yourself to rebuild an image, an explanation, or a concept from a few words.

■ Outline where possible. Preparing an outline helps you to discern the logical organization of information. As you take notes, organize them under main headings and subheadings.

■ Take the instructor at his or her word. If the instructor takes the time to make a list, then you should do so too. If he or she writes something on the board, it should be in your notes. If the instructor's voice indicates special concern, take special notes.

■ Come to class and take your own notes. Notes trigger the memory, but only if they are your notes.

## Doing Well on Tests

Follow these important study techniques to make the most of your time and effort preparing for tests.

■ Practice distillation. Do not try to reread but skim the assignments carefully, taking notes in your own words that record as economically as possible the important definitions, descriptions, and explanations. Do the same with any supplementary readings, handouts, and laboratory exercises. It takes practice to use this technique, but it is much easier to remember a few key phrases that lead to ever-increasing amounts of organized information than it is to memorize all of your notes. And the act of distillation in itself is a splendid memory device.

■ Combine and reorganize. Merge all your notes into a coherent study outline.

■ Become familiar with the type of questions that will be asked. Knowing whether the questions will be objective, short-answer, essay, or related to diagrams and other illustrations can help in your preparation. Some instructors make old tests available so that you can examine them or discover their evaluation styles if you inquire. If not, then turn to former students; there are usually some around the department or residence halls who have already experienced the instructor's tests.

■ Anticipate the questions that will likely be on the test. The really successful students almost seem to be able to predict the test items before they appear. Take your educated guesses and turn them into real questions.

■ Try cooperative study. This can best be described as role-playing and consists very simply of serving temporarily as the instructor. So go ahead and teach. If you can demonstrate a technique, illustrate an idea, or explain a process or theory to other students so that they can understand it, there is little doubt that you can answer test questions over the same material.

■ Avoid the "all-nighter." Use the early evening hours the night before the test for a final unhurried review of your study outline. Then get a good night's sleep.

## The Importance of Maps

Like graphs, tables, and diagrams, maps are excellent reference tools. Familiarize yourself with the maps in your textbook to better judge when it is appropriate to seek information from these important sources.

Maps are especially useful for comparison purposes and to illustrate relationships or the possible associations of two features, areas, or processes. But the map reader must beware. Only a small portion of the apparent associations of phenomena in space (areal associations) are actually cause-and-effect relationships. In some instances the similarities in a distribution result from another factor that has not been considered or mapped. For instance, a map of worldwide volcano distribution is almost exactly congruent with one of incidence of earthquakes, yet volcanoes are not the cause of major earthquakes, nor is the obverse true. A third factor, the location of tectonic plate boundaries, explains the first two phenomena.

Finally, remember that a map is the most important statement of the professional geographer. It is useful to all natural and social scientists, engineers, politicians, military planners, road builders, farmers, and countless others, but it is the essential expression of the geographer's primary concern with location, distribution, and spatial interaction.

## About Your Textbook

This textbook has been written for you, the student. It has been written so that the text can be read and understood easily. Explanations are as clear, concise, and uncomplicated as possible. Illustrations have been designed to complement the text and to help you visualize the processes, places, and phenomena being discussed. In addition, the authors do not believe it is sufficient to offer you a textbook that simply provides information to pass a course. We urge you to think critically about what you read in the textbook and hear in class.

As you learn about the physical aspects of Earth environments, ask yourself what they mean to you and to people throughout the world. Make an honest attempt to consider how what you are learning relates to the problems and issues of today and tomorrow. Practice using your geographic skills and knowledge in new situations so that you will continue to use them in the years ahead. Your textbook includes several special features that will support learning, encouraging you to go beyond memorization and to reason geographically.

**Chapter Activities** At the end of each chapter, Consider and Respond and Apply and Learn questions require you to go beyond a routine chapter review. The questions are designed so that you may apply your knowledge of physical geography and on occasion personally respond to critical issues in society today. Locate and Explore activities (found at the end of many chapters) teach you how to use the Google Earth application as an exploratory learning tool. Check with your instructor for answers to the problems.

**Caption Questions** With most illustrations and photographs in your textbook, a caption links the image with the chapter text it supports. Read the captions carefully because they explain the illustrations and may also contain new information. Wherever appropriate, questions at the ends of captions have been designed to help you seize the opportunity to consider your own personal reaction to the subject under consideration.

**Map Interpretation Series** It is a major goal of your textbook to help you become an adept map reader, and the Map Interpretation Series in your text has been designed to help you reach that goal.

**Environmental Systems** Viewing Earth as a system comprising many subsystems is a fundamental concept for researchers, instructors, and students in physical geography. The concept is introduced in Chapter 1 and reappears frequently throughout your textbook. The interrelationships and dependencies among the variables and components of Earth systems are important. Many of the illustrations included in the text will help you visualize how systems work. There are also diagrams designed to help you understand how human activities can affect the delicate balance that exists within many Earth systems.

As authors of your textbook, we wish you well in your studies. It is our fond hope that you will become better informed about our home—Earth and its varied environments—and that you will enjoy the study of physical geography.

# Brief Contents

# Contents

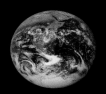

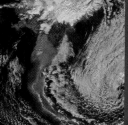

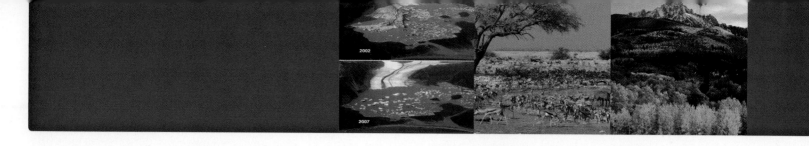

# List of Major Maps

## The World

# The Ocean

# The Contiguous United States

# North America

# Author Biographies

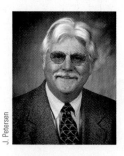

J. Petersen

**James F. Petersen** James F. Petersen is Professor of Geography at Texas State University, in San Marcos, Texas. A broadly trained physical geographer with strong interests in geomorphology and Earth Science education, he enjoys writing about topics related to physical geography for the public. Also involved in environmental interpretation, he has written a landform guidebook for Enchanted Rock State Natural Area in Texas and produced a number of field and trail guides. He is a past President of the National Council for Geographic Education and a recipient of the Distinguished Teaching Award from that organization. He has written or served as a senior consultant for nationally published educational materials at levels from middle school through university. Believing in the value of education in the field, every year since he began teaching he has taken students on extended field excursions to learn about Earth environments, locations, processes, and features through first-hand experience.

J. Petersen

**Dorothy Sack** Dorothy Sack, Professor of Geography at Ohio University in Athens, Ohio, is a physical geographer who specializes in geomorphology. Her research emphasizes arid region landforms, including geomorphic evidence of paleolakes, which contributes to paleoclimate reconstruction. She has published research results in a variety of professional journals, academic volumes, and Utah Geological Survey reports. She also has research interests and publications on the history of geomorphology and the impact of off-road vehicles. Her work has been funded by the National Geographic Society, NSF, Association of American Geographers (AAG), American Chemical Society, and other organizations. She is active in professional organizations, having served as chairperson of the AAG Geomorphology Specialty Group, and several other offices for the AAG, Geological Society of America, and History of Earth Sciences Society. She enjoys both teaching and research, has received the Outstanding Teacher Award from Ohio University's College of Arts and Sciences, and was recently named one of the university's transformative faculty in recognition of excellence in teaching.

J. Petersen

**Robert E. Gabler** During his nearly five decades of professional experience, Professor Gabler has taught geography at Hunter College, City of New York, Columbia University, and Western Illinois University, in addition to 5 years in public elementary and secondary schools. At times in his career at Western he served as Chairperson of the Geography and Geology Department, Chairperson of the Geography Department, and University Director of International Programs. He received three University Presidential Citations for Teaching Excellence and University Service, served two terms as Chairperson of the Faculty Senate, edited the *Bulletin of the Illinois Geographical Society,* and authored numerous articles in state and national periodicals. He is a Past President of the Illinois Geographical Society, former Director of Coordinators and Past President of the National Council for Geographic Education, and the recipient of the NCGE George J. Miller Distinguished Service Award.

# PHYSICAL GEOGRAPHY

# 1

# PHYSICAL GEOGRAPHY: EARTH ENVIRONMENTS AND SYSTEMS

## ■ OBJECTIVES

WHEN YOU COMPLETE THIS CHAPTER YOU SHOULD BE ABLE TO:

■ Discuss physical geography as a discipline and profession that considers both the natural world and human interaction with natural processes and areas.
■ Understand how geographic information and techniques are directly applicable in many career fields.
■ Describe the three major perspectives of physical geography: the spatial perspective, the physical science perspective, and the environmental perspective.
■ Conceptualize Earth as a system of interacting parts that respond to both natural and human-induced processes.
■ Explain how interactions between humans and their environments can be both advantageous and hazardous.
■ Recognize how physical geography invites better understanding of our environment.
■ Understand why every physical environment offers an array of advantages and challenges to human life and living conditions.
■ Explain why physical geography has relevance in your everyday life and the career applications of this discipline.

ABOUT 40 YEARS AGO, AS the last astronaut mission to the moon was on its way, the three crew members looked back at a spectacular view of Earth and took a photograph. This view of Earth, surrounded by the emptiness of space, illustrated the fact that life on our planet is dependent on self-contained environmental and natural resources that are not limitless. Today it is said that more people have seen this photograph than any other image in history. This image of Earth continues to be an internationally known symbol for environmental awareness, and it has increased our concern for conserving Earth's resources and environments.

Viewed from space, Earth is beautiful and intriguing—a life-giving planetary oasis. From this perspective we can begin to appreciate "the big picture," a global view of our planet's physical geography through its display of environmental diversity. Characteristics of the oceans, the atmosphere, the land masses, and evidence of life as revealed by vegetated regions are apparent. Looking carefully, we can also recognize geographic patterns, shaped by processes that make our world dynamic and ever-changing.

"The Blue Marble." Lunar astronauts' view of Earth: An oasis of life in the vastness of space. NASA Headquarters—Greatest Images of NASA

From a human perspective, Earth may seem immense and almost limitless. In contrast, viewing the "big picture" reveals Earth's fragile nature—a spherical island of life existing in the vast, dark emptiness of space. Except for the external addition of energy from the sun, our planet is a self-contained system that has all the requirements to sustain life. The natural characteristics of Earth and its environments provide the life-support systems for all living things. It is important to gain an understanding of the planet that sustains us by learning about the components and processes that change or regulate Earth's environmental conditions. Learning the relevant questions to ask is an important step toward finding answers and explanations. Understanding how Earth's features and processes interact to develop the environmental diversity on our planet is a major goal of a course in physical geography.

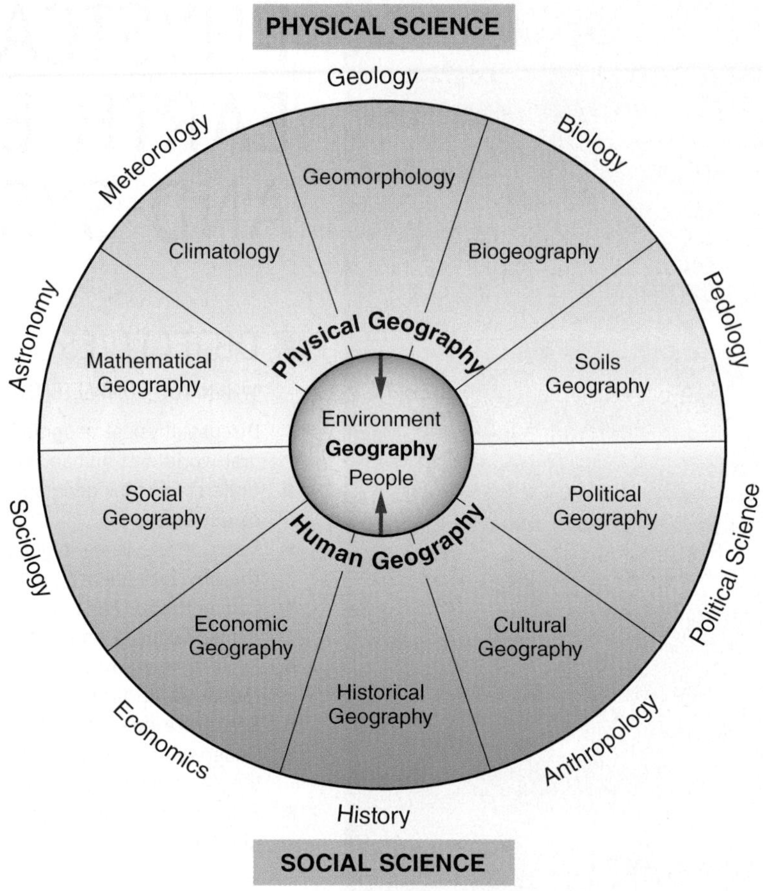

# The Study of Geography

**Geography** refers to the examination, description, and explanation of Earth—its variability from place to place, how places and features change over time, and the processes responsible for these variations and changes. The word *geography* comes from the Greek language. *Geo* refers to "Earth," and *graphy* means "picture or writing." Geography is often called the **spatial science** (the science of locational space) because it includes analyzing and explaining the locations, distributions, patterns, variations, and similarities or differences among phenomena on Earth's surface.

> Geography has much more to do with asking questions and solving problems than it does with rote memorization of facts. So what exactly is geography? It is an integrative discipline that brings together the physical and human dimensions of the world in the study of people, places, and environments. Its subject matter is the Earth's surface and the processes that shape it, the relationships between people and environments, and the connections between people and places.
>
> *Geography for Life,* Geography Education Standards Project

Geographers study processes that influenced Earth's landscapes in the past, how they affect them today, how a landscape may change in the future, and the significance of these changes. Geography is distinctive among the sciences by virtue of its definition and central purpose and may involve studying any topic related to the scientific analysis of human or natural processes on Earth. Because geography embraces the study of virtually any global phenomena, it is not surprising that the subject has many

● **FIGURE 1.1** Geography has many subdivisions that are related to other disciplines and share some of their interests. Geographers apply their own unique perspectives and approaches to these areas of study.

*What advantage might a geographer have when working with other physical scientists seeking a solution to a problem?*

subdivisions and typically geographers specialize in one or more subfields of the discipline ( ● Fig. 1.1). Geography is also subdivided academically; some geographers are social scientists and some are natural scientists. Most geographers are concerned with physical and/or human processes, especially how they affect our planet and its natural or human environments. **Human geography** deals with people's activities and their spatial nature. Human geographers are concerned with population distributions, cultural patterns, cities and urbanization, industrial and commercial location, natural resource use, and transportation networks.

Geographers are interested in how to divide areas into meaningful **regions,** which are areas identified by distinctive characteristics that distinguish them from other areas. Physical, human, or a combination of factors can define a region. *Regional geography* concentrates on the characteristics of a region (or of multiple regions). Geographers often seek to explain characteristics that two or more locations may have in common and why places vary in their geographic attributes. Geographers gather, organize, and analyze many kinds of data and information, yet a unifying factor is a focus on explaining spatial locations, distributions, and relationships.

# Physical Geography

The focus of **physical geography** is on the processes and features of Earth, including human activities where they interface with the environment. Geographers generally take a **holistic approach,** meaning that they often consider both the human and natural phenomena relevant to understanding aspects of our planet. Physical geographers are concerned with nearly all aspects of Earth and are trained to view a natural environment in its entirety ( ● Fig. 1.2). Yet most physical geographers focus their expertise on one or two specialties. For example, many *meteorologists* and *climatologists* have studied geography. Meteorologists are interested in the processes that affect daily weather and in forecasting weather conditions. **Climatology** deals with regional climates, the averages and extremes of weather data, understanding climate change and climatic hazards, and the impact of climate on human activities and the environment.

**Geomorphology** is the study of the characteristics of landforms and how interactions between Earth processes and surface materials contribute to their development and modification. *Geo-* *morphologists* are interested in understanding and explaining variations in landforms and landscapes, the processes involved, and the nature of Earth's surface features. The factors involved in landform development include gravity, running water, forces in the Earth's crust, flowing ice in glaciers, volcanic activity, and the erosion or deposition of Earth's surface materials. **Biogeography** involves studying the geographic ranges and patterns of vegetation or animal species, seeking to discover the environmental factors that limit or facilitate their distributions. *Biogeographers* examine natural and human-modified environments and the ecological processes that influence environmental characteristics and distributions, including vegetation change over time. Some geographers serve as *soil scientists,* mapping and analyzing soil types, determining the suitability of soils for certain uses such as agriculture, and working to conserve soil as a natural resource.

Water is critical to life on Earth, and concerns about the volume and quality of water available to meet human and environmental needs are increasing. Physical geographers are widely involved in the study of water bodies and water resources, including their processes, movements, impact, quality, and conserva-

● **FIGURE 1.2** Physical geographers study the elements and processes that affect natural environments. These include rock structures, landforms, soils, vegetation, climate, weather, and human impacts. This is in the White River National Forest in Colorado.

*What physical geography characteristics can you recognize in this scene?*

Martha Moran, Aspen Ranger District, White River National Forest

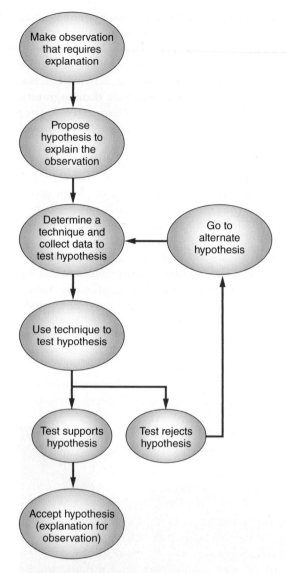

● **FIGURE 1.3** The scientific method, widely applicable in physical geography, involves following steps shown here.

**Making an observation that requires an explanation.** On a trip to the mountains, you notice that it gets colder as you go up in elevation. Is that just a result of local conditions on the day you were there, or is it a universal relationship?

**Restating the observation as a hypothesis.** Here is an example: As we go higher in elevation, the temperature gets cooler. (The answer may seem obvious, but although it is generally true, there are exceptions depending on environmental conditions, which will be discussed in later chapters.)

**Determining a technique for testing the hypothesis and collecting necessary data.** The next step is finding a technique for evaluating data (numerical information) and facts that relate to the hypothesis. In this case, you would gather temperature and elevation data (taken at about the same time for all data points) in the study area.

**Applying the technique or strategy to test the validity of the hypothesis.** Here we discover if the hypothesis is supported by adequate evidence, which must be collected under similar conditions to minimize bias. The technique will recommend either accepting or rejecting the hypothesis. If the hypothesis is rejected, we can test an alternate hypothesis, or we may just discover that our hypothesized relationship is not valid.

tion. They may serve as *hydrologists, oceanographers,* or *glaciologists.* Many geographers also function as *water resource managers,* who work to ensure that lakes, watersheds, springs, and groundwater sources are suitable to meet human or environmental needs, provide an adequate water supply, and are as free of pollution as possible.

Like other scientists, physical geographers typically apply the **scientific method** as they seek to learn about aspects of Earth. The scientific method involves finding the answers to questions and determining the validity of new ideas by objectively testing all pertinent evidence and facts related to the issue being studied ( ● Fig. 1.3). Using the scientific method, new ideas or proposed answers to questions are only accepted as valid if they are clearly supported by the evidence.

# Geographic Technologies and Tools

The technologies that physical geographers and other scientists use in their efforts to learn about Earth are rapidly changing. The current abilities of computer systems to capture, analyze, and display spatial data—functions that can be performed on a personal computer—were only a dream 30 years ago. Today, the Internet provides access to information and images on virtually any topic. The amounts of data, information, and imagery available for studying Earth and its environments have exploded. Continuous satellite imaging of Earth has been ongoing for about 40 years, which has given us a better perspective on environmental changes. Using images from space, we are able to see, measure, monitor, and map processes and the effects of certain processes, including many that are invisible to the naked eye. Displays of environmental data and information are becoming more vivid and striking as a result of sophisticated methods of data processing and visual representation. Increased computer power allows the presentation of high-resolution images, three-dimensional scenes, and even animated images of Earth's features, changes, and processes ( ● Fig. 1.4).

Satellite technology for determining the location of a positioning receiver on Earth's surface has many geographic and mapping applications. These systems are also widely used by the public, even on cell phones, for finding locations or navigating while hiking or traveling in vehicles. Today, most map making **(cartography)** and many aspects of map analysis are computer-assisted operations, although the ability to visually interpret a map, a landscape, or an environmental image remains an important geographic skill.

Physical geographers make observations and gather data in the field, but they must also keep up with new technologies that support traditional fieldwork. Technology may generate maps, images, and data, but a person who is knowledgeable about the geographic aspects of the subject being studied is essential to the processes of map analysis and problem solving. Many geographers are gainfully employed in positions that apply technology to the problems of understanding our planet and its environments, and their numbers are certain to increase in the future.

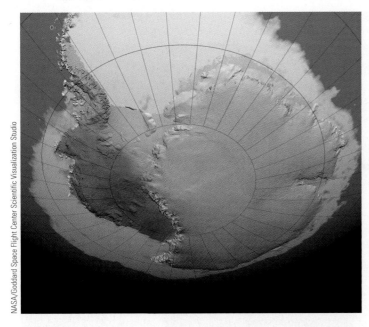

NASA/Goddard Space Flight Center Scientific Visualization Studio

● **FIGURE 1.4** This computer–generated, three-dimensional model of Antarctica was made by combining a 50-year history of temperature records from locations on the continent with modern satellite images of the ice surface. The red area shows the region that has experienced the largest temperature increase in response to global warming.

*What global warming impacts have caused concern in recent years?*

# Major Perspectives in Physical Geography

Your textbook will demonstrate three major perspectives that physical geography emphasizes: the *spatial perspective, physical science perspective,* and *environmental perspective.* Although the focus on each of these perspectives may vary from chapter to chapter, take note of how each perspective relates to the unique nature of geography as a discipline.

## The Spatial Perspective

Physical geographers have varied interests, but they share the common goals of understanding and explaining spatial variations on Earth's surface. The following five examples illustrate spatial factors that geographers typically consider and the problems they address. In keeping with the quote from *Geography for Life* that geography is about asking questions and solving problems, some questions have been included for each topic.

**Location** Geographic knowledge and studies often begin with locational information. Describing a location usually uses one of two methods: **absolute location,** which is expressed by a coordinate system (or address), and **relative location,** which identifies where a feature is in relation to something else, usually a fairly well-known location. For example, Pikes Peak in the Rocky

Mountains of Colorado, with an elevation of 4302 meters (14,115 ft), has a location of latitude 38°51′ N and longitude 105′03″ W. This kind of global address is an absolute location. However, another way to report its location would be to state that it is 36 kilometers (22 mi) west of Colorado Springs ( ● Fig. 1.5). This is an example of relative location (its position in relation to Colorado Springs). Typical spatial questions involving location include the following:

*Where is a certain type of Earth feature found, and where is it not found? Why is a certain feature located where it is? What methods can we use to locate a feature on Earth? How can we describe its location? What is the most likely or least likely location for a certain Earth feature?*

**Characteristics of Places** Physical geographers are interested in the environmental features and processes that make a place unique and in the shared characteristics between places. For example, what physical geographic features make the Rocky Mountains appear as they do? (See Figs. 1.2 and 1.5.) Further, how are the Appalachian Mountains different from the Rockies, and what characteristics are common to both mountain ranges? Another aspect of the characteristics of places is analyzing the environmental advantages and challenges that exist in a place. Questions might include: *How does an Australian desert compare to the Sonoran Desert of the southwestern United States? How do the grasslands of the North American Great Plains compare to the grasslands of Argentina? What are the environmental conditions at a particular site? How do places on Earth vary in their environments, and why? In what ways are places unique, and in what ways do they share similar characteristics with other places?*

**Spatial Distribution and Spatial Pattern** When studying how features are arranged in space, geographers are usually interested in two spatial factors. **Spatial distribution** is a locational characteristic that refers to the extent of the area or areas where a feature exists. For example, where on Earth do we find the tropical rainforests? What is the distribution of rainfall in the United States on a particular day? Where on Earth do major earthquakes occur? **Spatial pattern** refers to the arrangement of features in space—are they regular or random, clustered together or widely spaced? The distribution of population can be either dense or sparse ( ● Fig. 1.6). The spatial pattern of earthquakes may be aligned on a map because earthquake faults display similar linear patterns. *Where are certain features abundant, and where are they rare? How are particular factors or elements of physical geography arranged in space, and what spatial patterns exist, if any? What processes are responsible for these distributions or patterns? If a spatial pattern exists, what does it signify?*

**Spatial Interaction** Few processes on Earth operate in isolation because areas on our planet are interconnected (linked to conditions elsewhere on Earth). A condition, an occurrence, or a process in one place generally has an impact on other places. Unfortunately, the exact nature of a **spatial interaction**— whether one event actually causes another—is often difficult to

Pike's Peak                                                    Colorado Springs

● **FIGURE 1.5** A three-dimensional digital model shows the relative location of Pikes Peak to Colorado Springs, Colorado. Because this is a perspective view, the 36-km (22-mi) distance appears to be shorter than its actual ground distance. A satellite image was merged with elevation data gathered by radar from the space shuttle to create this scene.

*What can you learn about the physical geographic characteristics of this place from the image?*

● **FIGURE 1.6** A nighttime satellite image provides good illustrations of distribution and pattern, shown here on most of North America. Spatial distribution is where features are located (or perhaps, absent). Spatial pattern refers to their arrangement. Geographers seek to explain why these spatial relationships exist.

*Can you locate and propose possible explanations for two patterns and two distributions in this scene?*

# Natural Regions

The term *region* has a precise meaning and special significance to geographers. Simply stated, a region is an area that is defined by a certain shared characteristic (or a set of characteristics) existing within its boundaries. The concept of a region is a tool for thinking about and analyzing logical divisions of areas based on their geographic characteristics. Geographers not only study and explain regions, including their locations and characteristics, but also strive to delimit them—to outline their boundaries on a map. An unlimited number of regions can be derived for each of the four major Earth subsystems.

Regions help us understand the arrangement and nature of areas on our planet. Regions can also be divided into subregions. For example, North America is a region, but it can be subdivided into many subregions. Examples of subregions based on natural characteristics include the Atlantic Coastal Plain (similarity of landforms, geology, and locality), the Prairies (ecological type), the Sonoran Desert (climate type, ecological type, and locality), the Pacific Northwest (general locality), and Tornado Alley (region of high potential for these storms).

There are three important points to remember about natural regions. Each of these points has endless applications and adds considerably to the questions that the process of defining regions based on spatial characteristics seeks to answer.

■ **Natural regions can change in size and shape over time in response to environmental changes.** An example is desertification, the expansion of desert regions that has occurred in recent years. Using images from space, we can see and monitor changes in the areas covered by deserts and other natural regions.

■ **Boundaries separating different natural regions tend to be indistinct or transitional rather than sharp.** For example, on a climate map, lines separating desert from nondesert regions do not imply that extremely arid conditions instantly appear when the line is crossed. If we travel to a desert, it is likely to get progressively more arid as we approach our destination.

■ **Regions are spatial models, devised by humans, for geographic analysis, study, and understanding.** Natural regions are conceptual models that help us comprehend and organize spatial relationships and geographic distributions. Learning geography is an invitation to think spatially, and regions provide an essential, extremely useful conceptual framework in that process.

Understanding regions, through an awareness of how areas can be divided into geographically logical units and why it is useful to do so, is essential in geography. Regions help us to reason about, make sense of, and comprehend the spatial aspects of our world.

USDA Forest Service

Natural Regions: The Great Basin of the Western United States is a landform region that is clearly defined based on an important physical geographic characteristic. No rivers flow to the ocean from this arid and semiarid region of mountains and topographic basins. The rivers and streams that exist flow into enclosed basins where the water evaporates away from temporary lakes or they flow into lakes like the Great Salt Lake, which has no outlet to the sea. Topographic features called drainage divides (mountain ridges) form the outer edges of the Great Basin, defining and enclosing this natural region.

establish with complete confidence. It is much easier to determine that certain changes seem to be associated with each other without knowing if one event is the actual cause of the other. For example, the presence of abnormally warm ocean waters off South America's west coast, a condition called El Niño, seems to be related to unusual weather in other parts of the world. Clearing the tropical rainforest may have a widespread impact on world climates. Human activities that can contribute to global warming are increasing in conjunction with climate changes. Interconnections are one reason for considering interactions, impacts, and their potential links at various geographic scales: local, regional, and global. *What are some examples of relationships among places and features on Earth? How do they affect one another? What important interconnections link the oceans to the atmosphere and the atmosphere to the land surface?*

**The Changing Earth** Earth's features and landscapes are continuously changing in a spatial context. Weather conditions change from day to day, over the seasons, and from year to year. Storms, earthquakes, landslides, and stream processes modify the landscape. Coastlines may change position because of storm waves, tsunamis, human impacts, or changes in sea level. Areas that were once forested have been clear-cut, changing the nature of the environment there. Desertlike conditions seem to be expanding in many arid regions of the world. Vegetation and wildlife are becoming re-established in areas that were devastated by wildfires or volcanic eruptions. A new Hawaiian island is now forming beneath the waters of the Pacific Ocean. Volcanic islands have been destroyed and created in historic times ( ● Fig. 1.7).

⊕ In Google Earth fly to: 63.202°N, 20.601°W (Surtsey, Iceland).

● **FIGURE 1.7** Surtsey, Iceland, an island in the North Atlantic, did not exist until about 45 years ago when eruptions formed this new volcanic island. Since the 1960s, when the volcanic eruptions stopped, erosion by waves and other processes have reduced the island by half of its original size.

*Once the island formed and cooled, what other environmental changes appear to have taken place?*

World climates have changed throughout Earth's history, with attendant shifts in the distribution of plant and animal life. Today changes in Earth's climates and environments are complicated by the impact of human activities. Most of Earth's glaciers are shrinking in response to global warming. Earth and its environments are always changing, although at different time scales, so the impact and direction of certain changes can be difficult to determine. *How are Earth features changing in ways that can be recorded in a spatial sense? What processes contribute to the change? What is the rate of change? Does change occur in a cycle? Can humans witness this change as it is taking place, or is a long-term study required to recognize the change? Do all places on Earth experience the same levels of change, or is there spatial variation?*

The previous five topics illustrate geography's emphasis on the spatial perspective. Learning the relevant questions to ask is the first step toward finding answers and explanations, and it is a major objective of your physical geography course.

## The Physical Science Perspective

Physical geographers observe phenomena and compile data to find solutions to problems and the answers to questions that are also of interest to researchers in other physical sciences. Physical geographers who specialize in climatology share many ideas and information with atmospheric physicists. Soil geographers study some of the same elements and compounds analyzed by chemists. Biogeographers are concerned about environments that support the same plants and animals that are classified by biologists. However, physical geographers also bring distinctive points of view to scientific study—a holistic perspective and a spatial perspective. Physical geographers are concerned with the processes that affect Earth's physical environments at scales from global to regional to local. By examining the factors, features, and processes that influence an environment, and how these elements work together, we can better understand our planet's physical geography. We can also appreciate the importance of viewing Earth as a constantly functioning system.

**The Earth System** The analysis of systems provides physical geographers with an ideal method for studying relationships that affect Earth's features and environments. A **system** is any entity that consists of interrelated parts or components. Our planetary environment, the **Earth system,** relies on the interactions among a vast combination of factors that enable it to support life. The individual components of a system, termed *variables,* change by interacting with one another as parts of a functioning unit. A change in one aspect of the Earth system affects certain other parts. Systems can be divided into **subsystems,** which are functioning units of a major system that demonstrate strong internal connections. For example, the human body is a system that is composed of many subsystems (for example, the respiratory system, circulatory system, and digestive system). Examining the Earth system as consisting of a set of interdependent subsystems is a major concept in understanding physical geography.

Atmosphere

Biosphere

Earth System

Hydrosphere

Lithosphere

● **FIGURE 1.8** Earth's four major subsystems. Studying Earth as a system is central to understanding changes in our planet's environments and adjusting to or dealing with these changes. Earth consists of many interconnected subsystems.

*How do these systems overlap? For example, how does the atmosphere overlap with the hydrosphere or the biosphere?*

**Earth's Major Subsystems** Earth has four major subsystems ( ● Fig. 1.8). The **atmosphere** is the gaseous blanket of air that envelops, shields, and insulates Earth. Variations in atmospheric components and processes create the changing conditions that we know as weather and climate. The **hydrosphere** includes the waters of Earth—oceans, lakes, rivers, and glaciers. The **lithosphere** makes up the solid Earth—landforms, rocks, soils, and minerals. The **biosphere** is composed of all living things: people, other animals, and plants.

These major subsystems and the interactions among them nurture the conditions necessary for life. The hydrosphere provides the water supply for life on Earth, including humans, and provides a home environment for aquatic plants and animals. The hydrosphere affects the lithosphere as water moving in streams, waves, and currents shapes landforms. It also influences the atmosphere through evaporation, condensation, and the effects of ocean temperatures on climate. Many other examples of overlap exist among these four Earth subsystems. Soil can be examined as part of the lithosphere, the biosphere, or the hydrosphere because soils typi-

cally contain minerals, organisms, and water (and gases). The water stored in plants and animals is part of both the biosphere and the hydrosphere, and the water in clouds is a component of the atmosphere and the hydrosphere. The fact that we cannot draw sharp boundaries between these divisions underscores the interrelatedness among various components of the Earth system.

The presence of mountains influences the distribution of rainfall, and variations in rainfall affect the density, type, and variety of vegetation. Plants, moisture, and the underlying rock affect the kind of soil that forms in an area. Characteristics of vegetation and soils influence the runoff of water from the land, leading to completion of the circle because the amount of runoff is a major factor in stream erosion, which eventually can reduce the height of mountains. Cycles such as this operate to change our planet, but the Earth system is complex, and these cycles and processes operate at widely varying rates and over widely varying time spans.

The impact and intensity of interactions within and between Earth's subsystems are not equal everywhere. This inequality leads to our planet's environmental diversity and produces the wide variety of geographic patterns on Earth. However, like a machine, a computer, or the human body, planet Earth is a system that functions well only when all of its parts (and its subsystems) work together harmoniously.

**Earth Impacts** The Earth system is *dynamic,* continually responding to changes. We can directly observe some of these changes—the seasons, the ocean tides, earthquakes, floods, volcanic eruptions. Other environmental changes may take years, or more than a lifetime, to accumulate enough modification so that humans can recognize their impact. Long-term changes in our planet are often difficult to understand or predict with certainty. The evidence must be carefully studied to determine what is really occurring and what the potential consequences might be ( ● Fig. 1.9). Changes of this type include shifts in world climates, drought cycles, the spread of deserts, worldwide rise or fall in sea level, erosion of coastlines, and major changes in river systems.

Changes on Earth may be naturally caused or human induced, or they may result from a combination of these factors. Today there is much concern about environmental changes, such as the potential impacts of global warming, which centers on the increasing impact that human activities are exerting on Earth's natural systems. To understand our planet, therefore, we must learn about its components and the processes that operate to change or regulate the Earth system. Such knowledge is in the best interest for humankind as they interact with and influence Earth's natural systems, which form the habitat for all living things. Understanding changes in our planet is critical to human existence. We are, after all, a part of the Earth system.

1913

2005

● **FIGURE 1.9** Photographs taken 92 years apart in Montana's Glacier National Park show that Shepard Glacier, like other glaciers in the park, has dramatically receded during that time. This glacial retreat is in response to climate warming and droughts. The U.S. Geological Survey estimates that if this climate trend continues, the glaciers in the park will disappear by 2030.

*What other kinds of environmental change might require long-term observation and recording of evidence?*

## The Environmental Perspective

Today we are increasingly concerned about how the environment and ecology are affected by human activities. The news media often devote entire sections to these and other environmental issues. But what do we really mean when we use words like *environment, ecology,* or *ecosystem?* In the broadest sense, our **environment** can be defined as our surroundings; including all physical, social, and cultural aspects of our world that affect our growth, our health, and our way of living.

Environments are systems composed of a wide variety of elements and processes that involve interconnections among weather, climate, soils, rocks, terrain, plants, animals, water, and humans. Physical geography's holistic approach is well suited to environmental understanding because important factors are considered individually and also as integral parts of an environmental system.

**Ecosystems** The study of relationships between organisms and their environments is the science of **ecology.** The word **ecosystem** (a contraction of *ecological system*) refers to a community of organisms and the relationships of those organisms to each other and to their environment ( ● Fig. 1.10). An ecosystem is dynamic in that its various parts are always changing. For instance, plants grow, rain falls, animals eat, and soils develop—all affecting the environment of a particular ecosystem. Because each member of an ecosystem belongs to the environment of other parts of that system, a change in one often affects the living conditions for others. As those components react to the alteration, they transform the environment for the others. For example, a change in the weather from sunshine to rain affects plants, soils, and animals. Heavy rain, however, may carry away soils and plant nutrients so that vegetation may not be able to grow as well and thus, many animals may not have as much to eat. In contrast, the addition of moisture to the soil may help some

plants grow, increasing the amount of shade beneath them, which may keep other plants from growing.

The ecosystem concept can be applied on almost any scale, from local to global, in virtually any geographic location. Your backyard, a farm pond, a grassy field, a marsh, a forest, a coastal

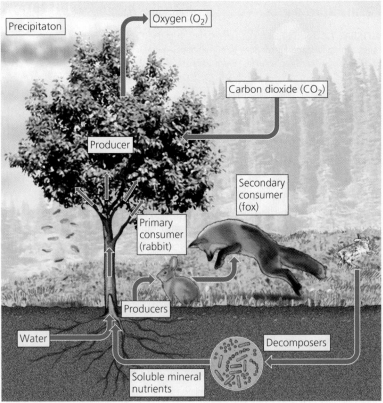

● **FIGURE 1.10** Ecosystems are an important aspect of natural environments, which are affected by the interaction of many processes and components.
Miller and Spoolman, *Living in the Environment.*

*How do ecosystems illustrate the interactions in the environment?*

shore, or a desert mountain can be viewed as an ecosystem. Ecosystems exist wherever there is an exchange of materials among living organisms and functional relationships between organisms and their natural surroundings. Certain ecosystems, such as a lake or a desert oasis, may have relatively clear-cut boundaries, but the limits of many others are not so precisely defined. Typically, the change from one ecosystem to another is transitional, occurring gradually over distance.

Physical geographers are keenly interested in environmental processes and interactions, and they consider relationships that involve humans and their activities. Much of human existence through time has been based on the adaptations people have made and the modifications they have imposed on their natural surroundings. Human activities will always affect the environment in some way, but if we understand the factors and processes involved, we can work to minimize the negative impacts.

**A Life-Support System** Earth's most critical characteristic is that it is a **life-support system.** On Earth, natural processes produce an adequate supply of oxygen; the sun interacts with the atmosphere, oceans, and land to maintain tolerable temperatures; and photosynthesis or other processes provide food supplies for living things. If a critical part of a life-support system is significantly changed or fails to operate properly, living organisms may no longer be able to survive.

About 45 years ago, Buckminster Fuller, a distinguished scientist and inventor, coined the notion of Spaceship Earth— the idea that our planet is a life-support system, transporting us through space. Spacecraft can provide a life-support system for astronauts, but they are dependent on Earth for sustenance, maintenance, and supplies of necessities ( ● Fig. 1.11). The Earth system, however, provides all the environmental constituents and conditions to support life. The only external input Earth requires is energy from the sun. Fuller said that knowing how Earth works may be required for human survival—but humans are only slowly learning the processes involved. He compared this information to the contents of an automobile owner's manual.

> If you own an automobile, you realize that you must put oil and gas into it, and you must put water in the radiator and take care of the car as a whole. You know that you are going to have to keep the machine in good order or it's going to be in trouble and fail to function.
>
> We have not been seeing our Spaceship Earth as an integrally designed machine which to be persistently successful must be comprehended and serviced in total ... there is one outstandingly important fact regarding Spaceship Earth, and that is that no instruction book came with it.
>
> R. Buckminster Fuller, *Operating Manual for Spaceship Earth*

**Sustainable Living** Today we realize that critical parts of our planet's life-support system, **natural resources,** can be abused, wasted, or exhausted, potentially threatening Earth's ability to support human life. We are rapidly depleting nonrenewable natural resources, like coal and oil, which, once

● **FIGURE 1.11** The International Space Station can function as a life-support system and astronauts can venture out on a spacewalk, but they remain dependent on resources such as air, food, and water that are shipped in from Earth.

*What do the limited resources on space vehicles suggest about our environmental situation on Earth?*

exhausted, will not be replaced. Other resources have natural limits on regeneration and are undergoing levels of human use, impact, or reduction, which on a worldwide basis currently exceed the natural productivity of Earth's environments. These overtaxed resources include forests, agricultural soils, water resources, and food sources from the oceans. This problem of using more resources in a year than their annual renewal, growth, or replacement is called **environmental overshoot.**

Ultimately, over the long term, the size of the human population cannot continually exceed the environmental resources necessary it requires for sustenance. The world population is at 7 billion, and United Nations' estimates indicate a population of more than 9 billion people by 2050 if current growth rates continue. More than half the world's people now tolerate substandard living conditions and insufficient food. It is estimated that it would take 1.5 planet Earths to continually support the world's population at current levels of human resource consumption ( ● Fig. 1.12). It is obvious that this trend of overshoot, which continues to rise, cannot go on forever.

## GEOGRAPHY'S ENVIRONMENTAL PERSPECTIVE

# Human–Environment Interactions

Interactions between humans and the Earth's environments work both ways and can be beneficial or detrimental. The effects of natural processes on humans and the impacts of humans on the environment have become topics of increasing concern. Certain environmental processes can be hazardous to human life and property, and certain human activities threaten to cause major, and possibly irrevocable, damage to Earth environments.

## Environmental Hazards

The environment becomes a hazard to people and other living things when, occasionally and often unpredictably, a natural process operates in an unusually intense or violent fashion. Molten rock and gases move upward toward the surface and suddenly trigger massive eruptions that can blow apart volcanic mountains. The powerful earthquake that hit Japan in 2011 and the massive tsunami wave it generated are examples of the potential for occurrences of natural processes to far exceed our expectable "norm," often with devastating results. Rain showers may become torrential rains that occur for days or weeks and cause flooding. Some tropical storms gain strength and reach coastlines with great intensity, such as Hurricane Katrina in 2005.

In September 2008, after Hurricane Ike became a powerful storm in the Atlantic Ocean, it passed over several islands in the Caribbean Sea, causing great damage, and continued into the Gulf of Mexico. Moving northwest, Ike made landfall near Galveston, Texas, a coastal city that had been rebuilt after being almost completely destroyed by a hurricane in 1900. Ike brought violent winds, high waves, and a massive 4.5- to 6.5-meter high (15–22 ft) surge of

U.S. Navy photo/Alexander Tidd

Environmental hazards: a tsunami's destructive force. This aerial view shows the virtually total devastation of the port town of Wakuya, Japan, after being battered by powerful tsunami waves that were generated by a massive earthquake in 2011. The earthquake was among the largest ever recorded.

seawater that swept low-lying coastal areas for several kilometers inland.

When a natural process operates in an extraordinary fashion it is a noteworthy environmental event, but it is not considered to be a natural hazard unless people or their properties are affected. Many natural hazards exist because people live where po-

tentially catastrophic environmental events may occur. Nearly every populated area of the world is associated with a natural hazard or perhaps several hazards. Forested regions are subject to fire; earthquakes, landslides, and volcanic activities plague mountain regions; violent storms threaten interior plains; and many coastal regions experience periodic hurricanes or severe winter storms.

## Human Impacts

Just as natural processes can pose a potential danger to people, humans and their activities can constitute a serious threat to the environment. Issues such as global warming, acid precipitation, deforestation and the extinction of biological species, damage to the atmospheric ozone layer, and desertification have been given high priority in the agendas at international conferences and when world leaders meet. Environmental concerns are recurring subjects in the news, on television, in books, and on the Internet.

Much environmental damage has resulted from atmospheric pollution associated with industrialization, particularly in support of wealthy, developed nations. But as population pressures mount and developing nations struggle to industrialize, human activities are exacting an increasing toll on the air, water, soils, and forests. Environmental deterioration is a problem of worldwide concern, and effective solutions must involve international cooperation.

Examining environmental issues from a physical geographic perspective requires that characteristics of the environment and the humans involved in those issues be given strong consideration. Physical environments are changing constantly and

human activities can often result in negative environmental consequences. In addition, worldwide, many people live in constant threat from earthquakes, fires, floods, and storms—environmental hazards that vary in their geographic distribution. These natural processes are a part of the physical environment, but the causes of problems and their solutions are involved with human–environmental interactions that include the economic, political, and social characteristics of the people who are affected. Geography's holistic approach is beneficial to understanding these issues because it takes both natural and human factors into consideration in dealing with environmental concerns.

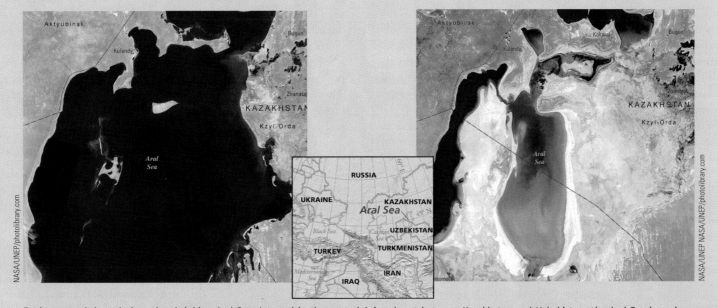

Environmental degradation: the shrinking Aral Sea. Located in the central Asian desert between Kazakhstan and Uzbekistan, the Aral Sea is an inland lake that does not have an outlet stream. Water that flows in is eventually lost by evaporation to the air. Before the 1960s, rivers flowing into the lake supplied enough water to maintain what was the fourth largest body of inland water in the world. Since that time, agricultural diversion of river water has caused the Aral Sea to dramatically shrink. The result has been the disappearance of many species that relied on the lake for survival, frequent dust storms, and an economic disaster for the local economy. Without the waters of the lake to moderate temperatures, the winters have become colder and the summers are hotter. Today, efforts are under way to restore at least part of the lake and its environments.

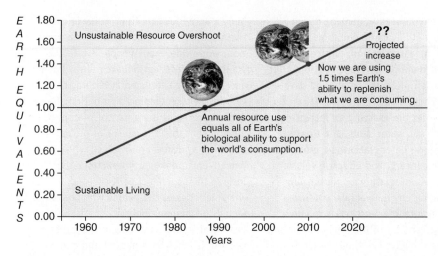

**FIGURE 1.12** Environmental overshoot. Based on human use of resources today, it would take 1.5 Earths to generate an adequate supply to meet our current level of consuming natural resources. This means that we are using 50% more than our planet can supply and sustain over the long term.
*Source:* Modified with permission from Worldwide Fund for Nature, Global Footprint Network. http://www.footprintnetwork.org/en/index.php/GFN/page/world_footprint

*What can we do about this situation that will help future generations?*

In recent years, there have been increasing movements by governments and environmental agencies toward encouraging **sustainable development.** The United Nations provided the following definition: *"Sustainable Development is development that meets the needs of the present without compromising the ability of future generations to meet their own needs."* What this means is that it will be impossible for humans to use, abuse, or destroy natural resources at their current rate, and it is irresponsible for humans to exhaust the resources that will be needed by our descendents. We should understand the impact of our individual and collective actions on the complex environmental systems of our planet. Geography has much to offer in understanding the factors involved in meeting this responsibility and in helping us learn more about environmental changes that are associated with human activities. We have the responsibility of helping to maintain our present and future habitat—the Earth System.

**Human–Environment Interactions** As we consider sustaining acceptable living standards for generations to come, it is essential to realize that environments do not change their nature to accommodate humans. Interactions between humans and environments are two-way relationships because the environment influences human behavior and humans affect the environment. The capability of different environments to adequately support a human population or absorb human impact varies widely.

Not all of the land on Earth is suitable as living space; thus usable land is yet another limited resource ( ● Fig. 1.13). The geographic distribution of population densities around the world, varying from uninhabited to dense settlements, reflects this disparity. As the world population grows, productive lands continue to diminish. Crops are sometimes grown in areas that are poorly suited to agriculture, whereas prime agricultural lands are built upon or paved over for other uses. We must use our lands wisely.

**Natural Hazards** In our search for livable space, we may live and work in locations that are subject to the potentially hazardous action of natural processes. News reports depict disasters as people are exposed to violent natural processes, such as floods and tornadoes. In 2011, we witnessed the consequences of an earthquake in New Zealand and another in Japan, which generated a devastating tsunami wave. The term **natural hazards** refers to natural processes, typically of unusual intensity, that put environments and human life or property at risk of damage or destruction (for example, constructing buildings in areas subject to flooding, severe storms, coastal erosion, landslides, or earthquakes). Natural processes, such as the peak flow of a river or the power of an earthquake, may vary considerably in intensity over time. Occasionally an extremely large flood may occur, or a particularly intense hurricane, tornado, or storm hits, or a very severe earthquake strikes. These are naturally occurring processes, but because a long time may transpire between severe events, governments and people may be ill prepared to adequately deal with the consequences. It is important to be aware of what natural hazards may affect the area where you live and how to either avoid or respond to their potential effects. Many geographers are interested in the factors involved with natural hazards, along with the potential intensity and frequency that a place might experience. They also work to find, classify, and map hazard zones, with recommendations about how to be best prepared for safeguarding life and property in an emergency resulting from a natural hazard.

**Human Impacts on the Environment** Human activities can have negative impacts on environmental conditions. One example is **pollution,** an undesirable or unhealthy environmental contamination. Critical resources, such as air, water, and even

● **FIGURE 1.13** The percentages of land and water areas on Earth. Habitable land is a limited resource on our planet.

*What options do we have for future settlement of Earth's lands?*

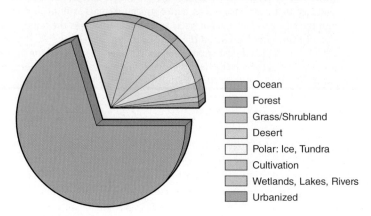

- Ocean
- Forest
- Grass/Shrubland
- Desert
- Polar: Ice, Tundra
- Cultivation
- Wetlands, Lakes, Rivers
- Urbanized

MODIS Rapid Response Team, NASA Goddard Space Flight Center

● **FIGURE 1.14** Smog over eastern China is carried to the Korean peninsula (right) and beyond. Occasionally winds carry air pollution from this region across the Pacific to North America.

land areas, can be polluted to the point where they become unusable or even lethal to some life forms. By polluting the oceans, we may be killing off important fish species, which may allow less desirable species to increase in number. Acid rain, caused by pollutants from industries and power plants, is damaging forests and killing fish in freshwater lakes. Air pollution has become a serious environmental problem for urban centers throughout the world. What some people do not realize, however, is that pollutants are often transported by winds and waterways hundreds or even thousands of kilometers from their source ( ● Fig. 1.14). Lead from automobile exhausts has been found in the ice of Antarctica, as has the insecticide DDT. Pollution is a worldwide problem that does not stop at political, or even continental, boundaries.

In modern times, the ability of humans to alter the landscape has been increasing. For example, a century ago the interconnected Kissimmee River–Lake Okeechobee–Everglades ecosystem was one of the most productive wetland regions in the world, but marshlands and slow-moving water stood in the way of urban and agricultural development. Intricate systems of ditches and canals were built, and since 1900 half of the original 1.6 million hectares (4 million acres) of the Everglades have disappeared. The Kissimmee River was channelized into an arrow-straight ditch, and wetlands were drained ( ● Fig. 1.15a). Levees have prevented water in

Lake Okeechobee from flowing to the Everglades, and highway construction disrupted the natural drainage patterns.

Fires have been more frequent and more destructive (Fig. 1.15b), and biotic communities have been eliminated by lowered water levels. When the area receives too much rain, portions of the Everglades are deliberately flooded to prevent drainage canals from overflowing. As a result, animals drown and birds cannot rest and reproduce. South Florida's wading bird population has decreased by 95% in the past hundred years. Without the natural purifying effects of wetland systems, water quality in south Florida has deteriorated; with lower water levels, saltwater encroachment is a serious problem in coastal areas.

Government agencies and scientists continue to work on restoring south Florida's ecosystems. The goals are to allow the Kissimmee River to flow naturally across its former flood plain (Fig. 1.15c), to return agricultural land to wetlands, and to restore water flow through the Everglades. The problems of south Florida should serve as a useful lesson. Alterations of the natural environment should not be undertaken without serious consideration of the potential consequences.

Poised at the interface between Earth and human existence, geography has much to offer in helping us understand how environments affect our lives and how our ways of living affect the

● **FIGURE 1.15** The Kissimmee River originally flowed in sweeping bends on its floodplain for 100 miles from Lake Kissimmee to Lake Okeechobee. (a) In the 1960s and early 1970s, the Kissimmee River was artificially straightened, disrupting the previously existing ecosystem at the expense of plants, animals, and human water supplies. (b) An ongoing problem is the invasion of weedy plants that causes a serious fire hazard during the dry season. Controlled burns by the U.S. Fish and Wildlife Department are necessary to avoid catastrophic fires and to help restore the natural vegetation. (c) After much work on habitat restoration that continues today, the Kissimmee is re-establishing its wetland environments and the form of its natural channel.

*What factors should be considered prior to any attempts to return rivers and wetland habitats to their original condition?*

environment. Scientific monitoring helps us learn more about environmental changes associated with human activities and what we should do to minimize or eliminate problems. People should make greater attempts to alter their behaviors and consumption of resources to accommodate the limitations, risks, and potentials of Earth's environments.

# Models and Systems

As they work to explain the features of Earth and its environments, geographers support these efforts, as other scientists do, by developing representations of the real world called *models*. A **model** is a useful simplification of a more complex reality that permits prediction, and every model is designed with a specific purpose in mind. As examples, maps and globes are models—simplified representations that provide us with useful information. Today many models are computer generated because computers can handle great amounts of data and perform the mathematical calculations that are often necessary to construct and display certain types of information.

There are many kinds of models ( ● Fig. 1.16). **Pictorial/graphic models** include pictures, maps, graphs, diagrams, drawings, and computer-generated visualizations. **Physical models** are solid three-dimensional representations, such as a world globe or a replica of a mountain or stream. **Mathematical/statistical models** are used to understand processes and predict possibilities such as river floods or the influence of climate change on daily weather. Words, language, and the definitions of terms or ideas also serve as models.

Another important type is a **conceptual model**—the mind imagery that we use for understanding our surroundings and experiences. Focus for a moment on the image that the word *mountain* (or *waterfall, cloud, tornado, beach, forest, desert*) generates in your mind. Most likely what you "see" (conceptualize) in your mind is sketchy rather than detailed, but enough information is there to convey a mental idea of the feature. This image is a conceptual model. For geographers, a particularly important conceptual model is the **mental map,** which we use to think about places, travel routes, and the distribution of features in space. How could we even begin to understand our world without conceptual models and, in terms of spatial understanding, without mental maps?

## Systems Analysis

If you try to think about Earth in its entirety, or to understand how a part of the Earth system works, often you will discover that there are just too many factors to envision. Our planet is too complex to permit a single model to explain all of its environmental components and how they affect one another. Yet it is often said that to be responsible citizens of Earth, we should "think globally, but act locally." To begin to comprehend Earth as a whole or to understand most of its environmental components, physical geographers use a powerful strategy for analysis called **systems analysis.** Systems analysis suggests that the way to understand how anything works is to use the following strategy:

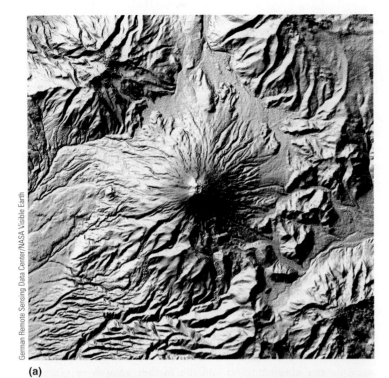

(a)

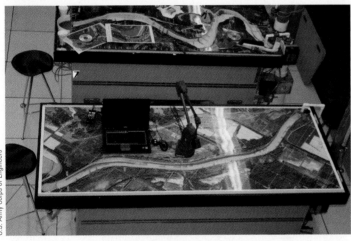

(b)

● **FIGURE 1.16** Models help us understand Earth and its subsystems by focusing our attention on major features or processes.

(a) A computer-generated elevation model of Cotopaxi, a large volcano in the Andes Mountains of Ecuador. The terrain and the drainage pattern of stream channels are well illustrated in this model. (b) This table-sized physical model of part of the Illinois River uses a channel and water flow to understand the behavior of the river under various conditions.

1. Clearly define the system that you are studying.
   *What are the boundaries (limits) of the system?*
2. Break the defined system down into its component parts (variables). The variables in a system are either matter or energy.
   *What important parts and processes are involved in this system?*
3. Attempt to understand how these variables are related to (or affect, react with, or have an impact on) one another.
   *How do the parts interact with one another to make the system work? What will happen in the system if a part changes?*

The human body provides an example of a system. ● Figure 1.17 shows a systems model where you can trace the movement of energy and matter into the system **(inputs)** and their movements out of the system **(outputs),** the storage of energy and matter, and interactions within the system. Of course, the human body is also composed of many subsystems (for example, the respiratory system, circulatory system, and digestive system). Subsystems can also be divided into subsystems and so on.

The systems approach is a beneficial strategy for studying environmental conditions on Earth. Subsystems, the interacting divisions of the Earth system, are also important to consider. The atmosphere, hydrosphere, lithosphere, and biosphere each function as a subsystem of Earth. The Earth system and its four major subsystems are typically further divided into smaller subsystems to focus attention on understanding a particular part of the whole. Examples of subsystems examined by physical geographers include the water cycle, climatic systems, storm systems, stream systems, the systematic heating of the atmosphere, and ecosystems. A great advantage of systems analysis is that it can be applied to environments at virtually any spatial scale from global to microscopic.

## How Systems Work

The Earth and its subsystems "work" by the movement (or transfer) of matter and energy and the processes involved with these transfers. For example, as shown in ● Figure 1.18, sunlight *(energy)* warms *(process)* a body of water *(matter)* and the water evaporates *(process)* into the atmosphere. Later, the water condenses *(process)* back into a liquid and the rain *(matter)* falls *(process)* on the land and runs off *(process)* downslope back to the sea. In a systems model, geographers can trace the input of energy or matter into the system, their storage in the system, their output from the system, and the interactions between components within the system. There are two basic kinds of systems.

Most Earth subsystems are **open systems** ( ● Figure 1.19a) because both energy and matter move freely across subsystem boundaries as inputs and outputs. A stream system like the one shown in Figure 1.18 is an excellent illustration of an open subsystem: Matter and energy in the form of soil particles, rock fragments, solar energy, and precipitation enter the stream, and water and sediments leave the stream where it empties into the ocean or some other body of water.

A **closed system** (Figure 1.19b) is one in which no substantial amount of *matter* crosses its boundaries, although *energy* can go in and out. Planet Earth is essentially a closed system. Except for meteorites that reach Earth's surface, the escape of atmospheric gas molecules to space, and a few moon rocks brought back by astronauts, the Earth system is essentially closed to the input or output of matter.

When we describe Earth as a system or as a set of interrelated systems, we are using conceptual models to help us organize our thinking about what we are observing. Throughout the following chapters, we will use the systems concept, and many other kinds of models, to simplify and illustrate complex features of the physical environment.

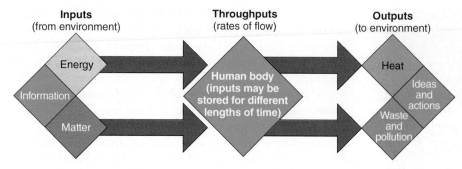

**Inputs** (from environment)

Energy

Information

Matter

**Throughputs** (rates of flow)

Human body (inputs may be stored for different lengths of time)

**Outputs** (to environment)

Heat

Ideas and actions

Waste and pollution

● **FIGURE 1.17** The human body is an example of a system, with inputs, storage, and outputs of energy and matter. Interactions between energy and matter drive the system.

*What characteristics of the human body as a system are similar to the Earth as a system?*

## Equilibrium in Earth Systems

We often hear about the "balance of nature." What this means is that natural systems have built-in mechanisms that tend to counterbalance, or accommodate, change without affecting the system dramatically. If the inputs entering the system are balanced by outputs, the system is said to have reached a state of **equilibrium.** Most systems are continually shifting slightly one way or another as they react to external conditions. This change within a range of tolerance is called **dynamic equilibrium**—that is, the balance is not static but in the long-term changes may be accumulating. A reservoir contained by a dam is a good example of dynamic equilibrium in a system ( ● Fig. 1.20).

The interactions that cause changes or adjustments between parts of a system are called **feedback.** Two kinds of feedback relationships operate in a system. **Negative feedback,** in which one change tends to offset another (an inverse relationship), creates a counteracting effect that is generally beneficial because it tends to maintain equilibrium in a system. Systems can also exhibit **positive feedback** sequences for a while—that is, changes that reinforce the direction of an initial change (a direct relationship). For example, several times in the past 2.6 million years Earth has experienced significant decreases in global temperatures. This cooling of the atmospheric system led to the growth of great glacial ice sheets that covered large areas. The highly reflective surfaces of these massive ice sheets increased the reflection of solar energy off of Earth and back to space, thus increasing the cooling trend and the further growth of glaciers. The result over a considerable period was positive feedback. Ultimately, however, the climate became so cold that evapora-

● **FIGURE 1.18** The water cycle provides examples of interactions between energy and matter, their storage in the system, and the processes involved. Being aware of energy and matter and the interactive processes that link them is important to understanding how environmental systems operate.

*Can you think of another environmental system and break it down into its components of energy, matter, storage in the system, and processes?*

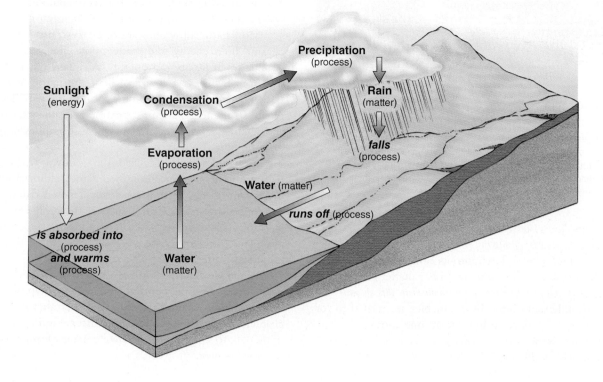

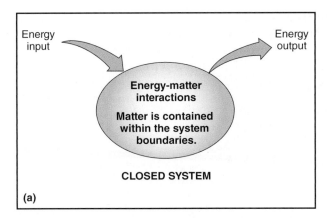

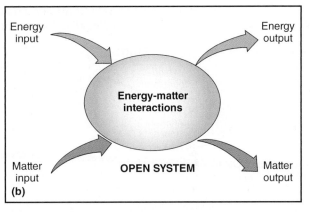

● **FIGURE 1.19** Two basic types of systems. (a) Earth's subsystems are open systems. Open systems involve the inputs and outputs of both energy and matter. (b) Earth is basically a closed system. Closed systems allow only energy to pass in and out. Solar energy (input) enters the Earth system, and that energy is dissipated (output) to space mainly as heat. External inputs of matter are virtually nil. Because Earth is a closed system, humans face limits to their available natural resources.

*What kind of system is illustrated in Fig. 1.18?*

tion from the oceans decreased and the cover of sea ice expanded, cutting off the supply of moisture to storms that fed the snow, which supported growth of the glaciers. The conditions that caused a reduction of snowfall is an example of what is called a **threshold,** a condition that causes a system to change dramatically, in this case bringing the positive feedback to a halt or completely reversing the effect of the feedback. A decrease in snowfall caused the glaciers to shrink and less solar energy was reflected back to space, so the climate began to warm, thus beginning another cycle.

● **FIGURE 1.20** A reservoir serves as an example of dynamic equilibrium in systems. The amount of water coming in may increase or decrease over time, but it must equal the water going out, or the level of the lake will rise or fall. If the input–output balance is not maintained, the reservoir will get larger or smaller as the system adjusts to storing more or less water. A state of equilibrium (balance) will exist between inputs, outputs, and storage in the system.

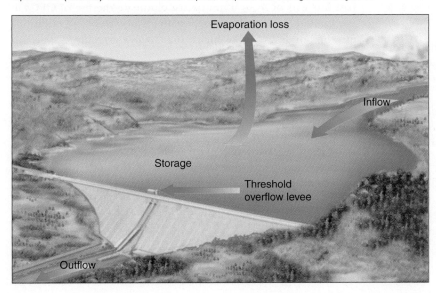

Thresholds are important regulators of systems processes. Threshold conditions, if reached or exceeded (or not met), can cause a fundamental change in a system and the way that it behaves. For example, earthquakes will not occur until the built-up stress reaches a threshold level that overcomes the strength of the rocks to resist breaking. As another example, fertilizing a plant will help it to grow larger and faster, but if more and more fertilizer is added, will this positive feedback relationship continue forever? Too much fertilizer may actually reach a threshold level, poisoning the plant. Either exceeding or not meeting certain critical conditions (thresholds) can change a system dramatically. With environmental systems, an important question that we often try to answer is how much change a system can tolerate without becoming drastically or irreversibly altered, particularly if the change has negative consequences.

Systems analysis allows us to see how these processes will affect the variables and helps us answer "what if?" questions. For example, if chlorofluorocarbons (CFCs) continue to deplete the ozone layer, what will happen? ● Figure 1.21 shows a **feedback loop**—a circular set of feedback operations that can be repeated as a cycle. Generally in natural systems, the eventual result of a feedback loop is negative feedback because the sequence of changes tends to reverse the direction of change in the initial element. Figure 1.21 also illustrates six of the most important factors related to damage to the ozone layer by CFCs. Each of these factors is linked by a feedback interaction to the next variable in the loop. To illustrate a feedback loop, consider a simplified example—a hypothetical scenario of what might happen if human-caused damage to the atmosphere's ozone layer continues unimpeded. The example will show you how

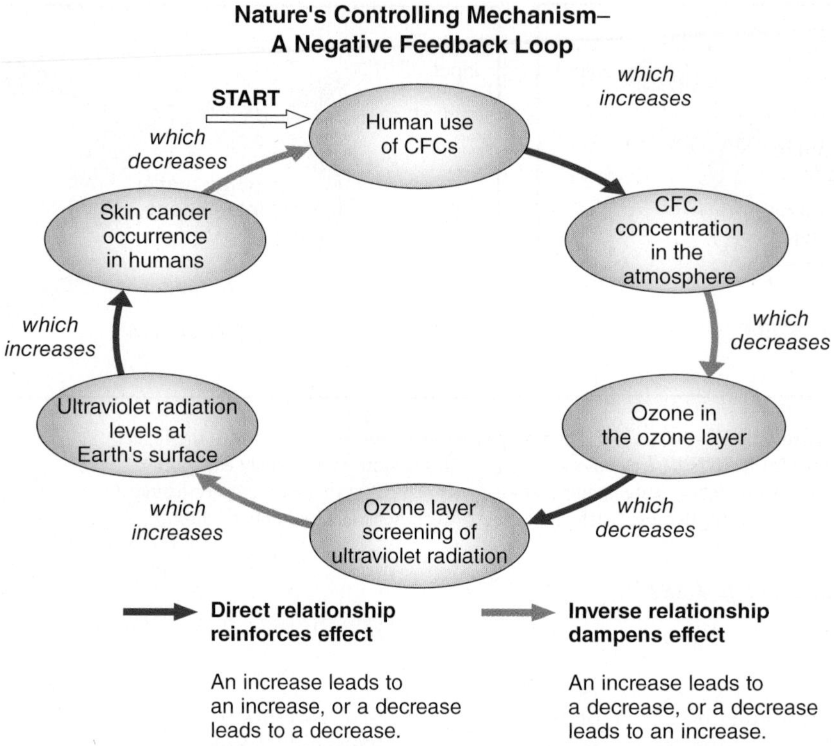

## Nature's Controlling Mechanism–
## A Negative Feedback Loop

● **FIGURE 1.21** A feedback loop illustrates how negative feedback tends to maintain system equilibrium. This example shows relationships between the ozone layer, which screens out harmful (cancer-related) ultraviolet (UV) energy from the sun, chlorofluorocarbons (CFCs), and potential impacts on life on Earth. CFCs have been used in air conditioning/ refrigeration systems and, if leaked to the atmosphere, they cause ozone depletion. A direct (positive feedback) relationship means that either an increase or a decrease in the first variable will lead to the same effect on the next. Inverse (negative feedback) relationships mean that a change in a variable will cause an opposite change in the next. After one pass through the negative-feedback loop, all subsequent changes in the next cycle will be reversed. A second feedback loop pass (reversing each increase or decrease interaction) illustrates how this works. The last link between skin cancer and human use of CFCs would likely result in people acting to reduce the problem.

*What might be the potential (extreme) alternative resulting from a lack of corrective action by humans?*

to think about Earth processes and interactions functioning as a system. First, consider some facts:

1. We know that the ozone layer in the upper atmosphere protects us by blocking harmful ultraviolet (UV) radiation from space, radiation that could cause harmful skin cancers and cell mutations.
2. We also know that CFCs, and some related chemicals that have been widely used in air conditioners, can migrate to the upper atmosphere and cause chemical reactions that destroy ozone.

Follow Figure 1.21 starting with the human use of CFCs at the top of the diagram, and trace the feedback links. Keep in mind that this systems example is simplified and presents an extreme scenario. In fact, efforts have been undertaken in the

past 30 years or so to minimize or eliminate the use of CFCs in the United States and internationally. Today new air conditioners for buildings and vehicles use "ozone-friendly," units for cooling.

Finally, some important questions remain: What is likely to happen to the amount of CFCs used if skin cancer continues to increase? Will humans act to correct the problem, or not? What would be the potential outcome in each case? Ironically, negative feedback loop operations are beneficial to the environment because they regulate a system through a tendency toward balance. Feedback loops in nature normally do not operate for extended periods on positive feedback because environmental limiting factors (thresholds) act to return the process to a state of equilibrium. What are some other examples of feedback operations in natural systems?

It is essential to remember that systems are models, so they are not the same as reality. They are products of the human mind and are only one way of looking at the real world. Examining various Earth subsystems helps us understand the natural processes involved in the development of the atmosphere, lithosphere, hydrosphere, and biosphere. Models also help us to simulate past events or predict future changes. But we must be careful not to confuse simplified models with the complexities of the real world.

## Physical Geography and You

The physical environment affects our everyday lives. The principles and perspectives of physical geography help us be environmentally aware, assess environmental situations, analyze the factors involved, and make informed decisions about courses of action. What are the environmental advantages and disadvantages of a particular home site? What season would be the best time to plant a tree? What environmental impacts might be expected from a proposed development? What potential natural hazards—flooding, landslides, earthquakes, hurricanes, and tornadoes—should you be aware of where you live? What can you do to minimize potential damage to your family, friends, and household from a natural hazard?

You may have wondered what people with a background in physical geography do in the workplace and what kinds of jobs do they hold. By applying their knowledge, skills, and techniques to real-world problems, physical geographers make major contributions to human well-being and to environmental stewardship. A recent publication about geography-related jobs by the U.S. Department of Labor stated that people in any career field that deals with maps, location, spatial data, or the environment would benefit from an educational background in geography ( ● Fig. 1.22).

Geography is a way of looking at the world and of observing its features. It involves asking questions about the nature of

© Ashley Cooper/Corbis

● **FIGURE 1.22** A geographer uses computer technology to work with data, maps, and imagery in the field.

*How is a geography background and an understanding of mapping and spatial concepts valuable in the workplace?*

those features and appreciating their beauty and complexity. It encourages you to seek explanations; gather information; and use geographic skills, tools, and knowledge to solve problems. Just as you see a painting differently after an art course, after this course you should see sunsets, waves, storms, deserts, valleys, rivers, forests, prairies, and mountains with a geographically "educated eye." You will see greater variety in the landscape because you will have been trained to observe Earth differently, with greater awareness and a deeper understanding.

# CHAPTER 1 ACTIVITIES

## ■ DEFINE AND RECALL

geography
spatial science
human geography
region
physical geography
holistic approach
climatology
geomorphology
biogeography
scientific method
cartography
absolute location
relative location
spatial distribution
spatial pattern
spatial interaction
system

Earth system
variable
subsystem
atmosphere
hydrosphere
lithosphere
biosphere
environment
ecology
ecosystem
life-support system
natural resource
environmental overshoot
sustainable development
natural hazards
pollution
model

pictorial/graphic model
physical model
mathematical/statistical model
conceptual model
mental map
systems analysis
input
output
open system
closed system
equilibrium
dynamic equilibrium
feedback
negative feedback
positive feedback
threshold
feedback loop

## ■ DISCUSS AND REVIEW

1. What are some of the subfields of physical geography, and what do geographers study in those areas of specialization?
2. Why is geography known as the spatial science? What are some topics in physical geography that illustrate the role of geography as the spatial science?
3. What does a holistic approach mean in terms of thinking about an environmental problem?
4. How do physical geography's three major perspectives make it unique among the sciences?
5. What are the four major divisions of the Earth system, and how do the divisions interact with one another?
6. How does the study of systems relate to the role of geography as a physical science?

7. How does the examination of human–environment relationships in ecosystems serve to illustrate the role of geography as an environmental science?
8. How do open and closed systems differ? Why would Earth be best described as a closed system but its subsystems are open systems?
9. What is a threshold in a system? What are some examples that relate to physical geography?
10. How does negative feedback maintain a tendency toward balance in a system?

## ■ CONSIDER AND RESPOND

1. Give examples from your local area that demonstrate each of the five topics listed concerning spatial science.
2. List some potential sources of pollution in your city or town. Could these kinds of pollution affect your life? What are some potential solutions to this problem?
3. Give one example of an environment in your local area that has been affected by human activity. In your opinion, was the change good or bad? What values are you using in making such a judgment?

4. There are advantages and disadvantages to the use of models and the study of systems by scientists. List and compare the advantages and disadvantages from the point of view of a physical geographer.
5. How can an understanding of physical geography be of value to you now and in the future? What steps should you take if you wish to seek employment as a physical geographer? What advantages might you have when applying for a job?

# APPLY AND LEARN

1. From memory, draw on paper the "mental map" that you envision when you think about the geography of your local natural environment (or neighborhood). Try to maintain some reasonable geographic (spatial) relationships for the sizes of areas and distances between features.

2. Draw a simple, circular feedback loop to illustrate the interactions between components and processes involved in some system that you are familiar with. Label the positive (direct) and negative (inverse) feedback relationships. What threshold conditions exist and how do they enact change?

 CourseMate – Make the most of your study time by accessing everything you need to succeed in one place. Read your textbook, take notes, review flashcards, watch videos, complete activities, take practice quizzes, and more online with CourseMate. Log in at **www.cengagebrain.com**.

# 2

# REPRESENTATIONS OF EARTH

A computer visualization based on actual satellite measurements of Earth's features and characteristics. Differences in ocean temperature are colorized and cloud heights are greatly exaggerated. R.B. Husar, Washington University; Data: NASA/ESA/NAVO/SSEC

## ■ OBJECTIVES

WHEN YOU COMPLETE THIS CHAPTER YOU SHOULD BE ABLE TO:

■ Explain the ways that Earth and its regions, places, and locations can be represented on a variety of visual media—maps, aerial photographs, and other imagery.

■ Assess the nature and importance of maps and maplike presentations of the planet, or parts of Earth, citing some examples.

■ Find and describe the locations of places using coordinate systems, use topographic maps to find elevations, and understand the three types of map scales.

■ Demonstrate knowledge of techniques that support geographic investigations, including mapping, spatial analysis, global positioning systems (GPS), geographic information systems (GIS), and remote sensing.

■ Evaluate the advantages and limitations of different kinds of representations of Earth and its areas.

■ Understand how the proper techniques, images, and maps can be used to best advantage in solving geographic problems.

PERHAPS AS SOON AS PEOPLE began to communicate with each other, they began to develop a language of location, using landscape features as directional cues. Today we still use landmarks to help us find our way, along with reading maps. The earliest known maps were drawn on rock surfaces, clay tablets, metal plates, papyrus, linen, or silk, or they were constructed of sticks ( ● Fig. 2.1). Ancient maps were a fundamental contribution to the beginnings of geography. Many of the basic principles for solving locational problems have been known for centuries, but the technologies applied to these tasks are rapidly improving and changing. Through history, maps have become increasingly more common as a result of the appearance of paper, followed by the printing press, the computer, and the Internet.

Digital and satellite technologies are widely used for locating, describing, storing, and accessing spatial data. Knowing where certain features are located and being able to convey that information to others is a first step toward describing and analyzing aspects of the Earth system.

Computer systems can be used to produce complex maps and three-dimensional displays of geographic features that would have been nearly impossible or extremely time-consuming to produce two decades ago. Geographers use these technologies to help them understand spatial relationships and for locational problem solving. Because maps are so commonly used to convey information, it is important to be able to read

● **FIGURE 2.1** In France cave paintings made between 17,000 and 35,000 years ago depict the migration routes of animals. This view shows detail of stags crossing a river, and experts suggest that the artwork represents a rudimentary map with marks that represent locational information. If so, this is the earliest known example of humans recording their spatial knowledge.

*Why would prehistoric humans want to record locational information?*

and interpret them correctly. An informed knowledge of spatial representation and the ability to communicate locational information is not only an integral part of physical geography, but it is also important in our daily lives.

# Maps and Location on Earth

*Cartography* is the science and profession of map making. Geographers who specialize in cartography design maps and globes to ensure that mapped information and data are accurate and effectively presented. Most cartographers would agree that the primary purpose of a map is to communicate spatial information. Maps and globes convey spatial information through graphic symbols, a "language of location," that must be understood to appreciate and comprehend the rich store of information that they display (see Appendix B).

Computer technology has revolutionized cartography. Maps that once had to be hand drawn ( ● Fig. 2.2) are now produced digitally and printed in a short amount of time. Computer-assisted

● **FIGURE 2.2** When maps had to be hand drawn, artistic talent was required in addition to knowledge of cartographic principles. Erwin Raisz, a famous and talented cartographer, drew this map of U.S. landforms in 1954.

*Are maps like this still valuable for learning about landscapes, or are they obsolete?*

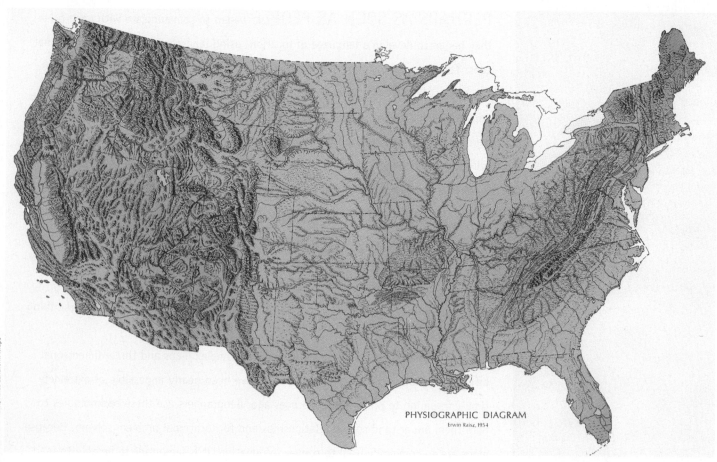

PHYSIOGRAPHIC DIAGRAM
Erwin Raisz, 1954

mapping allows easy map revision, which was a time-consuming process when maps were drawn by hand. Artistic talent is still an advantage in cartography, but it is most important for today's cartographers to be skilled in using computer-mapping systems and understand the principles of geography, cartography, and map design.

Cartographers can now gather spatial data and make maps faster than ever before—within hours—and the accuracy of these maps is excellent. Moreover, digital mapping enables mapmakers to experiment with a map's basic characteristics (for example, scale, projections), to combine and manipulate map data, to transmit entire maps electronically, and to produce unique maps on demand.

*Exploring Maps* United States Geological Survey (USGS)

Information that was once gathered little by little from ground observations and field surveys can now be collected instantly by satellites that flash recorded data back to Earth at the speed of light. Maps are an essential resource in navigation, political science, community planning, meteorology, surveying, history, geology, and many other career fields. Maps communicate important information for travel, recreation, education, the media, business, and entertainment.

Maps are ever-present in newspapers, on news and weather broadcasts, on the Internet, and, increasingly, on our cell phones and in vehicle navigation systems. How many maps do you see in a typical day? How many would that equal in a year? How do these maps affect your daily life?

## Earth's Shape and Size

We were first able to image our planet's shape from space in the 1960s, but Greek philosophers theorized that our planet was a sphere as early as 540 B.C. About 240 B.C. Eratosthenes, a philosopher with an interest in geography, not only understood that the Earth was round, but he also estimated the planet's circumference fairly accurately (how he accomplished this will be illustrated in the next chapter).

Mapping and describing global locations both require an understanding of our planet's form and features. Earth can generally be considered as a sphere, with an equatorial circumference of 39,840 kilometers (24,900 mi), but the centrifugal force caused by Earth's daily *rotation* bulges the equatorial region outward and slightly flattens the polar regions, forming an overall shape called an **oblate spheroid.** Yet, on a planetary scale, Earth's deviations from a true sphere are relatively minor. Earth's diameter at the *equator* is 12,758 kilometers (7927 mi), whereas from pole to pole it is 12,714 kilometers (7900 mi). On a 30.5-centimeter (12-in.) globe, this difference of 44 kilometers (27 mi) is about as thick as the wire in a paper clip. This variation from a spherical shape is less than one third of 1% and is not noticeable in views of Earth from space ( ● Fig. 2.3). Nevertheless, people working in very precise navigation, surveying, aeronautics, and cartography must consider Earth's deviations from a perfect sphere.

Landforms also cause deviations from true sphericity. Mount Everest in the Himalayas is the highest point on Earth at 8850 meters (29,035 ft) above sea level. The lowest point is the Chal-

● **FIGURE 2.3** Earth's spherical shape is clearly shown in this image. The flattening of the polar regions and the bulging of the equatorial region are too minor to be visible from space.

*Rettckli, Nazmi El Saleous, and Marit Jentoft-Nilsen/NASA GSFC*

*What does this suggest about the degree of "sphericity" of Earth?*

lenger Deep in the Mariana Trench of the Pacific Ocean, southwest of Guam, at 11,033 meters (36,200 ft) below sea level. The difference between these two elevations, 19,883 meters, or just more than 12 miles (19.2 km), would also be insignificant when reduced in scale on a 12-inch (30.5 cm) globe.

When ancient peoples began to sail the oceans, they recognized the need for ways of finding directions and describing locations. Long before the first compass was developed, humans understood that the positions of the sun and the stars—rising, setting, or circling in the sky—could provide locational information. Observing relationships between the sun and stars to find a position on Earth is a basic skill in **navigation,** the science of location and wayfinding. Navigation is basically the process of getting from where you are to where you want to go. High-tech navigational and mapping technologies are now in widespread public use on personal computers and hand-held systems that display locations and directions for use in hiking, traveling, and virtually any means of transportation. Today even our cell phones use these technologies, with the capability to display locations, routes, and maps.

## Globes and Great Circles

Because world globes have essentially the same spherical geometry as our planet, they represent geographic features and spatial relationships virtually without distortion. Globes correctly display the relative shapes, sizes, and comparative areas of Earth features, landforms, water bodies, and distances between locations, and they also preserve true compass directions. If we want to view the

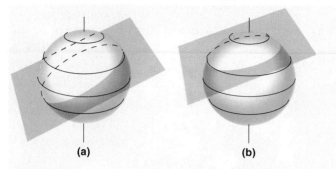

**(a)** **(b)**

● **FIGURE 2.4** Any imaginary geometric plane that passes through Earth's center, dividing it into two equal halves, forms a great circle where the plane intersects the planet's surface. This plane can be oriented in any direction as long as it defines two (equal) hemispheres (a). The plane shown in (b) slices the globe into unequal parts, so the line of intersection of such a plane with Earth is a small circle.

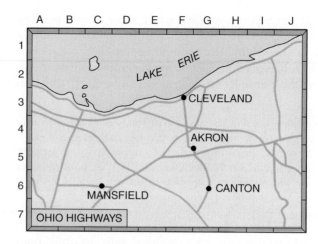

● **FIGURE 2.5** Using a simple rectangular coordinate system to locate a position. This map uses an alphanumeric location system, similar to that used on many road maps and campus maps.

*What are the rectangular coordinates of Mansfield? What is at location F-3?*

entire world, a globe provides the most accurate representation. Being familiar with the characteristics of a globe helps us to understand maps and how they are constructed.

A globe would not help us find our way on a hiking trail, however. It would be awkward to carry, and our location would appear as a tiny pinpoint with very little, if any, local information. We would need a *map* that clearly showed elevations, trails, and rivers and could be folded to easily carry it along.

An imaginary circle drawn in any direction on Earth's surface and whose plane passes through the center of Earth is a **great circle** ( ● Fig. 2.4a). It is called "great" because this is the largest circle that can be drawn around Earth while connecting any two points on the surface. Great circles divide Earth into equal halves called **hemispheres.** An important example of a great circle is the *circle of illumination,* which divides Earth into light and dark halves—a day hemisphere and a night hemisphere. Great circles are useful to navigation because following the trace of any great circle means taking shortest travel route between two locations on Earth's surface. Any circle on Earth's surface that does not divide the planet into equal halves is called a **small circle** (Fig. 2.4b). The planes of small circles do not pass through the Earth's center.

The shortest route between two places can be located by finding the great circle that connects them. You can put a rubber band (or string) around a globe to visualize this spatial relationship. Connect any two cities, such as Beijing and New York, San Francisco and Tokyo, New Orleans and Paris, or Kansas City and Moscow, by stretching a large rubber band around the globe so that it touches both cities and divides the globe in half. The rubber band then marks the shortest route between these two cities. Navigators chart *great circle routes* for aircraft and ships because traveling the shortest distance saves time and fuel. The farther away two points are on Earth, the greater the travel distance savings will be by following the great circle route that connects them.

## Latitude and Longitude

Imagine that you are traveling by car and you want to visit the Pro Football Hall of Fame in Canton, Ohio. Using the Ohio road map, you look up Canton in the map index and find that it is

located at "G-6." The letter G and the number 6 meet in a box marked on the map. In box G-6, you locate Canton ( ● Fig. 2.5). What you have used is a **coordinate system** of intersecting lines, a system of *grid cells* on the map. Without a locational coordinate system, it would be difficult to describe a location. A problem that a sphere presents, however, is deciding where the starting points should be for a grid system. Without reference points, either natural or arbitrary, a sphere is a geometric form that looks the same from any direction and has no natural beginning or end points. Earth's coordinate system of *latitude and longitude* is based on a set of reference lines that are naturally defined based on its planetary *rotation* and another set that was arbitrarily defined by international agreement.

**Measuring Latitude** The **North Pole** and the **South Pole** provide two natural reference points because they mark the opposite positions of Earth's *rotational* axis, around which it turns in 24 hours. The **equator,** halfway between the poles, forms a great circle that divides the planet into the Northern and Southern Hemispheres. The equator is the reference line for measuring **latitude** in degrees north or degrees south—the equator is 0° latitude. North or south of the equator, the angles and their arcs increase until we reach the North or South Pole at the maximum latitudes of 90° north or 90° south. A **sextant** can be used to determine latitude by celestial navigation ( ● Fig. 2.6). This instrument measures the angle between our *horizon,* the visual boundary line between the sky and Earth, and a celestial body, such as the noonday sun or the North Star (Polaris).

To find the latitude of Los Angeles, for example, imagine two lines that radiate out from the center of Earth. One goes straight to Los Angeles and the other goes to the equator at a point directly south of the city. These two lines form an angle that is the latitudinal distance (in degrees) that Los Angeles lies

© Bernie Bernard, TDI-Brooks International, Inc.

● **FIGURE 2.6** Finding latitude by celestial navigation. A traditional way to determine latitude is to measure the angle between the horizon and a celestial body with a sextant. Today the satellite-based global positioning system (GPS) supports most air and sea navigation (and land travel).

*With high-tech location systems like the GPS available, why would it still be important to know how to use a sextant?*

north of the equator ( ● Fig. 2.7a). The angle made by these two lines is just over 34°—so the latitude of Los Angeles is about 34°N (north of the equator). Because Earth's circumference is approximately 40,000 kilometers (25,000 mi) and there are 360 degrees in a circle, we can divide (40,000 km/360°) to find that 1° of latitude equals about 111 kilometers (69 mi).

A single degree of latitude covers a relatively large distance, so degrees are further divided into minutes (′) and seconds (″) of arc. There are 60 minutes of arc in a degree. Actually, Los Angeles is located at 34°03′N (34 degrees, 3 minutes north latitude). We can get even more precise: 1 minute is equal to 60 seconds of arc. We could locate a different position at latitude 23°34′ 12″S, which we would read as 23 degrees, 34 minutes, 12 seconds south latitude. A minute of latitude equals 1.85 kilometers (1.15 mi), and a second is about 31 meters (102 ft).

The latitude of a location, however, is only half of its global address. Los Angeles is located approximately 34° north of the equator, but an infinite number of points exist on the same line of latitude.

**Measuring Longitude** To accurately describe the location of Los Angeles, we must also determine where it is situated along the line of 34°N latitude. However, to describe an east or west position, we must have a starting line, just as the equator provides our reference for latitude. To find a location east or west, we use longitude lines, which run from pole to pole, each one forming half of a great circle. The global position of the 0° east–west line for longitude is arbitrary, and this is the grid reference line that was established by international agreement. The longitude line passing through Greenwich, England (near London) was accepted as the **prime meridian,** or 0° longitude, in 1884. **Longitude** is the angular distance east or west of the prime meridian.

Longitude is also measured in degrees, minutes, and seconds. Imagine a line drawn from the center of Earth to the point where the north–south running line of longitude that passes through Los Angeles crosses the equator. A second imaginary line will go

● **FIGURE 2.7** Finding a location by latitude and longitude. (a) The geometric basis for the latitude of Los Angeles, California. Latitude is the angular distance in degrees north or south of the equator. (b) The geometric basis for the longitude of Los Angeles. Longitude is the angular distance in degrees east or west of the prime meridian, which passes through Greenwich, England. (c) The location of Los Angeles is 34°N, 118°W.

*What is the latitude of the North Pole, and does it have a longitude?*

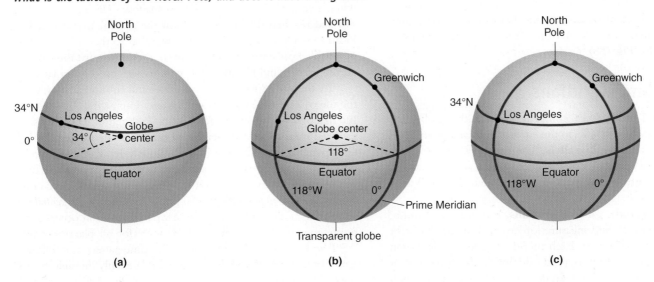

(a)                    (b)                    (c)

from the center of Earth to the point where the prime meridian crosses the equator (this location is 0°E or W and 0°N or S). Figure 2.7b shows that these two lines drawn from Earth's center define an angle, the arc of which is the angular distance that Los Angeles lies west of the prime meridian (118°W longitude). Figure 2.7c provides the location of Los Angeles by latitude and longitude.

Our longitude increases as we go farther east or west from 0° at the prime meridian. Traveling eastward from the prime meridian, we will eventually be halfway around the world from Greenwich, in the middle of the Pacific Ocean at 180°. Longitude is measured in degrees up to a maximum of 180° east or west of the prime meridian. Along the prime meridian (0°E–W) or the 180° meridian, the E–W designation does not matter, and along the equator (0°N–S), the N–S designation does not matter and is not needed for indicating location.

**Decimal Degrees** Instead of using minutes and seconds of arc, **decimal degrees** of longitude and latitude are also used to describe the location of a point. Decimal degrees, expressed to many decimal places, are very precise in pinpointing a location. Also, computer mapping systems (GPS, GIS, Google Earth) handle decimals much more readily than degrees, minutes, and seconds, and some systems require decimal locations. Many computer systems use decimal degrees without the N, S, E, W letters to indicate cardinal directions. Instead, north latitudes are designated as positive numbers, and south latitudes are designated as negative numbers (with a minus sign). For longitudes, east is positive and west is negative. For example, the Statue of Liberty in New York is located at 40.6894, −74.0447. Latitude is always listed first and the degree symbol (°) is generally not necessary. Some of the latitude/longitude coordinates found in this book are in decimal degrees with letters to indicate cardinal direction.

# The Geographic Grid

Every point on Earth's surface can be located by its latitude north or south of the equator in degrees and its longitude east or west of the prime meridian in degrees. Lines that run east and west around the globe to mark latitude and lines that run north and south from pole to pole to indicate longitude form the **geographic grid** ( ● Fig. 2.8).

## Parallels and Meridians

The east–west latitude lines completely circle the globe, are evenly spaced, and are parallel to the equator and each other. This is why they are known as **parallels.** The equator is the only parallel that is a great circle; all other lines of latitude are small circles. One degree of latitude equals about 111 kilometers (69 mi) anywhere on Earth.

Lines of longitude, called **meridians,** run north and south, converge at the poles, and measure longitudinal distances east or west of the prime meridian. Each meridian of longitude, when joined with its mate on the opposite side of Earth, forms a great circle. Meridians at any given latitude are evenly spaced, although

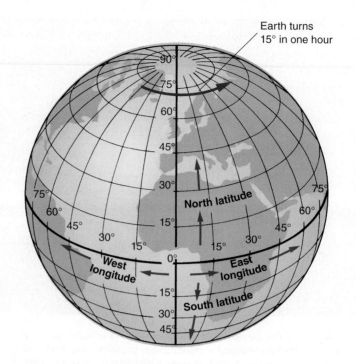

● **FIGURE 2.8** A globelike representation of Earth, which shows the geographic grid with parallels of latitude and meridians of longitude at 15° intervals.

*How do parallels and meridians differ?*

meridians get closer together as they move toward the poles and away from the equator. At the equator, meridians separated by 1° of longitude are about 111 kilometers (69 mi) apart, but at 60°N or 60°S latitude, they are only half that distance apart, about 56 kilometers (35 mi).

## Longitude and Time

The world's **time zones** were established based on the relationships between longitude, Earth's *rotation,* and time. Until about 125 years ago, each town or area used what was known as *local time.* **Solar noon** was determined by the precise moment in a day when a vertical stake cast its shortest shadow. This meant that the sun had reached its highest angle in the sky for that day at that location—noon—and local clocks were set to that time. Because of Earth's *rotation* on its axis, noon in a town toward the east occurred earlier, and towns to the west experienced noon later.

By the late 1800s advances in travel and communication made the use of local time by each community impractical. In 1884, the International Meridian Conference in Washington, D.C., set standardized time zones and established the longitude passing through Greenwich as the prime meridian (0° longitude). Earth was divided into 24 time zones, one for each hour in a day because Earth turns 15° of longitude in an hour (360° ÷ 24 hours). Ideally, each time zone spans 15° of longitude. The prime meridian is the *central meridian* of its time zone, and the time when solar noon occurs at the prime meridian was established as noon for all places between 7.5°E and 7.5°W of that meridian. The same pattern was followed around the world. Every line of longitude evenly divisible by 15° is the central meridian for a time zone extending 7.5° of longitude on either side. However, as shown in ● Figure 2.9, time zone boundaries do not follow meridians exactly. In the United States, time zone boundaries commonly follow state lines. It would be inconvenient and confusing to have a time zone boundary dividing a city or town into two time zones, so jogs in the lines were established to avoid most of these problems.

The time of day at the prime meridian, known as *Greenwich Mean Time* (GMT, but also called Universal Time, UTC, or Zulu Time), is used as a worldwide reference. Times to the east or west can be easily determined by comparing them with GMT. A place 90°E of the prime meridian would be 6 hours later (90° ÷ 15° per hour), whereas in the Pacific Time Zone of the United States and Canada, whose central meridian is 120°W, the time would be 8 hours earlier than GMT.

For navigation, longitude can be determined with a *chronometer,* an extremely accurate clock. Two chronometers are used, one set on GMT and the other on local time. The number of hours between them, earlier or later, determines longitude (1 hour = 15°

of longitude). Until the advent of electronic navigation by ground and satellite-based systems, the sextant and chronometer were a navigator's basic tools for determining location.

## The International Date Line

On the opposite side of Earth from the prime meridian is the **International Date Line.** It is a line that generally follows the 180th meridian, except for jogs to separate Alaska and Siberia and to skirt some Pacific island groups ( ● Fig. 2.10). At the International Date Line, we turn our calendar back a full day if we are traveling east and forward a full day if we are traveling west. Thus, if we are going east from Tokyo to San Francisco and it is 4:30 p.m. Monday just as we cross the International Date Line, it will be 4:30 p.m. Sunday on the other side. If we are traveling west from Alaska to Siberia and it is 10:00 a.m. Wednesday when we reach the International Date Line, it will be 10:00 a.m. Thursday once we cross it. As a way of remembering this relationship, many world maps

● **FIGURE 2.9** World time zones reflect the fact that Earth rotates through 15° of longitude each hour. Thus, time zones are approximately 15° wide. Political boundaries usually prevent the time zones from following a meridian perfectly.

*How many hours of difference are there between the time zone where you live and Greenwich, England, and is it earlier in England or later?*

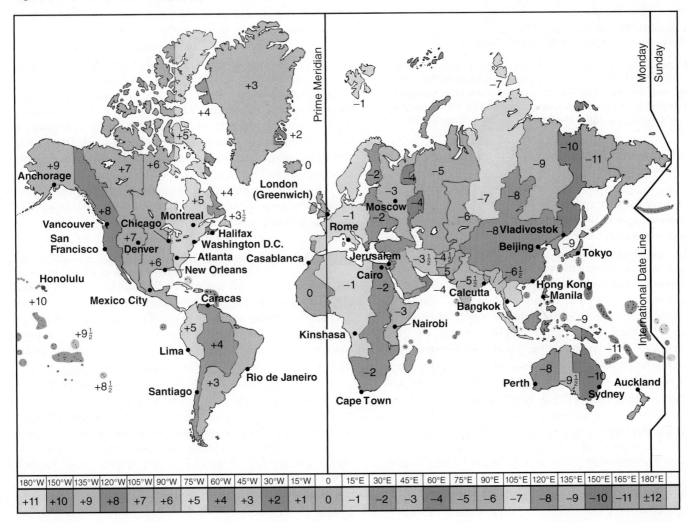

| 180°W | 150°W | 135°W | 120°W | 105°W | 90°W | 75°W | 60°W | 45°W | 30°W | 15°W | 0 | 15°E | 30°E | 45°E | 60°E | 75°E | 90°E | 105°E | 120°E | 135°E | 150°E | 165°E | 180°E |
|---|---|---|---|---|---|---|---|---|---|---|---|---|---|---|---|---|---|---|---|---|---|---|---|
| +11 | +10 | +9 | +8 | +7 | +6 | +5 | +4 | +3 | +2 | +1 | 0 | −1 | −2 | −3 | −4 | −5 | −6 | −7 | −8 | −9 | −10 | −11 | ±12 |

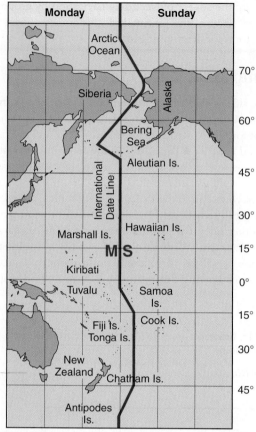

**Monday | Sunday**

135°E  150°E  165°E  180°  165°W  150°W  135°W

● **FIGURE 2.10** The International Date Line. The new day officially begins at the International Date Line (IDL) and then sweeps westward around the Earth to disappear when it again reaches the IDL. Maps and globes often have either "Monday | Sunday" or "M | S" shown on opposite sides of the line to indicate the direction of day change. This is the IDL as it is officially accepted by the United States.

*Why does the International Date Line deviate from the 180° meridian in some places?*

and globes have Monday and Sunday (M | S) labeled in that order on the opposite sides of the International Date Line. To find the correct day, you just substitute the current day for Monday or Sunday, and use the same relationship.

The International Date Line was not established officially until the 1880s, but the need for such a line on Earth to adjust the day was inadvertently discovered by Magellan's crew who, from 1519 to 1521, were the first to circumnavigate the world. Sailing westward from Spain when they returned from their voyage, the crew noticed that 1 day had apparently been missed in the ship's log. What actually happened is that in going around the world in a westward direction, the crew had experienced one less sunset and one less sunrise than had occurred in Spain during their absence.

## The U.S. Public Lands Survey System

The longitude and latitude system was designed to locate the *points* where those lines intersect. A different system is used in much of the United States to define and locate land *areas.* This is

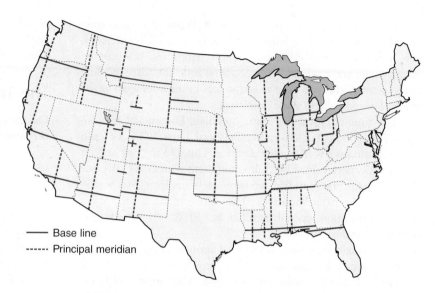

● **FIGURE 2.11** Principal meridians and base lines of the U.S. Public Lands Survey System (Township and Range System).

*Why wasn't the Township and Range System applied throughout the eastern United States?*

the **U.S. Public Lands Survey System,** or the *Township and Range System,* developed for parceling public lands west of Pennsylvania. Lands in the eastern United States had already been surveyed into irregular parcels at the time this system was established. The Township and Range System divides land areas into parcels based on north–south lines called **principal meridians** and east–west lines called **base lines** ( ● Fig. 2.11). Base lines were surveyed along parallels of latitude. The north–south meridians, although perpendicular to the base lines, had to be adjusted (jogged) along their length to accommodate Earth's curvature. If these adjustments were not made, the north–south lines would tend to converge and land parcels defined by this system would be smaller in northern regions of the United States.

The Township and Range System forms a grid of nearly square parcels called *townships* laid out in horizontal *tiers* north and south of the base lines and in vertical *columns* ranging east and west of the principal meridians. A **township** is a square plot 6 miles on a side (36 sq mi, or 93 sq km). As illustrated in ● Figure 2.12, townships are first labeled by their position north or south of a base line; thus, a township in the third tier south of a base line will be labeled Township 3 South, which is abbreviated T3S. However, we must also name a township according to its *range*—its location east or west of the principal meridian for the survey area. Thus, if Township 3 South is in the second range east of the principal meridian, its full location can be given as T3S/R2E (Range 2 East).

The Public Lands Survey System divides townships into 36 **sections** of 1 square mile, or 640 acres (2.6 sq km, or 259 ha). Sections are designated by numbers from 1 to 36 beginning in the northeastern most section with section 1, snaking back and forth across the township, and ending in the southeast corner with section 36. Sections are divided into four *quarter sections,* named by their location within the section—northeast, northwest, southeast, and southwest, each with 160 acres (65 ha). Quarter sections are also

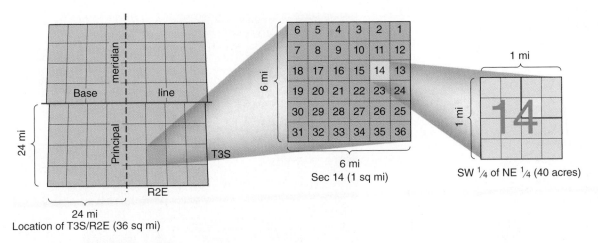

Location of T3S/R2E (36 sq mi)

**● FIGURE 2.12** The method of location for areas of land according to the Public Lands Survey System.

*How would you describe the extreme southeastern 40 acres of section 20 in the middle diagram?*

subdivided into four *quarter-quarter sections*, sometimes known as *forties*, each with an area of 40 acres (16.25 ha). These quarter-quarter sections, or 40-acre plots, are also named after their position in the quarter: the northeast, northwest, southeast, and southwest forties. Thus, we can describe the location of the 40-acre tract that is shaded in Figure 2.12 as being in the SW ¼ of the NE ¼ of Sec. 14, T3S/R2E, which we can find if we locate the principal meridian and the base line. The order is consistent from smaller division to larger, and township location is always listed before range (T3S/R2E).

The Township and Range system has exerted an enormous influence on landscapes in many areas of the United States and gives most of the Midwest and West a checkerboard appearance from the air or from space (● Fig. 2.13). Road maps in states that use this survey system strongly reflect its grid, and many roads follow the regular and angular boundaries between square parcels of land.

## The Global Positioning System

The **Global Positioning System (GPS)** is a technology for determining a location on Earth. This high-tech system was originally created for military applications but today is being adapted to many public uses, from surveying to navigation. The global positioning system uses radio signals, transmitted by a network of satellites orbiting 17,700 kilometers (11,000 mi) above Earth (● Fig. 2.14). By accessing signals from several satellites, a GPS receiver calculates the distances from those satellites to its location on Earth. GPS is based on the principle of *triangulation*, which means that if we can find the distance to our position, measured from three or more different locations (in this case, satellites), we can determine our location. GPS receivers vary in size, and handheld units are common (● Fig. 2.15). Small GPS

**● FIGURE 2.13** Rectangular field patterns resulting from the Public Lands Survey System in the Midwest and western United States. Note the slight jog in the field pattern to the right of the farm buildings near the lower edge of the photo.

*How do you know this photo was not taken in the Midwestern United States?*

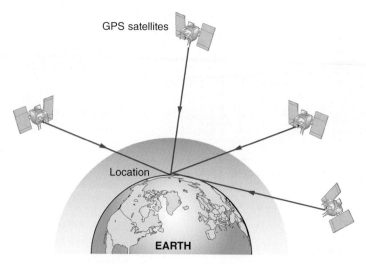

GPS satellites

Location

EARTH

● **FIGURE 2.14** The global positioning system (GPS) uses signals from a network of satellites to determine a position on Earth. A GPS receiver on the ground calculates the distances from several satellites (a minimum of three) to find its location by longitude, latitude, and elevation. With the distance from three satellites, a position can be located within meters, but with more satellite signals and sophisticated GPS equipment, the position can be located precisely.

● **FIGURE 2.15** A global positioning system receiver provides a readout of its latitudinal and longitudinal position. Small handheld units provide an accuracy that is acceptable for many uses, and many can also display locations on a map. This receiver was mounted on a motorcycle for navigation on a trip to Alaska; the latitude shown is at the Arctic Circle.

*What other uses can you think of for a small GPS unit like this that displays its longitude and latitude as it moves from place to place?*

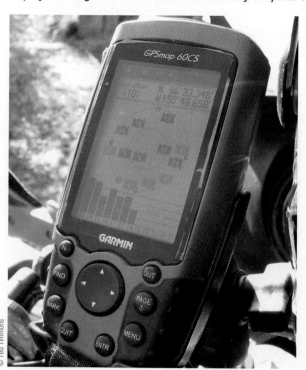

© Ted Timmons

NPS/Cape Cod National Seashore

● **FIGURE 2.16** An intern at Cape Cod National Seashore in Massachusetts helps with collecting accurate global positioning system locations to support studies of beach changes. Many geography students serve in summer intern positions with federal government agencies.

receivers are very useful to travelers, hikers, and backpackers who need to keep track of their location. GPS systems are integrated into cell phones so that a caller can be located in case of an emergency. The distances from a receiver to the satellites are calculated by measuring the time it takes for a satellite radio signal, broadcast at the speed of light, to arrive at the receiver. A GPS receiver performs these calculations and displays a locational readout in latitude, longitude, and elevation or on a map display. Map-based GPS systems—where GPS data is translated to a map display—not only are becoming popular for hikers, but also larger units are widely used in vehicles and on boats and aircraft. GPS systems also support map making from data gathered in the field. With sophisticated GPS equipment and techniques, it is possible to find and describe locational coordinates within small fractions of a meter ( ● Fig. 2.16).

# Maps and Map Projections

Maps are extremely versatile—they can be reproduced easily, can depict the entire Earth or show a small area in great detail, are easy to handle and transport, and can be displayed on a computer monitor. Yet, it is impossible for one map to fit all uses. The

● **FIGURE 2.17** This map uses colors to show elevations and the topography on Mars. The area shown is about 160 km (100 mi) wide.

*How were we able to map Mars in such detail?*

many different varieties of maps all have qualities that can be either advantageous or problematic, depending on the application. Knowing some basic concepts concerning maps and cartography will greatly enhance a person's ability to effectively use a map and to select the right map for a particular task.

## Advantages of Maps

If a picture is worth a thousand words, then a map is worth a million. Because they are graphic representations and use symbolic language, maps show spatial relationships and portray geographic information with great efficiency. As visual representations, maps can supply an enormous amount of information that would take many pages to describe in words, most likely with less success.

Imagine trying to tell someone about all of the information that a map of your city, county, state, or campus provides: sizes, areas, distances, directions, street patterns, railroads, bus routes, hospitals, schools, libraries, museums, highway routes, business districts, residential areas, population centers, and so forth. Maps can display true courses for navigation and accurate shapes of Earth features. They can be used to measure areas or distances, and they can show the best route from one place to another. The potential applications of maps are practically infinite, even "out of this world," because our space programs have produced detailed maps of the moon and Mars ( ● Fig. 2.17).

Cartographers can produce maps to illustrate almost any relationship in the environment. For many reasons, whether it is presented on paper, on a computer screen, or in the mind, the map is the geographer's most important tool.

## Limitations of Maps

On a globe we can directly compare the size, shape, and area of Earth features and we can measure distance, direction, shortest routes, and true directions. Yet, because of the distortion inher-

ent in maps, we can never compare or measure all of these properties on a single map. It is impossible to present a spherical planet on a flat (two-dimensional) surface and accurately maintain all of its geometric properties. The problem of representing a sphere on a flat map has been described as trying to flatten out an eggshell.

On maps that show large regions or the whole world, Earth's curvature causes apparent and pronounced distortion, but when a map depicts only a small area, the distortion should be negligible. If we use a map of a state park for hiking, the distortion will be too small to affect us. To be skilled map users, we must know which properties a certain map depicts accurately, which features it distorts, and for what purpose a map is best suited. If we are aware of these map characteristics, we can make accurate comparisons and measurements on maps and better understand the information that the map conveys.

## Examples of Map Projections

Transferring a spherical grid onto a flat surface produces a **map projection.** Although maps are not actually made this way, certain projections can be demonstrated by putting a light inside a transparent globe so that the grid lines are projected onto a flat surface (plane), a cone, a cylinder, or other geometric forms that are flat or can be cut and flattened out ( ● Fig. 2.18). Today map projections are developed mathematically, using computers to fit the geographic grid to a surface. Map projections will always distort the shape, area, direction, or distance of map features, or some combination thereof, so it is important for mapmakers to choose the best projection for the task.

**Planar Projections** Projecting the grid lines onto a plane, or flat surface, produces a map called a **planar projection** (Fig. 2.18a). These maps are most often used to show the polar regions, with the pole centrally located on the circular map, which displays one hemisphere.

**Conic Projections** Maps of middle-latitude regions, such as the contiguous United States, are typically based on a **conic projection,** which portrays these latitudes with minimal distortion. In a simple conic projection, a cone is fitted over the globe with its pointed top centered over a pole (Fig. 2.18b). Parallels of latitude on a conic projection are arcs that become smaller toward the pole, and meridians appear as straight lines radiating toward the pole.

**Cylindrical Projections** A well-known example of a **cylindrical projection** (Fig. 2.18c) is the **Mercator projection,** once commonly used in schools and textbooks, although much less so in recent years. The Mercator world map is a mathematically adjusted cylindrical projection on which meridians appear as parallel lines instead of converging at the poles. Obviously, there is enormous east–west distortion of areas in the high latitudes because the distances between meridians are stretched to the same

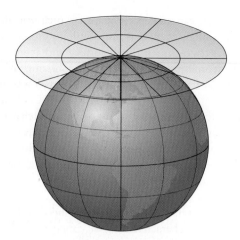

**(a) Planar projection**

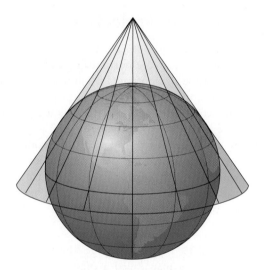

**(b) Conical projection**

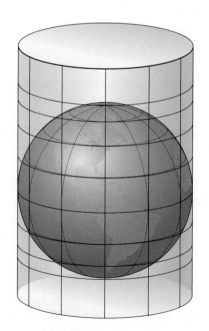

**(c) Cylindrical projection**

● **FIGURE 2.18** The concept behind the development of (a) planar, (b) conic, and (c) cylindrical projections. Although projections are not actually produced this way, they can be demonstrated by projecting light from a transparent globe.
From Petersen/Sack/Gabler. *Fundamentals of Physical Geography,* 1e. Copyright © 2011 Brooks/Cole, a part of Cengage Learning, Inc. Reproduced by permission. www.cengage.com/permissions.

### Why do we use different map projections?

width that they are at the equator ( ● Fig. 2.19). The spacing of parallels on a Mercator projection is also not equal, as they are on Earth. This projection does not display all areas accurately, and size distortion increases toward the poles.

Gerhardus Mercator devised this map in 1569 to provide a property that no other world projection has. A straight line drawn anywhere on a Mercator projection is a line of true compass direction, called a *rhumb line,* which was very important in navigation (see Fig. 2.19). On Mercator's map, navigators could draw a straight line between their location and the place where they wanted to go, and then follow a constant compass direction to get to their destination.

● **FIGURE 2.19** The Mercator projection was designed for navigation but has often been misused as a general-purpose world map. Its most useful property is that lines of constant compass directions are straight lines. The Mercator projection was developed from a cylindrical projection.

### Compare the sizes of Greenland and South America on this map to their sizes on a globe. Is the distortion great or small?

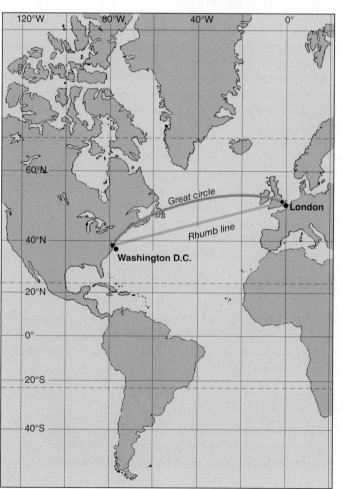

# Properties of Map Projections

The geographic grid has four important geometric properties: (1) Parallels of latitude are always parallel, (2) parallels are evenly spaced, (3) meridians of longitude converge at the poles, and (4) meridians and parallels always cross at right angles. There are thousands of ways to transfer a spherical grid onto a flat surface to make a map projection, but no map projection can maintain all four of these properties at once. Because it is impossible to have all these properties on the same map, cartographers must decide which properties to preserve at the expense of others. Closely examining a map's grid system to determine how these four properties are affected will help us discover areas of greatest and least distortion.

Today map projections are developed mathematically, using computers to fit the geographic grid to a surface. Distortions in the geographic grid that are required to make a map can affect the geometry of several characteristics of the areas and features that a map portrays.

**Shape** Flat maps cannot depict large regions of Earth without distorting either their shape or their comparative sizes in terms of area. However, using the proper map projection will depict the true shapes of continents, regions, mountain ranges, lakes, islands, and bays. Maps that maintain the correct shapes of areas are **conformal maps.** To preserve the shapes of Earth features on a conformal map, meridians and parallels always cross at right angles just as they do on the globe.

The Mercator projection does present correct shapes, so it is a conformal map, but areas away from the equator are exaggerated in size. Because it was once commonly used as a world map in schools and textbooks, the Mercator projection's distortions led generations of students to believe incorrectly that Greenland is as large as South America. On Mercator's projection, Greenland is shown as being about equal in size to South America (see Fig. 2.19), but South America is more than eight times larger.

**Area** Cartographers are able to create a world map that maintains correct area relationships—that is, areas on the map have the same size proportions to each other as they have in reality. Thus, if we cover any two parts of the map with, let's say, a quarter, no matter where the quarter is placed it will cover equivalent areas on Earth. Maps drawn with this property, called **equal-area maps,** should be used if size comparisons are being made between two or more areas. The property of equal area is also essential when examining spatial distributions. As long as the map displays equal area and a symbol represents the same quantity throughout the map, we can get a good idea of the distribution of any feature—for example, people, earthquakes, hurricanes, rainfall, or volcanoes. However, equal-area maps distort the shapes of map features ( ● Fig. 2.20) because it is impossible to show both equal areas and correct shapes on the same map.

● **FIGURE 2.20** An equal-area world projection map. This map preserves area relationships but distorts the shape of landmasses.

*Which world map would you prefer, one that preserves area or one that preserves shape, and why?*

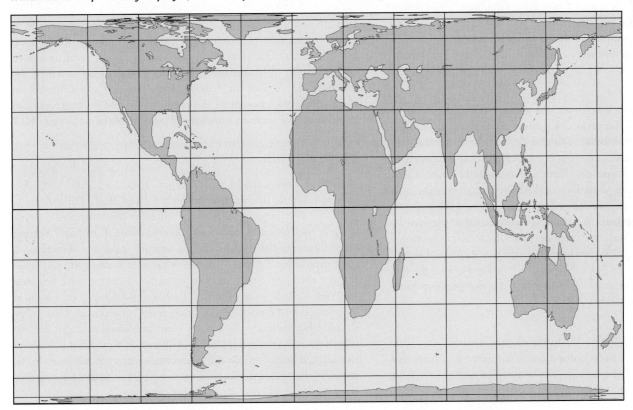

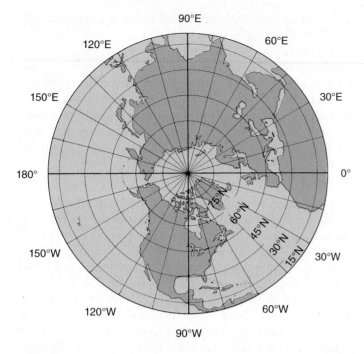

● **FIGURE 2.21** An azimuthal map centered on the North Pole. Although a polar view is the conventional orientation of such a map, it could be centered anywhere on Earth.

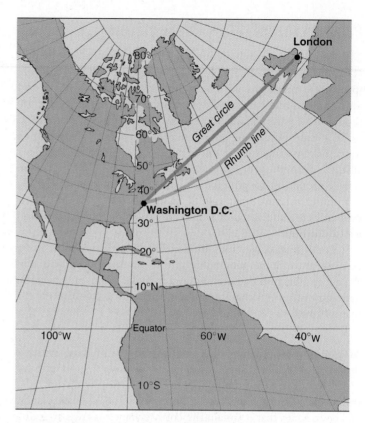

● **FIGURE 2.22** The gnomonic projection produces extreme distortion of distances, shapes, and areas, yet it is valuable for navigation because it is the only projection that shows all great circles as straight lines. It is developed from a planar projection.

*Compare this figure with Figure 2.19. How do these two projections differ?*

**Distance** No flat map can maintain a constant distance scale over Earth's entire surface. The scale on a map that depicts a large area cannot be applied equally everywhere on that map. On maps of small areas, however, distance distortions will be minor, and the accuracy will usually be sufficient for most purposes. Maps can be made with the property of **equidistance** in specific instances. That is, on a world map, the equator may have equidistance (a constant scale) along its length, and all meridians may have equidistance, but not the parallels. On another map, all straight lines drawn from the center may have equidistance, but the scale will not be constant unless lines are drawn from the center.

**Direction** Because longitude and latitude directions run in straight lines, not all flat maps can show true directions as straight lines. Thus, lines of latitude or longitude that curve on maps are not drawn as true compass directions. **Azimuthal projections** produce maps that show true directions as straight lines. These projections are drawn with a central focus, and all straight lines drawn from that center are true compass directions ( ● Fig. 2.21). **Gnomonic projections** are planar projections that present the grid with great distortion. Despite their distortion, gnomonic projections ( ● Fig. 2.22) are the only maps that display all arcs of great circles as straight lines. Navigators can draw a straight line between their location and this line will be the shortest route between the two places.

**Compromise Projections** In developing a world map, one cartographic strategy is to compromise by creating a map that shows both area and shape fairly well but is not really correct for either property. These world maps are **compromise projections**

that are neither conformal nor equal area, but an effort is made to balance distortion to produce an "accurate looking" global map ( ● Fig. 2.23a). An *interrupted projection* can also be used to reduce the distortion of landmasses (Fig. 2.23b) by moving much of the distortion to the oceanic regions. If our interest was centered on the world ocean, however, the projection could be interrupted in the continental areas to minimize distortion of the ocean basins.

## Map Basics

Maps not only contain spatial information and data; they also display essential information about the map itself. This information and certain graphic features (often in the margins) are intended to facilitate using and understanding the map. Among these items are the map title, date, legend, scale, and direction. A map should have a title that tells what area is depicted and what subject the map concerns. For example, a hiking and camping map for Yellowstone National Park should have a title such as "Yellowstone National Park: Trails and Camp Sites." Most maps should also indicate when they were published and the date to which its information applies. For instance, a population map of the United States should tell when the census was taken to let us know if the map information is current or outdated or whether the map is intended to show historical data.

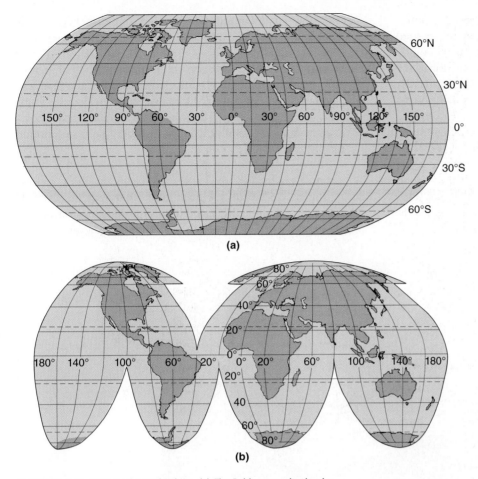

**(a)**

**(b)**

● **FIGURE 2.23** Compromise projections. (a) The Robinson projection is considered a compromise projection because it departs from equal area to show both area and shape reasonably well, although neither properties are truly accurate. Distortion in projections can be also reduced by using an interrupted projection (b)—that is, by having a central meridian for each segment of the map.

*Compare the distortion of these maps with the Mercator projection (Fig. 2.19). What is a disadvantage of (b) in terms of usage?*

**Legend** A map should have a **legend**—a key to symbols used on the map. For example, if one dot represents 1000 people or the symbol of a pine tree represents a roadside park, the legend should explain this information. If color shading is used on the map to represent elevations, different climatic regions, or other factors, then a key to the color coding should be provided. Map symbols can be designed to represent virtually any feature (see Appendix B).

**Scale** Obviously, maps depict features smaller than they actually are. If the map is used for measuring sizes or distances, or if the size of the area represented might be unclear to the map user, it is essential to know the map scale ( ● Fig. 2.24). A map **scale** is an expression of the relationship between a distance on Earth and the same distance as it appears on the map. Knowing the map scale is essential for accurately measuring distances and for determining areas. Map scales can be conveyed in three basic ways.

A **verbal scale** is a statement on the map that indicates, for example, "1 centimeter to 100 kilometers" (1 cm represents 100 km) or "1 inch to 1 mile" (1 in on the map represents 1 mi on the ground). Stating a verbal scale tends to be how most of us would refer to a map scale in conversation. A verbal scale, however, will no longer be correct if the original map is reduced or enlarged. When stating a verbal scale, it is acceptable to use different map units (centimeters, inches) to represent another measure of true length it represents (kilometers, miles).

A **representative fraction (RF) scale** is a ratio between a unit of distance on the map to the distance that unit represents in reality (expressed in the same units). Because a ratio is also a fraction, the units of measure, being the same in the numerator and denominator, cancel each other out. An RF scale is therefore free of units of measurement and can be used with any unit of linear measurement—meters, centimeters, feet, inches—as long as the same unit is used on both sides of the ratio. As an example, a map may have an RF scale of 1:63,360, which can also be expressed 1/63,360. This RF scale can mean that 1 inch on the map repre-

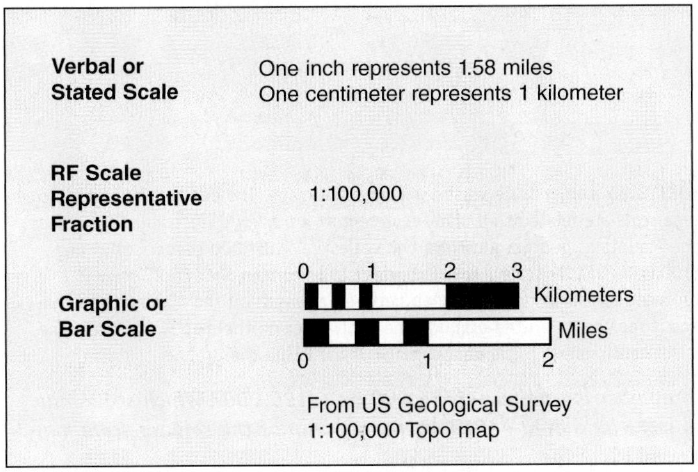

| Verbal or Stated Scale | One inch represents 1.58 miles<br>One centimeter represents 1 kilometer |
|---|---|
| RF Scale Representative Fraction | 1:100,000 |
| Graphic or Bar Scale | From US Geological Survey<br>1:100,000 Topo map |

● **FIGURE 2.24** Map scales. A *verbal scale* states the relationship between a map measurement and the corresponding distance that it represents on the Earth. Verbal scales generally mix units (centimeters/kilometer or inches/mile). A *representative fraction* (RF) scale is a ratio between a distance on a map (1 unit) and its actual length on the ground (here, 100,000 units). An RF scale requires that measurements be in the same units both on the map and on the ground. A *graphic scale* is a device used for measuring distances on the map in terms of distances on the ground.

From Petersen/Sack/Gabler. *Fundamentals of Physical Geography*, 1e. Copyright © 2011 Brooks/Cole, a part of Cengage Learning, Inc. Reproduced by permission. www.cengage.com/permissions.

sents 63,360 inches on the ground. It also means that 1 centimeter on the map represents 63,360 centimeters on the ground. Knowing that 1 inch on the map represents 63,360 inches on the ground may be difficult to conceptualize unless we realize that 63,360 inches is equal to 1 mile. Thus, the representative fraction 1:63,360 means the map has the same scale as a map with a verbal scale of 1 inch to 1 mile.

A **graphic scale,** or **bar scale,** is useful for making distance measurements on a map. Graphic scales are graduated lines (or bars) marked with map distances that are proportional to distances on the Earth. To use a graphic scale, take a straight edge of a piece of paper and mark the distance between any two points on the map. Then use the graphic scale to find the equivalent distance on Earth's surface. Graphic scales have two major advantages:

1. It is easy to determine distances on the map because the graphic scale can be used like a ruler to make measurements.
2. They are applicable even if the map is reduced or enlarged because the graphic scale (on the map) will also change proportionally in size. This is particularly useful because maps can be reproduced easily in a reduced or enlarged scale using computers or photocopiers. The map and the graphic scale, however, must be enlarged or reduced together (the same amount) for the graphic scale to be applicable.

Maps are often described as being of small, medium, or large scale ( ● Fig. 2.25). *Small-scale* maps show large areas in a relatively small size, include little detail, and have large denominators in their representative fractions. *Large-scale* maps show small areas of Earth's

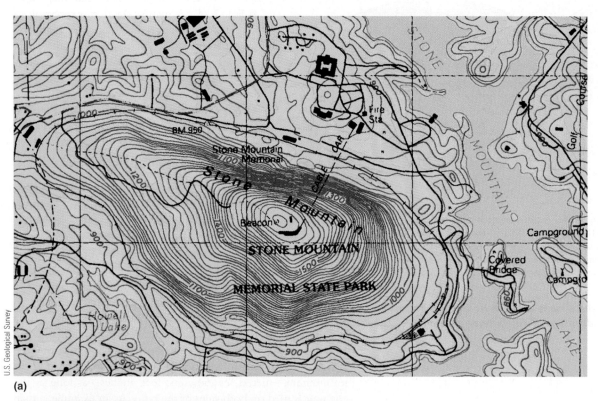

(a)

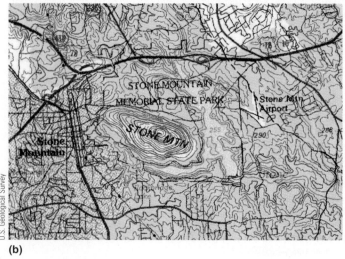

(b)

U.S. Geological Survey

● **FIGURE 2.25** Larger-scale versus smaller-scale maps. The designations *small scale* and *large scale* are related to a map's *representative fraction* (RF) scale. These maps of Stone Mountain, Georgia illustrate two scales: (a) 1:24,000 (larger scale) and (b) 1:100,000 (smaller scale). It is important to remember that an RF scale is a *fraction* that represents the proportion between a length on the map and the true distance it represents on the ground. One centimeter on the map would equal the number of centimeters in the denominator of the RF on the ground.

*Which number is smaller—1/24,000 or 1/100,000? Which scale map shows more land area—the larger-scale map or the smaller-scale map?*

surface in greater detail and have smaller denominators in their representative fractions. To avoid confusion, remember that 1/2 is a *larger* fraction than 1/100 and small scale means that Earth features are shown very small. A large-scale map would show the same features larger. Maps with representative fractions larger than 1:25,000 are large scale. Medium-scale maps have representative fractions between 1:25,000 and 1:250,000. Small-scale maps have representative fractions less than 1:250,000. This classification follows the guidelines of the U.S. Geological Survey (USGS), publisher of many maps for the federal government and for public use.

## Direction

**Direction** The orientation and geometry of the geographic grid give us an indication of direction because parallels of latitude are east–west lines and meridians of longitude run directly north–south. Many maps have an arrow pointing to north as displayed on the map. A north arrow may indicate either *true north* or *magnetic north*—or two north arrows may be given, one for true north and one for magnetic north. Generally if there is only one north arrow, it represents the direction to true north.

Earth has a magnetic field that makes the planet act like a giant bar magnet, with a magnetic north pole and a magnetic south pole,

each with opposite charges. Although the magnetic poles shift position slightly over time, they are located in the Arctic and Antarctic regions and do not coincide with the geographic poles. Aligning itself with Earth's magnetic field, the north-seeking end of a compass needle will point toward the magnetic north pole. If we know the **magnetic declination,** the angular difference between magnetic north and true geographic north (for our location), we can compensate for this difference ( ● Fig. 2.26a). Thus, if our compass points north and we know that the magnetic declination for our location is 20°E, we can adjust our course knowing that our compass is pointing 20°E of true north. To do this, we should turn 20°W from the direction indicated by our compass to face true north. Magnetic declination varies from place to place and changes through time. For this reason, magnetic declination maps are revised periodically, so if you are navigating by compass, it is important to have a recent map. A map of magnetic declination is called an *isogonic map* (Fig. 2.26b), and *isogonic lines* connect locations that have equal declination.

Compass directions can be given by either the azimuth system or the bearing system. In the **azimuth** system, direction is given in degrees of a full circle (360°) clockwise from north. That is, if we imagine the 360° of a circle with north at 0° (and at 360°)

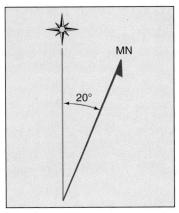

(a)

● **FIGURE 2.26** (a) Map symbol showing true north, symbolized by a star representing Polaris (the North Star), and magnetic north, symbolized by an arrow. The example indicates 20°E magnetic declination. (b) Isogonic map of the conterminous United States, showing the magnetic declination that must be added or subtracted from a compass reading to determine true directions.

*What is the magnetic declination of your hometown to the nearest degree?*

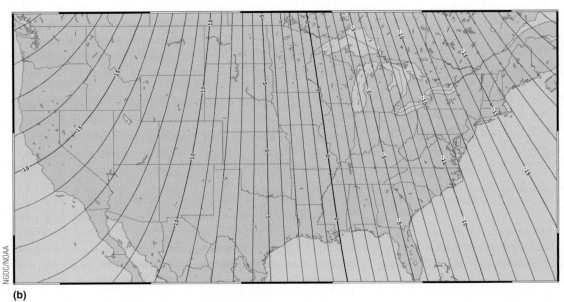

(b)

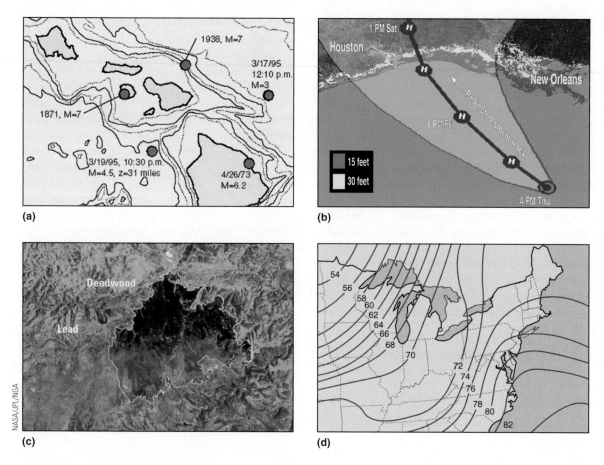

● **FIGURE 2.27** Discrete and continuous spatial data (variables). *Discrete variables* represent features that are present at certain locations but do not exist everywhere. The locations, distributions, and patterns of discrete features are of great interest in understanding spatial relationships. Discrete variables can be (a) *points* representing, for example, locations of large earthquakes in Hawaii (or places where lightning has struck or locations of water-pollution sources), (b) *lines* as in the path taken by Hurricane Rita (or river channels, tornado paths, or earthquake fault lines), or (c) *areas* like the land burned by a wildfire (or clear-cuts in a forest, or the area where an earthquake was felt). *A continuous variable* means that every location has a certain measurable characteristic—for example, everywhere on Earth has an elevation, even if it is zero (at sea level) or below (a negative value). Changes in a continuous variable over an area can be represented by isolines, shading, or colors or with a 3-D appearance. The map (d) shows the continuous distribution of temperature variation in part of eastern North America.

*Can you name other environmental examples of discrete and continuous variables?*

and read the degrees clockwise, we can describe a direction by its number of degrees away from north. For instance, straight east would have an azimuth of 90°, and due south would be 180°. The **bearing** system divides compass directions into four quadrants of 90° (N, E, S, W), each numbered by directions in degrees away from either north or south. Using this system, an azimuth of 20° would be north, 20° east (20° east of due north), and an azimuth of 210° would be south, 30° west (30° west of due south). Both azimuths and bearings are used for mapping, surveying, and navigation for both military and civilian purposes.

## Thematic Maps

**Thematic maps** are designed to focus attention on the spatial extent and distribution of one feature (or a few related ones). Examples include maps of climate, vegetation, soils, earthquake epicenters, or tornadoes.

**Discrete and Continuous Data** There are two major types of spatial data, discrete and continuous. **Discrete data** mean that a phenomenon is located at a particular place—for example, hot springs, tropical rainforests, rivers, tornado paths, or earthquake faults. Discrete data are represented on maps by point, area, or line symbols to show their locations and distributions ( ● Figs. 2.27a–c). *Regions* are discrete areas that exhibit a common characteristic or set of characteristics within their boundaries and are typically represented by different colors or shading to differentiate one region from another. Physical geographic regions include areas of similar soil, climate, vegetation, landform type, or many other characteristics (see the world and regional maps throughout this book).

**Continuous data** mean that a measurable numerical value for a certain characteristic exists everywhere on Earth (or within the area of interest displayed); for example, every location on Earth has a measurable elevation (or temperature, or air pressure, or population density). The distribution of continuous data is

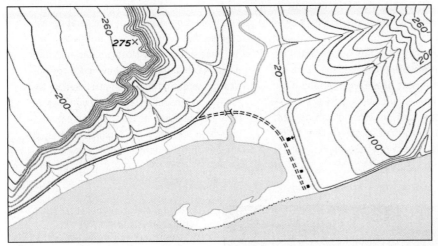

often shown using **isolines**—lines on a map that connect points with the same numerical value (Fig. 2.27d). Isolines that we will be using later on include *isotherms,* which connect points of equal temperature; *isobars,* which connect points of equal barometric pressure; *isobaths* (also called bathymetric contours), which connect points with equal water depth; and *isohyets,* which connect points receiving equal amounts of precipitation.

## Topographic Maps

**Topographic contour lines** are isolines that connect points on a map that are at the same elevation above mean sea level (or below sea level, such as in Death Valley, California). For example, if we walk around a hill along the 1200-foot *contour line* shown on the map, we would always be 1200 feet above sea level, maintaining a constant elevation by walking on a level line. Contour lines are an excellent means for showing elevation changes and the form of the land surface on a map. The arrangement, spacing, and shape of contours give a map reader a mental image of what the topography (the "lay of the land") is like ( ● Fig. 2.28).

    ● Figure 2.29 illustrates how contour lines portray the land surface. The bottom portion of the diagram is a simple *contour map* of an asymmetrical hill. Note that the elevation difference between adjacent contour lines on this map is 20 feet. The constant difference in elevation between adjacent contour lines is called the **contour interval.**

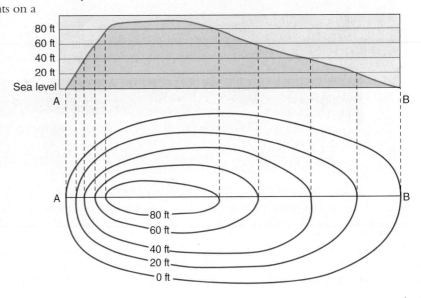

# Using Vertical Exaggeration to Portray Topography

**M**ost maps present a landscape as if viewed from directly overhead, looking straight down. This perspective is referred to as a map view or plan view (like architectural house plans). On a flat map, illustrating terrain and showing elevation differences require some kind of symbol to display elevations. Topographic maps use contour lines, which can also be enhanced by relief shading.

For many purposes, either a side view (profile) view, or an oblique (perspective) view of the terrain helps us visualize the landscape. Block diagrams, 3-D models of Earth's surface, show the topography from a perspective view and information about the

subsurface can be included. They provide a perspective similar to looking out an airplane window or from a high vantage point. Such diagrams are not always intended for making accurate measurements, however, and many block diagrams represent hypothetical or stylized, rather than actual, landscapes.

A topographic profile (see Fig. 2.29) illustrates the shape of a land surface as if viewed directly from the side. Profiles are graphs of elevation changes over distance along a transect line. Elevation and distance information collected from a topographic map or from other elevation data can be used to draw a topographic profile to show the terrain. If the geology of the

subsurface also is represented, such profiles are called geologic cross-sections.

Block diagrams, profiles, and cross-sections are most often drawn with *vertical exaggeration*, which enhances changes in elevation. This makes mountains appear taller, valleys deeper, slopes steeper, and the terrain more rugged. Vertical exaggeration is used so that subtle changes in the terrain are more noticeable. Most profiles and block diagrams should indicate how much the vertical presentation has been stretched so that there is no misunderstanding. Two times vertical exaggeration means that the features presented are two times higher than they really are, but the horizontal scale is correct.

This digital elevation model (DEM) of Anatahan Island (16° 22″ N, 145° 40′ E) and the surrounding Pacific Ocean floor is presented in 3-D and colorized according to elevation and seafloor depth relative to sea level. This image of Anatahan has three times vertical exaggeration. The vertical scale has been stretched three times compared to the horizontal scale. The vertical scale bar represents a distance of 3800 meters, so taking the vertical exaggeration into account, what horizontal distance would the same scale bar length represent in meters?

Compare the vertically exaggerated elevation model to this natural scale (not vertically exaggerated) version. This is how the island and the seafloor actually look in terms of slope steepness and relief.

If we hiked from point A to point B, what kind of terrain would we cover? We start from sea level at point A and immediately begin to climb. We cross the 20-foot contour line, then the 40-foot, then the 60-foot, and near the top of our hill the 80-foot contour level. After walking over a relatively broad summit that is above 80 feet but not as high as 100 feet (or we would cross another contour line), we once again cross the 80-foot contour line, which means we must be starting down. During our descent, we cross each lower level in turn until we arrive back at sea level (point B).

In the top portion of Figure 2.29, a **profile** (side view) helps us to visualize the topography we covered in our walk. We can see why the trip up the mountain was more difficult than the trip down. Closely spaced contour lines near point A represent a steeper slope than the more widely spaced contour lines near point B. Actually, we have discovered something that is true of all isoline maps: The closer together the lines are on the map, the steeper the **gradient** (the greater the rate of vertical change per unit of horizontal distance). When studying a contour map, we should understand that the slope between contours almost always changes gradually, and it is unlikely that the land drops off in steps downslope as the contour lines might suggest.

Topographic maps use symbols to show many other features in addition to elevations (see Appendix B)—for instance, water bodies such as streams, lakes, rivers, and oceans or cultural features such as towns, cities, bridges, and railroads. The USGS produces topographic maps of the United States at several different scales. Some of these maps—1:24,000, 1:62,500, and 1:250,000—use English units for their contour intervals. Many recent maps are produced at scales of 1:25,000 and 1:100,000 and use metric units. Contour maps of undersea topography, called *bathymetric charts,* are produced in the United States by the National Ocean Service.

# Modern Map Making

Today the vast majority of mapmakers use computer technologies. For most mapping projects, computer systems are faster, more efficient, and less expensive than the hand-drawn cartographic techniques they have replaced. Computer systems make it easy to make scale changes or metric conversions and changes in projection, contour interval, symbols, colors, and directions of view (rotating the orientation). The ability to interact with a map and make on-the-spot modifications is essential when representing changing phenomena such as weather systems, air pollution, ocean currents, volcanic eruptions, and forest fires. Digital maps can be instantly disseminated and shared via the Internet, which is a great advantage when spatial information is rapidly changing or when it is important to communicate mapped information as soon as possible.

# Geographic Information Systems

A **geographic information system (GIS)** is a versatile map-based computer technology that stores geographic databases, supports spatial data analysis, and facilitates the production of digital maps. The database for a map may include information on coastlines, political boundaries, city locations, river systems, map projections, and coordinate systems. A GIS assists the user in the entry, analysis, manipulation, and display of geographic information derived from combining any number of **digital map layers,** each composed of a specific *thematic map* ( ● Fig. 2.30). Spatial data representing elevations, depths, temperatures, or populations can be stored in a digital database, accessed, and displayed. A GIS can be used to make the scale and map projection of these map layers identical, thus allowing the information from several or all layers to be combined into new and more meaningful composite maps. GIS is especially useful to geographers as they work to address problems that require large amounts of spatial data from a variety of sources.

● **FIGURE 2.30** Geographic information systems (GIS) store different information and data as individual map layers. GIS technology is widely used in geographic and environmental studies in which several different variables need to be assessed and compared spatially to solve a problem. No longer limited to specialized fields, there is an abundance of mapping applications found on the World Wide Web, powered by GIS.

*Can you think of some ways in which GIS has influenced your life?*

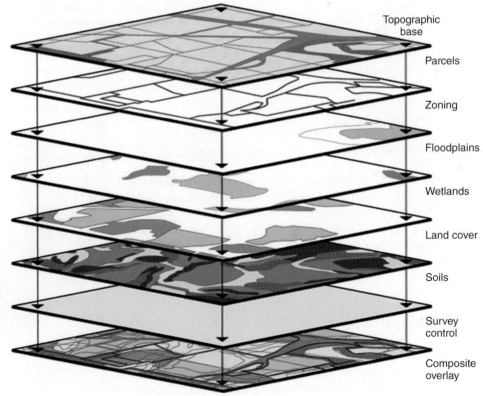

Topographic base

Parcels

Zoning

Floodplains

Wetlands

Land cover

Soils

Survey control

Composite overlay

**What a GIS Does** Imagine that you are in a giant map library with hundreds of paper maps, all of the same area, but each map shows a different aspect of the same location: One map shows roads, another highways, another trails, another rivers (or soils, or vegetation, or slopes, or rainfall, and so on almost to infinity). The maps were originally produced at different sizes, scales, and projections (including some maps that do not preserve shape or area). These cartographic factors make it difficult to *visually* overlay and compare the spatial information among these different maps. You also have digital terrain models and satellite images that you would like to be able to produce at the same scale and projection making spatial comparisons to the maps. Further, because few aspects of the environment involve only one factor and seldom exist in spatial isolation, you want to be able to combine a selection of these geographic aspects on a single composite map. You have a spatial–geographic problem, and to solve that problem, you need a way to make several representations of a part of Earth directly comparable. What you need is a GIS and the knowledge of how to use this system.

**Data and Attribute Entry** The first step is to enter map and image data into a geographic information system that stores them in digital files as individual map layers (see Fig. 2.30). Each category of spatial data and information is fitted to a common set of mapped grid coordinates such as latitude and longitude. The thematic data and information entered for each map layer, called *attributes,* can then be easily expanded and revised as needed.

**Registration, Display, and Analysis** A GIS can display any layer or any combination of layers, geometrically registered (fitted) to any map projection and at any scale that you specify. The maps, images, and data sets can now be directly compared at the same size, based on the same map projection and map scale. A GIS can digitally *overlay* any set of thematic map layers that are needed. If you want to see the locations of *homes* on a *river floodplain,* a GIS can be used to provide an instant map by combining and displaying the *home* and *floodplain* map layers simultaneously. If you want to see *earthquake faults* and *artificial landfill areas* in relation to locations of *fire stations* and *police stations,* that composite map will require four layers, but this is no problem for a GIS to display. And if you want to display new data in your map, such as the locations of earthquake-related calls each station receives, that can be done easily by creating a new layer using software tools.

**Digital Elevation Models** Computer-generated three-dimensional (3-D) views of elevation data, called **digital elevation models (DEMs),** are particularly effective at displaying topography in a way that simulates a 3-D view ( ● Fig. 2.31). Digital elevation data can be used to make many types of terrain

*NASA/GSFC/and U.S./Japan ASTER Science Team*

● **FIGURE 2.31** A digital elevation model (DEM) of the Grand Canyon, in Arizona, presented in 3-D and colorized according to elevation.

displays and maps, including color-scaled contour maps, where areas between contours are assigned a certain color; conventional topographic maps; and shaded-relief maps. Digital terrain models may be designed to show **vertical exaggeration** by stretching the vertical scale of the display to enhance the relief of an area (see the feature on vertical exaggeration in this chapter). Actually, any geographic factor represented by continuous spatial data can be displayed either as a two-dimensional contour map or as a 3-D surface to enhance the visibility of the spatial variation that it conveys ( ● Fig. 2.32).

● **FIGURE 2.32** Earthquake hazard in the conterminous United States: a continuous variable displayed as a continuous surface in 3-D perspective. Here it is easy to develop a mental map of how potential earthquake danger varies across this part of the United States.

*Are you surprised by any of the locations that are shown to have substantial earthquake hazard?*

*Can you think of other applications for geographic information systems?*

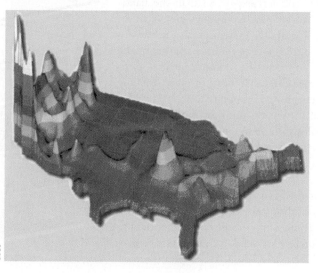

*USGS*

**Visualization Models** Also referred to as *visualizations*, **visualization models** are computer-generated image models designed to illustrate and explain complex processes and features. Many visualizations are presented as 3-D images and/or as animations. For example, the Earth image shown in the chapter opening image is a visualization model. Many visualizations, however, use vertical exaggeration to emphasize the characteristics they display (such as the cloud heights in the chapter opening image).

Visualization models can combine and present several components of the Earth system in stunning 3-D views, based on actual environmental data and satellite images or air photos. An example is shown in ● Figure 2.33, in which a satellite image and a DEM are map layers in a GIS that can be combined in a 3-D view to produce a landscape visualization model, the Rocky Mountain front at Salt Lake City, based on real image and elevation data. This process is called **draping** (like draping cloth over some ob-

U.S. Geological Survey

**(a)**

NASA

**(b)**

NASA/JPL/NIMA

**(c)**

● **FIGURE 2.33** A visualization model created with a geographic information system. (a) A digital elevation model (DEM) of Salt Lake City, Utah, and the Rocky Mountain front displays elevations in a grid with a three-dimensional appearance. (b) A satellite image of the same area. (c) These two data layers can be combined with a GIS to produce a digital 3-D model of the landscape. Digital models like this can be rotated on a computer screen to be viewed from any angle or direction. The examples here are enlarged to show the pixel resolution.
⊕ In Google Earth fly to: 40.760°N, 111.890°W (Salt Lake City).

● **FIGURE 2.34** The physical environment of Grand Teton National Park, Wyoming is presented in this landscape visualization. Satellite imagery and elevation data were combined to produce this scene and were computer enhanced to show a 3-D perspective. The steepness and height of the Grand Tetons are so great that no vertical exaggeration was necessary to enhance the relief. This image, unlike the example in Figure 2.33, is not greatly enlarged so the pixels are less visible.

*Does the terrain in this landscape look vertically exaggerated to you, or does the scene look fairly natural?*

ject), but the scale and the perspective are accurately registered among the map layers. Visualization models help us understand and conceptualize many environmental processes and features.

Today the products and techniques of cartography are different from their beginning forms and they continue to be improved, but the goal of making a representation of Earth remains the same—to effectively communicate geographic and spatial knowledge in a visual format. An example of a digital landscape visualization produced by combining elevation data and a satellite image is shown in ● Figure 2.34.

### GIS in the Workplace

A simple example will help illustrate the utility of a GIS. Suppose you are a geographer working for the Natural Resources Conservation Service. Your current problem is to control erosion along the banks of a reservoir. You know that erosion is a function of many environmental variables, including soil types, slope steepness, vegetation characteristics, and others. Using a GIS, you would enter map data for each of these variables as a separate layer. You could analyze these variables individually, or you could integrate information from individual layers (soils, slope, vegetation, and so on) to identify the locations most susceptible to erosion. Your resources and personnel could then be directed toward controlling erosion in those target areas. In physical geography and the Earth sciences, GIS is being used to analyze potential coastal flooding from sea-level rise, areas in need of habitat restoration, flood hazard potential, and earthquake distributions, just to list a few examples. The

spatial analysis capabilities of a GIS are nearly unlimited and are applicable in almost any career field.

Many geographers are employed in careers that apply GIS along with other digital mapping and imaging technologies. The capacity of a GIS to integrate and analyze a wide variety of geographic information, from census data to landform characteristics, makes it useful to both human and physical geographers. With nearly unlimited applications in geography and other disciplines, GIS will continue to be an important tool for understanding our environment and making important decisions based on spatial information.

# Remote Sensing of the Environment

**Remote sensing** is the collection of information and data about distant objects or environments. Remote sensing involves the gathering and interpretation of aerial and space imagery, images that have many maplike qualities. Using remote sensing systems, we can also detect objects and scenes that are not visible to humans and can display them on images that we can visually interpret.

## Digital Imaging and Aerial Photography

Remote sensing is commonly divided into photographic techniques and digital imaging, which may use equipment similar to digital cameras or more sophisticated technologies. Today, with the recent widespread use of digital cameras, the use of the term

● **FIGURE 2.35** A digital satellite image of part of the San Francisco Bay Area. Vegetated areas are shown in red because this is a near-infrared image (see Fig. 2.37). The pixel resolution is 15 × 15 meters, which shows good detail for regional analysis. The inset illustrates the mosaic-like pixel sizes that display the runways of the former Alameda Naval Air Station.

"photograph" is changing in common usage, but in technical terms, **aerial photographs** are made by using cameras to record a picture on *film*. Digital cameras or image scanners produce a **digital image**—a picture that is converted into numerical data. A digital image is similar to a mosaic, made up of grid cells with varying colors or tones that form a picture. Each cell (pixel) has a locational address within the grid and a value that represents the brightness or color of the picture area that the pixel represents. The digital values in an array of grid cells (pixels) are translated into an image by computer technology.

Most images returned from space are digital because digital data can be easily broadcast back to Earth. Digital imagery also offers the advantages of computer-assisted data processing, image enhancement, and image sharing and can provide a thematic layer in a GIS. Digital images consist of **pixels,** short for "picture element," the smallest area resolved in a digital picture. A key factor in digital images is **spatial resolution,** expressed as how small an area (on the Earth) each pixel represents. For example, a satellite image with a resolution of 15 × 15 or 30 × 30 meters provides discernable detail about a city or small region ( ● Fig. 2.35). Satellites that image an entire hemisphere at once, or large continental areas, typically use resolutions that are much more coarse (e.g., 1 km × 1 km) to produce a more generalized scene. Satellites that image an entire hemisphere at once, or large continental areas, use coarse resolutions to produce a more generalized scene.

Digital cameras for personal use express resolution in *megapixels,* or how many million pixels make an image. The more megapixels a camera or digital scanner can image, the better the resolution and the sharper the image will be, but this also depends on how much an image is enlarged or reduced in size while maintaining the same spatial resolution. If the pixel size is small enough on the finished image, the mosaic effect will be either barely noticeable or invisible to the human eye.

## Aerial Photography and Image Interpretation

Aerial photographs have provided us with "bird's-eye" views of our environment via kites and balloons even before airplanes were invented, but aircraft led to a tremendous increase in the availability of aerial photography. Both air photos and digital images may be *oblique* ( ● Fig. 2.36a), taken at an angle other than

**(a)**

**(b)**

● **FIGURE 2.36** (a) Oblique photos provide a "natural view," like looking out of an airplane window. This oblique aerial photograph in natural color shows farmland, countryside, and forest. (b) Vertical photos provide a maplike view that is more useful for mapping and making measurements (as in this view of Tampa Bay, Florida).

*What are the benefits of an oblique view compared to a vertical view?*

perpendicular to Earth's surface, or *vertical* (Fig. 2.36b), looking straight down. Image interpreters use aerial photographs and digital imagery to examine and describe relationships among objects on Earth's surface. A device called a *stereoscope* allows overlapped pairs of images (typically aerial photos) taken from different positions to allow viewing of features in three dimensions.

**Near-infrared (NIR)** energy, light energy at wavelengths that are too long for our eyes to see, cuts though atmospheric haze better than visual light does. Natural-color photographs taken from

very high altitudes or from space can have low contrast or appear hazy. Natural-color photographs (or digital images) and color near-infrared pictures of the same scene each display different visual information about a landscape. Comparing these two different kinds of images is typically better than using a single format for understanding the environment that is depicted ( ● Fig. 2.37a and b). Photographs and digital images that use NIR tend to provide very clear images when taken from high altitude or space. Color NIR photographs and digital images are sometimes referred to as "false

● **FIGURE 2.37** Southern Florida and part of the Florida Keys. (a) A natural-color digital image compared to (b) a false color, near-infrared image. In the color infrared image, red tones indicate vegetation density and health; dark blue to black is clear, deep water; and light blue—shallow or muddy water. These images are comparable in color and tone to natural and color infrared photographs.

*If you were asked to make a map of vegetation or to map the coastal outline, which image would you prefer to use and why?*

**(a)**

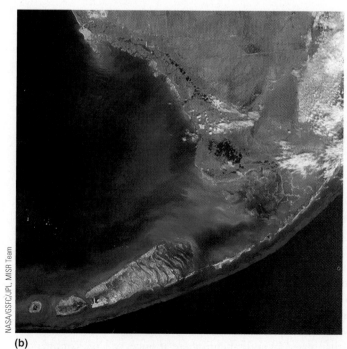

**(b)**

color" pictures because on NIR, healthy grasses, trees, and most plants will show up as bright red, rather than green (Fig. 2.37b). Near-infrared photographs and images have many applications for environmental study, particularly for water resources, vegetation, and crops. An incorrect, but widely held, notion of NIR techniques is that they image heat or temperature variations. Near-infrared energy is *light*. NIR images are produced by measuring reflections off of surfaces, and not by imaging radiated heat.

## Specialized Remote Sensing Techniques

Many different remote sensing systems are in use, each designed for specific imaging applications. Remote sensing may use ultraviolet (UV) light, visible light, NIR light, thermal infrared energy (heat), lasers, and microwave energy *(radar)* to produce images.

**Thermal infrared (TIR)** images show patterns of heat and temperature instead of light and can be taken either day or night by TIR sensors. Spatially recorded heat patterns are digitally converted into a visual image. Hot objects show up in light tones, and cool objects will be dark, but typically a computer is used to emphasize heat differences by colorizing the image. Some TIR applications include finding volcanic hot spots and geothermal sites, locating forest fires through dense smoke, finding leaks in building insulation or in pipelines, and detecting thermal pollution in lakes and rivers.

Weather satellites also use thermal infrared imaging, and we often see thermal images of cloud patterns and storms on television broadcasts. With the current concern about climate change and knowing that ocean temperatures have a strong influence on climate, monitoring ocean currents is also important. Ocean currents flow like "rivers" of sea water in the oceans, and they are can be tracked on thermal images because currents are warmer or cooler than the parts of the ocean that they flow through ( ● Fig. 2.38).

**Radar** (*RAdio Detection And Ranging*) transmits microwave energy (radio waves) and produces an image by reading the energy signals that are reflected back. Radar systems can operate day or night and can see through clouds. **Imaging radar** systems produce an image of the surface (topography, rock, water, ice, sand dunes, and so forth) by converting radar reflections into a maplike image. Data from radar mapping systems are often used to generate digital elevation models because radar can measure changes in elevation. **Side-looking airborne radar (SLAR)** was designed to image areas located to the side of an aircraft or from orbit. Excellent for mapping terrain and water features ( ● Fig. 2.39), SLAR is used most often to map remote, inhospitable, inaccessible, cloud-covered regions or heavily forested areas.

Radar systems are also used to monitor and track thunderstorms, hurricanes, and tornadoes. **Weather radar** systems produce maplike images of precipitation. These radar penetrate most clouds (day or night) but the transmitted signal reflects off of raindrops and other precipitation, producing an image on the radar screen.

**Lidar** (*LIght Distance And Ranging*) is a relatively new technology that uses an airborne scanning laser system to produce image surveys in high resolution with elevation data that offer excellent precision. Lidar is being used in determining amounts of beach erosion after storms, in vegetation and forestry studies, for mapping floodplains, and in urban planning, and it can even monitor certain types of air pollution.

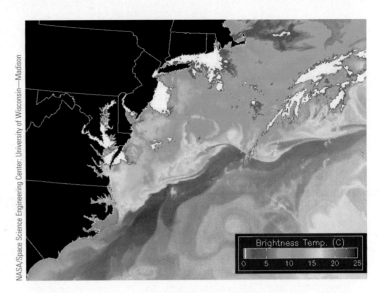

*NASA/Space Science Engineering Center: University of Wisconsin—Madison*

● **FIGURE 2.38** Thermal infrared images show patterns of heat—basically, warmer and cooler conditions. The Gulf Stream, a warm ocean current flowing along the east coast of North America, is shown here. This thermal image shows warm temperatures in red and cooler in blue, etc. Ocean currents and changes in ocean temperature affect weather and climate, so it is important to monitor ocean currents and the regional temperature patterns of the ocean.

*Which way do you think the Gulf Stream current is flowing— roughly northwest or southwest—and why?*

*NASA/JPL-Caltech*

● **FIGURE 2.39** Imaging radar reflections can produce an image of a landscape. This radar scene taken from Earth orbit shows the topography near Sunbury, Pennsylvania, where the West Branch River flows into the Susquehanna River (north is at the top of the image). Parallel ridges, separated by linear valleys, form the Appalachian Mountains in Pennsylvania. River bridges provide a sense of scale.

# Polar versus Geostationary Satellite Orbits

Satellite systems that return images from orbit are designed to produce many different kinds of Earth imagery. Some of these differences are related to the type of orbit the satellite system is using while scanning the surface. There are two distinctively different types of orbits: the *polar orbit* and the *geostationary orbit* (sometimes called a *geosynchronous orbit*), each with a different purpose.

The polar orbit was developed first; as its name implies, the satellite orbits Earth from pole to pole. This orbit has some distinct advantages. It is typically a low orbit for a satellite, usually varying in altitude from 700 kilometers (435 mi) to 900 kilometers (560 mi). At this height, but also depending on the equipment used, a polar-orbiting satellite can produce clear, close-up images of Earth. However, at this distance the satellite must move at a fast orbital velocity to overcome the gravitational pull of Earth. This velocity can vary, but for polar orbiters it averages around

27,400 kilometers/hour (17,000 mph), traveling completely around Earth in about 90 minutes. While the satellite orbits from pole to pole, Earth rotates on its axis below, so each orbit views a different path along the surface. Thus, polar orbits will at times cover the dark side of the planet. To adjust for this, a slightly modified polar orbit was developed, called a *sun synchronous orbit*. If the polar orbit is tilted a few degrees off the vertical, then it can remain over the sunlit part of the globe at all times. Most modern polar-orbiting satellites are sun synchronous (a near-polar orbit).

The geostationary orbit, developed later, offered some innovations in satellite image gathering. A geostationary orbit must have three characteristics: (1) It must move in the same direction as Earth's rotation; (2) it must orbit exactly over the equator; and (3) the orbit must be perfectly circular. The altitude of the orbit must be exact, at 35,900 kilometers (22,300 mi). At this greater height, the orbital velocity is less

than that for a polar orbit—11,120 kilometers/hour (6900 mph). When these conditions are met, the satellite's orbit is perfectly synchronized with Earth's rotation, and the satellite is always located over the same spot above Earth. This orbit offers some advantages. First, at its great distance, a geostationary satellite can view an entire Earth hemisphere in one image (that is, the half it is always facing—a companion satellite images the other hemisphere). Another great advantage is that geostationary satellites can send back a continuous stream of images for monitoring changes in our atmosphere and oceans. A film loop of successive geostationary images is what we see on TV weather broadcasts when we see motion in the atmosphere. Geostationary satellite images give us broad regional presentations of an entire hemisphere at once. Near polar–orbiting satellites take image after image in a swath and rely on Earth's rotation to cover much of the planet over a time span of about a week and a half.

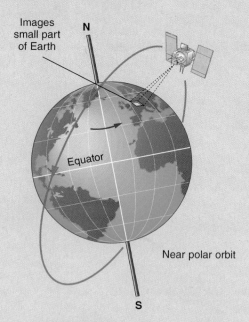

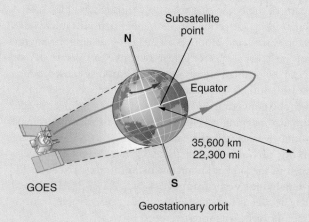

*Polar orbits* circle Earth approximately from pole to pole and use the movement of Earth as it turns on its axis to image small areas (perhaps 100 × 100 km) to gain good detail of the surface. This orbital technique yields nearly full Earth coverage in a mosaic of images, and the satellite travels over the same region every few days, always at the same local time. (Not to scale.)

*Geostationary orbits* are used with satellites orbiting above the equator at a speed that is synchronized with Earth rotation so that the satellite can image the same location continuously. Many weather satellites use this orbit at a height that will permit imaging an entire hemisphere of Earth. (Not to scale.)

● **FIGURE 2.40** This color-composite satellite image of the New Orleans area along the Mississippi River in Louisiana was taken 5 years before the disastrous impact of Hurricane Katrina, which devastated much of the city in 2005.

*What features and geographic patterns can you recognize on this natural-color image?*

**Sonar** (*SO*und *NA*vigation and *R*anging) uses the reflection of emitted sound waves to probe the ocean depths. Much of our understanding of sea floor topography, and mapping of the sea floor, has been a result of sonar applications.

**Multispectral remote sensing** uses and compares more than one kind of image of the same place using some combination of different remote sensing systems—for example, radar and TIR images or NIR and normal color photos/images. Common on satellites, *multispectral scanners* sense many kinds of energy simultaneously and relay them to receiving stations as separate digital images. Each part of the energy spectrum yields different information about aspects of the environment. The separate images, like thematic map layers in a GIS, can be combined, depending on which ones are needed for analysis.

Many types of images can be generated from multispectral data, but the most familiar is the **color composite image** (● Fig. 2.40). Blending three images of the same location, by overlaying pixel data from three different *wavelengths* of reflected light, creates a digital color composite. A common color composite im-age resembles a false-color NIR photograph with a color assignment that resembles color NIR photos. On a standard NIR color composite, red is healthy vegetation; pinkish tones may represent vegetation that is under stress; barren areas show up as white or brown; clear, deep-water bodies are dark blue or black; and muddy water appears light blue. Clouds and snow are bright white. Urbanized areas are blue–gray. Although the colors are visually important, the greatest benefit of digital multispectral imagery is that computers can be used to identify, classify, and map (in a first approximation) these kinds of areas automatically, based on color and tone differences. These digital images can be input as thematic layers for integration into a GIS, and geographic information systems are often closely linked to the analysis of remotely sensed images.

The use of digital technologies in mapping and imaging our planet and its features continues to provide us with data and information that contribute to our understanding of the Earth system. Through continuous monitoring of the Earth system, global, regional, and even local changes can be detected and mapped. Geographic information systems have the capability to match and combine thematic layers of any sort, instantly accessing any combination of layers that we need to solve complex spatial problems.

Maps and various kinds of representations of Earth continue to be essential tools for geographers and other scientists, whether they are on paper, displayed on a computer monitor, hand drawn in the field, or stored as a mental image. Digital mapping, GPS, GIS, and remote sensing have revolutionized the field of geography, but the fundamental principles concerning maps and cartography remain basically unchanged.

# CHAPTER 2 ACTIVITIES

## ■ DEFINE AND RECALL

oblate spheroid
navigation
great circle
hemisphere
small circle
coordinate system
North Pole
South Pole
equator
latitude
sextant
prime meridian
longitude
decimal degrees
geographic grid
parallel
meridian
time zone
solar noon
International Date Line
U.S Public Lands Survey System
principal meridian
base line
township
section

Global Positioning System (GPS)
map projection
planar projection
conic projection
cylindrical projection
Mercator projection
conformal map
equal-area map
equidistance
azimuthal map
gnomonic projection
compromise projection
legend
scale
verbal scale
representative fraction (RF) scale
graphic (bar) scale
magnetic declination
azimuth
bearing
thematic map
discrete data
continuous data
isoline
topographic contour line

contour interval
profile
gradient
geographic information system (GIS)
digital map layers
digital elevation model (DEM)
vertical exaggeration
visualization models
draping
remote sensing
aerial photograph
digital image
pixel
resolution (spatial resolution)
near-infrared (NIR)
thermal infrared (TIR)
radar
imaging radar
side-looking airborne radar (SLAR)
weather radar
lidar
sonar
multispectral remote sensing
color composite image

## ■ DISCUSS AND REVIEW

1. Why is a great circle useful for navigation?
2. What are the latitude and longitude coordinates of the place (town, city) where you live?
3. Approximately how precise in meters could you be if you tried to locate a building in your city to the nearest second of latitude and longitude? What about using a GPS?
4. What time zone are you in? What is the time difference between Greenwich mean time and your time zone?
5. If you fly across the Pacific Ocean from the United States to Japan, how will the International Date Line affect the day change?
6. How has the use of the Public Lands Survey System affected the landscape of the United States? Has your local area been affected by its use? If so, how?
7. Why is it impossible for maps to provide a completely accurate representation of Earth's surface? What is the

difference between a conformal map and an equal-area map?
8. What is the difference between a representative fraction and a verbal map scale?
9. What does a small-scale map show in comparison with a large-scale map?
10. What does the concept of *thematic map layers* mean in a geographic information system?
11. What specific advantages do computers offer to the mapmaking process?
12. What is the difference between a photograph and a digital image?
13. What does a weather radar image show to help us understand weather patterns?

# CONSIDER AND RESPOND

1. Select a place within the United States that you would most like to visit for a vacation. You have a highway map, a USGS topographic map, and a satellite image of the area. What kinds of information could you get from one of these sources that is not displayed on the other two? What spatial information do they share (visible on all three)?

2. If you were an applied geographer and wanted to use a geographic information system to build an information database about the environment of a park (pick a state or national park near you), what are the five most important layers of mapped information that you would want to have? What combinations of two or more layers would be particularly important to your purpose?

# APPLY AND LEARN

1. If it is 2:00 a.m. Tuesday in New York (EST), what time and day is it in California (PST)? What time is it in London (GMT)? What is the date and time in Sydney, Australia (151° East)?

2. A topographic profile has a linear scale of 1:2400, and a vertical scale of 1 inch equals 100 feet. How many feet does 1 inch equal on the linear scale? If there is vertical exaggeration, what is it?

3. If 10 centimeters (3.94 in.) on a map equal 1 kilometer (3281 ft) on the ground, what is the RF scale of the map? You can round the answer to the nearest thousand. This is the formula to use for scale conversions of this kind:

$$\frac{\text{Map distance}}{\text{Earth distance}} = \frac{1}{\text{RFD}}$$

where RFD = representative fraction denominator

# LOCATE AND EXPLORE

1. The coordinate system used on a globe is latitude and longitude, representing angular distance (in degrees) north and south of the equator and angular distance east and west of the prime meridian that passes through Greenwich, England. Using the Search window in Google Earth, fly to the heart of the following cities and identify the latitude and longitude. Measure the latitude and longitude using decimal degrees, and round off to two decimal places (for example, 41.89 N as opposed to 41°88′54.32″ N). Make sure that you correctly note whether the latitude is North (N or +) or South (S or −) of the equator and whether the longitude is East (E or +) or West (W or −) of the prime meridian.

   *Tip:* Change the latitude/longitude setting in the Tools > Options dialog box.

   | | |
   |---|---|
   | London, England | Cape Town, South Africa |
   | Paris, France | Moscow, Russia |
   | New York City | Beijing, China |
   | San Francisco, California | Sydney, Australia |
   | Buenos Aires, Argentina | Your hometown |

   Now reverse the latitude and longitude of your hometown (N to S, and W to E) and note where you are in the world.

2. To find your location on the surface of the Earth, you can use a global positioning system (GPS) device, which gives location in latitude and longitude. Enter the following coordinates into Google Earth to identify the location:

   | | |
   |---|---|
   | 41.89 N, 12.492 E | 27.175 N, 78.042 E |
   | 33.857 S, 151.215 E | 27.99 N, 86.92 E |
   | 29.975 N, 31.135 E | 40.822 N, 14.425 E |
   | 90.0 N, 0 E | 48.858 N, 2.295 E |
   | 90.0 S, 90.0 W | |

   *Tip:* Use the zoom, tilt, rotate, and elevation exaggeration functions of Google Earth to help view and interpret the landform object shown in the browser.

3. When looking at a topographic profile or using the terrain feature in Google Earth, you can control the elevation exaggeration, which is calculated as the ratio of the units on the horizontal *(x)* axis to the units on the vertical *(z)* axis. In Google Earth you can adjust the elevation exaggeration between 0.5 and 3, thereby making subtle objects more noticeable. (Go to Tools > Options to set the elevation exaggeration.) In Google Earth turn on the Elevation Exaggeration Layer and then Fly to Mount Everest, the Nebraska Sand Hills, and the Goosenecks of the San Juan River. Adjust the exaggeration from 0.5 to 3 and notice the change in the terrain. In which of these landscapes is a higher level of vertical exaggeration most useful in interpreting the natural terrain?

   *Tip:* Use the zoom, tilt, and rotate functions of Google Earth to help view and interpret the landform object shown in the browser.

4. Download the Stone Mountain Topographic Layer from this book's online resources found at www.cengagebrain.com. Using this layer as a reference, draw a topographic profile (similar to Fig. 2.29 in your text) from Point A to Point B along the line shown. Use the contour lines and Google Earth's elevation exaggeration and tilt features to decipher the landform.

   *Tip:* Adjust the transparency of the topographic layer to see the image below. Turn off unnecessary layers for better visibility.

 CourseMate – Make the most of your study time by accessing everything you need to succeed in one place. Read your textbook, take notes, review flashcards, watch videos, complete activities, take practice quizzes, and more online with CourseMate. Log in at **www.cengagebrain.com**.

# Map Interpretation

## ■ TOPOGRAPHIC MAPS

### The Map

Menan Buttes on the Snake River Plain in southern Idaho. Topographic maps show elevation changes with contour lines that connect points of equal elevation. The character of the landscape and the landforms can be interpreted from these maps if you understand how to interpret contour lines. Elevations, hiking routes, slope steepness, and the direction of stream flow can be determined by reading a topographic map. Standard symbols are used on USGS topographic maps to represent mapped features and information (a guide is in Appendix B).

The elevation difference between adjacent contour lines, called the *contour interval,* depends on the map scale and how much elevation change exists in the mapped area (the local relief). Elevations represented by topographic contours are typically divisible by ten. Mountainous areas are shown with greater intervals to keep contours from visually merging. Low relief areas use a smaller contour interval to show subtle relief features. It is important to note the map scale and the contour interval when first examining a topographic map.

Keep in mind several rules when interpreting contours:

- Closely spaced contours indicate a steep slope, and widely spaced contours indicate a gentle slope.
- Evenly spaced contours indicate a uniform slope.
- Closed contour lines represent a hill or a depression (if there are tick marks on the downslope side).
- Contour lines never cross but may converge along a vertical cliff.
- Contour lines bend upstream when they cross a valley.

### Interpreting the Map

1. The topographic map scale is now 1:12,000. What would 1 inch on the map represent in feet on the Earth's surface? What was the representative fraction scale of the original map scale before enlargement? What is the contour interval on this map?
2. What are the highest and lowest elevations on the map? Where are they located? What is the crater depth from the highest point on the rim to the lowest point in the crater?
3. Slope ratio is calculated by dividing elevation change by horizontal distance. For example, a 1000-meter-high elevation rise over a horizontal distance of 3000 meters would have a slope ratio of 1:3. What is the slope ratio for the slope of South Menan Butte from the 5289 elevation point down to the Snake River (below the "R" in river)?
4. Why would it be useful to have both a map and an aerial photograph when studying landforms? What is the chief advantage of each? If you were using a compass with this map, what is the magnetic declination as shown here?

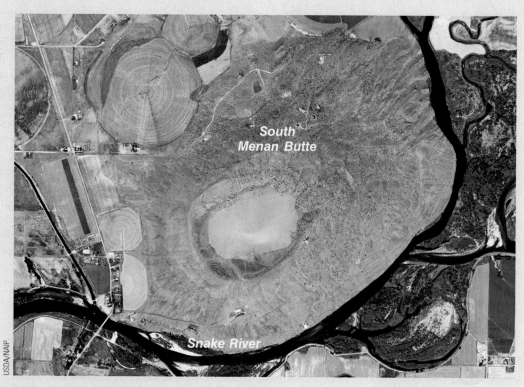

A natural color aerial photograph of South Menan Butte.

 In Google Earth fly to: 43.76°N, 111.97°W.

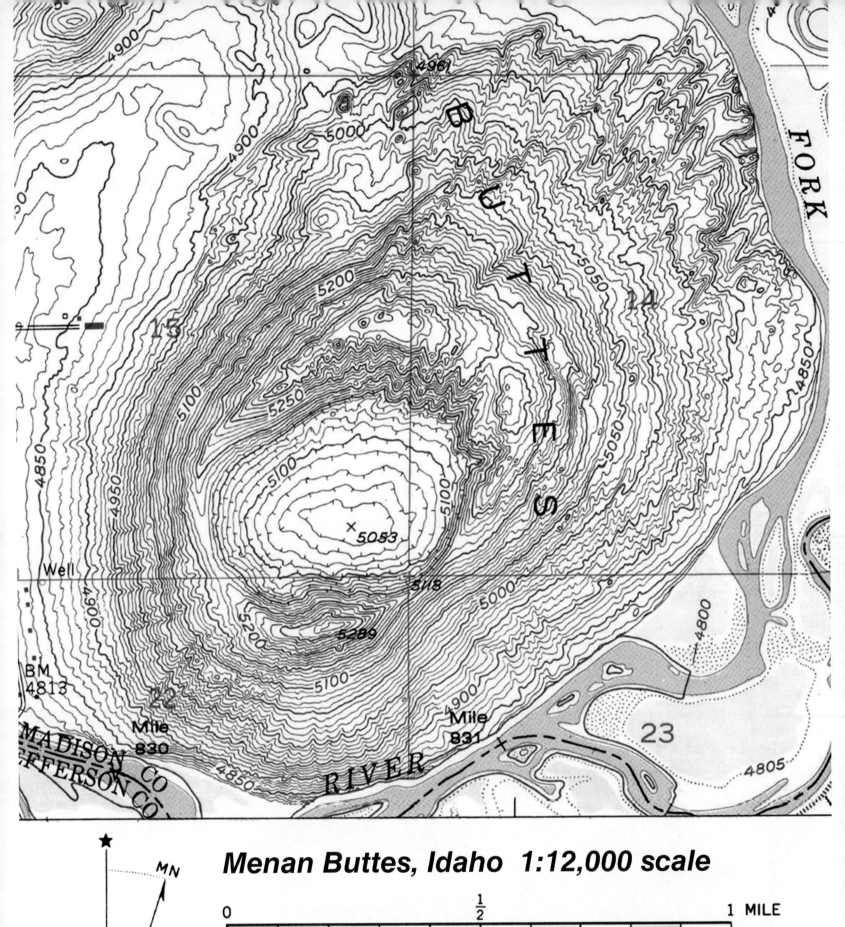

**Menan Buttes, Idaho 1:12,000 scale**

MN

17°

| 0 | | | | | | 1/2 | | | | | | 1 MILE |

| 0 | 1000 | 2000 | 3000 | 4000 | 5000 FEET |

| 0 | | .5 | | 1 KILOMETER |

# 3

# SOLAR ENERGY AND EARTH–SUN RELATIONSHIPS

## ■ OBJECTIVES

WHEN YOU COMPLETE THIS CHAPTER YOU SHOULD BE ABLE TO:

■ Explain why the sun is the ultimate source of energy that drives the various components of the Earth system.

■ Discuss the types of energy emitted by the sun as they are represented in the electromagnetic spectrum and their use or impact on Earth.

■ Describe how Earth–sun relationships affect the receipt and distribution of solar energy on our planet.

■ List each of the four astronomical seasons and their occurrence in terms of Earth–sun relationships.

■ Diagram the positions of Earth in its annual orbit around the sun during the solstice and equinox days including the orientation of the axis.

■ Explain how and why the length of daylight hours varies by latitude and by hemisphere.

■ Outline the significant relationships to solar radiation and the seasons of the Tropics of Cancer and Capricorn and the Arctic and Antarctic Circles.

THE SUN HAS A RADIUS 110 times that of Earth, has a mass that is 330,000 times greater, and reigns as the center of our solar system. The gravitational pull of this star, a fierce and stormy ball of gas, holds Earth in orbit, and its energy emissions power the Earth–atmosphere systems on which our lives depend. As the source of almost all the energy in our world, the sun holds the key to many of our questions about the Earth, the sky, and our weather and climate.

Everyone has wondered about environmental changes that occur throughout the year and from place to place on our planet's surface. Perhaps when you were young you wondered why summer was warmer than it was in winter and why the summertime hours of daylight were long but in other seasons they were shorter. These questions and many like them are probably as old as the earliest human thoughts, and the answers to them provide us with a better understanding of our world's physical geography and the daily and seasonal changes that directly influence our lives.

Sunset over the Pacific Ocean near San Diego, California. Sunrise and sunset are good times for us to think about the importance of the sun to life on Earth. U.S. Geological Survey/Guy DeMeo

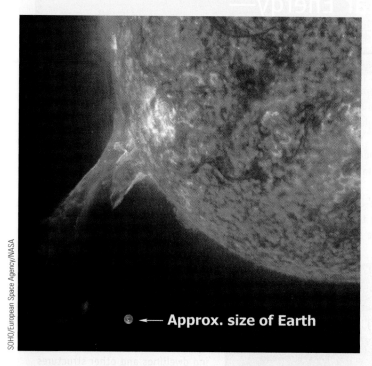

**← Approx. size of Earth**

● **FIGURE 3.4** The explosive thermonuclear fusion reactions that power the sun are similar to the explosion of a hydrogen bomb. This is a massive erupting solar prominence that expelled charged particles and electrons from the sun and into space.

*What elements drive a nuclear fusion reaction?*

these solar winds reach Earth, they are prevented from harming the surface by Earth's magnetic field and are confined to the upper atmosphere. During these times, they can disrupt radio and television communication and may disable orbiting satellites. The Earth's magnetic field tends to direct the solar wind into the outer atmosphere and to the areas surrounding our planet's magnetic poles (see Fig. 3.5). When this happens, the solar wind energizes ions in the outer atmosphere, resulting in an amazing light show known as an **aurora** ( ● Fig. 3.6). The **aurora borealis,** often called the northern lights ( ● Fig. 3.7), and the **aurora australis,** the southern lights, happen simultaneously in the northern latitudes and southern latitudes.

Sunspots, the best-known solar feature, influence the intensity of solar winds. Visible on the photosphere, **sunspots** are dark regions that are about 1500°C to 2000°C (2700°F–3600°F) cooler than the temperatures of the surrounding areas of the sun ( ● Fig. 3.8). During the 1600s Galileo began recording sunspots, and for many years they have been used as an indicator of solar activity. Sunspots seem to occur on an 11-year cycle from one maximum (where 100 or more may be visible) to the next. The next sunspot cycle should peak around the year 2012. An individual sunspot may last for less than a day or as long as 6 months. Just how sunspots might affect Earth's atmosphere is still a matter of controversy. Proving direct connections between sunspot numbers and weather or climate is difficult, but such relationships have been proposed. Times of minimum sunspots have been known to be times of cooler temperatures.

● **FIGURE 3.5** Earth's magnetic field protects the surface from the harmful effects of the solar wind.

*What impacts can the solar wind still have on our lives?*

**FIGURE 3.10**
the sun. Earth's r
to rise in the east

velocity for all
or 15° per hou
tance (not the
velocity at the
with a postage
360° but cover
you place the
Poles, howeve
rotation of the
where the dist
Earth. At Kam
460 meters (15
(1038 mi) per I
Russia (60°N
complete rotati
rotates at appro

We are u
angular velocit
(2) the atmospl
objects, either :
to Earth, to wl
these reference

Our altern
ing a light at a
east. You can s
while the othe
are continually
while others a
rotation and th
mean that half
radiation, while
2, the great ci
**circle of illur**

**FIGURE 3.6** Solar wind, when directed toward Earth, forms ring-shaped auroras over and around the magnetic poles in each hemisphere. This is the Northern Hemisphere's aurora borealis.

*What is the Southern Hemisphere's aurora called?*

## Solar Energy and Atmospheric Dynamics

As we have previously noted, our sun is the major source of energy, either directly or indirectly, for the Earth system. Earth does receive certain amounts of heat from its interior (volcanoes, hot springs, and geysers). When compared with the amount of

**FIGURE 3.7** The aurora borealis, seen here from an area in Alaska, interrupts the darkness of the winter sky with a beautiful and amazing light show. Energy interactions between the solar wind and ions in the atmosphere produce the auroras.

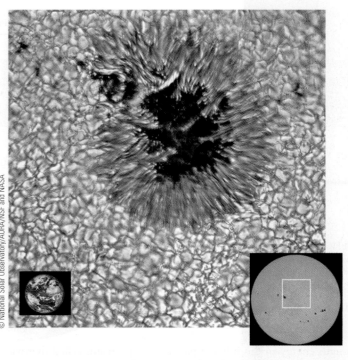

**FIGURE 3.8** Sunspots as they appear on the solar surface. Insets show the sun's area that is illustrated here and the relative size of Earth.

*About how many Earth diameters can fit east to west across this sunspot?*

energy received from the sun, however, these other sources are insignificant.

The sun emits **electromagnetic energy,** which travels at the speed of light in a range of varying wavelengths, called the **electromagnetic spectrum** ( ● Fig. 3.9). It takes about 8.3 minutes for this energy to reach Earth. Approximately 9% of solar energy consists of radiation with wavelengths that are shorter than visible light, which includes gamma rays, X-rays, and ultraviolet radiation. These short wavelengths cannot be seen but can affect tissues of the human body. It is well known that absorbing too much X-ray energy can be dangerous. Exposure to excessive ultraviolet energy gives us sunburned skin and is a primary cause of skin cancer. About 41% of the solar energy spectrum comes in the form of visible light rays, where each color is distinguishable by its specific wavelength band. Wavelengths longer than visible light are other parts of the electromagnetic spectrum, which are not visible to the human eye. About 49% of the sun's radiant energy is emitted in wavelengths that are longer than the range of human vision. Although these wavelengths are invisible, they can sometimes be sensed by human skin. Near-infrared energy, emitted in wavelengths of light that are longer than the range of human vision, are harmless to living organisms. Longer wave-

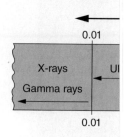

66

0.01

X-rays | UI

Gamma rays

0.01

0.4

● **FIGURE 3.9** Solar ener
wavelengths, which are m
of a meter). Visible light (
0.7 micrometers.

*Are radio signals cons*

length energy in the
infrared, can be felt a
falls into the very lon;
waves and those used
lectively, gamma ray:
near-infrared energy a
energy including the
lengths (microwaves,
**radiation** (see Fig. 3.
Through advanc
electromagnetic wave
tions. In the field of
television and radio l
crowaves are also u
gamma rays are used
tion security checks. I
reflected light energy
microwaves to detec
ships and to map elev:
make images of heat
temperature. Many o:
energy were discussed
The sun radiates
The rate of a planet's :
**constant** and has be
Earth's atmosphere by
at the outer edge of I
square meter (in SI un
meter). These both ar
has also been defined
square centimeter per
required to raise the t
all expressions of Eart
usage varies dependin
tific study. The atmos

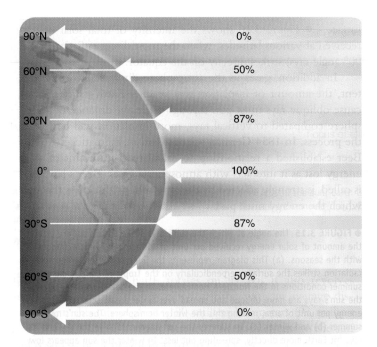

● **FIGURE 3.16** The percentage of incoming solar radiation (insolation) striking various latitudes during an equinox date according to Lambert's law.

*How much less solar energy is received at 60° latitude than that received at the equator?*

Because no insolation is received at night, the duration of solar energy is related to the daylight hours at a particular location on Earth, which varies over the seasons (Table 3.2). We all know that the days are longer in the summer and shorter in the winter. Of course, every day is 24 hours long, but what we are referring to is the length of time between sunrise and sunset. Obviously, the more hours of daylight that are experienced at a location, the greater the amount of solar radiation that should be received at that location. In the next section, we will discuss why periods of daylight vary in length through the seasons and from place to place.

## The Seasons

Many people erroneously believe that the changing distance between Earth and the sun must cause the seasons during Earth's annual revolution. As noted earlier, in terms of percentage, the change in this distance is very small. Further, for people in the Northern Hemisphere, Earth is actually closest to the sun in January and farthest away in July (see Fig. 3.13). This is exactly opposite of that hemisphere's seasonal temperature variations. As we will see, seasons are caused by the 23½° tilt of Earth's axis away from being perpendicular to the plane of the ecliptic (see Fig. 3.14) and the parallelism of the axis that is maintained as Earth orbits the sun.

**TABLE 3.2**
**Duration of Daylight for Certain Latitudes**

| Length of Day (Northern Hemisphere) (read down) | | | |
|---|---|---|---|
| LATITUDE (IN DEGREES) | MAR. 20/SEPT. 22 | JUNE 21 | DEC. 21 |
| 0.0 | 12 hr | 12 hr | 12 hr |
| 10.0 | 12 hr | 12 hr 35 min | 11 hr 25 min |
| 20.0 | 12 hr | 13 hr 12 min | 10 hr 48 min |
| 23.5 | 12 hr | 13 hr 35 min | 10 hr 41 min |
| 30.0 | 12 hr | 13 hr 56 min | 10 hr 4 min |
| 40.0 | 12 hr | 14 hr 52 min | 9 hr 8 min |
| 50.0 | 12 hr | 16 hr 18 min | 7 hr 42 min |
| 60.0 | 12 hr | 18 hr 27 min | 5 hr 33 min |
| 66.5 | 12 hr | 24 hr | 0 hr |
| 70.0 | 12 hr | 24 hr | 0 hr |
| 80.0 | 12 hr | 24 hr | 0 hr |
| 90.0 | 12 hr | 24 hr | 0 hr |
| LATITUDE | MAR. 20/SEPT. 22 | DEC. 21 | JUNE 21 |
| Length of Day (Southern Hemisphere) (read up) | | | |

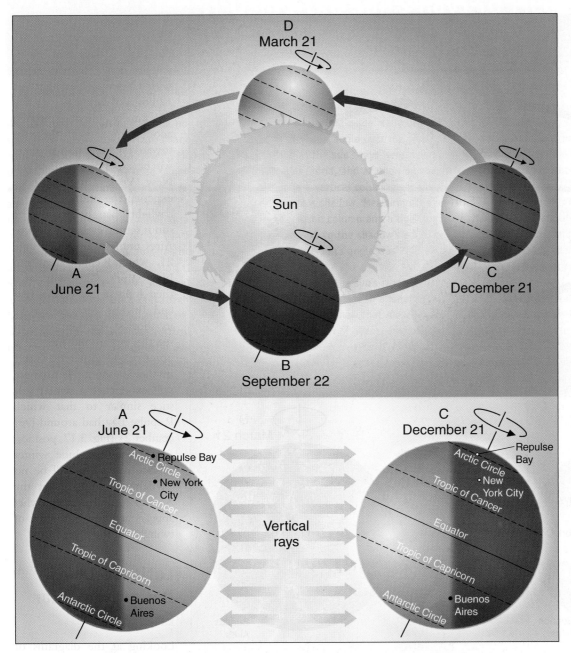

on the June solstice. On the same day, someone in New York City will experience a longer period of daylight than of darkness. However, someone in Buenos Aires, Argentina, will have a longer period of darkness than daylight on that day. June 21st is the winter solstice in the Southern Hemisphere. Thus, in the Southern Hemisphere, June 21 is the shortest day with the lowest sun angles of the year, and in the Northern Hemisphere, it is the longest day with the highest sun angles of the year.

Imagine the movement of Earth from its position at the June solstice toward a position a quarter of a year later, in September. As Earth moves toward that new position, we can imagine the changes that will be taking place in our three cities. In Repulse Bay the time of darkness will increase through July, August, and September. In New York sunset will be arriving earlier. In Buenos Aires the situation will be reversed; as Earth moves toward its position in September, the daylight hours in the Southern Hemisphere will begin to get longer and the nights will become shorter.

Finally, on or about September 22 Earth will reach a position known as an **equinox** (Latin: aequus, equal; nox, night). On this date (the **autumnal equinox** in the

● **FIGURE 3.17** The geometric relationships between Earth and the sun during the June and December solstices. Note the differing day lengths at the summer and winter solstices in the Northern and Southern Hemispheres.

About June 21, Earth is in a position in its orbit so that the northern tip of its axis is inclined toward the sun at an angle of 23½°. In other words, the plane of the ecliptic (the 90° sun angle) is directly on 23½° N latitude. This day during Earth's orbit is called the Northern Hemisphere summer **solstice** (from Latin: sol, sun; sistere, to stand). We can best see what is happening if we refer to ● Figure 3.17, position A. In that diagram, we can see that the Northern and Southern Hemispheres receive unequal amounts of light from the sun. That is, as we imagine the Earth rotating under these conditions, a larger area of the Northern Hemisphere than the Southern Hemisphere remains in daylight. Conversely, a larger area of the Southern Hemisphere than the Northern Hemisphere remains in darkness. Thus, a person living at Repulse Bay, Canada, north of the Arctic Circle, experiences a full 24 hours of daylight

Northern Hemisphere) day and night will be of equal length at all locations on Earth. Thus, on the equinox, daylight hours are identical for both hemispheres and at all latitudes. As you can see in ● Figure 3.18, position B, Earth's axis points neither toward nor away from the sun (imagine the axis is pointed at the reader); the circle of illumination passes through the poles, and it cuts Earth in half along its axis.

Imagine again the revolution and rotation of Earth while moving from around September 22 toward a new position another quarter of a year later in December. We can see that in Repulse Bay, Canada the nights will be getting longer until, on the winter solstice, which occurs on or about December 21, this northern town will experience 24 hours of darkness (Fig. 3.17, position C). The only natural light at all in Repulse Bay will be

NASA satellites, in an effort to understand the geographical distribution of places in terms of their potential for harnessing solar power as an energy resource.

The amount of insolation received by the Earth system is an important factor in understanding atmospheric and oceanic dynamics and the geographic distribution of climate, soils, and vegetation. Climate elements such as temperature, precipitation, and winds are controlled in part by the amount of insolation received at various locations on Earth. People depend on certain levels of insolation for physical comfort, and plant life is especially sensitive to the availability and seasonality of solar radiation. You may have noticed that plants can wilt if they receive too much sunshine or they may become brown in a dark corner away from a window. Over a longer period, certain plants have an annual cycle of budding, flowering, leafing, and losing their leaves. The increase and decrease in solar radiation that accompanies the changing seasons apparently determine this cycle. Animals also respond to seasonal changes. Some animals hibernate; many North American birds fly south toward warmer weather as winter approaches; and many animals breed at such a time that their offspring will be born in the spring, when warm weather is approaching.

Ancient civilizations around the world (from China to Mexico) realized the incredible influence of the sun's energy, and many societies worshiped the sun as chief among their gods ( ● Fig. 3.22). This should not be too surprising—they were well aware that the sun was vital to their survival. Today

most people do not worship the sun as a god, but it is extremely important to understand and appreciate its role as the ultimate energy source for the vast majority of Earth's natural subsystems.

● **FIGURE 3.22** El Castillo, a Mayan Pyramid at Chichén Itzá, Mexico, was oriented to the annual change in sun angle because the Maya worshiped the sun. Each side has 91 steps—the number of days between the solstice and equinox days. On the vernal equinox, the afternoon sun casts a shadow that makes it appear like a giant snake is crawling down the pyramid's north side.

*Why would ancient civilizations worship the sun?*

# CHAPTER 3 ACTIVITIES

## ■ DEFINE AND RECALL

| | | |
|---|---|---|
| galaxy | sunspot | insolation |
| light-year | electromagnetic energy | solstice (summer and winter) |
| solar system | electromagnetic spectrum | equinox |
| gravity | shortwave radiation | autumnal equinox |
| mass | longwave radiation | vernal equinox |
| planet | solar constant | Arctic Circle |
| satellite | calorie | Antarctic Circle |
| asteroid | galactic movement | vertical (direct) rays |
| comet | rotation | Tropic of Cancer |
| meteor | revolution | Tropic of Capricorn |
| meteorite | circle of illumination | solar declination |
| terrestrial planet | perihelion | analemma |
| giant planet (gas planet) | aphelion | zenith |
| fusion (thermonuclear) reaction | plane of the ecliptic | tropical zone |
| solar wind | angle of inclination | middle-latitude zone |
| aurora | parallelism of the axis | polar zone |
| aurora borealis | North Star (Polaris) | |
| aurora australis | | |

# DISCUSS AND REVIEW

1. What is a solar system? What bodies constitute our solar system?
2. How is the energy emitted from the sun produced?
3. Name the terrestrial planets. What do they have in common? Name the giant planets. What do they have in common?
4. Which planets are capable of maintaining a gaseous atmosphere?
5. The electromagnetic spectrum displays various types of energy by their wavelengths. Where is the division between longwave and shortwave energy? In what ways do humans use electromagnetic energy?
6. Is the amount of solar energy reaching Earth's outer atmosphere constant? What might make it change?
7. Describe briefly how the influence of Earth's rotation and revolution on insolation affect life on Earth.
8. If the sun is closest to Earth on January 3, why isn't the winter in the Northern Hemisphere warmer than winter in the Southern Hemisphere?
9. Identify the two major factors that cause regular variations in insolation throughout the year. How do they combine to cause the seasons?
10. On the world map of average annual insolation in Figure 3.21, what are some reasons why the patterns do not always follow exactly belts of latitude?

# CONSIDER AND RESPOND

1. Given what you know of the sun's relation to life on Earth, explain why the solstices and equinoxes have been so important to cultures all over the world.
2. Use the discussion of solar angle, including Figure 3.15, to explain why we can look directly at the sun at sunrise and sunset but not at the noon hour.
3. Describe in your own words the relationship between insolation and latitude.
4. Use the analemma presented in Figure 3.19 to determine the latitude where the noon sun will be directly overhead on February 12, July 30, November 2, and December 30.

# APPLY AND LEARN

1. Imagine you are at the equator on March 20. The noon sun would be directly overhead. However, for every degree of latitude that you travel to the north or south, the noon solar angle would decrease by the same amount. For example, if you travel to 40°N latitude, the solar angle would be 50°. Explain this relationship.

   Develop a formula or set of instructions to generalize this relationship.

   What would be the solar angle at 40°N on June 21? On December 21?

   Lambert's law can be expressed as: $I = I_0 \cos g$

   Where: $I =$ −Intensity of solar radiation received at the surface.

   $I_0 =$ −Intensity of solar radiation received from a 90° angle.

   $\cos =$ Cosine of g

   $g =$ the sun's zenith angle

   The zenith angle is measured from 90° straight above, down to the sun's position in the sky.

2. Using Lambert's law, calculate the intensity of the insolation on September 22 at the following locations:

   0° latitude

   27°S latitude

   40°N latitude

   65°S latitude

   83°N latitude

 CourseMate – Make the most of your study time by accessing everything you need to succeed in one place. Read your textbook, take notes, review flashcards, watch videos, complete activities, take practice quizzes, and more online with CourseMate. Log in at **www.cengagebrain.com**.

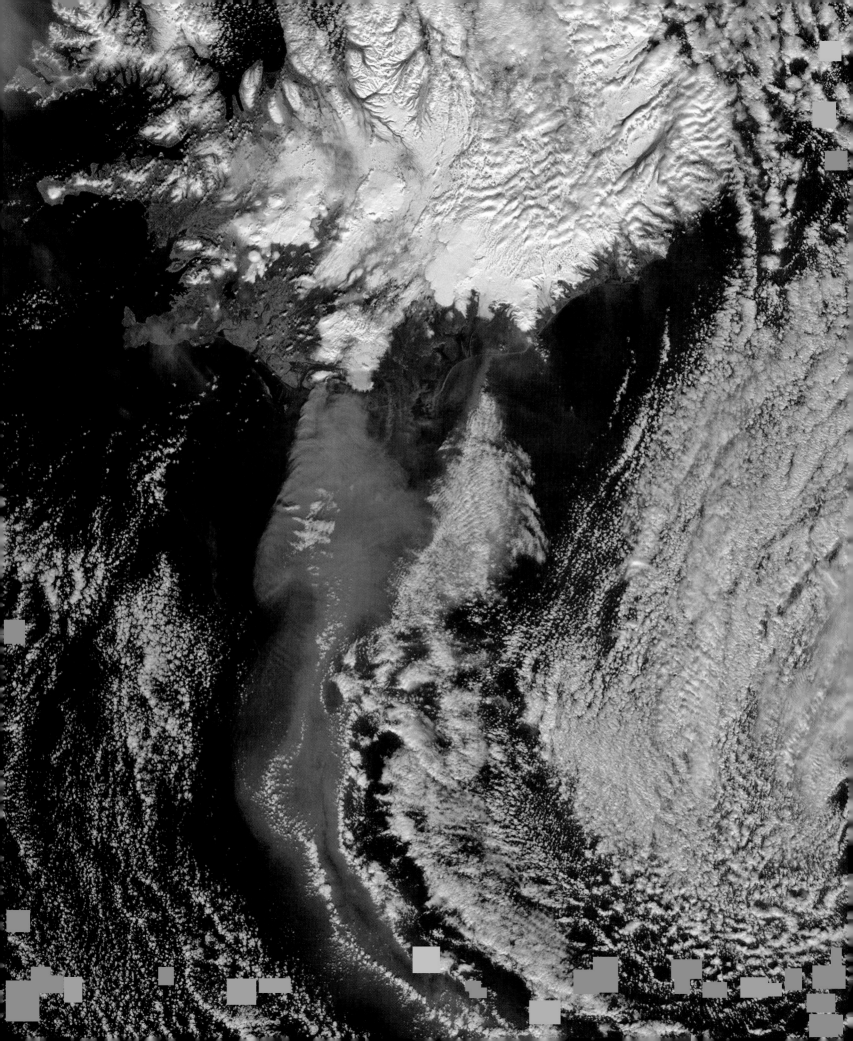

# 4

# THE ATMOSPHERE, TEMPERATURE, AND EARTH'S ENERGY BUDGET

## ■ OBJECTIVES

WHEN YOU COMPLETE THIS CHAPTER YOU SHOULD BE ABLE TO:

■ List several reasons why our planet's atmosphere is essential to life on Earth.
■ Outline the major gases found in the atmosphere.
■ Conceptualize the ways that solar energy interacts with Earth's atmosphere and surface.
■ Explain how the solar energy that reaches Earth's surface is transferred to the atmosphere.
■ Discuss the processes and the important roles that water plays in heat transfer.
■ Outline the characteristics of the temperature-based layers of the atmosphere.
■ Describe the controls on horizontal distribution of Earth's surface temperatures.
■ Explain the major inputs, outputs, and processes in Earth's energy budget and why it is in a state of energy equilibrium.

The 2010 eruption of Eyjafjallajökull volcano in Iceland emitted a huge plume of volcanic dust into the atmosphere. Swept by winds to Europe, the abrasive dust disrupted air travel in the region. Large volcanic eruptions can have a significant effect on the atmosphere and even influence climate. NASA/GSFC/Jeff Schmaltz/ MODIS Land Rapid Response Team

✛ In Google Earth fly to: 63.629°N, 19.619°W

THE ATMOSPHERE SURROUNDS OUR PLANET with an envelope of air and supplies the gases that support life. The atmosphere also profoundly affects the solar energy that heats Earth and powers its environmental systems. An awareness of the atmosphere's composition, structure, and dynamics is essential for understanding the critical roles that the atmosphere plays in the Earth system and for life on our planet. One description of the atmosphere is that it is like an ocean of air that surrounds Earth. This description reminds us that the atmosphere has dynamic air movements—currents and circulations—that create the changing conditions that we know as weather.

The atmosphere provides oxygen and water that are vital for animals and humans to survive and the water and carbon dioxide that plants require. Compared to our planet's size, the atmosphere is a rather thin film of air ( ● Fig. 4.1), yet it serves as an effective insulator, maintaining the viable temperatures we find on Earth. Life on Earth exists within rather narrow ranges of temperature. Without the atmosphere, our planet would experience temperature extremes of as much as 260°C (500°F) between day and night. Serving as a filter or shield, the atmosphere blocks

● **FIGURE 4.1** The atmosphere is a relatively thin envelope of gases that surrounds the Earth. Compared to the size of an apple, the thickness of its skin is an approximate comparison.

*What does this say about the potentially fragile nature of the atmosphere?*

much of the sun's ultraviolet (UV) radiation and protects us from meteor showers.

We can consider the moon—a celestial body with virtually no atmosphere—to see the importance of our atmosphere. A person standing on the moon without a space suit that supplies air would immediately die for lack of oxygen. Our lunar astronauts recorded temperatures of up to 204°C (400°F) on the hot, sunlit side of the moon, and on the dark side, they recorded temperatures approaching −121°C (−250°F). These temperature extremes would kill an unprotected human.

An astronaut on the moon would also notice an "unearthly" silence. On Earth we are able to hear noises and speech because sound waves move by vibrating molecules in the air. Because the moon has no atmosphere and no molecules to transmit the sound waves, a lunar visitor cannot hear any sounds or speak with an audible voice; only radio communications are possible. Also, because there is no atmosphere, it is not possible to fly airplanes or helicopters and it would be fatal to use a parachute. In addition, the moon's lack of atmosphere means that it receives no

protection from being bombarded by meteors. In contrast, most meteors burn up before reaching Earth's surface because of friction with the atmosphere and oxygen, which encourages burning. Without a protective atmosphere, ultraviolet rays from the sun would also burn a lunar visitor. On Earth, we are protected to a large degree from UV radiation because the ozone layer in the upper atmosphere absorbs a major portion of this harmful solar radiation.

We can see that, compared with our stark, silent, and lifeless moon, Earth harbors a hospitable environment for life almost solely because of the nature of our atmosphere. For example, many plants reproduce by pollen and spores that are carried by winds. Birds can fly only because of the air. Atmospheric processes maintain the cycling of water from the oceans to the land and back. The atmosphere is also an important medium for moving thermal energy from one place to another. The gas molecules and particles in the air interact with sunlight in a process called **scattering** (see the box in this chapter), the diffusion of various wavelengths of light, which gives us blue skies and the beautiful red and orange sunrises and sunsets. Without this scattering of light, which

● **FIGURE 4.2** The moon's lack of an atmosphere would be deadly to an unprotected astronaut.

*Why are shadows and the sky so dark on the moon?*

diffuses light in many directions, the sky would appear black, as it does from the moon ( ● Fig. 4.2).

Further, the atmosphere provides a means by which many Earth systems are able to tend toward equilibrium conditions. Changes in the weather ultimately result from atmospheric

processes that tend to equalize differences in air temperature and pressure by transferring energy and moisture and through interactions with oceanic circulation systems.

# Characteristics of the Atmosphere

The atmosphere extends to approximately 480 kilometers (300 mi) above Earth's surface. Certain characteristics related to Earth's atmosphere and interactions with energy continue beyond this altitude (such as Earth's magnetic field), but the density of gas molecules above that level is minimal. Gas molecule density decreases rapidly with altitude, and 97% of the air is concentrated in about the first 25 kilometers (16 mi). Because air has mass, the atmosphere exerts pressure on Earth's surface. At sea level, this pressure is about 1034 grams per square centimeter (14.7 lb/sq in.), but as the altitude *increases*, atmospheric pressure decreases. In Chapter 5, we will examine relationships between atmospheric pressure and altitude in more detail.

## Atmospheric Composition

Numerous gases make up the composition of the atmosphere (Table 4.1) and because of mixing by winds, most of these gases remain in the same proportions regardless of the atmospheric density. A bit more than 78% of the atmosphere's volume is nitrogen and nearly 21% consists of oxygen. Argon comprises most of the remaining 1%. The percentage of carbon dioxide in the atmo-

**TABLE 4.1**
**Composition of the Atmosphere Near the Earth's Surface**

| Permanent Gases | | | Variable Gases | | | |
|---|---|---|---|---|---|---|
| GAS | SYMBOL | PERCENT (BY VOLUME) DRY AIR | GAS (AND PARTICLES) | SYMBOL | PERCENT (BY VOLUME) | PARTS PER MILLION (PPM)* |
| Nitrogen | $N_2$ | 78.08 | Water vapor | $H_2O$ | 0 to 4 | |
| Oxygen | $O_2$ | 20.95 | Carbon dioxide | $CO_2$ | 0.0392 | 392* |
| Argon | Ar | 0.93 | Methane | $CH_4$ | 0.00017 | 1.7 |
| Neon | Ne | 0.0018 | Nitrous oxide | $N_2O$ | 0.00003 | 0.3 |
| Helium | He | 0.0005 | Ozone | $O_3$ | 0.000004 | 0.04† |
| Hydrogen | H | 0.00006 | Particles (dust, soot, etc.) | | 0.000001 | 0.01–0.15 |
| Xenon | Xe | 0.000009 | Chlorofluorocarbons (CFCs) | | 0.00000002 | 0.0002 |

*For $CO_2$, 392 parts per million means that out of every million air molecules, 392 are $CO_2$ molecules.

†Stratospheric values at altitudes between 11 km and 50 km are about 5 to 12 ppm.

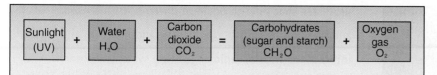

| Sunlight (UV) | + | Water $H_2O$ | + | Carbon dioxide $CO_2$ | = | Carbohydrates (sugar and starch) $CH_2O$ | + | Oxygen gas $O_2$ |

● **FIGURE 4.3** The equation of photosynthesis shows how plants use solar energy to manufacture sugars and starches from atmospheric carbon dioxide and water, liberating oxygen in the process. The stored food energy is then eaten by animals, which also breathe the oxygen released by photosynthesis.

sphere has risen over time, but it is a little less than 0.04% by volume. There also are traces of other gases: ozone, hydrogen, neon, xenon, helium, methane, nitrous oxide, krypton, and others.

**Nitrogen, Oxygen, Argon, and Carbon Dioxide** Of these four most abundant atmospheric gases, nitrogen gas ($N_2$) makes up the largest proportion of air, and nitrogen is very important for plant growth. In addition, several other atmospheric gases are vital to the development and maintenance of life. One of the most critical of these atmospheric gases is, of course, oxygen ($O_2$), which people and other animals use to breathe and oxidize (burn) the food they consume. Oxidation, which is the chemical combination of oxygen with other substances to create new products, also occurs in many other situations. Rapid oxidation takes place when we burn fossil fuels, wood, or refuse, which releases large amounts of heat. The decay of certain minerals or organic debris and the development of rust are examples of slow oxidation. These processes depend on the presence of oxygen in the atmosphere. The third most abundant gas in our atmosphere is Argon (Ar). It is not a chemically active gas and therefore neither helps nor hinders life on Earth. Argon is inert, is heavy, and is a stable product from the radioactive decay of minerals.

Carbon dioxide ($CO_2$) is the fourth most abundant atmospheric gas. The involvement of carbon dioxide in a system known as the *carbon cycle* has been studied for generations. Plants, through the process of **photosynthesis,** use carbon dioxide and water to produce carbohydrates (sugars and starches), in which energy, derived originally from the sun, is stored and used by vegetation ( ● Fig. 4.3). Oxygen is given off as a by-product of photosynthesis. Animals oxidize the carbohydrates, thus releasing the stored energy. A by-product of this process by animals is the release of carbon dioxide, which in turn is used by plants in photosynthesis and completes the cycle.

**Water Vapor, Liquids, Particulates, and Aerosols** Water in the gaseous state is called **water vapor** and is always mixed in some proportion with the air of the lower atmosphere. Water vapor is the most variable of the atmospheric gases and can range from 0.02% by volume in cold, dry climates to more than 4% in the humid tropics. Water vapor in the air will be discussed in some detail later under the broad topic of humidity, but it is important to note that variations in the percentage of water vapor over time and place are important to consider in understanding weather and climate.

Water vapor absorbs and stores heat in the atmosphere, preventing its rapid escape from Earth. Thus, like carbon dioxide, water vapor plays a large role in the insulating properties of the atmosphere. Water occurs naturally in all three states in the atmosphere—as a solid, a liquid, and a gas. Liquid water occurs as rain and as fine droplets in clouds, mist, and fog. Solid water (ice) is found in the atmosphere as ice crystals, snow, sleet, and hail.

**Particulates** are solid particles suspended in the atmosphere. **Aerosols** are solids or liquids suspended in the atmosphere—they include particulates, but they also include tiny liquid droplets and/or ice crystals composed of chemicals other than water. For example, sulfur dioxide crystals ($SO_2$) are atmospheric aerosols. Particulates can be considered as aerosols, but all aerosols are not necessarily particulate matter. Particulates and aerosols can be pollutants from transportation and industry, but particles from natural sources have always existed in our atmosphere ( ● Fig. 4.4). Particulates such as dust, smoke, pollen and spores, volcanic ash, bacteria, and salts from ocean spray can all play important roles in the absorption or blocking of energy and in the formation of rain and snow.

● **FIGURE 4.4** Volcanic eruptions, like this one at Mount Etna in Italy, add a variety of gases, particulates, aerosols, and water vapor into our atmosphere. This view is from the orbiting International Space Station.

*What other processes add particles to the atmosphere?*

⊕ In Google Earth fly to: 37.747°N, 14.998°E (Mt. Etna, Italy)

# Atmospheric Environmental Issues

Two gases in our atmosphere are particularly relevant to environmental issues. One is carbon dioxide, which is related to increases in global temperatures. The other is ozone, which comprises a layer in the upper atmosphere that protects Earth from excessive UV radiation but is endangered by other gases associated with industrialization. In recent years, scientists have directly linked carbon dioxide with long-term changes in Earth's atmospheric temperatures and climates. The major concern is with the relationship between carbon dioxide and rising global temperatures.

**The Greenhouse Effect** We are all familiar with what happens to the inside of a parked car left in the sun with the windows closed. Shortwave radiation (mainly visible light) from the sun penetrates the glass windows with ease ( ● Fig. 4.5). When this insolation strikes the interior of the car, it is absorbed, heating exposed surfaces. Energy, emitted from these surfaces as longwave thermal radiation (heat), cannot escape through the glass as freely. The result is that the vehicle's interior gets hotter throughout the day. Many drivers recognize this as a blast of hot air as the car door opens and seats that are very hot. In extreme cases, windows in some cars have cracked under the heat as a result of thermal expansion (many substances expand when heated). Even more serious is that temperatures can become so high in vehicles with closed windows (or even if they are only slightly open) as to pose a deadly threat to small children and pets left behind.

Termed the **greenhouse effect,** this process is the primary reason for the moderate temperatures observed on Earth. A greenhouse (glass structure that houses plants) will behave in a somewhat similar manner to a closed vehicle parked in the sun ( ● Fig. 4.6). Insolation (shortwave radiation) goes through the transparent glass roof and walls of the greenhouse and helps the plants inside to thrive, even in a cold outdoor environment. After being absorbed by the materials in the greenhouse, the shortwave energy is re-radiated as longwave thermal energy, which cannot escape rapidly, thus warming the greenhouse interior.

Like the glass of a greenhouse, carbon dioxide, water vapor, and other greenhouse gases in the atmosphere are largely transparent to incoming solar radiation but can impede the escape of longwave radiation by absorbing it and then radiating it back to Earth. Carbon dioxide emits about half of its absorbed thermal energy back to Earth's surface. Of course, although there are some similarities, the processes involving the glass of a car or a greenhouse and the atmosphere are significantly different. The heat of a closed car, or a greenhouse, increases because the air is trapped and cannot circulate to the outside air. Our atmosphere is free to circulate and radiated thermal energy is not so confined, so some loss to space is always ongoing.

The greenhouse effect is a natural and much needed process because without greenhouse gases in our atmosphere, our planet

● **FIGURE 4.5** (a) Greenhouse gases in our atmosphere allow solar radiation (sunlight) to penetrate Earth's atmosphere relatively unhampered, but they also store heat energy and retard the loss of terrestrial radiation (heat) from escaping into outer space. (b) A similar sort of heat buildup can occur in a closed car. Solar radiation streaming through the car windows is absorbed into the interior and is re-radiated as heat energy, but the glass does not allow thermal radiation to escape.

*How might you prevent your car's interior from becoming very hot on a summer day?*

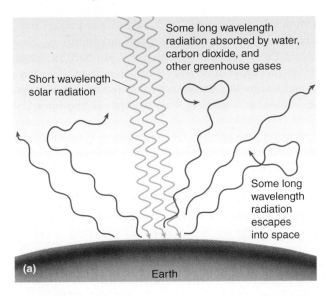

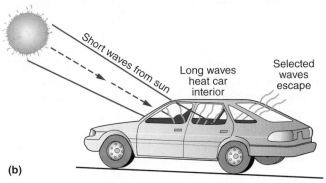

● **FIGURE 4.6** Greenhouses used by horticulturalists at Western Kentucky University. Notice the shaded roof on the greenhouse in the background, which, coupled with open windows and exhaust fans, bears witness to the fact that greenhouses can actually get too hot for the plants.

M. Trapasso

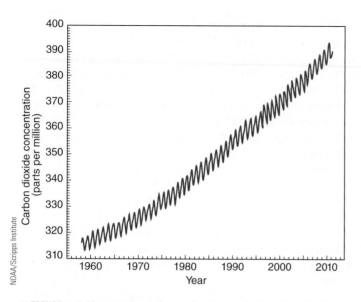

● **FIGURE 4.7** Since 1958, measurements of atmospheric carbon dioxide recorded at Mauna Loa, Hawaii have shown an upward trend.

*What factors seem to be responsible for this trend?*

would be too cold to sustain human life (or perhaps any life). However, a serious environmental issue has arisen in recent years, as growing concentrations of greenhouse gases have been related to measurable increases in global temperatures.

Since the Industrial Revolution began in the late 1800s, burning of fossil (carbon) fuels has added increasing amounts of carbon dioxide to the atmosphere. During this time there has also been massive deforestation, the removal of forests and other vegetation for agriculture and urban development. Vegetation uses large amounts of carbon dioxide in photosynthesis (see Fig. 4.3), and the extensive removal of vegetation permits more $CO_2$ to remain in the atmosphere. ● Figure 4.7 shows the increase of atmospheric carbon dioxide over time as a result of human contributions to this increase. Because carbon dioxide absorbs thermal infrared radiation emitted from Earth's surface and restricts its loss to space, increasing amounts of atmospheric $CO_2$ increase the greenhouse effect and produce a global rise in temperatures. Human influences on increasing carbon dioxide in the atmosphere have resulted in what has been referred to as the **enhanced greenhouse effect.**

The environmental impacts of global warming are many and varied. It is feared that continued global warming will significantly raise sea levels. Higher average temperatures will cause increased melting of ice sheets and mountain glaciers worldwide, and the meltwaters will add to rising ocean levels. Rising sea levels are of great concern to virtually all coastal regions but especially to areas that lie at or below sea level, like the Netherlands and such cities as New Orleans. Increasing temperatures can cause shifts in vegetation regions as plants begin to grow in new locations that previously lacked suitable growing conditions or die out in other places that have become too warm or dry for growth. With temperature shifts, scientists also expect changes in the migration patterns of birds and other animals and the move-

ment of insect-borne diseases to new areas. Also, major global temperature changes will cause a parallel shift in weather systems. Temperature and rainfall patterns may change worldwide, but the impact will be to differing degrees and with varying consequences among the world's regions. Although certain places may benefit and other locations may experience detrimental change, the process adapting to the impacts of significant climate change may be difficult worldwide. Currently, numerous researchers are closely monitoring the trends and amounts of climate change associated with Earth's temperatures and are studying the impacts of global warming related to the *enhanced greenhouse effect.* The issue of climate change will be discussed in more detail in Chapter 8.

**The Ozone Layer** Another vital gas in Earth's atmosphere is **ozone.** The ozone molecule ($O_3$) is related to the oxygen molecule ($O_2$), except it is made up of three oxygen atoms, whereas oxygen gas consists of only two. Ozone is formed in the upper atmosphere when an oxygen molecule is split into two oxygen atoms (O) by the sun's ultraviolet radiation. Then, the free unstable atoms join two oxygen gas molecules to form two molecules of ozone gas consisting of three oxygen atoms each:

$$(1)\ 2O_2 + 2O^- \rightarrow 2O_3 \text{ (using UV radiation)}$$

In the lower atmosphere, ozone is formed by electrical discharges (like high-tension power lines and lightning strokes) and incoming shortwave solar radiation. It is a toxic pollutant and a major component of urban smog, which can cause sore and watery eyes, soreness in the throat and sinuses, and difficulty breathing. Near the surface of Earth, ozone is a menace that can hurt life forms. However, in the upper atmosphere, ozone is essential to both terrestrial and marine life. Ozone is vital to living organisms because it is capable of absorbing large amounts of the sun's UV radiation.

In the upper atmosphere, UV radiation is consumed as it breaks the chemical bonds of ozone ($O_3$) to form an oxygen gas molecule ($O_2$) and an oxygen atom (O).

$$(2)\ 2O_3 \rightarrow 2O_2 + 2O \text{ (using UV radiation)}$$

After this process, more UV radiation is consumed to recombine the oxygen gas and the oxygen atom back into ozone, as in Formula 1. This process is repeated over and over again, thereby involving large amounts of UV energy that would otherwise reach Earth's surface. The chemistry of Formulas 1 and 2, repeating back and forth, creates a very efficient UV filter.

Without the ozone layer of the upper atmosphere, excessive UV radiation reaching Earth would severely burn human skin, increase the incidence of skin cancer and optical cataracts, destroy certain microscopic forms of marine life, and damage plants. Ultraviolet radiation is also responsible for suntans and painful sunburns, depending on an individual's skin tolerance and the length of exposure to the sun.

For many years there has been concern that chlorofluorocarbons (CFCs) and nitrogen oxides ($NO_x$), added to the atmosphere by human activities, may permanently damage Earth's fragile **ozone layer.** CFCs have been used extensively in refrigeration and air conditioning. As refrigeration became more common in around the world, CFCs entered the upper atmosphere; removed ozone through chemical processes; and threatened the ozone layer, our natural UV filter. Nitrogen oxide compounds, emitted in automobile and jet engine exhausts, also have the ability to enter and deplete our ozone shield.

The reaction that destroys ozone works in the following way: chlorine atoms (Cl), found in CFCs, split off and enter the stratosphere. There they bond to oxygen atoms (O) to form chlorine monoxide and oxygen gas.

$$(3)\ Cl + O_3 \rightarrow ClO + O_2$$

Oxygen atoms that have now bonded with the chlorine cannot be used to replenish the original amount of ozone ($O_3$) as in Formula 1. This and other chemical reactions attack the ozone layer and threaten our natural UV filter. Nitrogen oxide compounds ($NO_x$), emitted with exhaust from vehicles and jet engines, also have the ability to enter the stratosphere and destroy our ozone shield.

The UV radiation that the ozone layer allows to reach Earth does serve useful purposes. For instance, it has a vital function in the process of photosynthesis (see Fig. 4.3). It is important in the production of certain vitamins (especially vitamin D), it can help treat certain types of skin disorders, and it helps the growth of some beneficial viruses and bacteria. However, increasing amounts of UV radiation reaching the surface can become a serious problem for Earth's environments, and the ozone layer must be protected from the pollutants that threaten its existence.

## The Ozone "Hole"

Actually, there is no hole in our protective ozone layer. This term refers to the roughly circular zones of stratospheric ozone depletion that form in over the poles. From Earth's surface to the outer reaches of the atmosphere, there is always some ozone present (keep in mind that ozone is only a trace gas). There have been years when ozone was missing from specific levels above the ground, but there is always some ozone above us. If all the ozone in our atmosphere were forced down to sea level, the atmospheric pressure there would compress the ozone into a worldwide layer ranging from 3 to 4 millimeters (<1/16 of an inch) thick. So the ozone hole is really an area where the amount of ozone is considerably less than it should be. ● Figure 4.8 displays the ozone hole centered over the South Pole (the area of maximum ozone loss).

Atmospheric ozone levels are measured in Dobson units (du), established by G. M. B. Dobson along with his Dobson spectrometer in the late 1920s. A range of 300 to 400 du indicates a sufficient amount of ozone to prevent damage to Earth's life forms. To understand these units better, consider that 100 du equals 1 millimeter of thickness that ozone would have at sea level. Ozone measured inside the "hole" has dropped as low as 95 du in recent years, and the area of the ozone-deficient atmosphere (a more accurate description than a "hole") has exceeded that of the North American continent.

Ozone-destroying pollutants enter a circulation pattern in the stratosphere that transports them over the tropics, over the

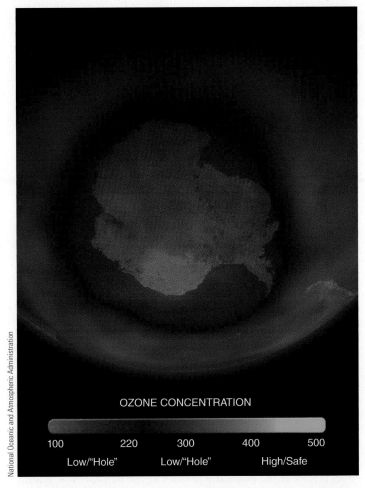

OZONE CONCENTRATION

| 100 | 220 | 300 | 400 | 500 |
| Low/"Hole" | | Low/"Hole" | | High/Safe |

National Oceanic and Atmospheric Administration

● **FIGURE 4.8** Satellite sensors monitor changes in the ozone hole *(shaded in red)* over Antarctica. This image shows the spatial extent of the ozone hole in September 2010. A value of 300 du is approximately the worldwide average for the ozone layer.

*What are the potential effects of ozone depletion on the world's human population?*

middle latitudes, and toward the poles. Near the South Pole, ozone and its destroyers (CFCs and $NO_x$) become trapped together in what is called the Southern Hemisphere polar vortex. This is a closed flow of extremely cold air that circles around the South Pole during the dark winter months (June, July, and August). During the South Polar spring (September, October, and November), incoming solar radiation starts to dissolve this circulation but also energizes the ozone-destroying Formula 3. The Southern Hemisphere ozone hole usually reaches its greatest extent each year within the first two weeks of October. Formula 3 works most effectively on ice crystal surfaces, and ice clouds in the stratosphere provide the perfect laboratory. When taking into account the stratospheric circulation, the Southern Hemisphere polar vortex, and the polar stratospheric clouds, the area of maximum ozone destruction would first occur at the poles and then would appear as a swirling elliptical or circular region like those seen in total ozone mapping spectrometer (TOMS) satellite images (see Fig. 4.8).

# Vertical Layers of the Atmosphere

Although people live and function primarily in the lowest level of the atmosphere, there are times, such as when we fly in an aircraft, climb a mountain, or travel to highland areas, when we leave our normal elevation. The thinner atmosphere may affect us if we are not accustomed to the atmospheric conditions at higher elevations. Visitors to Inca ruins in the high Andes or climbers in the Himalayas may experience altitude sickness, and even skiers in the Sierra Nevada or in the Rocky Mountains near mile-high Denver may need time to adjust. The air at these levels is much *thinner* than what most of us are used to experiencing. Thin air means that there is more empty space between air molecules so there is less oxygen and other gases in each breath of air.

Ascending from the Earth's surface to outer space, it is necessary to consider the changing conditions of the vertical layers of the atmosphere. Different systems have been used to divide the atmosphere into vertical layers. One system uses temperatures and the rates of temperature change. Another deals with the functions of these various layers.

## Atmospheric Layering by Temperature Characteristics

Based on temperature and rates of temperature change, the atmosphere can be divided into four layers, as shown in ● Figure 4.9. The first of these layers, lying closest to Earth's surface, is the **troposphere** (from Greek: tropo, turn—the turning or mixing zone), which extends about 8 to 16 kilometers (5–10 mi) above Earth. Its thickness, which varies with latitude, is least at the poles and greatest at the equator. It is within the troposphere that virtually all weather takes place.

Two distinct characteristics differentiate the troposphere from the other atmospheric layers. One is that water vapor and particulates are concentrated in this layer, so these constituents are rarely found above the troposphere. The other characteristic of the troposphere is that temperature normally decreases with increased altitude. The average rate of temperature decrease in the troposphere with altitude is called the **normal lapse rate** (or the *environmental lapse rate*); it amounts to 6.5°C per 1000 meters (3.6°F/1000 ft).

The altitude at which the temperature ceases to drop with increased altitude is called the **tropopause.** It is the boundary that separates the troposphere from the **stratosphere**—the layer of the atmosphere directly above the troposphere. Temperatures in the lower stratosphere remain fairly constant (about −57°C, or −70°F) to an altitude of about 32 kilometers (20 mi). It is in the stratosphere that we find the ozone layer and its conditions that protect Earth from excessive UV radiation. As the ozone layer absorbs UV radiation, this absorbed energy results in a release of heat; thus temperatures increase in the upper parts of the stratosphere. Some water exists in the stratosphere, but it appears as stratospheric ice clouds. These thin veils of ice clouds have no effect on weather as we experience it. Temperatures at the stratopause, the upper limit of the stratosphere, which is at an altitude of about 50 kilometers (30 mi), are about the same as temperatures on the surface, but almost none of that heat can be transferred because the air is so thin.

Above the stratosphere is the **mesosphere,** where temperatures tend to drop with increased altitude; and higher still is the

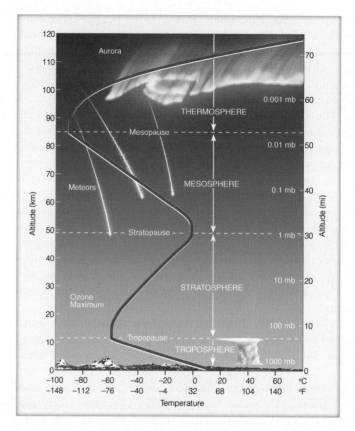

● **FIGURE 4.9** Vertical temperature changes in Earth's atmosphere are the basis for its subdivision into the troposphere, stratosphere, mesosphere, and thermosphere.

From PETERSEN/SACK/GABLER. *Fundamentals of Physical Geography, 1E.* Copyright © 2011 Brooks/Cole, a part of Cengage Learning, Inc. Reproduced by permission. www.cengage.com/permissions

*At what altitude is our atmosphere the coldest?*

**thermosphere,** where temperatures increase until they approach 1100°C (2000°F) at noon. Again, the air is so thin at this altitude that there is practically a vacuum and little heat can be transferred.

## Atmospheric Layering by Functional Characteristics

Astronomers, geographers, and communications experts sometimes use a different method for separating the atmosphere above the troposphere into layers based on their protective functions. This system divides the atmosphere into two distinct layers, the lowest being the **ozonosphere,** approximately the zone between 15 and 50 kilometers (10–30 mi) above the surface. The ozonosphere is another name for the ozone layer mentioned previously. Here, ozone effectively filters the UV energy from the sun and gives off heat (as thermal energy) instead.

From about 60 to 400 kilometers (40–250 mi) above the surface is the **ionosphere.** This name denotes the ionization of molecules and atoms in this layer, mostly as a result of UV energy, X-rays, and gamma radiation. Ionization refers to the process whereby atoms are changed to ions through the removal or addition of electrons, giving them an electrical charge. The ionosphere helps to shield Earth from harmful shortwave radiation. This electrically charged layer also aids in transmitting communication and broadcast signals to distant regions on Earth. It is in the ionosphere

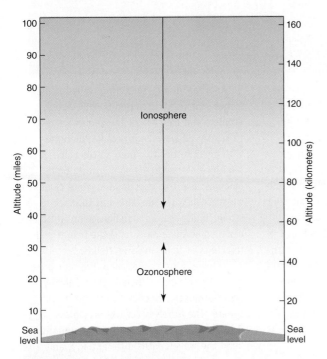

● **FIGURE 4.10** Vertical changes based on functions of the gases subdivide the atmosphere into two layers: the ozonosphere and the ionosphere.

*How do these layers protect life on Earth?*

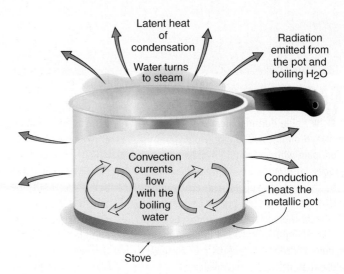

● **FIGURE 4.11** Mechanisms of heat transfer. Conduction by contact with a heat source heats the pot and then the pot heats the water. Convection occurs as hotter water flows upward, and cooler water sinks, forming a convective current in boiling water. Radiation, emitted as heat (thermal energy), flows outward into the surrounding air from the boiling water, the pot, and the heat source. Last, latent heat of condensation is released as water vapor turns back into a liquid as steam.

*How is advection involved in this small system?*

that the auroras (defined in Chapter 3) occur. The ionosphere gradually gives way to interplanetary space ( ● Fig. 4.10).

It is interesting to note that, when watching television reports from the International Space Station or the NASA space shuttle missions, the orbiting astronauts are still in the atmosphere. The background appears black, making it look like interplanetary space, but in reality these missions still take place within the realm of the outer atmosphere. One should also keep in mind that these different layering systems can focus on the same regions. For example, note that the thermosphere and the ionosphere occupy the same altitudes above Earth—that is, from 80 kilometers (50 mi) and outward. The names are different because of the criteria used in the differing systems.

# Energy Transfer Processes

The various forms of energy interact with the atmosphere and the Earth's surface in ways that are important to understanding our planet's weather and climate. Several important processes are responsible for the transfer of energy, from one form to another and from one place to another. These are radiation, conduction, convection, advection, and latent heat exchange ( ● Fig. 4.11).

## Radiation

Electromagnetic energy is transferred from the sun through space to Earth through the process of **radiation.** All objects with a temperature above absolute zero emit electromagnetic radiation. The characteristics of that radiation depend on the temperature of the radiating body. The warmer the object, the more energy it

will emit and the shorter the wavelengths at peak emission output. Because the sun's absolute temperature is 20 times that of Earth, the sun emits a great deal more energy, and radiates shorter wavelengths, compared to Earth. The sun's energy output per square meter is approximately 160,000 times that of our planet. The majority of solar energy is emitted at wavelengths shorter than 0.7 micrometers, whereas most of Earth's energy is radiated at much longer wavelengths of around 10 micrometers (see Fig. 3.9). Thus, *shortwave solar radiation (light)* reaches the surface and is absorbed, which heats the Earth, and because the Earth is cooler than the sun, it radiates energy in the form of longwave radiation *(heat)*. This *longwave thermal energy*, **terrestrial radiation,** emitted from Earth's surface (including the oceans) heats the atmosphere and accounts for the heat of the day. Basically the source of atmospheric heat is from terrestrial radiation, so the atmosphere mainly heats from the bottom upward.

## Conduction

The transfer of energy between two objects in contact with each other is called **conduction.** Heat flows from warmer objects or substances to the cooler ones they are in contact with, which tends to equalize temperatures. Conduction actually occurs as heat is passed from one molecule to another in a chainlike fashion. It is conduction that makes the bottom of a soup bowl too hot to touch.

Atmospheric conduction occurs at the interface (contact zone) between the atmosphere and Earth's surface. However, it is actually a relatively minor heat transfer process in terms of atmospheric warming because it affects only the layers of air closest to the surface. This is because air is a poor conductor of heat. In fact, air is the op-

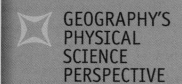
# Blue Skies, Red Sunsets, and Rainbows

When we look at the sky we are used to seeing certain colors: brilliant blue during a clear day, vibrant reds and oranges at sunset, clouds ranging from pure white to ominous gray–black. The process of *atmospheric scattering* explains why these colors appear. This scattering is explained in the text, but let's examine how it affects the colors of our atmosphere.

As sunlight is scattered in our atmosphere, certain wavelengths are scattered more than others. Each wavelength of visible light corresponds to one of the spectrum of colors. For example, wavelengths in the range of 0.45 to 0.50 micrometers (1 micrometer = 1/1,000,000 of a meter) correspond to the color blue and 0.66 to 0.70 micrometers correspond to the color red. These scattered wavelengths, when viewed by the human eye, are transmitted to the brain and translated into the colors we see. The particular wavelengths that are being scattered into the sky at our location are the colors that we see in the atmosphere.

In 1871 Lord Rayleigh, a radiation scientist, explained that gas molecules in the atmosphere (the smallest atmospheric particles) tend to scatter the shorter wavelengths of visible light (and also UV energy). Named after him, this process is called *Rayleigh scattering,* which makes a clear sky appear blue because blue light is being scattered preferentially by gas molecules in the air. In 1908 another radiation specialist, Gustav Mie, explained that oblique (or sharp) sun angles coming through the atmosphere tend to cause the scattering of longer wavelengths of visible light. Also, larger particles in the atmosphere, like dust, smoke, and pollutants, scatter the longer visible wavelengths, unless the accumulation of these particles is so thick that they block visibility. Mie scattering explains why sunsets bring about red- and orange-colored skies. Longer wavelengths in the visible spectrum also penetrate the atmosphere better than shortwave energy does, even in fog and haze. This is one reason why stoplights and brake lights are red because they are easier to see under various weather conditions.

Another form of scattering, called *nonselective scattering,* is caused by tiny water droplets and ice crystals in clouds that tend to scatter all wavelengths of visible light with equal intensity. When all the wavelengths are scattered equally, white light emerges, and this explains the white clouds we see. When we see dark-gray to black storm clouds, we are actually seeing white clouds plus shadows cast by the height and thickness of these massive clouds.

Rainbows are caused by a different process: the refraction (bending) of light by water droplets suspended in the atmosphere. The droplets act like tiny glass prisms, which take the incoming sunlight, bend the wavelengths differentially, and project the light as the colors of the spectrum. The factors necessary for seeing a rainbow include the sun angle and direction, the altitude of the water droplets, their density, and the observation angle. When you see a rainbow from a moving vehicle, it is only a matter of time until your observation angle changes and the rainbow disappears.

---

posite of a good conductor; it is a good insulator. This property is why an air layer is sometimes put between two panes of glass to help insulate a window. Insulation that contains a great deal of air-filled cavities is also used in sleeping bags and ski parkas. In fact, if air conducted heat well, our kitchens would become unbearable every time we turned on the stove or oven.

## Convection

In the atmosphere, as parcels of air near the surface are heated, they expand in volume, becoming less dense than the surrounding air, and therefore rise. This vertical transfer of heat through the atmosphere is called **convection;** it is much the same process that circulates boiling water in a pot. The water at the bottom, nearest to the heat source, becomes lighter and less dense as it warms. As the hot water rises, cooler, denser surface water replaces it by flowing downward. The water that flows down to the bottom is then warmed, so it flows upward to be replaced by cooler water moving downward. This vertical, circular movement of a fluid is called a convective current. These currents, set into motion by the heating of a fluid (liquid or gas), make up a convectional system. Convection currents account for much of the vertical transfer of heat within the atmosphere and the oceans and they have a great impact on the development of clouds and precipitation.

## Advection

The transfer of heat in a horizontal direction is called **advection.** Winds and ocean currents are the two major mechanisms of advective heat transfer. Advection is extremely important because it

This beautiful sunset over the ocean illustrates the role of scattering in producing the colorful skies. The blue sky aloft results from Rayleigh scatter because the air is thin at that altitude and gas molecules are scattering blue light. The sunset is red–orange because of longer wavelengths from Mie scattering caused by the thicker and denser atmosphere at that level. Part of the clouds are dark and in shadow, and the other parts are white because of nonselective scattering.

A double rainbow colors the sky behind the National Weather Service Forecast Office at Hanford in the San Joaquin Valley of California. Refracting as it passes through water droplets in the atmosphere, light separates into the colors of the spectrum to project a rainbow. The colors are in wavelength order, with shorter wavelengths on the inside of the arc. If a secondary rainbow forms, the color order reverses.

moves heat from areas where it is warm or hot to colder places, with a moderating effect on both of these areas. Winds and ocean currents help to distribute heat horizontally from the equatorial–tropical regions to the higher latitude and polar regions. Advection has a great influence on weather conditions and on the world's climates.

## Latent Heat Exchange

Water is the only substance that exists naturally in all three states of matter—as a solid, a liquid, and a gas—within the normal temperature range of Earth's surface and the troposphere. The energy transfer that occurs with the changes in water from one state to another is called **latent heat exchange.** This involves heat changing from *sensible heat* (which we can feel) to latent heat, which is stored in a form that we cannot feel, and the reverse—from latent to sensible heat. These exchanges from one state to another, which occur under natural conditions, are illustrated in ● Figure 4.12.

Molecules of a gas move faster than those of a liquid. During the process of **condensation,** when water vapor changes to liquid water, its molecules slow down and release some of their energy (590 cal/g, or 2470 joules/g) as sensible heat. Molecules in a solid move more slowly than those of a liquid, and when water undergoes **freezing,** changing into ice, less energy is released (80 cal/g, or 333 joules/g). When these processes are reversed, sensible heat becomes stored as latent heat. **Melting** takes up 80 calories per gram (333 joules/g), and **evaporation** takes up 590 calories per gram (2470 joules/g); the latter is called **evaporative cooling.**

When water evaporates, energy is stored in water vapor as latent heat. We are cooled by perspiration evaporating from our

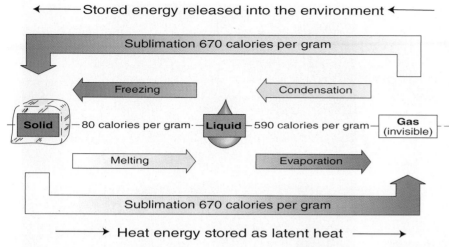

Stored energy released into the environment

Sublimation 670 calories per gram

Freezing          Condensation

Solid —80 calories per gram— Liquid —590 calories per gram— **Gas** (invisible)

Melting          Evaporation

Sublimation 670 calories per gram

Heat energy stored as latent heat

● **FIGURE 4.12** Energy exchanges between the three physical states of water. As you read the diagram, consider where heat comes from and where it goes. For example, when water freezes into ice, heat must come out of the water as it freezes and thus enters the environment. To evaporate water into vapor, the heat must go into the water for it to evaporate and therefore must come from the surrounding environment.

*Why are some of these energy exchanges referred to as "latent heat"?*

skin, thereby lowering the temperature. When water condenses from a gas to a liquid, the stored latent energy is released as sensible heat.

All of these major energy transfer processes play important roles in moving heat within the Earth system. Light energy from the sun is absorbed by the Earth and re-radiated as thermal infrared energy (heat). Conduction transfers heat by contact from warmer to colder substances. Convection moves heat in fluids (air, water) upward and moves cooler fluids downward. Advection moves heat to cooler areas and moves cool (air or water) to warmer areas.

Evaporation helps cool the atmosphere, whereas the release of latent heat of condensation helps warm the atmosphere. The massive amounts of heat released from condensation in the atmosphere provide a source of energy that drives many kinds of storms. These are the major processes that drive Earth's dynamic weather systems and on climates.

# Earth's Energy Budget

As the sun's energy passes through the atmosphere, a little more than half of its intensity is lost through various processes. However, as discussed in Chapter 3, the amount of insolation actually received at a particular location depends on the processes involved and also latitude, time of day, and time of year (all of which are related to the angle at which the sun's rays strike Earth). The transparency of the atmosphere (or the amount of cloud cover, moisture, carbon dioxide, and solid particles in the air) also plays a vital role.

Several processes affect the sun's energy as it interacts with the atmosphere. A determination (and inventory) of the incoming energy, its storage, and the outgoing energy processes is

called the **Earth's energy budget.** The following figures represent approximate averages for Earth for an average year ( ● Fig. 4.13):

27% reaches Earth's surface as direct radiation.

20% reaches the surface as diffuse radiation after being scattered.

26% of the energy is reflected directly back to space by clouds and the surface (*land and oceans*).

8% is *scattered* by minute atmospheric particles and returned to space as diffuse radiation.

19% is *absorbed* by the ozone layer and water vapor in the clouds.

In other words, on a worldwide average, 47% of the incoming solar radiation eventually reaches the surface, 19% is retained in the atmosphere, and 34% is returned to space. Because Earth's energy budget is in equilibrium, the 47% received at the surface is ultimately returned to the atmosphere by processes that we now examine.

## Heating the Atmosphere

The 19% of direct solar radiation retained by the atmosphere is stored in the clouds and the ozone layer and then eventually dissipated back to space. Other sources must be found to explain how the atmosphere is heated. The explanation lies in the 47% of incoming solar energy that reaches Earth's surface (land and water) and in the transfer of energy from Earth back to the atmosphere. This is accomplished through the energy transfer processes of conduction, convection, and latent heat exchange. These processes, however, are driven by the ultimate source of atmospheric heat, which is *terrestrial radiation,* the thermal infrared energy (heat) converted from the insolation received and re-radiated by Earth.

**The Energy Budget at Earth's Surface**  Now that we are familiar with the various means of heat transfer, we can examine what happens to the 47% of solar energy that reaches, and is absorbed into, Earth's surface (see Fig. 4.13). Approximately 14% of this energy is re-radiated by the Earth in the form of longwave *terrestrial radiation.* This 14% includes a net loss of 6% (of the total originally received by the atmosphere) directly to outer space; the other 8% is absorbed into the atmosphere. In addition, there is a net transfer back to the atmosphere (by conduction and convection) of 10% of the 47% that reached Earth. The remaining 23% is returned to the atmosphere by the release of latent heat of condensation. Thus, the 47% of the sun's original insolation that reached Earth's surface is returned to other components of the energy system and ultimately is lost back to space. There has been no long-term gain or loss.

**The Energy Budget in the Atmosphere**  At one time or another, about 60% of the solar energy intercepted by the Earth system is temporarily retained by the atmosphere. This includes

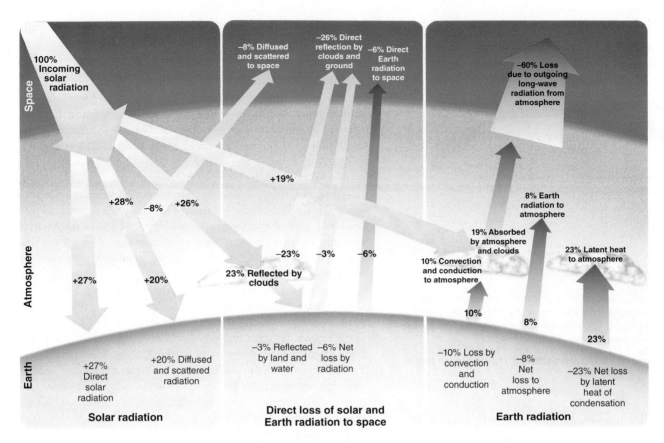

**Space**

100% Incoming solar radiation

−8% Diffused and scattered to space

−26% Direct reflection by clouds and ground

−6% Direct Earth radiation to space

−60% Loss due to outgoing long-wave radiation from atmosphere

**Atmosphere**

+28%  −8%  +26%

+19%

+27%  +20%

−23%  −3%  −6%

23% Reflected by clouds

8% Earth radiation to atmosphere

19% Absorbed by atmosphere and clouds

23% Latent heat to atmosphere

10% Convection and conduction to atmosphere

10%  8%  23%

**Earth**

+27% Direct solar radiation

+20% Diffused and scattered radiation

−3% Reflected by land and water

−6% Net loss by radiation

−10% Loss by convection and conduction

−8% Net loss to atmosphere

−23% Net loss by latent heat of condensation

**Solar radiation**

**Direct loss of solar and Earth radiation to space**

**Earth radiation**

● **FIGURE 4.13** Earth's energy budget indicates that a long-term global balance, or equilibrium, exists. Only 47% of the incoming solar radiation reaches and is absorbed by Earth's surface. Eventually, solar radiation received is re-radiated away by the Earth system (terrestrial radiation). However, the energy budget is dynamic. There is concern that increasing greenhouse gases from human activities are causing the atmosphere to absorb more terrestrial radiation, thus raising global temperatures.

19% of direct solar radiation absorbed by the clouds and the ozone layer, 8% emitted by *longwave radiation* from the Earth's surface, 10% transferred from the surface by *conduction* and *convection,* and 23% released by the *latent heat of condensation.* Some of this energy is recycled back to the surface, but eventually it is lost to outer space as more solar energy is received. Hence, just as was the case at Earth's surface, the energy budget in the atmosphere is in balance over long periods—a dynamic but generally stable system. However, many scientists believe that the enhanced greenhouse effect is causing an imbalance in the energy budget, with the impact of a global warming trend.

## Energy Balance

Examining the Earth's energy budget helps us understand the *open energy system* involved in heating the atmosphere. The *input* to the system is the incoming shortwave solar radiation (light) that reaches Earth's surface; this is balanced by the *output* of longwave terrestrial radiation (as heat) back to the atmosphere and lost to space. As these functions adjust to remain in balance, we can say that there is an overall **energy balance** in the energy budget of Earth, but it is in a state of *dynamic equilibrium.* Therefore, on Earth, over the long term, the energy budget is in balance. What

this means is that, because of this tendency toward balance, Earth has neither gotten continually warmer nor continually colder year after year through the entire history of the planet.

Of course, it should be noted that the percentages mentioned earlier are generalized and that they refer to *net* losses that occur over a long period. Earth also stores heat energy, part of it is re-radiated to the atmosphere, and then some of this energy is circulated back to Earth in a chain of cycles before it is finally lost into space.

**Variations in the Energy Budget** Remember that the figures we have seen for the energy budget are averages for the whole Earth system over many years. For any particular location, the energy budget is most likely not balanced. Some places have a surplus of incoming solar energy compared to outgoing energy loss, and others have a deficit. The main causes of these variations are differences in latitude and seasonal fluctuations.

As we know, the amount of insolation received is directly related to latitude ( ● Fig. 4.14). In the tropical and subtropical zones, where insolation is high throughout the year, more solar energy is received at Earth's surface and in the atmosphere than can be emitted back into space. In contrast, the Arctic and Antarctic

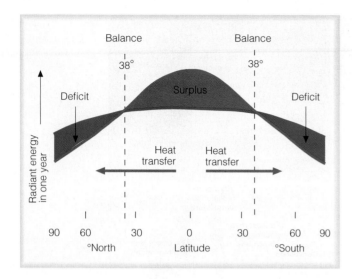

Balance       Balance

38°       38°

Surplus

Deficit       Deficit

Heat transfer    Heat transfer

Radiant energy in one year

90   60    30    0    30    60   90

°North      Latitude      °South

● **FIGURE 4.14** The energy budgets for specific places are not necessarily balanced. Low latitudes receive more solar energy than they lose by re-radiation back to space (energy surplus). High latitudes receive less energy from the sun than they radiate back to space (energy deficit).

*How is surplus energy from the low latitudes transferred to higher latitudes?*

zones receive so little insolation during the winter, when Earth is still emitting thermal energy, that there is a large deficit for the year. Locations in the middle-latitude zones have lower deficits or surpluses. The latitudes of 38° North and South mark the general division between equatorward zones that receive a surplus of solar energy and the poleward zones that run an annual deficit. If it were not for the heat transfers within the atmosphere, the land, and the oceans, particularly through advection, the tropical zones would get hotter and the polar zones would get colder over time.

During the year at any particular location, the energy budget varies over the seasons, with a tendency toward a surplus in the summer or high-sun season and a tendency toward a deficit 6 months later. Seasonal differences may be small near the equator, but they are generally large in the middle-latitude and polar zones.

# Air Temperature

## Temperature and Heat

Heat and temperature are highly related, but they are not the same. **Heat** is energy being transferred from one substance or medium to another as a result of their temperature differences. All substances consist of molecules that are constantly in motion (vibrating and colliding) and therefore possess kinetic energy—the energy of motion. **Temperature,** in comparison, is a measure of the average kinetic energy of individual molecules in a substance. When a substance is heated, its atoms and molecules vibrate faster, increasing its temperature. It is important to know that the amount of heat energy also depends on the mass of a substance, whereas the temperature refers to the energy of individual molecules. A burning match has a

high temperature but minimal heat; the oceans have moderate temperatures but a high content of heat—thermal energy.

**Temperature Scales** Three different scales are generally used in measuring temperature. The one that Americans are most familiar with is the **Fahrenheit scale,** devised in 1714 by Daniel Fahrenheit, a German scientist. On the Fahrenheit scale, the boiling temperature of water at sea level is 212°F and the temperature freezing temperature is 32°F. This temperature scale is used in the English system of measurements.

Anders Celsius, a Swedish astronomer, devised the **Celsius scale** (also called centigrade) in 1742. Celsius temperatures are used as a part of the International System of Units (the metric system). The freezing temperature of water at sea level on this scale was arbitrarily set at 0°C, and the boiling temperature of water was designated as 100°C.

The Celsius scale is used nearly everywhere but in the United States. Even in the United States the majority of the scientific community uses the Celsius scale to record and refer to temperatures. By this time, you have noticed that throughout this book comparable figures in both the Celsius (centigrade) and Fahrenheit scales are provided for temperatures. Similarly, whenever important figures for distance, area, weight, or speed are given, we use metric system units followed by measurements that use the English system. Appendix A in the back of your text may be used for comparison and conversion between the two systems.

● Figure 4.15 can help you compare the Fahrenheit and Celsius scales as you encounter temperatures that you wish to convert. In addition, the following formulas can be used for conversion from Fahrenheit to Celsius or vice versa:

$$C = (F - 32) \div 1.8$$

$$F = (C \times 1.8) + 32$$

A third temperature scale, used primarily by scientists, is the **Kelvin scale.** Lord Kelvin, a British scientist, felt that negative temperatures were not proper and should not be used. In his mind, no temperature should ever go below zero. This scale is based on the fact that temperature is related to the molecular movement. As the temperature is reduced, molecular motion slows. There is a temperature at which molecular motion stops and no further cooling is possible. This temperature, approximately −273°C (−460°F), is termed **absolute zero.** The Kelvin scale uses absolute zero as its starting point.

Thus, 0K equals −273°C. Conversion of Celsius to Kelvin is expressed by the following formula:

$$K = C + 273$$

## Short-Term Temperature Variations

Local changes in atmospheric temperature can result from a wide range of causes. These are related to the receipt and dissipation of energy from the sun and to various properties of Earth's surface and the atmosphere.

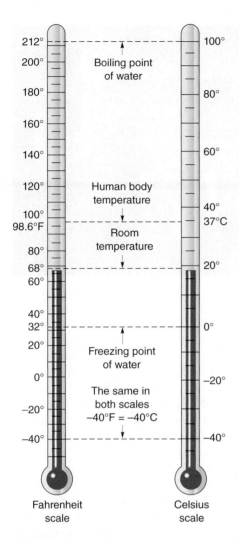

● **FIGURE 4.15** The Fahrenheit and Celsius temperature scales. The scales are aligned to permit direct conversion of temperature readings from one system to the other.

*When it is 70°F, what is the temperature in Celsius degrees?*

**The Daily Effects of Insolation**  As we noted earlier, the amount of insolation at any particular location varies both throughout the year (annually) and throughout the day (diurnally). Annual fluctuations are associated with the sun's changing declination with the seasons. **Diurnal** changes are related to the daily rotation of Earth about its axis. Diurnal typically refers to some cycle of day-to-night change. Each day insolation receipt begins at sunrise, reaches a maximum at noon (local solar time), and returns to zero at sunset. The exceptions to this occur in the polar regions during times of the year when the sun never rises above the horizon or during the times when it never sets.

Although insolation is greatest at noon, you may know that temperatures usually do not reach their maximum until 2 to 4 p.m. ( ● Fig. 4.16); this is called a **temperature lag.** This is because the insolation that is received by Earth between sunrise and the afternoon hours exceeds the energy being lost through terrestrial radiation. Hence, during that period, as Earth and atmosphere continue to gain energy, temperatures normally show a gradual increase.

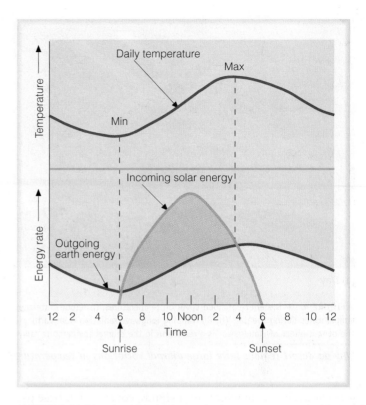

● **FIGURE 4.16** Diurnal and hourly changes in air temperature are strongly controlled by insolation and outgoing Earth radiation. When incoming energy exceeds outgoing energy (orange), the air temperature rises. When outgoing energy exceeds incoming energy (blue), air temperature drops.

*Why does temperature continue to rise even after the input of solar energy declines?*

Sometime around 3 to 4 p.m., when outgoing Earth radiation begins to exceed insolation, temperatures begin to decrease. The **daily temperature lag** of a few hours between the time of the highest sun angle (solar noon) and the warmest time of day results from the time it takes to heat Earth's surface to its daily maximum and for this energy to be radiated to the atmosphere.

Insolation receipt ends at sunset, and the heat that has been stored in Earth's surface layer during the day continues to be lost at night, so the ability to heat the atmosphere decreases as the night continues. The lowest temperatures typically occur just before dawn, when the maximum amount of energy has been radiated back to space and before insolation begins at sunrise. Thus, disregarding (for now) other factors that can cause temperature changes, we can see a predictable hourly temperature change over the day called the **daily march of temperature.** There is a decline from mid-afternoon until dusk and a rapid increase in the hours from dawn until the next maximum temperature is reached in the afternoon.

**Cloud Cover**  The extent of cloud cover is another factor that affects the temperatures of the surface and the atmosphere ( ● Fig. 4.17). Obviously, a heavy cover of clouds reduces the amount of insolation a place receives, causing daytime temperatures to be lower on a cloudy day. In contrast, we also have the

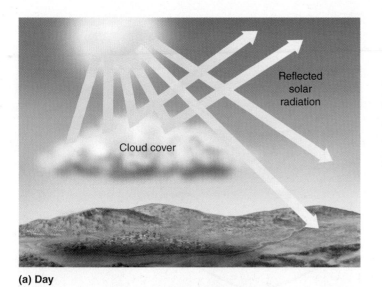

**(a) Day**

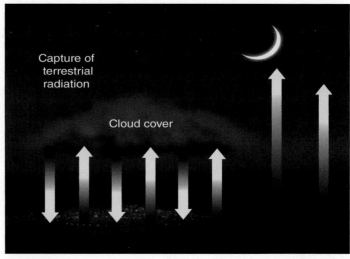

**(b) Night**

● **FIGURE 4.17** The effect of cloud cover on temperatures. (a) Clouds intercept solar radiation and lower air temperatures during the day. (b) By trapping longwave radiation from Earth, clouds increase air temperatures at night. The overall effect of cloud cover is a reduction in the diurnal temperature range.

*Why do desert regions have large diurnal variations in temperature?*

greenhouse effect, in which most clouds, composed in large part of water droplets, are capable of absorbing heat radiating from Earth. Clouds therefore can keep temperatures near Earth's surface warmer than they would otherwise be, especially at night. The general effect of cloud cover is to moderate temperature by lowering the potential maximum and raising the potential minimum temperatures. In other words, cloud cover makes for cooler days and warmer nights. Satellite images of global weather indicate that, at any time, about 50% of Earth is covered by clouds ( ● Fig. 4.18).

**Differential Heating of Land and Water** Bodies of water heat and cool slowly in comparison to the rates of heating and cooling of land surfaces. The air above the surface is heated or cooled in part by the surface characteristics beneath the air. Therefore, temperatures over water bodies or on land that is subjected to oceanic winds (**maritime** locations) tend to be more moderate than those of inland places at the same latitude. In contrast, the greater the **continentality** of a location (the distance removed from a large body of water), the less its temperature pattern will be modified. Areas deep within a continent and distant from the

● **FIGURE 4.18** This composite of several satellite images shows a variety of cloud cover and storm systems across Earth.

*In general, which are the cloudiest latitude zones and which are the zones with the clearest skies?*

ocean or winds from the ocean, particularly in regions outside the tropics, tend to experience larger annual and diurnal temperature ranges, compared to what coastal locations experience.

**Reflection** The ability of a surface to reflect solar energy (expressed as a percentage) is called its **albedo;** a high albedo surface reflects a high percentage of the energy that it receives. Albedo and absorption of energy are related because energy that is not reflected from a surface is absorbed. The higher the albedo, the lower the amount of energy absorbed and vice versa. The solar energy that the Earth system reflects back to space directly reduces the amount of insolation that will be absorbed for heating the atmosphere, oceans, and land surfaces.

As you may know from experience, snow and ice are good reflectors; they have an albedo of 90% to 95%. This is one reason why glaciers on high mountains do not melt away in the summer or why there may still be snow on the ground on a sunny day in the spring—the solar radiation is reflected away. Forests, however, have albedos of about 10% to 12%, which is good for the trees because they need solar energy for photosynthesis. The albedo of cloud cover varies, from 40% to 80%, according to the cloud thickness. The high albedo of cloud tops is why much solar radiation is reflected directly back to space from the tops of clouds.

The albedo of a water body varies greatly, depending on the water depth and the angle of the sun's rays. If the sun's rays are at a high angle, smooth water will reflect very little and absorb a great deal of energy. In fact, if the sun is vertical over a calm lake or ocean, the albedo will be only about 2%. However, low sun angles during sunrise and sunset, in winter, or at high latitudes can cause an albedo of more than 90% from the same water body ( ● Fig. 4.19). Likewise, a snow surface, in winter when solar angles are lower, can reflect up to 95% of the energy striking it, and skiers must be aware of the danger of severe sunburns and possible snow blindness from reflected solar radiation.

**Horizontal Air Movement** We have already seen that advection is the major mechanism of horizontal heat transfer over Earth's surface. Any air movement as a result of the wind, whether on a large or small scale, can have a significant short-term effect on temperatures. A wind blowing from an ocean to land will generally bring cooler temperatures in summer and warmer temperatures in winter. Large masses of air moving from polar regions into the middle latitudes can cause sharp drops in temperature, whereas tropical air moving poleward will usually bring warmer temperatures.

## Vertical Temperature Distributions

**Normal Lapse Rates** We have learned that Earth's atmosphere is primarily heated from the ground up as a result of longwave terrestrial radiation, conduction, and convection. Thus, temperatures in the troposphere are usually highest at low elevations and decrease as the altitude increases. As noted earlier in the chapter, this *average* decrease in air temperature with altitude is approximately 6.5°C per 1000 meters (3.6°F/1000 ft) and is referred to as the *normal lapse rate.* In this case, "normal" refers to the expected, average conditions in the troposphere.

U.S. Navy photo, Nathanael Miller

● **FIGURE 4.19** At low sun angles, water reflects most of the solar radiation that strikes it.

***Why is it so difficult to assign a single albedo value to a water surface?***

The actual lapse rate at a particular place can vary. Smaller lapse rates (which means less change with altitude) can exist if cold, dense air drains into a valley from a higher elevation or if advectional winds bring air in from a cooler region at the same elevation. In each case, the surface is cooled so that its temperature is closer to that at higher altitudes directly above it. However, if the surface is heated strongly on a hot summer afternoon, the air near the surface will be disproportionately warm and the lapse rate will be steep (great change with altitude).

**Temperature Inversions** Under certain circumstances, the normal decrease of temperature with increased altitude might be reversed and temperature may actually *increase* with altitude for several hundred meters. This atmospheric condition is called a **temperature inversion,** where a layer of warmer air aloft interrupts the altitudinal decrease in temperature. The altitudinal boundary between the cooler air below and the warmer air aloft can be abrupt and appear as a horizontal line across the sky, marking the **inversion layer.** Inversions tend to stabilize the air, causing less turbulence, which discourages the development of precipitation and storms. There are two main kinds of inversions, which form in different ways and at different altitudes: upperlevel and surface.

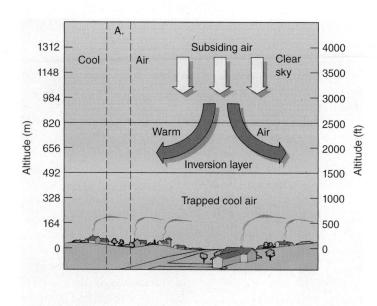 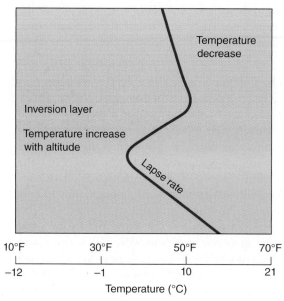

● **FIGURE 4.20** (Left) Upper-level inversions are caused by air subsiding from aloft, warming as it sinks to form a stable inversion layer over cooler air below (note column of air A). (Right) Lapse rate associated with the column of air (A) during a temperature inversion.

*How are the temperatures trending in the lowest layer, in the inversion layer, and above the inversion layer?*

**Upper-Level Inversions** Air settling slowly from the upper atmosphere can cause an **upper-level inversion** ( ● Fig. 4.20). This kind of atmospheric condition involves an inversion layer about 1000 or 2000 meters above the ground. Descending air is compressed as it sinks, causing increases in temperature and atmospheric stability. Inversions caused by descending air are most common at about 30° to 35° north and south latitudes.

An upper-level inversion, common to the coastal area of California, results when cool marine air drifting in from the Pacific Ocean moves under stable, warmer, and lighter air aloft created by subsidence and compression. Such an inversion layer tends to maintain itself—that is, the cold underlying air is heavier and cannot rise through the warmer air above. Not only does the cold air resist rising, moving, or mixing with the warmer air aloft, but pollutants, such as smoke, dust particles, and automobile exhaust, created at Earth's surface also fail to disperse. During these conditions, pollution and smoke will accumulate in the stable air below the inversion layer. Upper-level inversions are particularly acute in the Los Angeles area, which is located in a topographic basin surrounded by mountainous areas ( ● Fig. 4.21). Cooler air drifts into the basin from the ocean and then cannot escape either horizontally, because of the landform barriers, or vertically, because of the inversion.

**Surface Inversions** Some of the most noticeable temperature inversions are those that occur close to the surface, when the ground is cold because of heat loss from terrestrial radiation and then cools the lowest layer of air. These are called **surface inversions** ( ● Fig. 4.22). In this situation,

the coldest air is close to the surface, the inversion layer is much lower then those of an upper-level inversion, and the temperature rises with altitude. Surface inversions most often occur on clear nights with calm conditions in the middle and high latitudes. Snow cover or the recent advection of cool, dry air into an area may enhance the strength of an inversion. These conditions produce extremely rapid cooling of Earth's surface at night as it loses the heat accumulated from daytime insolation through ter-

● **FIGURE 4.21** Conditions producing smog-trapping upper-level inversions in the Los Angeles area.

*Why is the air clear above the inversion?*

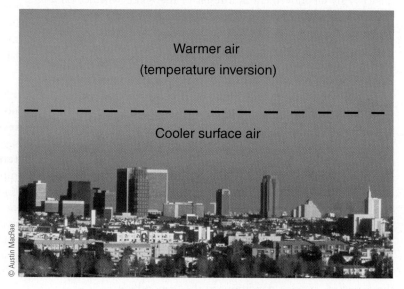

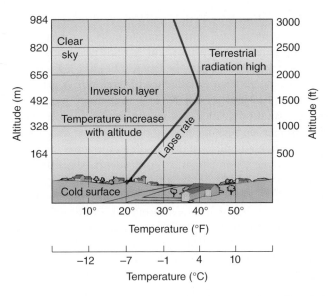

● **FIGURE 4.22** Surface inversions are caused by the rapid cooling of air above a cold land surface at night.

*What is the significance of an inversion in terms of temperatures in the atmosphere?*

restrial radiation. The air layer that is closest to Earth is cooled by conduction more than the air aloft. Calm air conditions near the surface are a result of these temperature inversions. As the sun rises in the morning and heats the ground, surface inversions tend to break up, but they may form again the next night if the air is clear and calm.

Because they occur when it is cold, surface inversions often cause the formation of dew, fog, and frost. Where the topography is hilly or mountainous, cold, dense air tends to flow downslope and accumulate in the valleys. Cold air on the valley floors and other low-lying areas sometimes produces fog or, if temperatures are cold enough, a killing frost. A variety of methods are used to prevent frosts from destroying crops during a surface inversion. For example, fruit trees are often planted on warmer hillsides at elevations that tend to be above the inversion layers instead of in the valleys. Farmers may also put blankets of straw, cloth, or other insulating materials over their plants. This prevents the escape of heat to space, insulates the plants from the cold air, and keeps the plants warmer. Large fans are often used in an effort to mix colder surface air with the warmer air aloft, thus disturbing the inversion ( ● Fig. 4.23). Huge orchard heaters that warm the air can also be used to disturb the temperature layers. Smudge pots, an older

● **FIGURE 4.23** Powerful wind machines can break up a surface inversion to avoid crop damage from frost. (a) These fan towers blow warm air downward from above an inversion layer, which moderates the temperature below to protect these vineyards from frost. (b) This large, horizontally mounted fan protects crops from frost by blowing cold air upward as much as 91 meters (300 ft) high and moderating the temperature by pulling in warmer air from aloft.

method of preventing frost, used oil fires to put heat into the air and emit particulates that encourage condensation, which releases some latent heat. However, smudge pot use has declined in favor because of their air-pollution potential.

## Controls of Earth's Surface Temperatures

There are several major **controls of temperature**: (1) latitude, (2) land and water distribution, (3) ocean currents, (4) altitude or elevation, (5) landform barriers, and (6) human activities.

**Latitude** The most important control of temperature variations involved in weather and climate is latitude. Because of changes in sun angle and length of daylight hours, distinct latitudinal patterns are influenced by the seasonal and annual receipt of solar energy. In general, annual insolation tends to decrease from the lower latitudes to the higher latitudes. Table 4.2 shows the average annual temperatures for several Northern Hemisphere locations. We can see that, responding to insolation (with one exception), a poleward decrease in temperature exists for these locations. The exception is the place near the equator. Because of heavy cloud cover in equatorial regions, annual temperatures there tend to be lower than at bordering regions to the north and south, where skies tend to be clearer.

**Land and Water Distribution** The oceans, seas, and large lakes serve as storehouses of water that store tremendous amounts of heat. The widespread extent and distribution of the oceans make them a major factor that exerts a great influence on atmospheric and weather conditions. Different substances heat and cool at varying rates, and land heats and cools much faster than water does. There are three reasons for this phenomenon. First, the **specific heat** of water is greater than that of land. Specific heat refers to the amount of heat necessary to raise the temperature of 1 gram of any substance 1°C. Water has a specific heat of 1 calorie per gram per degree Celsius (4.19 joules/gram °C) and must absorb more thermal energy than land (which has specific heat values of about 0.2 calories/gram degrees Celsius, or 0.84 joules/g °C) to be warmed by the same number of degrees in temperature.

Second, water is transparent, so solar energy passes through the surface and into the water below, yet the energy is concentrated on the surface of opaque materials like soil and rock. Thus, a given amount of heat will spread through a greater volume of water in comparison to land. Third, liquid water circulates and mixes, which transfers heat among the various depths within its mass. The result is that as summer changes to winter, land cools more rapidly than bodies of water, and as winter becomes summer, the land heats more rapidly. Air gets much of its heat from the surface below, so the differential heating of land and water produces temperature differences in the atmosphere above these two surfaces.

Seattle, Washington, has a mean temperature in July of 18°C (64°F), whereas the mean temperature during the same month in Minneapolis, Minnesota, is 21°C (70°F). Because these two cities are at similar latitudes, their annual pattern and receipt of solar energy should also be similar. Therefore, their different July temperatures must be related to a factor other than latitude. Much of this temperature difference can be attributed to the fact that Seattle is near the Pacific coast, whereas Minneapolis is in the heart of a large continent, far from moderating oceanic influences. Seattle stays cooler than Minneapolis in the summer because the ocean warms up slowly, keeping the air relatively cool. Minneapolis, in comparison, is in the center of a large landmass that warms very quickly and to a greater degree. In the winter, the opposite is true because Seattle is warmed by its nearness to ocean water, but Minneapolis is distant from the moderating influence of a large water body. The mean temperature in January is 4.5°C (40°F) in Seattle and −15.5°C (4°F) in Minneapolis.

Not only do water and land heat and cool at different rates, but so do various land surface materials. Soil, forest, grass, and rock surfaces all heat and cool differentially and thus affect the temperatures of overlying air.

**Ocean Currents** Surface ocean currents are large movements of water flowing like "rivers" in the ocean. They may flow from a place of warm temperatures to one of cooler temperatures and vice versa. These movements are another example of processes that tend to keep Earth systems in equilibrium—in this instance, a balance of temperature and density.

Earth rotation affects the movements of winds, which in turn affect the flow of ocean currents. In general, ocean currents move in a clockwise direction in the Northern Hemisphere and in a counterclockwise direction in the Southern Hemisphere ( ● Fig. 4.24). The ocean temperature greatly affects the air temperature above it. An ocean current that moves warm tropi-

## TABLE 4.2
### Average Annual Temperatures in the Northern Hemisphere

| Location | Latitude | (°C) | (°F) |
|---|---|---|---|
| Libreville, Gabon | 0°23′N | 26.5 | 80 |
| Ciudad Bolivar, Venezuela | 8°19′N | 27.5 | 82 |
| Bombay, India | 18°58′N | 26.5 | 80 |
| Amoy, China | 24°26′N | 22.0 | 72 |
| Raleigh, North Carolina | 35°50′N | 18.0 | 66 |
| Bordeaux, France | 44°50′N | 12.5 | 55 |
| Goose Bay, Labrador, Canada | 53°19′N | −1.0 | 31 |
| Markova, Russia | 64°45′N | −9.0 | 15 |
| Point Barrow, Alaska | 71°18′N | −12.0 | 10 |
| Mould Bay, NWT, Canada | 76°17′N | −17.5 | 0 |

cal water toward the poles (a **warm current**) or cold polar water toward the equator (a **cold current**) can significantly modify the air temperatures of regions near to where these currents flow. Ocean currents that pass close to land and are accompanied by onshore winds can have a significant impact on coastal climates.

The Gulf Stream, with its extension the North Atlantic Drift, is an example of an ocean current that moves warm water poleward. This warm water keeps the coasts of Great Britain, Iceland, and Norway ice free in wintertime and moderates the climates of nearby land areas ( ● Fig. 4.25). We can see the Gulf Stream's impact if we compare winter conditions in the British Isles with those of Labrador in northeastern Canada. Although these locations are at the same latitude, the average temperature in Glasgow, Scotland, in January is 4°C (39°F), whereas during the same month it is −21.5°C (−7°F) in Nain, Labrador.

The California Current off the west coast of North America helps moderate the climate of that coastal region as it brings cold water south. As this current forces ocean water southwest and away from the California coast, cold bottom water is drawn to the surface, causing further chilling of the air above. San Francisco's cool summers (July average: 14°C, or 58°F) show the effect of this current and the associated upwelling of cold waters.

**Altitude** As you know, temperatures in the troposphere decrease as the altitude or elevation increases. Anyone who has hiked upward 500, 1000, or 1500 meters in midsummer has experienced a decline in temperature with increasing elevation ( ● Fig. 4.26). Even if it is hot on the valley floor, you may need warm clothes once you climb a few thousand meters. In Southern California you can find snow for skiing if you go to an elevation of 2400 to 3000 meters (8000–10,000 ft). The summits of Mauna Loa and Mauna Kea, the high volcanoes on the island of Hawaii, receive snow in the winter, and some people actually snowboard and ski there. Mount Kenya, 5199 meters (17,058 ft) high and located at the equator, is cold enough to have glaciers on its summit. The city of Quito, Ecuador, only 1° south of the equator, has an average annual temperature of only 13°C (55°F) because it is located about 2900 meters (9500 ft) in elevation.

**Landform Barriers** Landform barriers, especially large mountain ranges, can block air movements from one place to another

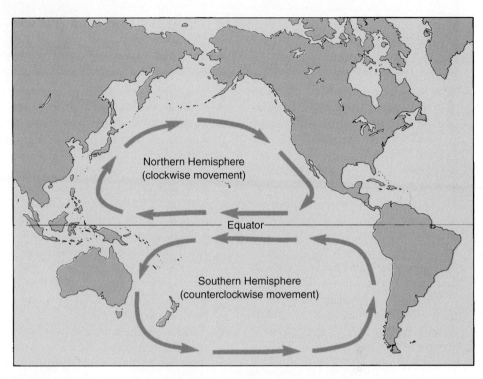

● **FIGURE 4.24** A simplified map of currents in the Pacific Ocean shows their basic rotary pattern. Major currents move clockwise in the Northern Hemisphere and counterclockwise in the Southern Hemisphere. A similar pattern exists in the Atlantic Ocean.

*What direction would a hurricane forming off of west Africa typically take as it approaches the United States?*

● **FIGURE 4.25** The Gulf Stream (the North Atlantic Drift farther north and eastward) is a warm current that moderates the climate of northern Europe.

*Use this figure and the information gained in Figure 4.24 to determine the route that sailing ships would likely follow from the United States to England and back.*

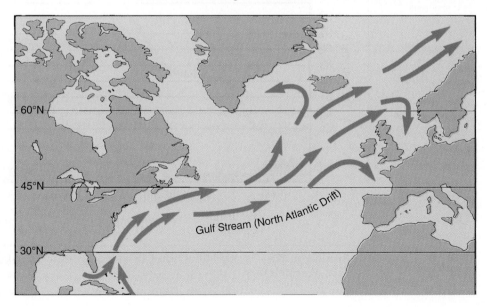

# The Urban Heat Island

In 1820 Sir Luke Howard, an amateur meteorologist, observed that temperatures in London, England were warmer than those of the surrounding countryside. This is an early reference to what has become known as an *urban heat island,* a measurable dome of warm air produced by a large urban area. Research on urban heat islands in the United States began in the mid-1950s, but serious urban climate monitoring started in the early 1970s. Since then, urban heat island has been the standard title for this phenomenon.

Several factors contribute to the increase in the temperatures of cities, which are generally 2 to 10°F (1–6°C) warmer than the surrounding rural areas.

1. Heating and cooling: In the summer air conditioners pump heat into the environment, and in the winter heated buildings lose heat to the outside. In a rural area, these factors also exist but in a much lower density.
2. Automobiles: In urban areas, great numbers of automobile and truck engines generate heat and expel hot exhaust. In the country fewer vehicles—cars, trucks, and farm machinery—are operating at any given time.

3. Industry: Factories, which operate large machinery that give off heat, tend to be located in or near urban areas. Fewer industries exist in the countryside.
4. Materials and buildings: The built environment of cities is dominated by concrete, asphalt, glass, and metal, which can heat up quickly during the day (with low specific heat and low albedos), as opposed to the grass, trees, and croplands of rural areas. Although urban building materials may cool down rather quickly at night, buildings, roads, and highways also continue to radiate heat until well after sunset.
5. Pollution levels: Particulates, aerosols, and greenhouse gas levels are higher in cities than in the surrounding countryside. A dome of pollution and haze can form over large cities. Certain pollutants reflect away some of the incoming solar radiation, shading the city somewhat by day, but this pollution may also absorb the heat trying to escape and re-radiate it back to the surface, especially at night.
6. Surface water: In the countryside, evaporation from lakes, ponds, streams, and water loss from vegetation can

help cool the surroundings by the latent heat of evaporation (590 cal/g, or 2470 joules/g). Urban areas are designed to drain surface water quickly and efficiently with gutters, drains, and sewers.

It has been estimated that by the year 2025, about 80% of the world's population will reside in urban areas. As world population grows and urbanization increases, these six factors will likely intensify, and the urban heat island will directly affect more people than the general climatic conditions of the region.

What is the solution? Planting more trees, shrubs, and grass along with adding park space to cities can help moderate urban heat island conditions. Roofs of buildings, surfaces of paved areas, and roads can be made from materials that are more reflective of solar radiation, reducing absorption and re-radiation of heat. The practice of constructing "green roofs," gardens planted on the tops of buildings, is also increasing. This is another way to help reduce the impact of an urban heat island.

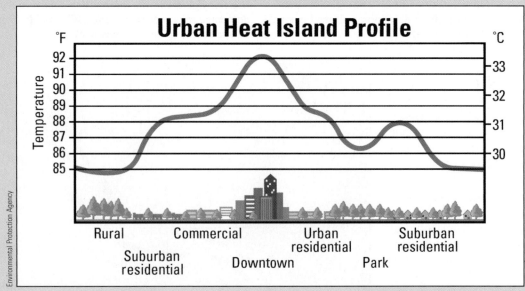

A temperature profile of an urban "heat island" shows the increase in temperature with increasing development and urbanization.

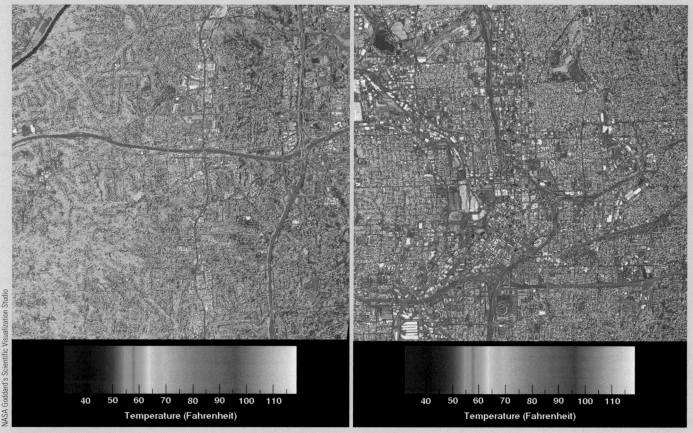

The urban heat island in Atlanta, Georgia. The daytime thermal infrared (heat) image on the right shows the heat retention of buildings, roads, and pavement in densely urbanized areas of Atlanta. The hottest areas are in red and the coolest areas are in blue. Red tones dominate the urban image.

The daytime suburban image on the left is dominated by yellow and green—areas that include residential neighborhoods, with lawns and trees, and vegetated areas.

Comparing the temperature values that the colors represent on the image of the cooler suburban area with those on the image of the urbanized city center provides a good illustration of the urban heat island effect.

In Google Earth fly to: 35.356°N, 138.734°W (Mt. Fuji).

● **FIGURE 4.26** Snow-capped mountains show visual evidence of the temperature decrease with altitude. This is the summit of Mount Fuji in Japan (3776 m/12,388 ft) during the fall. Note the fall colors on the slope and the elevation where trees cannot grow.

*At what rate per 1000 meters do temperatures decrease with height in the troposphere?*

and thus can also affect temperatures. During the winter the Himalayas keep cold, Northern Asiatic air from flowing south, giving most of the Indian subcontinent a year-round tropical climate. The geographic orientation of mountains is an important factor. In North America, for example, southern slopes face the sun and tend to be warmer than shady north-facing slopes. Snow on Northern Hemisphere mountains, which accumulates on south-facing slopes, experiences more melting and is typically restricted to higher elevations than the snow cover on north-facing slopes. North-facing slopes retain more snow for a longer season and the snow cover typically extends to lower elevations in comparison to south-facing slopes. In the Southern Hemisphere, the same is true but for south-facing slopes.

**Human Activities** Various human activities may also have an impact on the temperature. Congregations of people in cities and their built environments can result in what is called an *urban heat island* (see Geography's Spatial Perspective box). What this means is that urbanized areas tend to be warmer in both summer and in winter in comparison to the surrounding countryside. In cities, many thousands of automobiles add heat to the air, whereas rural areas experience much fewer vehicles. Air conditioning units and industries also expel warmth into the urban atmosphere and environment. Even the building material in cities (concrete, asphalt, glass, and metal) can heat up quickly during the day (with low specific heat and low albedo values), as opposed to the grass, trees,

and croplands of the surrounding countryside, which react more slowly to insolation. In addition, destroying forests, draining swamps, or creating large reservoirs can significantly affect local climatic patterns. The massive extent of deforestation also has an impact on global climate.

## Temperature Distribution at the Surface

One method for illustrating the temperature distribution over the Earth is to use a graphic mapping line symbol called an isotherm. **Isotherms** (from Greek: **isos,** equal; **therm,** heat) are lines that connect points of equal temperature. When constructing worldwide isothermal maps of temperature, the effects of elevation on temperatures may be accounted for by adjusting temperature readings to what they would be at sea level. This adjustment means adding 6.5°C for every 1000 meters of elevation (the normal lapse rate). The rate of temperature change over distance on an isothermal map is the **temperature gradient.** Closely spaced isotherms indicate a steep temperature gradient (high temperature change with distance), and widely spaced lines indicate a weak gradient (low temperature change with distance).

● Figures 4.27a and 4.27b show the horizontal, worldwide temperature distributions during January and July, when the seasonal extremes of high and low temperatures exist in the Northern and Southern Hemispheres. The easiest temperature distribution to recognize on these two maps is the general orientation of the isotherms; they run nearly east to west, illustrating the general relationship between temperature and latitude.

A comparison of Figures 4.27a and 4.27b reveals some additional temperature relationships. The Southern Hemisphere has the highest temperatures in January, and in July they are in the Northern Hemisphere. Note that on the July map, Portugal in the Northern Hemisphere is nearly on the 20°C isotherm, whereas in South Australia in the Southern Hemisphere the average July temperature is around 10°C, even though these locations are about the same distance from the equator. These temperature differences between the two hemispheres illustrate seasonal variation in insolation, changing as the 90° rays of the sun shift north and south across the equator between its positions at the June and December solstices.

Note that the greatest deviation from east–west trending temperatures occurs where the isotherms change from crossing landmasses to cross the oceans. As these isotherms leave the land, they usually bend rather sharply poleward in the hemisphere experiencing winter and toward the equator in the summer hemisphere. This pattern of isotherms is a direct reaction to the differential heating and cooling of land and water. The continents are warmer than the oceans in the summer and colder than the oceans in the winter. Another interesting geographic pattern can be seen on the January and July maps. Note that the isotherms poleward of 40° latitude have a more regular east–west orientation in the Southern Hemisphere compared to the Northern Hemisphere. This is because the Southern Hemisphere (often called the "water hemisphere") has very little land south of 40°S latitude, other than Antarctica, so there are few land and water contrasts. Note also that the temperature gradients are much

January °C

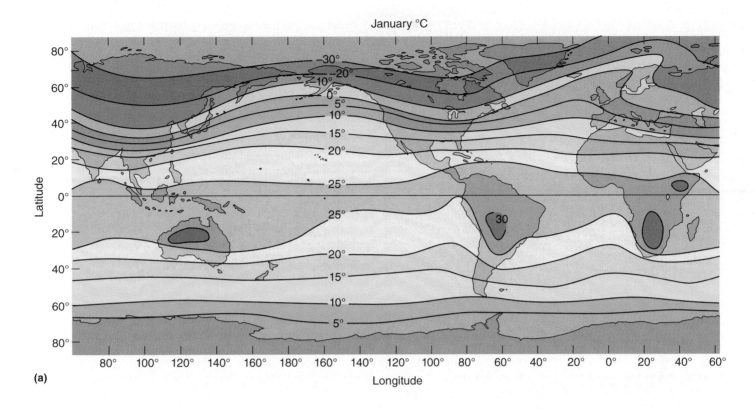

(a)

July °C

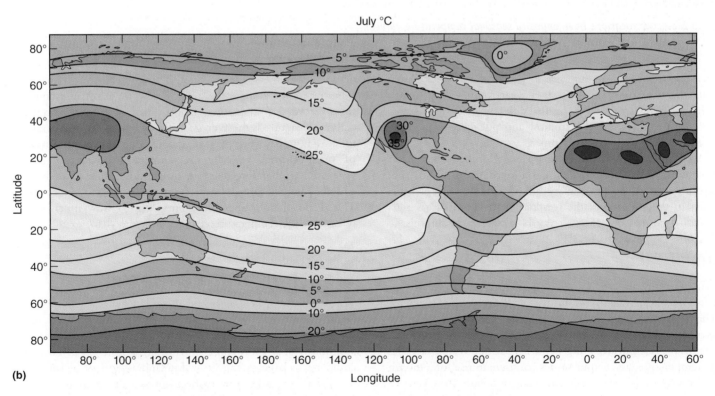

(b)

● **FIGURE 4.27** (a) Average sea-level temperatures in January (°C). (b) Average sea-level temperatures in July (°C).

*Observe the temperature gradients between the equator and northern Canada in January and July. Which is greater and why?*

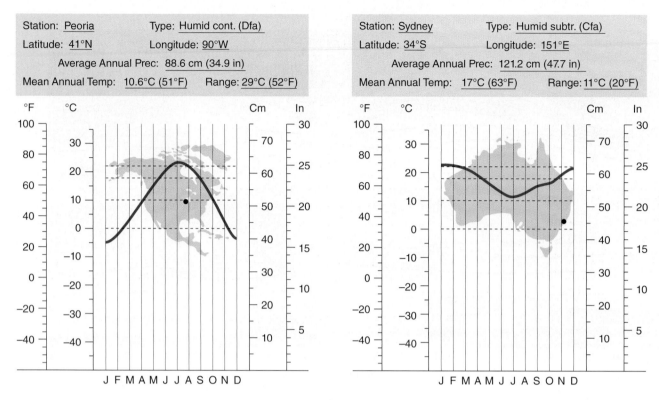

| Station: Peoria | Type: Humid cont. (Dfa) |
| Latitude: 41°N | Longitude: 90°W |

Average Annual Prec: 88.6 cm (34.9 in)

Mean Annual Temp: 10.6°C (51°F)  Range: 29°C (52°F)

| Station: Sydney | Type: Humid subtr. (Cfa) |
| Latitude: 34°S | Longitude: 151°E |

Average Annual Prec: 121.2 cm (47.7 in)

Mean Annual Temp: 17°C (63°F)  Range: 11°C (20°F)

● **FIGURE 4.28** The annual march of temperatures at Peoria, Illinois and Sydney, Australia.

*Why do these two locations have opposite seasonal temperature curves?*

steeper in winter than in summer in both hemispheres. This is because the tropical zones have high year-round temperatures, but the polar zones have large seasonal differences. The difference in temperature between the tropical and polar zones is much greater in winter than it is in summer.

As a final point, observe the sharp swing of the isotherms off the coasts of eastern North America, southwestern South America, and southwestern Africa in January and off of Southern California in July. In these locations, the bending of the isotherms is augmented by the presence of either warm or cool ocean currents.

## Annual Temperature Changes

January and July are commonly used for global isothermal maps as these are warmest and coldest times of year because of temperature lags of about 30 to 40 days after the solstices, when insolation is at a minimum or maximum (depending on the hemisphere). This **annual temperature lag** behind insolation is similar to the daily temperature lag. Temperatures continue to rise for a month or more after the summer solstice because insolation continues to exceed radiation loss. Temperatures continue to fall after the winter solstice until the increase in insolation finally matches Earth's radiation. In short, these lags exist because it takes time for Earth to heat or cool and transfer those temperature changes to the atmosphere.

Annual temperature changes for a location can be plotted in a graph (● Fig. 4.28). The mean temperature for each month in a place such as Peoria, Illinois, is recorded and a curved line is drawn to connect the 12 monthly temperatures. The mean monthly temperature is the average of the mean daily temperatures recorded at a weather station during a month. The mean daily temperature is the average of the high and low temperatures for a 24-hour day. Temperature graphs depict the **annual march of temperature,** basically the temperature changes over the year—the temperature decrease from midsummer to midwinter and the temperature increase from midwinter to midsummer. The shape of the temperature curve shows how much and at what rates the temperature changes over the year.

## Weather and Climate

We use the words *weather* and *climate* in general conversation, but it is important to carefully distinguish between these terms (● Fig. 4.29a and 4.29b). **Weather** refers to the atmospheric conditions at a given time and for a specific area. That area could be as large as the New England region or as small and specific as a weather observation station. The analysis of many weather observations for a place over a period of at least 30 years provides us with a description of its climate. **Climate** describes an area's average weather, but it also includes deviations from the normal or average that are likely to occur and the atmospheric processes involved. Climatic study is also concerned with the potential for extreme weather or climatic events, which can be significant. Thus, we could describe the climate of the southeastern United States in terms of average

(a)

(b)

● **FIGURE 4.29** (a) Weather, such as a snowy day, is a short-term meteorological event. (b) These extensive glaciers in Alaska's Kenai Fjords National Park, however, have formed as a result of a cold climate over the long term.

*Can you make the same weather versus climate comparisons using rainfall?*

annual temperatures and precipitation, but we would also have to include the likelihood of severe events such as hurricanes, tornadoes, or blizzards during certain periods of the year.

**Meteorology** involves monitoring, analyzing, and forecasting the weather. **Climatology** is the study and classification of climates, both past and present; their distribution on Earth; and the processes that influence climate. Weather and climate are of prime interest in physical geography because they have an impact on, and interrelate with, all of Earth's environments. The changing conditions of atmospheric elements such as temperature, wind, rainfall, and other forms of precipitation affect soils and vegetation, modify landforms, cause flooding, and in a multitude of other ways influence the function of Earth's physical environments.

Five fundamental atmospheric **elements** serve as the indicators of weather and climate: (1) solar energy, (2) temperature, (3) atmospheric pressure, (4) wind, and (5) precipitation. We must examine these elements to understand and categorize weather and climate. A weather forecast will generally include the current temperature, the probable temperature range, a description of the cloud cover, the chance of precipitation, wind speed and direction, and air pressure.

In Chapter 3 we noted that the amount of solar energy received at one place on Earth's surface varies during a day and throughout the year. The amount of insolation a place receives is the most important weather element; the other four elements depend in part on the intensity and duration of solar energy.

The temperature of the atmosphere at a given place on or near Earth's surface is largely a function of insolation, but it is also influenced by many other factors, such as elevation and the distribution of land and water. Unless precipitation is occurring, the air temperature may be the first weather element that we describe when someone asks us what it is like outside.

However, if it is raining, foggy, or snowing, we will probably mention that condition first. We tend to be less aware of the amount of water vapor or moisture in the air (except in very arid or very humid areas). However, moisture in the air is another vital weather element, and its variations play an important role in the likelihood of precipitation.

We all know that weather varies in both time and place. At some locations or during certain times of the year the weather can change almost daily. At other locations there may be weeks of uninterrupted sunshine, blue skies, and moderate temperatures followed by weeks of persistent rain. A few places experience only minor differences in the weather throughout the year. The language of the original people of Hawaii is said to have no word for weather because conditions there varied so little.

# Complexity of Earth's Energy Systems

This chapter was designed to show Earth's energy systems and dynamic balances. These systems and processes vary as a result of complex interrelationship between the Earth and its atmosphere, especially the energy gained and lost by the Earth system. The geographic distributions of energy and temperature variations exist horizontally across Earth's surface and vertically through the atmosphere. They also vary over both daily and seasonal time frames.

Earth's energy systems impose both diurnal and annual rhythms on virtually all environments and on our agriculture, recreational pursuits, clothing styles, architecture, and energy bills. Human activities are constantly influenced by temperature changes, which reflect the input–output patterns of Earth's energy systems.

## DEFINE AND RECALL

scattering
photosynthesis
water vapor
particulates
aerosols
greenhouse effect
enhanced greenhouse effect
ozone
ozone layer
troposphere
normal lapse rate (environmental lapse rate)
tropopause
stratosphere
mesosphere
thermosphere
ozonosphere
ionosphere
radiation
terrestrial radiation
conduction

convection
advection
latent heat exchange
condensation (release of heat)
freezing
melting
evaporation
evaporative cooling
Earth's energy budget
energy balance
heat
temperature
Fahrenheit scale
Celsius (centigrade) scale
Kelvin scale
absolute zero
diurnal
temperature lag
daily temperature lag
daily march of temperature

maritime
continentality
albedo
temperature inversion
inversion layer
upper-level inversion
surface inversion
controls of temperature
specific heat
warm current
cold current
isotherm
temperature gradient
annual temperature lag
annual march of temperature
weather
climate
meteorology
climatology
element (atmosphere)

## DISCUSS AND REVIEW

1. Why is it useful to think of the atmosphere as a thin film of air? As an ocean of air?
2. Name the gases of the atmosphere and give the percentage of the total that each supplies.
3. What function does ozone play in supporting life on Earth? Where and how is ozone formed?
4. How is the atmosphere subdivided? In what levels do you live? Have you been in any of the other levels?
5. How does Earth's atmosphere affect incoming solar radiation (insolation)? What processes prevent insolation from reaching Earth's surface? What percentages are involved in a generalized situation? What percentages reach the surface and by what processes?
6. Discuss the role of water in energy exchange. What characteristics of water make it so important?
7. How is the atmosphere heated from Earth's surface? What processes and percentages are involved in a generalized situation?
8. What is meant by Earth's energy budget? List and define the important energy exchanges that keep it in balance.
9. What is the temperature in Fahrenheit degrees today in your area? In Celsius degrees?
10. At what time of day does insolation reach its maximum and its minimum? Compare this to the typical times when the daily high and low temperatures occur.
11. How would albedo affect your selection of outdoor clothes on a hot, sunny day? On a cold, clear winter day?
12. What is a temperature inversion? Explain the two major types of temperature inversions.
13. Why do vineyards and orchards use wind machines and heaters? What methods are used to prevent frost damage to plants in your area, including landscaping plants?
14. Would you expect places like Seattle to have a milder or a harsher winter than Grand Forks, North Dakota? Why?
15. Describe the behavior of the isotherms in Figures 4.27a and 4.27b. What factors cause the greatest deviation from an east–west trend? What factors cause the greatest temperature shifts between the January and July maps (for the same location)?
16. What is the difference between meteorology and climatology?
17. What are the basic *atmospheric elements* of weather and climate?
18. What factors cause variations in the elements of weather and climate?

# CONSIDER AND RESPOND

1. Convert the following temperatures to Fahrenheit: 20°C, 30°C, and 15°C.
2. Convert the following temperatures to Celsius: 60°F, 15°F, and 90°F.
3. Convert the temperatures in items 1 and 2 to Kelvin.
4. Refer to Figure 4.13. List the major means by which the atmosphere gains heat and loses heat.
5. What are the major controls of weather and climate that operate in your area?
6. The normal lapse rate is 6.5°C/1000 meters. If the surface temperature is 25°C, what is the air temperature at 10,000 meters above Earth's surface? Convert your answer to degrees Fahrenheit.
7. Refer to Figures 4.27a and 4.27b. What location on Earth's surface exhibits the greatest annual range of temperature? Why?

 CourseMate – Make the most of your study time by accessing everything you need to succeed in one place. Read your textbook, take notes, review flashcards, watch videos, complete activities, take practice quizzes, and more online with CourseMate. Log in at **www.cengagebrain.com**.

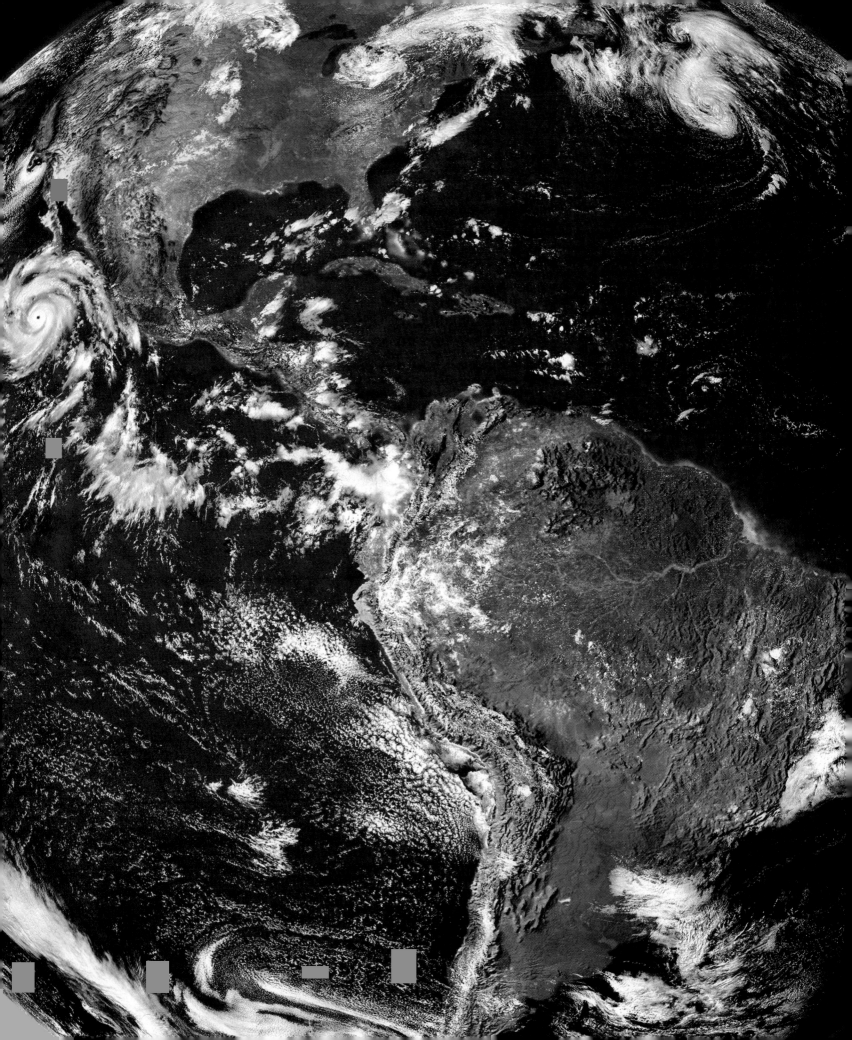

# 5

# ATMOSPHERIC PRESSURE, WINDS, AND CIRCULATION PATTERNS

## OUTLINE

**Variations in Atmospheric Pressure**
- Air Pressure, Altitude, and Elevation
- Horizontal Pressure Variations

**Basic Pressure Systems**
- Convergent and Divergent Circulation
- Mapping Pressure Distributions

**Wind**
- Winds and Pressure Gradients
- The Coriolis Effect and Wind
- Cyclones, Anticyclones, and Winds
- Geography's Environmental Perspective: Harnessing the Wind's Energy

**Global Pressure Systems**
- A Model of Global Pressure Belts
- Seasonal Pressure Differences

**Global Wind Systems**
- A Model of Atmospheric Circulation
- Winds in Latitudinal Zones
- Latitudinal Migration with the Seasons
- Longitudinal Differences in Winds

**Upper Air Winds and Jet Streams**

**Regional and Local Wind Systems**
- Monsoon Winds
- Local Winds
- Geography's Spatial Perspective: The Santa Ana Winds and Wildfires

**Ocean–Atmosphere Interactions**
- Ocean Currents
- El Niño
- North Atlantic Oscillation

The swirling circulation patterns in Earth's atmosphere are created by changes in atmospheric pressure and winds. NASA

## OBJECTIVES

WHEN YOU COMPLETE THIS CHAPTER YOU SHOULD BE ABLE TO:

- Explain why atmospheric pressure declines with altitude and generally varies with latitude because of air temperature differences and vertical air movement.
- Associate precipitation, clouds, and windy conditions with the rising air of low pressure systems and clear, calm conditions with the subsiding air of high pressure systems.
- Answer why latitudinal changes in sun angle and duration of daylight hours with the seasons cause temperature variations that influence atmospheric pressure.
- Provide examples of how differences in atmospheric pressure affect wind velocity and direction.
- Understand why the Coriolis effect apparently causes winds and ocean currents to bend to the right of the direction of motion in the Northern Hemisphere and to the left in the Southern Hemisphere.
- Outline the major latitudinal pressure systems and wind belts and their influence on the circulation of global winds and ocean currents.
- Discuss examples of how interchanges between the atmosphere and oceans influence weather systems.

AN INDIVIDUAL GAS MOLECULE IN the atmosphere weighs almost nothing. So it may be surprising to learn that as huge numbers of air molecules collide with anything, they exert an average pressure of 1034 grams per square centimeter (14.7 lb/sq in.) at sea level. People are not crushed by this atmospheric pressure because the air and water inside us—in our blood, tissues, and cells—exert an equal outward pressure that balances the atmospheric pressure.

Changes in atmospheric pressure exert a major influence on our weather and climate. Differences in atmospheric pressure circulate the air and create our winds, which cycle water from the oceans to the landmasses. Air pressure and winds are also important factors in driving the world's ocean currents. Wind movement also disperses seeds and pollen in the biosphere and affects the lithosphere by transporting dust, soil particles, and sand.

In 1643, Evangelista Torricelli, a student of Galileo, performed an experiment that was the basis for inventing the **mercury barometer,** an instrument that measures atmospheric (also called barometric) pressure. Torricelli filled a long glass tube, closed on one end, with mercury and inverted it in an open pan of mercury. The mercury inside the tube fell until it was at a height of about 76 centimeters (29.92 in.) above the mercury in the pan, leaving a vacuum bubble at the closed, upper end of the tube. The pressure exerted by the atmosphere on the mercury in the open pan was equal to the pressure from the mercury trying to drain from the tube. As the atmospheric pressure increased, it pushed the mercury to a higher level in the tube, and as the air pressure decreased, the mercury level in the column dropped proportionately.

In the strictest sense, a mercury barometer does not actually measure the pressure exerted by the atmosphere, but instead measures the response to atmospheric pressure. That is, when the atmosphere exerts a specific pressure, the mercury will respond by rising to a specific height ( ● Fig. 5.1a). An **aneroid barometer** (Fig. 5.1b), another device for measuring air pressure, uses a hollow metal cell that is sealed after removing some of the air inside. The cell, typically accordion shaped,

expands and contracts with pressure changes, and a needle pointer indicates the atmospheric pressure on a calibrated scale. Although we usually hear atmospheric pressure reported in inches of mercury, meteorologists typically work with actual pressure units, most often the millibar (mb). **Standard sea-level pressure** is equal to 1013.2 millibars and will support a level of 76 centimeters (29.92 in.) of mercury in a barometer. These values are important because they provide the cutoff measurements used to define areas (often called cells) of low pressure and areas of high pressure.

Our study of the atmospheric elements that produce weather and climate has considered variations in the receipt of solar energy and their influence on temperature patterns from the equator to the poles. In this chapter we learn that these temperature differences are one of the major causes of the development of higher and lower pressure areas that are often linked to latitude. In addition, we examine patterns of another kind—movement—the circulations of the atmosphere and oceans, which move energy and matter through Earth's subsystems.

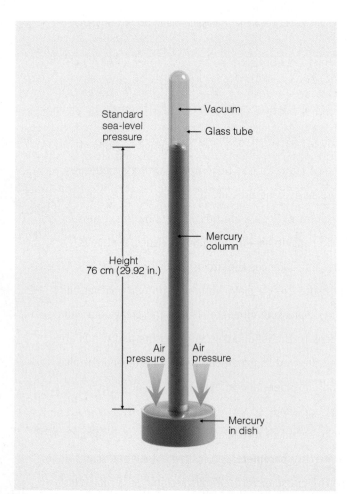

(a)

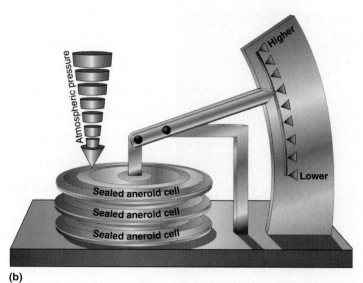

● **FIGURE 5.1** (a) A mercury barometer. Standard sea-level pressure of 1013.2 millibars will cause the mercury to rise 76 centimeters (29.92 in.) in the tube. (b) An aneroid barometer works by the expansion or contraction of the accordion-shaped cell in response to changes in air pressure, which moves an indicator needle up or down.
From Federal Aviation Administration Pilot's Handbook of Aeronautical Knowledge, Figure 11-7. http://www.faa.gov/library/manuals/aviation/pilot_handbook/
Copyright © 2011 Brooks/Cole, Cengage Learning

*What are some advantages of an aneroid barometer?*

(b)

Geographers are particularly interested in circulation patterns because they illustrate spatial interactions, one of geography's major themes and introduced in Chapter 1. Patterns of movement from one place to another reveal shared relationships, prompting efforts to understand the nature and effects of the interactions involved. It is also important to understand the causes of the spatial interactions taking place. As we examine the circulation patterns featured later in this chapter, you should make an effort to trace each pattern back to the fundamental influence of solar energy.

# Variations in Atmospheric Pressure

## Air Pressure, Altitude, and Elevation

Air pressure decreases with increasing elevation on Earth and with altitudes above it because the higher we go, the more widely spaced and diffused air molecules become. The increased space between gas molecules in the atmosphere results in lower air density and lower air pressure ( ● Fig. 5.2). In fact, at the top of Mount Everest (elevation 8848 m, or 29,028 ft), the air pressure is only about one third of the pressure at sea level. Note that altitude usually refers to a height in the air (above sea level), and elevation refers to height on the surface above (or below) sea level.

People are usually not sensitive to small, gradual changes in air pressure. However, when we climb to high elevations or fly

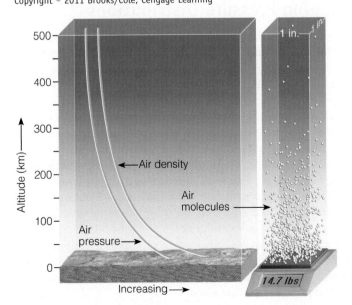

● **FIGURE 5.2** Both air pressure and air density decrease rapidly with increasing altitude.
Copyright © 2011 Brooks/Cole, Cengage Learning

to altitudes significantly above sea level, we become aware of the effects of air pressure on our bodies. While flying at altitudes around 10,000 to 12,000 meters (33,000–35,000 ft) commercial airliners are pressurized to maintain air pressure at safe levels for the passengers and crew. At cruising altitude, aircraft cabins are typically pressurized to an equivalent pressure that would be experienced at an elevation of about 2100 to 2400 meters (7000–8000 ft.). Still, pressurization may vary, so our ears may pop as they adjust to rapid pressure changes when ascending or descending. Hiking or skiing in locations that are a few thousand meters in elevation will affect us if we are used to sea-level pressure. Reduced air pressure means that there are fewer air molecules in a given volume of air, so less oxygen will be contained in each breath. Thus, we get out of breath much more easily at high elevations until our bodies adjust to the reduced air pressure and corresponding drop in oxygen level.

Altitude and elevation, however, are not the only factors that cause air pressure to change. On Earth's surface, variations in pressure are also related to the intensity of heating from insolation, local humidity, and global or regional air circulations. Changes in atmospheric pressure at a given location often indicate a change in the weather is occurring or coming. Weather systems themselves can be classified by their structure and tendency toward change of their pressure.

## Horizontal Pressure Variations

There are two main causes of horizontal variations in air pressure. One is thermal (determined by air temperature) and the other is dynamic (related to atmospheric air motion). In reality, although one of these processes may dominate, the two operate together.

Thermally induced changes in air pressure are relatively easy to understand. Both air movement and air density are related to temperature differences that result from the unequal distribution of insolation, differential heating of land and water, and the varying albedos that exist on Earth's surface. A scientific law states that the pressure and density of a gas will vary inversely with its temperature. Thus, as daytime heating warms the air in contact with Earth's surface, the air expands in volume and decreases in density. As the heated air decreases in density, it will rise and consequently lower the atmospheric pressure at the surface. Thermally induced rising of warm air contributes to the typically low pressures that dominate the equatorial regions and is a major cause of thunderstorms. When air becomes cold, the volume will decrease and the density will increase, causing the air to sink, and increasing the atmospheric pressure. For these reasons, polar regions regularly experience high pressure. The low pressure in the equatorial zone and the high pressures at the polar regions are considered thermally induced because the air temperature plays a dominant role in creating these pressure conditions.

Given this knowledge, we might expect that a gradual increase in air pressure would accompany the general decline in temperature from the equator to the poles. However, sea-level barometer readings indicate that atmospheric pressure does not

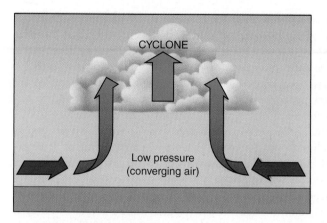 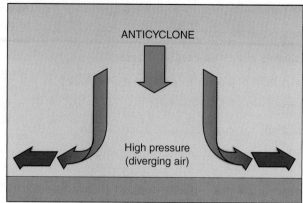

● **FIGURE 5.3** Winds converge and ascend in cyclones (low pressure centers) and descend and diverge from anticyclones (high pressure centers).

*What happens to the temperature of air if it rises, and what happens if it descends?*

increase in a regular pattern poleward from the equator. Instead, regions of high pressure in the subtropics and low pressure in the subpolar regions develop through dynamic air movements.

The dynamic processes that affect air pressure are related to broad patterns of atmospheric circulation. For example, at the equator when rising air encounters the tropopause it splits into two air currents flowing in opposite poleward directions (north and south). When these high-level air currents reach the subtropical regions, they encounter similar air currents flowing equatorward from the middle latitudes. As these opposing upper-air currents merge, the air aloft will "pile up" and descend, producing high pressure in the subtropical regions. As air sinks downward, it forms a high pressure area and flows outward from the high into an area with lower pressure, where it rises, which tends to reinforce the low air pressure at that location.

At the surface in the subpolar regions, air flowing out of the polar high pressure zones encounters air flowing out of the subtropical high pressure regions. The collision of these merging winds causes the air to rise, creating a dynamically induced zone of low pressure that is common in the subpolar regions. High pressure in the subtropical regions and low pressure in the subpolar regions are dominantly the result of dynamic air movement. These examples describe horizontal pressure variations on a global scale, which we will investigate further later in this chapter.

# Basic Pressure Systems

There are two basic types of pressure systems. An area where the pressure is lower than standard sea-level pressure is often simply called a **low,** but a **cyclone** is also a general term for a low pressure area. A high-pressure area is called a **high,** or an **anticyclone.** Low and high pressure areas are often referred to as cells and are represented by the capital letters **L** and **H** on weather maps that we see on TV, on the Internet, and in the newspaper.

A low, or cyclone, is an area where air is rising. As air moves upward away from the surface, it relieves pressure on that surface and barometer readings will fall. The situation in a high, or anti-cyclone, is just the opposite. Under conditions of high pressure, air descends and barometer readings rise, indicating an increased atmospheric pressure on the surface. Lows and highs are illustrated in ● Figure 5.3.

## Convergent and Divergent Circulation

Winds blow toward the center of a cyclone—that is, they converge on a low pressure cell. The center of a low pressure system serves as the focus for **convergent wind circulation.** In contrast, the winds blow outward and away from the center of an anticyclone, diverging from the high pressure cell. The center of a high pressure cell, where the pressure is highest, serves as the source for **divergent wind circulation.** Given existing pressure differences, we would expect converging and diverging winds to move in straight paths, as shown in Figure 5.3. Yet, wind motions are much more complex because they are influenced by other factors, which will be explained in the section on wind.

## Mapping Pressure Distributions

Maps are an excellent medium for geographers and meteorologists when they analyze the existing and changing spatial patterns that influence our weather. Air pressure is also strongly influenced by elevation in addition to spatial variations in air pressure. When atmospheric pressure is mapped or presented in weather broadcasts, the air pressure reported for every surface location is adjusted to reflect what it would be if that place was at sea level. Adjusting air pressure to its sea-level equivalent is important because the variations resulting from altitude are far greater than those caused by changes in the weather or in the air temperature. Without this adjustment, elevation differences would mask the regional differences, which are more important to understanding the weather. For example, if meteorologists did not adjust barometer readings to the sea level equivalent, Denver, Colorado (the "Mile High City") would always be reporting low pressure conditions, yet the barometric pressure there varies up and down, just like it does at every other place.

**Isobars** (from Greek: isos, equal; baros, weight) are lines on maps that connect points of equal pressure. When isobars are closely spaced, they portray a significant difference in pressure over a short distance, a strong **pressure gradient.** Isobars that are widely spaced indicate a weak pressure gradient. When depicted on a map, high and low pressure cells are outlined by concentric isobars that form a closed cell around centers of high or low pressure. In map view, cells of high and low pressure vary in shape from roughly circular to elongated. If pressure cells are somewhat elongated, they may be referred to as either a trough (low) or a ridge (high).

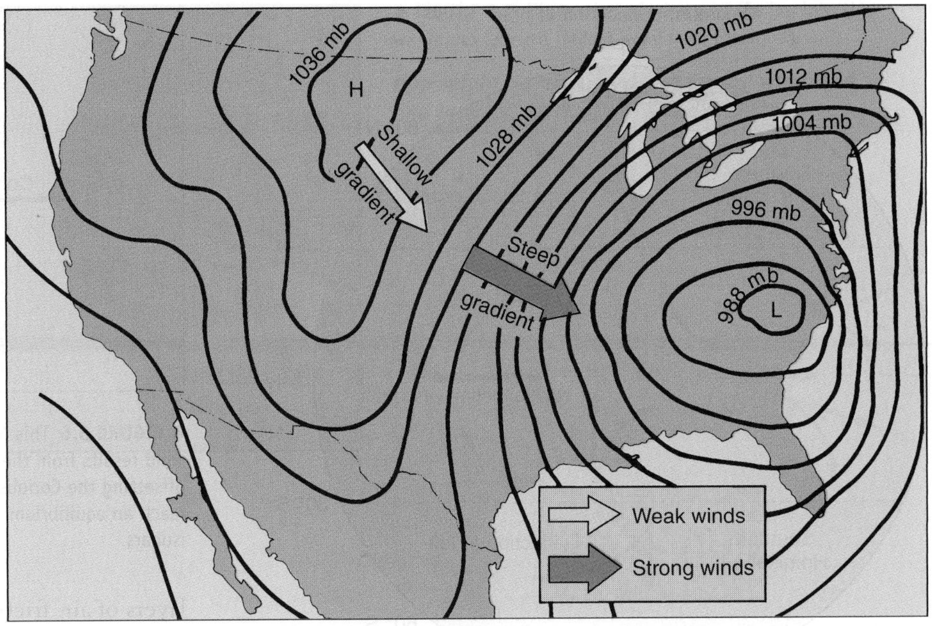

● **FIGURE 5.4** The relationship of wind to the pressure gradient: The steeper the pressure gradient, the stronger the resulting wind.

*What is the highest pressure shown on this map, and what is the lowest?*

# Wind

**Wind** is the movement of air in response to differences in atmospheric pressure. Winds also play a major role in correcting the temperature imbalances that occur over Earth's surface. On average, locations below 38° latitude receive more radiant energy than they lose, whereas locations poleward of 38° lose more than they gain (see Fig. 4.14). Through advection, Earth's planetary wind system transports heat energy poleward to help moderate temperatures by moving heat energy away from hot regions to colder regions. The global wind system also strongly influences the ocean currents, which also transport great quantities of heat energy from areas that receive a surplus to regions where a deficit exists. Thus, without winds and ocean currents, the equatorial regions would continually get hotter and the polar regions would continually get colder through time.

In addition to the advectional (horizontal) transport of heat energy, winds also transport water vapor from the air over water bodies where it has evaporated, to land surfaces where it condenses and precipitates. Without moisture-laden winds, land areas would be arid and barren. In addition, winds influence evaporation rates. Furthermore, as we become more aware of our energy needs, harnessing the power of the wind is becoming more important, along with other natural sources such as water power and solar energy. These are clean, abundant, and renewable energy sources.

## Winds and Pressure Gradients

Winds are the atmosphere's means of attempting to balance uneven pressure distributions, and they vary in velocity, duration, and direction. The velocity and strength of a wind depends on the intensity of the pressure gradient that produces the wind. As we noted previously, a pressure gradient is the rate of change in atmospheric pressure between two points, typically indicated on weather maps by the spacing of isobars. When the pressure gradient is steep with a large pressure change over a

short distance, the winds will be fast and strong ( ● Fig. 5.4). Winds tend to flow down a pressure gradient from high pressure to low pressure, just as water flows downslope from a high point to a low one. A useful little rhyme, "Winds always blow from high to low," is a reminder of the direction of winds. Steep pressure gradients produce faster and stronger winds, yet the wind does not generally flow in a straight line directly from high to low.

## The Coriolis Effect and Wind

Two factors, both related to Earth rotation, greatly influence wind direction. First, our fixed-grid system of latitude and longitude is constantly rotating with the planet. Thus, our frame of reference for tracking the path of any free-moving object—whether it is an aircraft, a missile, or the wind—is constantly changing its position. Second, the rotational speed of Earth's surface increases as we move equatorward and decreases as we move toward the poles (see Fig. 3.11). Referring to our previous example, someone in St. Petersburg (60°N latitude), where the distance around a parallel of latitude is about half of that at the equator, moves at about 840 kilometers per hour (525 mph) as Earth rotates, whereas someone in Kampala, Uganda, near the equator, moves at about 1680 kilometers per hour (1050 mph).

Because of Earth rotation, anything moving horizontally appears to be deflected to the right of its direction of travel in the Northern Hemisphere and to the left in the Southern Hemisphere. This apparent deflection is the **Coriolis effect.** The amount of deflection, or apparent curvature of the travel

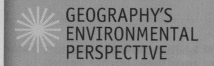
# Harnessing the Wind's Energy

For centuries, windmills provided the power to pump water and grind grain throughout the world. But the widespread availability of inexpensive electricity changed the role of most windmills to that of a nostalgic tourist attraction. Today the mounting needs for electricity and increasing problems from atmospheric pollution associated with fossil fuels encourage the use of wind power to generate electricity.

Wind power is an inexhaustible source of clean energy. Although the cost of electrical energy produced by the wind depends on favorable sites for locating wind turbines, wind power is already cost competitive with power produced from fossil fuels. One expert calls wind generation the fastest-growing electricity-producing technology in the world. During the past year, power production from the wind increased about 30%. The wind provides an estimated 175,000 megawatts of electricity worldwide.

There are two particularly important criteria in locating wind turbines. The site must have persistent strong winds, and it must be in an area where the electricity from the turbines can be linked to an existing electrical grid system. In the United States individual wind turbines can be found producing electricity (such as those located on farms in the Midwest and Great Plains), but most wind power is generated from *wind farms*. These are long rows or concentrated groups of many turbines. Each turbine can economically extract up to 60% of the wind's energy at minimum wind speeds of 20 kilometers (12 mi) per hour, although higher wind speeds are desirable. Because the power generated is proportional to the cube of the wind speed, a doubling of wind velocity increases energy production eight times.

Although North America currently lags behind Europe in the production of energy from the wind, the continent has great potential. Excellent sites for wind farms exist throughout the open plains of North America's interior and along its coasts from the Maritime Provinces of Canada to Texas and from California to the Pacific Northwest. The sites are available, the technology has been developed, the costs are competitive, and the resolve to shift from fossil fuels is growing. Between 2009 and 2010 the production of wind power in the United States increased about 30%, but this supplies only around 2% of the nation's current usage. The potential for expanding the generation capacity of this natural energy resource is great.

Wind turbines like this one in Germany generate electricity as the winds spin the propellers. The field of yellow flowers is planted with rapeseed, an agricultural crop that is harvested to make biofuels.

Windmills like this have been used for years to pump well water on ranches and farms in the semiarid and arid regions of North America.

In response to the pressure gradient, air from all directions tends to flow toward the center of a low pressure area (cyclone). Despite the fact that winds are apparently deflected to the right in the Northern Hemisphere, strong pressure gradients in a low pressure cell cause the winds to flow into the center of the low in a counterclockwise spiral. These spirals are reversed in direction in the Southern Hemisphere where winds and currents are apparently deflected to the left. Thus, in the Southern Hemisphere, winds moving away from an anticyclone do so in a counterclockwise spiral, and winds moving into a cyclone move in a clockwise spiral.

## Global Pressure Systems
### A Model of Global Pressure Belts

Using what we have learned about pressure on Earth's surface, we can construct a simplified model of the pressure belts of the world ( • Fig. 5.9). Later, we will see how conditions depart from our

model and why these differences occur by examining global pressure in more detail.

Centered approximately over the equator in our model is a belt of low pressure, or a **trough.** Because this is the region on Earth of greatest annual heating, we can conclude that the low pressure of this area, the **equatorial low (equatorial trough),** is determined primarily by thermal factors, which cause the air to rise.

North and south of the equatorial low, centered on about 30°N and 30°S, cells of relatively high pressure dominate. These are the **subtropical highs,** which result from dynamic air motion related to the sinking of convectional cells initiated at the equatorial low.

Poleward of the subtropical highs in both hemispheres are large belts of low pressure that extend through the upper-middle latitudes. Pressure decreases through these **subpolar lows** until about 65° latitude. Dynamic factors dominate the formation of the subpolar lows, as opposing winds collide and cause air to rise.

The Arctic and Antarctic regions are dominated by high pressure systems called the **polar highs.** Extremely cold tem-

● **FIGURE 5.9** Idealized world pressure belts. Note the arrows on the perimeter of the globe that illustrate the vertical flows of air associated with the surface pressure belts.

*Is the latitudinal pattern of these pressure belts the same for the Southern Hemisphere and the Northern Hemisphere?*

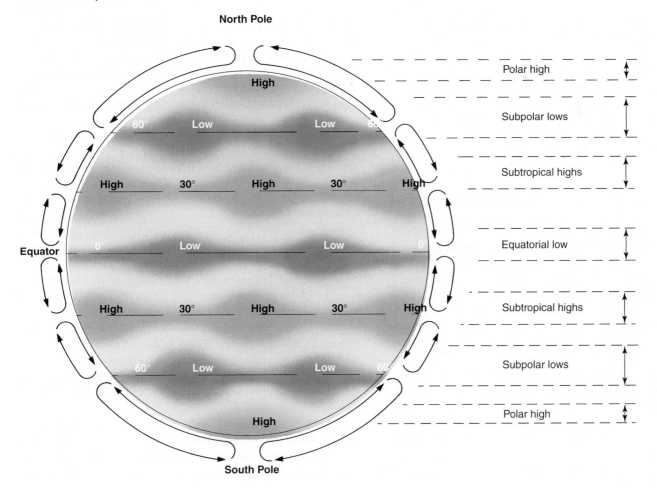

peratures and the consequent sinking of dense cold air create the higher pressures found in the polar regions.

This system of pressure belts is a generalized model, but it is useful for understanding global wind circulations and dominant pressure patterns. Yet, to some degree temperatures and atmospheric pressures change from month to month, day to day, and hour to hour at any location. This global model of pressure belts does not reflect these local changes, but it does give a general idea of the latitudinal patterns of surface atmospheric pressure that influence Earth's weather and climate regions.

As our idealized model indicates, processes in the atmosphere tend to form latitudinal belts of high and low pressure, but this model does not consider the alternating influences of ocean basins and continents that these belts traverse. In latitudes where continental landmasses are separated by ocean basins, pressure belts tend to break up to form cellular pressure systems. These cells of high and low pressure develop because the belts are affected by the differential heating of land and water. Landmasses also affect air movement and the development of pressure systems through surface friction and air flowing up and down mountains.

## Seasonal Pressure Differences

In general, the global atmospheric pressure belts shift northward in July and southward in January, following the changing position of the sun's direct rays as they migrate between the Tropics of Cancer and Capricorn. Thus, thermally induced seasonal variations affect the pressure patterns, as seen in ● Figure 5.10. Seasonal differences tend to be minimal at low latitudes, where little temperature variation occurs. At high latitudes, seasonal temperature differences are larger because of greater annual variations in daylight length and angle of the sun's rays. Landmasses also alter the pattern of seasonal pressure variations for certain latitudes. This is a particularly important factor in the Northern Hemisphere, where land accounts for 40% of surface area, as opposed to less than 20% in the Southern Hemisphere.

January During the Northern Hemisphere winter, middle and high latitude continents become much colder than the surrounding oceans. Figure 5.10a shows that in the Northern Hemisphere this variation leads to the development of high pressure cells over the land areas. In contrast, the subpolar lows develop over the oceans because they are comparatively warmer. Over eastern Asia, there is a strongly developed anticyclone during the winter months, known as the **Siberian High.** Its equivalent in North America, the **Canadian High,** is not as strong because North America is smaller than the Eurasian continent.

In addition to the Canadian High and the Siberian High, two low pressure centers develop: one in the North Atlantic, called the **Icelandic Low,** and the other in the North Pacific, called the **Aleutian Low.** These low pressure cells result from the clash of winds flowing out of the polar high to the north with winds from the subtropical highs to the south. The Aleutian and Icelandic lows are associated with cloudy, unstable weather and are a major source of winter storms. In contrast, the Canadian and Siberian Highs are

associated with clear, blue-sky days; calm, starry nights; and cold, stable weather. Therefore, during the winter months, cloudy and sometimes dangerously stormy weather tends to be associated with the two oceanic lows, and clear, but cold, weather is associated with the continental highs.

We can also see that in January the polar high in the Northern Hemisphere is well developed, primarily because of thermal cooling during the coldest time of the year. The subpolar lows have developed into the Aleutian and Icelandic Low pressure cells described earlier. The subtropical highs of the Northern Hemisphere have moved slightly south of their average annual position as the sun's rays migrate toward the Tropic of Capricorn. The equatorial low also shifts south of its average annual position at the equator.

In the Southern Hemisphere during January (where it is summer), the subtropical high pressure belt breaks into three cells centered over the oceans because the warmer continents produce lower pressures compared to those over the oceans. Because there is virtually no land between 45°S and 70°S latitude, the subpolar low circles Earth over ocean waters as a belt of low pressure rather than being divided into cells. There is little seasonal change in this belt of low pressure other than in the Southern Hemisphere summer (January) when it shifts a few degrees north of its July (winter) position.

July The high pressure over the North Pole weakens during the summer, primarily because of the lengthy (24-hour daylight) heating in that region (Fig. 5.10b). The Aleutian and Icelandic Lows also weaken and shift poleward from their winter position. The large landmasses of North America and Eurasia, which developed high pressure cells during the cold winter months, develop extensive low pressure cells slightly to the south during the summer.

In Asia a low pressure system develops, but it is divided into two separate cells by the Himalayas. The low pressure cell over northwest India is strengthened because it combines with the equatorial low, which has moved north of the position it had 6 months earlier. The subtropical highs in the Northern Hemisphere are stronger over the oceans than over the landmasses, and in the summer they have migrated northward from their winter position. In this more northerly location they have a strong influence on the climate of the nearby landmasses. The North Pacific subtropical high is termed the **Pacific High** (also called the Hawaiian High), a pressure system that greatly affects the climates of the west coast of North America. North Americans call the corresponding high pressure cell in the North Atlantic the **Bermuda High,** but it is the **Azores High** to Europeans and West Africans. The equatorial low moves north in July, following the seasonal migration of the sun's rays, and the subtropical highs of the Southern Hemisphere move equatorward of their January (summer) locations.

There are essentially seven belts of pressure (two polar highs, two subpolar lows, two subtropical highs, and one equatorial low). These pressure belts, however, break into pressure cells in some places primarily because of alternating continents and

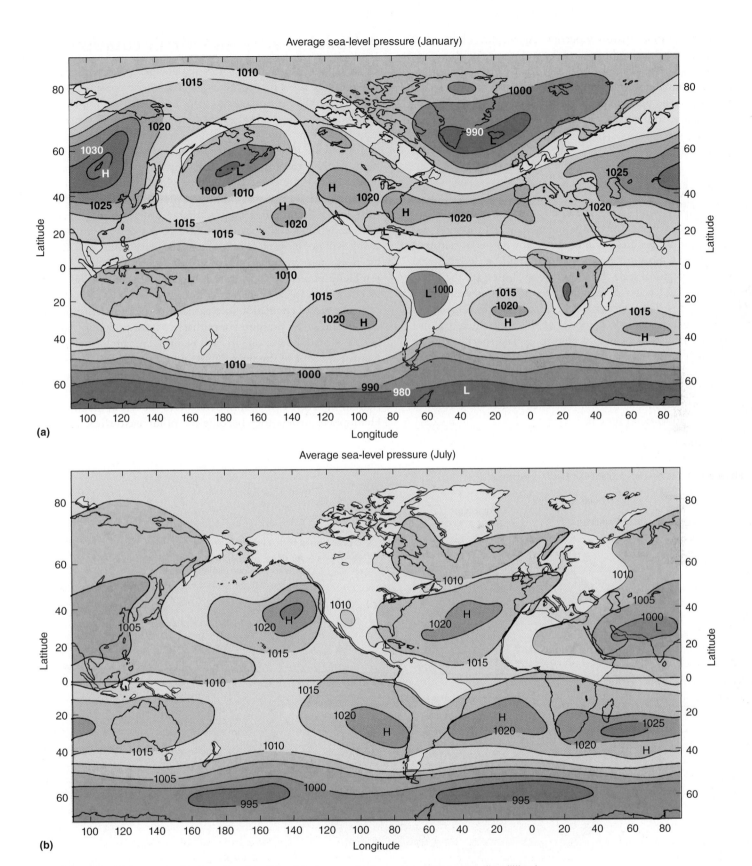

● **FIGURE 5.10** (a) Average sea-level pressure (in millibars) in January. (b) Average sea-level pressure (in millibars) in July.

*What is the difference between the January and July average sea-level pressures at your location? Why do they vary?*

oceans at the latitudes that they occupy. Again, this is because of the differences in heating between land and water. We have also seen that these belts and cells vary in size, intensity, and location with the seasons and with the migration of the sun's vertical rays. These global-scale pressure systems form a fairly regular latitudinal distribution, so they are sometimes referred to as semipermanent pressure systems.

# Global Wind Systems

Earth's planetary wind system is a response to the global pressure patterns and also plays a role in the maintenance of those regions of high and low pressures. Winds, the major means of transport for energy and moisture through the atmosphere, can also be examined on a global scale. For now, we will disregard the influences of land and water differences, variations in elevation, and seasonal changes. These factors will be addressed later. This allows us to construct a generalized model of the atmosphere's global circulation. This model can also help us explain specific climate features such as the rain and snow of the mountains of western North America and the arid regions directly to the east of those mountains. A basic wind model will provide insights for understanding the movement of ocean currents, which are driven, in part, by the global wind systems.

## A Model of Atmospheric Circulation

Because winds are caused by pressure differences, a system of global winds can be based on the model of atmospheric pressures (see Fig. 5.9). The convergence and divergence of winds are important to our understanding of global wind patterns. Surface air diverges from zones of high pressure and converges on areas of low pressure.

Knowing that surface winds originate from and flow out of highs to flow toward low pressure areas, we can use the global model of pressure belts to develop a global wind system model ( ● Fig. 5.11). This global circulation model, although simplified, takes into account differential heating, Earth rotation, and atmospheric dynamics. The Coriolis effect is also considered because winds will not blow along straight north–south paths. The Coriolis effect causes an apparent deflection to the right in the Northern Hemisphere and to the left in the Southern Hemisphere, and its effect increases with higher latitudes.

The global circulation model includes six wind belts in addition to the seven pressure zones that we have previously identified. Two wind belts, one in each hemisphere, are located where winds blow out of the polar highs and down the pressure gradients toward the subpolar lows. These winds in the high latitudes are very strongly affected by apparent Coriolis deflection to the right in the Northern Hemisphere and to the left in the Southern Hemisphere, where they become the **polar easterlies.**

● **FIGURE 5.11** The general circulation of Earth's atmosphere.

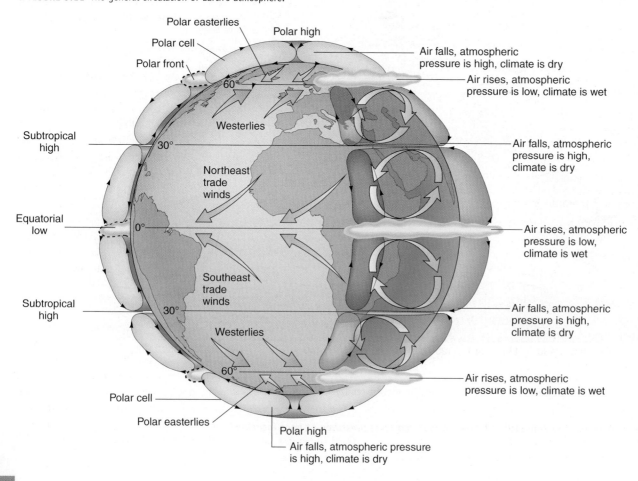

Divergent circulation out of the subtropical highs, flowing both equatorward and poleward, provides the source for the remaining four wind belts. In each hemisphere, winds flow out of the poleward sides of the subtropical highs into the subpolar lows. Because of the strong apparent bending of winds flowing from the higher latitude sides of the subtropical highs, the general wind movement is from the west. These winds of the upper-middle latitudes are the **westerlies.** The winds blowing from the subtropical highs toward the equator are the **trade winds.** Because the bending from the Coriolis effect in lower latitudes is minimal, they are the **northeast trades** in the Northern Hemisphere and the **southeast trades** south of the equator. While examining this model, it is important to remember that winds are named after the direction from which they come.

This model provides a basic concept of the global atmospheric circulation, although many other factors, local and regional, also influence the winds. The pressure systems, and consequently the winds, move in response to heating differences that change with the seasonal position of the sun. The winds are also influenced by continent and ocean differences between the Northern and Southern Hemisphere that affect high and low pressure zones.

## Winds in Latitudinal Zones

**Trade Winds** A good place to begin a more detailed examination of winds and associated weather patterns is in the vicinity of the subtropical highs. On Earth's surface the trade winds, which blow out of the subtropical highs toward the equatorial low in both the Northern and Southern Hemispheres, can be identified between latitudes 5° and 25°. Because of the Coriolis effect, the northern trade winds move away from the subtropical high in a clockwise direction out of the northeast. In the Southern Hemisphere, the trades diverge out of the subtropical high toward the equatorial low from the southeast in a counterclockwise direction. Because the trades tend to blow out of the east, they are also known as the **tropical easterlies.**

The trade winds tend to be constant, steady winds, consistent in their direction. The location of the trade winds, however, does vary somewhat during the seasons as they move a few degrees of latitude north and south with the sun's rays. Near their source in the subtropical highs, the weather of the trades is clear and dry, but after crossing large expanses of warm ocean waters, the trades have a high potential for stormy weather. Early Spanish sailing ships depended on the northeast trade winds to drive their galleons from Europe to destinations in Central and South America in search of gold, spices, and new lands. Going eastward toward home, navigators usually tried to plot a course using the westerlies to the north. The trade winds are one of the reasons why the Hawaiian Islands are popular with tourists; the steady winds help keep temperatures pleasant, even though Hawaii is located south of the Tropic of Cancer.

**The Intertropical Convergence Zone** The equatorial low, where the trade winds converge, coincides with the latitudinal belt with the world's heaviest precipitation and most persistent cloud cover. This area is called the **Intertropical Conver-**

gence Zone (ITCZ or ITC) because it is where the trade winds from the tropics in both hemispheres converge on the equatorial region. In the ITCZ humid air heated by the sun expands and rises, which maintains low pressure in the area and a high potential rainfall.

Close to the equator, roughly between 5°N and 5°S, is where the strongest belt of convergence is located, with rising air, heavy rainfall, and calm winds that have no prevailing direction. This zone is known as the **doldrums.** Without adequate winds for sailing, ships often remained stranded in the doldrums for days. It is interesting to note that the word doldrums in English means a listless condition or a depressed state of mind. The sailors were in the doldrums in more ways than one.

**Subtropical Highs** The areas of subtropical high pressure, generally located between latitudes 25° and 35°N and S, are the source of winds blowing generally poleward as the westerlies and equatorward as the trade winds. These subtropical belts of variable or calm winds have been called the **"horse latitudes."** This name comes from sailors having to resort to eating their horses or throwing them overboard to conserve drinking water and lighten the weight when their sailing ships became stranded in these latitudes. The centers of the subtropical highs, like the doldrums, are areas where there are no strong prevailing winds. The doldrums are characterized by convergence, rising air, and heavy rainfall. In contrast, the subtropical highs are areas of air descending from higher altitudes. Weather conditions there are typically clear, sunny, dry, and rainless, especially over the eastern sides of the ocean basins (along west coasts) where the high pressure cells are strongest.

**Westerlies** The winds that flow poleward out of the subtropical highs in the Northern Hemisphere are strongly deflected to the right and thus blow from the southwest. These winds in the Southern Hemisphere are strongly deflected to the left and blow out of the northwest. Thus, these winds are correctly labeled the westerlies. They tend to be less consistent in direction in comparison to the trade winds, but they are usually stronger winds and may be associated with stormy weather. The westerlies occur between about 35° and 65°N and S latitudes. In the Southern Hemisphere, where there is less land than in the Northern Hemisphere to affect wind development, the westerlies attain their greatest consistency and strength. Much of Canada and most of the United States, except Florida, Hawaii, and Alaska, are influenced by weather brought by the westerlies.

**Polar Winds** Accurate surface observations of pressure and wind are sparse in the polar regions. Much of our information is derived from satellite data and imagery. Atmospheric pressure tends to be consistently high throughout the year at the poles. The polar highs feed prevailing winds (the polar easterlies) that circle the polar regions and blow easterly toward the subpolar low pressure systems. This circulation around the poles expands outward in winter and contracts in summer in response to seasonal temperatures.

**Polar Front** The polar wind systems can be highly variable, blowing at times with great speed and intensity. When the cold air flowing out of the polar regions and the warmer air moving in the path of the westerlies meet, they do so like two warring armies. One does not absorb the other, but the denser, heavier cold air pushes warm air upward, forcing it to rise rapidly. The line along which these two great wind systems battle is appropriately known as the **polar front.** The weather that results from the meeting of cold polar air and warmer air from the subtropics can be very stormy. In fact, most of the storms that move slowly through the middle latitudes following the path of the prevailing westerlies originate along the polar front.

## Latitudinal Migration with the Seasons

Just as insolation, temperature, and pressure systems migrate north and south, Earth's wind systems also migrate with the seasons. During the Northern Hemisphere summer months, maximum insolation is received north of the equator. This condition causes the pressure belts to move north also, and the wind belts of both hemispheres shift accordingly. Six months later, when maximum heating is taking place south of the equator, the wind systems shift south in response to the migration of the pressure systems. Thus, seasonal variations in wind and pressure conditions are important factors of the atmosphere's circulation patterns that differ from our idealized model.

The boundary zones between two wind or pressure systems are the regions that are most strongly influenced by these seasonal migrations. During the winter months, these boundary zones will be subject to the impact of one system. As summer approaches, the system that dominated in the winter will migrate poleward and the equatorward system will move in to influence the region. Two of these boundary zones in each hemisphere experience these distinctive summer-to-winter fluctuations in climate. The first lies between latitudes 5° and 15°, where the wet equatorial low of the high-sun season (summer) alternates with the dry subtropical high of the low-sun season (winter). The second occurs between 30° and 40°, where the subtropical high dominates in summer but is replaced by the wetter westerlies and the polar front in winter.

California is an example of a region located within a transition zone between two wind and pressure systems ( ● Fig. 5.12). During the winter this region is under the influence of the westerlies blowing out of the Pacific high. These winds, turbulent and full of moisture from the ocean, bring winter rains and storms to "sunny" California. As summer approaches, however, the subtropical high and its associated westerlies move north. As California comes under the influence of the calm and steady subtropical high pressure system, it experiences the climate for which it is famous: day after day of warmth, with clear, blue skies. This alternation between moist winters and dry summers is typical of the western sides of all landmasses between about 30° and 40° latitude, north and south of the equator.

● **FIGURE 5.12** Winter and summer positions of the Pacific High in relation to California. In the winter, this anticyclone lies well to the south and feeds the westerlies that bring storms and rain from the North Pacific to California. The influence of high pressure dominates during the summer. The high pressure cell blocks storms and produces warm, sunny, and dry conditions.

*In what ways would the seasonal migration of the Pacific High affect agriculture in California?*

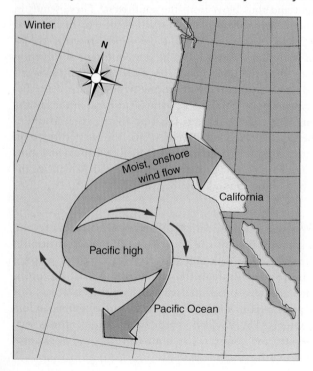
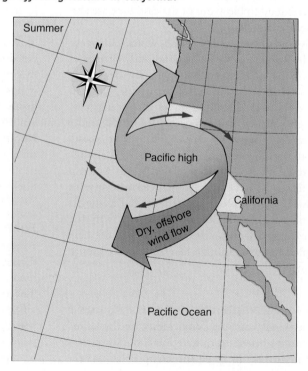

## Longitudinal Differences in Winds

We have already examined the important latitudinal differences in pressure and winds. Significant longitudinal variations also occur, especially in the region of the subtropical highs.

As was previously noted, the subtropical high pressure cells, which are generally centered over the oceans, are much stronger on their eastern sides than on their western sides. Thus, over the eastern portions of the oceans (west coasts of the continents) in the subtropics, subsidence and divergence are especially noticeable. Upper-level temperature inversions, associated with the subtropical highs, develop air that is clear and calm. Air flowing equatorward from this eastern side of the subtropical high produces steady trade winds with clear, dry weather.

Over the western portions of the oceans (eastern coasts), conditions are markedly different. During its passage over the ocean, the diverging air gradually warms and picks up moisture, so turbulent and stormy weather conditions are likely to develop. As indicated in ● Figure 5.13, winds in the western portions of the subtropical anticyclones, influenced by the Coriolis effect, bend poleward and toward landmasses. The trade winds in these areas are especially weak or nonexistent much of the year.

Figures 5.10a and b illustrate that there are great land–sea contrasts in temperature and pressure throughout the year in the higher latitudes, especially in the Northern Hemisphere. During the cold winters the land is associated with pressures that are higher than those over the oceans, and thus there are strong, cold winds from the land to the sea. In the summer the situation changes, with relatively low pressure existing over the continents because of higher temperatures. Wind directions change and the pattern is reversed so that in summer the winds affected by these conditions tend to flow from the sea toward the land.

# Upper Air Winds and Jet Streams

We have examined the atmosphere's surface wind patterns, but upper-level winds are also important—in particular, the winds at altitudes above 5000 meters (16,500 ft) and at higher levels in the upper troposphere. The formation, movement, and decay of cyclones and anticyclones in the middle latitudes depend to a great extent on high altitude flows of air in the atmosphere.

The upper air wind circulation is less complex compared to the circulation of surface winds. In the upper troposphere, an average westerly flow, the upper air westerlies, is maintained poleward of about 15° to 20° latitude in both hemispheres. Because of the reduced frictional drag high above the surface, the upper air westerlies blow much faster than their surface counterparts. This characteristic of upper air winds became apparent during World War II

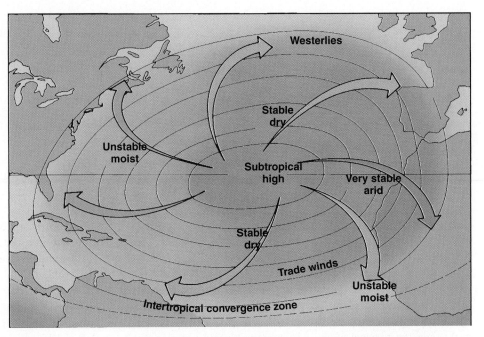

● **FIGURE 5.13** Circulation pattern in a Northern Hemisphere subtropical anticyclone. Subsidence of air is strongest in the eastern part of the anticyclone, producing calm air and arid conditions over adjacent land areas.

*What wind system is fed by the northern margin of this high pressure area, and what wind system blows generally toward the south?*

when high-altitude aircraft flying eastward covered distances faster than when they flew westward. Pilots had encountered the upper air westerlies or perhaps even the jet streams—very strong air currents embedded within the upper air westerlies. The **jet streams** are high altitude examples of geostrophic winds, flowing parallel between isobars in response to a balance between the Coriolis effect and the pressure gradient.

The upper air westerlies form as a response to temperature difference between warm tropical air and cold polar air. Air in the equatorial latitudes is warmed, rises by convection to high altitudes, and then flows toward the polar regions. At first, this seems to contradict our previous statement, relative to surface winds, that air flows from cold areas (high pressure) toward warm areas (low pressure). This apparent discrepancy disappears, however, if you recall that the pressure gradient, down which the flow takes place, must be assessed between two points at the same elevation. A column of cold air will exert a higher pressure at Earth's surface compared to a column of warm air. Consequently, the pressure gradient established at the surface will result in a flow from the cooler air toward the warmer air. Cold air, however, is denser and more compact than warm air, and pressure decreases with height more rapidly in cold air than in warm air. As a result, at a specific height above Earth's surface, a lower pressure will be encountered above cold surface air than above warm surface air. This will result in an air flow from the warmer surface air toward the colder surface air at that height. ● Figure 5.14 illustrates this concept.

As the upper air winds flow from the equator toward the poles (down the pressure gradient), they are turned eastward because of the Coriolis effect. The net result is a broad circumpolar flow of westerly winds throughout most of the upper atmosphere

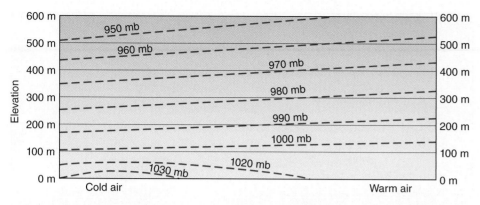

● **FIGURE 5.14** Variation of pressure surfaces with height. Note that the horizontal pressure gradient is from cold to warm air at the surface and in the opposite direction at higher elevations (such as 400 m).

( ● Fig. 5.15). Because the upper air westerlies form in response to the temperature gradient between tropical and polar areas, it is not surprising that they are strongest in winter (the low-sun season) when the thermal contrast is greatest. During summer (the high-sun season) when the temperature contrast over the hemisphere is much reduced, the upper air westerlies move more slowly. The temperature gradient between tropical and polar air, especially in winter, is concentrated where the warm tropical air meets cold polar air. This boundary is the polar front, which has a strong pressure gradient.

The best-known jet stream is the **polar jet stream,** which flows in the tropopause above the polar front, the area of the subpolar low. Ranging from 40 to 160 kilometers (25–100 mi) in width and up to 2 or 3 kilometers (1–2 mi) in depth, the polar jet stream is a faster, internal current of air within the upper air westerlies. Whereas the polar jet stream flows over the middle latitudes, another westerly **subtropical jet stream** flows above the subtropical highs in the lower-middle latitudes ( ● Fig. 5.16). Both jet streams are best developed in winter when temperatures exhibit their steepest gradient, and in the summer they weaken in intensity. During the summer, the subtropical jet stream frequently disappears completely and the polar jet stream migrates poleward. ● Figure 5.17 shows the position of these jet streams as they relate to the vertical and surface circulation of the atmosphere.

● **FIGURE 5.15** The upper air westerlies form a broad circumpolar flow throughout most of the upper atmosphere.

*What would you call these kinds of winds based on their relationship to the pressure gradient?*

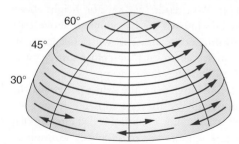

In general, the upper air westerlies and the associated polar jet stream flow in a fairly smooth pattern ( ● Fig. 5.18a). At times, however, the upper air westerlies develop a wave form, called long waves or **Rossby waves,** named after the Swedish meteorologist who discovered their existence (Fig. 5.18b). Rossby waves result from cold polar air pushing into the lower latitudes, forming troughs of low pressure, while warm tropical air moves into higher latitudes, forming ridges of high pressure. It is when the upper air circulation is in this configuration that middle latitude surface weather is most influenced. We will examine this influence in more detail in Chapter 7.

Eventually, the upper air Rossby waves become so elongated that "tongues" of air are cut off, forming warm and cold cells in the upper air (Fig. 5.18c and d). This process helps maintain a net poleward exchange of heat energy from equatorial and tropical areas. The cells eventually dissipate, and the normal pattern returns (Fig. 5.18a). A complete cycle takes approximately 4 to 8 weeks. Although it is not completely clear why the upper atmosphere goes into these oscillating patterns, we are currently gaining additional insights. One possible cause is variations in temperatures of the ocean surface. If the oceans in the northern Pacific or near the equator become unusually warm or cold (for example, during El Niño or La Niña, discussed later in this chapter), this apparently

● **FIGURE 5.16** Approximate location of the subtropical jet stream and area of activity of the polar front jet stream *(shaded)* in the Northern Hemisphere winter.

*Which jet stream is most likely to affect your home state?*

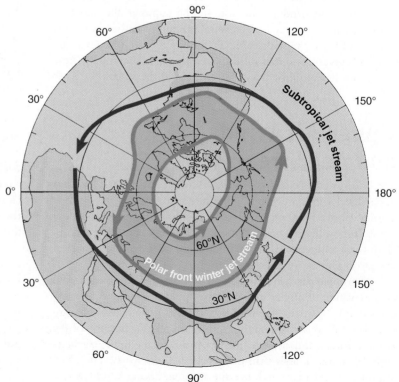

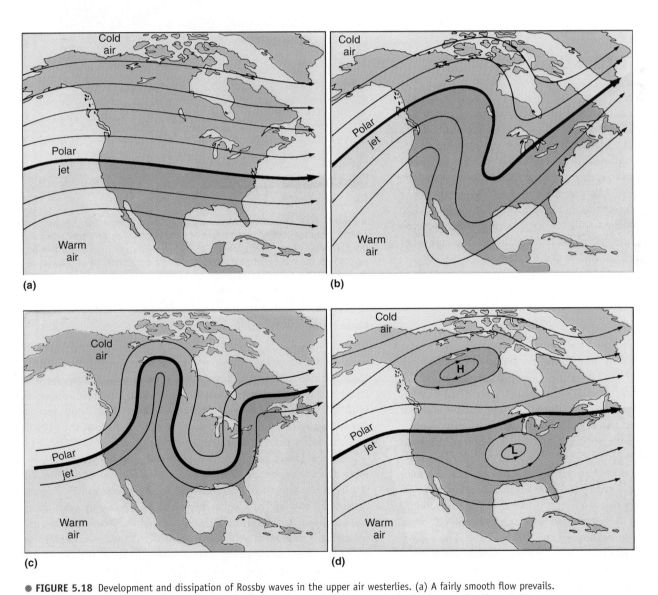

• **FIGURE 5.18** Development and dissipation of Rossby waves in the upper air westerlies. (a) A fairly smooth flow prevails. (b) Rossby waves form, with a ridge of warm air extending into Canada and a trough of cold air extending down to Texas. (c) The trough and ridge may begin to elongate and turn back on themselves. (d) The trough and ridge are cut off into cells that will eventually dissipate. The flow will then return to a pattern similar to (a).

*How are Rossby waves associated with the changeable weather of the central and eastern United States?*

triggers oscillations, which continue until the ocean-surface temperature returns to normal conditions. Other causes are also being investigated.

In addition to their influence on weather, jet streams are important for other reasons. They can carry pollutants, such as radioactive or volcanic dust (called ash), over great distances and at relatively rapid rates. It was the polar jet stream that carried volcanic ash from the 1980 eruption of Mount St. Helens eastward across the United States and Southern Canada. The westerly flow of the jet stream also carried volcanic ash in 2010 from an eruption in Iceland to Europe, disrupting air travel. Nuclear fallout from the Chernobyl incident in the former Soviet Union could be monitored in succeeding days as it crossed the Pacific, and later the United States, in the jet streams. Because of a westerly tailwind from the jet stream flying times on commercial airplanes from North America to Europe may be significantly shorter than those in the reverse direction.

# Regional and Local Wind Systems

The global wind system illustrates general circulation patterns that reflect latitudinal imbalances in pressure and temperature. On a regional or local scale, additional wind systems develop in response to similar conditions. Monsoon winds provide an example, sub-continental in size, that is related to seasonal temperature and pressure variations. Many regions and localities experience differences in wind directions and conditions that are related to the seasons. At the smallest scale are local winds, which develop in response to the diurnal (daily) variations in heating and other local effects upon pressure and winds.

## Monsoon Winds

The term *monsoon* comes from the Arabic word *mausim,* meaning season. Arab sailors have used this word for many centuries to describe seasonal reversals in wind direction across the Arabian Sea between Arabia and India. As a meteorological term, **monsoon** refers to the directional reversal of winds from one season to another. Usually, a monsoon occurs when a wind blowing warm humid air from the ocean toward the land in summer shifts to a dry, cooler wind blowing toward the sea from the land in winter. Monsoon conditions imply a complete reversal of seasonal wind directions.

The monsoon is most characteristic of southern Asia, although it also occurs on other continents. As the large Asian landmass cools to winter temperatures that are much colder than the surrounding oceans, the continent develops a strong high pressure cell from which there must be an outflow of air ( ● Fig. 5.19). These cold, dry winds blow southward from the continent toward the tropical low that exists over the warmer oceans. During this time, the winter monsoon is a dry season as air is coming from a dry continental area.

In summer the Asian continent becomes very warm and develops a large low pressure center. This development is reinforced by a poleward shift of warm, moist tropical air to a position over southern Asia. Warm, humid, tropical air from the

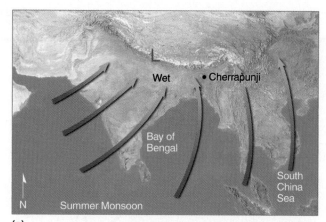

**(a)**

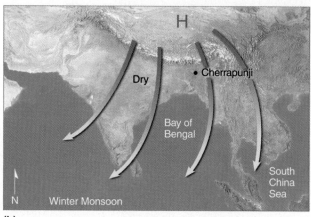

**(b)**

● **FIGURE 5.19** The Asiatic monsoon's seasonal wind reversals. (a) The wet summer monsoon with an onshore flow of humid tropical air from the Indian Ocean results in heavy precipitation. (b) The dry winter monsoon with a flow of dry air from the Asian continent produces drought conditions. Despite having a dry season, Cherrapungi, India, is one of the world's rainiest places.
Copyright © 2011 Brooks/Cole, Cengage Learning

*How do the seasonal changes of wind direction in Asia compare to those of the southern United States?*

ocean is attracted into this low. Full of water vapor, any turbulence, convection, or landform barrier that makes this air rise will bring about the heavy precipitation and wet summers associated with the monsoon. Extremely heavy rainfall and flooding can occur during the summer monsoon in the foothills of the Himalayas, in other parts of India, and in areas of Southeast Asia.

In the lower latitudes, a monsoonal shift in winds can be related to the migration of the direct rays of the sun. For example, winds in the equatorial zone migrate during the summer months northward toward the southern coast of Asia, bringing with them warm, moist, turbulent air. The winds of the Southern Hemisphere also migrate north with the sun, with some crossing the equator. These winds bring warm, moist air (from the ocean) to the southern and especially the southeastern coasts of India. In the winter months, the equatorial and tropical winds migrate south, leaving southern Asia under the influence of the dry, calm winds of the tropical Northern Hemisphere. Asia and northern Australia are true monsoon areas, with a full 180° wind shift with changes from summer to winter. Other regions, like the southern United

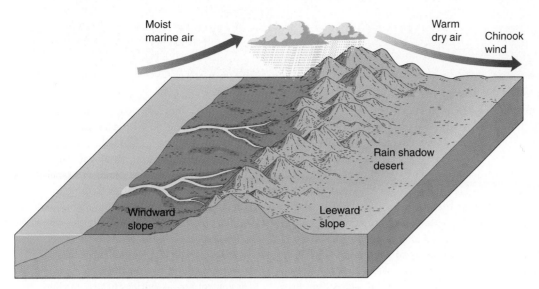

Moist marine air

Warm dry air

Chinook wind

Rain shadow desert

Windward slope

Leeward slope

● **FIGURE 5.20** Chinook (or Foehn) winds result when air ascends a mountain barrier, becoming cooler as it expands and losing some of its moisture through condensation and precipitation. As the air descends the leeward side of the range, its relative humidity decreases lower as the air compresses and warms. This produces the relatively warm, dry conditions with which Chinook winds are associated.

*The term Chinook means "snow eater." Can you offer an explanation for how this name came about?*

States and West Africa, have "monsoonal tendencies" but are not monsoons in the true meaning of the term.

The phenomenon of monsoon winds and their characteristic seasonal shifting cannot be fully explained, however, by the differential heating of land and water or the seasonal shifting of tropical and subtropical wind belts. Some aspects of the monsoon system—for example, its "burst" or sudden transition between dry and wet in southern Asia—must have other causes. Meteorologists looking for a more complete explanation of the monsoon are examining the role played by the jet stream and other wind movements of the upper atmosphere.

## Local Winds

The major circulation patterns of the atmosphere are important to understanding the climate regions on Earth and the climatic differences between those regions. Yet, we are all aware that many of the winds that affect weather function on a local scale. Despite affecting much smaller areas, local winds are also important. Much like larger wind systems, local winds are also affected by local terrain features and by land–water temperature differences.

**Chinooks and Other Warming Winds** One type of local wind is known by several names in different parts of the world—for example, **chinook** in the Rocky Mountain area and **foehn** (pronounced "fern") in the Alps. Chinook-type winds occur when air passes over a mountain range. After crossing the mountains, as these winds flow down the leeward slope, the air compresses and heats at a greater rate compared to the degree to which it cooled when it ascended the windward slope ( ● Fig. 5.20). Thus, the air flows into the valley below as warm, dry winds. The rapid temperature rise brought about by such winds has been known to damage crops, increase forest-fire hazard, and set off avalanches.

An especially hot and dry wind is the **Santa Ana wind** of Southern California. It forms when high pressure develops over the desert regions of Southern California and Nevada. The clockwise circulation out of the high pressure cell drives warm dry air from the desert over the mountains of Eastern California, and the air becomes

warmer and more arid as it moves down the western slopes. The hot, dry Santa Ana winds are notorious for fanning wildfires, which plague the southwestern United States, especially in California.

**Drainage Winds** Also known as **katabatic winds, drainage winds** are local to mountainous regions and occur under cold, calm, and clear conditions. Cold, dense air can accumulate in a high valley, plateau, or snowfield within a mountainous area. Because cold air is very dense, it tends to flow downward, escaping through passes, flowing down valleys, and pouring out onto the land below. The katabatic winds that result from cold air accumulating over ice sheets in Greenland and Antarctica can be extremely cold and strong. Drainage winds in other regions are known by many local names; for example, on the Adriatic coast, they are called the *bora;* in France, the *mistral;* and in Alaska, the *Taku.*

**Land Breeze–Sea Breeze** The **land breeze–sea breeze** is a diurnal (daily) cycle of local winds that occurs in response to pressure differences caused by the differential heating of land and water ( ● Fig. 5.21). During the day the land and the air above it heats quickly, warming to a higher temperature than that of the nearby ocean. The warm daytime air above the land expands and rises, creating a local area of low pressure, and the rising air is replaced by denser, cooler air from over the body of water. Thus, a sea breeze of cool, moist air blows in over the land during the day. Alleviation of the heat by a cooling sea breeze explains one reason why seashores are so popular in summer. In some situations, however, sea breezes are responsible for afternoon cloud cover and light rain, which could spoil an otherwise sunny day at the shore. A sea breeze can cause a 5°C to 9°C (9°F–16°F) reduction in temperature along the coast and a lesser influence on land perhaps as far from the ocean as 15 to 50 kilometers (9–30 mi). Large lakes and their shorelines can also experience this kind of local wind to cool cities like Chicago, Milwaukee and Cleveland during hot summer days. At night, the land and the air above it cools more quickly and to a lower temperature than the nearby water body and the air above it. Consequently, the pressure builds higher over the land and air flows out to the lower pressure over the water, creating a

**W**ildfires require three factors to occur: *oxygen, fuel,* and an *ignition source.* The conditions for all three factors vary in their spatial distribution. Where all three factors have the potential to exist, wildfire danger is high. Atmospheric oxygen is a constant, but winds, which supply more oxygen as a fire consumes it, vary with location, weather, and terrain. High winds cause fires to spread faster and make them difficult to extinguish. Fuel for wildfires is usually supplied by dry vegetative litter (leaves, branches, and dry annual grasses), and certain environments have more of this fuel than others. Dense vegetation supports the spread of fires. Vegetation can also become desiccated (dried out) during a drought or a dry season. In addition, once a fire becomes large, extreme heat along the edges of the burning areas causes vegetation to lose its moisture through evaporation. Ignition sources are the means by which a fire is started. Lightning and human causes, such as campfires and trash fires, are the main ignition sources for wildfires.

Southern California offers a regional example where these three factors combine with the local physical geography to create an environment conducive to wildfire hazard. This is a region where many people live in forested or scrub-covered locales along the urban–wildland fringe—areas that are very susceptible to fire. High pressure, warm weather, and low relative humidity dominate the Mediterranean climate of Southern California for much of the year. When these conditions occur, this region experiences high fire potential because of warm dry air and vegetation that has dried out during the arid summer season.

The most dangerous conditions for wildfires in Southern California occur as high winds sweep the region. When a strong high pressure cell forms to the east of Southern California, the clockwise (anticyclonic) circulation directs winds from the north and east toward the coast. These warm, dry winds (the Santa Ana winds) blow down from nearby high-desert regions, becoming warmer and drier as

they descend into the coastal lowlands. Santa Ana winds are most common in the fall and winter, with wind speeds that can be 50 to 90 kilometers per hour (30–50 mph). Stronger wind gusts can reach 160 kilometers per hour (100 mph). Just like using a bellows or blowing on a campfire to get it started, the Santa Ana winds produce *fire weather* that can cause the spread of a wildfire to be extremely rapid after ignition. Most people take great care during these times to avoid or control any activities that could cause a fire, but occasional accidents, acts of arson, or lightning strikes ignite a wildfire.

Ironically, although the Santa Ana winds create dangerous fire conditions, they also provide some benefits because the winds tend to blow air pollutants offshore and out of the urban region. In addition, because these strong winds flow in the opposite direction of ocean waves, experienced surfers enjoy higher than normal waves when Santa Ana winds are present.

Geographic setting and wind directions for the Santa Ana winds, generated by a high pressure cell to the east of Southern California.

Residents in an area near Los Angeles, California, had to flee their hillside homes because of this brush fire, which was driven by the Santa Ana winds in October of 2008.

Land breeze

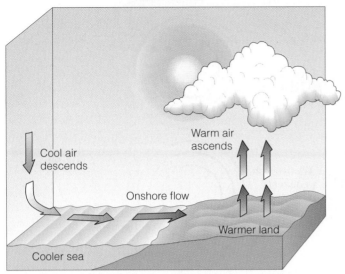

Sea breeze

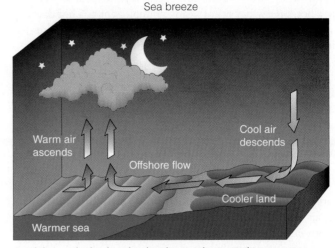

Cool air descends

Warm air ascends

Onshore flow

Warmer land

Cooler sea

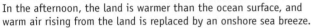

Warm air ascends

Offshore flow

Cool air descends

Cooler land

Warmer sea

In the afternoon, the land is warmer than the ocean surface, and warm air rising from the land is replaced by an onshore sea breeze.

At night, as the land cools, the air over the ocean is now warmer than the air over the land. The ocean air rises. Air flows offshore to replace it, generating an offshore flow (a land breeze).

● **FIGURE 5.21** Land and sea breezes. This day-to-night reversal of winds is a consequence of the different rates of heating and cooling of land and water areas. The land becomes warmer than the sea during the day and colder than the sea at night; the air flows from the cooler to the warmer area.

*How does the relative difference in air pressure over the land change from day to night compared to the air pressure over the water?*

land breeze. For thousands of years, sailboats have left their coasts at dawn, when there is still a land breeze, and have returned with the sea breeze of the late afternoon. The land–sea breeze cycle is a classic example of a diurnal wind shift.

**Mountain Breeze–Valley Breeze** Under the calming influence of a high pressure system, there is a daily **mountain breeze–valley breeze** cycle ( ● Fig. 5.22), which is somewhat similar in mechanism to the land breeze–sea breeze. During the day, when the sun heats the valleys and mountain slopes, the high

exposed slopes heat faster than the lower and shadier valley. Air on the slope expands and rises, drawing air from the valley and up the mountain sides. This warm daytime breeze is the *valley breeze,* named, like most winds, for its place of origin. Clouds, which can often be seen obscuring mountain peaks, can actually be the visible evidence of condensation in the warm air rising from the valleys. At night the valley and mountain slopes cool considerably because of radiation loss of heat energy to space through the clear air, yet they receive no solar radiation. The air cools, sinks, and flows down into the valley as a cool *mountain breeze.*

● **FIGURE 5.22** Mountain and valley breezes. This daily reversal of winds results from the heating of mountain slopes during the daytime and their cooling at night. Warm air is drawn up slopes during the day, and cold air drains down the slopes at night.

*How might a green, shady valley floor and a bare, rocky mountain slope contribute to these changes?*

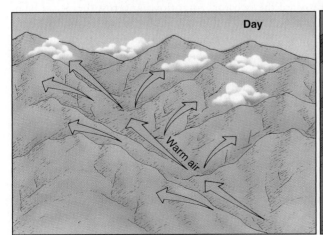

Day

Warm air

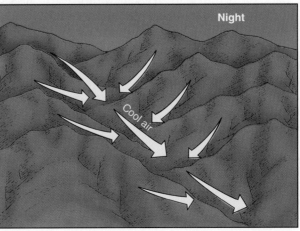

Night

Cool air

There is no question that winds, both local and global, are effective elements of atmospheric conditions and dynamics. We know that a hot, breezy day is not nearly as unpleasant as a hot day without any wind. This difference exists because winds increase evaporation, resulting in the removal of heat from our bodies, the land, the air, animals, and plants. In contrast, wind on a cold day increases our discomfort, mainly by conducting heat away from our bodies and advecting heat away from our surroundings.

# Ocean–Atmosphere Interactions

The majority of Earth's surface area acts as an interface between two fluids, air in the atmosphere and water in the oceans. Because they are both characterized as fluids (because they can flow), the dynamics of gases and waters follow the same laws of physics. The major difference is in their densities. Water molecules are much more closely packed together, so water's density is 800 times higher than the density of air at sea level. Yet, air motion in the atmosphere can cause or strongly affect movements in the oceans, and the oceans influence the atmosphere in many ways. Some of these interactions only affect small or local areas, some are regional, and others can have a global impact. In recent decades, oceanographers, geographers, and atmospheric scientists have combined their expertise to research some of these ocean–atmosphere relationships.

One of the best-known ocean–atmosphere interactions is the constant wind motion that creates waves and affects surface ocean currents. Because of the high density of water compared to air, faster movements in the atmosphere are reflected as much slower movements in the oceans. Ocean–atmosphere interactions exist at many scales with respect to both time and geographic area, and it will take many years of study before they are fully understood. The remainder of this chapter will discuss some of these very important relationships.

## Ocean Currents

Like the global wind system, ocean currents play a significant role in helping to moderate latitudinal imbalances in temperature. **Surface ocean currents** are fairly steady flows of seawater that move in a prevailing direction, and they behave somewhat like rivers in the ocean. The surface water temperatures of ocean currents flowing along the coasts have a great influence on the climate of coastal locations.

Earth's surface wind system is a primary factor in the flow of these ocean currents. Other factors include the Coriolis effect and the size, shape, and depth of the basin of a sea or ocean. Ocean currents are also influenced and driven by differences in density that result from temperature, evaporation, and salinity and the action of waves and tides.

The major ocean currents move in broad circulatory patterns, called **gyres,** which flow around the subtropical highs. Because of the Coriolis effect and the direction of flow around a cell of high pressure, the gyres follow a clockwise direction in the Northern Hemisphere and a counterclockwise direction in the Southern

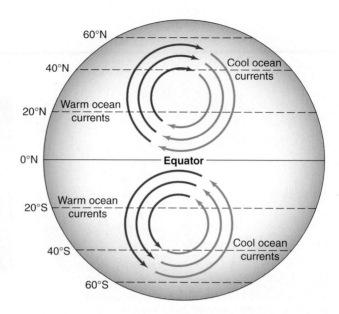

● **FIGURE 5.23** The major ocean currents flow in broad gyres in opposite directions in the Northern and Southern Hemispheres. In the middle latitudes cold currents flow equatorward along west coasts and warm currents flow poleward along east coasts.

***What is the main control on the direction of these gyres?***

Hemisphere ( ● Fig. 5.23). As a general rule, surface currents do not cross the equator where the impact of Coriolis effect is minimal.

Waters near the equator in both hemispheres are driven westward by the tropical easterlies or the trade winds. The related ocean current is called the Equatorial Current. At the western margin of the ocean, its warm tropical waters are deflected poleward along the coastline. As these warm waters move into higher latitudes, they move through waters cooler than the water flowing in the current so they are identified as *warm currents* ( ● Fig. 5.24).

In the Northern Hemisphere, warm currents, such as the Gulf Stream and the Kuroshio (Japan) Current, are deflected strongly to the right (or east) because of the increasing apparent impact of Coriolis that occurs with higher latitudes. At about 40°N, the westerlies begin to drive these warm waters eastward across the ocean, forming the North Atlantic Drift and the North Pacific Drift. Eventually, these currents encounter landmasses at the eastern margin of the ocean, and most of the waters are deflected toward the equator. After flowing across the ocean basin in higher latitudes, the currents have lost much of their warmth and are now are cooler than the adjacent waters as they flow equatorward into the subtropical latitudes. They now have become *cold currents.* Nearing the equator, these currents complete the circulation pattern when they rejoin the westward-flowing Equatorial Current.

On the eastern side of the North Atlantic, the North Atlantic Drift flows into the seas north of the British Isles and around Scandinavia, keeping those areas warmer than their latitudes would suggest. Some Norwegian ports north of the Arctic Circle remain ice free because of the moderating effect of relatively warmer water. Cold polar water in the Labrador Current and Oyashio (Kurile) Current flows southward into the ocean basins of the Atlantic and Pacific and along the margins of the continents.

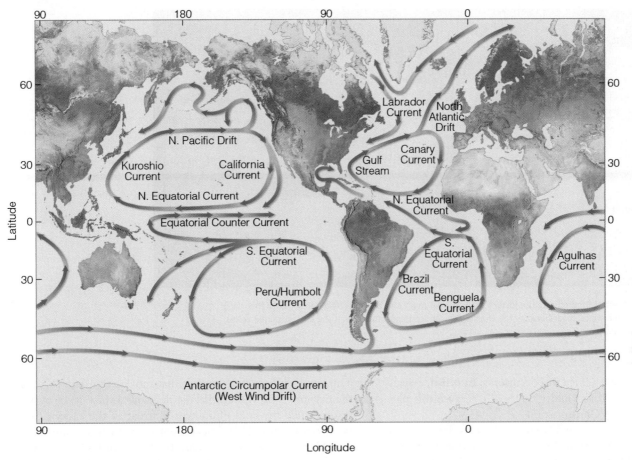

● **FIGURE 5.24** Map of the major world ocean currents showing the locations of warm and cold currents.
Copyright © 2011 Brooks/Cole, Cengage Learning

*How does this map of ocean currents help explain the mild winters in London, England?*

Oceanic circulation in the Southern Hemisphere is comparable to that in the Northern Hemisphere except that the gyres flow in a counterclockwise direction. Also, because in the Southern Hemisphere there is little land poleward of 40°S, the Antarctic Circumpolar Current (sometimes called the West Wind Drift) circles around Antarctica as a cool (cold) current across the Southern Ocean almost without interruption. It is cooled by the influence of its high latitude location and cold air from the Antarctic ice sheet.

In general, warm currents flow poleward as they carry tropical waters into the cooler waters of higher latitudes, as in the case of the Gulf Stream or the Brazil Current. Cold currents flow equatorward along west coasts of continents (for example, the California Current and the Humboldt Current). Warm currents tend to bring humidity and warmth to the east coasts of continents along which they flow, and cool currents tend to have a drying and cooling effect on the west coasts. The contact between the atmosphere and ocean currents is one reason why subtropical highs have a strong side and a weak side. Subtropical highs on the west coast of continents are in contact with cold ocean currents, which cool the air and stabilize and strengthen the eastern side of a subtropical high. On the east coasts of continents, contacts with warm ocean currents cause the western sides of subtropical highs to be less stable and weaker.

The general circulation patterns of oceanic surface currents are consistent throughout the year. However, the position of the currents may respond to seasonal changes in atmospheric heating and circulation. In addition, in the North Indian Ocean, the direction of circulation reverses seasonally according to the monsoon winds.

**Upwelling** frequently reinforces cold currents flowing along west coasts in subtropical latitudes. As the trade winds in these latitudes drive the surface waters offshore, the wind's frictional drag on the ocean surface displaces water to the west. Surface waters are dragged away and replaced by deeper, colder water that rises to the surface. This upwelling of cold waters reinforces the strength and effect of the California, Humboldt (Peru), Canary, and Benguela Currents.

## El Niño

As you can see in Figure 5.24, the cold Humboldt Current flows equatorward along the coasts of Ecuador and Peru. When the current approaches the equator, the westward-flowing trade winds cause upwelling of nutrient-rich cold water along the coast. Fishing, especially for anchovies, is a major local industry.

Every year, usually during the months of November and December, a weak warm countercurrent replaces the normally cold coastal waters. Without the upwelling of nutrients from below to feed the fish, fishing comes to a standstill. Fishermen in this region have known this phenomenon for hundreds of years. In fact, this is the time of year they traditionally set aside to tend to their equipment and await the return of cold water. The resi-

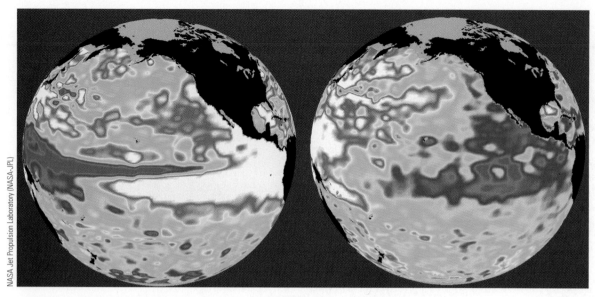

● **FIGURE 5.25** These thermal infrared satellite images show El Niño *(left)* and La Niña *(right)* episodes in the tropical Pacific. The red and white shades display the warmer sea surface temperatures, and the blue or purple colors mark areas of cooler temperatures.

*What are some impacts of an El Niño?*

dents of the region have named this occurrence **El Niño,** a Spanish reference to the Christ child ("the boy") because this phenomenon occurs about Christmastime.

The warm-water countercurrent usually lasts for 2 months or less, but occasionally it can last for several months. In these situations, water temperatures are raised not just along the coast, but also for thousands of kilometers offshore ( ● Fig. 5.25). Over the past decade, the term *El Niño* has come to describe these exceptionally strong episodes, and not just the annual event. During the past 50 years, approximately 18 of those years experienced El Niño conditions (with sea-surface temperatures 0.5°C warmer than normal for 6 consecutive months). Not only do the El Niño events affect the temperature of the equatorial Pacific, but the strongest of them also have an impact on global weather patterns.

**El Niño and the Southern Oscillation** Understanding the processes that interact to produce an El Niño requires studying conditions across the Pacific, not just the waters off of South America. In the 1920s Sir Gilbert Walker, a British scientist, discovered a connection between surface-pressure readings at weather stations on the eastern and western sides of the Pacific basin. He noted that a rise in pressure in the eastern Pacific is usually accompanied by a fall in pressure in the western Pacific and vice versa. He called this seesaw pattern the **Southern Oscillation.** The link between El Niño and the Southern Oscillation is so great that they are often referred to jointly as ENSO (El Niño/Southern Oscillation). Today the atmospheric pressure values from Darwin, Australia, are compared to those recorded on the Island of Tahiti, and the relationship between these two values is used to define the Southern Oscillation.

During a typical year, the eastern Pacific has a higher pressure than the western Pacific. This east-to-west pressure gradient strengthens the trade winds over the equatorial Pacific waters.

This results in a surface current that moves from east to west at the equator. The western Pacific develops a thick layer of warm water, while the cold Humboldt Current in the eastern Pacific (west coast of South America) is enhanced by upwelling.

Then, for unknown reasons, the Southern Oscillation swings in the opposite direction, dramatically changing the usual conditions described earlier, with pressure increasing in the western Pacific and decreasing in the eastern Pacific. This pressure gradient change causes the trade winds to weaken or, in some cases, they may reverse. This causes warm water in the western Pacific to flow eastward, increasing sea-surface temperatures in the central and eastern Pacific. This eastward shift signals the beginning of El Niño.

In contrast, at times and for reasons not fully understood, the trade winds intensify. These more powerful trade winds will cause even stronger upwelling than usual to occur. As a result, sea-surface temperatures will be colder than normal off the South American west coast. This condition is known as **La Niña** (in Spanish, "little girl," but scientifically simply the opposite of El Niño). La Niña episodes will at times, but not always, bring about the opposite effects of an El Niño episode (see Fig. 5.25).

**El Niño and Global Weather** Cold ocean waters impede cloud formation, except for coastal fogs. Thus, under normal conditions, clouds tend to develop over the warm waters of the western Pacific but not over the cold waters of the eastern Pacific ocean basin ( ● Fig. 5.26a). However, during an El Niño, when warm water migrates eastward, clouds develop over the equatorial region of the Pacific (Fig. 5.26b). These clouds can build to heights of 18,000 meters (59,000 ft). Clouds of this magnitude can disrupt the high-altitude wind flow above the equator. As we have seen, a change in the upper air wind flow in one portion of the atmosphere will trigger wind flow changes in other portions of the atmosphere. Alterations in the upper air winds result in changes to surface weather.

Scientists have tried to document as many past El Niño events as possible by piecing together historical evidence, such as sea-surface temperature records, daily observations of atmospheric pressure and rainfall, fishery records from South America, and even the writings of Spanish colonists living along the coasts of Peru and Ecuador dating back to the 15th century. Additional evidence comes from the growth patterns of coral and trees. Researchers are constantly discovering new techniques to identify the El Niños of the past by studying historical evidence.

Based on this historical evidence, El Niños seem to have occurred as far back as the records go. One disturbing fact is that they appear to be occurring more often. Records indicate that during the 16th century an El Niño occurred, on average, every 6 years. Evidence gathered over the past few decades indicates that El Niños are now occurring, on average, every 2.2 years. Even more alarming is the fact that they appear to be getting stronger. The one in 1997–1998 surpassed the record-setting El Niño of 1982–1983.

The 1997–1998 El Niño brought copious and damaging rainfall to the southern United States, from California to Florida. Snowstorms in the northeast portion of the United States were more frequent and stronger than in most years. The warm El Niño waters fueled Hurricane Linda, which devastated the western coast of Mexico. Linda was the strongest hurricane ever recorded in the eastern Pacific.

In recent years, scientists have become better able to monitor and forecast El Niño and La Niña events. An elaborate network of ocean-anchored weather buoys plus satellite observations provide an enormous amount of data that can be analyzed by computer to help predict the formation and strength of El Niño and La Niña events. The knowledge gained from these scientific efforts can help us prepare for weather events that are associated with El Niño and La Niña conditions.

## North Atlantic Oscillation

Our improved observation skills have led to the discovery of the **North Atlantic Oscillation (NAO)**—a relationship between the Azores (subtropical) High and the Icelandic (subpolar) Low. The east-to-west, seesaw motion of the Icelandic Low and the Azores High control the strength of the westerly winds and the direction of storm tracks across the North Atlantic. There are two recognizable phases associated with the established NAO index.

A positive NAO index phase is identified by higher than average pressures in the Azores High and lower than average pressures

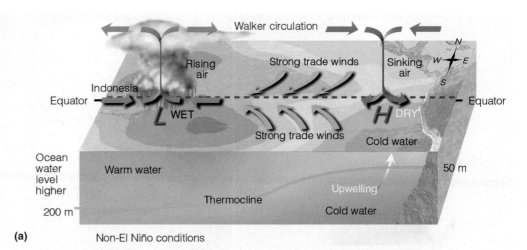

(a) Non-El Niño conditions

(b) El Niño conditions

● **FIGURE 5.26** (a) In a non-El Niño (normal) year, upwelling of cold sea water and the trade winds are both strong along the west coast of South America and bring rains to Southeast Asia. (b) During El Niño, the easterly trade winds weaken, allowing the central Pacific to warm and the rainy area to migrate eastward.

Copyright © 2011 Brooks/Cole, Cengage Learning

***Near what country or countries does El Niño begin?***

in the Icelandic Low. The increased pressure difference between the two systems results in stronger winter storms, occurring more often and following a more northerly track ( ● Fig. 5.27a). This promotes warm and wet winters in Europe but cold, dry winters in Canada and Greenland. The eastern United States may experience a mild and wet winter.

The negative NAO index phase occurs with a weak Azores High and higher pressure in the Icelandic Low. The smaller pressure gradient between these two systems will weaken the westerlies resulting in fewer and weaker winter storms (Fig. 5.27b). Northern Europe will experience cold air with moist air moving into the Mediterranean. The East Coast of the United States will experience more cold air and snowy winters. This index varies from year to year but also has a tendency to stay in one phase for periods lasting several years in a row.

The NAO is not as well understood as ENSO. Truly, both oscillations require more research in the future if scientists are to better understand how these ocean phenomena affect weather and climate. Will scientists ever be able to predict the occurrence of such

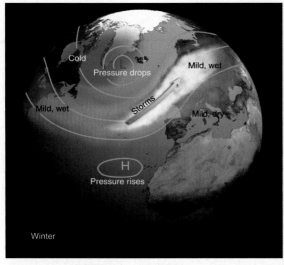

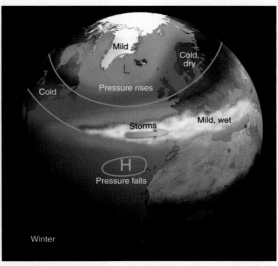

(a) Positive phase

(b) Negative phase

● **FIGURE 5.27** Positions of the pressure systems and winds involved with the (a) positive and (b) negative phases of the North Atlantic Oscillation (NAO).

*Which two pressure systems are used to establish the NAO phases?*

phenomena as ENSO or the NAO? No one can answer that question, but as our technology improves, our forecasting ability will also increase. We have made tremendous progress. In the past few decades, we have come to recognize the close association between the atmosphere and hydrosphere and to better understand the complex relationships between these Earth systems.

This chapter began with an examination of the behavior of atmospheric gases as they interact with solar radiation and dynamic forces. This information enabled a definition and detailed discussion of global pressure systems and their accompanying winds. This discussion in turn permitted a description of atmospheric circulation patterns on global, regional, and local scales.

Once again, we can recognize the workings of interactions among Earth's systems. Earth's energy budget influences movements in our atmosphere, which in turn affect ocean circulation, and ocean currents affect atmospheric processes and conditions.

Solar radiation → Atmosphere → Hydrosphere → Back to the atmosphere

The many interconnections between the atmosphere and the oceans directly influence conditions of weather and climate. In the following chapters, we will examine the role of atmospheric systems in controlling variations in weather and climate and, later, weather and climate systems as they affect surface landforms.

# CHAPTER 5 ACTIVITIES

## ■ DEFINE AND RECALL

| | | |
|---|---|---|
| mercury barometer | subpolar lows | jet stream |
| aneroid barometer | polar highs | polar front jet stream |
| standard sea-level pressure | Siberian High | subtropical jet stream |
| cyclone (low) | Canadian High | Rossby wave |
| anticyclone (high) | Icelandic Low | monsoon |
| convergent wind circulation | Aleutian Low | chinook |
| divergent wind circulation | Pacific High | foehn |
| isobar | Bermuda High (Azores High) | Santa Ana wind |
| pressure gradient | polar easterlies | drainage wind (katabatic wind) |
| wind | westerlies | land breeze–sea breeze |
| Coriolis effect | trade winds | mountain breeze–valley breeze |
| friction | northeast trades | surface ocean current |
| geostrophic wind | southeast trades | gyre |
| windward | tropical easterlies | upwelling |
| leeward | Intertropical Convergence Zone (ITCZ) | El Niño |
| prevailing wind | doldrums | Southern Oscillation |
| trough | horse latitudes | La Niña |
| equatorial low (equatorial trough) | polar front | North Atlantic Oscillation (NAO) |
| subtropical highs | | |

# DISCUSS AND REVIEW

1. What is standard atmospheric pressure at sea level? How do you suppose Earth's gravity is related to atmospheric pressure?
2. Horizontal variations in air pressure are caused by thermal or dynamic factors. How do these two factors differ? How are they related?
3. What kind of pressure (high or low) would you expect to find in the center of an anticyclone? Describe and diagram the wind patterns of anticyclones in the Northern and Southern Hemispheres.
4. What is the circulation pattern around a center of low pressure (cyclone) in the Northern Hemisphere? In the Southern Hemisphere? Draw diagrams to illustrate these circulation patterns.
5. Explain how water and land surfaces affect the pressure overhead during summer and winter. How does this relate to the afternoon sea breeze?
6. How do landmasses affect the development of belts of atmospheric pressure over Earth's surface?
7. Why do the global wind systems and pressure belts migrate with the seasons?
8. How are the land breeze–sea breeze and monsoon circulations similar? How are they different?
9. What effect on valley farms could a strong drainage wind have?
10. What are monsoons? What causes them? Name some nations that are concerned with the arrival of the "wet monsoon."
11. What is the relationship between ocean currents and global surface wind systems? How does the gyre in the Northern Hemisphere differ from the one in the Southern Hemisphere?
12. Where are the major warm and cool ocean currents located in respect to Earth's continents? Which currents have the greatest effect on North America?
13. What is an El Niño? What are some of its impacts on global weather?

# CONSIDER AND RESPOND

1. Look at the January (Fig. 5.10a) and July (Fig. 5.10b) maps of average sea-level pressure. Answer the following questions:
   a. Why is the subtropical high pressure belt more continuous (linear, not cellular) in the Southern Hemisphere than in the Northern Hemisphere in July?
   b. During July what area of the United States exhibits the lowest average pressure? Why?
2. How did the trade winds of the Atlantic Ocean affect the sailing routes of early traders?
3. Describe the movements of the upper air. How have pilots applied their experience of the upper air to their flying patterns?
4. Is the polar front jet stream stronger in the summer or the winter? Why?
5. What effect would Chinook-type winds have on farming, forestry, and ski resorts?

# APPLY AND LEARN

1. The amount of power that can be generated by wind is determined by the equation:

$$P = \tfrac{1}{2}D \times S^3$$

where $P$ is the power in watts, $D$ is the density, and $S$ is the wind speed in meters per second (m/sec). Because $D = 1.293$ kg/m$^3$, we can rewrite the equation as

$$P = 0.65 \times S^3$$

   a. How much power (in watts) is generated by the following wind speeds: 2 meters per second, 6 meters per second, 10 meters per second, 12 meters per second?
   b. Because wind power increases significantly with increased wind speed, very windy areas are ideal locations for "wind farms." Cities A and B both have average wind speeds of 6 meters per second. However, city A tends to have very consistent winds; in city B, half of its winds tend to be at 2 meters per second and the other half is at 10 meters per second. Which site would be the better location for a wind generation plant?
2. Atmospheric pressure decreases at the rate of 0.036 millibar per foot as one ascends through the lower portion of the atmosphere.
   a. The Sears Tower in Chicago, Illinois, is one of the world's tallest buildings at 1450 feet. If the street-level pressure is 1020.4 millibars, what is the pressure at the top of the Sears Tower?
   b. If the difference in atmospheric pressure between the top and ground floor of an office building is 13.5 millibars, how tall is the building?
   c. A single story of a building is 12 feet. You enter an elevator on the top floor of the building and want to descend five floors. The elevator has no floor markings—only a barometer. If the initial reading was 1003.2 millibars, at what pressure reading would you want to get off?

 CourseMate - Make the most of your study time by accessing everything you need to succeed in one place. Read your textbook, take notes, review flashcards, watch videos, complete activities, take practice quizzes, and more online with CourseMate. Log in at **www.cengagebrain.com**.

# 6

# HUMIDITY, CONDENSATION, AND PRECIPITATION

## ■ OBJECTIVES

WHEN YOU COMPLETE THIS CHAPTER YOU SHOULD BE ABLE TO:

■ Explain why Earth's surface is dominated by water but freshwater remains a limited and precious resource.

■ Outline the processes in the hydrologic cycle, including how water circulates among and interacts with the lithosphere, atmosphere, hydrosphere, and biosphere.

■ Apply adiabatic lapse rates to determine temperature changes in air that rises, expands, and cools compared with air that sinks, compresses, and warms.

■ Understand that relative humidity is a percentage of moisture saturation in the air, and know why it is dependent on air temperature and moisture content.

■ Explain why if no change in moisture content occurs, as air warms the relative humidity will fall and when air cools the relative humidity will rise.

■ Determine what processes cause the air to reach the dew point temperature and attain a relative humidity of 100%, a condition that can produce condensation and precipitation.

■ Describe the temperature, humidity, pressure, and winds that influence the potential for precipitation and the kinds of precipitation that may result.

■ Provide examples of why precipitation, evaporation, and water availability all vary in terms of their geographic distribution on Earth.

**EARTH IS OFTEN REFERRED TO** as "the Blue Planet." The abundance of water on our planet is unique in the solar system and vital to life on Earth. The physical and chemical characteristics of water are fundamentally responsible for supporting life, but water also supports and maintains our planet's environments. Although some living organisms can survive without air, no living thing can survive without water. Water is necessary for photosynthesis, cell growth, protein formation, soil formation, and the absorption of nutrients by plants and animals.

Water is the only substance on Earth that occurs naturally as a liquid, a solid, and a gas. The structure of water molecules is such that water can dissolve an enormous number of substances—so many, in fact, that it has been called the *universal solvent*. As a result of its solvent properties, water is almost never found in a pure state. Even rainwater is filled with impurities picked up in the atmosphere, many of which facilitate the development of clouds and precipitation. Most rain absorbs some carbon dioxide from the air, making rainwater a very weak form of carbonic acid. The action of rainwater includes

This massive thunderstorm with great vertical uplift produces intense rainfall. *VORTEX II/Sean Waugh NOAA/NSSL* dissolving and washing away certain minerals in the rocks and landforms, and although this process typically acts slowly, in certain areas its influence on landforms and the landscape is significant (discussed in Chapter 16). The weak acidity of rainwater should not be confused with the environmentally damaging *acid rain,* which is at least ten times more acidic.

Not only can water dissolve and transport many minerals, but it also can transport solid particles in suspension. These characteristics make water an important transportation process to many aspects of the Earth system. Water supplies nutrients that would not otherwise be available to plants. Water carries minerals and nutrients down rivers and streams, through the soil, through openings in subsurface rocks, and through living organisms. It deposits solid materials on stream floodplains, in river deltas, and on the ocean floor.

The surface tension of water and the behavior of water molecules in drawing together cause **capillary action**—the ability of water to pull itself upward through small openings. Capillary action transports dissolved material in an upward direction through rock and soil against the pull of gravity. This process also moves water into the stems and leaves of plants—even to the uppermost needles of the great California redwoods and to the tops of tall rainforest trees. Capillary action is also important in the movement of blood through our bodies. Without it, many of our cells could not receive the necessary nutrients carried by the blood.

Another important property of water is that it expands when it freezes into a solid. Most substances contract when cooled and expand when heated. Water does contract until it is cooled below 4°C (39°F), but then it expands as the crystal structure of ice develops. Expansion from crystallization during freezing makes ice about 9% less dense than liquid water, which is why ice floats on water. This is also why about 90% of ice cubes, sea ice, and icebergs extend below

● **FIGURE 6.1** These huge icebergs off the coast of Antarctica float because ice is about 9% less dense than water.

*If ice floating in a beverage glass melts completely before you can drink from it, will the liquid level rise, fall, or remain the same as before? Why?*

the water surface and about 10% is exposed above the water level ( ● Fig. 6.1).

Finally, water is slow to heat and slow to cool. Bodies of water are reservoirs that store heat and can moderate the cold of winter, yet the same water body will have a cooling effect in summer. This moderating temperature effect is typical in the vicinity of lakes and on seacoasts.

Earth's water—the *hydrosphere* (from Latin: *hydros,* water)—exists in all other major subsystems of the planet, occurring naturally in all three states: as a liquid in rivers, lakes, oceans, and rain; as a solid as snow and ice; and as water vapor as an atmospheric gas. Even the water temporarily stored in living things could be considered part of the hydrosphere. About 73% of Earth's surface is covered by water, with the largest proportion contained in the world's oceans. The water content of the Earth system is about 1.33 billion cubic kilometers (326 million mi$^3$) including all forms of water on our planet ( ● Fig. 6.2).

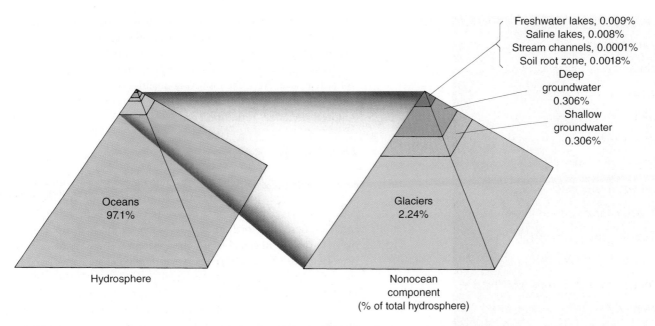

Freshwater lakes, 0.009%
Saline lakes, 0.008%
Stream channels, 0.0001%
Soil root zone, 0.0018%
Deep groundwater 0.306%
Shallow groundwater 0.306%

Oceans 97.1%

Glaciers 2.24%

Hydrosphere

Nonocean component (% of total hydrosphere)

● **FIGURE 6.2** Earth's water sources. The vast majority of water in the hydrosphere is seawater in the world's oceans. The freshwater supply stored in polar ice sheets is relatively unavailable for use.

*How might global warming or cooling alter the data shown in this figure?*

# The Hydrologic Cycle

The system that circulates water from one part of the overall Earth system to another is known as the **hydrologic cycle.** The atmosphere gains water vapor by evaporation from water bodies and land surfaces. In the hydrologic cycle, water is continually transferred from one state to another—as a vapor, liquid, or solid. When water vapor condenses into a liquid and falls as precipitation, several things may happen. First, the precipitation may go directly into a body of water, where it is immediately available for evaporation back into the atmosphere. Alternatively, it may fall on the land surface, where it runs off into rivers, streams, ponds, or lakes. Rainwater may be absorbed into the ground, where it can either be stored in the soil or flow through open spaces, called *interstices;* in loose surface materials like sand, gravel, and silt; or through voids in solid rock. Most of the water in the ground or on the surface will ultimately reach the oceans, returning to the original source of those waters. Some of the water that falls as snow may accumulate to become part of the massive ice cover in the polar regions, or in mountain glaciers, whereas some snow will also thaw into feed streams. In fact, glaciers in the polar regions store most of Earth's freshwater, mainly on Greenland and Antarctica ( ● Fig. 6.3), where the average ice thickness is more than 2000 m (1.25 mi). Water temporarily becomes a part of living things when it is used and stored by plants and animals. In short, the hydrologic cycle contains six major water storage areas: the atmosphere, the oceans, bodies of freshwater, plants and animals, snow and glacial ice, and the groundwater beneath the surface.

Evaporation returns liquid water to the atmosphere as a gas, and ice can revert directly to water vapor (gas) through sublimation. Water evaporates from all bodies of water, from plants and animals, and from soils. Water can evaporate from falling precipitation, and if it falls through very dry air below, rain can completely evaporate before it hits the surface. Once evaporation returns liquid water to the atmosphere as gaseous water vapor, the cycle can be repeated through condensation and precipitation.

The hydrologic cycle is basically a continuous cycle of evaporation, sublimation, condensation, precipitation, and transportation of water over the land, in water bodies, and in the ground ( ● Fig. 6.4). All of these processes are constantly operating. The hydrologic cycle for *the Earth system as a whole* can be considered a *closed system* because energy flows into and out of the system but there is no gain or loss of water (matter). Although Earth's hydrologic cycle is a closed system, its subsystems (like the hydrologic cycle of a region or water body) operate as open systems with energy and matter flowing both in and out. Earth's hydrosphere and its subsystems operate through many processes and interactions that change water from one state to another and transport water from the atmosphere to the surface and back.

The hydrologic cycle is an extremely important part of the Earth system. It is linked to numerous other subsystems that rely on water as a medium of movement and to sustain life. The movement of water and its transfer from one state to another play major roles in redistributing energy in the atmosphere, hydrosphere, biosphere, and lithosphere.

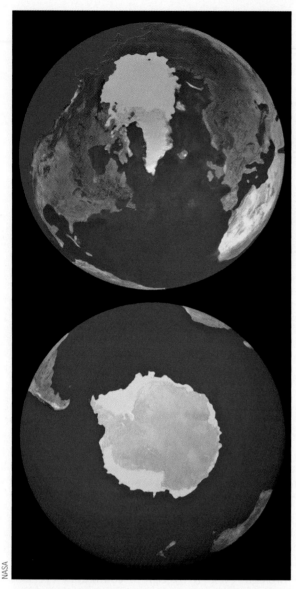

NASA

● **FIGURE 6.3** Earth's surface is mainly covered by the oceans. Most of Earth's freshwater is held in the glacial ice of the polar regions, as seen in these images.

*Can you distinguish between the Greenland and Antarctic glaciers compared to the ice shelves and seasonal sea ice that float on the ocean surface?*

# Water in the Atmosphere

In the atmosphere water exists as ice (snow, hail), as tiny liquid droplets that form clouds and fog, and as water vapor. Water vapor is a tasteless, odorless, and transparent gas that is mixed with the other atmospheric gases in varying proportions. The troposphere contains 99% of atmospheric water vapor, mainly within an altitude of about 5500 meters (18,000 ft). Water vapor makes up a small but highly variable percentage of the atmosphere by volume, yet atmospheric water is the source of all condensation and precipitation. Through the exchange of water among its states through evaporation, condensation, precipitation, melting, and freezing, water plays a significant role in regulating and modifying tempera-

tures locally and globally. In addition, as was noted in Chapter 3, atmospheric water vapor reflects and absorbs significant portions of incoming solar energy and outgoing terrestrial radiation, and it reradiates some of that absorbed energy from the atmosphere back to Earth as heat. Also, the insulating qualities of water vapor reduce the loss of heat from the surface, helping to maintain the moderate temperature ranges found on Earth.

## The Water Budget

Because Earth's hydrosphere is basically a *closed system,* virtually no water is received from outside the Earth system nor lost from it. Although water cycles in and out of the atmosphere, lithosphere, and biosphere, the total amount of water in the hydrosphere remains virtually constant. Despite the apparent vastness of the oceans, large lakes, and huge glaciers, however, the comparative volume of all of Earth's waters is relatively small compared with the planet's size ( ● Fig. 6.5). Water is an essential, but limited, resource.

The Earth operates on a **water budget** in which the total quantity of water remains basically the same and deficits balance gains throughout the entire system. An increase in water within one component of the hydrologic subsystem must be accounted for by a loss in another. The balance in the water budget is dynamic, and the amount of water associated with any one component of the hydrosphere changes continuously over time and place, particularly where water is stored. For example, about 18,600 years ago, during the last major ice age, glaciers expanded, sea level dropped, and evaporation and precipitation decreased. Less water was stored in the oceans, but this was balanced by increased storage on the land as glacial ice. Certain changes in the water budget are human induced—for example, the damming of a river stores more water on the land surface. Further, in an arid location, high evaporation rates from an artificial reservoir will cause water loss to the atmosphere and diminish river flow.

The atmosphere gives up a great deal of water by condensation into fog, dew, frost, and various forms of precipitation (rain, snow, hail, sleet). If the quantity of water in the atmosphere remains at the same level through time, this means that the atmosphere must be absorbing an equal amount of water to that which it give up from other parts of the system. In a single minute, through precipitation or condensation, the atmosphere gives up more than a billion tons of water, while an equal amount of water evaporates and sublimates back to the atmosphere as water vapor.

If we look again at our discussion of the energy budget in Chapter 4, we can see that a part of that budget is latent heat exchange, particularly release of latent heat through condensation. This energy, of course, is originally derived from the sun. Solar energy is used in evaporation, stored in water vapor, and released during condensation. Although the energy transfers involved in evaporation and condensation account for a small portion of the total energy budget, the actual amount of energy is significant. Imagine how much energy is released every minute when a billion tons of water condense out of the atmosphere. This vast storehouse of energy, the latent heat of condensation, is a major energy source that powers storms, particularly hurricanes and thunderstorms.

The amount of water vapor that can be held by a parcel of air is limited, and air temperature is a very important determinant

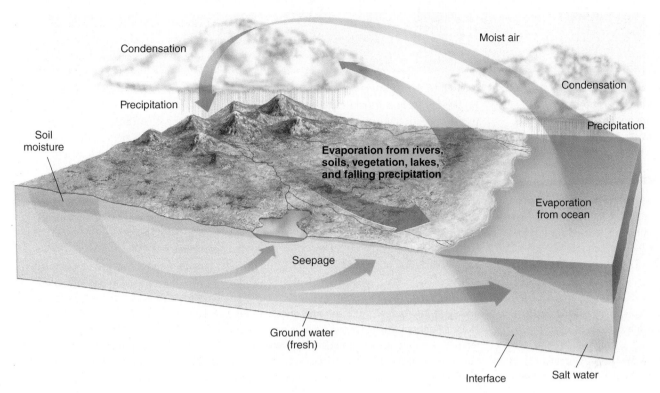

**FIGURE 6.4** The hydrologic cycle involves the circulation of water from one part of the Earth system to another. Largely through condensation, precipitation, and evaporation, water is cycled continually between the atmosphere, the soil, lakes and streams, plants and animals, glacial ice, subsurface storage, and back to the principal reservoir—the oceans.

*Is the hydrologic cycle a closed system or an open system? Why?*

of that amount. Warmer air can hold a greater quantity of water vapor compared with what can be held in colder air. Therefore, we can make a generalization that air in the cold polar regions can hold far less water vapor (approximately 0.2% by volume) than the air of the hot tropical and equatorial regions, which can contain as much as 5% by volume.

## Saturation and the Dew Point Temperature

When air of a given temperature holds all the water vapor that it possibly can, it is said to be in a state of **saturation** and to have reached its moisture **capacity.** If a constant temperature is maintained in a quantity of air, there will come a point, as more water vapor is added, when the air will be saturated and unable to hold any more water vapor. This situation often occurs when taking a shower. The air in the room becomes increasingly humid until a point is reached at which it cannot contain more water. The excess water vapor will then condense onto the colder mirrors and walls.

The air's capacity to hold water vapor, as seen in ● Figure 6.6, increases with rising temperatures. Some examples from reading this figure will illustrate the relationship between air temperature and water vapor capacity. If a parcel of air at 30°C is saturated, it will contain 30 grams of water vapor in each cubic meter of air (30 g/m³). Now, suppose the air temperature increases to 40°C *without* increasing the water vapor content. The parcel will no longer be saturated because air at 40°C can hold more than

**FIGURE 6.5** Compared to Earth's size, the total amount of water in the oceans, lakes, and rivers, as well as in the atmosphere and the ground would fit into the blue sphere shown here. If Earth was the size of a basketball, all of the planet's water would fit into the size of a ping-pong ball.

*What does this relationship mean in terms of water as a resource on Earth?*

# The Wettest and Driest Places on Earth

Although there may be disagreement as to which exact locations on Earth hold the distinction of being the wettest and driest, there are two geographical locations that lay claim to these titles.

Mount Waialeale (Hawaiian: for rippling waters) stands at 1569 meters (5148 ft) above sea level and is the second highest peak on the island of Kauai in the Hawaiian Island Chain. This location averages 11.68 meters (460 in.) of rain per year, with a record of 17.34 meters (683 in.) in 1982.

There are several reasons why this peak consistently accumulates huge amounts of rainfall: (1) Being northernmost of the inhabited Hawaiian Islands allows some exposure to winter time frontal systems; (2) its shape is nearly conical and thus all sides are exposed to the moisture-laden winds; and (3) its very steep cliffs cause the humid air to rise dramatically—over 1000 meters (3000 ft) in less than a half mile distance. This allows the rainfall to concentrate more on one spot instead of being spread out over a gentler mountain slope.

Although some frontal activity brings about winter rainfall, most of it comes in the form of orographic precipitation. The trade winds flow mainly out of the east, but any direction of the winds will waft the moisture-laden ocean air up the sides of this mountain. The precipitation falls mainly in the form of light rain and drizzle. The area around this mountain peak is perennially covered in fog and mist. The rainfall is rarely heavy, but it is consistent. The small amounts of rain each day accumulate through the year to reach these huge totals.

The driest spot on Earth is located in northern Chile, South America. Two locations can claim some fame in this discussion. Arica, Chile, claims the lowest average annual precipitation at 0.5 millimeters (considered a *trace*) of precipitation. However, the location shown in the image here is located between Arica and the city of Iquique, Chile. This valley nested in the Andes mountain range has not received rain according to anyone's living memory. No data have been collected because as far as anyone knows, it has never rained there. No doubt, if it ever rains at this location it will be recorded and likely celebrated! This location contains no vegetation and is likely not to contain many living microbes either.

In Google Earth fly to: 22.076°N, 159.499°W.

Northern Chile is dominated by two mechanisms that keep it rain free. The first is the dominance of a subtropical high pressure system. This semi-permanent anticyclone with its stabilizing force remains in close proximity to this location for long periods. Second, this valley is on the leeward (rain shadow) side of the Andes Mountains. Thus, with both of these processes working together, this location simply does not receive precipitation.

In Google Earth fly to: 19.500°S, 69.852°W.

Although both the wettest and driest locations are unique in their own ways, by applying our knowledge of pressure systems, winds, and precipitation processes it is not difficult to understand how these geographical distinctions can exist.

30 g/m³ of water vapor (actually 50 g/m³). Conversely, if the temperature of saturated air decreases from 30°C (containing 30 g/m³ of water vapor) to 20°C, which at that temperature has the capacity of only 17 g/m³, some 13 grams of water vapor will condense out of the air because of the reduced moisture capacity.

If an unsaturated parcel of air is continually cooled, it will eventually reach a temperature at which the air will become saturated. This critical temperature is known as the **dew point**—the temperature at which condensation takes place. For example, we know that if a parcel of air at 30°C contains 20 grams per cubic meter of water vapor, it is not saturated because it can hold 30 g/m³. However, if that parcel of air cools to 21°C, it would become saturated because the capacity of air at 21°C is 20 grams per cubic meter. Thus, that air parcel at 30°C has a dew point temperature of 21°C. The cooling of air to below its *dew point temperature* brings about the condensation that precedes precipitation.

Because the water vapor capacity of air increases with rising temperatures, warm air in the equatorial regions has a higher dew point temperature than the cold air in the polar regions. Thus, because the atmosphere can hold more water in the equatorial

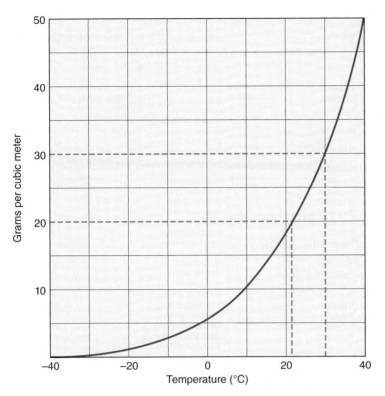

**FIGURE 6.6** Maximum absolute humidity by temperature. This graph shows the maximum amount of water vapor that can be contained in a cubic meter of air and how it varies over a range of temperatures.

*Compare the change in capacity if the temperature rises from 0°C to 10°C if the temperature rises from 20°C to 30°C. How does the relationship between temperature and capacity change?*

regions, there is a greater potential for large quantities of precipitation compared with the precipitation of the polar regions. Also, although there are some exceptions in places that have a dry summer, middle latitude locations generally have a greater potential for heavy precipitation during the warmer summer months rather than the colder winter months.

## Humidity

The amount of water vapor in the air at a time and place is called **humidity.** Humidity is commonly expressed by one of three ways. These quantitative methods, each of which provides different information that contributes to our understanding of weather and climate.

**Absolute and Specific Humidity**   The measure of the mass of water vapor that exists within a given volume of air is called **absolute humidity.** It is expressed either by the metric system as the number of grams of water vapor per cubic meter of air (g/m³) or by the English system as grains per cubic foot (gr/ft³). **Specific humidity** is the mass of water vapor (given in grams) per mass of air (given in kilograms). These two methods quantify the actual amount of water vapor contained in the air. Because most water vapor gets into the air by the evaporation of water from the surface, both absolute and specific humidity tend to decrease with altitude and elevation.

This relationship also occurs because the air is colder aloft and cannot hold as much moisture as the warmer air at lower altitudes.

We have learned that air compresses as it sinks and expands as it rises. Thus, a parcel of air will change in volume as it moves vertically, but there may be no change in the amount of water vapor in that quantity of air. What this means is that absolute humidity, which measures the amount of water vapor by volume of air, can vary as the volume of the air parcel also changes as it moves up or down. Specific humidity, which compares the mass of water vapor to the mass of the air, changes *only* as the quantity of the water vapor changes. For this reason, when assessing changes in water vapor content for large masses of air, which often undergo vertical movement, specific humidity is the preferred measurement among climatologists and meteorologists.

**Relative Humidity**   The best known expression of the water vapor in the air is **relative humidity,** which is the one commonly given in weather reports. Relative humidity is the ratio between the actual amount of water vapor in the air and the maximum amount of water vapor that the air could hold at that temperature expressed as a percentage. The stated percentage expresses how close the air is to being fully saturated with water vapor. Air that is at its moisture-holding capacity, saturated with water vapor, has a relative humidity of 100%. Using this method to describe humidity has both strengths and weaknesses. Its strength lies in how easily it is communicated and understood by the public. People easily understand the meaning of "percent," but they may not conceptualize the significance of "grams per cubic meter." However, relative humidity percentages can vary widely through the day, even when the water vapor content of the air remains the same. Two basic factors cause the relative humidity to rise or fall—changes in air temperature and changes in the amount of water vapor that the air contains. The unsteady nature of relative humidity is its weakness, particularly for scientific purposes.

If the temperature and absolute humidity of an air parcel are known, its relative humidity can be determined by using Figure 6.6. For instance, if a parcel of air has a temperature of 30°C and an absolute humidity of 20 grams per cubic meter, the graph indicates that if it were saturated, its absolute humidity would be 30 grams per cubic meter. To determine relative humidity (RH), divide 20 grams (actual content) by 30 grams (content at capacity) and multiply by 100 (to get an answer in percent):

(actual content in air ÷ maximum air can hold at that temperature) × 100 = RH%

(20 grams ÷ 30 grams) × 100 = 67%

The relative humidity in this case is 67%. The air is holding only two thirds (0.67) of the water vapor it could contain at 30°C; the air is at 67% of its maximum moisture holding capacity.

Two important factors are involved in the geographic distribution and spatial variations of relative humidity. One of these is availability of moisture. For example, the air above a water body is apt to contain more moisture than air (of similar temperature) over land because there is more water available for evaporation. Conversely, the air over a region like the central Sahara Desert is usually very dry in part because it is far from the oceans and little water is

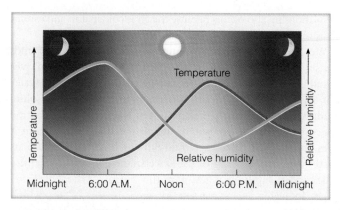

● **FIGURE 6.7** A graph of air temperature and relative humidity changes during a 24-hour day illustrates the relationship between air temperature and relative humidity. Even with no change in the moisture content of the air, as the day warms, the relative humidity drops. At night when it is cooler, the relative humidity rises.
Copyright © 2011 Brooks/Cole, Cengage Learning.

*How can the relationship between air temperature and relative humidity be applied when using a hair dryer?*

available to be evaporated. The second factor that causes variations in relative humidity is temperature. In regions of higher temperature, relative humidity for air containing the same amount of water vapor will be lower than it would be in a cooler region.

The relative humidity will vary if the amount of water vapor increases as a result of evaporation *or* if the temperature increases or decreases. Thus, although the quantity of water vapor may not change during the day or overnight, the relative humidity will vary with the daily temperature cycle. As air temperature increases from sunrise to its maximum in mid-afternoon, the relative humidity will decrease as the air warms and becomes capable of holding greater quantities of water vapor. When the air cools, decreasing toward its minimum temperature around sunrise, the relative humidity increases with the declining nighttime temperature ( ● Fig. 6.7).

Relative humidity affects our comfort levels on hot days because of its relationship to the evaporation rate. Evaporation is a cooling process because the thermal energy used to change perspiration to water vapor becomes locked in the water vapor as latent heat and is removed from our skin. On a hot day when it is 35°C (95°F), it is more uncomfortable in Atlanta, Georgia, with the relative humidity at 90%, than it is in Tucson, Arizona, where the relative humidity is only 15% at the same temperature. At a relative humidity of 15%, perspiration would evaporate faster, so you would feel much cooler. When the relative humidity is 90%, the air is nearly saturated, much less evaporation would take place, and less heat would be drawn from our skin.

# Sources of Atmospheric Moisture

Although water evaporates into the atmosphere from many different places, the most important are the Earth's bodies of water. Water also evaporates from wet ground surfaces and soils; from droplets of moisture on vegetation; from city pavements, building roofs, cars, and other surfaces; and even from falling precipitation.

Vegetation provides another source of water vapor to the atmosphere. Plants give up water in a process called **transpiration,** which can be a significant source of atmospheric moisture. A mature oak tree, for instance, can transpire 400 liters (105 gal) of water per day, and a cornfield may add 11,000 to 15,000 liters (2900–4000 gal) of water to the atmosphere per day for each acre under cultivation. In some parts of the world—notably the tropical rainforests with their heavy, lush vegetation—transpiration accounts for a significant amount of atmospheric humidity. The combined effects of evaporation and transpiration are referred to as **evapotranspiration,** which accounts for the majority of the water vapor in the atmosphere.

## Evaporation Rates

Several factors affect the rates of evaporation. First, evaporation is affected by the amount and temperature of accessible water. As Table 6.1 shows, evaporation rates tend to be greater over the oceans when compared with the evapotranspiration rates over the continents, even though the data for the continents includes transpiration by plants. The only regions where this generalization does not fit is in the equatorial areas between 0° and 10°N and S. Here the land vegetation is so lush that transpiration provides the air with huge quantities of moisture.

Second is relative humidity, the degree to which the air is saturated with water vapor. The drier the air and the lower the relative humidity, the greater the evaporation rate will be. Most people have had some direct experience with this relationship. Compare the length of time it takes a swimsuit and wet towel to dry on a hot day when the air is dry with how long it takes on a day when the air is hot and humid.

The wind is a third factor that affects evaporation rates. If there is no wind, the air that overlies a water surface will approach saturation as more and more molecules of water change to water vapor. Once saturation is reached, however, evaporation will cease. If the conditions are windy, the wind will blow the saturated or nearly saturated air away from the evaporating water source, replacing it with lower humidity air. This allows evaporation to continue as long as the wind keeps blowing saturated air away and bringing in drier air. Anyone who has gone swimming on a windy day has experienced the chilling effects of rapid evaporation.

Air temperature also strongly affects evaporation rates by influencing the first and second factors listed earlier. As air temperature increases, so does the water temperature at the evaporation source. Temperature increases ensure that more energy is available to the water molecules, enhancing the transition from a liquid to a gas, and more evaporation takes place. As the temperature of the air increases, its moisture holding capacity also increases. As air becomes warmer its molecules are increasingly energized, moving farther apart, and the air density decreases. With more energy and wider spacing between air molecules, additional water molecules can enter the atmosphere, thus increasing evaporation.

## Potential Evapotranspiration

So far we have discussed actual evaporation and transpiration (evapotranspiration). However, geographers and meteorologists are also concerned with **potential evapotranspiration** ( ● Fig. 6.8).

**TABLE 6.1**
**Distribution of Actual Mean Evapotranspiration**

| | Latitude | | | | | |
|---|---|---|---|---|---|---|
| Zone | 60°–50° | 50°–40° | 40°–30° | 30°–20° | 20°–10° | 10°–0° |
| **Northern Hemisphere** | | | | | | |
| Continents | 36.6 cm (14.2 in.) | 33.0 (13.0) | 38.0 (15.0) | 50.0 (19.7) | 79.0 (31.1) | 115.0 (45.3) |
| Oceans | 40.0 (15.7) | 70.0 (27.6) | 96.0 (37.8) | 115.0 (45.3) | 120.0 (47.2) | 100.0 (39.4) |
| Mean | 38.0 (15.0) | 51.0 (20.1) | 71.0 (28.0) | 91.0 (35.8) | 109.0 (42.9) | 103.0 (40.6) |
| **Southern Hemisphere** | | | | | | |
| Continents | 20.0 cm (7.9 in.) | NA | 51.0 (20.1) | 41.0 (16.1) | 90.0 (35.4) | 122.0 (48.0) |
| Oceans | 23.0 (9.1) | 58.0 (22.8) | 89.0 (35.0) | 112.0 (44.1) | 119.0 (47.2) | 114.0 (44.9) |
| Mean | 22.5 (8.8) | NA | NA | 99.0 (39.0) | 113.0 (44.5) | 116.0 (45.7) |

What this term refers to is that for various environmental reasons, the actual evapotranspiration that occurs does not match the *potential* (maximum) *evapotranspiration* that would take place with an unlimited supply of water available. In a desert region with hot dry air, for example, the potential for evapotranspiration may greatly exceed the water available to evaporate or transpire. Various formulas have been derived for estimating the potential evapotranspiration at a location because it is difficult to measure directly. These formulas commonly use temperature, latitude, vegetation, and soil character (permeability, water-retention ability) as factors that could affect the potential evapotranspiration.

In places where precipitation exceeds potential evapotranspiration, there is a surplus of water for storage in the ground and in water bodies, allowing water to flow in streams and rivers away

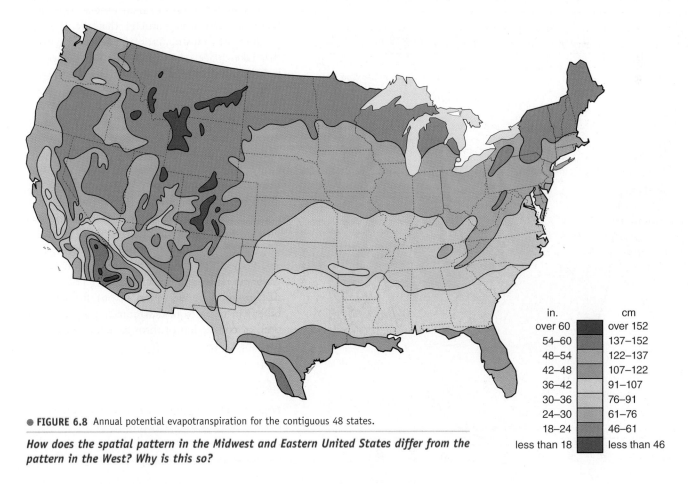

| in. | cm |
|---|---|
| over 60 | over 152 |
| 54–60 | 137–152 |
| 48–54 | 122–137 |
| 42–48 | 107–122 |
| 36–42 | 91–107 |
| 30–36 | 76–91 |
| 24–30 | 61–76 |
| 18–24 | 46–61 |
| less than 18 | less than 46 |

● **FIGURE 6.8** Annual potential evapotranspiration for the contiguous 48 states.

*How does the spatial pattern in the Midwest and Eastern United States differ from the pattern in the West? Why is this so?*

from those areas. Water can also be exported from these wetter places to drier locations if systems of canals or pipelines are feasible. When potential evapotranspiration exceeds precipitation, as it does during the dry summer months in California and in the arid West, no water is available for storage. In fact, the water stored during previous rainy months evaporates quickly into the warm, dry air or is transpired into the air by vegetation ( ● Fig. 6.9). The soil dries out and the vegetation turns brown as the available water is absorbed into air that could hold much more moisture. For this reason, wildfires are a potential hazard during the late summer and early fall in California and can occur almost any time in desert areas.

An understanding of potential evapotranspiration can be used to determine how much water will be lost from crops or a reservoir in environments that experience dry seasons. By assessing the daily or weekly relationship between potential evapotranspiration and precipitation, agriculturalists can determine when and how much irrigation water is needed for their crops.

# Condensation, Fog, and Clouds

Condensation, the process by which water vapor changes into liquid water, occurs when air saturated with humidity is cooled. Once the air temperature cools until it has a relative humidity of 100%

(meaning the air has reached the dew point temperature), condensation will begin and continue with additional cooling. Whether condensation will occur depends on the relative humidity of the air and the degree of cooling. In the very dry air of Death Valley, California, a great amount of cooling must take place before the dew point is reached. In contrast, on a humid summer afternoon in Biloxi, Mississippi, a minimal amount of cooling may bring on condensation.

Condensation is what causes the formation of water droplets on the side of a glass containing a cold drink on a warm afternoon. The temperature of the air in contact with the cold glass is lowered and the air's capacity to hold water vapor decreases. If air touching the glass is cooled sufficiently, it will reach the dew point temperature, which means that its relative humidity will be 100%. Further cooling will result in condensation that forms water droplets on the glass.

## Condensation Nuclei

An atmospheric factor that reinforces the condensation process is the presence of **condensation nuclei.** These are minute particles in the atmosphere that provide a surface upon which condensation can take place. Sea-salt particles in the air are common condensation nuclei that come from the evaporation of saltwater spray from ocean waves. Other common nuclei include dust, smoke, pollen, and volcanic material. Some particles are more *hygroscopic* (having the property to attract moisture) than others. Typically, these are chemical particles that are the by-products of industrialization. The condensation that takes place on certain chemical nuclei can be corrosive and hazardous to human health; when it is, we know it as *smog* (a term that was derived by combining the words *smoke* and *fog)*.

Theoretically, if all particles were removed from a volume of air, that air could be below its dew point without condensation occurring. Conversely, if there is a superabundance of particles, condensation may take place at a relative humidity that is less than 100%. For example, ocean fogs, an accumulation of condensation droplets formed on particles of sea salt, can form when the relative humidity is as low as 92%.

Fog, clouds, and dew all result from water vapor condensing in the atmosphere. The cooling that produces condensation in one form or another can result from radiational cooling, advection, or convection or through a combination of these processes.

## Fog

Fogs and clouds appear when water vapor condenses and forms a large number of droplets. Not being transparent to light in the way that water vapor is, these masses of condensed water

● **FIGURE 6.9** The water budget for San Francisco, California. This graph illustrates the water budget system, which "keeps score" of the balance between water input by precipitation and water loss to evaporation and transpiration, permitting month-by-month estimates of both runoff and soil moisture.

*When would irrigation be necessary in areas with this kind of a seasonal water budget?*

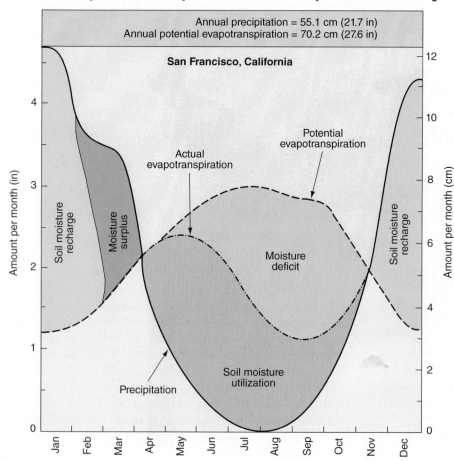

Annual precipitation = 55.1 cm (21.7 in)
Annual potential evapotranspiration = 70.2 cm (27.6 in)

San Francisco, California

droplets appear to us as fog or clouds, in a great variety of shapes and forms, usually in shades of white or gray.

On a worldwide basis, fog is a minor form of condensation, but in certain regions it has important climatic effects. The "drip factor" from fog helps sustain vegetation along some desert coastlines where fog occurs. Fog also plays havoc with transportation systems. Navigation on the seas is more difficult in fog. Air travel can be greatly impeded by fog; dense fog can cause airports to delay flights until visibility improves. Highway travel can also be greatly hampered by heavy fogs, which can lead to huge, chain-reaction vehicle pileups.

**Radiation Fog** Cold ground surfaces that result from radiational cooling can produce **radiation fog,** also called **ground fog,** or **temperature-inversion fog.** This kind of fog typically occurs on cold, clear, calm nights and lasts until morning. Clear, calm atmospheric conditions allow massive amounts of terrestrial radiation to be lost from the surface. With no incoming radiation at night, the ground becomes cold as it gives up much of the heat that it received during the day. The air directly above the surface is cooled by conduction through contact with the cold ground. Because the cold surface can cool only the lower few meters of the atmosphere, a temperature inversion is created as air near the surface becomes colder than the warmer air above. If this cold layer of air at the surface cools to below its dew point temperature, condensation will occur, often in the form of a low-lying fog. However, winds strong enough to disturb the inversion layer can prevent the formation of fog by not allowing the air to stay at the surface long enough to become cooled below its dew point.

The chances of a temperature-inversion fog occurring are increased in valleys and depressions, where cold air drains down from higher areas into lowland areas. During a cold night this air can be cooled below its dew point, resulting in a fog that forms like a pond in the valley bottom ( ● Fig. 6.10). It is common in mountain areas to see early morning radiation fog in the valleys while snow-capped mountaintops shine against a clear blue sky. Radiation fog has a diurnal cycle, forming during the night and becoming densest around sunrise when temperatures are lowest. It typically "burns off" during the day when solar energy slowly penetrates the fog, warming the ground surface. The ground in turn warms the air directly above it, increasing its temperature and its capacity to hold water vapor, which causes the fog to evaporate.

Radiation fog can become very dense in industrial areas where high concentrations of chemical particles in the air provide abundant condensation nuclei. These fogs are usually thicker than "natural" radiation fogs and less easily dissipated by winds or the daytime sun.

**Advection Fog** A common type of fog, called **advection fog,** occurs when warm, moist air moves over a colder land or water surface. When the overriding warm air cools below its dew point temperature through heat loss by conduction to the colder surface below, condensation occurs to produce fog. Advection fog is usually less localized than radiation fog. It is also less likely to have a

NOAA

● **FIGURE 6.10** Radiation fog forms when intense nighttime cooling by loss of terrestrial radiation causes cold temperatures near the surface on clear, calm nights. These fogs can also be reinforced when cold air drains into a valley. This inversion-related fog is in Reno, Nevada.

diurnal cycle of forming and burning off, although if the fog is not too thick, it can be burned off during the day to return again in the early evening. More common, however, advection fog is persistent and spreads over a large area for days at a time.

During the winter advection fog forms in many land areas of the middle and high latitudes. In the United States, for example, these fogs occur when warmer, moist air from the Gulf of Mexico flows northward over the cold, frozen, and snow-covered upper Mississippi Valley. Advection fog can also develop over a lake when warm air from the land flows over colder water. These fogs commonly occur when a warm air mass passes over the cool surface of Lake Michigan or another one of the Great Lakes.

During the summer advection fog can also form over the oceans or large lakes. Widespread advection fogs develop offshore and along coasts when a warm air mass moves over a cold ocean current, cooling the air sufficiently to bring about condensation. These advection fogs are also known as coastal fogs or sea fogs. These are common fogs in the summertime along the West Coast of the United States and are related to cold ocean currents. In the summer months the Pacific subtropical high moves north and air drifting toward the coast passes over the cold California Current. Condensation occurs and forms fog that can flow into coastal areas, pushed from behind by the eastward air movement and pulled inland by the low pressure of the warmer land ( ● Fig. 6.11). Advection fogs also occur in New England, especially along the coasts of Maine and the Canadian Maritime Provinces, when warm, moist air over the Gulf Stream flows north over cold waters of the Labrador Current. Advection fog over the Grand Banks off Newfoundland has long been a hazard for marine navigation. This region has been referred to as the foggiest area in the world.

● **FIGURE 6.11** Advection fog is caused by warm, moist air passing over colder water or a colder coastal surface. Here on the Oregon coast the sea fog also drifts onshore.

## Upslope Fog

**Upslope Fog** Another type of fog, **upslope fog,** clings to windward sides of mountains. Its appearance has also been the source of geographic place names—for example, the Great Smoky Mountains, where this type of fog is very common. During early morning hours when a moist breeze ascends a slope, the air may cool to the dew point, leaving a blanket of fog behind ( ● Fig. 6.12). In wet, tropical areas mountain slopes may be covered in a misty fog at any time of day. Because of the very humid air in those regions, reaching the dew point may result from only a minor drop in temperature.

## Dew and Frost

A cover of tiny water droplets that form by condensation on cool surfaces, such as plants, buildings, and metal objects, is called **dew.** Dew collects on surfaces that are good radiators of heat (such as a vehicle or blades of grass) that give up large amounts of heat during the night. If air cools to the dew point temperature when it comes in contact with these cold surfaces, water droplets will form in beads on the surface. If the air temperature is below freezing, 0°C (32°F), **frost** forms. It is important to note that frost is not frozen dew but results from a *sublimation* process—water vapor changing directly from the gaseous to the frozen state.

Sometimes, although air temperatures may be below 0°C (32°F), the liquid droplets that make up clouds and fogs may not freeze into solid particles. These are *supercooled* water droplets and when they come in contact with a surface such as a tree branch, a window, a vehicle, or an airplane wing, ice crystals develop on that surface in a formation known as **rime** ( ● Fig. 6.13). Icing on the wings, nose, and tail of an airplane is extremely hazardous and can cause aviation disasters. The airlines use de-icing treatments to avoid this problem if the weather conditions are conducive to ice forming on aircraft during flight and before takeoff.

● **FIGURE 6.12** Upslope fog is caused by moist air adiabatically cooling as it rises up this mountain slope in the German Alps.

*What unique problems might coastal residents face as a result of fog?*

● **FIGURE 6.13** This automatic camera was placed to monitor volcanic activity at Mt. St. Helens in Washington state. A large accumulation of rime, however, put the camera temporarily out of commission.

*What does rime have to do with aircraft safety?*

# Clouds

Clouds consist of billions of tiny water droplets and/or ice crystals so small (some measured in thousandths of a millimeter) that they remain suspended in the atmosphere. Clouds are an intriguing part of our environment. Clouds, with their great variety of shapes and hues, and the colors of the sky provide us with an ever-changing backdrop to the natural scenery on Earth. Clouds can appear white, or as shades of gray, or even deep gray approaching black. The apparent color of clouds depends on how thick or dense they are and if the sun is shining on the cloud surface that we can see. Thick clouds absorb a great deal of the sunlight that strikes them, which blocks light from our view and makes the cloud appear dark. Clouds also seem dark when seen from their shaded side instead of their sunlit side.

Clouds are the most common form of condensation and are important for many reasons. First, they are the source of all precipitation. **Precipitation** refers to any liquid or solid water particles that fall from the sky. Obviously, not all clouds result in precipitation, but precipitation will not occur without the formation of a cloud. Clouds serve an important function in the energy budget because they absorb some of the incoming solar energy. They also reflect some of that energy back to space and scatter or diffuse other wavelengths of incoming energy before it strikes the surface. In addition, clouds absorb some of Earth's thermal radiation so that it is not directly lost to space and then re-radiate it back to the surface.

**Cloud Forms** Clouds result from various air movements in our atmosphere. In 1803 Sir Luke Howard, an Englishman, proposed the first method of cloud classification, which has been modified through time into the system that is in use today. Cloud names generally (but not always) consist of two parts. The first part refers to the cloud's height: low-level clouds, below 2000 meters (6500 ft), are called **strato;** middle-level clouds, from 2000 to 6000 meters (6500–19,700 ft), are named **alto;** and high-level clouds, above 6000 meters (19,700 ft), are termed **cirro.**

The second part of the name concerns the morphology, or shape, of the clouds. The three basic shapes are termed *cirrus, cumulus,* and *stratus.* Classification systems categorize these cloud formations into many subtypes; however, most subtypes are variations of these three basic shapes. ● Figure 6.14 illustrates the appearance and the general heights of common clouds, and ● Figure 6.15 provides images to help you recognize the major cloud types.

**Cirrus** clouds (from Latin: *cirrus,* a lock or wisp of hair) form at very high altitudes, normally 6000 to 10,000 meters (19,800–36,300 ft), and consist of ice crystals rather than droplets of water. They are thin, stringy, white clouds that trail like feathers across the sky. When associated with fair weather, cirrus clouds are scattered white patches in a clear blue sky.

**Cumulus** clouds (from Latin: *cumulus,* heap or pile) develop vertically rather than forming the more horizontal structures of the cirrus and stratus forms. Cumulus are massive piles of clouds, rounded in appearance, usually with a flat base, which can be

● **FIGURE 6.14** Clouds are named based on their height and their form.

*Observe this figure and those in Figure 6.15; what cloud type is present in your area today?*

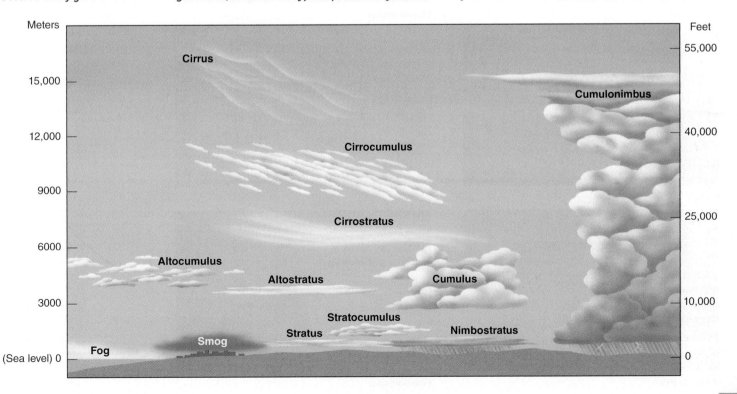

● **FIGURE 6.15** Types of clouds.

Cirrus

Cirrocumulus

Cirrostratus

Altocumulus

Altostratus

Stratocumulus

Stratus

Cumulus

Nimbostratus

Cumulonimbus

anywhere from 500 to 12,000 meters (1650–39,600 ft) above sea level. From this base, they develop into great rounded structures, often with tops like cauliflowers. Cumulus clouds provide visible evidence of an unstable atmosphere. Their base is the altitude where condensation begins in a rising column of air. Basically, this is the altitude where the dew point temperature is reached.

**Stratus** clouds (from Latin: *stratus,* layer) can appear anywhere from near the surface to almost 6000 meters (19,800 ft) and the varieties of stratus clouds are based in part on their altitude. The basic characteristic of stratus clouds is their horizontal sheet-like appearance, lying in layers with fairly uniform thickness. This horizontal configuration indicates that they form in relatively stable atmospheric conditions, which minimize vertical development.

Often stratus clouds cover the entire sky with a gray cloud layer. It is stratus clouds that make up the dull, gray, overcast sky common to winter days in much of the Midwestern and Eastern United States. A formation of stratus clouds may overlie an area for days and any precipitation will be light but steady and persistent.

Examine Figures 6.14 and 6.15 and become familiar with the basic cloud types and their names. Keep in mind that some cloud shapes exist in all three levels—for example, *stratocumulus* (*strato* = low level + *cumulus* = a rounded shape), *altocumulus,* and *cirrocumulus*. These three share the similar rounded or cauliflower appearance of cumulus clouds, which do exist at all three levels. You may notice that *altostratus* (*alto* = middle level + *stratus* = layered shape) and *cirrostratus* have two-part names, but low-level layered clouds are just called *stratus*. Thin, stringy *cirrus* clouds are found only as high-level clouds, so the term *cirro* (meaning high level cloud) is not necessary here.

Other terms used in describing clouds are **nimbo** or **nimbus,** meaning precipitation (rain is falling). Thus, the *nimbostratus* cloud may bring a long-lasting drizzle, and the cumulonimbus are the thunderstorm clouds. Cumulonimbus clouds can form a flat top called an anvil head and a relatively flat base, and they become darker as they grow higher and thicker, blocking the incoming sunlight. Cumulonimbus clouds are the source of many atmospheric concerns including high-speed winds, torrential rain, flash flooding, thunder, lightning, hail, and possibly tornadoes. This type of cloud develops from rapidly rising columns of air that ascend to towering heights.

# Adiabatic Heating and Cooling

Clouds typically develop from the cooling that results as a parcel of air rises. Rising parcels of air will expand as they encounter decreasing atmospheric pressure with altitude. Expansion of air allows the gas molecules to spread out, causing the parcel's air temperature to decrease. This is known as **adiabatic cooling,** a temperature decrease that occurs at a lapse rate of approximately 10°C per 1000 meters (5.6°F/1000 ft). Air descending through the atmosphere is compressed by increasing pressure and undergoes **adiabatic heating,** which increases air temperature by the same rate.

However, a rising parcel of air will eventually cool to its dew point temperature and reach its saturation capacity (relative humidity = 100%), and water vapor begins to condense, forming cloud droplets. After condensation occurs, the adiabatic cooling of rising air will decrease at a lower rate because of the release of latent heat of condensation into the air. To differentiate between these two adiabatic cooling rates, we refer to the pre-condensation rate (10°C/1000 m) as the **dry adiabatic lapse rate** and the lower rate post-condensation rate as the **wet adiabatic lapse rate.** The latter rate averages 5°C per 1000 meters (3.2°F/1000 ft) but varies according to the amount of water vapor that condenses out of the air.

Rising parcels of air will always cool at either the dry or the wet adiabatic rate. Which rate operates depends on whether condensation is occurring (wet adiabatic rate) or is not occurring (dry adiabatic rate). As air descends the temperature will warm by compression, increasing its capacity to hold water vapor and preventing condensation. Thus, the temperature of descending air that is being compressed always increases at the dry adiabatic rate. It is important to note that adiabatic temperature changes are the result of changes in air volume and density. Adiabatic heating and cooling do not involve the addition or subtraction of heat from external sources.

It is also extremely important to differentiate between the normal lapse rate, *where a temperature measuring device is moving up or down,* and the adiabatic lapse rates, *where the air is moving up or down.* In Chapter 4, we learned that, in general, atmospheric temperatures decrease with increasing altitude. This is the normal lapse rate (also called the environmental lapse rate), which is variable but averages 6.5°C per 1000 meters (3.6°F/1000 ft) and is measured by meteorological instruments sent aloft. The normal lapse rate reflects the vertical temperature structure of the atmosphere. Adiabatic lapse rates indicate temperature changes that result when air moves up or down and whether or not condensation is occurring ( ● Fig. 6.16).

## Instability and Stability

Although adiabatic cooling results in cloud development, most of the varying cloud forms are related to differing degrees of vertical air movement. Some clouds are associated with air that is rapidly rising and buoyant. Other cloud forms result when air resists vertical movement.

An air parcel will be buoyant and rise as long as it is warmer than the surrounding air. When it reaches a part of the atmosphere that has the same temperature, it will stop rising. Air parcels that rise because they are warmer than the surrounding atmosphere are said to be *unstable* (or *instable*). In contrast, air that is colder than the surrounding atmosphere tends to sink to lower levels. Sinking air is said to be *stable.*

Determining the stability or instability of an air parcel involves answering a fairly simple question. If an air parcel rises to a specific altitude (cooling at an *adiabatic* lapse rate), would it be warmer, colder, or the same temperature as the surrounding air (as determined by the normal lapse rate) at that same altitude?

If the air parcel is warmer than the atmospheric air at the selected elevation, then the parcel will be unstable and would continue to rise because warmer air is less dense and therefore buoyant. Thus, under conditions of **instability,** the normal lapse rate must be *greater than* the adiabatic lapse rate in operation. For example, if the normal lapse rate is 12°C per 1000 meters and the ground temperature is 30°C, then the atmospheric air tempera-

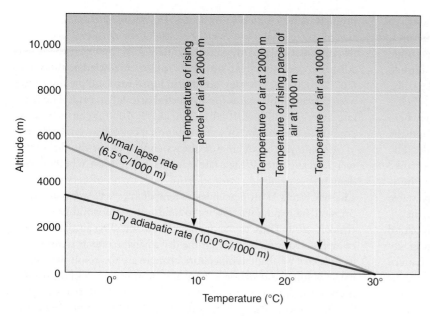

● **FIGURE 6.16** Comparison of the dry adiabatic lapse rate and the environmental, or normal, lapse rate. The normal lapse rate is the average vertical change in temperature. Air displaced upward will cool (at the dry adiabatic rate) because of expansion.

*In this example, using the normal lapse rate determine the temperature of the layer of air at 2000 meters.*

ture at 2000 meters would be 6°C. In comparison, an air parcel (assuming that no condensation occurs) lifted to 2000 meters would have a temperature of 10°C. Because the air parcel is warmer than the atmospheric air around it, it is unstable and will continue to rise ( ● Fig. 6.17).

● **FIGURE 6.17** Relationship between lapse rates and air mass stability. When an air parcel is forced to rise, it cools adiabatically. Whether it continues to rise or resists vertical motion depends on whether adiabatic cooling is greater or less than the prevailing vertical temperature lapse rate. If the adiabatic cooling rate exceeds the lapse rate, the lifted air will be colder than its surroundings and will tend to sink when the lifting force is removed. If the adiabatic cooling rate is less than the lapse rate, the lifted air will be warmer than its surroundings and will be buoyant, continuing to rise even after the original lifting force is removed.

*In these examples, what would be the temperature of the lifted air if it rose to 2000 meters?*

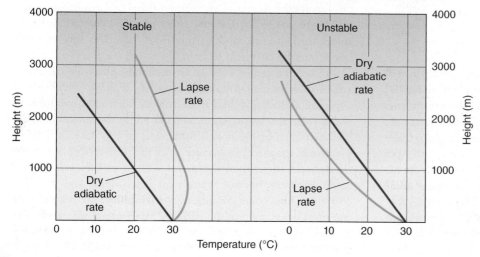

Now, assume that it is another day and all the conditions are the same except that measurements indicate the normal lapse rate on this day is 2°C per 1000 meters. Consequently, although our air parcel if lifted to 2000 meters would still have a temperature of 10°C, the temperature of the atmosphere at 2000 meters would now be 26°C. Thus, the air parcel would be colder and would sink back toward Earth as a result of its greater density (see Fig. 6.17). Under conditions of **stability** the normal lapse rate is *less than* the adiabatic lapse rate in operation. If an air parcel, lifted to a specific elevation, has the same temperature as the atmospheric air surrounding it, it is neither stable nor unstable. Instead, it is considered *neutral;* it will neither rise nor sink but will remain at that elevation.

Whether an air parcel will be stable or unstable is related to the amount of cooling and heating of air at Earth's surface. With air cooling through radiation and conduction on a cool, clear night, air near the surface will be relatively close in temperature to that aloft, and the normal lapse rate will be low, enhancing stability. With the rapid heating of the surface on a hot summer day, there will be a very steep normal lapse rate because the air near the surface is so much warmer than that above, and instability will be enhanced. Pressure zones can also be related to atmospheric stability. In areas of high pressure, stability is maintained by air slowly subsiding from aloft, and in low pressure regions instability is promoted by the tendency for air to rise.

## Precipitation Processes

The condensed water droplets that float around within clouds do not fall to Earth because they are so tiny (0.02 mm, or less than 1000th of an inch) that gravity does not overcome the buoyant effects of air and the currents and updrafts that exist in clouds. ● Figure 6.18 shows the relative sizes of a condensation nucleus, a cloud droplet, and a raindrop. It takes about a million tiny cloud droplets to form one raindrop.

Precipitation occurs when droplets of water or ice crystals become too large and heavy to be held aloft, so they fall as rain, snow, sleet, or hail. The type of precipitation depends largely on how it formed, the temperature in the cloud, and the air temperature below as it falls to Earth. The most widely accepted theories for how precipitation develops include the **collision–coalescence process** for warm clouds and the **Bergeron (or ice crystal) process** for cold clouds.

Precipitation in the tropics and in warm clouds is likely to form by collision–coalescence, a process that is well described by its name. Water is cohesive (able to stick to itself), so as water droplets collide while circulating in a cloud, they tend to coalesce (or grow together) until they become large

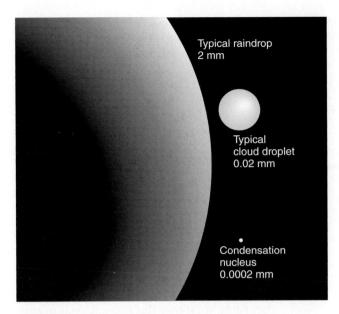

● **FIGURE 6.18** The relative sizes of raindrops, cloud droplets, and condensation nuclei.

and heavy enough to fall. In falling, larger droplets overtake smaller, more buoyant droplets and capture them to form larger raindrops. The mass of these growing raindrops eventually overcomes the updrafts of the cloud and fall to Earth under the pull of gravity. This process occurs in the warm section of clouds where all the moisture exists as liquid water ( ● Fig. 6.19).

At higher latitudes, storm clouds can possess three distinctive layers. The lowest is a warm layer where the temperatures are above the freezing point of 0°C (32°F) and water droplets are liquid. Higher up, the second layer is composed of some ice

● **FIGURE 6.19** Collision and coalescence. (a) Tiny cloud droplets falling at about the same speed are unlikely to collide and coalesce, and if they do collide, they tend to bounce off of each other because of the surface tension of water. (b) Large droplets falling rapidly can capture some of the smaller droplets.

*Why do these tiny droplets fall at different speeds?*

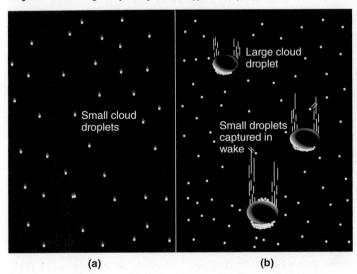

(a)                                   (b)

crystals but mainly **supercooled water** (liquid water cooler than 0°C). In the uppermost layer of these tall clouds, if temperatures are lower than or equal to −40°C (−40°F), ice crystals will dominate ( ● Fig. 6.20). It is in relation to these layered clouds that Scandinavian meteorologist Tor Bergeron presented his explanation.

The Bergeron (or ice crystal) process begins at great heights in the ice crystal and supercooled water layer of the clouds. Supercooled water has a tendency to freeze on any available surface. It is for this reason that aircraft flying through middle- to high-latitude storms run the risk of severe icing and potential disaster. The ice crystals can become *freezing nuclei* upon which the supercooled water can freeze to form growing ice crystals. This process can also create snow. If the frozen precipitation falls through lower cloud layers that have above-freezing temperatures, the ice crystals melt and fall as liquid rain. Finally, as raindrops fall through the warmer section of the cloud, the collision–coalescence process may cause the raindrops to grow larger. Therefore, according to Bergeron, rain in these clouds begins as frozen precipitation and melts into a liquid before reaching Earth.

## Forms of Precipitation

**Rain,** consisting of droplets of liquid water, is by far the most common form of precipitation. Raindrops vary in size but are generally about 2 to 5 millimeters (approximately 0.1–0.25 in.) in diameter (see Fig. 6.18). As we all know, rain can come in many ways: as a brief afternoon shower, a steady rainfall, or the deluge of a tropical rainstorm. When the temperature of an air mass is only slightly below the dew point, the raindrops may be very small (about 0.5 mm or less in diameter) and close together. The result is a fine mist called **drizzle.** Drizzle is so light that it is greatly affected by the direction of air currents and the variability of winds. Consequently, drizzle seldom falls vertically.

**Snow** is the second most common form of precipitation. When water vapor is frozen directly into a solid without first passing through a stage as liquid water (or sublimation), it forms minute ice crystals around the freezing nuclei (of the Bergeron process). These crystals develop into hexagonal ice crystals that make up the intricate six-sided form of snowflakes. Snow will reach the ground only if the cloud and the air below maintain sub-freezing temperatures (below 0°C or 32°F).

**Sleet** is rain that freezes as it falls through a thick layer of sub-freezing air near the surface. The cold air causes raindrops to freeze into small solid particles of clear or milky ice. In English-speaking countries outside the United States, sleet refers not to this phenomenon of frozen rain but rather to a mixture of rain and snow.

**Hail** is a less common form of precipitation than rain, snow, or sleet and occurs most often during the spring and summer months as a result of thunderstorm activity. Hail forms as lumps of ice, *hailstones,* which can vary in size from 5 millimeters (0.2 in) in diameter to larger than a baseball. The U.S. record for hailstone size fell in Vivian, South Dakota in July 2010 ( ● Fig. 6.21). The world record is a hailstone 30 centimeters (12 in.) in diameter that fell in Australia. Hail forms when ice crystals are lifted by strong updrafts in a cumulonimbus (thunderstorm) cloud. As these ice crystals circulate around the storm cloud, supercooled water droplets attach

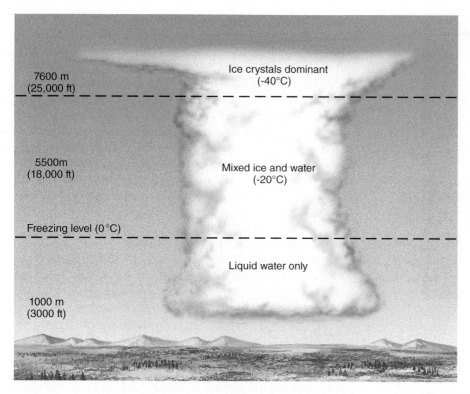

● **FIGURE 6.20** The distribution of water, supercooled water, and ice crystals in a towering cumulonimbus storm cloud according to the Bergeron process theory.

*What is the difference between water and supercooled water?*

● **FIGURE 6.21** A severe thunderstorm in South Dakota produced the largest hailstone (by diameter) ever recorded in the United States in July 2010. Shown here, its diameter was 20.3 cm (8 in) and it weighed 0.89 kilograms (1.94 pounds).

*What gives hail stones their generally spherical shape?*

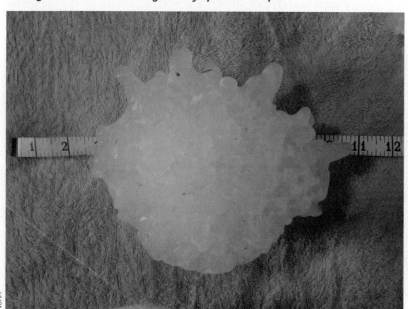

themselves as increasing thicknesses of frozen layers. Sometimes these pellets are lifted up into the cold layer of air and then dropped again and again. The resulting hailstone, made up of concentric layers of ice, has a frosty, opaque appearance when it finally breaks out of the strong updrafts of the cloud formation and falls to Earth. The larger the hailstone, the more times it is cycled through the freezing process and accumulated additional frozen layers. Dropping from the sky, hailstones can be highly destructive to livestock, crops and other vegetation, and vehicles and buildings. Although primarily destructive of property, hailstones have been known to injure or kill animals and humans.

On occasion, raindrops can also have a temperature below freezing and maintain liquid form. These supercooled raindrops will instantly freeze if they fall onto a surface that is also at a sub-freezing temperature. The resulting icy covering on trees, plants, and telephone and power lines is known as **freezing rain (or glaze).** People usually call this kind of precipitation an "ice storm" ( ● Fig. 6.22). Because of the weight of ice, glazing can break tree branches that bring down telephone and power lines. Surface ice accumulations cause extremely slippery conditions that make driving dangerous and some roads impassable. Yet ice storms can also produce a beautiful natural landscape. Sunlight glitters on the ice, reflecting and making a diamond-like surface that covers trees, plants, buildings, and vehicles.

## Factors Necessary for Precipitation

Three factors are necessary for precipitation to develop. The first is the presence of *moist air,* which provides a moisture source (for precipitation) and energy (as latent heat of condensation). Second are the *condensation nuclei* around which the water vapor can condense. Third is an **uplift mechanism** that forces the air to rise enough to cool (by the dry adiabatic rate) to the dew point temperature and then continue to cool at the wet adiabatic lapse rate.

Precipitation results from one of four major uplift mechanisms that force parcels of air to rise. All precipitation can be traced to one of these four uplift mechanisms ( ● Fig. 6.23):

- **Convectional precipitation** results from the vertical displacement of warm air upward in a convectional system.
- **Frontal precipitation** takes place when a warmer air mass rises after encountering a colder, denser air mass.
- **Cyclonic (or convergence) precipitation** occurs when air converges upon and is lifted upward into a low pressure system.
- **Orographic precipitation** results when a moving air mass encounters a land barrier, usually a mountain, and is forced to rise.

**Convectional Precipitation** Convection occurs as air, heated near the surface, expands, becomes lighter, and rises. **Convectional precipitation** is most common in the hot, humid tropical and equatorial areas and during the summer in many middle-latitude locations. Once condensation begins in a convectional column of air, further lifting will be encouraged by additional energy from the release of latent heat of condensation.

To reinforce our understanding of convectional precipitation, let's apply what we have learned about instability and stability. ● Figure 6.24 illustrates two different cases where air rises as a result of convection. In both, the lapse rate in the free atmosphere is the same; it is especially high during the first few thousand meters but slows after that (as on a hot summer day).

In the first case (Fig. 6.24a), the air parcel is not very humid and thus the dry adiabatic rate applies throughout its ascent. By the time the air reaches 3000 meters (9900 ft), its temperature and density are the same as those of the surrounding atmospheric air. At this point, convectional uplift stops.

In the second case (Fig. 6.24b) we have introduced the latent heat of condensation. Here again the unsaturated rising column of air cools at the dry adiabatic rate of 10°C per 1000 meters (5.6°F/1000 ft) for the first 1000 meters (3300 ft). However, because the air parcel is humid, the rising air column soon reaches the dew point, condensation takes place, and cumulus clouds begin to form. As condensation occurs, the latent heat locked up in the water vapor is released, heating the moving parcel of air and retarding the adiabatic cooling so that the rising air now cools at the wet adiabatic rate (5°C/1000 meters). Here the temperature of the rising air parcel remains warmer than that of the atmospheric layer it is passing through and the air parcel will continue to rise. In this case, which incorporates the latent heat of condensation, massive amounts of condensation occur, reinforcing the towering cumulus clouds and their thunderstorm potential.

Convectional uplift can cause the heavy precipitation, thunder, lightning, and tornadoes that can accompany thunderstorms, especially in the spring and summer afternoons. The strong convectional updrafts that occur in towering cumulonimbus clouds frequently produce hail along with the thundershowers.

**Frontal Precipitation** The zone of contact between relatively warm and relatively cold bodies of air is known as a **front.** As was mentioned in the previous chapter, the concept of a weather *front* comes from the line of contact between oppos-

*FEMA Photo/Michael Raphael*

● **FIGURE 6.22** The coating of ice on the sign and the branches are from freezing rain during an ice storm. A glazing of ice like this also makes roads and sidewalks extremely slippery and hazardous. Tree branches can also break under the extra weight of the ice cover.

*Why are power failures a common occurrence with ice storms?*

● **FIGURE 6.23** Four uplift mechanisms for air. The principal cause of precipitation is upward movement of moist air resulting from convectional, frontal, cyclonic, or orographic lifting.

*What kind of air movement is common to all four diagrams?*

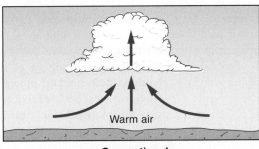

**Convectional**

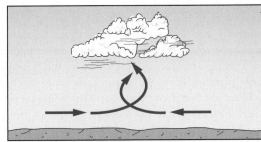

**Cyclonic (convergence)**

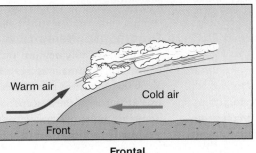

**Frontal**

**Orographic**

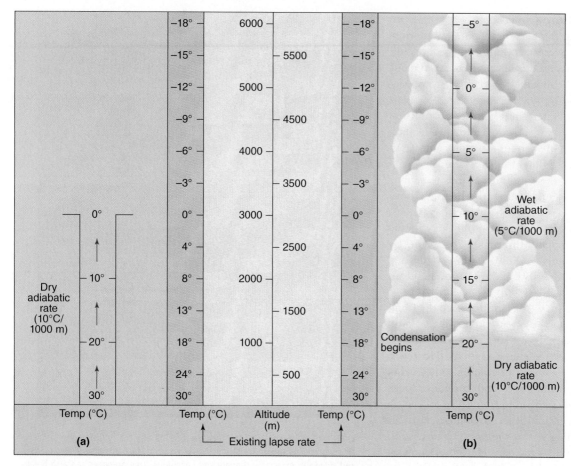

● **FIGURE 6.24** Effect of humidity on air mass stability. (a) Warm, dry air rises and cools at the dry adiabatic rate, soon becoming the same temperature as the surrounding air, at which point convectional uplift terminates. Because the rising dry air did not cool to its dew point temperature by the time convectional lifting ended, no cloud formed. (b) Rising warm, moist air soon cools to its dew point temperature. The upward moving air subsequently cools at the wet adiabatic rate, which keeps the air warmer than the surrounding atmosphere so that the uplift continues. Only when all moisture is removed by condensation will the air cool rapidly enough at the dry adiabatic rate to become stable.

*What would be necessary for the cloud in (b) to stop its upward development at 4500 meters?*

ing armies. When two large bodies of air that differ in temperature, humidity, and density collide, the warmer, less dense air mass is lifted above the colder body of air. Collision causes uplift and uplift results in cooling—producing condensation and precipitation. **Frontal precipitation** develops as warm, moisture-laden air collides with cold air and rises above the front. Depending on many factors, especially the temperature and moisture content of the clashing *air masses,* frontal precipitation can produce many kinds of weather, from cloudy overcast conditions to rain, snow, or ice storms.

Understanding fronts requires knowing what causes unlike bodies of air to collide and the weather conditions associated with different kinds of fronts. This will be discussed in Chapter 7, where we will take a more detailed look at frontal disturbances and precipitation.

## Cyclonic (Convergence) Precipitation

Cyclonic uplift (convergence) was introduced in Chapter 5 and involves air interacting with a cell of low pressure *(cyclone).* Air will flow from all directions into a low pressure system that is roughly circular, upward, and in a counterclockwise manner in the Northern Hemisphere. When air *converges* on a cyclone, it is pulled into the rising air of the low pressure cell. Therefore, cloudy conditions and precipitation are common around the center of a cyclone. Hurricanes and the rain associated with these storms **(cyclonic precipitation)** are fed by air flowing into the convergent uplift around a circular low pressure system and the energy resulting from the release of latent heat of condensation.

## Orographic Precipitation

When land barriers—such as a mountain, a hilly region, or the escarpment (steep edge) of a plateau or mountain range—lie in the path of prevailing winds, air is forced to rise above these barriers. Masses of air are cooled by expansion as they ascend over a topographic barrier and condensation takes place. The resultant precipitation is termed **orographic precipitation** (from Greek: *oros,* mountains).

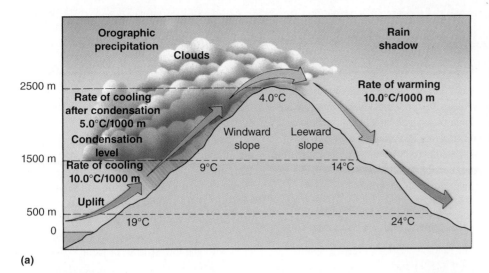

**(a)**

Orographic precipitation

Rain shadow

Clouds

2500 m

Rate of cooling after condensation 5.0°C/1000 m

Condensation level

4.0°C

Windward slope

Leeward slope

Rate of warming 10.0°C/1000 m

1500 m

Rate of cooling 10.0°C/1000 m

9°C

14°C

500 m

Uplift

0

19°C

24°C

R. Gabler

**(b)**

R. Gabler

**(c)**

● **FIGURE 6.25** Orographic precipitation and the rain shadow effect. (a) Orographic uplift over the windward (western) slope of the Sierras produces condensation, cloud formation, and precipitation, resulting in (b) dense stands of forest. (c) Semiarid or rain-shadow conditions occur on the leeward (eastern) slope of the Sierras.

*Can you identify a mountain range in Eurasia in which the leeward side of that range is in the rain shadow?*

As long as the air parcel rising up the mountainside remains stable (cooling at a greater rate than the normal lapse rate), any resulting cloud cover will be a type of stratus cloud. However, the situation can be complicated by the same circumstances illustrated in Figure 6.24b. A potentially unstable air parcel may need only the initial lift provided by an orographic barrier to set it in motion. In this case, it will continue to rise of its own accord (no longer forced) as it seeks air of its own temperature and density. Once the land barrier provides the initial thrust, it has performed its function as a lifting mechanism.

As orographic precipitation falls on the windward side of a mountain, the air parcel and the clouds lose some of their moisture content (the *absolute* or *specific humidity* declines). However, continued cooling as the air rises can maintain the relative humidity at 100%, so precipitation continues as long as the air contains adequate moisture and continues to rise. As the air descends the leeward slope, its temperature warms (at the dry adiabatic rate) and condensation ceases. The leeward side of an orographic barrier is called the **rain shadow** ( ● Fig. 6.25a). Just as being in a shadow means that you are not receiving any direct sunlight, being in the rain shadow means that an area does not receive much rain (or other precipitation). The effects of orographic precipitation and the rain shadow effect on the vegetation patterns (Figs. 6.25b and c) can be observed on a mountain range. The windward side of mountains (for example, the Sierra Nevada in California) will be heavily forested. The opposite slopes in the rain shadow will be drier, usually with a sparse cover of vegetation.

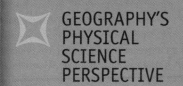
# The Lifting Condensation Level

When we look at clouds in the atmosphere, it is often easy to see their relatively flat bases. Cloud tops may appear irregular, but the bases of clouds are often flat. Even if the cloud bases do not seem flat, it will be obvious that the clouds you see all seem to have formed at the same level above the surface. This level represents the altitude to which the air must be lifted (and cooled at the dry adiabatic rate) before saturation is reached. Basically, this is the altitude at which rising air reaches its dew point temperature and condensation begins. Any additional lifting and clouds will form and build upward. Therefore, the height at which clouds form from lifting is called the lifting condensation level (LCL) and can be estimated by the equation:

$$\text{LCL (in meters)} = 125 \text{ meters} \times$$
$$(\text{Celsius temperature} - \text{Celsius dew point})$$

For example, if the surface temperature is 7.2°C (45°F) and the dew point temperature is 4.4°C (40°F), then the LCL is estimated at 350 meters (1148 ft) above the surface.

*Caution:* Keep in mind that different layers of clouds may exist at the same time. Low, middle, and high clouds as defined in this chapter may all appear on the same afternoon. These clouds may have formed in other regions and be only passing overhead. The formula presented here is best used with the lowest level of cloud cover that appears overhead.

J. Petersen

# Distribution of Precipitation
## Distribution Over Time

To understand the significance of a location's precipitation, consider *average annual precipitation* to get an impression of the moisture that a region gets during a year. Also examine the annual or monthly number of *rain days*—days when 1.0 millimeter (0.01 in) or more of rain were received during a 24-hour period. Less than this amount is known as a **trace** of rain.

By dividing the number of rain days in a month or year by the total number of days in that period, the resulting figure represents the probability of rain. Such a measure is important to farmers and to ski or summer resort owners whose incomes may depend on precipitation or the lack of it.

Another factor to be considered is the *average monthly precipitation.* By examining precipitation data for all 12 months, the seasonal variations in precipitation for a location can be determined ( ● Fig. 6.26). For instance, in describing the climate of California, average annual precipitation would not give the full story because an annual average would not show the distinct wet and dry seasons that characterize this region, but monthly averages would give us that information.

## Horizontal Precipitation Distribution

Great geographic variability exists in the global distribution of precipitation. The map in ● Figure 6.27 shows the average annual precipitation of Earth's land areas. Although there is a zonal distribution of precipitation related to latitude, this distribution is obviously not the only factor involved in the amount of precipitation an area receives.

The likelihood and amount of precipitation are based on two factors. First, precipitation depends on the degree of lifting that occurs in the air of a particular region. This lifting may be a result of the collision of different air masses (frontal), the convergence of air into a low pressure system (cyclonic or convergence), differential heating of Earth's surface (convection), the lifting that results when an air mass encounters a rise in Earth's surface (orographic), or a combination of these processes. The second factor

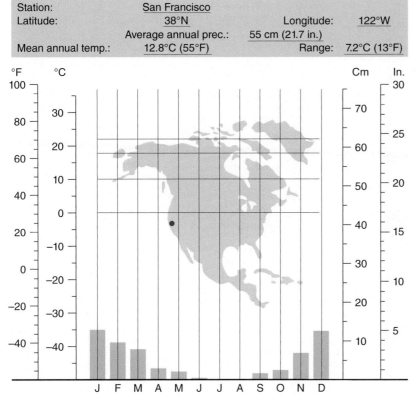

| Station: | San Francisco | | |
|---|---|---|---|
| Latitude: | 38°N | Longitude: | 122°W |
| | Average annual prec.: | 55 cm (21.7 in.) | |
| Mean annual temp.: | 12.8°C (55°F) | Range: | 7.2°C (13°F) |

● **FIGURE 6.26** Average monthly precipitation in San Francisco, California is represented by colored bars. A graph of monthly precipitation gives a good impression of the seasonal weather in a place. If we had only the annual precipitation total, we would not know that nearly all the precipitation occurs in only half of the year.

*How would this rainfall pattern affect agriculture?*

affecting the likelihood of precipitation depends on the internal characteristics of the air itself, including its degree of instability, its temperature, and its humidity.

Because higher temperatures allow air masses to hold greater amounts of water vapor and because, conversely, cold air masses can hold less water vapor, we can expect a general decrease of precipitation from the equator to the poles that is related to the unequal zonal distribution of incoming solar energy discussed in Chapter 3. Further study of the map in Figure 6.27 will reveal a great deal of variability in average annual precipitation beyond the general pattern of a decrease with increased latitude. The following discussion examines some of these variations and the reasons for them. We also apply what we have already learned about temperature, pressure systems, wind belts, and precipitation.

## Latitudinal Distribution

Latitudinal position has a strong impact on the distribution of precipitation because the occurrence, absence, and variations of many weather and climate factors are related to latitude. For example, because warmer air can hold more water vapor and colder air can hold less, there is a general decrease of precipitation from the equator to the poles.

The equatorial zone is generally an area of high precipitation—typically more than 200 cm (79 in) annually. High temperatures and instability in the equatorial region lead to a general pattern of rising air, which generates precipitation. This tendency is reinforced by the convergence of the trade winds as they flow toward the equator from opposite hemispheres. In fact, the *intertropical convergence zone* is one of the two great zones where air masses converge. The other is along the *polar front,* between air from the subtropical highs and from the polar regions.

In general, the air of the trade wind zones is stable compared to the instability in the equatorial region. Under the influence of these steady winds, few atmospheric disturbances would lead to convergent or convectional lifting. However, because the trade winds have an easterly flow, when they move onshore along east coasts or islands with high elevations they carry moisture from the oceans. Thus, within the trade wind belt, east coasts tend to be wetter than west coasts. In fact, where the air of the equatorial and trade wind regions—with its high temperatures and vast amounts of moisture—moves onshore from the ocean and meets a landform barrier, record rainfalls can be measured. The windward slope of Mount Waialeale on Kauai, Hawaii, at approximately 22°N latitude, holds the world's record for greatest average annual rainfall—1198 cm (471 in).

Moving poleward from the trade wind belts, we would encounter the subtropical high pressure zones where air is subsiding. As it sinks, it is warmed adiabatically, increasing its moisture-holding capacity and reducing the precipitation potential in this area. In fact, ● Figure 6.28 indicates that a drop in precipitation amounts corresponds to the latitudes of the subtropical high pressure cells. These zones of subtropical high pressure are where most of the great deserts of the world are located—in northern and southern Africa, Arabia, North America, and Australia. The exceptions to this subtropical aridity occur along the east sides of continents, where the subtropical high pressure cells are weak and wind direction is onshore. This exception is especially true of the monsoon regions.

In the zones of the westerlies, from about 35° to 65°N and S latitude, precipitation occurs largely from the collision of cold, dry polar air masses and warm, humid subtropical air masses along the polar front. Thus, there is much frontal precipitation in this zone.

In the middle latitudes, the continental interiors are drier than the coasts because they are farther away from the oceans. However, where air in the prevailing westerlies is forced to rise, as it does when it crosses the Cascades Range and Sierra Nevada of the Pacific Northwest and California, especially during the winter months, there is heavy orographic precipitation. Thus, in the middle latitudes, continental west coasts tend to be wet and precipitation decreases eastward toward the continental interiors. Along eastern coasts within the westerlies, precipitation usually increases once again because of proximity to humid air from the oceans. Here convection and the convergence associated with hurricanes bring precipitation, mainly in the summer months.

In the United States the interior lowlands are not as dry as we might expect within the prevailing westerlies. This is because

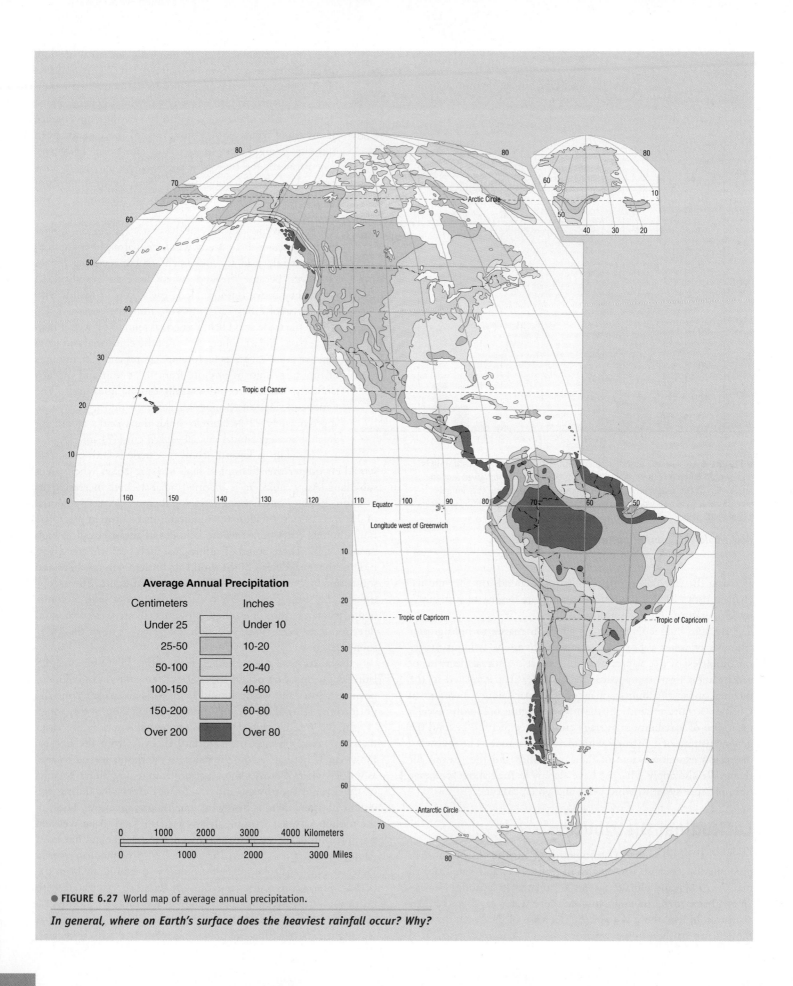

● **FIGURE 6.27** World map of average annual precipitation.

*In general, where on Earth's surface does the heaviest rainfall occur? Why?*

CHAPTER 6 • HUMIDITY, CONDENSATION, AND PRECIPITATION

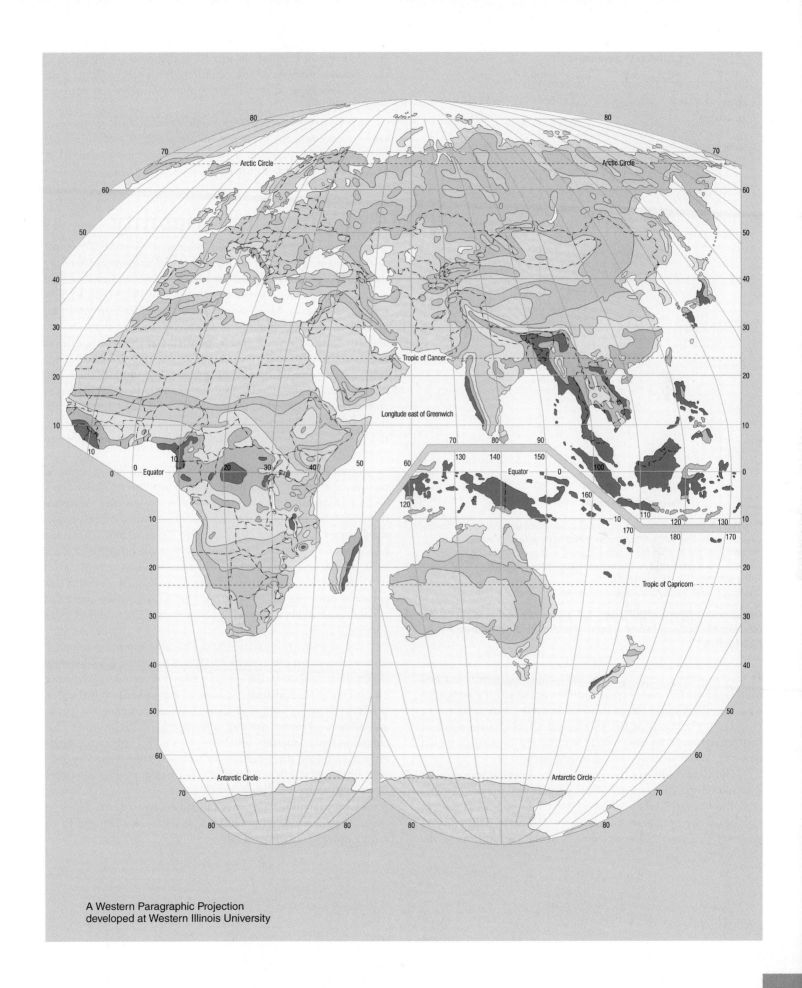

A Western Paragraphic Projection
developed at Western Illinois University

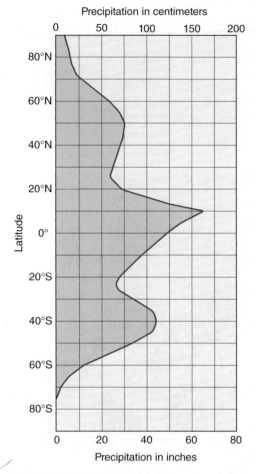

Precipitation in centimeters

● **FIGURE 6.28** Latitudinal distribution of average annual precipitation. Earth has four distinctive precipitation zones: the tropics, with high precipitation caused by air converging; the middle latitudes, with precipitation associated with the polar front; and two zones of low precipitation caused by subsiding air in the subtropical and polar regions.

Copyright © 2011 Brooks/Cole, Cengage Learning.

*Compare this figure with Figures 5.9 and 5.10a and b. What is the relationship between world rainfall patterns and world pressure distribution?*

of frontal activity resulting from the conflicting northward and southward movements of polar and subtropical air. If a high east–west mountain range extended from central Texas to northern Florida, the lowlands of the continental United States north of that range would be much drier because the mountains would block the moist air from the Gulf of Mexico.

Also characteristic of the westerlies are desert areas in the rain shadows of mountain ranges. This is one reason for the extreme aridity of California's Death Valley; the deserts of Nevada; eastern California; the mountain-ringed deserts of eastern Asia; and Argentina's Patagonian Desert, which is in the rain shadow of the

Andes. Note in Figure 6.28 that there is greater precipitation in the middle latitudes of the Southern Hemisphere, where the oceans cover more area than the continents, unlike the northern middle latitudes.

Moving poleward, low temperatures lead to low evaporation rates. In addition, the polar regions are generally areas of subsiding air and high pressure. These factors combine to cause low precipitation amounts in the polar zones.

# Precipitation Variability

The rainfall amounts depicted on the map in Figure 6.27 are annual averages. However, in many parts of the world there are significant variations in precipitation, both within any one year and between years. For example, areas like the Mediterranean region, California, Chile, South Africa, and Western Australia, located on the west sides of the continents, roughly between 30° and 40° latitude, receive much more rain in the winter than in the summer. Also some areas between 10° and 20° latitude receive much more of their precipitation in the summer (high-sun season) than in the winter (low-sun season). Rainfall totals can change markedly from one year to the next and tragically for many of the world's people. The drier a place is on the average, the greater will be the statistical variability in its precipitation (compare ● Fig. 6.29 with Fig. 6.27).

To make the situation worse for people in arid or semiarid regions, a year with a particularly high amount of rainfall may be more than offset by several years of below-average precipitation. This situation has occurred recently in West Africa's Sahel, the Russian steppe, and the American Great Plains. There can be years of drought and years of flood, each bringing its own kind of disaster. Farmers, construction workers, people who work in resorts, and others whose economic well-being depends in one way or another on weather are at the mercy of variable probabilities of rainfall on an annual, monthly, or even a seasonal basis.

We all know from watching the weather reports that rainfall cannot always be predicted with 100% accuracy. This inability results from the interaction of many factors involved in producing precipitation—temperature, moisture, atmospheric disturbances, landform barriers, fronts, air mass movement, upper air winds, and differential surface heating, among others. Variations in precipitation types, amounts, reliability, seasonality, and spatial distributions not only affect patterns of weather, climate, and vegetation, but also the soils and landforms of our planet. These weather factors also affect our daily lives. By using our knowledge of why certain precipitation types develop and where they are likely to occur, the patterns of environmental diversity that exist on Earth can be more easily understood.

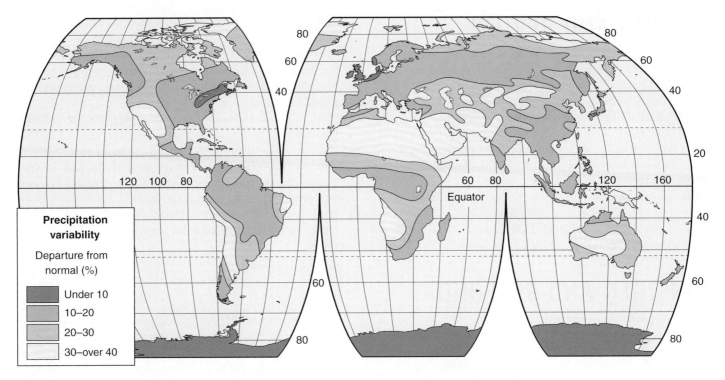

● **FIGURE 6.29** World map of precipitation variability. The greatest variability in annual precipitation totals occurs in the dry regions, accentuating the critical problem of moisture supply in those parts of the world.

*Compare this figure with the map in Figure 6.27. What are some of the similarities and differences?*

# CHAPTER **6** ACTIVITIES

## ■ DEFINE AND RECALL

| | | |
|---|---|---|
| capillary action | dew | collision–coalescence process |
| hydrologic cycle | frost | Bergeron (ice crystal) process |
| water budget | rime | supercooled water |
| saturation | precipitation | rain |
| capacity (moisture) | strato | drizzle |
| dew point | alto | snow |
| humidity | cirro | sleet |
| absolute humidity | cirrus | hail |
| specific humidity | cumulus | freezing rain (glaze) |
| relative humidity | stratus | uplift mechanism |
| transpiration | nimbus (nimbo) | convectional precipitation |
| potential evapotranspiration | adiabatic cooling | front |
| condensation nuclei | adiabatic heating | frontal precipitation |
| radiation fog (ground or temperature-inversion fog) | dry adiabatic lapse rate | cyclonic (convergence) precipitation |
| | wet adiabatic lapse rate | orographic precipitation |
| advection fog | instability | rain shadow |
| upslope fog | stability | trace (of rain) |

# DISCUSS AND REVIEW

1. How is the hydrologic cycle related to Earth's water budget?
2. What is the difference between absolute and specific humidity? What is relative humidity?
3. Imagine that you are deciding when to water a lawn or garden. What time of day would be best for conserving water? Why?
4. What factors must be taken into consideration when calculating potential evapotranspiration? How is evapotranspiration related to the water budget of a region?
5. What factors affect the formation of temperature inversion fogs?
6. What causes adiabatic cooling? Differentiate between the normal lapse rate and adiabatic lapse rates.
7. How and why does the wet adiabatic lapse rate differ from the dry adiabatic lapse rate?
8. How is atmospheric stability related to the adiabatic lapse rates?
9. What atmospheric conditions are necessary for precipitation to occur?
10. Find out how many inches of precipitation have fallen in your area this year. Is that average or unusually high or low?
11. Compare and contrast convectional, orographic, cyclonic (convergence), and frontal precipitation.
12. How is rainfall variability related to total annual rainfall? How might this relationship be considered a double problem for people?

# CONSIDER AND RESPOND

1. Refer to Figure 6.6.
   a. What is the water vapor capacity of air at 0°C? 20°C? 30°C?
   b. If a parcel of air at 30°C has an absolute humidity (actual water vapor content) of 20.5 grams per cubic meter, what is the parcel's relative humidity?
   c. If the relative humidity of a parcel of air is 33% and the air temperature is 15°C, what is the absolute humidity of the air in grams per cubic meter?
   d. A major concern of northern-climate residents is the low relative humidity within their homes during the winter. Low relative humidity is not healthy, and it has an adverse effect on the homes' furnishings. The problem results when cold air, which can hold little water vapor, is brought indoors and heated up. The following example will illustrate the problem: Assume that the air outside is 5°C and has a relative humidity of 60%. What is the actual water vapor content of this air? If it is brought indoors (through the doors, windows, and cracks in the home) and heated to 20°C, with no increase in water vapor content, what is the new relative humidity?
2. Recall that as a parcel of air rises, it expands and cools. The rate of cooling, termed the dry adiabatic lapse rate, is 10°C per 1000 meters. (A descending parcel of air will always warm at this rate.) In addition, the dew point temperature decreases about 2°C per 1000 meters within a rising parcel of air. At the height at which the dew point temperature is reached, condensation begins and, as discussed in the text, the wet adiabatic lapse rate of 5°C per 1000 meters becomes operational. When the wet adiabatic lapse rate is in operation, the dew point temperature will be the same as the air temperature. When a parcel descends through the atmosphere, its dew point temperature increases 2°C per 1000 meters. The height at which condensation begins, termed the lifting condensation level (LCL), can be determined by using the formula found in Geography's Physical Science Perspective: The Lifting Condensation Level box.
   a. A parcel of air has a temperature of 25°C and a dew point temperature of 14°C. What is the height of the LCL? If that parcel were to rise to 4000 meters, what would be its temperature?
   b. A parcel of air at 6000 meters has a temperature of −5°C and a dew point of −10°C. If it descended to 2000 meters, what would be its temperature and dew point temperature?

# APPLY AND LEARN

1. Using the following data set, for each month of the year calculate: (a) a running total of precipitation month to month, (b) the departure from the mean value (in surplus or deficit) for each month, and (c) the annual departure from the mean value (in surplus or deficit) for the whole year.
   How does the year begin with respect to surplus or deficit? How does the year end?
   Which month accumulated the greatest deficit for the year? Which month is the first to show a surplus?

| Month | Recorded Rainfall (inches) | Mean Rainfall (inches) |
|---|---|---|
| January | 4.94 | 4.94 |
| February | 2.40 | 4.10 |
| March | 3.02 | 5.26 |
| April | 3.21 | 4.36 |
| May | 3.69 | 4.22 |
| June | 2.69 | 4.17 |
| July | 5.08 | 4.28 |
| August | 5.41 | 3.62 |
| September | 7.89 | 3.19 |
| October | 4.10 | 2.68 |
| November | 4.27 | 3.72 |
| December | 2.77 | 4.36 |

2. Using the following data set, answer the following questions: (a) Following the procedure described in the section Distribution of Precipitation, calculate the probability of precipitation for each month of the year. (b) Which month has the highest probability? Which has the lowest probability? (c) What is the average probability for rainfall for the year?

| Month | Number of Days with >1 mm (0.01 in.) of Rain |
|---|---|
| January | 12 |
| February | 10 |
| March | 7 |
| April | 11 |
| May | 16 |
| June | 11 |
| July | 8 |
| August | 9 |
| September | 10 |
| October | 10 |
| November | 11 |
| December | 9 |

# LOCATE AND EXPLORE

1. Using Google Earth, fly to the Aral Sea (45.2°N, 59.9°E) on the Uzbekistan–Kazakhstan border. Once you arrive at your coordinates, zoom out to view the extent of the Aral Sea, which was once one of the largest lakes in the world. As a result of irrigation projects and stream diversions, much of the water flowing into the lake was cut off and the lake has been significantly reduced in size. You can see the outline of the lake before it was reduced in size. Assuming that the lake can be characterized as a rectangle (area = length × width), what has been the change in the lake's area (in square miles and as a percentage of the original lake area)? Tip: Use the ruler tool to measure the width and length.

 CourseMate - Make the most of your study time by accessing everything you need to succeed in one place. Read your textbook, take notes, review flashcards, watch videos, complete activities, take practice quizzes, and more online with CourseMate. Log in at **www.cengagebrain.com**.

# 7

# AIR MASSES AND WEATHER SYSTEMS

## ■ OBJECTIVES

WHEN YOU COMPLETE THIS CHAPTER YOU SHOULD BE ABLE TO:

■ Outline and explain the major air mass types, their characteristics, and their source regions.
■ Describe all four types of fronts and the types of weather that occur with their passage or presence.
■ Distinguish between the general atmospheric conditions associated with anticyclones and cyclones.
■ Discuss the characteristics of a middle-latitude cyclone, the factors that influence its movement, and its stages of development.
■ Understand the potentially serious weather conditions, possible damage, and hazards associated with hurricanes, thunderstorms, and tornadoes.
■ Explain the difference between weather and climate, and be aware of the factors that make weather forecasting a complex process.
■ Interpret a weather map and understand the symbols that are used to show atmospheric conditions, including pressure, winds, and storm types.

IN THIS CHAPTER WE WILL apply much of what we have learned about insolation, heat, temperature, pressure, wind, and moisture conditions as we examine weather systems and the kinds of storms (atmospheric disturbances) that accompany them. We will explore storm types and characteristics, origins, and development. Understanding Earth's varied weather systems—their general characteristics, when they may occur, how they form, and how they may have an impact on affected regions—is important in the study of physical geography. In addition to causing temperature changes and producing precipitation, weather systems are a major means of energy exchange. Weather systems can also present hazards, such as floods, damaging winds, lightning, and violent thunderstorms. We will begin the chapter with a detailed study of air masses and fronts, two atmospheric elements that not only affect weather systems, but also strongly influence regional climates and environmental diversity. These two important topics will be considered in forthcoming chapters.

Hurricane Norbert swirls toward the Baja Peninsula of Mexico. In the United States we tend to think most often about hurricanes along the coasts of the Atlantic Ocean and the Gulf of Mexico, but they affect many other coastal regions of the world. *NASA/GSFC MODIS Rapid Response Team, Jeff Schmaltz*

# Air Masses

An **air mass** is a large body of air, subcontinental in size, that is relatively homogeneous in terms of temperature and humidity. The regional extent of an air mass, however, may be 20 or 30 degrees of latitude, so temperature and humidity variations exist from its poleward edges to its equatorward edges. The characteristics, locations, and movement of an air mass exert considerable impacts on the weather. Changes caused by contact with differing land and ocean surfaces also affect the characteristics of air masses. As a result of this temperature and moisture uniformity, the density of air will be much the same throughout any one level within an air mass.

The characteristics of temperature and humidity within an air mass are determined by the nature of its **source region**—the place where the air mass originates. Good source regions for generating air masses share several characteristics. For an air mass to have similar temperature and humidity characteristics, the source region must have a nearly homogeneous surface. For example, source regions include deserts, low-relief continental areas, ice sheets, or oceans but not mountainous areas or locations with a combination of surfaces. In addition, an air mass must have sufficient time to acquire the characteristics of the source region. Hence, gently settling, slowly diverging air will mimic a source region, whereas air that is converging and rising will not.

On weather maps used by climatologists and meteorologists, air masses are identified by a simple letter code. The first part is a lowercase letter. There are two choices: The letter *m* refers to **maritime,** which means that the air mass originated over the ocean and therefore these air masses are relatively moist. The letter *c* refers to **continental,** which means that the air mass originated over land and therefore these air masses are relatively dry. The second letter is always a capital letter that helps to locate the latitudinal area of the source region. *E* stands for equatorial and the air is very warm. The letter *T* identifies a tropical origin and is a warm air mass. The letter *P* represents polar, and this air can be quite cold, depending on its maritime or continental origin. Last, an *A* identifies arctic air (AA for antarctic), which is very cold. These six letters can be combined to give us the classification of air masses that is in use today: **maritime equatorial (mE, or sometimes just E), maritime tropical (mT), continental tropical (cT), continental polar (cP), maritime polar (mP),** and **continental arctic (cA).** The origin and typical movement of the five air mass types that affect North America are shown in ● Figure 7.1, and all six types are described in some detail in Table 7.1. From now on we will frequently use these symbols to refer to the different air mass types rather than using the full names.

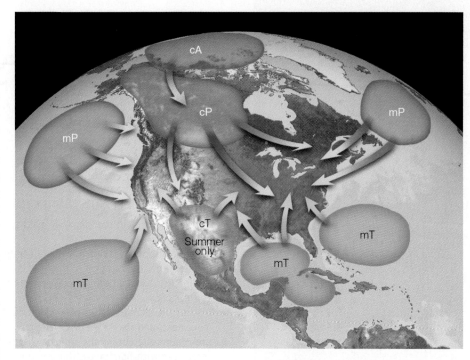

● **FIGURE 7.1** Source regions of North American air masses. Air mass movements import the temperature and moisture characteristics of these source regions into far distant areas.

*Use Table 7.1 and this figure to determine which air masses affect your location. Are there seasonal variations?*

## Modification and Stability of Air Masses

Air masses require sufficient time to develop, basically acquiring the temperature and humidity characteristics of their source region. Yet air masses do not remain stationary over their source regions indefinitely as a result of the general atmospheric circulation. As an air mass travels over Earth's surface, it generally retains its distinct characteristics. Temperature and humidity modifications, however, occur as the air mass gains or loses thermal energy or moisture by interacting with landmasses or water bodies. This gain or loss of thermal energy, humidity, or both can either make an air mass more stable or cause it to become unstable.

Air mass modification can also involve moisture content. For example, in North America during the early-winter to midwinter seasons, cold, dry cP or cA air from Canada can move southeastward across the Great Lakes region. While passing over the lakes, the air mass can pick up moisture, thus increasing its humidity. This modified cP or cA air reaches the frigid leeward shores of the Great Lakes and the cooling causes precipitation, generally referred to as **lake effect snow,** which can be heavy ( ● Fig. 7.2). The areas of snow cover may appear as *snow belts,* bands of snow that extend downwind from the lakes. The chances for lake effect snow events diminish in late winter as the lakes freeze over, thus cutting off the moisture supply to the air masses flowing across them.

## TABLE 7.1
### Types of Air Masses

| Source | Region | Usual Characteristics at Source | Accompanying Weather |
| --- | --- | --- | --- |
| Maritime Equatorial (mE) | Equatorial oceans | Ascending air, very high | High temperature and humidity, heavy moisture content rainfall; never reaches the United States |
| Maritime Tropical (mT) | Tropical and subtropical oceans | Subsiding air; fairly stable but some instability on western side of oceans; warm and humid | High temperatures and humidity, cumulus clouds, convectional rain in summer; mild temperatures, overcast skies, fog, drizzle, and occasional snowfall in winter; heavy precipitation along mT/cP fronts in all seasons |
| Continental Tropical (cT) | Deserts and dry plateaus of subtropical latitudes | Subsiding air aloft; generally stable but some local instability at surface; hot and very dry | High temperatures, low humidity, clear skies, rare precipitation |
| Maritime Polar (mP) | Oceans between 40° and 60° latitude | Ascending air and general instability, especially in winter; mild and moist | Mild temperatures, high humidity; overcast skies and frequent fogs and precipitation, especially during winter; clear skies and fair weather common in summer; heavy orographic precipitation, including snow, in mountainous areas |
| Continental Polar (cP) | Plains and plateaus of subpolar and polar latitudes | Subsiding and stable air, especially in winter; cold and dry | Cool (summer) to very cold (winter) temperatures, low humidity; clear skies except along fronts; heavy precipitation, including winter snow, along cP/mT fronts |
| Continental Arctic (cA) | Arctic Ocean, Greenland, and Antarctica | Subsiding very stable air; very cold and very dry | Seldom reaches United States, but when it does, bitter cold, subzero temperatures, clear skies, often calm conditions |

## North American Air Masses

Five types of air masses *(cA, cP, mP, mT,* and *cT)* influence the weather of North America. Because the middle latitudes are located between several source regions, a great many storms and precipitation events in that latitudinal zone involve the collision of unlike air masses (see Fig. 7.1). Further, as the source regions change with the seasons, primarily because of changing insolation, the air masses also will vary according to the season.

Most of us are familiar with the weather of the United States or Canada; therefore, this chapter will concentrate on the air masses of North America and their impacts on weather conditions. The processes involving North American air masses are also generally applicable worldwide and are important to understanding the global climate regions that will be addressed in the following chapters.

**Continental Arctic Air Masses (cA)** The frozen Arctic Ocean in winter and the frigid land surface in far northern Canada and Alaska serve as source regions for this air mass type. These air masses are extremely cold, very dry, and very stable. Although *cA* air masses will affect parts of Canada, during the winter if these air masses become extremely well-developed, they will occasionally travel far enough south to affect the United States. When continental arctic air extends down into the midwestern or even the southeastern United States from Texas to

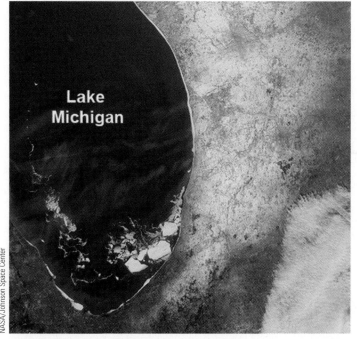

● **FIGURE 7.2** This lake effect snow accumulated on the eastern shore of Lake Michigan as a storm moved west to east.

*NASA/Johnson Space Center*

*What two main factors contribute to the increased precipitation caused by the lake effect?*

Florida, record-setting cold temperatures can be experienced. If a *cA* air mass remains in regions that are not accustomed to extreme cold for extended periods, vegetation can be severely damaged or killed. Water pipes may also freeze and break because they often are not as well insulated compared to those in regions that expect hard freezes in the winter.

### Continental Polar Air Masses (cP)

At their source region in northern-central North America, *cP* air masses are cold, dry, and stable, resulting in clear, cold weather. Because North America has no major east–west trending landform barriers, *cP* air can migrate to southern Canada and as far south as the Gulf of Mexico or southern Florida. The movement of a continental polar air mass into the midwestern or the southern United States brings a cold wave characterized by clear, dry air and colder-than-average temperatures. Although the air temperature in an air mass warms as it moves south, continental polar air masses can bring freezing conditions as far south as Texas and Florida. The westerly circulation of the middle latitudes, along with mountain barriers, rarely allows a *cP* air mass to move westward to the West Coast. When a *cP* air mass does reach Washington, Oregon, or California, however, it brings with it unusual freezing temperatures that can cause great agricultural damage.

### Maritime Polar Air Masses (mP)

During winter months, when the oceans tend to be warmer than the land, *maritime polar* air masses, although they are damp and cool, tend to be warmer than the air masses on land (*cP* air masses). Maritime polar air masses develop in the northern Pacific Ocean and move with the westerly circulation to affect the weather of southwestern Canada and the western United States, particularly in the Northwest. When *mP* air is affected by an uplift mechanism (such as a mass of colder, denser air, or a mountain range), the result is usually cloudy weather and precipitation (snow during the winter in inland and higher elevation areas). Maritime polar air masses may also continue eastward, becoming the source of moisture for snowstorms after crossing the western mountain ranges.

Generally, the *mP* air masses that develop over the northern Atlantic Ocean do not affect the weather of the United States because the westerlies push those air masses toward Europe. On some occasions, however, a strong low pressure cell may stall off the north Atlantic Coast. Cyclonic winds from the poleward side of the cyclonic system cause a northeasterly onshore flow, with cool or cold, damp winds, and rain or heavy snows. Known as **nor'easters** because the storms come from the northeast, these weather systems can bring serious winter conditions to the New England states. The weather during a nor'easter typically includes strong winds, heavy snowfall, and coastal damage from high waves that are driven by winds blowing onshore.

### Maritime Tropical Air Masses (mT)

The Gulf of Mexico and the subtropical areas of the Atlantic and Pacific Oceans are source regions for *mT* air masses. Long days and intense insolation during summer produce air masses in the *mT* source regions that are warm and humid. During the summer, however, the land is even warmer than the *mT* air masses. This difference in temperature results in convective uplift that causes precipitation and strong thunderstorms on hot, humid days. Maritime tropical air masses are responsible for much of the hot, humid summertime weather in wide areas of the United States that are east of the Rocky Mountains, particularly in the southern and eastern parts of this region.

During the wintertime, the tropical and subtropical oceans and the Gulf of Mexico remain warm with warm and humid air aloft. As this warm, moist air moves northward into the south-central United States, it travels over increasingly cooler land surfaces. The lower layers of air are chilled and the drop in temperature increases the relative humidity, often resulting in fog. If a tropical air mass reaches *cP* air migrating southward from Canada, the warm *mT* air will be forced to rise over the colder, drier *cP* air and frontal precipitation will occur.

Maritime tropical air masses also form in the subtropical latitudes over the Pacific Ocean. These air masses tend to be slightly cooler than those that form over the Gulf of Mexico and the Atlantic, partly because of their passage over the cooler California Current. A Pacific *mT* air mass is also more stable because of the strong subsidence associated with the eastern portion of the Pacific subtropical high. This air mass contributes to the dry summers of Southern California but occasionally brings moisture in winter as it rises over the mountains of the Pacific Coast.

### Continental Tropical Air Masses (cT)

A fifth type of air mass affects North America, but it is mainly important to the weather of the southwestern United States. This is the *cT* air mass that develops over large, homogeneous land surfaces in the subtropics. The weather typical of the *cT* air mass is usually very hot and dry, with clear skies and major heating from the sun during the daytime.

The Sahara Desert of North Africa is a prime example of a source region for this type of air mass. In North America there is little land in the latitudes to serve as a source region for a *cT* air mass of a great proportion compared to the other air mass types that affect the continent. A *cT* air mass can form over the deserts of the southwestern United States and central Mexico in the summer. In the source region, a *cT* air mass provides clear weather that is hot and dry (sometimes very hot and very dry). When *cT* air masses move eastward, however, they are usually modified by contact with larger, stronger air masses with lower temperature and higher humidity or by passing over bodies of water. At times, *cT* air from Mexico and Texas meets with *mT* air from the Gulf of Mexico. This boundary is known as a *dry line*. Here, the drier air is denser and will lift the moister air over it. This mechanism of uplift may act as a trigger for precipitation episodes and perhaps thunderstorm activity.

# Fronts

Air masses migrate with the general atmospheric circulation. The middle latitudes are where the clash of unlike air masses is most common and frequent. Air masses differ primarily in their temperature and in their moisture content. When different air masses come together, they do not mix easily but instead come in contact along sloping boundaries called *fronts*.

The sloping surface of a front develops as a warmer, lighter air mass is lifted, rising above a cooler, denser air mass. This forced rising of air, known as *frontal uplift,* is a major source of precipitation in middle-latitude countries, where contrasting air masses are most likely to converge. The United States and southern Canada are located between the source regions for five different air masses, all of which migrate seasonally.

The steepness of the frontal surface is governed primarily by the degree of difference and relative rate of advancement of the two converging air masses. When two strongly contrasting air masses converge—for example, when a warm and humid *mT* air mass meets a cold, dry *cP* air mass—the frontal surface tends to be steep, with strong frontal uplift. Given similar temperature conditions and moisture content, a steep slope, with greater frontal uplift, will produce heavier precipitation than will a gentler slope.

Fronts are differentiated based on whether a colder air mass is moving in on a warmer one or vice versa. The weather that occurs along a front also depends on which air mass is the "aggressor." Clashing air masses form a frontal zone that can cover an area from 2 to 3 kilometers (1–2 mi) wide to as wide as 150 kilometers (90 mi). Although weather maps use one-dimensional line symbols to separate two different air masses, a front is actually a three-dimensional surface with length, width, and height. Generally, then, it is more accurate to speak of a frontal *zone* rather than a frontal *line*.

## Cold Fronts

A **cold front** occurs when a cold air mass actively moves in on a warmer air mass, pushing it upward. Because the colder air is denser than the warm air it displaces, it stays at the surface and forces the warmer air to rise. As we can see in ● Figure 7.3, cold fronts usually have a relatively steep slope. For example, the warm air may rise 1 meter vertically for every 40 to 80 meters of horizontal distance. If a warm air mass is unstable and has a high moisture content, heavy precipitation can result, sometimes in the form of violent thunderstorms. **Squall lines** result when several storms align themselves along a cold front. Cold fronts are usually associated with strong weather disturbances and sharp changes in temperature, air pressure, and wind.

## Warm Fronts

When a warmer air mass is the aggressor, invading a region occupied by a colder air mass, a **warm front** forms. At a warm front, warmer air slowly pushes against the cold air and rises over the colder, denser air mass at the surface. The slope of a warm front is much gentler than that of a cold front. For example, the warm air may rise only 1 meter vertically for every 100 or even 200 meters of horizontal distance. Thus, the uplift along a warm front will not be as strong as that occurring along a cold front. The result is that the weather associated with the passage of a warm front tends to be less violent and the changes are less abrupt than those associated with cold fronts.

● Figure 7.4 shows why an advancing warm front affects the weather of areas well in advance of the surface location of the front. The cloud types that precede a front typically indicate the weather changes that can be expected as a front approaches.

## Stationary and Occluded Fronts

When a frontal boundary fails to move in any appreciable direction as air masses converge, this forms a **stationary front.** Under the influence of a stationary front, locations may experience clouds, drizzle, and rain (or possible thunderstorms), sometimes for several days. A stationary front and its accompanying weather will remain until it dissipates as the contrasts between the two air masses diminish or the atmospheric circulation finally causes one of the air masses to move.

An **occluded front** occurs when a faster moving cold front overtakes a warm front, pushing the warmer air aloft. This frontal situation usually occurs in the latter stages of a storm and on the poleward side of a middle-latitude cyclone, a storm type that will be discussed next. Areas under an occluded front tend to experience gray, overcast skies and perhaps light precipitation. Map symbols for the four frontal types are shown in ● Figure 7.5.

● **FIGURE 7.3** Cross-section of a cold front. Cold fronts generally move rapidly, with a blunt forward edge that drives adjacent warmer air upward. This rapid uplift can produce intense precipitation from moisture supplied by the warmer air.

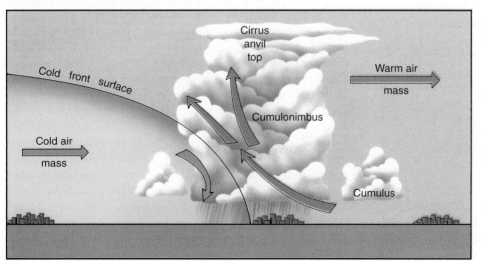

Cirrus anvil top

Warm air mass

Cold front surface

Cold air mass

Cumulonimbus

Cumulus

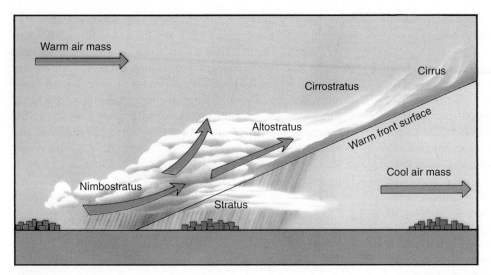

● **FIGURE 7.4** Cross-section of a warm front. Warm fronts advance more slowly compared with cold fronts as the warm air slides upward over a wedge of colder air. The gentle rise of the warm air produces stratus clouds and gentle rain.

*Compare Figures 7.3 and 7.4. How are they different? How are they similar?*

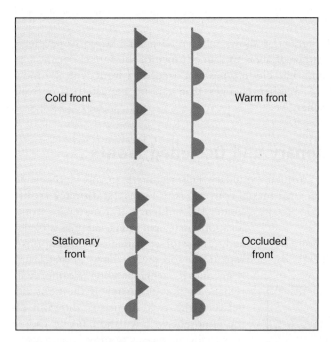

● **FIGURE 7.5** The four major frontal symbols used on weather maps. The "teeth" point in the direction that an airmass on the other side of the line is moving or attempting to move.

# Atmospheric Disturbances

## Anticyclones and Cyclones

We have previously distinguished anticyclones and cyclones according to differences in pressure and wind direction. We also have identified large areas of semi-permanent cyclonic and anti-

cyclonic circulation in Earth's atmosphere (the subtropical high, for example). Secondary circulations, storms, and other atmospheric disturbances are embedded within the wind belts of the general atmospheric circulation. The term **atmospheric disturbance** is used because it is a more general term than "storm" and includes atmospheric conditions that cannot be classified as storms. Now, when examining middle-latitude atmospheric disturbances, we can use the terms *anticyclone* and *cyclone* to refer to moving cells of high and low pressure, which change with the seasons and drift along the path of the prevailing winds. It is important to remember that in a cyclone pressure decreases toward the center and in an anticyclone pressure increases toward the center. We should note here that the atmospheric pressures in these two systems (highs and lows) are relative. As systems of higher pressure, anticyclones are usually characterized by clear skies, gentle winds, and a general lack of precipitation. Cyclones are centers of converging, rising air, often with cloudy, windy conditions, with good chances for precipitation. Cyclones include hurricanes, tornadoes, and the storms of the middle latitudes with their associated fronts of various types. The wind intensities involved in cyclonic or anticyclonic systems will depend on the steepness of the *pressure gradient*, which is determined by how much change in air pressure occurs over a horizontal distance.

**Mapping Pressure Systems** Cells of high and low pressure are easy to visualize if we imagine these pressure systems as if they were land surfaces. A cyclone is shaped like a basin ( ● Fig. 7.6). Winds converging toward the center of a cyclone will move rapidly if the pressure gradient is great, just as water will flow faster into a natural basin if the sides are steep and the depression is deep. If we visualize an anticyclone as a hill or mountain, then we can see that air diverging from an anticyclone will flow at speeds directly related to how high the pressure is at the center of the cell, in a manner similar to water moving down mountain slopes at speeds related to the landform's height. Thus, if there is a steep pressure gradient in a cyclone, with the pressure much lower at the center than at the outer portions of the system, the winds will converge toward the center with considerable velocity.

On a surface weather map, cyclones and anticyclones are depicted by roughly concentric isobars of increasing pressure toward the center of a high and of decreasing pressure toward the center of a low. A high pressure cell will typically cover a larger area than a low, but both pressure systems are capable of covering and affecting extensive areas. There are times, for example, when nearly the entire midwestern United States is under the influence of the same system. The average diameter of an anticyclone is about 1500 kilometers (900 mi) and a cyclone averages about 1000 kilometers (600 mi).

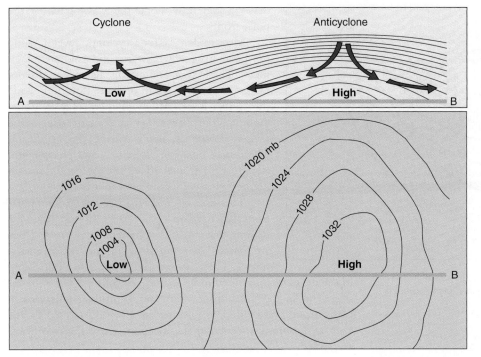

● **FIGURE 7.6** The horizontal and vertical structure of pressure systems. Close spacing of isobars around a cyclone or anticyclone indicates a steep pressure gradient that will produce strong winds. Wide spacing of isobars indicates a weaker system.

*Where would be the strongest winds in this figure? Where would be the weakest winds?*

**Anticyclones** An anticyclone is a high pressure area with subsiding air in the center, which displaces surface air with winds blowing outward, away from the center of the system. Hence, an anticyclone has diverging winds, with atmospheric pressure decreasing toward the outer limits of the system. Anticyclones tend to be fair weather systems because of the temperature and stability increase that occurs in subsiding air, reducing the possibility of condensation.

Air subsidence in the center of an anticyclone encourages stability because the air is warmed adiabatically, increasing its capacity to hold moisture. Although the weather resulting from the high pressure influence of an anticyclone is often clear, with no rainfall, there are certain conditions under which some precipitation can occur within a high pressure system. When such a system passes near or crosses a large body of water, the resulting evaporation can cause variations in humidity significant enough to result in some precipitation.

The relatively high pressures associated with anticyclones in the middle latitudes of North America have two sources. Some anticyclones move into the middle latitudes from northern Canada and the Arctic Ocean in what are called **polar outbreaks** of frigid continental polar or even continental arctic air. These outbreaks can be extensive, covering much of the midwestern and eastern United States, and they occasionally push into the sub-tropical Gulf Coast areas. The temperatures in an anticyclone that developed in a *cP* or *cA* air mass can be markedly lower than those expected for any given time of year, dipping far below freezing in the winter. Such an outbreak would typically be preceded by the squalls, clouds, and rain or snow associated with a cold front. A period of cold or cool, clear, and fair weather follows the frontal passage as the influences of modified polar air are felt. Other anticyclones are generated in the subtropical high pressure regions. When they move across the United States toward the north and northeast, they bring waves of hot, clear weather in summer and unseasonably warm days in the winter.

**Cyclones** A cyclone is a low pressure area that is typified by uplifting air and winds that tend to converge on the center of the low in an attempt to equalize pressure. As this air flows toward the center of a low pressure system, it feeds air into an upward spiral of air (convergent uplift, also known as cyclonic uplift), which can produce clouds and precipitation. Compared to anticyclones, cyclones are much more varied and complex in the ways that they form and in the weather that they generate. Low pressure systems generate storms and precipitation of all kinds through the adiabatic cooling of rising air and the resultant condensation that can occur. The types of cyclonic storms, their impacts, and their associated weather phenomena that will be discussed in detail in this chapter will focus on North American locations. The discussion of these storms, however, applies just as well to middle-latitude European locations.

**General Movement** The cyclones and anticyclones of the middle latitudes are steered, or guided, along paths influenced by the upper air westerlies (or the jet stream). Although the upper air flow can be variable and have large oscillations, a general west-to-east pattern prevails. As a result, people in the United States look at the weather occurring to the west to see what they might expect in the next few days. Most storms that develop in the Great Plains or on the West Coast move across the United States during a period of a few days at an average speed of 36 kilometers per hour (23 mph) and then travel into the North Atlantic.

Although neither cyclones nor anticyclones develop in exactly the same places at the same times each year, they do tend to arise in certain areas or regions more frequently than in others. Depending on the season, they also follow generally similar paths, known as **storm tracks** ( ● Fig. 7.7). In addition, atmospheric disturbances that develop in the middle latitudes during the winter are greater in number and intensity because the temperature variations between air masses are stronger during the winter months.

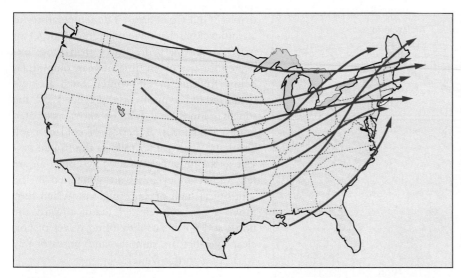

● **FIGURE 7.7** Common storm tracks for the United States. Virtually all middle-latitude cyclonic storms move from west to east with the prevailing westerlies and swing northeastward across the Atlantic coast.

*What storm tracks influence your location?*

## Middle-Latitude Cyclones

Because they are so important to the weather of North America, we will concentrate on examining **middle-latitude cyclones,** also known as **extratropical cyclones** ("extratropical" means outside the tropics). These migrating storms, typically formed of opposing cold, dry polar air and warm, humid tropical air, can cause significant variation in the day-to-day weather of the areas they affect. It is not unusual in some parts of the United States and Canada for people to experience a beautiful warm day in early spring followed by falling snow the next morning. Variability is common for middle-latitude weather, especially during fall and spring when conditions can also change from a period of cold, clear, dry days to a period of snow, only to be followed by one or two more moderate but humid days.

The weather associated with middle-latitude cyclones can vary widely with the seasons and with air mass conditions, and no two storms are ever identical. Storms vary in intensity, longevity, speed of travel, wind strength, amount and type of cloud cover, the quantity and type of precipitation, and the area they affect. Yet there is a useful model that describes and generalizes the typical characteristics of a middle-latitude cyclonic storm.

Shortly after World War I, Norwegian meteorologists Jacob Bjerknes and Halvor Solberg proposed the *polar front theory,* concerning the development, movement, and dissipation of middle-latitude storms. They recognized the middle latitudes as a region where different air masses, such as cold polar air and warm subtropical air, commonly meet at a boundary called the *polar front.* Although the polar front may be a continuous boundary that circles Earth, it is typically fragmented into several frontal segments. The polar front moves north and south with the seasons and is stronger in winter than in summer. The upper air westerlies (see Figs. 5.17 and 5.18), also known as the *polar jet stream,*

develop and flow along the wavy track of the polar front.

Most middle-latitude cyclones develop along the polar front where warm and cold air masses meet. These two contrasting air masses do not merge but may move in opposite directions along the frontal zone. Although there may be some slight uplift of the warmer air along the edge of the denser, colder air, uplift will not be significant. There may be some cloudiness and precipitation along such a frontal zone, but the conditions are not yet what we would refer to as a storm.

The line of convergence along the polar front may develop a wave form (in map view) for reasons that are not completely understood but are related to wind flow in the upper troposphere. These wave forms represent an initial step in the development of a middle-latitude cyclone ( ● Fig. 7.8). At this bend in the polar front, warm air is pushing poleward (a warm front) and cold air is pushing equatorward (a cold front), with a low pressure center at the location where the two fronts are joined.

As the contrasting air masses compete for position, clouds and precipitation increase along the fronts, spreading over a larger area. Precipitation along the cold front will be less widespread than that along the warm front, but it will be more intense. One factor that affects the kind of precipitation that occurs at the warm front is the stability of the warm air mass. If it is relatively stable, then its movement over the cold air mass may cause only a fine drizzle or a light, powdery snow if the temperatures are cold enough. In contrast, if the warm air mass is moist and unstable, uplift may set off heavier precipitation. As you can see by referring again to Figure 7.4, the precipitation that falls at a warm front may *appear* to be coming from the colder air. Although the weather may feel cool and damp, the precipitation is actually coming from the overriding warmer air mass above and then falling through the colder air mass to reach Earth's surface.

Because a cold front typically moves faster, it will eventually overtake the warm front. This produces an *occluded front.* Once this occurs, the system will soon die because the temperature, pressure, and humidity differences that powered the storm diminish at the occluded front. Occlusions are usually accompanied by cloudy overcast conditions and light rain (or snow). Occlusions represent the dissipation of a middle-latitude cyclone, at least in the poleward section of the storm.

**Middle-Latitude Cyclones and Weather** The various sections of a middle-latitude cyclone generate different weather conditions. Therefore the weather that a location experiences depends on which portion of the middle-latitude cyclone is over the location. Because the entire cyclonic system tends to travel as a unit from west to east, a specific sequence of weather can be expected at a given location as the cyclone passes.

Let's examine the typical passage of a middle-latitude cyclone during the late spring by following a westerly track (see

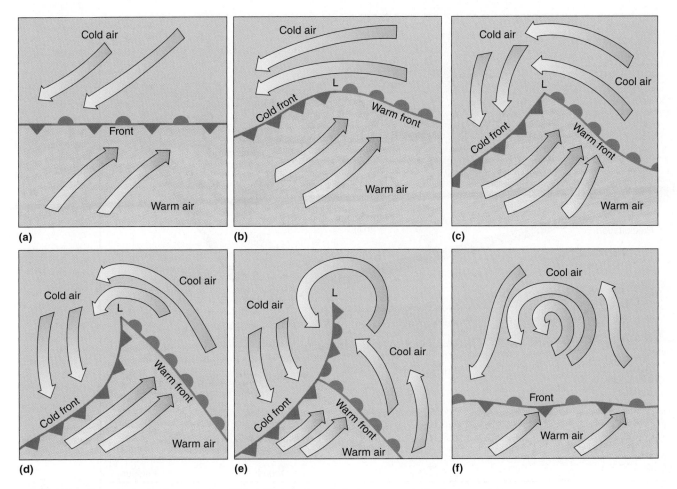

● **FIGURE 7.8** Stages in the development of a middle-latitude cyclone. Each view represents the development in a position somewhat eastward of the preceding view as the cyclone travels along its storm track. It is important to know that the storm and its associated fronts, wind directions, and air masses generally move as a unit. Note the occlusion in (e).

*In (c), where would you expect rain to develop? Why?*

Figure 7.7) that will take it across Illinois, Indiana, Ohio, and Pennsylvania and finally out over the Atlantic Ocean. A view of this storm on a weather map, at a specific time in its journey, is presented in ● Figure 7.9a. Figure 7.9b is a cross-sectional view of the storm, north of the cyclone's center, and Figure 7.9c is a cross-sectional view south of the cyclone's center. As the storm continues eastward, moving at 33 to 50 kilometers per hour (20–30 mph) the sequence of weather will be different for Detroit, where the warm and cold fronts will pass to the south, compared to Pittsburgh, which will experience the passage of both fronts. To illustrate these differing weather sequences, we will examine the changing weather in Pittsburgh, in contrast to that which Detroit experiences as the cyclonic system moves east. We will also examine the weather of other cities that are affected by the passage of this storm system.

A middle-latitude cyclonic storm is composed of two dissimilar air masses, usually with significant temperature contrasts. The sector of warm, humid $mT$ air between the two fronts of the cyclone is usually considerably warmer than the cold $cP$ air surrounding it. The temperature contrast between these two air masses is accentuated during the winter when the source region for $cP$ air is the very cold cell of high pressure in Canada. During the summer, the contrast between these air masses is greatly reduced.

Because of the temperature difference, atmospheric pressure in the warm sector is considerably lower than the atmospheric pressure in the cold sector behind the cold front. In advance of the warm front, the pressure is also high, but as the warm front approaches Pittsburgh, the pressure will decrease because of rising air along the front. After the warm front passes through Pittsburgh, the pressure will stop falling and the temperature will rise as warm $mT$ air invades the area from the south.

At this time, Indianapolis has already experienced the warm front's passage and is now awaiting the cold front. After the cold front passes Indianapolis, the pressure will rise rapidly, the clouds will clear, and the temperature will drop. Detroit, which is to the north of the cyclone's center, will miss the warm air sector entirely and experience a slight increase in pressure and a temperature change from cool to colder as the cyclone moves to the east.

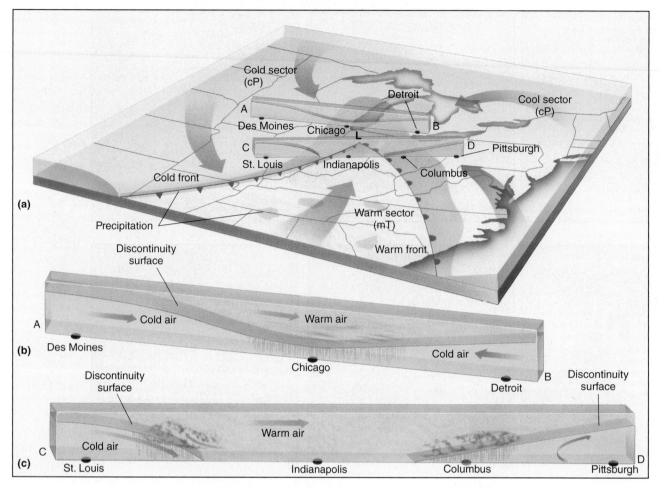

● **FIGURE 7.9** A classic middle-latitude cyclonic system. This diagram models a middle-latitude cyclonic storm positioned over the Midwest as the system moves eastward: (a) a map view of the weather system; (b) a cross-section along line A-B north of the center of low pressure; and (c) a cross-section along line C-D south of the center of low pressure.

*Why are the cross-sectional views important to understanding the different fronts and the weather associated with them?*

Changes in wind direction are one signal of the approach and passing of a cyclonic storm and its associated fronts. Because a middle-latitude cyclone is a low pressure system, the winds tend to flow counterclockwise (in the northern hemisphere) and toward its center. Winds associated with a cyclonic storm tend to be stronger in winter when the pressure and temperature differences between air masses are considerable.

Indianapolis, now located in the warm sector south of the center of the low, is receiving winds from the south. Pittsburgh, located to the east and ahead of the warm front, has southeast winds. As the cyclonic system moves eastward and the warm front passes, the winds in Pittsburgh will shift to the south-southwest. After the cold front passes, the winds in Pittsburgh will be from the north–northwest. St. Louis has already experienced the passage of the cold front with cold weather and winds from the northwest.

The changing direction of wind, clockwise around the compass from east to southeast to south to southwest to west and northwest, is called a **veering wind shift** and indicates that your position is south of the center of a low. Detroit, however, is also experiencing winds from the southeast but will undergo a completely different sequence of directional wind changes as the cyclonic storm moves eastward. Detroit's winds will shift, coming from the northeast as the center of the storm passes to the south. Chicago has just undergone this shift. Finally, after the storm has passed, the winds will generally blow from the northwest. Des Moines, to the west of the storm, currently has northwest winds. A change of wind direction like this, from east to northeast to north to northwest, is called a **backing wind shift,** as the wind "backs" counterclockwise around the compass. A backing wind shift indicates that you are north of the cyclone's center.

The type and intensity of precipitation and cloud cover also vary as a cyclonic storm moves through a location. In Pittsburgh the first sign of the approaching warm front will be high cirrus clouds; although the warm front is approaching, the front may still be 150 to 300 kilometers (90–180 mi) away. As the warm front gets closer, the clouds will thicken and lower, and in Pittsburgh light rain and drizzle may begin (or snow in winter) and stratus clouds will blanket the sky. After the warm front has passed, precipitation will stop and the skies will clear.

As the cold front passes, warm air in its path will be forced to move aloft rapidly. This may mean that there will be a cold, hard rain (or heavy snowfall in winter), but the band of precipitation normally will not be very wide because of the steeper angle of the surface along a cold front compared to the wedge-like warm front. In our springtime example, the cold front and the band of precipitation have just passed St. Louis. During the winter, however, a cold front is likely to bring snow, followed by the cold, clear conditions of the *cP* air mass, and increasing atmospheric pressure.

Located in the latitudinal center of the cyclonic system, Pittsburgh can expect three zones of precipitation as it passes over its location: (1) a broad area of overcast and drizzle in advance of the warm front (or light snow in winter); (2) a zone within the warm sector where clearing occurs; and (3) a narrow band of heavy precipitation associated with the cold front (rain or snow, depending on the season and temperatures) (see Fig. 7.9c). However, locations to the north of the center of the cyclonic storm, such as Detroit, will usually experience light precipitation and overcast cloudy conditions resulting from warm air being lifted above cold air from the north (see Fig. 7.9b).

As you can see, the various fronts and sectors of a middle-latitude cyclone are accompanied by different weather. If we know where the cyclone will pass relative to our location, we can make a fairly accurate forecast of what our weather will be like as the storm moves to the east along its track.

**Surface Weather and Upper Air Flow** Steering surface storm systems is only one of the ways that the upper air winds influence our surface weather. A less obvious influence is related to the undulating, wave-like flow so often exhibited by the upper air winds. As the air passes through these waves, it undergoes divergence or convergence because of the atmospheric dynamics associated with curved flow of winds. These flow dynamics produce alternating pressure areas of ridges (highs/divergent winds) and troughs (lows/convergent winds).

The region between a ridge and an adjacent trough (A–B in ● Fig. 7.10) is an area of upper-level convergence. Because any action taken in one part of the atmosphere is countered by an opposite reaction somewhere else, upper-air convergence is compensated for by divergence at the surface. In this area, the air is pushed downward, promoting anticyclonic circulation. This pattern will either inhibit the formation of a middle-latitude cyclone altogether, or cause an existing storm to weaken or dissipate. In contrast, the region between a trough and the next downwind ridge (B–C in Fig. 7.10) is an area of upper-level divergence, which in turn is compensated for by surface convergence. This is an area where air is drawn upward, causing cyclonic circulation at the surface that will enhance the prospects for storm development or strengthen an existing storm.

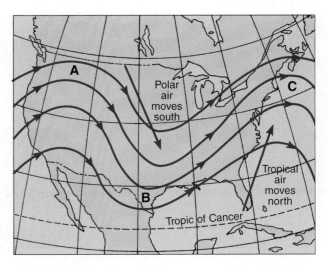

● **FIGURE 7.10** Wave forms in the jet stream. An upper-air wind pattern, long waves, such as those depicted here, can significantly influence the distribution of temperatures and precipitation at the surface.

*On which side of a wave (east or west) would you expect storms to develop?*

In addition to storm development or dissipation, temperatures are affected by upper air flow. If we assume that our "average" flow of upper air is from west to east, then any deviation from that pattern will cause either colder air from the north or warmer air from the south to be advected into an area. For example, after the atmosphere has been in a wave-like pattern for a few days, the areas in the vicinity of a trough (area B in Fig. 7.10) will be colder than normal as polar air from higher latitudes dips into that area. This intrusion of cold dry air is a *polar outbreak,* which occurs when continental polar air or arctic air is pushed by a major expansion of the polar high and drives a strong cold front into the lower middle latitudes of North America and sometimes to the subtropics.

Just the opposite effect on temperature and weather conditions will occur at locations near a ridge (area C in Fig. 7.10). Here, warmer air from more southerly latitudes will be drawn up toward the top of the ridge. As ● Figure 7.11 shows, the jet stream actually curves with less regularity than in our model. Compare Figures 7.10 and 7.11 to see the difference between the theoretical and the real waves in the polar jet stream.

## Hurricanes

A **hurricane** is a circular, cyclonic system with wind speeds in excess of 118 kilometers per hour (74 mph) and a diameter of 160 to 640 kilometers (100–400 mi). These tropical storms form and develop in the tropical oceans, but steered by pressure and wind systems, they are often directed toward the middle latitudes. Extending upward to heights of 12 to 14 kilometers (40,000–45,000 ft) or higher, the hurricane is a towering column of spiraling air ( ● Fig. 7.12). Although its diameter may be less than that of a *middle-latitude cyclone* with its extended *fronts,* a hurricane is essentially the largest storm on Earth. At its base, air is

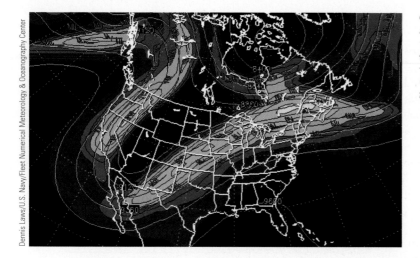

● **FIGURE 7.11** Polar front jet stream analysis. This map shows the winds at an altitude where the atmospheric pressure is 300 mb (approximately 10,000 m or 33,000 feet above sea level). At this altitude the patterns in the long waves that the jet stream follows are very apparent.

*Which country would experience the most weather changes as this wave pattern moves to the east?*

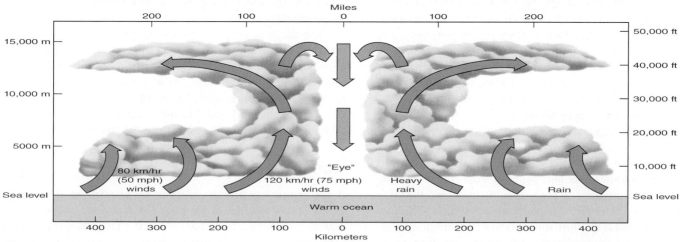

**(a)**

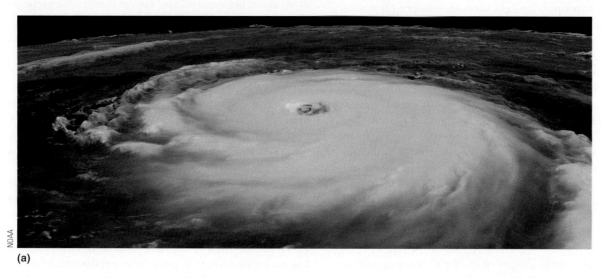

**(b)**

● **FIGURE 7.12** (a) Hurricane Katrina, as seen from space, can be directly compared with the diagram of a hurricane below. (b) Cross-section of a hurricane (with much vertical exaggeration) showing its circulation pattern: the inflow of air in the spiraling arms of the cyclonic system, rising air in the towering circular wall cloud, and outflow in the upper atmosphere. Subsidence of air in the storm's center produces the distinctive calm, cloudless "eye" of the hurricane.

*Which features in the Katrina image can you match with the cross-sectional diagram?*

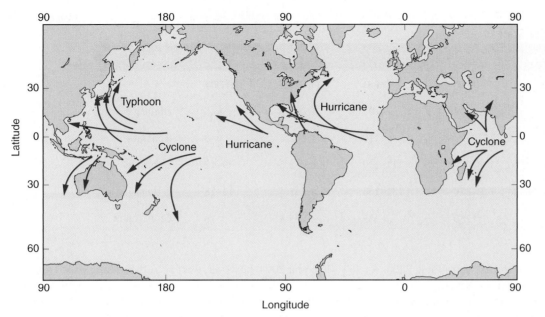

● **FIGURE 7.13** A world map showing major "Hurricane Alleys" and their regional names.

*Which coastlines seem unaffected by these tracks?*

sucked in by low pressure surrounding the hurricane's center and rises rapidly to the top where it spirals outward. The rapid upward movement of moisture-laden air produces enormous amounts of rain. Massive release of latent heat energy from condensation increases the power that drives the storm.

Hurricanes have extremely low pressure at their centers and very strong pressure gradients that produce powerful high-velocity winds. Unlike middle-latitude cyclones, hurricanes form from a single air mass and do not have the different sectors of strongly contrasting temperature that power a frontal system. In contrast, at the surface, hurricanes have a circular distribution of warm temperatures. At the center is the eye of a hurricane, an area of calm, clear, usually warm and humid, but rainless air. Sailors traveling through the eye have been surprised to see birds flying there. Unable to leave the eye because of the strong winds surrounding it, these birds will often alight on a passing ship as a resting spot.

Although a great deal of time, effort, and money has been spent on studying the development, growth, and paths of hurricanes, much is still not known. It is not yet possible to predict a hurricane's path with great certainty, even though it can be tracked with radar and studied by planes and weather satellites. As with tornadoes, there are also areas where hurricanes are most likely to develop and strike ( ● Fig. 7.13). For North America, the most susceptible areas are the Atlantic and Gulf Coastal regions, and hurricanes can continue to do damage as they dissipate and move inland.

The development of a hurricane requires a warm ocean surface of about 27°C (80°F) or higher and warm, moist overlying air. These factors explain why hurricanes occur most often in late summer and early fall when maritime air masses have maximum humidity and ocean surface temperatures are highest. Hurricane season in the Atlantic Basin officially runs from June 1 through November 30.

Hurricanes begin as weak tropical disturbances over the ocean, such as *easterly waves,* which are weak, trough-shaped, low pressure areas. Traveling slowly in the belt of the trade winds from east to west, if the pressure becomes lower and the winds strengthen, a tropical storm will develop.

Names are assigned to these storms once they reach *tropical storm* status, with wind speeds between 62 and 118 kilometers per hour (39–74 mph). Each year the names are selected from a different alphabetic list of alternating female and male names—one list for the North Atlantic and one for the North Pacific. If a hurricane is especially destructive and becomes a part of recorded history, its name is retired and never used again. Andrew, Carla, Hugo, and Katrina are just some of the nearly 70 names that have been retired since the naming of storms began in the 1950s.

Hurricanes do not last long over land because their source of moisture (and consequently their source of energy) is cut off, and friction with the land surface reduces wind speeds. North Atlantic hurricanes move first toward the west with the trade winds and then move north and northeast. Over land they become simple cyclonic storms. Even then, with their power significantly reduced, hurricanes can still do great damage.

Hurricanes can occur over most subtropical and tropical oceans and seas. The South Atlantic was once the exception, although it was not known why. However, on March 26, 2004 tropical Cyclone Catarina was the first known hurricane to attack the southern coast of Brazil, much to the amazement of atmospheric scientists. In the South Pacific, Australia, and South Asia hurricanes are called *cyclones.* Near the Philippines they are known as *bagyos,* but in most of East Asia they are called **typhoons.**

**Hurricane Intensities and Impacts** The results of a hurricane landfall can be devastating destruction of property and loss of life. The **Saffir–Simpson Hurricane Scale** provides a means of classifying hurricane intensity and potential damage by assigning a category from 1 to 5 based on a combination of central pressure,

TABLE 7.2
Saffir–Simpson Hurricane Intensity Scale

| Scale Number | Central Pressure | Wind Speed | | Storm Surge | | Damage |
|---|---|---|---|---|---|---|
| (CATEGORY) | (MILLIBARS) | (KPH) | (MPH) | (METERS) | (FEET) | |
| 1 | 980 | 119–153 | 74–95 | 1.2–1.5 | 4–5 | Minimal |
| 2 | 965–979 | 154–177 | 96–110 | 1.6–2.4 | 6–8 | Moderate |
| 3 | 945–964 | 178–209 | 111–130 | 2.5–3.6 | 9–12 | Extensive |
| 4 | 920–944 | 210–250 | 131–155 | 3.7–5.4 | 13–18 | Extreme |
| 5 | <920 | >250 | >155 | >5.4 | >18 | Catastrophic |

wind speed, and the potential height of its storm surge (Table 7.2). From one year to the next, the exact number and severity of *tropical cyclones* can vary dramatically. The continental United States usually averages three hurricane landfalls a year yet experienced no landfalls from September 2008 through 2010 ( ● Fig. 7.14).

The year 2004, however, was a record-breaking year. Typhoon Tokage struck the Japanese Coast near Tokyo, with significant loss of life. In all, 10 tropical cyclones pounded Japan in 2004. In the Caribbean and Gulf of Mexico, three hurricanes—Charley, Frances, and Jeanne—hit Florida directly. A fourth, Ivan, struck

● **FIGURE 7.14** This map shows the hurricane landfall sites along the U.S. Atlantic and Gulf Coasts from 1950 to 2009. The sites are labeled by the storm name, the year, and the Saffir–Simpson category. No hurricanes made landfall on the continental United States during 2009 or 2010.

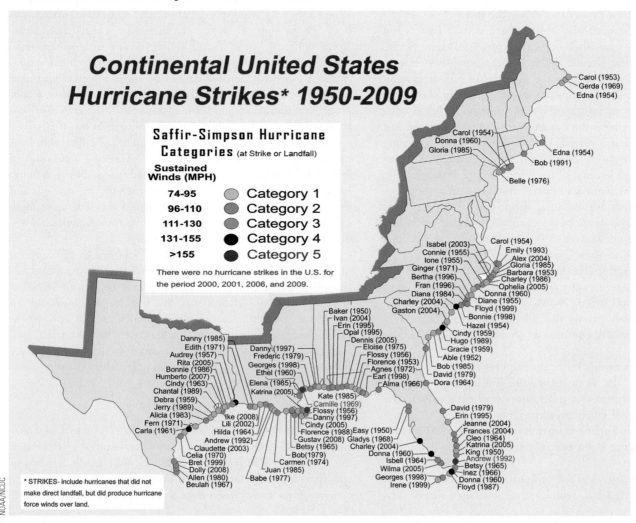

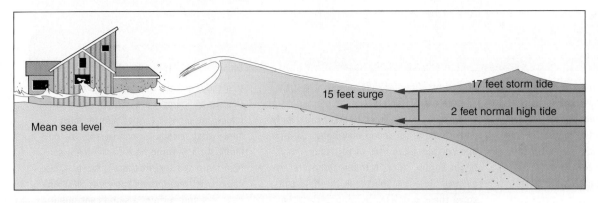

● **FIGURE 7.15** As a hurricane makes landfall, a storm surge may be generated. These surges are mounds of water, topped by battering waves, that can flow over low-elevation coastal areas with tremendous destructive force.

*What can people who live in such regions do to protect themselves when a serious storm surge is threatening?*

Gulf Shores, Mississippi and caused devastation in Florida's western panhandle. The damages from these storms were estimated at $23 billion. This amount exceeds even the $20 billion cost of damage by Hurricane Andrew, which, when it hit in 1992, was the costliest natural disaster in U.S. history.

Among their most serious hazards of hurricanes are the high seas pushed onshore by the strong winds. These **storm surges** can flood, and sometimes destroy, entire coastal communities ( ● Fig. 7.15). Hurricane Katrina, with 225-kilometer-per-hour (140 mph) winds and storm surges of more than 16 feet, struck the Louisiana, Mississippi, and Alabama coasts in August 2005. The storm surges breached the levee systems designed to protect the city of New Orleans, much of which is below sea level.

The subsequent flooding caused massive destruction. Responsible for the deaths of more than 2000 people, the winds and floodwaters of Katrina caused damages estimated to be in excess of $125 billion. Hurricane Rita, even more powerful, followed in

September with landfall near the Texas–Louisiana state line, devastating coastal areas in that region. Because Rita's eye missed the heavily populated Houston area, the estimated $10 billion damage was much less than for Katrina.

In 2006, the hurricane season was unusual because no hurricanes made landfall in the United States. The following year, two category 5 hurricanes struck Mexico and Central America, causing much destruction, but once again no powerful tropical storms affected the United States. In 2008, Hurricane Ike, a powerful storm with 230-kilometer-per-hour (145 mph) winds, made landfall on the Gulf Coast, with a storm surge that caused widespread damage to the coastal region near Galveston, Texas and in some cases complete destruction of coastal communities ( ● Fig. 7.16). The storm's path continued directly to Houston, causing billions of dollars of damage. Also in 2008, major hurricanes struck Haiti, Cuba, Jamaica, and other Caribbean islands, causing many deaths and serious destruction. In February

● **FIGURE 7.16** Extensive destruction and damage to a community along the Texas Gulf Coast caused by the storm surge generated by Hurricane Ike in 2008.

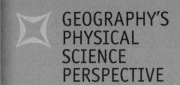

# Tornado Chasers and Tornado Spotters

It is amazing how many students choose to study meteorology because they want to chase tornadoes. This was especially true after the 1996 movie *Twister*. If this type of activity has ever interested you, you should know a few things from the start. First of all, there is a distinct difference between a *tornado spotter* and a *tornado chaser*.

Tornado spotters are trained by the National Weather Service (NWS) to serve their communities by watching and warning for severe weather. When severe weather is approaching, Doppler radar may not tell the complete story about what is happening on the ground. Sometimes Doppler radar only shows where a tornado may begin to form, and certain categories of tornadoes may begin before radar "tornado signature" is even detected. A spotter in the field can solve these problems by pinpointing the tornado touchdown and tracking the storm at a safe distance.

National Weather Service professionals often conduct these training sessions for state Emergency Management Agencies and amateur radio groups. Amateur radio operators (known as "hams") are well suited for this important service. A good spotter must know what to look for *and* be able to communicate a warning back to the NWS, so ham radio spotters fit the bill nicely. These volunteers form groups known as SKYWARN networks and perform a significant public service.

Tornado chasers, however, are not necessarily out to warn others of danger. Some of them are simply thrill seekers engaging in an exciting but potentially

NOAA/NSSL VORTEX II, Mike Coniglio

Scientists who "chase" tornadoes seek to observe them from as near as is safely possible to gain important information that helps us understand how these storms function. With a tornado in the background, this vehicle is mounted with special equipment for tornado tracking and data gathering.

2011, a massive category 5 hurricane, Cyclone Yasi, struck the coast of Queensland, Australia ( ● Fig. 7.17) with wind gusts as high as 290 kph (180 mph). The storm caused major flooding and considerable damage to structures and croplands in a wide area of northeastern Australia.

In spite of the damage they cause, hurricanes also have beneficial impacts. They are an important source of much-needed rainfall in regions of the southeastern United States such as Florida and in other areas of the world. Hurricanes are a natural and important means of moderating latitudinal temperatures by transferring surplus heat energy away from the tropics and into the cooler latitudes.

## Snowstorms and Blizzards

Snow-producing events mainly occur within the middle and higher latitudes because they are associated with colder winter temperatures. However, the areas affected by these storms may be extensive. Snowstorms are also triggered by the same uplift mechanisms that cause other types of precipitation events—orographic, frontal, and convergence (cyclonic). Convection is more of a warm-weather mechanism and is less likely to be involved in producing snowfall. The middle- to high-latitude winters experience snowfall events of varying severity. They can come as a heavy snowfall or as a *snow flurry*, a brief, light

dangerous hobby. Others chase tornadoes to study and understand them better in the interest of science. Meteorological scientists and their students collect data for research. Others who follow severe storms are professional photographers, photojournalists, and news reporters. Chasers are usually more mobile than spotters and will drive hundreds of miles to encounter a tornado. More often than not, a tornado chase results in a "bust," meaning a failed trip. Experienced tornado chasers may go on 50 or 60 trips over several years without seeing a tornado. Some spend their vacation time driving cross-country hoping to see a tornado—only to go home disappointed.

There are an estimated 1000 tornado chasers in the United States. Their average age is about 35 years, but ages range from 18 to 65. Women comprise about 2% of this group. Many tornado chasers have a college education. Although most live in the Great Plains or Midwest (where tornadoes are more frequent), tornado chasers now reside in all the lower 48 states. Regardless of who they are and where they live, most of them have one thing in common: a working knowledge of meteorology. Academic and government programs direct much effort and sophisticated equipment, mounted on vehicles, to advance our knowledge of these fascinating, powerful, and dangerous storms. Tornado formation is often unpredictable and they can appear so quickly that there will always be some risk of injury or property damage to residents of tornado prone regions. The knowledge gained through the hard work and dedication of scientists who sometimes risk their lives to closely monitor tornadoes makes a major contribution to minimizing the human impact of these storms.

NOAA/NSSL VORTEX II, J. J. Gourley

Lightning strikes light up the sky behind a VORTEX2 storm chase vehicle in Enid, Oklahoma. Tornadoes are not the only danger for storm chasers; intense thunderstorms also can produce lightning and large hail.

snowfall that may not produce any accumulation of snow on the ground. A **snowstorm** is a weather event where frozen precipitation falls as snow, typically with much accumulation, and the storm may be severe. Some snowstorms create enough turbulence to create lightning discharges, a phenomenon once thought to be impossible.

A **blizzard** is the most severe weather event involving snowfall. Blizzards are characterized as heavy snowstorms accompanied by very strong winds. At wind speeds of about 55 kilometers per hour (35 mph) or greater, a blizzard can reduce visibility to zero because of the falling and blowing snow. Here, the term *whiteout* can apply. Visibility can be reduced so much that all a person can see is white, and an individual can completely lose track of distance and direction. This is especially dangerous for people using any mode of transportation. Airport closings and traffic accidents are common during blizzards ( ● Fig. 7.18).

## Thunderstorms

Thunderstorms are common storms of the middle and lower latitudes, particularly in summer, and they are typically localized (rather than broad regional) storms. Very simply, thunderstorms are often intense atmospheric disturbances that are accompanied by

● **FIGURE 7.17** This massive hurricane, Cyclone Yasi, hit the coast of Queensland, Australia in February 2011. A Category 5 storm, Yasi had winds of nearly 300 kilometers per hour (186 mph), which demolished homes, caused severe crop damage, toppled or stripped large trees, and brought a half a meter of rain.

*Compare the circulation of Cyclone Yasi here with the image of Hurricane Katrina in Figure 7.12a. Explain any difference that you see.*

thunder and lightning. Lightning is a massive discharge of electricity within the atmosphere or between the atmosphere and the ground (including structures on the ground). Lightning is produced when strong positive and negative electrical charges are generated within a cloud. It is believed that the intense friction of air between air and moving ice particles within a cumulonimbus cloud generates these charges. Usually, but not always, positive charges tend to cluster in the upper portion of the cloud, with negative charges clustering at lower levels. When the potential difference between

● **FIGURE 7.18** A weather station in central Illinois during a blizzard in February 2000. With winds gusting to 45 miles per hour, this blizzard can greatly reduce visibility.

*How far would you estimate the visibility to be in this area?*

these charges becomes large enough to overcome the natural insulating effects of the air, an electrical discharge takes place producing a lightning flash. These discharges, often involving more than 1 million volts of electricity, can occur within the cloud, between two clouds, or from cloud to ground. The air immediately around the discharge is momentarily heated to temperatures in excess of 25,000°C (45,000°F), which is about four times hotter than the surface of the sun! The heated air expands nearly instantaneously and explosively, creating the sonic shock wave we call thunder ( ● Fig. 7.19).

Thunderstorms usually cover relatively small areas of a few square kilometers, although there may be a series of related thunderstorms covering a larger region. The intensity of a thunderstorm depends on the degree of instability of the air and the amount of water vapor it holds. A thunderstorm will die out when most of its water vapor has condensed and there will no longer be enough energy available for continued vertical movement. In fact, most thunderstorms last an hour or less. Thunderstorms result from the very strong uplift of moist air. The trigger mechanism causing that uplift can be thermal convection (warm unstable air rising on a warm afternoon; ● Fig. 7.20a), orographic uplift (warm moist air ramping up a mountainside, Fig. 7.20b), or frontal uplift (see Figs. 7.3 and 7.4).

**Convective thunderstorms,** on a worldwide basis, are most common in the lower latitudes during the warmer months and during the warmer times of the day. Solar heating encourages thunderstorm development because intense surface heating steepens the environmental lapse rate, which leads to increased instability of the air and allows for greater moisture holding capacity, adding to the buoyancy of the air.

**Orographic thunderstorms** occur when air is forced to rise over land barriers, providing the necessary initial trigger action that leads to the development of thunderstorm cells. Thunderstorms of orographic origin play a large role in the tremendous precipitation that occurs in some areas of South and Southeast Asia. In North America they occur over the mountains of the West (the Rockies and the Sierra Nevada), especially during summer afternoons when heating of south-facing slopes increases the air's instability. For this reason, pilots of small planes try to avoid flying in the mountains during summer afternoons for fear of getting caught in the turbulence generated during a thunderstorm.

**Frontal thunderstorms** are often associated with cold fronts where a cooler air mass forces a warm, humid air mass to rise. This kind of frontal activity can bring about the strong, vertical updrafts necessary for heavy precipitation. At times, cold fronts are immediately preceded by a line of thunderstorms (a squall line) resulting from uplift of the edge of a warm air mass (see Fig. 7.4).

● **FIGURE 7.19** A cross-sectional view of a thunderstorm showing the distribution of electrical charges.

*Where do you place a lightning bolt in this diagram?*

As was mentioned in the discussion of precipitation types, hail is a product of thunderstorms when the vertical updrafts of the cells are strong enough to repeatedly carry water droplets into a freezing layer of air that supports pellets of ice. Fortunately, only a very small percentage of storms around the world produce hail. In fact, thunderstorms only occasionally produce hail in the lower latitudes. In the United States there is little hail along the Gulf of Mexico where thunderstorms are most common.

## Tornadoes

A **tornado** is a small, intense cyclone with very low pressure, violent updrafts, and strong converging winds that extends from a thunderstorm cloud to the ground. Tornadoes are the most violent storms on Earth ( ● Fig. 7.21 and ● Fig. 7.22). The interior of North America experiences more tornadoes than anywhere else, particularly a region centered on Oklahoma and Kansas that is referred to as "Tornado Alley." The United States is the country that experiences the most tornadoes in the world, and Canada is second. In August 1976, a tornado touched down near the village of Kiana, Alaska, 47 km (29 mi) north of the Arctic Circle (the highest known latitude for a tornado).

Although only 1% of all thunderstorms produce a tornado, 80% of all tornadoes are associated with thunderstorms and middle-latitude cyclones. The remaining 20% of tornadoes are spawned by hurricanes that make landfall. In the past decade, more than 1000 tornadoes have occurred each year in the United States, most of them from March to July in the late afternoon or early evening in the central part of the country. ● Figures 7.23a and 7.23b present maps of where tornadoes are most likely to occur and the season when they are most likely to occur. It is important to know that these maps are generalized to show important spatial patterns. Tornadoes can occur in any of the United States and are not limited to the seasons that are shown. An EF–3 tornado struck an army base in Missouri on December 31, 2010.

Government records of tornado activity began in 1875. Accounts of tornadoes occurring prior to 1875 must be investigated by examining other sources. These early historic accounts are sometimes unverifiable and vague, but they do offer insights about early perceptions of tornadoes in North America. The following accounts describe a tornado that killed several people as it swept across several counties in western Illinois on May 21, 1859.

> It was "a violent storm or hurricane [which] did immense damage to houses, barns, fences, and also caused some destruction of life." It was described as having a "frightful ... balloon or funnel shape, and appeared ... peculiarly bright and luminous, not at all black or dark in any of its parts, except its base." A vivid account of what must be related to the output of static electricity is given in this account of the

● **FIGURE 7.20** Thermal convection and orographic uplift.

*What are the other mechanisms of uplift?*

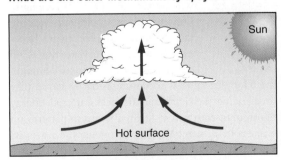

**(a) Convectional uplift**

**(b) Orographic uplift**

● **FIGURE 7.21** A powerful F4 tornado in central Illinois during July 2004.

● **FIGURE 7.22** Terrible destruction caused by an F5 tornado at Greensburg, Kansas on May 16, 2007.

of the few individuals to apparently have had a tornado pass directly over him and live to tell about it. He described the light in the tornado's center as "so brilliant that he could not endure it with his eyes open, and for the most part kept them shut ... Yet [inside the tornado] there was no wind, no thunder and no noise whatever...."

—Transactions of the Illinois Natural History Society Phillips Bros., 1861

Fortunately, tornadoes are relatively small and short-lived. Even in Tornado Alley, a tornado is likely to strike a given locale only once in 250 years. Because of their relatively small size and limited life span, tornadoes are extremely difficult to detect and forecast. However, relatively new radar technology, **Doppler radar,** improves tornado detection and forecasting significantly. Doppler radar has more power concentrated in a narrower beam than previous radar units. This allows meteorologists to assess the patterns of storms in much greater detail ( ● Fig. 7.24). Even more important is a Doppler radar's ability to measure wind speeds flowing toward the radar site and wind speeds blowing away from the radar site (by measuring the Doppler effect). When the energy emitted by radar strikes precipitation, a small portion is scattered back to the radar antenna. Depending on whether the precipitation is moving toward or away from the radar site, the wavelength of the returned energy is either compressed or elongated. The faster the winds flow, the greater will be the change in wavelength. Previous radars could not measure this change. Doppler radar, however, uses it to estimate the wind circulation and cyclonic rotation within the storm. Doppler radar is so sensitive that it can display wind patterns in clear air by detecting the backscattered energy from clouds, pollution, insects, and so forth. The storm patterns on Doppler radar images help meteorologists recognize the potential development of a tornado, thus increasing the warning time to the public (see Fig. 7.24). The U.S. government, through the NEXRAD program (NEXt-generation weather-RADar), has in-

same tornado as it swept Morgan county: "Mr. Cowell was plowing his field ... He saw the frightful cloud approaching ... and at once attempted to drive his horses and plow to the house.... The horses suddenly took fright ... their manes and tails and all their hair 'stood right out straight' as he expressed it, and ... the iron in the harness ... and plow, in his language 'seemed all covered with fire.' He felt a violent pulling of his hair which left 'his head sore for some days' and the hair itself rigid and inflexible." In addition, Mr. Cowell was one

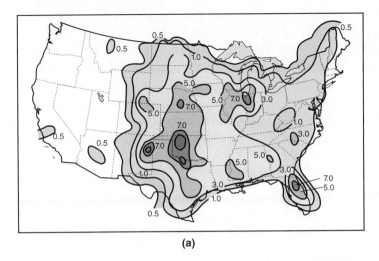

**(a)**

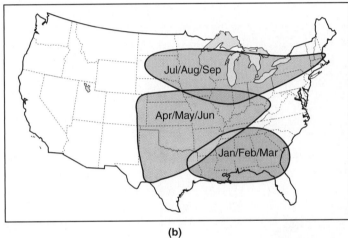

**(b)**

● **FIGURE 7.23** (a) Average annual number of tornadoes per 26,000 square kilometers (10,000 sq mi). (b) Seasonal times of peak tornado activity.

*How do tornadoes affect your geographic area?*

stalled more than 150 Weather Service Doppler Radar (WSR–88D) sites across the country.

Doppler radar observations indicate that most tornadoes (63%) are in a category considered as fairly weak with wind speeds of 162 kilometers per hour (100 mph) or less. About 35% of tornadoes can be classified as strong, with wind speeds reaching 320 kilometers per hour (200 mph). Nearly 70% of all tornado fatalities result from what are considered to be the most violent tornadoes. Although rare (only 2% of all tornadoes reach this stage), these may last for hours and have wind speeds in excess of 480 kilometers per hour (300 mph).

The late T. Theodore Fujita, a former professor at the University of Chicago, originated a scale of tornado intensity, based on the damage that a tornado produced, which is generally a function of wind velocity. The scale is termed the Fujita Intensity Scale or more commonly the F-scale (Table 7.3). Since Fujita's time, Doppler radar has provided the capability to accurately estimate the actual wind speed of a particular tornado. In 2007, the National

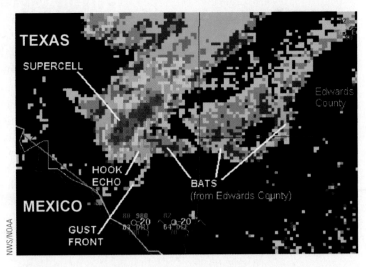

● **FIGURE 7.24** This NEXRAD Doppler radar image of thunderstorms shows details of a severe storm with a hook-shaped pattern that is associated with the development of tornadoes. Colors show rainfall intensity: green—light rainfall, yellow—moderate, and orange–red—heavy. This radar image also picked up reflections from huge groups of flying bats, among the millions that live in this region.

*How do patterns on Doppler radar images help us to understand weather conditions as they occur?*

Weather Service adopted a refined and modified version of the original F-scale, based on new data and observations that were not available to Fujita. This is the **Enhanced Fujita Scale,** the EF-scale, and the main difference is in how wind speeds are now estimated based on damage observations, and changes in the wind speeds were integrated for the new EF-scale.

A tornado first appears as a swirling, twisting funnel cloud that moves across the landscape at about 35 to 51 kilometers per hour (22–32 mph). Its narrow end may be only 100 meters (330 ft) across. The funnel cloud becomes a tornado when its lower, narrow end is in contact with the ground. The greatest damage is done often along a relatively linear track where the base of the tornado was in contact with the surface ( ● Fig. 7.25). When the base of a tornado is above the ground, the end can swirl and twist, but little or nothing is done to the surface below. Tornadoes can appear milky white or even black, depending on the brightness and direction of sunlight and the amount of dust and debris picked up by the storm as it travels across the land. Although most tornado damage is caused by the violent winds, most tornado injuries and deaths result from flying debris. The relatively small size and short duration of a tornado typically limit the number of deaths caused by tornadoes. In fact, more people die from lightning strikes each year than from tornadoes. At times, however, severe storms may spawn a **tornado outbreak,** where multiple tornadoes are produced by the same weather system. The worst outbreak in recorded history occurred on April 3 and 4, 1974, when 148 twisters touched down in 13 states, injuring almost 5500 people and killing 330 others.

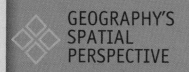
# Forecasting Hurricane Strikes and Wind Intensities

Hurricanes (also called typhoons or cyclones) are generated over tropical or subtropical oceans and build strength as they move over regions of warm ocean water. Ships and aircraft regularly avoid hurricane paths by navigating away from these huge, violent storms. People living in the path of an oncoming hurricane try to prepare their homes, other structures, and belongings because they may have to evacuate if the hazard potential of the impending storm is great enough. Landfall refers to the location where the eye of the storm encounters the coastline. Storm surges are the most dangerous hazard associated with hurricanes, where the ocean violently washes over and floods low-lying coastal areas. In 1900, 6000 residents of Galveston Island in Texas were killed when a hurricane pushed a 7-meter-high wall of water over the island. Much of the city was destroyed by this storm, the worst natural disaster ever to occur in the United States in terms of lives lost. In 2008 the same area was hit by Hurricane Ike, which caused serious and widespread damage by a storm surge of lesser intensity than the one in 1900. Early warning and evacuation saved many lives because the coastal residents had time to leave for safer areas.

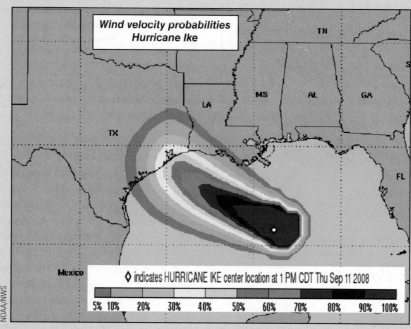

Hurricane Ike heads for the Gulf Coast in 2008. A probability map shows where hurricane velocity winds are occurring (dark red and purple) and the velocity probabilities for places located in the forecasted track. The center of the curved zones on the northwest side indicates the most likely place for landfall. The bay, with blue water, is very near to the site of the actual landfall, so this prediction was very good.

# Weak Tropical Disturbances

Until World War II the weather of tropical regions was described as hot and humid, generally fair, but basically rather monotonous. The only tropical disturbance that was given considerable attention was the tropical cyclone (also called a *hurricane* or, in other parts of the world, a *typhoon* or *cyclone*), a spectacular but relatively uncommon storm that affects only islands, coastal lands, and ships at sea.

Even a few decades ago, an aura of mystery remained about the weather of the tropics. One reason for this lack of information was that the few weather stations located in tropical areas were widely scattered and often poorly equipped. As a result, it was difficult to understand completely the passing weather disturbances in the tropics.

Largely through satellite technology and computer analysis, it is now known that there are different kinds of weak atmospheric disturbances affecting weather in the tropics, although it

Today we have sophisticated technologies for tracking and evaluating tropical storms. Computer models, developed from mapping the behavior of past storms, are used to indicate a hurricane's most likely path and landfall location and its potential wind speed. The closer a storm is to the coast, the more accurate the predicted landfall site should be. In some cases a hurricane may begin to move in a completely different direction, but with computer analysis and satellite data and images, the maps can be revised daily to incorporate any changes in the storm. In general, hurricanes that originate in the North Atlantic Ocean tend to move westward toward North America and then turn northward along the Atlantic Coast or toward the Gulf Coast.

Nature is unpredictable, so when potential hurricane paths and landfall sites are shown on maps, it is by the degree of probability for a hurricane's path and wind intensity. These maps help local authorities and residents decide on the best course of action in preparing for a hurricane's approach. Regions where a hurricane is considered likely to move next are represented on the map by color shadings that correspond to a reasonably probable path and wind intensity. In recent years, the National Weather Service has worked to develop computer models that will yield better predictions of hurricane paths, intensities, possible storm surges, and landfall areas. If you live in a coastal area affected by hurricanes and tropical disturbances, understanding these maps may be important to your safety and your ability to prepare for a coming storm.

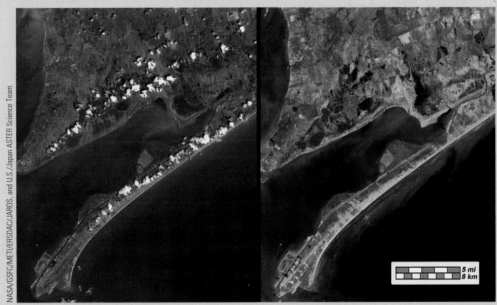

Before and after near-infrared satellite images along the Texas coast after the storm surge generated by Hurricane Ike in 2008. Refer to Fig. 7.16 to see a close view of the damage done right along the coast. The surge traveled up to 7 miles inland on this low-lying coastal area.

⊕ In Google Earth fly to: 29.680°N, 94.420°W.

is likely that the importance of these disturbances has not yet been fully recognized. The primary impact of weak atmospheric disturbances on the weather of tropical regions is not on the temperature but rather that they affect the cloud cover and the precipitation amounts. Temperatures in the tropics are largely unaffected during the passage of a tropical storm, except that as the cloud cover increases, temperature extremes are reduced.

The best-known weak tropical disturbance is the **easterly wave.** ● Figure 7.26 illustrates an easterly wave as a trough-shaped, weak, low pressure region, generally aligned on an approximate north–south axis. Traveling slowly as the converging trade winds transport them from east to west, they are preceded by fair, dry weather and followed by cloudy, showery weather. This occurs because air tends to converge into the low from its trailing end in the east, which causes lifting and convectional showers. Divergence and subsidence to the west account for the existing fair weather. Easterly waves, however, can also represent the beginnings of a tropical

TABLE 7.3
The Fujita Tornado Intensity Scale and the Enhanced Fujita Scale

| F-SCALE | Wind Speed | | EF-SCALE | Wind Speed | | EXPECTED DAMAGE |
| | KPH | MPH | | KPH | MPH | |
|---------|------|------|----------|------|------|-----------------|
| F-0 | <116 | <72 | EF-0 | <138 | <86 | **Light Damage**<br>Damage to chimneys and billboards; broken branches; shallow-rooted trees pushed over |
| F-1 | 116–180 | 72–112 | EF-1 | 138–177 | 86–110 | **Moderate Damage**<br>Surfaces peeled off roofs; mobile homes pushed off foundations or overturned; exterior doors blown off; windows broken; moving autos pushed off the road |
| F-2 | 181–253 | 113–157 | EF-2 | 178–217 | 111–135 | **Considerable Damage**<br>Roofs torn off houses; mobile homes demolished; boxcars pushed over; large trees snapped or uprooted; light-object missiles generated |
| F-3 | 254–332 | 158–206 | EF-3 | 218–265 | 136–165 | **Severe Damage**<br>Roofs and some walls torn off well-constructed houses; trains overturned; most trees in forest uprooted; heavy cars lifted off ground and thrown |
| F-4 | 333–419 | 207–260 | EF-4 | 266–322 | 166–200 | **Devastating Damage**<br>Well-constructed houses leveled; structures with weak foundations blown some distance; cars thrown and large missiles generated |
| F-5 | >419 | >260 | EF-5 | >322 | >200 | **Incredible Damage**<br>Strong frame houses lifted off foundations and carried considerable distance to disintegrate; automobile-sized missiles fly through the air farther than 100 meters; trees debarked; incredible phenomena occur |

● **FIGURE 7.25** The destructive track of a powerful tornado is visible on this satellite image as a linear swath of damage across the landscape of La Plata, Maryland.

Lawrence Ong, EO-1 Mission Science Office, NASA GSFC

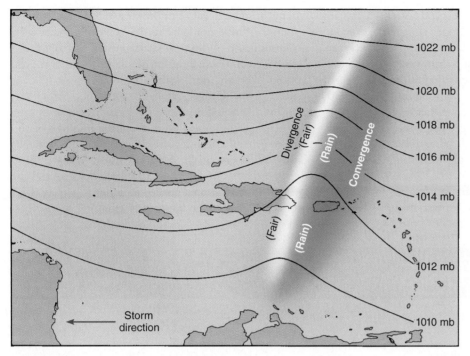

depression that can develop into a tropical cyclone (hurricane).

## Weather Forecasting

Weather forecasting, at least in the way it proceeds, is fairly straightforward. Meteorological observations are made, collected, and mapped to depict the current state of the atmosphere. From this information and data, the probable movement and any anticipated growth, movement, or decay of current weather systems is projected for a specific length of time into the future.

When a weather forecast goes wrong—which we all know occurs—it is usually because limited or erroneous information has been collected and processed or because of high potential variability in the anticipated path or growth of weather systems. Small errors can compound over time. For example, when recording the path of a storm, an error of a few degrees may cause the projected location of a storm to be off by several miles even in a 2-hour forecast. In a 48-hour forecast, the projected location could be off by hundreds of miles. Because of the complex uncertainties that are inherent in weather systems and atmospheric conditions, the farther into the future one tries to forecast, the greater the chance of error.

Weather forecasts are not perfect, but they are much better today than they were in the past. Much of this improvement can be attributed to increasingly sophisticated technology and equipment. Increased knowledge and surveillance of the upper atmosphere have also improved the accuracy of weather prediction. Weather satellites have helped tremendously by providing continuing images and data in real time of weather and weather systems locally, regionally, and globally ( ● Fig. 7.27). Satellite data and images have been of particular value to forecasters in coastal areas of the United States. Before the advent of weather satellites, forecasters had to rely on information relayed from ships, which left enormous areas of the oceans unobserved. Thus, forecasters could be caught off guard by unexpected weather events.

In addition, computers allow rapid processing and mapping of observed weather conditions. Computers can also be used to process numerical data and generate forecasts based on the physical laws and equations that govern our atmosphere. Numerical and long-term forecasts that are based on solving statistical relationships and complex equations would not be possible without computers, at least not within the short time span allotted to

● **FIGURE 7.26** A typical easterly wave in the tropics. Note that the isobars (and resulting winds) do not close in a circle but merely make a poleward "kink," indicating a low pressure trough rather than a closed cell. The resulting weather is a consequence of convergence of air coming into the trough, producing rains, and divergence of air coming out of the trough, producing clear skies.

*Why do easterly waves move toward the west?*

● **FIGURE 7.27** This weather satellite image shows the cloud and weather patterns of two continents and the adjacent ocean areas. Examine this image closely.

*Can you identify the equatorial low's belt of clouds, the clearer skies of the subtropical highs, and the cold and warm fronts of a middle latitude storm in the United States?*

weather forecasters. Today some of the world's largest and fastest computers are used to forecast the weather and model climatic conditions.

Although forecasters now possess a great deal of knowledge and access to a variety of highly sophisticated technologies that were previously unavailable, weather forecasting is not foolproof. If we are more aware of some of the problems that weather forecasters face, we should also be more understanding when a forecast fails. Weather forecasters combine science with art, fact with interpretation, and data with intuition to come up with some probabilities about future weather conditions.

In this chapter we combined some of the atmospheric elements that we learned in previous chapters to explain major weather systems in some detail. Temperature and moisture differences characterize air masses, and fronts develop along the convergence of unlike air masses. With their characteristics of temperature, humidity, pressure, and wind, air masses and fronts form the weather systems that we deal with over the seasons. Certain weather systems may be relatively small, like tornadoes, or they may cover large areas, like middle-latitude cyclones. Weather systems affect our lives in many different ways, from the joys of a perfect day, to the inconvenience of an incorrect forecast, to understanding how to prepare for a potentially disastrous storm. Knowledge of how weather systems vary and operate can enrich our lives and also provides some of the background needed to understand the world's climates, which are addressed in the next chapters.

# CHAPTER 7 ACTIVITIES

## ■ DEFINE AND RECALL

air mass
source region
maritime
continental
maritime equatorial *(mE)*
maritime tropical *(mT)*
continental tropical *(cT)*
continental polar *(cP)*
maritime polar *(mP)*
continental arctic *(cA)*
lake effect snow
nor'easters
cold front

squall line
warm front
stationary front
occluded front
atmospheric disturbance
polar outbreak
storm tracks
middle-latitude cyclone (extratropical cyclone)
veering wind shift
backing wind shift
hurricane
typhoon

Saffir–Simpson Hurricane Scale
storm surge
snowstorm
blizzard
convective thunderstorm
orographic thunderstorm
frontal thunderstorm
tornado
Doppler radar
Enhanced Fujita Scale
tornado outbreak
easterly wave

## ■ DISCUSS AND REVIEW

1. What is an air mass?
2. Do all areas on Earth produce air masses? Why or why not?
3. What letter symbols are used to identify air masses? How are these combined? What air masses influence the weather of North America? Where and at what time of the year are they most effective?
4. Use Table 7.1 and Figure 7.1 to find out what kinds of air masses are most likely to affect your local area. How do they affect weather in your area?
5. Why are air masses classified by whether they develop over water or over land?
6. What kind of air mass forms over the southwestern United States in summer? What weather conditions are associated with this air mass?
7. What is a front? How do they develop?
8. Compare warm and cold fronts. How do they differ in duration and precipitation characteristics?
9. What kind of weather often results from a stationary front? What kinds of forces tend to break up stationary fronts?
10. How does the westerly circulation affect air masses in your area? What kinds of weather result?
11. Draw a diagram of a fully developed middle-latitude cyclone that includes the center of the low with several isobars, the warm and cold fronts, wind direction arrows, appropriate labeling of warm and cold air masses, and zones of precipitation.
12. If the wind changes to a clockwise direction, what is the shift called? Where does it locate you in relation to the center of a low pressure system? Explain why this happens.
13. How does the configuration of the upper air winds play a role in the surface weather conditions?
14. Describe the sequence of weather events over a 48-hour period in St. Louis, Missouri if a typical low pressure system (cyclone) passes 300 kilometers (180 mi) north of that location in the spring.
15. List three major causes of thunderstorms. How might the storms that develop from each of these causes differ?

# CONSIDER AND RESPOND

1. Collect a 3-day series of weather maps from your local news-paper. Based on the migration of high and low pressure systems during that period, discuss the likely pattern of the upper air winds.
2. Look at Figure 7.9a. Assume you are driving from point A to point B. Describe the changes in weather (temperature, wind speed and direction, barometric pressure, precipitation, and cloud cover) you would encounter on your trip. Do the same analysis for a trip from point C to point D.

3. The location of the polar front changes with seasons. Why? In what way is Figure 7.23b related to the seasonal migration of the polar front?
4. Redraw Figure 7.8d so that it depicts a Southern Hemi-sphere example.
5. List the ideal conditions for the development of a hurricane.

# APPLY AND LEARN

1. Wind speeds are sometimes given in knots (nautical miles per hour) instead of statute miles per hours as most people use. A nautical mile (used mainly by sailors and pilots) is equal to 6080 feet, slightly longer than the statute mile at 5280 feet; therefore, a knot is a little faster than 1 statute mile per hour (mph). The conversion from miles per hour to knots is:

$$\text{KNOTS} = \text{MPH} \times 0.87$$

Using Table 7.2, convert the ranges of wind speeds for hurricane categories 1 through 5 from miles per hour to knots.

2. A cold front squall line moving 35 miles per hour has spawned tornadoes at 9:00 a.m. near Memphis, Tennessee. The remainder of Tennessee is covered by an unstable $mT$ air mass. Make a statement about what we might expect to happen in Nashville, Tennessee (some 200 miles away) and when.

# LOCATE AND EXPLORE

1. Using Google Earth and the weather layer provided by Google Earth (in the Layers window), track the daily varia-tion in weather (temperature, pressure, wind speed, wind direction, dew point, precipitation, and cloud type) at any weather station in the following cities. Can you see any re-lationship in the weather among these cities? Can you use the weather in Omaha to predict the weather later in the week in Louisville or Washington?
Portland, Oregon (45.53°N, 122.69°W)
Casper, Wyoming (42.86°N, 106.29°W)
Omaha, Nebraska (41.25°N, 95.88°W)
Louisville, Kentucky (38.27°N, 85.74°W)
Washington, D.C. (38.93°N, 77.03°W).
2. Using Google Earth and the weather layer provided by Google Earth (in the Layers window), look at the different weather systems across the United States and their associated cloud type. Over the next week, go outside every day at the same time and look at the sky. Record the cloud type

in your area using the examples in Figure 6.15 from your text as a guide. How do the cloud type and general weather conditions for your area (rain, sun, wind, and so on) relate to the regional weather systems (low pressures, high pres-sures, fronts, and so on) that are currently tracking across the United States?
3. Using Google Earth and the wind vector layer for Hurricane Katrina (download from CourseMate at www.cengagebrain.com), describe how the wind speed and direction changed in New Orleans, Louisiana (30.0°N, 90.1°W) and Biloxi, Missis-sippi (30.4°N, 88.9°W) as the storm passed. These layers show the speed of the wind (background color) and the direction (small arrows) every 2 hours as the storm passed. Provide an explanation for the observed wind speed and direction based on distance of each city from the center of the storm and whether the city was east or west of the storm. Tip: Create an $x$-$y$ scatter plot in a digital spreadsheet (for example, Excel) using time on the $x$-axis and wind speed on the $y$-axis.

 CourseMate – Make the most of your study time by accessing everything you need to succeed in one place. Read your textbook, take notes, review flashcards, watch videos, complete activities, take practice quizzes, and more online with CourseMate. Log in at **www.cengagebrain.com**.

# Map Interpretation

## ■ WEATHER MAPS

Weather maps that portray meteorological conditions over a large area at a given moment in time are important for current weather descriptions and forecasting. Simultaneous observations of meteorological data are recorded at weather stations across the United States (and worldwide). This information is electronically relayed to the National Centers for Environmental Prediction near Washington, D.C., where the data are analyzed and mapped and a U.S. map is published daily.

A different map from the same day (or any recent day) can be found here: http://www.hpc.ncep.noaa.gov/dailywxmap/index_20080903.html.

Meteorologists then use the information and data to depict a general weather picture over a larger area. For example, isobars (lines of equal atmospheric pressure) are drawn to reveal the locations of cyclones (L) and anticyclones (H) and to indicate frontal boundaries. Areas that were receiving precipitation at the time the map depicts are highlighted with shading. The result is a map of weather conditions that can be used to forecast changes in weather patterns. The following map is accompanied by a satellite image that was taken on the same date.

1. What is the pressure interval (in millibars) between adjacent isobars on this weather map?

2. What kind of front is stretching across eastern Canada to Maine?

3. What kind of air mass is dominating the midwestern part of the United States?

4. What are the highest and lowest values for isobars on this map in millibars. How did you know where to look for these maximum and minimum values? Where is the pressure gradient the steepest?

5. What kind of front is approaching the Pacific Northwest?

6. Do the fronts and cyclonic storms on the map accurately depict the cloud cover shown on the satellite image? (Disregard the fact that the map and the image are the same day but not the same time of day.)

7. Find the low pressure systems on the satellite image. Does the circulation look correct for the Northern Hemisphere?

8. List the three kinds of fronts on this map with a general statement of their location. The fourth kind of front is not occurring on this day; what kind is it?

9. Do the locations of the fronts and areas of precipitation depicted on the map agree with the idealized relationship represented in Figure 7.9 (Middle-Latitude Cyclonic Systems)?

10. On the map what kind of frontal symbol lies off the coast of Virginia and North Carolina?

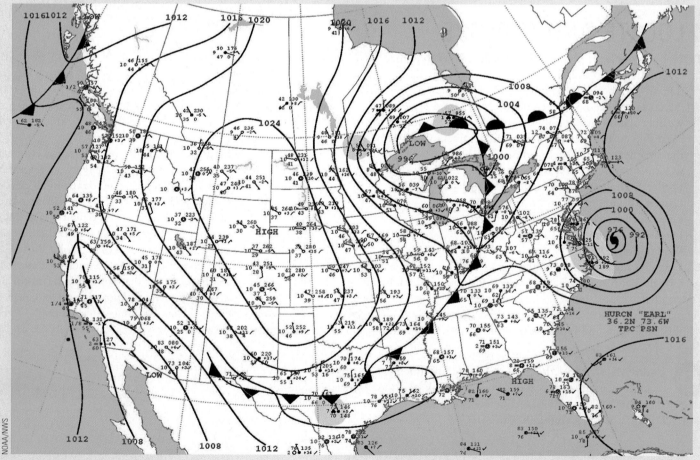

This weather map from September 3, 2010 illustrates the spatial distribution of quantitative weather elements (air pressure, temperatures, wind speed, wind directions) and the locations of fronts and areas of precipitation. Isobars define high and low pressure cells and the kinds of fronts are also identified. The weather conditions at meteorological stations are shown by numerical values and symbols.

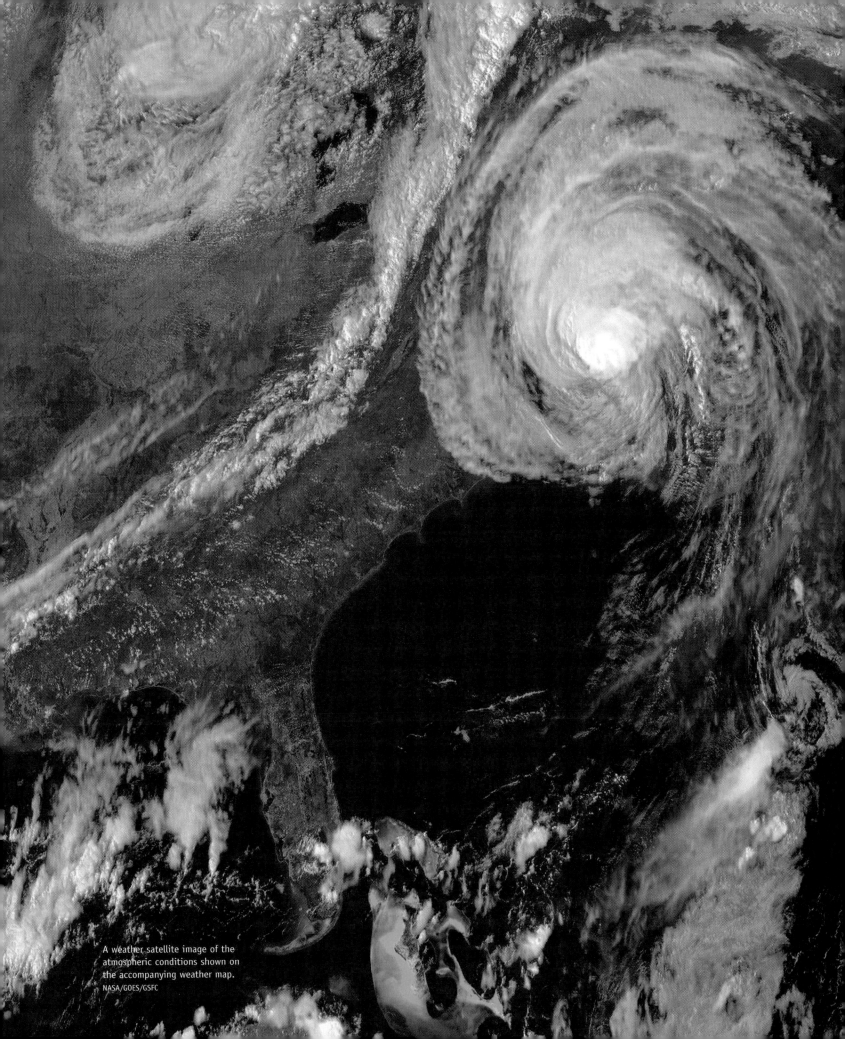

A weather satellite image of the atmospheric conditions shown on the accompanying weather map.

2002

2007

# 8

# GLOBAL CLIMATES AND CLIMATE CHANGE

## ▨ OBJECTIVES

WHEN YOU COMPLETE THIS CHAPTER YOU SHOULD BE ABLE TO:

■ Compare and contrast the advantages and limitations of the Thornthwaite and Köppen climate classification systems.

■ Apply temperature and precipitation statistics to classify climates using the Köppen system.

■ Recognize how the information on climographs is used to identify the climate of a place.

■ Understand why vegetation types are so closely related to climate regions in the Köppen system.

■ Explain why global climates have changed over the past few million years and how Earth scientists have documented these changes.

■ Appreciate why it is so difficult to determine the cause of an Ice Age and recall what hypotheses have been suggested.

■ Discuss the nature of recent global warming and the impact it would likely have on Earth environments.

■ Understand the results of extensive research into the role that human activity has played in causing climate change.

WEATHER AND CLIMATE WERE DEFINED briefly in Chapter 4, and in Chapter 7 we discussed the fundamentals of weather. In this chapter we begin the study of climate in greater detail. Because the study of both weather and climate focuses on the same atmospheric elements and involves the collection and examination of much the same data, the terms are often misunderstood or used interchangeably. However, science requires the student to clearly distinguish between weather and climate.

Weather describes specific atmospheric conditions "here and now." It also involves remarkably accurate forecasts of conditions a week or more in advance and memorable events that occurred at specific times in the past. In contrast, climate describes generalized atmospheric conditions, "past, present, and future," based on statistical averages, extremes, and patterns of change for the atmospheric elements throughout a year over as many years as records have been kept. Earlier climate conditions can be reconstructed from past evidence. In addition, because weather conditions change so often and vary so widely from place to place, weather descriptions refer to specific locations or limited areas, whereas descriptions of climate refer to atmospheric

*The Bear Glacier in Alaska, like many glaciers worldwide, has been shrinking in size under current climatic conditions. These photos show the dramatic change in a 5-year period.* USGS

conditions that are similar throughout large recurring areas that change slowly over periods of many years and are distributed in regular patterns over all Earth's surface.

As we begin the detailed study of climate, you should take special note again of the following illustrations located in previous chapters: average sea-level temperatures in January and July (Fig. 4.27); average sea-level pressures in January and July (Fig. 5.10); the general circulation of Earth's atmosphere (Fig. 5.11); the map of the major world ocean currents (Fig. 5.24); and the world map of average annual precipitation (Fig. 6.27). Each of these global representations of data affects atmospheric conditions that are essential to understanding one or more aspects of global climates and their distributions. We recommend that you review these figures now and refer to them when appropriate as the nature and characteristics of world climates are discussed in this chapter and the two chapters that follow.

In the first part of this chapter we will introduce the characteristics and classification of modern climates. Because climate can be defined at different scales, from a single hillside to a region as large as the Sahara, which stretches across much of northern Africa, two systems of describing and classifying climates are discussed. The Thornthwaite system is introduced because it is one of the most effective systems available to various scientists for the classification of climates on a more local scale. The Köppen system, however, has been widely adopted by physical geographers and other scientists; in a modified version it will be the basis for the worldwide regional study of present-day climates in Chapters 9 and 10.

The remainder of this chapter focuses on climate change. Climates in the past were not the same as they are today, and there is every reason to believe that future climates also will be different. It is now widely recognized that humans have an impact on Earth's climate.

For more than a century, scientists have realized that Earth has experienced major climate shifts during its history. It was believed that these shifts were gradual and could not be detected by humans during their lifetimes. However, recent research reveals that climate has shifted repeatedly between extremes over some exceedingly short intervals.

To predict future climates it is critical that we examine the details of past climate changes, including both the magnitude and rates of prehistoric climate change. Earth has experienced both ice ages and lengthy periods that were warmer than today. These fluctuations serve as indicators of the natural variability of climate in the absence of significant human impact. Using knowledge of present and past climates and models of how and why climate changes, we conclude this chapter with some predictions of future climate trends.

# Classifying Climates

Knowledge that climate varies from region to region dates to ancient times. The early Greeks (such as Aristotle, circa 350 B.C.) classified the known world into torrid, temperate, and frigid zones based on their relative warmth. It was also recognized that these zones varied systematically with latitude and that the flora and fauna also reflected these changes. With the further exploration of the world, naturalists noticed that the distribution of climates could be explained using factors such as sun angles, prevailing winds, elevation, and proximity to large water bodies.

The two weather variables used most often as indicators of climate are temperature and precipitation. To classify climates accurately, climatologists require a minimum of 30 years of data to describe the climate of an area. The invention of an instrument to reliably measure temperature—the thermometer—dates only to Galileo in the early 17th century. Sporadic collection of temperature and precipitation data at distant colonies by Europeans settlers began in the 1700s but was not routine until the mid-19th century. This was soon followed in the early 20th century by some of the first attempts to classify global climates using actual temperature and precipitation data.

As we have seen in earlier chapters, temperature and precipitation vary greatly over Earth's surface. Climatologists have worked to reduce the infinite number of worldwide variations in atmospheric elements to a comprehensible number of groups by combining elements with similar statistics (● Fig. 8.1)—that is, they can classify climates strictly on the basis of atmospheric elements, ignoring the causes of those variations (such as the frequency of air mass movements). This type of classification, based on statistical and mathematical parameters or physical characteristics, is called an **empirical classification.** A classification based on the causes, or *genesis,* of climate variation is known as a **genetic classification.**

Ordering the vast wealth of available climate data into descriptions of major climate groups, on either an empirical or a genetic basis, enables geographers to concentrate on the larger-scale causes of climatic differentiation. In addition, they can ex-

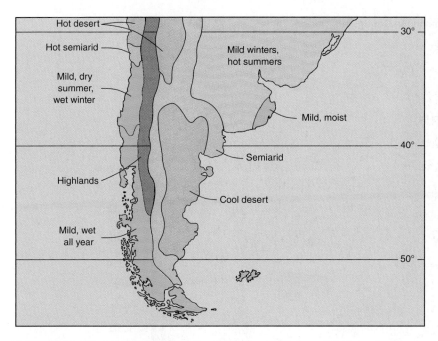

● **FIGURE 8.1** This map shows the diversity of climates possible in a relatively small area, including portions of Chile, Argentina, Uruguay, and Brazil. The climates range from dry to wet and from hot to cold, with many possible combinations of temperature and moisture characteristics.

*What can you suggest as the causes for the major changes in climate types as you follow the 40°S latitude line from west to east across South America?*

amine exceptions to the general relationships, the causes of which are often one or more of the other atmospheric controls. Finally, differentiating climates helps explain the distribution of other climate-related phenomena of importance to humans.

Despite their value, climate classifications are not without problems. Climate is a generalization about observed facts and data based on the averages and probabilities of weather. It does not describe a real weather situation; instead, it presents a composite picture of weather over the seasons. Within such a generalization, it is impossible to include the many variations that actually exist. Thus, classification systems must sometimes be adjusted to changes in climate. On a global scale, generalizations, simplifications, and compromises are made to distinguish among climate types and regions.

## The Thornthwaite System

One system for classifying climates concentrates on a local scale. This system is especially useful for soil scientists, water resource specialists, and agriculturalists. For example, for a farmer interested in growing a specific crop in a particular area, a system that classifies large regions of Earth is inadequate. Identifying the major vegetation type of the region and the annual range of both temperature and precipitation does not provide a farmer with information concerning the amounts and timing of annual soil moisture surpluses or deficits. From an agricultural perspective, it is much more important to know that moisture will be available in the growing season, whether it comes directly in the form of precipitation or from the soil.

Developed by an American climatologist, C. Warren Thornthwaite, the **Thornthwaite system** establishes moisture availability at the subregional scale ( ● Fig. 8.2). It is the system preferred by those examining climates on a local scale. Development of detailed climate classification systems such as the Thornthwaite system became possible only after temperature and precipitation data were widely collected at numerous locations beginning in the latter half of the 19th century.

The Thornthwaite system is based on the concept of **potential evapotranspiration (potential ET),** which approximates the water use of plants with an unlimited water supply. (Evapotranspiration, discussed in Chapter 6, is a combination of evaporation and transpiration, or water loss through vegetation.) Potential ET is a theoretical value that increases with increasing temperature, winds, and length of daylight and decreases with increasing humidity. In contrast, **actual evapotranspiration (actual ET)** reflects actual water use by plants. This water can be supplied during the dry season by soil moisture if the soil is saturated, the climate is relatively cool, and/or the day lengths are short. Thus, measurements of actual ET relative to potential ET and available soil moisture are the determining factors for most vegetation and crop growth.

Turn back to Figure 6.9. This graph shows a visual representation of the Thornthwaite system as it applies to the San Francisco, California area. You should quickly note the striking difference between the relationship of annual potential evapotranspiration in January with that in July. It becomes obvious how the Thornthwaite system helps California farmers determine the best planting, growing, and harvesting seasons for their crops. It also helps them decide when irrigation might be required and how much water might be needed.

The Thornthwaite system recognizes three climate zones based on potential ET values: low-latitude climates, with potential ET greater than 130 centimeters (51 in.); middle-latitude climates, with potential ET less than 130 but greater than 52.5 centimeters (20.5 in.); and high-latitude climates, with potential ET less than 52.5 centimeters. Climate zones may be subdivided based on how long and by how much actual ET is below potential ET. Moist climates have either a surplus or a minor deficit of less than 15 centimeters (6 in.). Dry climates have an annual deficit greater than 15 centimeters.

Thornthwaite's original equations for potential ET were based on analyses of data collected in the midwestern and eastern United States. The method was subsequently used with less success in other parts of the world. Over the past few decades, many attempts have been made to improve the accuracy of the Thornthwaite system for regions outside the United States.

## The Köppen System

The most widely used climate classification is based on regional temperature and precipitation patterns. It is referred to as the **Köppen system** after the German botanist and climatologist

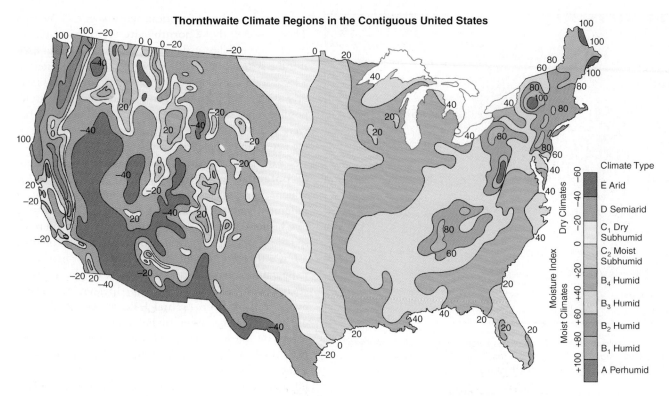

## Thornthwaite Climate Regions in the Contiguous United States

| Climate Type | |
|---|---|
| Dry Climates | E Arid |
| | D Semiarid |
| | C₁ Dry Subhumid |
| | C₂ Moist Subhumid |
| Moist Climates | B₄ Humid |
| | B₃ Humid |
| | B₂ Humid |
| | B₁ Humid |
| | A Perhumid |

● **FIGURE 8.2** Thornthwaite climate regions in the contiguous United States are based on the relationship between precipitation (P) and potential evapotranspiration (ET). The moisture index (MI) for a region is determined by this simple equation:

$$MI = 100 \times \frac{P - \text{Potential ET}}{\text{Potential ET}}$$

Where precipitation exceeds potential ET, the index is positive; where potential ET exceeds precipitation, the index is negative.

*What are the moisture index and Thornthwaite climate type for coastal southern California?*

who developed it. Wladimir Köppen recognized that major vegetation associations reflect the area's climate. Hence, his climate regions were formulated to coincide with well-defined vegetation regions, and each climate region was described by the natural vegetation most often found there. Evidence of the strong influence of Köppen's system is seen in the wide usage of his climatic terminology, even in nonscientific literature (for example, steppe climate, tundra climate, rainforest climate).

### Advantages and Limitations of the Köppen System

Not only are temperature and precipitation two of the easiest weather elements to measure, but they are also measured more often and in more parts of the world than any other variables. By using temperature and precipitation statistics to define his boundaries, Köppen was able to develop precise definitions for each climate region, eliminating the imprecision that can develop in verbal and sometimes in genetic classifications.

Moreover, temperature and precipitation are the most important and effective weather elements. Variations caused by the atmospheric controls will show up most obviously in temperature and precipitation statistics. At the same time, temperature and precipitation are the weather elements that most directly affect humans, other animals, vegetation, soils, and the form of the landscape.

Köppen's climate boundaries were designed to define the vegetation regions. Thus, Köppen's climate boundaries reflect

"vegetation lines." For example, the Köppen classification uses the 10°C (50°F) monthly isotherm because of its relevance to the timberline—the line beyond which it is too cold for trees to thrive. For this reason, Köppen defined the treeless polar climates as those areas where the mean temperature of the warmest month is below 10°C. Clearly, if climates are divided according to associated vegetation types and if the division is based on the atmospheric elements of temperature and precipitation, then the result will be a visible association of vegetation with climate types. The relationship with the visible world in Köppen's climate classification system is one of its most appealing features to geographers and other scientists.

Köppen's system has limitations, however. For example, Köppen considered only average monthly temperature and precipitation in making his climate classifications. These two elements permit estimates of precipitation effectiveness but do not measure it with enough precision to permit comparison from one specific locality to another. In addition, for the purposes of generalization and simplification, Köppen ignored winds, cloud cover, intensity of precipitation, humidity, and daily temperature extremes—much, in fact, of what makes local weather and climate distinctive.

**Simplified Köppen Classification** The Köppen system, as modified by later climatologists, divides the world into six major climate categories. The first four are based on the annual range of temperatures: humid tropical climates *(A)*, humid

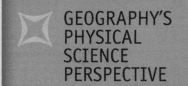

# GEOGRAPHY'S PHYSICAL SCIENCE PERSPECTIVE

# Using Climographs

As stated at the beginning of this chapter, weather and climate are different ways of looking at how our atmosphere affects various locations on Earth. *Weather* deals with the state of the atmosphere at one point in time or in the short term. It describes what is going on outside today or in the next few days. *Climate* deals with the conditions of the atmosphere in the long term—in other words, how the atmosphere behaves in a particular area through the months and years. Usually, a minimum record of 30 years is required to establish what an area's climate might bring. Therefore a climate can be described as, but is not restricted to, a compilation of average

values used to summarize atmospheric conditions of an area.

It is possible to summarize the nature of the climate at any point on Earth in graph form, as shown in the accompanying figure. Given information on mean monthly temperature and rainfall, we can express the nature of the changes in these two elements throughout the year simply by plotting their values as points above or below (in the case of temperature) a zero line. To make the pattern of the monthly temperature changes clearer, we can connect the monthly values with a continuous line, producing an annual temperature curve. Monthly precipitation amounts are usually shown as bars reaching to various

heights above the line of zero precipitation. Such a display of a location's climate is called a climograph. To read the graph, one must relate the temperature curve to the values given along the left side and the precipitation amounts to the scale on the right. Other information may also be displayed, depending on the type of climograph used.

The climograph shown here represents the type that we use in Chapters 9 and 10. This climograph can be used to determine the Köppen classification of the station and to show its specific temperature and rainfall measures. The climate classification abbreviations relating to all climographs are found in Table 8.1.

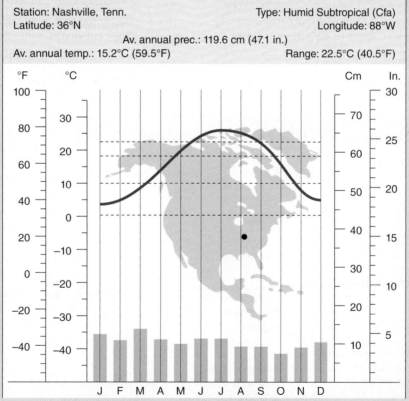

A standard climograph showing average monthly temperature (curve) and rainfall (bars). The horizontal index lines at 0°C (32°F), 10°C (50°F), 18°C (64.4°F), and 22°C (71.6°F) are the Köppen temperature parameters by which the station is classified.

mesothermal (mild winter) climates *(C)*, humid microthermal (severe winter) climates *(D)*, and polar climates *(E)*. Another category, the arid and semiarid climates *(BW and BS)*, identifies regions that are characteristically dry based on both temperature and precipitation values. Because plants need more moisture to survive as the temperature increases, the arid and semiarid climates include regions where the temperatures range from cold to very hot. The final category, highland climates *(H)*, identifies mountainous regions where vegetation and climate vary rapidly as a result of changes in elevation and exposure.

Within each of the first five major categories (all but the highland category), individual climate types and subtypes are differentiated from one another by specific parameters of temperature and precipitation. Table 1 in the "Graph Interpretation" exercise outlines the letter designations and procedures for determining the types and subtypes of the Köppen classification system. This table can be used with any Köppen climate type presented in this chapter and in Chapters 9 and 10.

## The Distribution of Climate Types

Five of the six major climate categories of the Köppen classification include enough differences in the ranges, total amounts, and seasonality of temperature and precipitation to produce the 13 distinctive climate types listed in Table 8.1. The tropical and arid climate types are discussed in some detail in the next chapter, and the mesothermal, micro-

thermal, and polar climates are presented in Chapter 10 along with a brief coverage of undifferentiated highland climates.

Before you continue reading, take the time to study the classification table (in the Graph Interpretation feature at the end of the chapter). Note how many terms in the table are already so familiar to you that they suggest a mental picture: humid, desert, rainforest, ice sheet, continental, marine, and so on. You may instinctively know more about climate than you realize. Then as you examine the photographs associated with the brief introduction to the Köppen system that follows (Figs. 8.3–8.16), discover how closely your mental image of each climate type based on terminology matches the photograph illustrating a location within each climate type. Even more important, what can you identify by studying each photograph that might help you determine the climate type without reading the figure caption?

**Tropical *(A)* Climates** Near the equator we find high temperatures year round because the noon sun is never far from 90° (directly overhead). Humid climates of this type with no winter season are Köppen's **tropical climates.** As his boundary for tropical climates, Köppen chose 18°C (64.4°F) for the average temperature of the coldest month because it closely coincides with the geographic limit of certain tropical palms.

Table 8.1 shows that there are three humid tropical climates, reflecting major differences in the amount and distribution of rainfall within the tropical regions. Tropical climates extend poleward to 30° latitude or higher in the continent's interior but to lower latitudes near the coasts because of the moderating influence of the oceans on coastal temperatures.

Regions near the equator are influenced by the *intertropical convergence zone (ITCZ)*. However, the convergent and rising air of the ITCZ, which brings rain to the tropics, is not anchored in one place; instead, it follows the 90° sun angle (see The Analemma in Chapter 3), migrating with the seasons. Within 5° to 10° latitude of the equator, rainfall occurs year-round because the ITCZ moves through twice a year and is never far away (see ● Fig. 8.3). Poleward of this zone the stabilizing influence of *subtropical high pressure systems* causes precipitation to become seasonal. When the ITCZ is over the region during the high-sun period (summer), there is adequate rainfall. However, during the low-sun period (winter), the subtropical highs and trade winds invade the area, bringing clear, dry weather.

We find the tropical rainforest climate *(Af)* in the equatorial region flanked both north and south by the dry-winter tropical savanna *(Aw)* climate

## Table 8.1
## Simplified Köppen Climate Classes

| Climates | Climograph Abbreviation |
|---|---|
| **Humid Tropical Climates (A)** | |
| Tropical Rainforest Climate | Tropical Rf. |
| Tropical Monsoon Climate | Tropical Mon. |
| Tropical Savanna Climate | Tropical Sav. |
| | |
| **Arid Climates (B)** | |
| Steppe Climate | Low-lat./Mid-lat. Steppe |
| Desert Climate | Low-lat./Mid-lat. Desert |
| | |
| **Humid Mesothermal (Mild Winter) Climates (C)** | |
| Mediterranean Climate | Medit. |
| Humid Subtropical Climate | Humid Subt. |
| Marine West Coast Climate | Marine W.C. |
| | |
| **Humid Microthermal (Severe Winter) Climates (D)** | |
| Humid Continental, Hot-Summer Climate | Humid Cont. H.S. |
| Humid Continental, Mild-Summer Climate | Humid Cont. M.S. |
| Subarctic Climate | Subarctic |
| | |
| **Polar Climates (E)** | |
| Tundra Climate | Tundra |
| Ice-sheet Climate | Ice-sheet |
| | |
| **Highland Climates (H)** | |
| Various climates based on elevation differences. | No single climograph can depict these varied (or various) climates |

● **FIGURE 8.3** Tropical rainforest climate: island of Jamaica.

● **FIGURE 8.4** Tropical savanna climate: East African high plains.

(see ● Fig. 8.4). Finally, along coasts facing the strong, moisture-laden inflow of air associated with the summer monsoon, we find the tropical monsoon *(Am)* climate (see ● Fig. 8.5).

Examine now ● Figure 8.6, the world map of climates. Note that it shows clearly the progression of the humid tropical climates from the equator toward the poles in the equatorial regions of Africa and South America and the major regions associated with the tropical monsoons of Asia. The atmospheric processes that produce the various tropical *(A)* climates are discussed in Chapter 9.

## Polar *(E)* Climates

Just as the tropical climates lack winters (cold periods), the polar climates—at least statistically—lack summers. **Polar climates,** as defined by Köppen, are areas in which no month has an average temperature exceeding 10°C (50°F). Poleward of this temperature boundary, trees cannot survive. The 10°C isotherm for the warmest month more or less coincides with the Arctic Circle, poleward of which the sun does not rise above

● **FIGURE 8.5** Tropical monsoon climate: Himalayan foothills, West Bengal, India.

the horizon in midwinter and, although the length of day increases during polar summers, the insolation strikes at a low angle.

The polar climates are subdivided into tundra and ice-sheet climates. The ice-sheet *(EF)* climate ( ● Fig. 8.7) has no month with an average temperature above 0°C (32°F). The tundra *(ET)* climate ( ● Fig. 8.8) occurs where at least 1 month averages above 0°C (32°F). Look at the far northern regions of Eurasia, North America, and Antarctica in Figure 8.6. The processes creating the polar *(E)* climates are explained in Chapter 10.

## Mesothermal *(C)* and Microthermal *(D)* Climates

Except where arid climates intervene, the lands between the tropical and polar climates are occupied by the transitional middle-latitude mesothermal and microthermal climates. Because they are neither tropical nor polar, the mild and severe winter climates must have at least 1 month averaging below 18°C (64.4°F) and 1 month averaging above 10°C (50°F). Although both middle-latitude climate categories have distinct temperature seasons, the **microthermal climates** have severe winters with at least 1 month averaging below freezing. Once again, vegetation reflects the climatic differences. In the severe-winter climates, all broadleaf and even some species of needle-leaf trees defoliate naturally during the winter (generally, needle-leaf trees do not defoliate in winter) because soil water is temporarily frozen and unavailable. Much of the natural vegetation of the mild-winter **mesothermal climates** retains its foliage throughout the year because liquid water is always present in the soil. The line separating mild from severe winters usually lies in the vicinity of the 40th parallel.

A number of important internal differences within the mesothermal and microthermal climate groups produce individual climate types based on precipitation patterns or seasonal temperature contrasts. The Mediterranean, or dry summer, mesothermal *(Csa, Csb)* climate (like Southern California or southern Spain in Fig. 8.6) appears along west coasts between 30° and 40° latitude ( ● Fig. 8.9). On the east coasts, in generally the same latitudes,

● **FIGURE 8.7** Ice-sheet climate: glaciers near the southern coast of Greenland.

J. Petersen

USGS

● **FIGURE 8.10** Humid subtropical climate: dense mixed forest with pine, oak, and palm species on a coastal island in Georgia.

Alaska Image Library/USFWS

● **FIGURE 8.8** Tundra climate: caribou in the Alaskan tundra during the summer.

the humid subtropical climate is found ( ● Fig. 8.10). This type of climate is found in regions like the southeastern United States and southeastern China in Figure 8.6.

The distinction between the humid subtropical *(Cfa)* and marine west coast *(Cfb, Cfc)* climates illustrates a second important criterion for the internal subdivisions of middle latitude climates: seasonal contrasts. Both mesothermal climates have year-round precipitation, but humid subtropical summer temperatures are much higher than those in the marine west coast climate. Therefore, summers are hot. In contrast, the mild summers of the marine west coast climate, located poleward of the Mediterranean climate along continental west coasts, often extend beyond 60° latitude ( ● Fig. 8.11). Some examples of marine west coast climates, shown in Figure 8.6, are along the northwest coast of the United States and extending into Canada, the west coast of Europe, and the British Isles.

Another example of internal differences is found among microthermal *(D)* climates, which usually receive year-round precipitation

R. Gabler

● **FIGURE 8.9** Mediterranean climate: village in southern Spain.

R. Gabler

● **FIGURE 8.11** Marine west coast climate: North Sea coast of Scotland.

● **FIGURE 8.12** Humid continental climate: fall color along Wagner Falls in the upper peninsula of Michigan.

● **FIGURE 8.14** Desert climate: Sonoran Desert of Arizona.

associated with middle-latitude cyclones traveling along the polar front. Internal subdivision into climate types is based on summers that become shorter and cooler and winters that become longer and more severe with increasing latitude and continentality. Microthermal climates are found exclusively in the Northern Hemisphere ( ● Fig. 8.12) because there is no land in the Southern Hemisphere latitudes that would normally be occupied by these climate types. In the Northern Hemisphere these climates progress poleward through the humid continental, hot-summer *(Dfa, Dwa)* climate to the humid continental, mild-summer *(Dfb, Dwb)* climate and, finally, to the subarctic *(Dfc, Dfd, Dwc, and Dwd)* climate ( ● Fig. 8.13). Microthermal regions can be seen in the eastern United States and Canada or northward through eastern Europe in Figure 8.6.

**Arid *(B)* Climates** Climates dominated by year-round moisture deficiency are called **arid climates.** These climates will penetrate deep into the continents, interrupting the latitudinal zonation of climates that would otherwise exist. The definition of climatic aridity is that precipitation received is less than potential ET. Aridity does not depend solely on the amount of precipitation received; potential ET rates and temperature must also be taken into account. In a low-latitude climate with relatively high temperatures, the potential ET rate is greater than in a colder, higher-latitude climate. As a result, more rain must fall in the lower latitudes to produce the same effects (on vegetation) that smaller amounts of precipitation produce in areas with lower temperatures and, consequently, lower potential ET rates. Potential ET rates also decrease with altitude, which helps to explain why higher altitude (highland) climates are distinguished separately.

Arid climates are concentrated in a zone from about 15°N and S to about 30°N and S latitude along the western coasts, expanding much farther poleward over the heart of each landmass. The correspondence between the arid climates and the belt of subtropical high pressure systems is unmistakable (like in the southwestern United States, central Australia, and north Africa in Fig. 8.6), and the poleward expansion is a consequence of remoteness from the oceanic moisture supply.

In desert *(BW)* climates, the annual amount of precipitation is less than half the annual potential ET ( ● Fig. 8.14). Bordering the deserts are steppe *(BS)* climates—semiarid climates that are transitional between the extreme aridity of the deserts and the moisture surplus of the humid climates ( ● Fig. 8.15). The definition of the steppe climate is an area where annual precipitation is less than potential ET but more than half the potential ET. *B* climates and the processes that create them are discussed in more detail in Chapter 9.

**Highland *(H)* Climates** The pattern of climates and extent of aridity are affected by irregularities in Earth's surface, such as the presence of deep gulfs, interior seas, or significant highlands. The climatic patterns of Europe and North America differ because of such variations.

Highlands can channel air mass movements and create abrupt climatic divides. Their own microclimates form an intricate pattern related to elevation, cloud cover, and exposure ( ● Fig. 8.16). One significant effect of highlands aligned at right angles to the

● **FIGURE 8.13** Subarctic climate: These trees endure a long winter in Alaska.

● **FIGURE 8.15** Steppe climate: sand hills of Nebraska.

● **FIGURE 8.16** Highland climate: Uncompahare National Forest, Colorado.

prevailing wind direction is the creation of arid regions extending tens to hundreds of kilometers leeward. Look at the mountain ranges in Figure 8.6; the Rockies, the Andes, the Alps, and the Himalayas show *H* climates. These undifferentiated **highland climates** are discussed in more detail in Chapter 10.

## Climate Regions

Each of our modified Köppen climate types is defined by specific parameters for monthly averages of temperature and precipitation; thus, it is possible to draw boundaries between these types on a world map. The areas within these boundaries are examples of one type of world region. The term **region,** as used by geographers, refers to an area that has recognizably similar internal characteristics that are distinct from those of other areas. A region may be described on any basis that unifies it and differentiates it from others.

As we examine the climate regions of the world in the chapters that follow, you should make frequent reference to the map

of world climate regions (see Fig. 8.6). It shows the patterns of Earth's climates as they are distributed over each continent. However, a word of caution is in order. On a map of climate regions, distinct lines separate one region from another. Obviously, the lines do not mark points where there are abrupt changes in temperature or precipitation conditions. Rather, the lines signify **zones of transition** between different climate regions. Furthermore, these zones or boundaries between regions are based on monthly and annual averages and may shift as temperature and moisture statistics change over the years.

The actual transition from one climate region to another is gradual, except in cases in which the change is brought about by an unusual climate control such as a mountain barrier. It would be more accurate to depict climate regions and their zones of transition on a map by showing one color fading into another. Always keep in mind as we describe Earth's climates that it is the core areas of the regions that best exhibit the characteristics that distinguish one climate from another.

Now, let's look more closely at Figure 8.6. One thing that is immediately noticeable is the change in climate with latitude. This is especially apparent in North America when we examine the East Coast of the United States moving north into Canada. Here the sun angles and length of day play an important role. We can also see that similar climates usually appear in similar latitudes and/or in similar locations with respect to landmasses, ocean currents, or topography. These climate patterns emphasize the close relationship among climate, the weather elements, and the climate controls. These elements and controls are more fully correlated with their corresponding climates in Chapters 9 and 10. There is an order to Earth's atmospheric conditions and so also to its climate regions.

A striking variation in these global climate patterns becomes apparent when we compare the Northern and Southern Hemispheres. The Southern Hemisphere lacks the large landmasses of the Northern Hemisphere; thus, no climates in the higher latitudes (in land regions) can be classified as humid microthermal, and only one small peninsula of Antarctica can be said to have a tundra climate.

## Scale and Climate

Climate can be measured at different scales (macro, meso, or micro). The climate of a large (macro) region, such as the Sahara, may be described correctly as hot and dry. Climate can also be described at mesoscale levels; for example, the climate of coastal Southern California is sunny and warm, with dry summers and wet winters. Finally, climate can be described at local scales, such as on the slopes of a single hill. This is termed a **microclimate.**

At the microclimate level, many factors will cause the climate to differ from nearby areas. For example, in the United States and other regions north of the Tropic of Cancer, south-facing slopes tend to be warmer and drier than north-facing slopes because they receive more sunlight ( ● Fig. 8.17). This variable is referred to as *slope aspect*—the direction a mountain slope faces in respect to the sun's rays. Microclimatic differences such as slope aspect can cause significant differences in vegetation and soil moisture. In what is

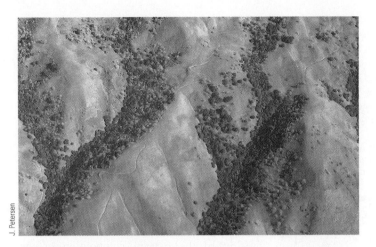

● **FIGURE 8.17** This aerial photograph, facing eastward over valleys in the Coast Ranges of California, illustrates the significance of slope aspect. South-facing slopes receive direct rays of the sun and are hotter and drier than the more shaded north-facing slopes. The south-facing slopes support grasses and only a few trees, whereas the shaded, north-facing slopes are tree covered.

*Why do the differing angles that the sun's rays strike the two opposite slopes affect temperatures?*

sometimes called *topoclimates,* tall mountains often possess vertical zones of vegetation that reflect changes in the microclimates as one ascends from the base of the mountain (which may be surrounded by a tropical-type vegetation) to higher slopes with middle latitude–type vegetation to the summit covered with ice and snow.

Human activities also can influence microclimates. Recent research indicates that the construction of a large reservoir leads to greater annual precipitation immediately downwind of this impounded water. This occurs because the lake supplies additional water vapor to passing storms, which intensifies the rainfall or snows immediately downwind of the lake. These microclimatic effects are similar to the *lake-effect snows* that occur downwind of the Great Lakes in the early winter when the lakes are not frozen (discussed in Chapter 7). Another example of human impact on microclimates is the urban heat-island effect (discussed in Chapter 4), which leads to changes in temperature (urban centers tend to be warmer than their outlying rural areas), rainfall, wind speeds, and many other phenomena.

# Climates of the Past

To try to predict future climates, it is critical to understand the magnitude and frequency of previous climate changes. Knowledge that Earth experienced major climate changes in the past is not new. In 1837 Louis Agassiz, a European naturalist, proposed that Earth had experienced major periods of *glaciation,* periods known as ice ages, when large areas of the continents were covered by huge sheets of ice. He presented evidence that glaciers (flowing ice) had once covered most of England, northern Europe, and Asia and the foothill regions of the Alps. Agassiz arrived in the United States in 1846 and found similar evidence of widespread glaciation throughout North America. Subsequent re-

search by geologists who have studied rock structures that recorded hundreds of millions of years of Earth history indicates that there were several periods of past major glaciations in addition to the one Agassiz described.

## The Recent Ice Ages

Until the 1960s it was widely believed that during its most recent history Earth had experienced four major glacial advances followed by warmer interglacial periods. These glacial cycles occurred during the geologic epoch known as the *Pleistocene* (from about 2.6 million years up to 12,000 years ago). In Europe these glacial epochs were termed the Günz (oldest), Mindel, Riss, and Würm. Likewise in North America, evidence of four glacial periods was recognized; these were termed the Nebraskan (oldest), Kansan, Illinoian, and Wisconsinan glaciations ( ● Fig. 8.18).

A major problem with studying the advance and retreat of glaciers on land is that each subsequent advance of the glaciers tends to destroy, bury, or greatly disrupt the sedimentary evidence of the previous glacial period. The evidence of the fourfold record of glacial advances was largely recognized on the basis of glacial deposits lying beyond the limit of the more recent glaciations. Evidence of "average" glacial advances that were subsequently overridden by more recent glaciers was rarely recognized.

Before the advent of radiometric dating techniques (mineral and organic material can be dated by measuring the extent to which radioactive elements in the material have decayed through time), the timing of the glacial advances in both Europe and the United States was only crudely known. For example, estimates of the age of the last interglacial period were based on the rates at which Niagara Falls had eroded headward after the areas were first exposed when the glaciers retreated. Calculations ranging from 8000 to 30,000 years ago were produced.

## Modern Research

Two major advances in scientific knowledge about climate change occurred in the 1950s. First, radiometric techniques, such as radiocarbon dating, that measured the absolute ages of landforms produced by the glaciers began to be widely used. Radiocarbon dating conclusively showed that the last ice advance peaked about 18,000 years ago. This ice sheet covered essentially all of Canada and the northern United States, extending down to the Ohio and Missouri Rivers. It flowed over modern-day city sites such as Boston, New York, Indianapolis, and Des Moines.

The second major discovery was that evidence of detailed climate changes has been recorded in the sediments on the ocean floors. Unlike the continental record, the deep-sea sedimentary record had not been disrupted by subsequent glacial advances. Rather, the slow, continuous sediment record provides a complete history of climate changes during the past several million years. The most important discovery of the deep-sea record is that Earth has experienced numerous major glacial advances during the Pleistocene period, not just the four that had been identified previously. Today, the names of only two of the North American glacial periods, the Illinoian and Wisconsinan, have been retained.

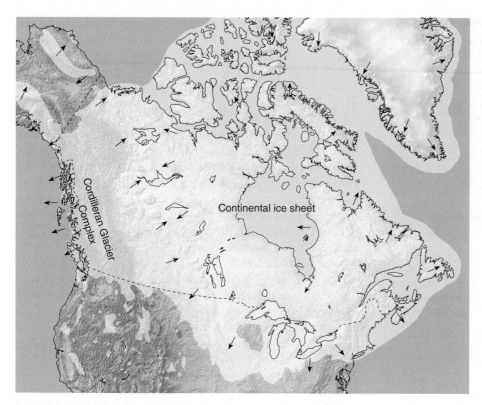

● **FIGURE 8.18** This map identifies the extensive areas of Canada and the northern United States that were covered by moving sheets of ice as recently as 18,000 years ago.

*Why does the ice move in various directions in different regions of the continent?*

Because the deep-sea sedimentary record is so important to climate-change studies, it is important to understand how the record is deciphered. The deep-sea mud contains the microscopic record of innumerable surface-dwelling marine animals that built tiny shells for protection. When they died, these tiny shells sank to the seafloor, forming the layers of mud. Different species thrive in different surface-water temperatures; therefore, the stratigraphic record of the tiny fossils produces a detailed history of water-temperature fluctuations.

These tiny seashells are composed of calcium carbonate ($CaCO_3$); therefore, the analyses also record the oxygen composition of the seawater in which they were formed. One common measurement technique for determining oxygen composition is known as **oxygen-isotope analysis** (further explained in the next section). Modern seawater has a fixed ratio of the two oxygen isotopes. The $^{18}O/^{16}O$ ratios will indicate changes in ocean temperatures relating to glacial cycles. A review of the oxygen-isotope record indicates that the last glacial advance, which peaked about 18,000 years ago, was only one of many major glacial advances during the past 2.6 million years. Evidence has suggested there may have been as many as 28 glacial-type climatic episodes.

Turn now to a close examination of ● Figure 8.19. This graph shows a record of average global temperatures for the past 110,000 years, based on oxygen-isotope analyses of ice cores. Note

● **FIGURE 8.19** Analyses of oxygen isotope ratios in ice cores taken from the glacial ice of Antarctica and Greenland provide evidence of surprising shifts of climate over short periods.

*Has the general trend of temperatures on Earth been warmer or colder during the Holocene?*

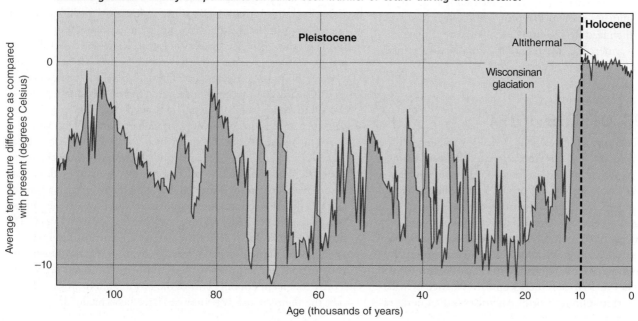

that temperatures have varied as much as 10°C (18°F) or more during this brief period of Earth history. More significantly, climatologists believe that the modern climate epoch, known as the *Holocene,* is a time of relatively stable, warm temperatures when compared with most of the last 2.6 million years of intermittent glaciation. Based on the deep-sea record, it appears that global climates tend to rest at one of two extremes: a very cold interval characterized by major glaciers and lower sea levels and shorter intervals between the glacial advances marked by unusually warm temperatures and high sea levels. With the realization that global climates have changed dramatically numerous times, two obvious questions arise: What causes global climate to change, and how quickly does global climate change from one extreme to the other?

## Methods for Revealing Climates of the Past

Many methods are used to uncover clues to the climates of the past. They are too numerous to mention in this one chapter. However, there are a few reliable methods that have been used for many years in reconstructing paleoclimates (ancient climates). Radiocarbon dating, mentioned earlier, is a means of determining how old an object (which contained carbon) may be. This is a helpful tool but in itself does not indicate the climate of the past. *Oxygen isotope analysis,* however, is a means of reconstructing paleoclimates.

To understand this method, it is helpful to review some basic definitions used in physics and chemistry. *Isotopes* are defined as atoms with the same atomic number but different atomic mass. The atomic number is equal to the positive charge of the nucleus—essentially, the number of *protons* in the nucleus. The atomic mass (or atomic weight) is equal to the number of *protons* and *neutrons* that comprise the nucleus of the atom. *Electrons,* which orbit the atom, are negatively charged particles that possess no appreciable mass (or weight). In a neutral atom, the number of electrons should equal the number of protons. When an electron is gained or lost to the atom, then a net (−) or (+) charge, respectively, will result, and the particle is then classified as an *ion.*

When dealing with isotopes, the atomic number (proton number) must remain the same (giving the atom its identity in the periodic table of elements), but the number of neutrons can vary. In the example of oxygen isotopes, the atomic number is always 8. This is necessary to identify the atom as oxygen. However, the atom may contain 8 neutrons ($^{16}O$, the lighter isotope) or 10 neutrons ($^{18}O$, the heavier isotope). In oxygen-isotope analysis, the ratio of $^{18}O$ to $^{16}O$ is measured and compared to normal values. We have already discussed the $^{18}O/^{16}O$ ratios of seafloor ($CaCO_3$) sediment; however, when dealing with yearly layers within Greenland and Antarctic ice cores, we must use them in a different way. When water evaporates from the ocean, slightly more of the $^{16}O$ than $^{18}O$ evaporates because water containing the lighter-weight oxygen evaporates more readily. During an ice age, the evaporated water is stored in the form of glacial ice rather than returned to the oceans and the $^{18}O/^{16}O$ ratio in the ocean changes slightly to reflect the $^{16}O$-enriched water being stored in the glaciers. In this way $^{18}O/^{16}O$ ratios can help reconstruct climates of the past from glacial ice layers ( ● Fig. 8.20).

Through time paleoclimatologists have discovered new and different ways to determine climates of the past. The oxygen-isotope analysis is one of the most widely accepted methods, but there also are other long-established methods. Two are worth a brief discussion: dendrochronology and palynology.

**Dendrochronology** (or tree-ring dating) has been used for decades. This analysis calls for the examination of tree rings exposed by cores taken through the middle of certain species of trees. The core (small enough so as not to harm the tree) will reveal each yearly tree ring. The rings are counted back through time to establish a time scale for the analysis. Each ring, by its thickness, color, and texture, can reveal the climate conditions (temperature and precipitation characteristics) during that particular year of the tree's growth. Thus, a short-term climate record can be determined through careful examination of tree rings ( ● Fig. 8.21).

**Palynology** (pollen analysis) is also a well-established way of reconstructing past climates and environments. Although pollen samples can be recovered from a variety of environments, lakes and organic bogs are the best places to extract the samples to be analyzed. A core is drilled and removed to show the layers of sediment and organic material all the way to the bottom layer of the lake or bog. Then each layer with organic material can be radiocarbon dated to identify its age. Afterward, all the pollen is

● **FIGURE 8.20** Paleoclimates can be reconstructed using $^{18}O/^{16}O$ ratios found in ice cores gathered from Greenland and Antarctica.

*How can they distinguish the age of the various layers of ice?*

© Kendrick Taylor, Desert Research Institute

Finally, a **precession cycle** has been recognized with a periodicity of 21,000 years. The precession cycle determines the time of year that perihelion occurs. Today Earth is closest to the sun on January 3 and, as a result, receives about 3.5% greater insolation than the average in January. When aphelion occurs on January 3 in about 10,500 years, the Northern Hemisphere winters should be somewhat colder (Fig. 8.24c and d).

These cycles operate collectively, and the combined effect of the three cycles can be calculated. The first person to examine all three of these cycles in detail was the mathematician Milutin Milankovitch, who completed the complex mathematical calculations and showed how these changes in Earth's orbit would affect insolation. Milankovitch's calculations indicated that numerous glacial cycles should occur during 1 million-year intervals.

By the late 1970s most paleoclimate (ancient climate) scientists were convinced that an unusually good correlation existed between the deep-sea record and Milankovitch's predictions. This suggests that the primary driving force behind glacial cycles is regular orbital variations, and it indicates that long-term climate cycles are entirely predictable. Unfortunately for humans, the Milankovitch theory indicates that the warm Holocene interglacial will soon end and that Earth is destined to experience full glacial conditions (glacial ice possibly as far south as the Ohio and Missouri Rivers) in about 20,000 years.

## Changes in the Ocean

Oceans cover more than 70% of Earth's surface. Their enormous volume and high heat capacity make the oceans the single largest buffer against changes in Earth's climate. Whenever changes occur in oceanic temperatures, chemistry, or circulation, significant changes in global climate are certain to follow.

Surface oceanic currents are driven mostly by winds. However, a much slower circulation deep below the surface moves large volumes of water between the oceans. A major driving force of the deep circulation appears to be differences in water buoyancy caused by differences in salinity (salt content). Where surface evaporation is rapid, the rising salinity content causes the seawater density to increase, inducing subsidence. However, when major influxes of freshwater flow from adjacent continents or concentrations of melting icebergs flood into the oceans, the salinity is reduced, thereby increasing the buoyancy of the water. When the surface water is buoyant, deepwater circulation slows. In many cases, the freshwater influx is immediately followed by a major flow of warm surface waters into the North Atlantic, causing an abrupt warming of the Northern Hemisphere. Subsurface ocean currents are also affected by water temperature. Extremely cold Arctic and Antarctic waters are dense and tend to subside, whereas tropical water is warmer and may tend to rise. Therefore, salinity and temperature taken together bring about complex subsurface flows deep within our ocean basins.

In modern times, short-term changes in Pacific circulation are primarily responsible for El Niño and La Niña events (discussed in Chapter 5). The onset of El Niño/Southern Oscillation (ENSO) climatic events is both rapid and global in extent, and it is widely believed that changes in oceanic circulation may be responsible for similar rapid climate changes during the last 2.6 million years.

## Changes in Landmasses

The third category of climate change theories involves changes in Earth's surface to explain lengthy periods of cold climates. A number of ice ages, some with multiple glacial advances, occurred during Earth's history. To explain some of the previous glacial periods, scientists have proposed several factors that might be responsible. For example, one characteristic that all of these glacial periods have in common with the Pleistocene is the presence of a continent in polar latitudes. Polar continents permit glaciers to accumulate on land, which results in lowered sea levels and consequent global effects.

Another geologic factor sometimes invoked as a cause of climate change is the formation, disappearance, or movement of a landmass that restricts oceanic or atmospheric circulation. For example, eruptions of volcanoes and the formation of the Isthmus of Panama severed the connection between the Atlantic and Pacific, thereby closing a pathway of significant ocean circulation. This redirection of ocean water created the Gulf Stream Current/North Atlantic Drift (see again Fig. 5.24). Another example is the uplift of the Himalayas, altering atmospheric flows, and monsoonal effects in Asia. Both of these events, and several other significant changes, immediately predate the onset of the modern series of glaciations. Which events caused climate changes and which are simply coincidences have yet to be determined.

Shifting in landmasses (in both latitude and altitude) will affect changes in the types and distributions of vegetation. These changes would further affect atmospheric composition and atmospheric circulation patterns.

A final group of theories involves changes in albedo, caused either by major snow accumulations on high-latitude landmasses or by large oceanic ice sheets drifting into lower latitudes. The increased reflection of sunlight starts a positive feedback cycle of cooling that may end when the polar oceans freeze, shutting off the primary moisture source for the polar ice sheets.

## Impact Events

As described in Chapter 3, *asteroids* are small, rocky, or metallic solar system bodies, usually less than 800 kilometers (500 mi) in diameter. They may break apart into smaller pieces called meteoroids. These objects orbit our sun, along with comets. *Comets* are small objects of rocky or iron material held together by ice. A comet's ice will vaporize in sunlight, leaving a distinguishable tail of dust or gas. Through time these objects have struck Earth, some with devastating impact. There is no doubt such impacts will occur again. It is not a matter of if but when another impact will take place ( ● Fig. 8.25).

On a daily basis Earth is bombarded with tons of this material; most are so small that they burn up in our atmosphere before hitting the ground. At night these small objects are seen as shooting stars. Objects that are smaller than 40 meters in diameter will incinerate with the friction encountered in our atmosphere. Objects ranging from 40 meters to about 1 kilometer in diameter can do tremendous damage on a local scale when they reach Earth's surface. This size impact can be expected every 100 years or so. The last occurred in 1908 near Tunguska, Siberia, and devastated a huge area of the Siberian wilderness with a blast estimated at 15 megatons (a megaton explosion is equal to 1 million tons of TNT).

● **FIGURE 8.25** Barringer Crater, Arizona, shows the results of an impact with an iron-nickel meteorite of about 50 meters (165 ft) in diameter.

*What effect should these changes in receipt of insolation have on global climates?*

© Bob Llewellyn/Medioimages/jupiterimages

Every few hundred thousand years or so, an object with a diameter of greater than 1.6 kilometers (1 mi or greater) will hit Earth, producing severe environmental damage and climate change on a global scale. The power of the blasts from such impacts could equal a million megatons of energy. The likely effect would be an "impact winter," characterized by skies darkened with particulates, thereby blocking insolation and causing a drastic drop in temperatures. Firestorms would result from heated impact debris raining back down on Earth, and large amounts of acid rain would precipitate. A catastrophe of this kind would result in loss of crops worldwide, followed by starvation and disease. The largest known impacts on Earth in its history, like the one that may have contributed to the extinction of the dinosaurs 65 million years ago, have been estimated to be about 15 kilometers (about 10 mi) in diameter and may have exploded with a force of 100 million megatons.

A vigilant group of professional and amateur astronomers are constantly watching the night skies for **near Earth objects (NEOs)** or objects whose trajectory may bring them into a collision course with our planet. If we know far enough in advance that an NEO will collide with Earth, we may be able to avert a disaster by modifying the object's course so as to avoid a collision. At this point, however, whether anything can be done about such a collision remains mere speculation.

## Changes in Earth's Atmosphere

Many theories attribute climate changes to variations in atmospheric constituents—some of which enter the atmosphere through natural processes and others that are a product of human activities. Certain gases tend to retain atmospheric heat energy and tend to cause warming trends, if their levels increase. Solid particles, such as dust and volcanic emissions, block solar radiation, thus increasing amounts tend to cause cooling. Volcanic eruptions can pump enormous quantities of particulates and aerosols (especially sulfur dioxide) into the stratosphere, where strong winds can spread it around the world ( ● Fig. 8.26).

● **FIGURE 8.26** In 1991 the massive eruptions of Mount Pinatubo in the Philippines expelled gases and particulates 35 km (22 mi) into the atmosphere. Satellites tracked the cloud of volcanic emissions as it traveled around the world several times.

*Besides affecting the climate, what other hazards result from volcanic explosions?*

⊕ In Google Earth fly to: 15.145°N, 120.349°E (Mt. Pinatubo, Philippines). Examine the landform from multiple angles.

USGS/Richard P. Hoblitt

# Graph Interpretation

## THE KÖPPEN CLIMATE CLASSIFICATION SYSTEM

The key to understanding any classification system is found by personally practicing use of the system. This is one reason why the Consider and Respond review sections of both Chapters 9 and 10 are based on the classification of data from sites selected throughout the world. Correctly classifying these sites in the modified Köppen system may seem complicated at first, but you will find that, after applying the system to a few of the sample locations, the determination of a correct letter symbol and associated climate name for any other site data should be routine.

Before you begin, take the time to familiarize yourself with Table 1. You will note that there are precise definitions in regard to

temperature or precipitation that identify a site as one of the five major climate categories in the Köppen system (*A*, tropical; *B*, arid; *C*, mesothermal; *D*, microthermal; *E*, polar). Furthermore, you will note that the additional letters required to identify the actual climate type also have precise definitions or are determined by the use of the graphs. In other words, Table 1 is all you need to classify a site if monthly and annual means of precipitation and temperature are available. Table 1 should be used in a systematic fashion to determine first the major climate category and, once that is determined, the second and third letter symbols (if needed) that complete the classification.

### TABLE 1
**Simplified Köppen Classification of Climates**

| First Letter | Second Letter | Third Letter |
|---|---|---|
| *E*  Warmest month less than 10°C (50°F) | *T*  Warmest month between 10°C (50°F) and 0°C (32°F) | NO THIRD LETTER (with polar climates) |
| POLAR CLIMATES | | SUMMERLESS |
| *ET*–Tundra | *F*  Warmest month below 0°C (32°F) | |
| *EF*–Ice Sheet | | |
| | | |
| *B*  Arid or semiarid climates | *S*  Semiarid climate (see Graph 1) | *h*  Mean annual temperature greater than 18°C (64.4°F) |
| ARID CLIMATES | | |
| *BS*–Steppe | *W*  Arid climate (see Graph 1) | *k*  Mean annual temperature less than 18°C (64.4°F) |
| *BW*–Desert | | |

## Graph 1  Humid/Dry Climate Boundaries

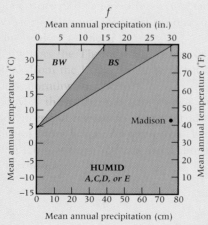

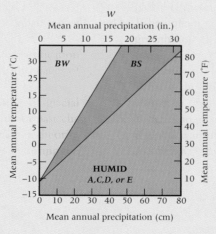

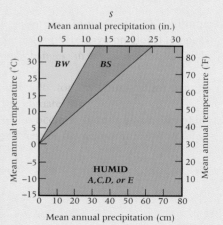

TABLE 1
**Simplified Köppen Classification of Climates** *(Continued)*

| First Letter | Second Letter | Third Letter |
|---|---|---|
| **A** Coolest month greater than 18°C (64.4°F) TROPICAL CLIMATES *Am*—Tropical monsoon *Aw*—Tropical savanna *Af*—Tropical rainforest | **f** Driest month has at least 6 cm (2.4 in.) of precipitation **m** Seasonally, excessively moist (see Graph 2) **w** Dry winter, wet summer (see Graph 2) | NO THIRD LETTER (with tropical climates) WINTERLESS |
| **C** Coldest month between 18°C (64.4°F) and 0°C (32°F); at least 1 month over 10°C (50°F) MESOTHERMAL CLIMATES *Csa, Csb*—Mediterranean *Cfa, Cwa*—Humid subtropical *Cfb, Cfc*—Marine west coast | **s** (DRY SUMMER) Driest month in the summer half of the year, with less than 3 cm (1.2 in.) of precipitation and less than one third of the wettest winter month | **a** Warmest month above 22°C (71.6°F) **b** Warmest month below 22°C (71.6°F), with at least 4 months above 10°C (50°F) |
| **D** Coldest month less than 0°C (32°F); at least 1 month over 10°C (50°F) MICROTHERMAL CLIMATES *Dfa, Dwa*—Humid continental, hot summer *Dfb, Dwb*—Humid continental, mild summer *Dfc, Dwc, Dfd, Dwd*—Subarctic | **w** (DRY WINTER) Driest month in the winter half of the year, with less than one tenth the precipitation of the wettest summer month **f** (ALWAYS MOIST) Does not meet conditions for *s* or *w* above | **c** Warmest month below 22°C (71.6°F), with 1 to 3 months above 10°C (50°F) **d** Same as *c*, but coldest month is below −38°C (−36.4°F) |

## Graph 2

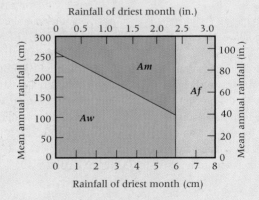

As you begin to classify climates, it is strongly recommended that you use the following procedure. (After a few examples you may find that you can omit some steps with a glance at the statistics.)

Step 1. Ask: Is this a polar climate (E)? Is the warmest month less than 10°C (50°F)? If so, is the warmest month between 10°C (50°F) and 0°C (32°F) (ET) or below 0°C (32°F) (EF)? If not, move on to Step 2.

Step 2. Ask: Is there a seasonal concentration of precipitation? Examine the monthly precipitation data for the driest and wettest summer and winter months for the site. Take careful note of temperature data also because you must determine whether the site is located in the Northern or Southern Hemisphere. (April to September are summer months in the Northern Hemisphere but winter months in the Southern Hemisphere. Similarly, the Northern Hemisphere winter months of October to March are summer south of the equator.) As the table indicates, a site has a dry summer (s) if the driest month in summer has less than 3 centimeters (1.2 in.) of precipitation and less than one third of the precipitation of the wettest winter month. It has a dry winter (w) if the driest month in winter has less than one tenth the precipitation of the wettest summer month. If the site has neither a dry summer nor a dry winter, it is classified as having an even distribution of precipitation (f). Move on to Step 3.

Step 3. Ask: Is this an arid climate (B)? Use one of the small graphs (included in Graph 1) to decide. Based on your answer in Step 2, select one of the small graphs and compare mean annual temperature with mean annual precipitation. The graph will indicate whether the site is an arid (B) climate or not. If it is, the graph

will indicate which one (BW or BS). You should further classify the site by adding h if the mean annual temperature is above 18°C (64.4°F) and k if it is below. If the site is neither BW nor BS, it is a humid climate (A, C, or D). Move on to Step 4a.

Step 4a. Ask: Is this a tropical climate (A)? The site has a tropical climate if the temperature of the coolest month is higher than 18°C (64.4°F). If so, use Graph 2 in Table 1 to determine which tropical climate the site represents. (Note that there are no additional lowercase letters required.) If not, move on to Step 4b.

Step 4b. Ask: Which major middle latitude climate group does that site represent, mesothermal (C) or microthermal (D)? If the temperature of the coldest month is between 18°C (64.4°F) and 0°C (32°F), the site has a mesothermal climate. If it is below 0°C (32°F), it has a microthermal climate. Once you have answered the question, move on to Step 5.

Step 5. Ask: What was the distribution of precipitation? This was determined back in Step 2. Add s, w, or f for a C climate or w or f for a D climate to the letter symbol for the climate. Then, move on to Step 6.

Step 6. Ask: What is needed to express the details of seasonal temperature for the site? Refer again to Table 1 and the definitions for the letter symbols. Add a, b, or c for the mesothermal (C) climates or a, b, c, or d for the microthermal (D) climates, and you have completed the classification of your climate. However, note that you may not have come this far because you might have completed your classification at Steps 1, 3, or 4a.

We should now be ready to try out the use of Table 1, following the steps we have recommended. Data for Madison, Wisconsin, is presented below for our example.

| | J | F | M | A | M | J | J | A | S | O | N | D | Year |
|---|---|---|---|---|---|---|---|---|---|---|---|---|---|
| T (°C) | −8 | −7 | −1 | 7 | 13 | 19 | 21 | 21 | 16 | 10 | 2 | −6 | 7 |
| P (cm) | 3.3 | 2.5 | 4.8 | 6.9 | 8.6 | 11.0 | 9.6 | 7.9 | 8.6 | 5.6 | 4.8 | 3.8 | 77.0 |

The correct answer is derived as follows:

Step 1. We must determine whether our site has an E climate. Because Madison has several months averaging above 10°C, it does not have an E climate.

Step 2. We must determine if there is a seasonal concentration of precipitation. Because Madison is driest in winter, we compare the 2.5 centimeters of February precipitation with the precipitation of June (1/10 of 11.0 cm, or 1.1 cm) and conclude that Madison has neither a dry summer nor a dry winter but instead has an even distribution of precipitation (f). (Note: The 2.5 cm of February precipitation is not less than 1/10 [1.1 cm] of June precipitation.)

Step 3. Next we assess, through the use of Graphs 1(f), 1(w), or 1(s), whether our site is an arid climate (BW, BS) or a humid climate (A, C, or D). Because we have previously determined that Madison has an even distribution of precipitation, we will use Graph 1(f). Based on Madison's mean annual precipitation (77.0 cm) and mean annual temperature (7°C), we conclude that Madison is a humid climate (A, C, or D).

Step 4. Now we must assess which humid climate type Madison falls under. Because the coldest month (−8°C) is below 18°C, Madison does not have an A climate. Although the warmest month (21°C) is above 10°C, the coldest month (−8°C) is not between 0°C and 18°C, so Madison does *not* have a *C* climate. Because the warmest month (21°C) is above 10°C and the coldest month is below 0°C, Madison *does* have a *D* climate.

Step 5. Because Madison has a *D* climate, the second letter will be *w* or *f*. Because precipitation in the driest month of winter (2.5 cm) is not less than one tenth of the amount of the wettest summer month ($1/10 \times 11.0$ cm = 1.1 cm), Madison does *not* have a *Dw* climate. Madison therefore has a *Df* climate.

Step 6. Because Madison is a *Df* climate, the third letter will be *a, b, c,* or *d*. Because the average temperature of the warmest month (21°C) is not above 22°C, Madison does *not* have a *Dfa* climate. Because the average temperature of the warmest month is below 22°C, with at least 4 months above 10°C, Madison is a *Dfb* climate.

# 9

# LOW-LATITUDE AND ARID CLIMATE REGIONS

## OUTLINE

**Humid Tropical Climate Regions**

Tropical Rainforest Climate

Geography's Environmental Perspective: Rainforest Deforestation

Tropical Monsoon Climate

Tropical Savanna Climate

**Arid Climate Regions**

Desert Climates

Geography's Spatial Perspective: Desertification

Steppe Climates

Waterholes are important to sustaining animals in Tropical Savanna climate regions during the dry season. This scene is in Etosha National Park in Namibia, southern Africa. dirkr/Shutterstock.com

## OBJECTIVES

WHEN YOU COMPLETE THIS CHAPTER YOU SHOULD BE ABLE TO:

- Explain the temperature basis used to designate a climate as tropical, the precipitation characteristics used to classify the different humid tropical climates, and the criteria for designating a climate as arid (desert and steppe).
- Locate the general areas of each tropical (rainforest, monsoon, and savanna) and arid climate (desert and steppe) on a world map, and explain the major factors that control their global distribution.
- Outline the major characteristics of each of the tropical and arid climates.
- Describe important vegetation types related to each of the humid tropical and arid climates.
- Recognize a specific humid tropical or arid climate in the Köppen classification system by examining a climograph or a set of data from a representative station.
- Explain some of the adaptations humans have made to living in the humid tropical and arid climate regions.
- Discuss the environmental challenges, hazards, and benefits to living in the humid tropical and arid climate regions.

THE ODDS ARE OVERWHELMING THAT within your lifetime you will travel to destinations far from where you live today, or perhaps you have done so already. You may move to a new home or place of employment, or you may travel on business or for pleasure. Whatever the reason, it is likely that you will ask the question almost every other traveler asks: "What will the weather be like at my destination?"

As you have learned from previous chapters, the question should probably be, "I wonder what the climate is like?" The constant variability associated with weather in many areas of the world may make it difficult to predict. However, the long-term averages and ranges on which climate is based allow geographers to provide travelers with a general idea of the atmospheric conditions likely to be experienced at specific world locations during different times of the year.

Of course, there are many reasons other than travel why knowledge of climate and its variation on Earth is a valuable asset. Being familiar with climates in other areas of the world helps us understand the adaptations to atmospheric conditions that have been made by the people who live there. We can better appreciate some

of their economic activities and certain aspects of their cultures. In addition, the climate of any place on Earth is a major controlling factor on native vegetation and animal life. It influences the rate and manner by which rock material is modified, contributing to the development of soil. Climate is an important factor in the processes that influence the development and modification of landforms and physical landscapes. In short, knowledge of climates provides endless clues, not only to atmospheric conditions, but also to numerous other aspects of the physical environment.

In this chapter and the next you will be provided with a broad survey of world climates: their locations, distributions, general characteristics, processes, associated features, and related human activities. The information contained in these two chapters can serve as a valuable knowledge base as you prepare for the future. Throughout these chapters we use the modified version of the Köppen climate classification that was introduced in Chapter 8. It is interesting to note that each of the climates discussed can be found in North America, so you will still find that your knowledge of climates provides a valuable preparation for traveling, vacationing, or moving to a new location, even if you never journey beyond this continent.

# Humid Tropical Climate Regions

We have already learned a good deal about the tropical (A) climate regions of the humid tropics through a preliminary discussion of these climate types in Chapter 8. In addition, we can review the location of these climates in relation to other climate regions by regularly examining the world map of climates (see Fig. 8.6). It now remains for us to identify the major characteristics of each humid tropical climate type in turn, along with its associated world regions.

As you read the detailed discussion of each of Köppen's five major climate categories described in this chapter and in Chapter 10, you should begin by taking careful note of two important aids to identifying and recalling the general location and distinguishing characteristics of each climate and its regional location and extent. The first aid is a world index map focused exclusively on locating the climates to be discussed. The second is a table that summarizes important information about each climate. As you study humid tropical climates, for example, begin by examining ● Figure 9.1 and

● **FIGURE 9.1** Index map of humid tropical climates.

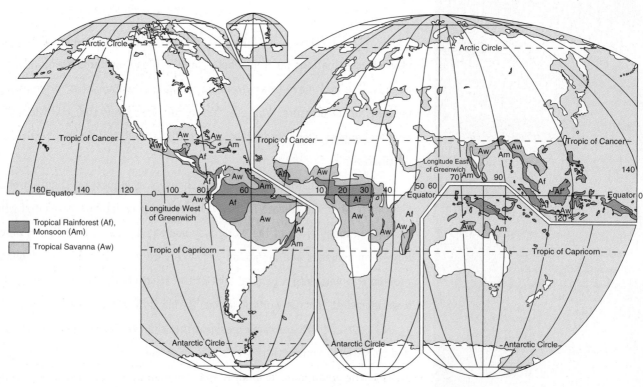

Humid Tropical Climates

TABLE 9.1
The Humid Tropical Climates

| Name and Description | Controlling Factors | Geographic Distribution | Distinguishing Characteristics | Related Features |
|---|---|---|---|---|
| **Tropical Rainforest** | | | | |
| Coolest month above 18°C (64.4°F); driest month with at least 6 cm (2.4 in.) of precipitation | High year-round insolation and precipitation of doldrums (ITCZ); rising air along trade wind coasts | Amazon R. Basin, Congo R. Basin, east coast of Central America, east coast of Brazil, east coast of Madagascar, Malaysia, Indonesia, Philippines | Constant high temperatures; equal length of day and night; lowest (2°C–3°C/3°F–5°F) annual temperature ranges; evenly distributed heavy precipitation; high amount of cloud cover and humidity | Tropical rainforest vegetation (selva); jungle where light penetrates; tropical iron-rich soils; climbing and flying animals, reptiles, and insects; slash-and-burn agriculture |
| **Tropical Monsoon** | | | | |
| Coolest month above 18°C (64.4°F); one or more months with less than 6 cm (2.4 in.) of precipitation; excessively wet during rainy season | Summer onshore and winter offshore air movement related to shifting ITCZ and changing pressure conditions over large landmasses; also transitional between rainforest and savanna | Coastal areas of southwest India, Sri Lanka, Bangladesh, Myanmar, southwest Africa, Guyana, Surinam, French Guiana, northeast and southeast Brazil | Heavy high-sun rainfall (especially with orographic lifting), short low-sun drought; 2°C–6°C (3°F–10°F) annual temperature range, highest temperature just prior to rainy season | Forest vegetation with fewer species than tropical rainforest; grading to jungle and thorn forest in drier margins; iron-rich soils; rainforest animals with larger leaf-eaters and carnivores near savannas; paddy rice agriculture |
| **Tropical Savanna** | | | | |
| Coolest month above 18°C (64.4°F); wet during high-sun season, dry during lower-sun season | Alternation between high-sun doldrums (ITCZ) and low-sun subtropical highs and trades caused by shifting winds and pressure belts | Northern and eastern India, interior Myanmar and Indo-Chinese Peninsula; northern Australia; borderlands of Congo R., south central Africa; llanos of Venezuela, campos of Brazil; western Central America, south Florida, and Caribbean Islands | Distinct high-sun wet and low-sun dry seasons; rainfall averaging 75–150 cm (30–60 in.); highest temperature ranges for humid tropical climates | Grasslands with scattered, drought-resistant trees, scrub, and thorn bushes; poor soils for farming, grazing more common; large herbivores, carnivores, and scavengers |

Table 9.1. Together they provide the locations and previews of the significant facts associated with each individual tropical climate. Table 9.1 even serves as a reminder that, although each of the humid tropical climates has high average temperatures throughout the year, the major differences among them are the annual amounts and distributions of precipitation for each climate. When studying a climate category, it is useful to refer to both map and table before, during, and after reading the text.

## Tropical Rainforest Climate

The **tropical rainforest climate** probably comes most readily to mind when someone says the word *tropical*. Hot and wet throughout the year, tropical rainforest climates have been the stage for many stories of both fact and fiction. If you are a fan of classic movies, you could not easily forget the life-and-death struggle with the elements portrayed by Humphrey Bogart and Katharine Hepburn in the 1951 film *The African Queen*. Or, more

# Rainforest Deforestation

Currently, an area of Earth's virgin tropical forest somewhat larger than a football field is being destroyed every second. During a recent decade, the rate of deforestation doubled, and in the Amazon Basin, where more than half of the rainforest resources are located, the rate nearly tripled. Simple mathematics indicates that even at the present rate of deforestation, the Amazon rainforest will virtually disappear in 150 years.

From the point of view of a developing country, there are basic economic reasons for clearing the Amazon rainforest, and Peru, Ecuador, Colombia, and Brazil (where most of the forest is located) are developing countries. Tropical timber sales provide short-term income to finance national growth and repay staggering debts owed to foreign banks. In addition, Brazil in particular has viewed the Amazon rainforest as a frontier land available for agricultural development and for resettlement of the poor from overcrowded urban areas.

No one questions that the decisions concerning the future of the Amazon rainforest rest with the governments of nations that control these resources. So why has the rainforest become a serious international issue? The answer is found in the concerns of physical geographers and other environmental scientists throughout the world.

When the tropical rainforests are removed, the hydrologic cycle and energy budgets in the previously forested areas are dramatically altered, often irreversibly. In tropical rainforests the canopy shades the forest floor, thus helping to keep it cooler. In addition, the huge mass of vegetation provides a tremendous amount of water vapor to the atmosphere through transpiration. The water vapor condenses to form clouds, which in turn provide rainfall to nourish the forest. With the forests removed, transpiration is diminished, which leads to less cloud cover and less rainfall. With fewer clouds and no forest

A section of Amazon rainforest cleared by slash-and-burn techniques for potential farming or grazing.

Cattle grazing along the Rio Salimoes in Brazil in an area of former rainforest.

recently you may have seen newspaper stories or movies depicting challenging environmental conditions under which drug traffickers have held kidnapped victims in the forests of Colombia. Just by visiting this type of climate, one will readily experience the high temperatures, oppressive humidity, and the frequent heavy rains, which sustain the massive and distinctive vegetative growth for which the tropical rainforests are known ( ● Fig. 9.2).

**Constant Heat and Humidity** To obtain a more precise understanding of atmospheric conditions within a specific climate type, the first step should be to carefully examine the climographs that are representative of each climate region. ● Figure 9.3 illustrates climate data from weather stations at two tropical rainforest locations—Akassa, Nigeria, and Ocean Island, in the Gilbert Islands of the Pacific Ocean. Note that average monthly temperatures recorded for both stations are above 25°C (77°F), which is about the average for most rainforest regions. Because these stations are located within 5° of the equator, the sun's rays at noon are always close to being directly overhead. For typical rainforest stations, days and nights are of almost equal length, and the amount of insolation received remains nearly constant throughout the year. Consequently, no appreciable temperature variations can be linked to the sun angle and therefore be considered seasonal. In other words, the concept of summer and winter as being hot and cold seasons, respectively, does not exist in the tropical rainforest climate regions.

canopy, more solar energy reaches Earth's surface.

The unfortunate outcome is that areas that are deforested soon become hotter and drier, and any ecosystem in place is seriously damaged or destroyed. Once the rainforests are removed, the soils lose their source of plant nutrients, and this precludes the growth of any significant crops or plants. In addition, the rainforest, which was in harmony with the soil, cannot reestablish itself. Thus, the multitude of flora and fauna species indigenous to the rainforest is lost forever. It is impossible to calculate the true cost of this reduction in biodiversity (the total number of different plant and animal species in the Earth system).

The lost species may have held secrets to increased food production; a cure for AIDS, cancer, or other health problems; or a base for better insecticides that do not harm the environment. Similar services to humanity already have been provided by tropical forest species.

Tropical deforestation is also threatening the natural chemistry of the atmosphere. Rainforests are a major source of the atmospheric oxygen so essential to all animal life. And deforestation encourages global warming by enhancing the greenhouse effect because forests act as a major reservoir of carbon dioxide. It has been estimated that forest clearing since the mid-1800s has contributed more than 130 billion tons of carbon to the atmosphere, more than two thirds as much as has been added by the burning of coal, oil, and natural gases combined.

What can be done? The reasons for tropical deforestation and the solutions to the problem may be economic, but the issues are extremely complex. It is not sufficient for the rest of the world to point out to governments of tropical nations that their forests are a major key to human survival. It is unacceptable for scientists and politicians from nations where barely one fourth of the original forests remain to insist that the citizens of the tropics cease cutting trees and establish forest plantations on deforested land. These are desired outcomes, but it is first the responsibility of all the world's people to help resolve the serious economic and social problems that have prevented most tropical nations from considering their forests as a sustainable resource.

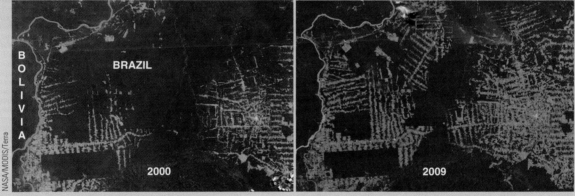

This pair of satellite images shows an example of tropical rainforest deforestation in the State of Rondônia, Brazil between 2000 and 2009. Over the past 30 years in this area of Brazil, a part of the Amazon Basin, the amount of deforested areas has become about equal to the size of the state of Kansas. The region is between Buritis, Brazil (15.63°S 46.43°W) and Nova Mamore, which is to the west of Buritis.

The **annual temperature range**—the difference between the average temperatures of the warmest and coolest months of the year—reflects the consistently high angle of the sun's rays. As indicated in Figure 9.3, the annual range is seldom more than 2°C or 3°C (4°F or 5°F). In fact, at Ocean Island in the central Pacific, the annual range is 0°C because of the additional moderating influence of the ocean and the nearly uniform pattern of insolation.

One of the most interesting features of the tropical rainforest climate is that the **daily (diurnal) temperature ranges**—the differences between the highest and lowest temperatures during the day—are usually far greater than the annual range. Highs of 30°C to 35°C (86°F–95°F) and lows of 20°C to 24°C (68°F–75°F) produce daily ranges of 10°C to 15°C (18°F–27°F). However, the drop in temperature at night is small comfort. High humidity in these regions causes even the cooler evenings to seem oppressive. (Recall that water vapor is a greenhouse gas and helps retain heat energy.)

The climographs of Figure 9.3 illustrate that significant variations in precipitation can occur even within rainforest regions. Although most rainforest locations receive more than 200 centimeters (80 in.) a year of precipitation and the average is in the neighborhood of 250 centimeters (100 in.), some locations record an annual precipitation of more than 500 centimeters (200 in.). Ocean locations, particularly if they are mountainous and near a great source of moisture, tend to receive the most rain. Mount Waialeale

● **FIGURE 9.2** (a) The typical vegetation in a tropical rainforest climate forms a cover of trees, growing at different heights to make a multilayered treetop canopy. This is the rainforest canopy in the Amazon region of Brazil. (b) Tall and massive hardwood trees with distinctive buttressed trunks thrive in a climate that is hot and wet all year but shady on the forest floor. Copyright © 2011 Brooks/Cole, Cengage Learning

*How many tree layers can you see in (a)?*

in the Hawaiian Islands receives a yearly average of 1168 centimeters (460 in.), making it the wettest spot on Earth. As a group, climate stations in the humid tropics experience much higher annual totals than typical humid middle-latitude stations. Compare, for example, the 365 centimeters (144 in.) in Akassa, Nigeria with the average 112 centimeters (44 in.) received annually in Portland, Oregon or the 61 centimeters (24 in.) received in London, England.

We should recall that the heavy precipitation received in tropical rainforest climates is associated with the warm, humid air of the equatorial low and the unstable conditions along the ITCZ (intertropical convergence zone). Both convection and convergence serve as uplift mechanisms, causing humid air to rise and condense, resulting in the heavy rains that are characteristic of this climate. These processes are enhanced on the east coasts of continents where warm ocean currents allow humid tropical climates to extend farther poleward. On west coasts where cold ocean currents flow, uplift mechanisms are somewhat inhibited. There is heavy cloud cover during the warmer, daylight hours when convection is at its peak, although the nights and early mornings can be quite clear. Variations in rainfall can usually be traced to the ITCZ and its low pressure cells of varying strength. Many tropical rainforest locations (Akassa, for example) exhibit two periods of maximum

precipitation during the year, one during each appearance of the ITCZ as it follows the migration of the sun's 90° rays (recall, the sun crosses the equator on equinox days in March 21 and September 22). In addition, although no season can be called dry, during some months it may rain on only 15 or 20 days.

**Cloud Forests** Highland areas near a seacoast both in the tropical regions (such as Costa Rica; ● Fig. 9.4) and in the middle latitudes (coastal Washington State) may contain **cloud forests.** Here, moisture-laden maritime air is lifted up the windward slopes of mountains. This orographic precipitation may not fall as heavy rain showers or thunderstorms but rather as an almost constant misty fog. Through time, this cloudy environment can precipitate enough moisture to qualify as a rainforest but without the oppressive heat found in the rainforests of the low-altitude tropics. One advantage of visiting the cloud forests is a scarcity of flying insects that cannot survive in the cooler temperatures experienced at higher elevations in the tropics.

**A Delicate Balance** The most common vegetation associations in the tropical rainforest climate regions are multistoried, broadleaf evergreen forests made up of many tree species whose

Akassa, Nigeria
5°N
Av. temp.: 25.5°C (78°F)

Tropical Rf (Af)
6°E
Precip.: 365 cm (143.8 in.)
Range: 2.2°C (4°F)

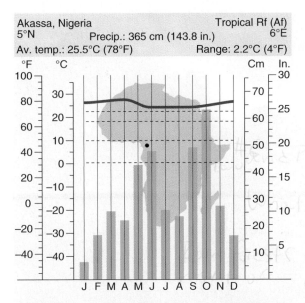

Ocean Is., Gilbert Is.
1°S
Av. temp.: 27°C (80.6°F)

Tropical Rf (Af)
170°E
Precip.: 213 cm (83.9 in.)
Range: 0°

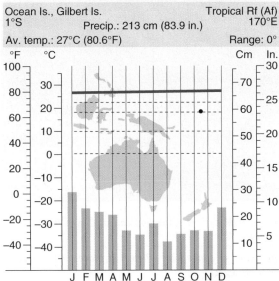

● **FIGURE 9.3** Climographs for tropical rainforest climate stations.

*Why is it difficult, without looking at the climograph keys, to determine whether each station is located in the Northern or Southern Hemisphere?*

● **FIGURE 9.4** A cloud forest in the highlands of Costa Rica.

*Can a constant, misty rain drop as much water as less frequent rainstorms?*

● **FIGURE 9.5** On the forest floor a tropical rainforest offers considerable open space.

*Why is this so?*

tops form a thick, almost continuous canopy cover that blocks out much of the sun's light. This type of rainforest is sometimes called a **selva.** Within the selva there is usually little undergrowth on the forest floor because sunlight cannot penetrate enough to support much low-growing vegetation ( ● Fig. 9.5). When a tree dies in a rainforest and the new opening in the canopy allows for sunlight to enter, another tree will soon fill that void to make use of the insolation that is made available by the forest opening. In a rainforest, sunlight is more important to vegetative growth than rainfall.

The relationship between the rainforest vegetation and the soil that supports that vegetation growth is so close that there exists a nearly perfect ecological balance between the two, threatened only by people's efforts to earn a living from the soil. The

trees of the selva supply the tropical soils with most of the nutrients that the trees themselves need for growth. As leaves, flowers, and branches fall to the ground or roots die, numerous soil-dwelling animals and bacteria act on them, transforming forest litter into organic matter with vital nutrients. However, if the trees are removed, there is no replenishment of these nutrients and no natural barrier (forest litter and root systems) to prevent large amounts of rain from percolating through the soil. This percolating water can dissolve and remove nutrients and minerals from the topsoil. Intense activities of microorganisms, worms, termites, ants, and other insects cause rapid deterioration of the remaining organic debris, and soon all that remains is an infertile mixture of insoluble aluminum, iron, and manganese compounds. In essence, without the vegetation to protect and nurture the soils, rainforest soils tend to be quite infertile.

In recent years there has been large-scale harvesting of the tropical rainforests by the lumber industry, and land has been cleared for agriculture and livestock production, especially in the Amazon River basin. Such deforestation can have a significant and potentially permanent impact on the delicate balance that exists among Earth's systems.

Environmental conditions vary from place to place within climate regions; therefore, the typical rainforest situation that we have just described does not apply everywhere in the tropical rainforest climate regions. Some areas are covered by true jungle, a term often misused when describing the rainforest. **Jungle** is a dense tangle of vines and smaller trees that develops where direct sunlight reaches the ground, as in clearings or along rivers and streams ( ● Fig. 9.6). Other regions have soils that remain fertile through deposition by stream flood deposits, and others provide nutrients through natural processes of soil development from bedrock. Examples of the former region are found along river floodplains; examples of the latter are volcanic regions such as in Indonesia or in the limestone areas of Malaysia and Vietnam. Only in such regions of continuous soil fertility can agriculture be intensive and continuous enough to support population centers in a tropical rainforest climate region.

**Human Activities** In most tropical rainforest climate regions, humans are far outnumbered by other forms of animal life. Although there are few large animal species, a great variety of smaller tree-dwelling and aquatic species live in the rainforest. Small predatory cats, birds, monkeys, bats, alligators, crocodiles, snakes, and amphibians such as frogs of many varieties abound. Animals that can fly or climb into the food-rich leaf canopy have become the dominant animals in this world of trees.

Most common of all, however, are the insects. Mosquitoes, ants, termites, flies, beetles, grasshoppers, butterflies, and bees live everywhere in the rainforest. Insects can breed continuously in this climate without danger from cold or drought.

In addition to the insects, there are genuine health hazards for human inhabitants of the tropical rainforest. Not only does the humid, sultry weather impose uncomfortable living conditions, but also any open wound would heal slowly in the steamy environment. The tropical rainforest climate also supports a variety of parasites and disease-carrying insects that threaten human survival. Malaria, yellow fever, dengue fever, and sleeping sickness are insect-borne (sometimes fatal) diseases of the tropics, but they are uncommon in the middle latitudes.

Whenever native populations have existed in the rainforest, subsistence hunting and gathering of fruits, berries, small animals, and fish have been important. As agriculture was introduced to these regions land was cleared and crops such as manioc, yams, beans, maize (corn), bananas, and sugarcane were grown. It has been the practice to cut down the smaller trees, burn the resulting debris, and plant the desired crops. With the forest gone, farming is possible for only 2 or 3 years before the soil is completely exhausted of its small supply of nutrients and the surrounding area

● **FIGURE 9.6** Jungle vegetation along the Rio Madre de Dios in Peru, a tributary to the Amazon River.

*Why is the vegetation so dense here when it is more open inside the forest at ground level?*

is depleted of game. At this point, the native population moves to another area of forest to begin the practice over again. This kind of subsistence agriculture is known as **slash-and-burn** or simply **shifting cultivation.** Its impact on the close ecological balance between soil and forest is obvious in many rainforest regions. Sometimes the damage done to the system is irreparable, and only jungle, thorn bushes, or scrub vegetation will return to cleared areas ( ● Fig. 9.7).

In terms of numbers of people supported, the most important agricultural use of the tropical rainforest climate is the wet-field (paddy) rice agriculture on river floodplains in southeastern Asia. However, this type of agriculture is best developed in the monsoon variant of this climate. Commercial plantation agriculture is also significant. The principal plantation crops are rubber, sugarcane, and cacao, all of which originally grew with abundance in the forests of the Amazon Basin but are now of greatest importance in other rainforest regions—rubber in Malaysia and Indonesia and sugarcane and cacao in West Africa and the Caribbean region.

● **FIGURE 9.7** This opening in the tropical rainforest of Jamaica is the result of land cleared for shifting (slash-and-burn) cultivation.

*What types of human activity might be responsible for clearings in the heart of a middle-latitude forest?*

## Tropical Monsoon Climate

As you examine ● Figure 9.8 to begin study of the **tropical monsoon climate,** the strikingly uneven distribution of monthly precipitation in tropical monsoon regions is immediately apparent. Although the index map of Figure 9.1 identifies minor costal

regions of South America, Africa, Australia, and Caribbean islands that match the statistical definitions for this climate, we associate tropical monsoon regions most closely with the peninsular lands of Southeast Asia. Here the alternating circulation of air (from sea to land in summer and from land to sea in winter) is strongly related to the shifting of the ITCZ. During the summer, the ITCZ moves north into the Indian subcontinent and adjoining lands to latitudes of 20° to 25°N. This is in part a result of the attracting force of the strong low pressure system on the Asian continent. However, as we have previously noted, the mechanism is complex and involves changes in upper air flow and in surface currents. During the winter months of the Northern Hemisphere, the moisture-laden summer monsoon is replaced by an outflow of dry air from the massive Siberian high pressure system that develops in the winter season over central Asia. By this time, the ITCZ has shifted to its southernmost position (see Fig. 5.19).

Check the climograph in Figure 9.8 for Freetown, Sierra Leone in Africa. It confirms that climate regions outside of Asia also fit the simplified Köppen classification of tropical monsoon. A modified version of the monsoonal wind shift occurs at Freetown, Sierra Leone, but the climate there might also be described as transitional between the constantly wet rainforest climate and the sharply seasonal wet and dry conditions of the tropical savanna.

### Distinctions Between Rainforest and Monsoon Climates
Whatever the factors are that produce a tropical monsoon climate, these regions have strong similarities to those classified as tropical rainforest. In fact, although their core regions are distinctly different, the two climates are often intermixed over *zones of transition*. A major reason for the similarity between monsoon and rainforest climates is that monsoon regions have enough precipitation to allow continuous vegetative growth with no dormant period during the year. Rains are so abundant and in-

● **FIGURE 9.8** Climographs for tropical monsoon climate stations.

*What is the approximate range of precipitation between the highest and lowest months?*

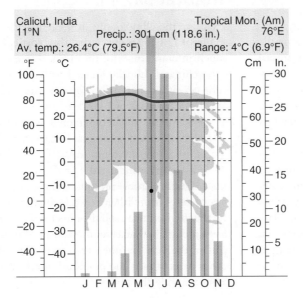

Calicut, India
11°N
Av. temp.: 26.4°C (79.5°F)
Tropical Mon. (Am)
76°E
Precip.: 301 cm (118.6 in.)
Range: 4°C (6.9°F)

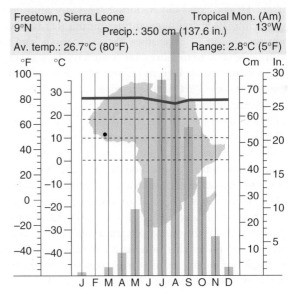

Freetown, Sierra Leone
9°N
Av. temp.: 26.7°C (80°F)
Tropical Mon. (Am)
13°W
Precip.: 350 cm (137.6 in.)
Range: 2.8°C (5°F)

(a)  (b)

● **FIGURE 9.9** Crop selection and agricultural production must be adjusted to the (a) wet and (b) dry seasons throughout Southeast Asia.

*Would it be beneficial to the people of Southeast Asia if the traditional rice farming methods were replaced by mechanized rice agriculture as practiced in the United States?*

tense and the dry season is so short that the soils usually do not dry out completely. As a result, this climate and its soils support a plant cover much like that of the tropical rainforests.

However, there are clear distinctions between rainforest and monsoon climate regions. The most important distinctions concern the seasonal distributions and amounts of precipitation. The monsoon climate has a short dry season, whereas the rainforest does not. Perhaps even more interesting, the average rainfall in monsoon regions varies more widely from place to place. It usually totals between 150 and 400 centimeters (60 to 150 in.) and may be massive where the onshore monsoon winds are forced to rise over mountain barriers. Mahabaleshwar at an elevation of 1362 meters (4467 ft), on the windward side of India's Western Ghats, averages more than 630 centimeters (250 in.) of rain during the 5 months of the summer monsoon.

The annual march of temperature of the monsoon climate also differs appreciably from the monotony of the rainforest climate. Heavy cloud cover during the rainy monsoon season reduces insolation and temperatures during that time of year. During the period of clear skies just prior to the onslaught of the rains, higher temperatures are recorded. As a result, the annual temperature range in a monsoon climate is 2°C to 6°C (compared with 2°C–3°C in the tropical rainforest).

Some additional distinctions between tropical monsoon and rainforest regions can be found in vegetation and animal life. Toward the wetter margins, the monsoon forest resembles the tropical rainforest, but fewer species are present and certain ones become dominant. Seasonality of rainfall in the monsoon regions narrows the range of species that will prosper. Toward the drier margins of the climate, trees grow farther apart and monsoon forest often gives way to jungle or a dwarfed thorn forest. The composition of the animal kingdom here also changes. Climbing and flying species that dominate the forest are joined by larger, hooved leaf eaters and by larger carnivores such as the famous tigers of Bengal (northeast India and Bangladesh).

**Effects of Seasonal Change** Seasonal precipitation of the tropical monsoon climate is of major importance for economic reasons, especially to the people of Southeast Asia and India. Most of the people living in those areas are farmers, and their major crop is rice, which is the staple food for millions of Asians. Rice is most often an irrigated crop, so the monsoon rains are very important to its growth. Harvesting, however, must be done during the dry season ( ● Fig. 9.9).

Each year an adequate food supply for much of south and Southeast Asia depends on the arrival and departure of monsoon rains. The difference between famine and survival for many people in these regions is very much associated with the climate and the significant variation of rainfall from year to year, which can be typical in monsoon climate regions.

## Tropical Savanna Climate

Examine the climographs in ● Figure 9.10 and again the map in Figure 9.1. They indicate that the **tropical savanna climate** has much in common with the tropical rainforest and tropical monsoon climates. Usually located between 5° and 20° or more latitude on either side of the equator, the sun's vertical rays at noon are never far from overhead, the receipt of solar energy is nearly at a maximum, and temperatures remain constantly high. Days and nights are of nearly equal length throughout the year, as they are in other tropical regions.

However, as indicated in the climographs, the tropical savanna has a distinct seasonal precipitation pattern that identifies it and differentiates it from other tropical climates. As the wind and pressure belts shift in latitude with the direct angle of the sun, savanna regions are under the influence of the rain-producing ITCZ (doldrums) for part of the year and the rain-suppressing subtropical highs for the other part. In fact, the poleward limits of the savanna climate are approximately the poleward limits of migration of the ITCZ, and

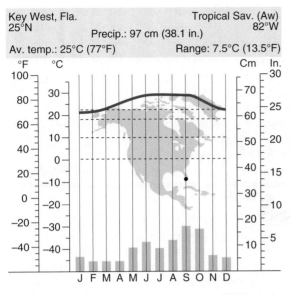

Key West, Fla.
25°N
Tropical Sav. (Aw)
82°W
Precip.: 97 cm (38.1 in.)
Av. temp.: 25°C (77°F)    Range: 7.5°C (13.5°F)

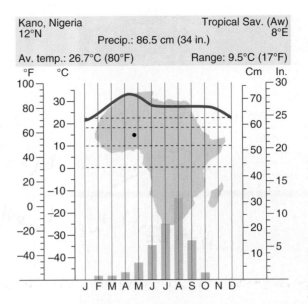

Kano, Nigeria
12°N
Tropical Sav. (Aw)
8°E
Precip.: 86.5 cm (34 in.)
Av. temp.: 26.7°C (80°F)    Range: 9.5°C (17°F)

● **FIGURE 9.10** Climographs for tropical savanna climate stations.

*Consider the differences in climate and human use of the environment between Key West and Kano. Which are more important in the geography of the two places, the physical or the human factors?*

the equatorward limits of this climate are the equatorward limits of movement by the subtropical high pressure systems.

As you can see in Figure 9.1 and Table 9.1, the greatest areas of savanna climate are found on the poleward sides of rainforest climates in Central and South America and in Africa. Other savanna regions occur in India, peninsular Southeast Asia, and Australia. In some instances, the climate extends poleward of the tropics, as it does in the southernmost portion of Florida.

## Transitional Features of the Savanna

Of particular interest to the geographer is the transitional nature of the tropical savanna. Often situated between humid tropical rainforest climates on one side and the rain-deficient (semi-arid) steppe climates on the other, the savanna experiences some of the characteristics of both of these climates. During the rainy, high-sun season, atmospheric conditions resemble those of the rainforest, whereas the low-sun season can be as dry as nearby arid lands are all year. The gradational nature of the climate causes precipitation patterns to vary considerably (see Fig. 9.10). Savanna locations close to the rainforest may receive rain during every month, and total annual precipitation may exceed 180 centimeters (70 in.). In contrast, drier margins of the savanna have longer and more intensive periods of drought and lower annual rainfalls, less than 100 centimeters (40 in.).

Other characteristics of the savanna demonstrate its transitional nature. The higher temperatures just prior to the arrival of the ITCZ produce annual temperature ranges 3°C to 6°C

(5°F–11°F) wider than those of the rainforest but still not as wide as those of the steppe and desert. Although the typical savanna vegetation (known as *llanos* in Venezuela and *campos* in Brazil) is a mixture of grassland and trees, scrub, and thorn bushes, there can be considerable variation. Near the equatorward margins of savanna regions, grasses are taller and trees, where they exist, grow fairly close together. Toward the drier, poleward margins, trees are more widely scattered and smaller and the grasses are shorter ( ● Fig. 9.11). Soils also are affected by the climatic transition as the iron-rich, reddish

● **FIGURE 9.11** Giraffes have always been a majestic sight in the savanna climate of East Africa.

*How is a giraffe's height so well adapted to the savanna environment?*

Courtesy of Norman Bettis

soils of the wetter sections are replaced by darker-colored, more organic-rich soils in the drier regions.

Both vegetation and soils have made special adaptations to the alternating wet–dry seasons of the savanna. During the wet (high-sun) period, the grasslands are green and the trees are covered with foliage. During the dry (low-sun) period, the grass turns brown, dry, and lifeless and most of the trees lose their leaves as an aid in reducing moisture loss through transpiration. Certain trees develop deep roots that can reach down to water in the soil during the dry season. Trees in savanna regions are also fire resistant, an advantage for survival where dry grasses may burn during the winter drought.

## Savanna Potential

Conditions within tropical savanna regions are not well suited to agriculture, although many of our domesticated grasses (grains) are presumed to have originated as wild savanna grasses. Rainfall is far less predictable than in the rainforest or even the monsoon climate regions. For example, Nairobi, Kenya has an average rainfall of 86 centimeters (34 in.). Yet from year to year the amount of rain received may vary from 50 to 150 centimeters (20–60 in.). As a rule, the drier the savanna location, the more unreliable the rainfall becomes. However, the high-sun rains are essential for human and animal survival in savanna regions. When they are late or deficient, as they have been in West Africa in recent years, severe drought and famine result. When the rains last longer than usual or are excessive, however, they can cause major floods, often followed by outbreaks of disease.

Savanna soils (except in areas of recent stream deposits) also tend to limit productivity. During the wet season rains, they may become sticky as wet clay; during the dry season, they are hard and almost impenetrable. Consequently, people in the savannas have often found the soils better suited to grazing than to farming. The Masai, a tribe of cattle herders of East Africa, are world-famous examples ( ● Fig. 9.12). However, even animal grazing has its problems. Many savanna regions make poor pasturelands, at least during the dry part of the year.

The savannas of Africa are the best known of the world's savanna regions. They have been veritable zoological gardens for the larger tropical animals—to such an extent that the popularity of classic photo safaris has made the African savannas a major center for tourism. The grasslands support many different herbivores (plant eaters), such as the elephant, rhinoceros, giraffe, zebra, and wildebeest. The herbivores are eaten by carnivores (flesh eaters), such as the lion, leopard, and cheetah. Scavengers, such as hyenas, jackals, and vultures, devour what remains of the carnivore's kill. During the dry season, the herbivores find grasses and water along stream banks and forest margins and at isolated water holes ( ● Fig. 9.13).

● **FIGURE 9.12** The grasslands of the East African savanna are well suited to supporting large numbers of grazing animals. These Masai cattle herders in Kenya count their wealth by the numbers of animals they own.

*What environmental problems may be created as cattle herds grow?*

● **FIGURE 9.13** In tropical climate regions that experience a lengthy dry season, like this site in East Africa, water holes are important resources for sustaining the wildlife population.

The carnivores follow the herbivores to the water, and a few human hunters and scavengers follow them both.

# Arid Climate Regions

As you begin your study of arid climates by examining ● Figure 9.14, the map of the world's arid regions, and Table 9.2, you will find ample documentation for something you undoubtedly already realized: Arid climate regions *(B)* in the simplified Köppen system are widely distributed over Earth's surface. They are found from the vicinity of the equator to more than 50°N and S latitude. There are two major concentrations of deserts, and each illustrates one of the important causes of climatic aridity. The first is centered on the Tropics of Cancer and Capricorn (23½°N and S latitudes) and extends 10° to 15° poleward and equatorward from there. This region contains the most extensive areas of arid climates in the world. The second is located poleward of the first and occupies continental interiors, particularly in the Northern Hemisphere.

The concentration of deserts in the vicinity of the two tropic lines is directly related to the subtropical high pressure systems. Although the boundaries of the subtropical highs may migrate north and south with the direct rays of the sun, their influence remains constant in these latitudes. We have already learned that the subsidence and divergence of air associated with these systems is strongest along the west coasts (recall that cold ocean currents off the western coasts of continents help stabilize the atmosphere). Hence, clear weather and dry conditions associated with subtropical high pressure extend inland from the western coasts of

each landmass in the subtropics. The Atacama, Namib, and Kalahari Deserts and the desert of Baja California are restricted in their development by the small size of the landmass or by landform barriers to the interior. However, the western part of North Africa and the Middle East comprises the greatest stretch of desert in the world and includes the Sahara, Arabian, and Thar Deserts. Similarly, the Australian Desert occupies most of the interior of the Australian continent.

The second concentration of deserts is located within continental interiors that are remote from moisture-carrying winds. Such arid lands include the vast cold-winter deserts of inner Asia and the Great Basin of the western United States. The dry conditions of the latter region extend northward into the Columbia Plateau and southward into the Colorado Plateau and are increased by mountain barriers that restrict the movement of rain-bearing air masses from the Pacific. Similar *rain-shadow* conditions help to explain the Patagonia Desert of Argentina and the arid lands of western China.

Both wind direction and ocean currents can accentuate aridity in coastal regions. When prevailing winds blow parallel to a coastline instead of onshore, desert conditions are likely to occur because little moisture is brought inland. This seems to be the case in eastern Africa and in northeastern Brazil. Where a cold current flows next to a coastal desert, foggy conditions may develop. Warm, moist air from the ocean may cool to its dew point as it passes over the cooler current. A temperature inversion is created, increasing stability and preventing the upward movement of air required for precipitation. The unique, fog-shrouded coastal deserts in Chile (the Atacama), southwest Africa (the

● **FIGURE 9.14** A map of the world's arid lands.

*What does a comparison of this map with the Map of World Population Density (inside back cover) suggest?*

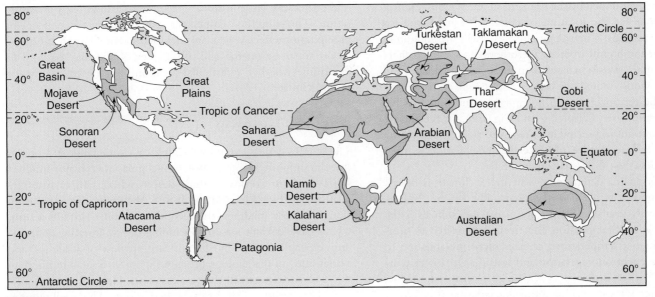

Low-latitude and middle-latitude deserts (BWh, BWk)

Low-latitude and middle-latitude steppes (BSh, BSk)

TABLE 9.2
The Arid Climates

| Name and Description | Controlling Factors | Geographic Distribution | Distinguishing Characteristics | Related Features |
|---|---|---|---|---|
| **Desert** | | | | |
| Precipitation less than half of potential evapotranspiration; mean annual temperature above 18°C (64.4°F) (low-lat.), below (mid-lat.) | Descending, diverging circulation of subtropical highs; continentality often linked with rainshadow location | Coastal Chile and Peru, southern Argentina, southwest Africa, central Australia, Baja California and interior Mexico, North Africa, Arabia, Iran, Pakistan and western India (low-lat.); inner Asia and western United States (mid-lat.) | Aridity; low relative humidity; irregular and unreliable rainfall; highest percentage of sunshine; highest diurnal temperature range; highest daytime temperatures; windy conditions | Xerophytic vegetation; often barren, rocky, or sandy surface; desert soils; excessive salinity; usually small, nocturnal burrowing animals; nomadic herding |
| **Steppe** | | | | |
| Precipitation more than half but less than potential evapotranspiration mean annual temperature above 18°C (64.4°F) (low-lat.), below (mid-lat.) | Same as deserts; usually transitional between deserts and humid climates | Peripheral to deserts, especially in Argentina, northern and southern Africa, Australia, central and southwest Asia, and western United States | Semiarid conditions, annual rainfall distribution similar to nearest humid climate; temperatures vary with latitude, elevation, and continentality | Dry savanna (tropics) or short grass vegetation; highly fertile black and brown soils; grazing animals in vast herds; predators and small animals; ranching and dry farming |

Namib), and Baja California have the lowest precipitation of any regions on Earth.

Figure 9.14 shows deserts of the world to be core areas of aridity, usually surrounded by the semiarid steppe regions. Hence, our explanations for the location of deserts also hold true for steppe climate regions. The steppe climates either are subhumid borderlands of the humid tropical, mesothermal, and microthermal climates or are transitional between these humid climates and deserts. As previously noted, we classify both steppe and desert on the basis of the relation between precipitation and potential ET (evapotranspiration, see Chapter 6 and Fig. 6.8). In the **desert climate** the amount of precipitation received is less than half the potential ET. In the **steppe climate** the precipitation is more than half but less than the total potential ET.

The criterion for determining whether a climate is desert, steppe, or humid is *precipitation effectiveness*. The amount of precipitation actually available for use by plants and animals is the *effective precipitation*. Precipitation effectiveness is related to temperature. At higher temperatures, it takes more precipitation to have the same effect on vegetation and soils than at lower temperatures. The result is that areas with higher temperatures that promote greater ET can receive more precipitation than cooler regions and yet have a more arid climate.

Because of the temperature influence, precipitation effectiveness depends on the season in which an arid region's meager precipitation is concentrated. Precipitation received during the low-sun (cooler) period will be more effective than that received during the high-sun period when temperatures are higher because less moisture will be lost through evapotranspiration. The simplified Köppen graphs based on the concept of precipitation effectiveness are included in the "Graph Interpretation" exercise (at the end of Chapter 8) and may be used to determine whether a particular location has a desert, steppe, or humid climate.

## Desert Climates

The world's deserts extend through such a wide range of latitudes that the simplified Köppen system recognizes two major subdivisions. The first are low-latitude deserts where temperatures are relatively high year-round and frost is absent or infrequent even along poleward margins; the second are middle-latitude deserts, which have distinct seasons, including below-freezing temperatures during winter. Your preliminary preview of Table 9.2 and a brief study of the climographs in ● Figure 9.15 serve to illustrate the differences between the two divisions. However, the significant characteristic of all deserts is their aridity. The small number of occasions on which we will distinguish between low-latitude and middle-latitude deserts in the discussion that follows emphasizes the relative unimportance of temperature.

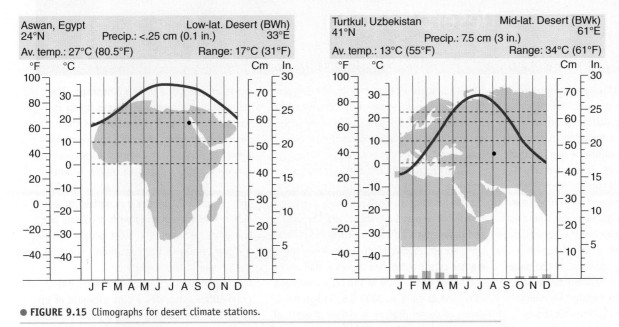

Aswan, Egypt                  Low-lat. Desert (BWh)
24°N          Precip.: <.25 cm (0.1 in.)        33°E
Av. temp.: 27°C (80.5°F)          Range: 17°C (31°F)

Turtkul, Uzbekistan          Mid-lat. Desert (BWk)
41°N          Precip.: 7.5 cm (3 in.)            61°E
Av. temp.: 13°C (55°F)            Range: 34°C (61°F)

● **FIGURE 9.15** Climographs for desert climate stations.

*If you consider the serious limitations of desert climates, how do you explain why some people choose to live in desert regions?*

**Land of Extremes** By definition, deserts are associated with a minimum of precipitation, but they also represent the extremes in other atmospheric conditions. With few clouds and little water vapor in the air, as much as 90% of insolation reaches the surface in desert regions. This is why the highest insolation and highest temperatures are recorded in low-latitude desert areas and not in the more humid tropical climates that are closer to the equator. Again, because the relative humidity of desert air is so low, and with little or no cloud cover, there is little atmospheric effect and much of the energy received during the day is radiated back to the atmosphere at night. Consequently, nighttime temperatures in the desert drop far below their daytime highs. This rather extreme cycle of day-to-night heating and cooling give low-latitude deserts the greatest diurnal temperature ranges in the world, and middle-latitude deserts are not far behind. In the spring and fall, these ranges may be as great as 40°C (72°F) in a day. More common diurnal temperature ranges in deserts are 22°C to 28°C (40°F–50°F).

The sun's rays are so intense in the clear, dry desert air that temperatures in shade are much lower than those a few steps away in direct sunlight. (Keep in mind that all temperatures for official meteorological statistics are recorded in the shade.) Khartoum, Sudan (in the Sahara) has an *average* annual temperature of 29.5°C (85°F), which is a *shade* temperature. Temperatures in the intense desert sun under cloudless skies at Khartoum are often 43°C (110°F) or more. Soil temperatures rise close to 95°C (200°F) in midsummer in the Mojave Desert of Southern California.

During low-sun or winter months, deserts experience colder temperatures than humid areas at the same latitude, and in summer they experience hotter temperatures. Just as with the high diurnal ranges in deserts, these high annual temperature ranges can be attributed to the lack of moisture in the air.

Annual temperature ranges are usually greater in middle-latitude deserts, such as the Gobi in Asia, than in low-latitude deserts because of the colder winters experienced at higher latitudes. Compare, for example, the climograph for Aswan in south central Egypt—at 24°N, a low-latitude desert location—with the climograph for Turtkul, Uzbekistan—at 41°N, a middle-latitude desert location (see Fig. 9.15). The annual range for Aswan is 17°C (31°F); in Turtkul, it is 34°C (61°F).

Precipitation in the desert climate is irregular and unreliable, but when it comes, it may arrive in an enormous cloudburst, bringing more precipitation in a single rainfall than has been recorded in years ( ● Fig. 9.16). Recall from Chapter 6 that the *variability of precipitation* is greater in regions where precipitation totals are lowest. This happened in the extreme at the port of Walvis Bay, on the coast of the Namib Desert, a cold-current coastal desert of southwest Africa. The equivalent of 10 years' rainfall was received in one night when a freak storm dumped 3.2 centimeters (1.3 in.) of rain.

The actual amount of water vapor in the air may be high in desert areas. However, the hot daytime temperatures increase the capacity of the air to hold moisture so that the relative humidity during daylight hours is quite low (10%–30%). Desert nights are a different story. Radiation of energy is rapid in the clear air. As temperature drops, relative humidity increases and the formation of dew in the cool hours of early morning is common. Where measurements have been made, the amount of dew formed has sometimes considerably exceeded the annual rainfall for that location. It has been suggested, in fact, that dew may be of great importance to plant and animal life in many desert regions. Studies are now being carried out on the use of dew as a moisture source for certain crops, thereby minimizing the need for large-scale irrigation.

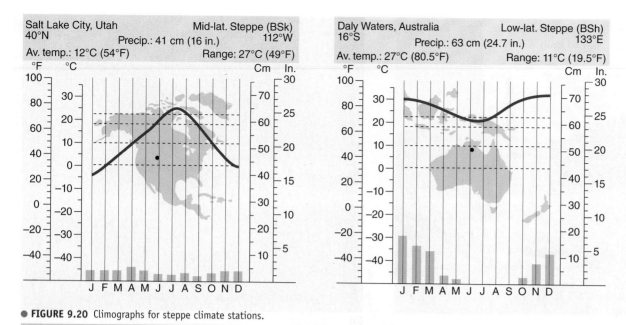

Salt Lake City, Utah      Mid-lat. Steppe (BSk)
40°N      Precip.: 41 cm (16 in.)      112°W
Av. temp.: 12°C (54°F)      Range: 27°C (49°F)

Daly Waters, Australia      Low-lat. Steppe (BSh)
16°S      Precip.: 63 cm (24.7 in.)      133°E
Av. temp.: 27°C (80.5°F)      Range: 11°C (19.5°F)

● **FIGURE 9.20** Climographs for steppe climate stations.

***What are the major differences between the climograph data for Salt Lake City and Daly Waters? What are the causes of these differences?***

nant. ● Figure 9.21 can be described as a *tropical scrub* or *tropical thorn forest* and can be a severely limiting environment for human activities.

Both low-latitude and middle-latitude steppe varieties are identified by mean annual temperature (Table 9.2). As in the

● **FIGURE 9.21** This tropical scrub or tropical thorn forest can be a hostile environment for humans and many animals, but these African warthogs thrive there.

***Why is it important not to wander off the well-worn paths and trails in this environment?***

desert, steppe temperatures vary throughout the climate type with latitude, distance from the sea, and elevation. The climographs of Figure 9.20 demonstrate that although summer temperatures are high in all steppe regions, the differences in winter temperatures can produce annual ranges in middle-latitude steppes that are two or three times as great as those in low-latitude steppes.

**A Dangerous Appeal** Although the surface vegetative cover is often incomplete, in the more humid areas of middle-latitude steppe regions the grasses have been excellent for pasture. In North America the semiarid steppe region was the realm of bison and antelope; in Africa it is the domain of wildebeests and zebra. Steppe soils are usually high in organic matter and soluble minerals. Attributes such as these have attracted farmers and herders alike to the rich grasslands—but not without penalty to both humans and land.

The semiarid steppe climate is problematic for agriculture, even in the middle-latitude steppe, and people take a sizable risk when they attempt to farm. Although dry-farmed wheat and drought-resistant barley and sorghum can be successfully raised because both farming methods and crops are adapted to the environment, the use of techniques employed in more humid regions can lead to serious problems. During dry cycles, crops can fail year after year, and as the land is stripped of its natural sod, the soil is exposed to wind erosion. Even using the grasses for grazing domesticated animals is not always a solution because overgrazing can just as quickly create "Dust Bowl" conditions (as occurred during the extreme droughts of the 1930s in the southern plains of the United States).

The difficulties in making steppe regions more productive point out again the sensitive ecological balance of Earth's systems. Natural rains in the steppe regions are usually sufficient to support a vegetation cover of short grasses that can feed roaming herbivores that graze on them. The herbivore population in turn is kept in check by carnivores that prey on them ( ● Fig. 9.22). When people enter the scene, however, introducing more grazing animals, plowing the land, or killing off the predators, the ecological balance is tipped, and sometimes the results are disastrous.

The climates discussed in this chapter represent some of the greatest extremes in atmospheric elements worldwide. These extremes vary from the heavy precipitation in tropical rainforest and tropical monsoon climates to the extreme dryness in subtropical desert regions. Daily temperatures vary greatest in the middle latitude desert regions, and precipitation variability is always greatest in the most arid, desert regions. Such climates may be harsh and delicate at the same time. These climates make survival difficult for the human populations that inhabit them, and in turn, the human intervention can have devastating effects on these environments.

UNEP/Peter Prorkosch

● **FIGURE 9.22** This zebra has more to worry about than an attack by a predator.

*What other types of dangers might threaten this environment?*

# CHAPTER 9 ACTIVITIES

## ■ DEFINE AND RECALL

tropical rainforest climate
annual temperature range
daily (diurnal) temperature range
cloud forest

selva
jungle
slash–and–burn (shifting cultivation)
tropical monsoon climate

tropical savanna climate
desert climate
steppe climate
xerophytic

## ■ DISCUSS AND REVIEW

1. Explain the role of insolation in the existence of three different tropical environments.
2. Describe the three tropical (A) climates.
3. How do ocean currents affect tropical and arid climates?
4. What are the controlling factors that explain the tropical rainforest climate? Give a brief verbal description of this climate type.
5. Describe the delicate balance between vegetative growth and soil fertility in a tropical rainforest environment.
6. What is the difference between a rainforest and a jungle? What factor differentiates these two forest types?

7. Where do you find the greatest difference in temperature between direct sunlight and shade, the rainforest or the savanna? Why are these two so different?
8. Explain the seasonal precipitation pattern of the tropical savanna climate. State some of the transitional features of this climate. How have vegetation and soils adapted to the wet–dry seasons?
9. What conditions give rise to desert climates?
10. How do steppes differ from deserts? Why might human use of steppe regions in some ways be more hazardous than use of deserts?

11. What exceptions are there to the general rule that the geographic pattern of climate regions is produced by Earth–sun relationships? How do the Northern and Southern Hemispheres differ in climate patterns, and why?

12. Do each of the following for the listed climates: tropical rainforest, tropical monsoon, tropical savanna, desert, and steppe.

a. Identify the climate from a set of data or a climograph provided in the chapter indicating average monthly temperature and precipitation for a representative station within a region typical of that climate.

b. Match the climate type with a written statement that includes one or more of the following: the statistical parameters of the climate in the modified Köppen classification; the particular climate controls (controlling factors) that produce the climate; and the geographic distributions of the climate, stated in terms of physical or political location.

c. Distinguish between the important subtypes (if any) of each climate by identifying the characteristics that separate them from one another.

# CONSIDER AND RESPOND

1. Based on the classification scheme presented in the "Graph Interpretation" exercise (end of Chapter 8), classify the following climate stations from the data provided.

| | | J | F | M | A | M | J | J | A | S | O | N | D | Yr. |
|---|---|---|---|---|---|---|---|---|---|---|---|---|---|---|
| a. | Temp. (°C) | 20 | 23 | 27 | 31 | 34 | 34 | 34 | 33 | 33 | 31 | 26 | 21 | 29 |
| | Precip. (cm) | 0.0 | 0.0 | 0.0 | 0.0 | 0.2 | 0.1 | 1.1 | 0.2 | 0.0 | 0.0 | 0.0 | 0.0 | 1.6 |
| b. | Temp. (°C) | 21 | 24 | 28 | 31 | 30 | 28 | 26 | 25 | 26 | 27 | 25 | 22 | 26 |
| | Precip. (cm) | 0.0 | 0.0 | 0.2 | 0.8 | 7.1 | 11.9 | 20.9 | 31.1 | 13.7 | 1.4 | 0.0 | 0.0 | 87.2 |
| c. | Temp. (°C) | 19 | 20 | 21 | 23 | 26 | 27 | 28 | 28 | 27 | 26 | 22 | 20 | 24 |
| | Precip. (cm) | 5.1 | 4.8 | 5.8 | 9.9 | 16.3 | 18.0 | 17.0 | 17.0 | 24.0 | 20.0 | 7.1 | 3.0 | 149.0 |
| d. | Temp. (°C) | 2 | 4 | 8 | 13 | 18 | 24 | 26 | 24 | 21 | 14 | 7 | 3 | 14 |
| | Precip. (cm) | 1.0 | 1.0 | 1.0 | 1.3 | 2.0 | 1.5 | 3.0 | 3.3 | 2.3 | 2.0 | 1.0 | 1.3 | 20.6 |
| e. | Temp. (°C) | 27 | 26 | 27 | 27 | 27 | 27 | 27 | 27 | 27 | 27 | 27 | 27 | 27 |
| | Precip. (cm) | 31.8 | 35.8 | 35.8 | 32.0 | 25.9 | 17.0 | 15.0 | 11.2 | 8.9 | 8.4 | 6.6 | 15.5 | 243.8 |
| f. | Temp. (°C) | 13 | 17 | 17 | 19 | 22 | 24 | 26 | 26 | 26 | 24 | 19 | 15 | 20 |
| | Precip. (cm) | 6.6 | 2.0 | 2.0 | 0.5 | 0.3 | 0.0 | 0.0 | 0.0 | 0.3 | 1.8 | 4.6 | 6.6 | 26.7 |

2. The following six locations are represented by the data in the previous table, although not in the order listed: Albuquerque, New Mexico; Belém, Brazil; Benghazi, Libya; Faya, Chad; Kano, Nigeria; Miami, Florida. Use an atlas and your knowledge of climates to match the climatic data with the locations.

3. Benghazi, Libya and Albuquerque, New Mexico both exhibit steppe climates, but the cause (or control) of their dry conditions is quite different. Describe the primary factor responsible for the dry conditions at each location.

4. Tropical B climates are located more poleward than A climates, yet their daytime high temperatures are often higher. Why?

5. Kano, Nigeria has copious amounts of rainfall during the summer season. What are the two sources, or causes, of this rainfall?

# APPLY AND LEARN

| | Jan | Feb | Mar | Apr | May | Jun | Jul | Aug | Sep | Oct | Nov | Dec |
|---|---|---|---|---|---|---|---|---|---|---|---|---|
| Average Max. Temperature (F) | 66.2 | 73.4 | 81.1 | 89.7 | 99.5 | 109.2 | 115.6 | 113.8 | 105.8 | 92.9 | 76.3 | 65.2 |
| Average Min. Temperature (F) | 39.3 | 46.0 | 53.8 | 61.5 | 71.5 | 80.6 | 87.2 | 85.3 | 75.2 | 61.6 | 47.8 | 38.1 |
| Average Total Precipitation (in.) | 0.32 | 0.50 | 0.31 | 0.13 | 0.07 | 0.05 | 0.13 | 0.14 | 0.19 | 0.10 | 0.19 | 0.16 |

1. The data set shows measurements collected from Death Valley, California between April 1, 1961 and December 31, 2005. For this data set, calculate the mean annual maximum temperature and annual maximum temperature range; the mean annual minimum temperature and annual minimum temperature range; and the mean annual precipitation values and the annual precipitation range.

2. The highest temperature ever recorded in the United States was set here in July 1913, at 134°F. How much higher was that record from the mean maximum temperature for July in the data set? Is July typically the month with the highest maximum temperatures?

3. The Amazon rainforest occupies some 5.5 million square kilometers. Between May 2000 and August 2005, Brazil lost more than 132,000 square kilometers of forest—an area larger than Greece. Assuming a constant rate★ of deforestation at approximately 132,000 square kilometers per 5-year time period, how many years will it take before the Amazon rainforest is completely destroyed?

★Actually, the destruction of the Amazon rainforest is occurring at an accelerating rate.

 CourseMate - Make the most of your study time by accessing everything you need to succeed in one place. Read your textbook, take notes, review flashcards, watch videos, complete activities, take practice quizzes, and more online with CourseMate. Log in at **www.cengagebrain.com**.

# 10

# MIDDLE-LATITUDE, POLAR, AND HIGHLAND CLIMATE REGIONS

Bull moose in a pond on the fringe of the tundra in Denali National Park, Alaska.
© Design Pics Inc./Alamy

## ▉ OBJECTIVES

WHEN YOU COMPLETE THIS CHAPTER YOU SHOULD BE ABLE TO:

■ Locate the general areas of each mesothermal, microthermal, and polar climate on a world map, and explain the major factors that control their global distribution.
■ Outline the major characteristics of each mesothermal (moderate winter), microthermal, (cold winter) and polar climate.
■ Describe the important vegetation types and human adaptations related to each mesothermal, microthermal, and polar climate.
■ State the significant differences between Mediterranean climate and the humid subtropical climate, and explain why the two are found on opposite sides of landmasses.
■ Identify the differences between humid continental hot-summer and mild-summer climate subtypes in North America, and understand why their differences are not as important on other continents.
■ Recognize a specific mesothermal, microthermal, or polar climate in the Köppen classification system by examining a climograph or a set of data from a representative station.
■ Explain why highland climates are so variable and what factors control the nature of a highland climate at a given time and location.

IN THIS CHAPTER WE WILL begin with a discussion of Köppen's *(C)* and *(D)* climates, the categories that dominate the middle-latitudes, and then follow with an examination of the *(E)* climates, the category that occupies the polar regions. Each of Köppen's climate categories has certain things in common. As we have already noted, the humid tropical *(A)* climates exhibit the constant characteristic of heat, whereas we instinctively know that the polar *(E)* climates exhibit the constant characteristic of cold. The arid *(B)* climates have inadequate precipitation in common. But what is the constant characteristic of the mesothermal *(C)* and microthermal *(D)* middle-latitude climates, which we are about to examine? If there is one, then it is the oxymoron "constant inconsistency." Each of the middle-latitude climates is dominated by the changing of the seasons and the variability of atmospheric conditions associated with migrating air masses or cyclonic activity along the polar front. If change is constant in the mesothermal and microthermal climate groups, then degree of change is what distinguishes one climate type from another. In one or another of the middle-latitude climates, summers vary from hot to cool and winters vary from mild to extremely cold. Some climates receive adequate monthly precipitation year-round; others experience winter drought or, even more challenging to humans who live there, dry months during the normal summer growing season. Despite the changing atmospheric conditions of middle-latitude climates, their regions are home to the majority of Earth's people and they are major factors in some of Earth's most attractive, interesting, and productive physical environments. As you read, it is once again advisable to follow a regular study pattern that makes full use of the illustrations associated with climate. First, before the climate is discussed in the text, you should become familiar with its global location and identify adjoining climates. This is best accomplished by studying the related index map and turning again to Figure 8.6, the world map of climates. Once locations have been identified, preview the significant information about the climate that is included in the table summarizing the climates of the category. In addition, we suggest that before reading the text you will gain a valuable overview of a new climate if you first examine one or two climographs from recording stations located in the climate's region.

When you read about the middle-latitudes, keep in mind two important climatic trends that help explain the differences between one climate type and another. Both trends have simple explanations but have major impacts on the Köppen classification system. The first trend—that average temperatures decrease as latitude increases towards the poles—is a direct response to decreasing insolation. It is obviously the basic reason for four Köppen climate categories between the equator and the poles, but it also is the basis for significant distinctions between individual middle-latitude climates. The second trend—that average precipitation decreases toward the interior of the continents—is explained by increasing distance from the oceans, Earth's major source of precipitation. It is most important as an explanation for the location of boundaries between humid and arid middle-latitudes climates. However, it also plays a role in distinguishing between individual climate types and subtypes.

# Humid Mesothermal Climate Regions

When we use the term *mesothermal* (from Greek: *mesos*, middle) in describing climates, we are usually referring to the moderate temperatures that characterize such regions. However, we could also be referring to their middle position between those climates that have high temperature throughout the year and those that experience severe cold. By definition, the mesothermal climates experience seasonality, with distinct summers and winters that distinguish them from the humid tropics. Their summers are long and their winters are mild, and this separates them climatically from the microthermal climates, which lie poleward.

The three distinct mesothermal climates introduced in Chapter 8 are displayed in Table 10.1. In all three, the annual precipitation exceeds the annual potential evapotranspiration, but the Mediterranean climate has a lengthy period of precipitation deficit in the summer season that distinguishes it from the humid subtropical and the marine west coast climates. The latter two are further differentiated by the fact that the humid subtropical regions have hot summers, whereas the marine west coast regions experience mild summers.

## Mediterranean Climate

The **Mediterranean climate** *(Csa, Csb)* is one of the best examples for organizing a study of the environment or developing an understanding of world regions based on climate classification. ● Figure 10.1 shows that the climate appears with remarkable regularity in the vicinity of 30° to 40° latitude along the west coasts of each landmass. ● Figure 10.2 indicates that the alternating controls of subtropical high pressure in summer and westerly wind movement in winter are so predictable that all Mediterranean regions have notably similar and easily recognized temperature and precipitation characteristics. The special appearance, combinations, and climatic adaptations of Mediterranean vegetation not only are unusual, but also are clearly distinguishable from those of other climates. Agricultural practices, crops, recreational activities, and architectural styles all exhibit strong similarities within Mediterranean lands.

**Characteristics** The major characteristics of the Mediterranean climate are a dry summer; a mild, moist winter; and abundant sunshine (90% of possible sunshine in summer and as much as 50%–60% even during the rainy winter season). Summers are warm throughout the climate, but there are enough differences between the monthly temperatures in coastal and interior locations to recognize two distinct subtypes. The moderate-summer subtype *(Csb)* has the lower summer temperatures associated with coastal fogs and a strong maritime influence. The hot-summer subtype *(Csa)* is located further inland and reflects an increased influence of *continentality*. The inland version has higher summer and daytime temperatures and slightly cooler winter and nighttime temperatures than its coastal counterpart, and it also has

**TABLE 10.1**
**The Mesothermal Climates**

| Name and Description | Controlling Factors | Geographic Distribution | Distinguishing Characteristics | Related Features |
|---|---|---|---|---|
| **Mediterranean** | | | | |
| Warmest month above 10°C (50°F); coldest month between 18°C (64.4°F) and 0°C (32°F); summer drought; hot summers (inland), mild summers (coastal) | West coast location between 30° and 40°N and S latitudes; alternation between subtropical highs in summer and westerlies in winter | Central California; central Chile; Mediterranean Sea borderlands, Iranian highlands; Cape Town area of South Africa; southern and southwestern Australia | Mild, moist winters and hot, dry summers inland with cooler, often foggy coasts; high percentage of sunshine; high summer diurnal temperature range; frost danger | Sclerophyllous vegetation; low, tough brush (chaparral); scrub woodlands; varied soils, erosion in Old World regions; winter-sown grains, olives, grapes, vegetables, citrus, irrigation |
| **Humid Subtropical** | | | | |
| Warmest month above 10°C (50°F); coldest month between 18°C (64.4°F) and 0°C (32°F); hot summers; generally year-round precipitation, winter drought (Asia) | East coast location between 20° and 40°N and S latitudes; humid onshore (monsoonal) air movement in summer, cyclonic storms in winter | Southeastern United States; southeastern South America; coastal southeast South Africa and eastern Australia; eastern Asia from northern India through south China to southern Japan | High humidity; summers like humid tropics; frost with polar air masses in winter; precipitation 62–250 cm (25–100 in.), decreasing inland; monsoon influence in Asia | Mixed forests, some grasslands, pines in sandy areas; soils productive with regular fertilization; rice, wheat, corn, cotton, tobacco, sugarcane, citrus |
| **Marine West Coast** | | | | |
| Warmest month above 10°C (50°F); coldest month between 18°C (64.4°F) and 0°C (32°F); year-round precipitation; mild to cool summers | West coast location under the year-round influence of the westerlies; warm ocean currents along some coasts | Coastal Oregon, Washington, British Columbia, and southern Alaska; southern Chile; interior South Africa; southeast Australia and New Zealand; northwest Europe | Mild winters, mild summers, low annual temperature range; heavy cloud cover, high humidity; frequent cyclonic storms, with prolonged rain, drizzle, or fog; 3- to 4-month frost period | Naturally forested, green year-round; soils require fertilization; root crops, deciduous fruits, winter wheat, rye, pasture and grazing animals; coastal fisheries |

greater annual and diurnal temperature ranges. Compare, for example, the annual range for Red Bluff, California, an inland station, with that of San Francisco, a coastal station about 240 kilometers (150 mi) farther south (see Fig. 10.2).

Whichever the subtype, Mediterranean summers clearly show the influence of the subtropical highs. [...] ks go by without a sign of rain, and evapotranspiration [...] are high. Effective precipitation is lower than actual p[...]tion, and the summer drought is as intense as that of th[...] t. Days are warm to hot, skies are blue and clear, and s[...] is abundant. The high percentage of insolation co[...] th nearly vertical rays of the noon sun may drive d[...] emperatures as high as 30°C to 38°C (86°F–100°F) [...] where moderated by a strong ocean breeze or coastal [...]

Fog is common throughout the year in coastal locations and is especially noticeable during the summer. As moist maritime air moves onshore, it passes over the cold ocean currents that typically parallel west coasts in Mediterranean latitudes. The air is cooled; condensation takes place; and fog regularly creeps in during the late afternoon, remains through the night, and "burns off" during the morning hours. As in the desert, radiation loss is rapid at night, and even in summer, nighttime temperatures are commonly only 10°C to 15°C (50°F–60°F).

Winter is the rainy season in the Mediterranean climate. The average annual rainfall in these regions is usually between 35 and 75 centimeters (15–30 in.), with 75% or more of the total rain falling during the winter months. The precipitation results primarily from the cyclonic storms and frontal systems common

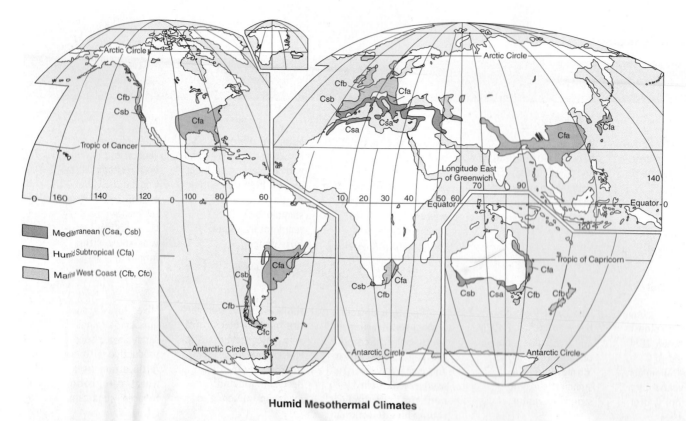

**Humid Mesothermal Climates**

● **FIGURE 10.1** Index map of humid mesothermal climates.

with the westerlies. Annual amounts increase with elevation and decrease with increased distance from the ocean. Only because the rain comes during the cooler months, when evapotranspiration rates are lower, is there sufficient precipitation to make this a humid climate.

Despite the rain during the winter season, there are often many days of fine, mild weather. Insolation is still usually above 50% and the average temperature of the coldest month rarely falls below 4°C to 10°C (40°F–50°F). Frost is uncommon and, because of its rarity, many less hardy tropical varieties of fruits and vegetables are grown in these regions. However, when frost does occur, it can do great damage.

**Special Adaptations** The summer drought, not frost, is the great challenge to vegetation in Mediterranean regions. The natural vegetation reflects the wet–dry seasonal pattern of the climate. During the rainy season, the land is covered with lush, green grasses that turn golden and then brown under the summer drought. Only with winter and the return of the rains does the landscape become green again. Much of the natural vegetation is **sclerophyllous** (hard-leafed) and drought resistant. Like xerophytes, these plants have tough surfaces; shiny, thick leaves that resist moisture loss; and deep roots to help combat aridity.

One of the most familiar plant communities is made up of many low, scrubby bushes that grow together in a thick tangle. In the western United States, this is called **chaparral** ( ● Fig. 10.3).

Chaparral areas have the potential for frequent wild fires that occur in this dry brush typically during the summer and fall. The fires help perpetuate the chaparral because the associated heat is required to open seedpods and allow many chaparral species to reproduce. People often remove chaparral as a preventive measure against fires, but the removal can have disastrous results because the chaparral acts as a check against erosion of soils during the rainy season. With the chaparral removed, soils wash or slide down hillsides during the heavy rains of winter, frequently taking homes with them.

Brush similar to the chaparral of California is called *mallee* in Australia, *mattoral* in Chile, and *maquis* in France. In fact, the "maquis" French Underground that fought against Nazi occupation in World War II literally meant "underbrush."

Trees in the Mediterranean climate also respond to the seasonally wet and dry moisture conditions. Because of their drought-resistant qualities, the needle-leaf pines are among the more common species. Groves of deciduous and evergreen oaks appear in depressions where moisture collects and on the shady north sides of hills where evapotranspiration rates are lower. Where the summer drought is more distinct, the scrub and woodlands open up to parklands of grasses and scattered oak trees. Even the great redwood forests of northern California probably could not survive without the heavy fogs that regularly invade the coastal lands in summer ( ● Fig. 10.4).

In response to the dry summer, most soils in Mediterranean regions are high in soluble nutrients, but the potential for agriculture differs widely from one region to another. The lime-rich soils around the Mediterranean Sea were highly pro-

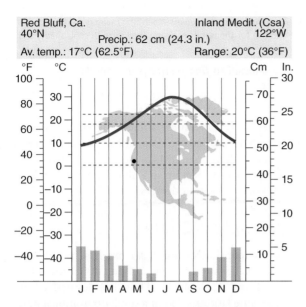

Red Bluff, Ca.                    Inland Medit. (Csa)
40°N                                              122°W
           Precip.: 62 cm (24.3 in.)
Av. temp.: 17°C (62.5°F)        Range: 20°C (36°F)

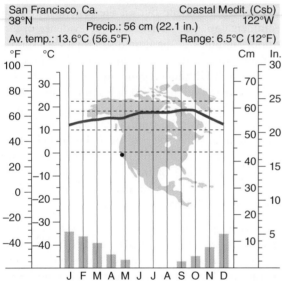

San Francisco, Ca.               Coastal Medit. (Csb)
38°N                                              122°W
           Precip.: 56 cm (22.1 in.)
Av. temp.: 13.6°C (56.5°F)      Range: 6.5°C (12°F)

● **FIGURE 10.2** Climographs for Mediterranean climate stations.

*In what way do these climographs differ? What causes the differences?*

ductive, but destructive agricultural practices and overgrazing over thousands of years of human use have caused serious erosion problems. Today, bare white limestone is widely visible on hillsides in Spain, Italy, Greece, Crete, Syria, Lebanon, Israel, and Jordan. In the Mediterranean regions of California, the soils are dense clays, gluelike when wet and hard as concrete when dry. During the Spanish period in California, the clay was formed into adobe bricks and used for building material. These clay soils are hazardous on slopes because they absorb so much water during the wet season that they can become mudflows destructive to roads and homes in areas where vegetation has been removed by fire or human interference.

In all Mediterranean regions, the most productive areas are the lowlands covered by stream deposits. Here farmers have made

● **FIGURE 10.3** Typical remnant chaparral vegetation as found in southern Spain.

*Why would you expect to find little "native" vegetation left in Old World Mediterranean lands?*

special adaptations to climatic conditions. There is sufficient rainfall in the cool season to permit fall planting and spring harvesting of winter wheat and barley. These grasses originally grew wild in the eastern Mediterranean region. Grapevines, fig and olive trees, and the cork oak, which undoubtedly were also native to Mediterranean lands, are especially well adapted to the dry summers because of their deep roots and thick, well-insulated stems or bark (● Fig. 10.5). Where water for irrigation is available, an incredible diversity of crops may be seen. These include, in addition to those already mentioned, oranges, lemons, limes, melons, dates, rice, cotton, deciduous fruits, various types of nuts, and countless vegetables. California, blessed both with fertile valleys for growing fruits, vegetables, and flowers and with snow meltwater for irrigation, is probably the most agriculturally productive of the Mediterranean regions.

Even the houses of these regions show people's adaptation to the climate. Usually white or pastel in color, they gleam in the brilliant sunshine against clear blue skies. Many have shuttered windows to reduce incoming sunlight and to keep the houses

Photo by Jeff Leonard, courtesy Humbolt County

● **FIGURE 10.4** California redwoods (*Sequoia sempervirens*) may reach heights of 100 meters (330 ft) and live for thousands of years. They grow in cool, shady, and often foggy areas.

*Why is it considered unusual to find redwoods growing in a Mediterranean climate?*

cool in summer. However, much less attention is paid to keeping homes warm during the cooler winter months. This is true even in the United States, where many Midwesterners and Easterners are surprised at the lack of insulation in California homes and the small number and size of heating devices.

## Humid Subtropical Climate

The **humid subtropical climate** *(Cfa)* extends inland from continental east coasts between 15° and 20° and 40°N and S latitude (see Fig. 10.1 and Table 10.1). Thus, it is located within approximately the same latitudes and in a similar transitional position as the Mediterranean climate but on the eastern instead of the western continental margins. There is ample evidence of this climatic transition. Summers in the humid subtropics are similar to the humid tropical climates farther equatorward. When the noon sun is nearly overhead, these regions are subject to the importation of moist tropical air masses. High temperatures, high relative humidity, and frequent convectional showers are all characteristics that they share with the tropical climates. In contrast, during the winter months, when the pressure and wind belts have shifted equatorward, the humid subtropical regions are more commonly under the influence of the cyclonic systems of the continental middle latitudes. Polar air masses can bring colder temperatures and occasional frost.

### Comparison with the Mediterranean Climate

The major similarities and differences between the humid subtropical and Mediterranean climates are most easily noted by comparing the climographs of humid subtropical New Orleans, Louisiana ( ● Fig. 10.6) with the climograph of Red Bluff, California (see Fig. 10.2), an inland version of the Mediterranean climate. The two climographs indicate that both climates have adequate precipitation, mild winters, and hot summers—but New Orleans has no dry season. To fully understand the reason for this significant difference between the climographs and the climates they represent, you should look once again at the index map of humid mesothermal climates (see Fig. 10.1). The map will remind you that the two climate stations are located in the lower-middle latitudes on opposite sides of a large landmass facing oceans with atmospheres influenced in the summer by the subtropical highs.

Now turn back to Chapter 5 and look again at Figure 5.13. Although the diagram shows the position of a subtropical high in the summer over the Atlantic Ocean, there is a similar high positioned over the Pacific Ocean during the same period. Whereas the Mediterranean regions, represented by Red Bluff (see Fig. 10.2), are under the stable, dry, drought-producing eastern flank of the

R. Gabler

● **FIGURE 10.5** Vineyards similar to this one in Tuscany, Italy are a common sight throughout Mediterranean regions.

*What characteristics make this climate especially suited for grape production?*

In Google Earth fly to: 43.567°N, 10.980°E (Tuscany, Italy).

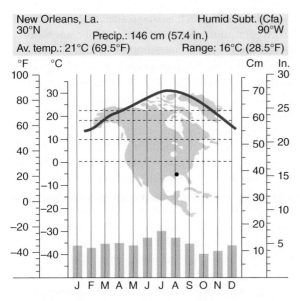

New Orleans, La.          Humid Subt. (Cfa)
30°N                                       90°W
Precip.: 146 cm (57.4 in.)
Av. temp.: 21°C (69.5°F)        Range: 16°C (28.5°F)

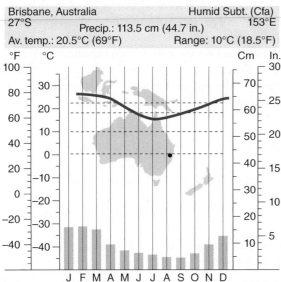

Brisbane, Australia         Humid Subt. (Cfa)
27°S                                       153°E
Precip.: 113.5 cm (44.7 in.)
Av. temp.: 20.5°C (69°F)       Range: 10°C (18.5°F)

● **FIGURE 10.6** Climographs for humid subtropical climate stations.

*What characteristics in these graphs suggest a location's hemisphere?*

Pacific high pressure system, the humid subtropical regions, represented by New Orleans (Fig. 10.6), are located on the weak western side of the Atlantic subtropical high. On the western side, subsidence and stability are greatly reduced or absent and precipitation is ample throughout the summer months. Here again, ocean temperatures play a significant role. The warm ocean currents that are commonly found along continental east coasts in these latitudes also moderate the winter temperatures and warm the lower atmosphere, thus increasing lapse rates, which enhances instability. Furthermore, a modified monsoon effect (especially in Asia and to some extent in the southern United States) increases summer precipitation as the moist, unstable tropical air is drawn in over the land.

As might be expected from the year-round rainfall, average annual precipitation for humid subtropical locations usually exceeds that for Mediterranean stations and may vary more

widely. Humid subtropical regions receive anywhere from 60 to 250 centimeters (25–100 in.) a year. Precipitation generally decreases inland toward continental interiors and away from the oceanic sources of moisture. It is not surprising that these regions are noticeably drier the closer they are to steppe regions inland toward their western margins.

Both the Mediterranean and humid subtropical climates receive winter moisture from cyclonic storms, which travel along the polar front. As we have noted, the great contrast occurs in the summer when the humid subtropics receive substantial precipitation from convectional showers, supplemented in certain regions by a modified monsoon effect. In addition, because of the shift in the sun and wind belts during the summer months, the humid subtropical climates are subject to tropical storms, some of which develop into hurricanes (or typhoons), especially in late summer. These three factors—the modified monsoon effect, convectional activity, and tropical storms—combine in most of these regions to produce a precipitation maximum in late summer. The climographs for New Orleans, Louisiana, and Brisbane, Australia, illustrate these effects (Fig. 10.6).

A subtype of the humid subtropical climate is found most often on the Asian continent, where the monsoon effect is most pronounced because of the magnitude of the seasonal pressure changes over this immense landmass. There the low-sun period, or winter season, is noticeably drier than the high-sun period. High pressure over the continent blocks the importation of moist air so that some months receive less than 3 centimeters (1.2 in.) of precipitation.

Temperatures in the humid subtropics are much like those of the Mediterranean regions. Annual ranges are similar, despite a greater variation among climate stations in the humid subtropical climate, primarily because the climate covers a far larger land area. Mediterranean stations record higher summer daytime temperatures, but summer months in both climates average around 25°C (77°F), increasing to as much as 32°C (90°F) as maritime influence decreases inland. Winter months in both climates average around 7°C to 14°C (45°F–57°F). Frost is a similar problem. The long growing season in the warmest humid subtropical regions enables farmers to grow such delicate crops as oranges, grapefruit, and lemons, but, as in the Mediterranean climates, farmers must be prepared with various means to protect their more sensitive crops from the danger of freezing. The growers of citrus crops in Florida are concentrated in the Central Lake District to take advantage of the moderating influence of nearby bodies of water.

Variation in humidity greatly affects the effective temperatures—the temperatures we feel—in the humid subtropical and Mediterranean climates. The summer temperatures in humid subtropical regions feel far warmer than they are because of the high humidity there. In fact, summers in this climate are oppressively hot, sultry, and uncomfortable. Nor is there the relief of lower night temperatures, as in the Mediterranean regions. The high humidity of the humid subtropical climate prevents much radiative loss of heat at night. Consequently, the air remains hot and sticky. Diurnal temperature ranges, in winter and in summer, are far smaller in the humid subtropical than in the drier Mediterranean regions.

Despite the relatively mild temperatures, humid subtropical winters seem cold and damp, again because of the high humidity.

## A Productive Climate

Vegetation generally thrives in humid subtropical regions, with their abundant rainfall, high temperatures, and long growing season. The wetter portions support forests of broad-leaf deciduous trees, pine forests on sandy soils, and mixed forests ( ● Fig. 10.7). In the drier interiors near the steppe regions, forests give way to grasslands, which require less moisture. There is an abundant and varied fauna. A few of the common species are deer, bears, foxes, rabbits, squirrels, opossums, raccoons, skunks, and birds of many sizes and species. Bird life in lake and marsh areas is incredibly rich. Alligators inhabit the American swamps in North Carolina, Georgia, and Florida.

As in the tropics, soils tend to have limited fertility because of rapid removal of soluble nutrients. However, there are excep-

tions in drier grassland areas, such as the pampas of Argentina and Uruguay, which constitute South America's "bread basket." Whatever the soil resource, the humid subtropical regions are of enormous agricultural value because of their favorable temperature and moisture characteristics. They have been used intensively for both subsistence crops, such as rice and wheat in Asia, and commercial crops, such as cotton and tobacco in the United States. When we consider that this climate (with its monsoon phase) is characteristic of south China and of the most densely populated portions of both India and Japan, we realize that this climate regimen contains and feeds far more human beings than any other type ( ● Fig. 10.8). The care with which agriculture has been practiced and the soil resources conserved over thousands of years of intensive use in eastern Asia are in sharp contrast to the agricultural exploitation of the past 200 years in the corresponding area of the United States. The traditional system of cotton and tobacco farming, in particular, devastated the land by exhausting the soil and triggering massive sheet and gully erosion. Over much of the old cotton and tobacco belt, extending from the Carolinas through Georgia and Alabama to the Mississippi Delta, all topsoil or a significant part of it has been lost, and we see only the red-clay subsoil and occasionally bare rock where crops formerly flourished. In the remaining areas, practices have had to change to conserve the soil that is left. Heavy applications of fertilizer, scientific crop rotation, and careful tilling of the land are now the rule.

Where forests still form the major natural vegetation, forest products, such as lumber, pulpwood, and turpentine, are important commercially ( ● Fig. 10.9). The long-leaf and slash pines of the southeastern United States are a source of lumber, as well as the resinous products of the pine tree (pitch, tar, resin, and turpentine). The absence of temperature and moisture limitations

● **FIGURE 10.7** Forest vegetation similar to this in Myakka State Park originally covered much of humid subtropical central Florida.

*How has the physical landscape of central Florida been changed by human occupancy?*

R. Gabler

● **FIGURE 10.8** The fields bordering the Yulong River in Yangshuo, China, are ideally suited for rice production.

*Why is rice a preferred crop in humid subtropical regions of China?*

James Feltz

● **FIGURE 10.9** A commercial tree farm in Alabama. Note that the pine trees are planted in orderly rows to expedite cultivation and harvest.

*Why are tree farms common in the U.S. Southeast and Pacific Northwest but not in New England or the upper Midwest?*

strongly favors forest growth. In Georgia, for example, trees may grow two to four times faster than in colder regions such as New England. This means that trees can be planted and harvested in much less time than in cooler forested regions, offering distinct commercial advantages.

The fact that living things thrive in the humid subtropical climate also presents certain problems because parasites and disease-carrying insects also thrive along with other life forms. African "killer bees," for example, have migrated through Mexico and now pose a new threat to humans and livestock in the southern United States.

Agriculture and lumbering are not the only important industries in the humid subtropical climates. In the southeastern United States, livestock raising has been greatly increasing in importance. Despite the commercial advantages of this climate, people often find it an uncomfortable one in which to live. The development and spread of air conditioning helps mitigate this problem. Where the ocean offers relief from the summer heat, as in Florida, the humid subtropical climate is an attractive recreation and retirement region. The beauty of its more unusual features, such as its cypress swamps and forests draped with Spanish moss, has to be experienced to be fully appreciated.

## Marine West Coast Climate

To begin your study of the **marine west coast climate** *(Cfb, Cfc),* it is even more important than it usually is with a new climate type to turn your attention to the climate's global loca-

tion. A quick review of Figures 10.1 and 8.6 reveals that marine west coast regions are found in latitudes that seem unusual for a mesothermal climate. These regions are invariably situated poleward of the other mesothermal climates. The northern hemisphere boundaries of the marine west coast climate even reach beyond the Arctic Circle. In those latitudes the climate is adjacent to the subarctic climate, the coldest of the microthermal climates. Mesothermal marine west coast regions seem to serve as an exception to the trend that temperatures decrease as latitudes increase toward the poles. Can you recall from your study of previous chapters why these regions can be located so close to the poles? If not, then, as is often the case with climate types, the answer to a question regarding climate location might be suggested in the title of the climate. Remember that oceans heat and cool more slowly than adjacent landmasses and therefore have a moderating effect on climates. Proximity to the sea and prevailing onshore winds make the marine west coast climate one of the most temperate in the world. Found in those middle-latitude regions (between 40° and 65°) that are continuously influenced by the westerlies, the climate receives ample precipitation and experiences no monthly temperatures averaging below 0°C (32°F). Summers are mild to cool. The climograph for Bordeaux, France in ● Figure 10.10 is representative of the mild-summer marine west coast climate *(Cfb),* and the climograph for Reykjavik, Iceland represents the cool-summer variety *(Cfc).*

**Oceanic Influences** As they travel onshore, the westerlies carry with them the moderating marine influence on temperature and much moisture. In addition, warm ocean currents, such as the North Atlantic Drift, bathe some of the coastal lands in the latitudes of the marine west coast climate, further moderating climatic conditions and accentuating humidity. This latter influence is particularly noticeable in Europe where the marine west coast climate extends along the coast of Norway to beyond the Arctic Circle (see Fig. 10.1).

In this climate zone, the marine influence is so strong that temperatures decrease little with poleward movement. Thus, the influence of the oceans is even stronger than latitude in determining these temperatures. This is obvious when we examine isotherms in the areas of marine west coast climates on a map of world temperatures (see Fig. 4.27). Wherever the marine west coast climate prevails, the isotherms swing poleward, parallel to the coast, clearly demonstrating the dominant *marine influence.*

Another result of the ocean's moderating effect is that the annual temperature ranges in the marine west coast climates are relatively small, considering the latitude. For an illustration of this, compare the monthly temperature graphs for Portland,

● **FIGURE 10.13** Reliable precipitation makes a diversified type of agriculture possible in the marine west coast climate areas, with emphasis on grain, orchard crops, vineyards, vegetables, and dairying. The village in the photograph is Iphofen, Germany.

*Although the climate is similar, why would a photo taken in a marine west coast agricultural region of the United States depict a scene that is significantly different from this one?*

**Resource Potential** There is little doubt that this climate offers certain advantages for agriculture. The small annual temperature ranges, mild winters, long growing seasons, and abundant precipitation all favor plant growth. Many crops, such as wheat, barley, and rye, can be grown farther poleward than in more continental regions. Although the soils common to these regions are not naturally rich in soluble nutrients, highly successful agriculture is possible with the application of natural or commercial fertilizers ( ● Fig. 10.13). Root crops (such as potatoes, beets, and turnips), deciduous fruits (such as apples and pears), berries, and grapes join the grains previously mentioned as important farm products. Grass in particular requires little sunshine, and pastures are always lush. The greenness of Ireland—the Emerald Isle—is evidence of these favorable conditions, as is the abundance of herds of beef and dairy cattle.

The magnificent forests that form the natural vegetation of the marine west coast regions have been a readily available resource. Some of the finest stands of commercial timber in the world are found along the Pacific coast of North America where pines, firs, and spruces abound, commonly exceeding 30 meters (100 ft) in height. Europe and the British Isles were once heavily forested, but most of those forests (even the famous Sherwood Forest of Robin Hood fame) have been cut down for building material and have been replaced by agricultural lands and urbanization.

# Humid Microthermal Climate Regions

Our definition of a humid microthermal climate includes one in which temperatures are high enough during part of the year to have a recognizable summer and cold enough 6 months later to have a distinct winter. In between are two periods, called spring and fall, when all life, and especially vegetation, makes preparations for the temperature extremes. Thus, in this section, we talk about climate regions that clearly display four readily identifiable seasons.

Seasonality is not the only reason we often use the word *variable* when describing the humid microthermal climates. As ● Figure 10.14 indicates, these climates are generally located between 35°N and 75°N on the North American and Eurasian landmasses. (Because there are no large landmasses at similar latitudes in the Southern Hemisphere, these climate types exist only in the Northern Hemisphere.) Thus, they share the westerlies and the storms of the polar front with the marine west coast climate. However, their position in the continental interiors and in high latitudes prevents them from experiencing the moderating influence of the oceans. In fact, the dominance of continentality in these climates is best demonstrated by the fact that they do not exist in the Southern Hemisphere where there are no large landmasses in the appropriate latitudes.

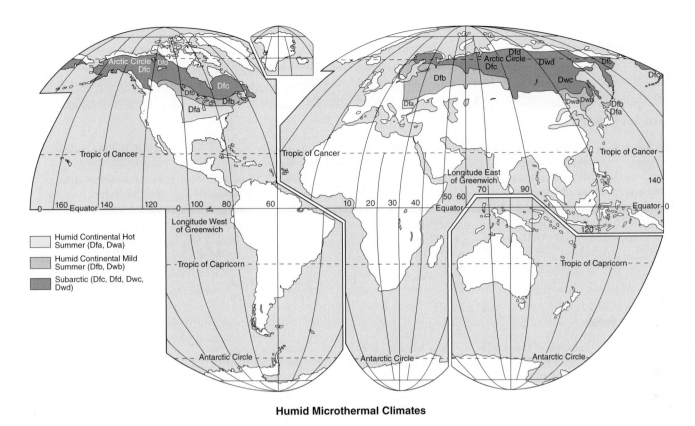

**Humid Microthermal Climates**

● **FIGURE 10.14** Index map of humid microthermal climates.

The recognition of three separate microthermal climates is based mainly on latitude and the resulting differences in the length and severity of the seasons (Table 10.2). Winters tend to be longer and colder toward the poleward margins because of latitude and toward interiors throughout the microthermal climates because of the continental influence. Summers inland are also inclined to be hotter, but they become progressively shorter as the winter season lengthens poleward. Thus, the three microthermal climates can be defined as humid continental, hot summer *(Dfa, Dwa)*; humid continental, mild summer *(Dfb, Dwb)*; and subarctic *(Dfc, Dfd, Dwc, Dwd)*, with a cool summer and, in extreme cases, a long, bitterly cold winter.

All microthermal climates have several features in common. By definition, they all experience a surplus of precipitation over potential evapotranspiration, and they have year-round precipitation. An exception to this rule lies in an area of Asia where the Siberian High causes winter droughts to occur. The greater frequency of maritime tropical air masses in summer and continental polar air masses in winter combined with the monsoonal effect and strong summer convection produce a precipitation maximum in the summer. Although the length of time that snow remains on the ground increases poleward and toward the continental interior ( ● Fig. 10.15), all three microthermal climates experience significant snow cover. This decreases the effectiveness of insolation and helps explain their cold winter temperatures. Finally, the unpredictable and variable nature of the weather is especially apparent in the humid microthermal climates.

With these generalizations in mind, let's compare the microthermal climates with the mesothermal climates we have just examined. Regions with microthermal climates have more severe winters, a lasting snow cover, shorter summers, shorter growing seasons, shorter frost-free seasons, a truer four-season development, lower nighttime temperatures, greater average annual temperature ranges, lower relative humidity, and much more variable weather than do the mesothermal climate regions.

## Humid Continental, Hot-Summer Climate

Unlike the other two microthermal climates, the **humid continental, hot-summer climate** *(Dfa, Dwa)* is relatively limited in its distribution on the Eurasian landmass (see Table 10.2). This is unfortunate for the people of Europe and Asia because it has by far the greatest agricultural potential and is the most productive of the microthermal climates. In the United States this climate is distributed over a wide area that begins with the eastern seaboard of New York, New Jersey, and southern New England and stretches continuously across the heartland of the eastern United States to encompass much of the American Midwest. It is one of the most densely populated, highly developed, and agriculturally productive regions in the world.

In terms of environmental conditions, the hot-summer variety of microthermal climate has some obvious advantages over its

**TABLE 10.2**
**The Microthermal Climates**

| Name and Description | Controlling Factors | Geographic Distribution | Distinguishing Characteristics | Related Features |
|---|---|---|---|---|
| **Humid Continental, Hot Summer** | | | | |
| Warmest month above 10°C (50°F); coldest month below 0°C (32°F); hot summers; usually year-round precipitation, winter drought (Asia) | Location in the lower middle latitudes (35°–45°); cyclonic storms along the polar front; prevailing westerlies; continentality; polar anticyclone in winter (Asia) | Eastern and midwestern U.S. from Atlantic coast to 100°W longitude; east central Europe; northern and northeast (Manchuria) China, northern Korea, and Honshu (Japan) | Hot, often humid summers; occasional winter cold waves; rather large annual temperature ranges; weather variability; precipitation 50–115 cm (20–45 in.), decreasing inland and poleward; 140- to 200-day growing season | Broad-leaf deciduous and mixed forest; moderately fertile soils with fertilization in wetter areas; highly fertile grassland and prairie soils in drier areas; "corn belt," soybeans, hay, oats, winter wheat |
| **Humid Continental, Mild Summer** | | | | |
| Warmest month above 10°C (50°F); coldest month below 0°C (32°F); mild summers; usually year-round precipitation, winter drought (Asia) | Location in the middle latitudes (45°–55°); cyclonic storms along the polar front; prevailing westerlies; continentality; polar anticyclone in winter (Asia) | New England, the Great Lakes region, and south central Canada; southeastern Scandinavia; eastern Europe, west central Asia; eastern Manchuria (China) and Hokkaido (Japan) | Moderate summers; long winters with frequent spells of clear, cold weather; large annual temperature ranges; variable weather with less total precipitation than farther south; 90- to 130-day growing season | Mixed or coniferous forest; moderately fertile soils with fertilization in wetter areas; highly fertile grassland and prairie soils in drier areas; spring wheat, corn for fodder, root crops, hay, dairying |
| **Subarctic** | | | | |
| Warmest month above 10°C (50°F); coldest month below 0°C (32°F); cool summers, cold winters poleward; usually year-round precipitation, winter drought (Asia) | Location in the higher middle latitudes (50°–70°); westerlies in summer, strong polar anticyclone in winter (Asia); occasional cyclonic storms; extreme continentality | Northern North America from Newfoundland to Alaska; northern Eurasia from Scandinavia through most of Siberia to the Bering Sea and the Sea of Okhotsk | Brief, cool summers; long, bitterly cold winters; largest annual temperature ranges; lowest temperatures outside Antarctica; low precipitation, 20–50 cm (10–20 in.); unreliable 50- to 80-day growing season; permafrost common | Northern coniferous forest (taiga); strongly acidic soils; poor drainage and swampy conditions in warm season; experimental vegetables and root crops |

poleward counterparts. Its higher summer temperatures and longer growing season permit farmers to produce a wide variety of crops. Those lands within the hot-summer region that were covered by ice sheets are far enough equatorward that there has been sufficient time for most negative effects of continental glaciation to be removed, and primarily positive effects remain. Soils are inclined to be more fertile, especially under forest cover, where the typical soil-forming processes are not as extreme and where deciduous trees are more common than the acid-associated pine. Of course, some advantages are matched by liabilities. The lower fuel bills of winter in the humid continental, hot-summer climate are often more than offset by the cost of air conditioning during the long, hot summers not found in other microthermal climates.

**Internal Variations** From place to place within the humid continental, hot-summer climate, there are significant differences in temperature characteristics. The length of the growing season is directly related to latitude, varying from 200 days equatorward to as little as 140 days along poleward margins of the climate.

*Continentality* has the greatest impact on temperature differences. The degree of continentality (distance from continental

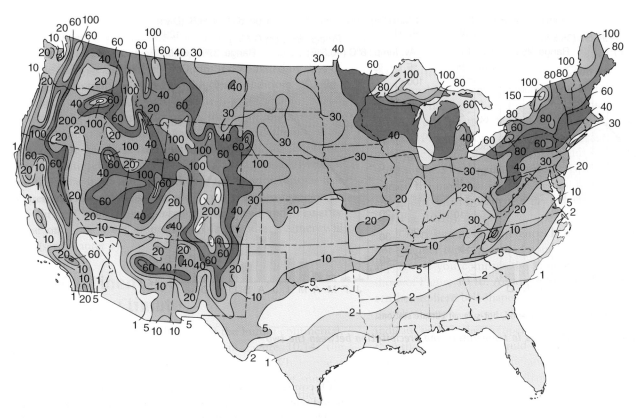

● **FIGURE 10.15** Map of the contiguous United States showing average annual number of days of snow cover.

*What areas of the United States average the greatest number of days of snow cover?*

coastlines) and size of a continent have significant effects on both summer and winter temperatures and, as a result, on temperature range. As an example, examine the climographs in ● Figure 10.16. Note the differences in both summer and winter temperatures recorded at two weather stations and compare the temperature ranges. Galesburg, Illinois, a typical station in North America, has slightly milder summers, less frigid winters, and a lower temperature range than Shenyang, northeast China (Manchuria), which is located at almost the same latitude but which experiences the greater seasonal contrasts of the Eurasian landmass. Throughout the humid continental, hot-summer climate, temperature ranges are consistently large and they become progressively larger toward continental interiors. As should be expected, near the coasts in this climate region, temperatures may be modified by a slight *maritime* influence so that they are milder, in both summer and winter, than those at inland locations at comparable latitudes. Large lakes may cause a similar effect. The amount and distribution of precipitation are also variable from station to station. The total precipitation received decreases both poleward and inland ( ● Fig. 10.17). A move in either direction is a move away from the source regions of warm maritime air masses that provide much of the moisture for cyclonic storms and convectional showers. This decrease can be seen in the average annual precipitation figures for the following cities, all at a latitude of about 40°N: New York (longitude 74°W), 115 centimeters (45 in.); Indianapolis, Indiana (86°W), 100 centimeters (40 in.);

Hannibal, Missouri (92°W), 90 centimeters (35 in.); and Grand Island, Nebraska (98°W), 60 centimeters (24 in.). Most stations have a precipitation maximum in summer when the warm, moist air masses dominate. In certain regions of Asia, not only does a summer maximum of precipitation exist, but also the monsoon circulation inhibits winter precipitation to such a degree that they experience winter drought (see again Shenyang in Fig. 10.16).

As might be expected, vegetation and soils vary with the climatic elements, especially precipitation. In the wetter regions, forests and forest soils predominate. At one time in certain sections of the American Midwest, tall prairie grasses grew where precipitation was insufficient to support forests; in all the drier portions of the climate, grasslands are the natural vegetation. The soils that developed under these grasslands are among the richest in the world.

**Seasonal Changes** The four seasons are highly developed in the humid microthermal, hot-summer climate. Each is distinct from the other three and has a character all its own. The winter is cold and often snowy; the spring is warmer, with frequent showers that produce flowers, budding leaves, and green grasses; the hot, humid summer brings occasional violent thunderstorms; and the fall has periods of both clear and rainy weather, with mild days and frosty nights in which the green leaves of summer turn to colorful reds, oranges, yellows, and browns before falling to the ground.

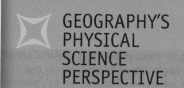
# Effective Temperatures

*Effective temperatures* (formerly known as *sensible temperatures*) are temperatures as they might be experienced by a person at rest, in ordinary clothing, in a motionless atmosphere. In other words, at any given time, how comfortable does a temperature feel to the individual experiencing it? This temperature value cannot be obtained by simply reading a thermometer.

Several factors come into play when considering effective temperatures. These factors can be divided into two categories: atmospheric factors and human factors. Of the atmospheric factors, four are most important. First is the actual temperature; thermometers will help distinguish between cold and warm days. Second is humidity; because the evaporation of sweat is a cooling process for the human body, humid days feel warmer than dry days. Third is

**Relative humidity (%)**

| Air temperature (°F) | 0 | 5 | 10 | 15 | 20 | 25 | 30 | 35 | 40 | 45 | 50 | 55 | 60 | 65 | 70 | 75 | 80 | 85 | 90 | 95 | 100 |
|---|---|---|---|---|---|---|---|---|---|---|---|---|---|---|---|---|---|---|---|---|---|
| 140 | 125 | | | | | | | | | | | | | | | | | | | | |
| 135 | 120 | 128 | | | | | | | | | | | | | | | | | | | |
| 130 | 117 | 122 | 131 | | | | | | | | | | | | | | | | | | |
| 125 | 111 | 116 | 123 | 131 | 141 | | | | | | | | | | | | | | | | |
| 120 | 107 | 111 | 116 | 123 | 130 | 139 | 148 | | | | | | | | | | | | | | |
| 115 | 103 | 107 | 111 | 115 | 120 | 127 | 135 | 143 | 151 | | | | | | | | | | | | |
| 110 | 99 | 102 | 105 | 108 | 112 | 117 | 123 | 130 | 137 | 143 | 150 | | | | | | | | | | |
| 105 | 95 | 97 | 100 | 102 | 105 | 109 | 113 | 118 | 123 | 129 | 135 | 142 | 149 | | | | | | | | |
| 100 | 91 | 93 | 95 | 97 | 99 | 101 | 104 | 107 | 110 | 115 | 120 | 126 | 132 | 138 | 144 | | | | | | |
| 95 | 87 | 88 | 90 | 91 | 93 | 94 | 96 | 98 | 101 | 104 | 107 | 110 | 114 | 119 | 124 | 130 | 136 | | | | |
| 90 | 83 | 84 | 85 | 86 | 87 | 88 | 90 | 91 | 93 | 95 | 96 | 98 | 100 | 102 | 106 | 109 | 113 | 117 | 122 | | |
| 85 | 78 | 79 | 80 | 81 | 82 | 83 | 84 | 85 | 86 | 87 | 88 | 89 | 90 | 91 | 93 | 95 | 97 | 99 | 102 | 105 | 108 |
| 80 | 73 | 74 | 75 | 76 | 77 | 77 | 78 | 79 | 79 | 80 | 81 | 81 | 82 | 83 | 85 | 86 | 86 | 87 | 88 | 89 | 91 |
| 75 | 69 | 69 | 70 | 71 | 72 | 72 | 73 | 73 | 74 | 74 | 75 | 75 | 76 | 76 | 77 | 77 | 78 | 78 | 79 | 79 | 80 |
| 70 | 64 | 64 | 65 | 65 | 66 | 66 | 67 | 67 | 68 | 68 | 69 | 69 | 70 | 70 | 70 | 70 | 71 | 71 | 71 | 71 | 72 |

Heat index (or apparent temperature)

A chart of effective temperatures can be used to determine the heat index, which combines temperature and humidity conditions to find an apparent temperature.

● **FIGURE 10.22** Climographs of the subarctic climate stations.

*Why would people settle in such severe-winter climate regions?*

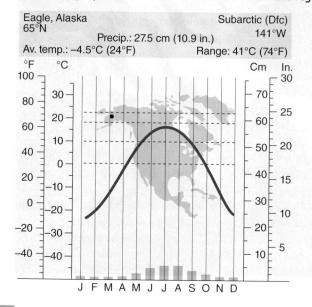

Eagle, Alaska
65°N
Subarctic (Dfc)
141°W
Precip.: 27.5 cm (10.9 in.)
Av. temp.: −4.5°C (24°F)
Range: 41°C (74°F)

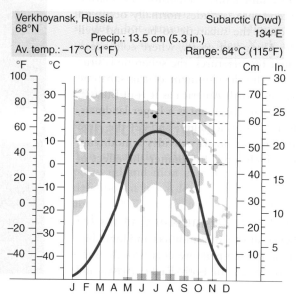

Verkhoyansk, Russia
68°N
Subarctic (Dwd)
134°E
Precip.: 13.5 cm (5.3 in.)
Av. temp.: −17°C (1°F)
Range: 64°C (115°F)

wind speed; winds not only carry heat from the body but can also accelerate the evaporation of sweat. Fourth is the percentage of clear sky; shady areas are cooler than sunny areas.

Of the many human factors, the following stand out as important. First is respiration; breathing in a lungful of cold air will make one feel colder. Second is perspiration; the evaporative-cooling process is quite efficient for the human body, but it differs from one individual to another. Third is the amount of activity involved; physical work or playing a physical sport can heat the body rapidly. Fourth is the amount of exposed skin; tank tops versus sweatshirts can make a world of difference.

Effective temperatures are established by considering the interplay between these two sets of factors. For example, the well-known heat index (also called apparent temperature) takes into account the temperature and humidity of a summer day and calculates how it might feel. The equally well-known wind chill index considers both temperature and wind speed to establish how cold it might feel on a winter day. Keep in mind that these are only theoretical values. No one can predict exactly how comfortable a particular person will feel on a given day. However, these temperature indices can help guide us when dealing with seasonally extreme days.

## Wind Chill Chart

Temperature (°F)

| Wind (mph) \ Temp | 40 | 35 | 30 | 25 | 20 | 15 | 10 | 5 | 0 | −5 | −10 | −15 | −20 | −25 | −30 | −35 | −40 | −45 |
|---|---|---|---|---|---|---|---|---|---|---|---|---|---|---|---|---|---|---|
| 5 | 36 | 31 | 25 | 19 | 13 | 7 | 1 | -5 | -11 | -16 | -22 | -28 | -34 | -40 | -46 | -52 | -57 | -63 |
| 10 | 34 | 27 | 21 | 15 | 9 | 3 | -4 | -10 | -16 | -22 | -28 | -35 | -41 | -47 | -53 | -59 | -66 | -72 |
| 15 | 32 | 25 | 19 | 13 | 6 | 0 | -7 | -13 | -19 | -26 | -32 | -39 | -45 | -51 | -58 | -64 | -71 | -77 |
| 20 | 30 | 24 | 17 | 11 | 4 | -2 | -9 | -15 | -22 | -29 | -35 | -42 | -48 | -55 | -61 | -68 | -74 | -81 |
| 25 | 29 | 23 | 16 | 9 | 3 | -4 | -11 | -17 | -24 | -31 | -37 | -44 | -51 | -58 | -64 | -71 | -78 | -84 |
| 30 | 28 | 22 | 15 | 8 | 1 | -5 | -12 | -19 | -26 | -33 | -39 | -46 | -53 | -60 | -67 | -73 | -80 | -87 |
| 35 | 28 | 21 | 14 | 7 | 0 | -7 | -14 | -21 | -27 | -34 | -41 | -48 | -55 | -62 | -69 | -76 | -82 | -89 |
| 40 | 27 | 20 | 13 | 6 | -1 | -8 | -15 | -22 | -29 | -36 | -43 | -50 | -57 | -64 | -71 | -78 | -84 | -91 |
| 45 | 26 | 19 | 12 | 5 | -2 | -9 | -16 | -23 | -30 | -37 | -44 | -51 | -58 | -65 | -72 | -79 | -86 | -93 |
| 50 | 26 | 19 | 12 | 4 | -3 | -10 | -17 | -24 | -31 | -38 | -45 | -52 | -60 | -67 | -74 | -81 | -88 | -95 |
| 55 | 25 | 18 | 11 | 4 | -3 | -11 | -18 | -25 | -32 | -39 | -46 | -54 | -61 | -68 | -75 | -82 | -89 | -97 |
| 60 | 25 | 17 | 10 | 3 | -4 | -11 | -19 | -26 | -33 | -40 | -48 | -55 | -62 | -69 | -76 | -84 | -91 | -98 |

Frostbite occurs in 15 minutes or less

A chart of effective temperatures can be used to determine wind chill, which shows what the combination of cold temperatures and wind speed will make the temperature outside feel like.

Verkhoyansk, Russia there are more months with average temperatures below freezing than above. Also note how steeply the temperatures rise and fall between summer and winter. The rapid heating and cooling associated with continental interiors in the higher latitudes allow little time for the in-between seasons of spring and fall. At Eagle, Alaska, a station in the Klondike region of the Yukon River Valley, the temperature climbs 8°C to 10°C (15°F–20°F) per month as summer approaches and drops just as rapidly prior to the next winter season. At Verkhoyansk in Siberia, the change between the seasonal extremes is even more rapid, averaging 15°C to 20°C (30°F–40°F) per month.

Because of the high latitudes of these regions, summer days are quite long and nights are short. The noon sun is as high in the sky during a subarctic summer as during a subtropical winter. The combination of a moderately high angle of the sun's rays and many hours of daylight means that some subarctic locations receive as much insolation at the time of the summer solstice as the equator does. As a result, temperatures during the 1 to 3 months of the subarctic summer usually average 10°C to 15°C (50°F–60°F), and on some days they may even approach 30°C (86°F). Thus, the brief summer in the subarctic climate is often pleasantly warm, even hot, on some days.

The winter season in the subarctic is bitter and intense and lasts for as long as 8 months. Eagle, Alaska has 8 months with average temperatures below freezing. In the Siberian subarctic the January temperatures regularly *average* −40°C to −50°C (−40°F to −60°F). The coldest temperatures in the Northern Hemisphere have been recorded there and are officially −68°C (−90°F) at both Verkhoyansk and Oymyakon, Russia and unofficially −78°C (−108°F) at Oymyakon. In addition, the winter nights, with an average 18 to 20 hours of darkness extending well into one's working hours, can be mentally depressing and can increase the impression of climatic severity.

As a direct result of the intense heating and cooling of the land, the subarctic has the largest annual temperature ranges of any climate. Average annual ranges in equatorward margins of the climate vary from near 40°C (72°F) to more than 45°C (80°F). The exceptions are near western coasts, where warm ocean currents and the marine influence may significantly modify winter temperatures. Average annual ranges for pole-ward stations are even greater. The climograph for Verkhoyansk, which indicates a range of 64°C (115°F), is an extreme example.

As with subarctic temperature, latitude and continentality influence subarctic precipitation. These climate controls combine to limit annual precipitation amounts to less than 50 centimeters (20 in.) for most regions and to 25 centimeters (10 in.) or less in northern and interior locations. The low temperatures in the subarctic reduce the moisture-holding capacity of the air, thus minimizing precipitation during the occasional passage of cyclonic storms. Location toward the center of large landmasses or near lee coasts increases distance from oceanic sources of moisture. Finally, the higher latitudes occupied by the subarctic climate are dominated by the polar anticyclone, especially in the winter season. The subsidence and divergence of air in the polar anticyclone limit the opportunity for lifting and hence for precipitation in the subarctic regions. This high pressure system also blocks the entry of moist air from warmer areas to the south.

National Park Service – Wrangell-St. Elias

● **FIGURE 10.23** Seemingly endless tracts of taiga (boreal forest) are typical of the vegetation throughout much of the American and Canadian subarctic. This photo was taken in Alaska's Wrangell-St. Elias National Park and Preserve.

*Why are these kinds of virgin forests currently of little economic value?*

Subarctic precipitation is cyclonic or frontal, and because the polar anticyclone is weaker and farther north during the warmer summer months, more precipitation comes during that season. The meager winter precipitation falls as fine, dry snow. Although there is not as much snowfall as in less severe climates, the temperatures remain so cold for so long that the snow cover lasts for as long as 7 or 8 months. During this time, there is almost no melting of snow, especially in the dark shadows of the forest.

**A Limiting Environment** The climatic restrictions of subarctic regions place distinct limitations on plant and animal life and on human activity. The characteristic vegetation is coniferous forest, adapted to the severe temperatures; the physiologic drought associated with frozen soil water; and the infertile soils. Seemingly endless tracts of spruce, fir, and pine thrive over enormous areas, untouched by humans ( ● Fig. 10.23). In Russia the forest is called the *taiga* (or *boreal forests* in other regions), and this name is sometimes given to the subarctic climate type itself.

The brief summers and long, cold winters severely limit the growth of vegetation in subarctic regions. Even the trees are shorter and more slender than comparable species in less severe climate regimens. There is little hope for agriculture. The growing season averages 50 to 75 days, and frost may occur even during June, July, or August. Thus, in some years a subarctic location may have no truly frost-free season. Although scientists are working to develop plant species that can take advantage of the long hours of daylight in summer, only minimal success has been achieved in southern parts of the climate

with certain vegetables such as cabbage and root crops such as potatoes.

A particularly vexing problem to people in subarctic (and tundra) regions is **permafrost,** a permanently frozen layer of subsoil and underlying rock that may extend to a depth of 300 meters (1000 ft) or more in the northernmost sections of the climate. Permafrost is present over much of the subarctic climate, but it varies greatly in thickness and is often discontinuous. Where it occurs the land is frozen completely from the surface down in winter. The warm temperatures of spring and summer cause the top few feet to thaw out, but because the land beneath this thawed top layer remains frozen, water cannot percolate downward, and the thawed soil becomes sodden with moisture, especially in spring, when there is an abundant supply of water from the melting snow. Permafrost poses a problem to agriculture by preventing proper soil drainage. The seasonal freeze and thaw of the surface layer above the permafrost also poses special problems for construction engineers. The cycle causes repeated expansion and contraction, heaving the surface up and then letting it sag down. The effects of this cycle break up roads, force buried pipelines out of the ground, cause walls and bridge piers to collapse, and even swivel buildings off their foundations. Landscapes called **patterned ground,** or **frost polygons,** are commonly found in regions of subarctic yearly freezing and thawing ( ● Fig. 10.24).

With agriculture being a questionable occupation at best, there is little economic incentive to draw humans to subarctic regions. Logging is unimportant because of small tree size. Even use of the vast forests for paper, pulp, and wood products is re-

● **FIGURE 10.24** Frost polygons, or patterned ground, as seen in the Arctic National Wildlife Refuge of Alaska. Repeated freezing and thawing cause the soil to produce polygonal shapes. When the ground freezes, it expands, and it shrinks when it thaws.

*Why are the edges heaved upward?*

# Polar Climate Regions

The polar climates are the last of Köppen's humid climate subdivisions to be differentiated on the basis of temperature. These climate regions are situated at the greatest distance from the equator, and they owe their existence primarily to the low annual amounts of insolation they receive. No polar station experiences a month with average temperatures as high as 10°C (50°F), and hence all are without a warm summer (Table 10.3). Trees cannot survive in such a regimen. In the regions where at least 1 month averages above 0°C (32°F), they are replaced by tundra vegetation. Elsewhere, the surface is covered by great expanses of frozen ice. Thus, there are two polar climate types, tundra *(ET)* and ice sheet *(EF)*.

Two important points should be kept in mind in the discussion of polar climate regions. First, these regions have a large net annual radiation loss; that is, they give up much more radiation or energy than they receive from the sun during a year, resulting in a major radiation deficiency. The transfer of heat from lower to higher latitudes to make up this deficiency is the driving force of the general atmospheric circulation. Without this compensating poleward transfer

stricted by their interior location far from world markets. Miners occasionally exploit rich ore deposits; others, many of them native to the subarctic regions, pursue hunting, trapping, and fishing of the relatively limited wildlife. Fur-bearing animals such as mink, fox, wolf, ermine, otter, and muskrat are of greatest value.

**TABLE 10.3**
**The Polar Climates**

| Name and Description | Controlling Factors | Geographic Distribution | Distinguishing Characteristics | Related Features |
|---|---|---|---|---|
| **Tundra** | | | | |
| Warmest month between 0°C (32°F) and 10°C (50°F); precipitation exceeds potential evapotranspiration | Location in the high latitudes; subsidence and divergence of the polar anticyclone; proximity to coasts | Arctic Ocean borderlands of North America, Greenland, and Eurasia; Antarctic Peninsula; some polar islands | At least 9 months average below freezing; low evaporation; precipitation usually below 25.5 cm (10 in.); coastal fog; strong winds | Tundra vegetation; tundra soils; permafrost; swamps and bogs during melting period; life most common in nearby seas; Inuit; mineral and oil resources; defense industry |
| **Ice Sheet** | | | | |
| Warmest month below 0°C (32°F); precipitation exceeds potential evaporation | Location in the high latitudes and interior of landmasses; year-round influence of the polar anticyclone; ice cover; elevation | Antarctica; interior Greenland; permanently frozen portions of the Arctic Ocean and associated islands | Summerless; all months average below freezing; world's coldest temperature; extremely meager precipitation in the form of snow, evaporation even less; gale-force winds | Ice- and snow-covered surface; no vegetation; no exposed soils; only sea life or aquatic birds; scientific exploration |

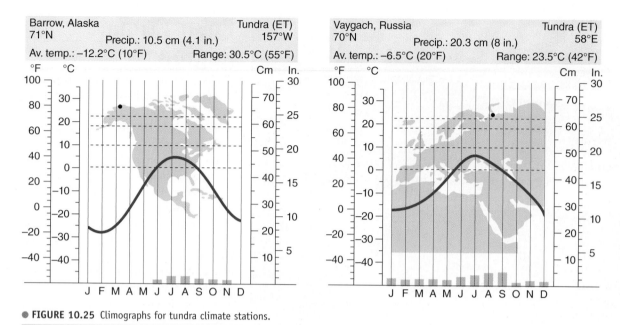

Barrow, Alaska
71°N
Precip.: 10.5 cm (4.1 in.)
Av. temp.: −12.2°C (10°F)
Tundra (ET)
157°W
Range: 30.5°C (55°F)

Vaygach, Russia
70°N
Precip.: 20.3 cm (8 in.)
Av. temp.: −6.5°C (20°F)
Tundra (ET)
58°E
Range: 23.5°C (42°F)

● **FIGURE 10.25** Climographs for tundra climate stations.

*Why is it not surprising that both stations are located in the Northern Hemisphere?*

of heat from the lower latitudes, the polar regions would become too cold to permit any form of life and the equatorial regions would heat to temperatures no organism could survive.

A second and equally important characteristic of polar climates is the unique pattern of day and night. At the poles, 6 months of relative darkness, caused when the sun is positioned below the horizon, alternate with 6 months of daylight during which the sun is above the horizon. Even when the sun is above the horizon, however, the sun's rays are at a sharply oblique angle, and little insolation is received for the number of hours of daylight. Moving outward from the poles, the lengths of periods of continuous winter night and continuous summer day decrease rapidly from 6 months at the poles to 24 hours at the Arctic and Antarctic Circles (66 1/2°N and S). Here the 24-hour night or day occurs only at the winter and summer solstices, respectively.

Antarctic Peninsula, it is everywhere adjacent to the Arctic Ocean, you should already know the answer. Tundra temperatures are directly influenced by the tundra's maritime location.

It almost seems inappropriate to call the unpleasantly chilly and damp conditions of the tundra's warmer season "summer." Temperatures average around 4°C (40°F) to 10°C (50°F) for the warmest month, and frosts occur regularly. The air does warm sufficiently to melt the thin snow cover and the ice on small bodies of water, but this only causes marshes, swamps, and bogs to form across the land because drainage is blocked by permafrost ( ● Fig. 10.26).

● **FIGURE 10.26** Permafrost regions, such as this area at the base of the Alaska Range, become almost impenetrable swampland during the brief Alaskan summer. Travel over land is feasible only in the winter season.

*What is the preferred means of travel in the summer?*

## Tundra Climate

Turn back to Figure 8.6 and compare the location of the **tundra climate** *(ET)* with that of the subarctic climate. You can see that the tundra climate is situated closer to the poles. Now compare the winter temperatures and temperature ranges of Barrow, Alaska and Vaygach, Russia, tundra climates in ● Figure 10.25, with those of Eagle, Alaska and Verkhoyansk, Russia, subarctic climates in Figure 10.22. Note that even though the temperature ranges in tundra climates are large, they are not as large as those in subarctic climates, and winter temperatures are not as severe. How can this be true, when the tundra climate is closer to the poles? If you noticed that the tundra climate is also situated along the periphery of landmasses and, with the exception of the

Alaska Image Library/USFWS

● **FIGURE 10.27** An island off the Antarctic Peninsula displays thick, bright patches of moss as its most complex vegetation.

*What weather elements might help to form this stark landscape?*

Clouds of black flies, mosquitoes, and gnats swarm in this soggy landscape, known as **muskeg** in Canada and Alaska. A bright note in the landscape is provided by the enormous number of migratory birds that nest in the arctic regions in summer and feed on the insects. However, as soon as the shrinking daylight hours of autumn approach, these birds depart for warmer climates.

Winters are cold and seem to last forever, especially in tundra locations where the sun may be below the horizon for days at a time. The climograph for Barrow, Alaska, illustrates the low temperatures of this climate. Note that average monthly temperatures are *below freezing* 9 months of the year. The average annual temperature is −12°C (10°F). The low-growing tundra vegetation survives despite the forbidding environment. It consists of lichens, mosses, sedges, flowering herbaceous plants, small shrubs, and grasses. In particular, the plants have adjusted to the conditions associated with nearly universal permafrost ( ● Fig. 10.27).

The tundra regions exhibit several other significant climatic characteristics. Diurnal temperature ranges are small because insolation is uniformly high during the long summer days and uniformly low during the long winter nights. Precipitation is generally low, except in eastern Canada and Greenland, because of exceedingly low absolute humidity and the influence of the polar anticyclone. Icy winds sweep across the open land surface and are an added factor in eliminating the trees that might impede their progress. Coastal fog is characteristic in marine locations, where cool polar maritime air drifts onshore and is chilled below the dew point by contact with the even colder land.

## Ice-Sheet Climate

The **ice-sheet climate** *(EF)* is the most severe and restrictive climate on Earth. As Table 10.3 indicates, it covers large areas in both the Northern and Southern Hemispheres, a total of about 16 million square kilometers (6 million sq mi)— nearly the same area as the United States and Canada combined. All average monthly temperatures are below freezing, and because most surfaces are covered with glacial ice, no vegetation can survive in this climate. It is a virtually lifeless region of perpetual frost.

Antarctica is the coldest place on Earth (although Siberia sometimes has longer and more severe periods of cold in winter). The world's coldest temperature, −88°C (−127°F), was recorded at Vostok, Antarctica. Consider the climographs for Little America, Antarctica, and Eismitte, Greenland, for a fuller picture of the cold ice-sheet temperatures ( ● Fig. 10.28).

The primary reason for the low temperatures of ice-sheet climates is the minimal insolation received in these regions. Not only is little or no insolation received during half the year, but also the sun's radiant energy that is received arrives at sharply oblique angles. In addition, the perpetual snow and ice cover of this climate reflects nearly all incoming radiation. A further factor, in both Greenland and Antarctica, is elevation. The ice sheets covering both regions rise more than 3000 meters (10,000 ft) above sea level ( ● Fig. 10.29). Naturally, this elevation contributes to the cold temperatures.

The polar anticyclone severely limits precipitation in the ice-sheet climate to the fine, dry snow associated with occasional cyclonic storms. Precipitation is so meager in this climate that regions within this regimen are sometimes incorrectly referred to as "polar deserts." However, because of the exceedingly low evaporation rates (evapotranspiration rates are not considered in this climate because there is no plant life to account for any transpiration) associated with the severely cold temperatures, precipitation still exceeds potential evaporation, and the climate can be classified as humid. The annual precipitation surplus produces glaciers, which export snowfall similar to the way rivers export rainfall.

The strong and persistent polar winds are another staple of the harsh ice-sheet climate. Mawson Base, Antarctica, for example, has approximately 340 days a year with gale-force winds of 15 meters per second (33 mph) or more. *Katabatic winds,* which are caused by the downslope drainage of heavy cold air accumulated over ice sheets, are common along the edges of the polar ice. The winds of these regions can result in whiteouts—periods of zero visibility as a result of blowing fine snow and ice crystals.

## Human Activity in Polar Regions

The climatic severity that limits animal life in polar regions to a few scattered species in the tundra is just as restrictive on human settlement. The celebrated Lapps of northern Europe migrate

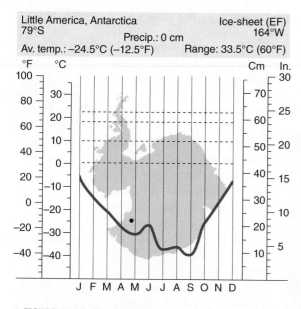

Little America, Antarctica  Ice-sheet (EF)
79°S  164°W
Precip.: 0 cm
Av. temp.: −24.5°C (−12.5°F)  Range: 33.5°C (60°F)

J F M A M J J A S O N D

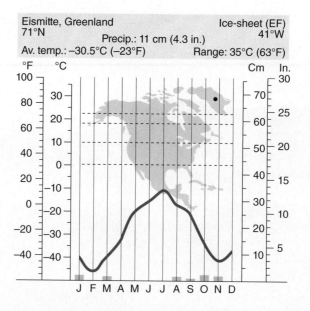

Eismitte, Greenland  Ice-sheet (EF)
71°N  41°W
Precip.: 11 cm (4.3 in.)
Av. temp.: −30.5°C (−23°F)  Range: 35°C (63°F)

J F M A M J J A S O N D

● **FIGURE 10.28** Climographs for ice-sheet climate stations.

*If you were to accept an offer for an all-expense-paid trip to visit either Greenland or Antarctica, which would you choose, and why would you go?*

with their reindeer to the tundra from the adjacent forest during warmer months. They join the musk ox, arctic hare, fox, wolf, and polar bear that manage to make a home there despite the prohibitive environment. Only the Inuit (Eskimos) of Alaska, northern Canada, and Greenland have in the past succeeded in developing a year-round lifestyle adapted to the tundra regimen, yet even this group relies less on the resources of the tundra than on the large variety of fish and sea mammals, such as cod, salmon, halibut, seal, walrus, and whale, that occupy the adjacent seas.

As their communication with the rest of the world has increased and they have become acquainted with alternative lifestyles, the permanent population of Inuit living in the tundra has greatly diminished, and life for those remaining has changed drastically. Some have gained new economic security

● **FIGURE 10.29** The Antarctic ice sheet resulting from the ice-sheet climate of the south polar regions is as much as 4000 meters (13,200 ft) thick locally. Where it is thinnest, it is floated by seawater to produce an ice shelf. The smaller Greenland ice sheet is about 3000 meters (10,000 ft) thick.

*What reasons might be given for the fact that more land in Greenland than in Antarctica is free of glacial ice?*

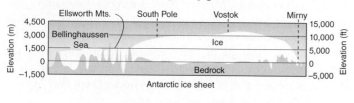

Antarctic ice sheet

Greenland ice sheet

through employment at defense installations or at sites where they join other skilled workers from outside the region to exploit mineral or energy resources. However, the new population centers based on the construction and maintenance of radar and missile defense stations or, as in the case of Alaska's North Slope, on the production and transportation of oil, cannot be considered permanent ( ● Fig. 10.30). Workers depend

● **FIGURE 10.30** Mile-marker zero on the Alaska pipeline, near Prudhoe Bay. This is one very profitable venture for humans in the North Slope oil fields. However, in 1989, the *Exxon Valdez* spilled 11 million gallons of crude oil into Prince William Sound, Alaska.

*Considering the vulnerability of Alaska's physical environment, should development of the North Slope oil fields have been permitted?*

M. Trapasso

● **FIGURE 10.31** Greenland's ice sheet covers about 85% of its surface. Here we see the ice sheet where it meets the ocean.

*What kind of activities might bring individuals from other regions to an ice-sheet climate?*

on other regions for support and often inhabit this region only temporarily.

The ice-sheet climate cannot serve as a home for humans or other animals. Even the penguins, gulls, leopard seals, and polar bears are coastal inhabitants. It is without question the harshest, most restrictive, most nearly lifeless climate zone on Earth ( ● Fig. 10.31), yet, especially in Antarctica, it is of strategic importance and of great scientific interest. Scientists study the oxygen-isotope ratios of the Antarctic ice cores and the gas bubbles trapped within them to help them reconstruct past climates. They have also noted that ozone concentrations have decreased over Antarctica every fall for several decades. The result is a hole in the ozone layer above Antarctica that is as large as the North American continent. This decrease in the ozone layer is a major environmental concern. Antarctica's strategic value is so widely recognized that the world's nations have voluntarily given up claims to territorial rights on the continent in exchange for cooperative scientific exploration on behalf of all humankind.

# Highland Climate Regions

As we saw in Chapter 4, temperature decreases with increasing altitude at the rate of about 6.5°C per 1000 meters (3.6°F per 1000 ft). Thus, you might suspect that highland regions exhibit broad zones of climate based on changes in temperature with elevation that roughly correspond to Köppen's climate zones based on change of temperature with latitude. This is indeed the case, with one important exception: Seasons only exist in high-

lands if they also exist in the nearby lowland regions. For example, although zones of increasingly cooler temperature occur at progressively higher elevations in the tropical climate regions, the seasonal changes of Köppen's middle-latitude climates are not present ( ● Fig. 10.32).

Elevation is only one of several controls of highland climates; **exposure** is another. Just as some continental coasts face the prevailing wind, so do some mountain slopes; others are lee slopes or are sheltered behind higher topography. The nature of the wind, its temperature, and its moisture content depend on whether the mountain is (1) in a coastal location or deep in a continental interior and (2) at a high or low latitude within or beyond the reaches of cyclonic storms and monsoon circulation. In the middle and high latitudes, mountain slopes and valley walls that face the equator receive the direct rays of the sun and are warm; poleward-facing slopes are shadowed and cool. West-facing slopes feel the hot afternoon sun, whereas east-facing slopes are sunlit only in the cool of the morning. This factor, known as **slope aspect**, affects where people live in the mountains and where particular crops will do best. The higher one rises in the mountains, the more important direct sunlight is as a source of warmth and energy for plant and animal life processes.

Complexity is the hallmark of highland climates. Every mountain range of significance is composed of a mosaic of climates far too intricate to differentiate on a world map or even on a map of a single continent. Highland climates are therefore undifferentiated, signifying climate complexity. Highland climates are indicated on Figure 8.6 wherever there is marked local variation in climate as a consequence of elevation, exposure, and slope aspect. We can see that these regions are distributed widely over Earth but are particularly concentrated in Asia, central Europe, and western North and South America.

The areas of highland climate on the world map are cool, moist islands in the midst of the zonal climates that dominate the areas around them. Consequently, highland areas are also biotic islands, supporting a flora and fauna adapted to cooler and wetter conditions than those of the surrounding lowlands. This coolness is part of the highland charm, particularly where mountains rise cloaked with forests above arid plains, as do the Canadian Rocky Mountains and California's Sierra Nevada.

Highlands stimulate moisture condensation and precipitation by forcing moving air masses to rise over them ( ● Fig. 10.33). Where mountain slopes are rocky and forest free, their surfaces grow warm during the day, causing upward convection, which often produces afternoon thundershowers. Mountains receive abundant precipitation and are the source area for multitudes of streams that join to form the great rivers of all of the continents.

There are few streams of significance whose headwaters do not lie in rugged highlands. Much of the stream flow on all continents

● **FIGURE 10.32** Note the terraced fields among the Inca ruins of Machu Picchu. Situated at 13 degrees south latitude and 2410 meters (8000 ft) above sea level, food production in the year-round spring-like climate of this tropical highland location supported an estimated 50,000 people.

*Why would the development of a similar ancient civilization at such high elevations be unlikely in the higher middle latitudes?*

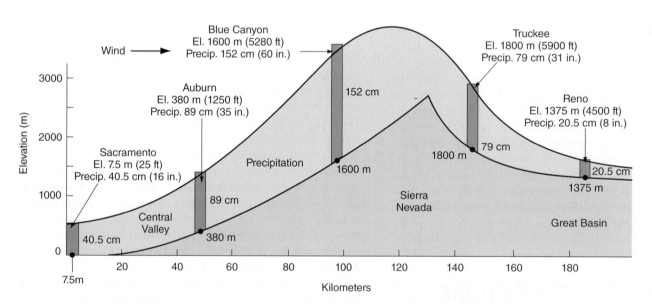

● **FIGURE 10.33** Variation in precipitation caused by uplift of air crossing the Sierra Nevada range of California from west to east. The maximum precipitation occurs on the windward slope because air in the summit region is too cool to retain a large supply of moisture. Note the strong rain shadow to the lee that gives Reno a desert climate.

*Taking into consideration the locations of the recording stations, during what season of the year does the maximum precipitation on the windward slope occur?*

is produced by the summer melting of mountain snowfields. Thus, the mountains not only draw moisture from the atmosphere, but also store much of it in a form that gradually releases it throughout summer droughts when water is most needed for irrigation and for municipal and domestic use.

## Peculiarities of Mountain Climates

A general characteristic of mountain weather is its variability from hour to hour and from place to place. Strong orographic flow over mountains often causes clouds to form very quickly, leading to thunderstorms and longer rains that do not affect surrounding cloud-free lowlands. Where the cloud cover is diminished, diurnal temperature ranges over mountains are far greater than those over lowlands. Because mountains penetrate upward beyond the densest part of the atmosphere, the greenhouse effect is less developed at those high elevation locations than anywhere else on Earth. The thinner layer of low-density air above a mountain site does not greatly impede insolation, thus allowing surfaces to warm dramatically during the daytime. By the same token, the atmosphere in these areas does little to impede longwave radiation loss at night. Consequently, air temperatures overnight are cooler than the elevation alone would indicate. Because the atmospheric shield is thinnest at high elevations, plants, animals, and humans receive proportionately more of the sun's shortwave radiation at high altitudes. Ultraviolet radiation is particularly noticeable; severe sunburn is one of the real hazards of a day in the high country.

In the middle and high latitudes, mountains rise from mesothermal and microthermal climates into tundra and snow-covered zones. The lower slopes of mountains are commonly forested with conifers, which become more stunted as one moves upward, until the last dwarfed tree is passed at the **tree line.** This is the line beyond which low winter temperatures and severe wind stress eliminate all forms of vegetation except those that grow low to the ground, where they can be protected by a blanket of snow ( ● Fig. 10.34). Where mountains are high enough, snow or ice permanently covers the land surface. The line above which summer melting is insufficient to remove all of the preceding winter's snowfall is called the **snow line**.

In tropical mountain regions, the vertical zonation of climate is even more pronounced. Both tree line and snow line occur at higher elevations than in middle latitudes. Any seasonal change is mainly restricted to rainfall; temperatures are stable year-round, regardless of elevation. Each climate zone has its own particular association of natural vegetation and has given rise to a distinctive crop combination where agriculture is practiced ( ● Fig. 10.35). In South America four vertical climate zones are recognized: *tierra caliente* (hot lands), *tierra templada* (temperate lands), *tierra fría* (cool lands), and *tierra helada* (frozen lands).

R. Gabler

● **FIGURE 10.34** As in this example in the Colorado Rocky Mountains, the last tree species found at the tree line are stunted, prostrate forms, which often produce an elfin forest. Where the trees are especially gnarled and misshapen by wind stress, the vegetation is called *krummholz* (crooked wood).

*What do you see in the photograph that indicates prevailing wind direction?*

## Highland Climates and Human Activity

In middle-latitude highlands soils are poor; the growing season is short; and the winter snow cover is heavy in the conifer zone, which dominates the lower and middle mountain slopes. Therefore, little agriculture is practiced and permanent settlements in the mountains are few. However, as the winter snow melts off the high ground just below the bare rocky peaks, grass springs into life and humans drive herds of cattle and flocks of sheep and goats up from the warmer valleys. The high pastures are lush throughout the summer, but in early fall they are once again vacated by the animals and their keepers, who return to the valleys. This seasonal movement of herds and herders between alpine pastures and villages in the valleys, termed *transhumance,* was once common in the European highlands (the Alps, Pyrenees, Carpathian Mountains, and mountains throughout Scandinavia) and is still practiced there on a reduced scale.

Otherwise, the middle-latitude highlands serve mainly as sources of timber and of minerals formed by the same geologic forces that elevated the mountains and as arenas for recreation—in both summer and winter. Recreational use of the highlands is a relatively recent phenomenon, resulting both from new interest in mountain areas and from new access routes by road, rail, and air that did not exist a century ago.

In contrast to poleward mountain regions, tropical highlands may actually experience more favorable climatic conditions and are often a greater attraction to human settlement (from ancient to modern times) than adjacent lowlands. In fact, large permanent populations are supported throughout the tropics where topogra-

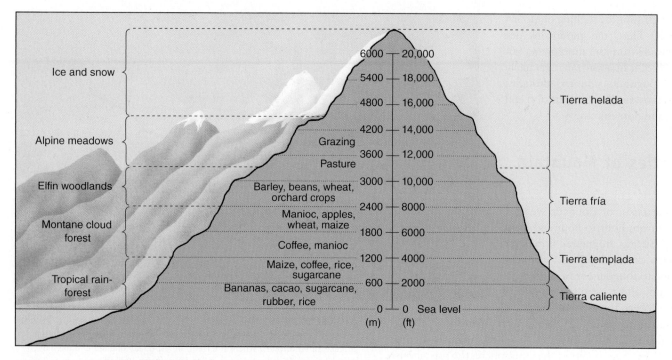

● **FIGURE 10.35** Natural vegetation, vertical climate zones, and agricultural products in tropical mountains. Note that this example extends from tropical life zones to the zone of permanent snow and ice. There is little seasonal temperature change in tropical mountains, which allows life forms sensitive to low temperatures to survive at relatively high elevations.

*When Europeans first settled in the highlands of tropical South America, in which vertical climate zone did they prefer to live?*

phy and soil favor agriculture in the vertical climate zones. Highland climates are at such a premium in many areas that steep mountain slopes have been extensively terraced to produce level land for agricultural use. Spectacular agricultural terraces can be seen in Peru, Yemen, the Philippines, and many other tropical highlands. Where the climate is appropriate and population pressure is high, people have created a topography to suit their needs, although they have had to hack it out of mountainsides.

This chapter has discussed the major climates that dominate the middle latitudes of Earth. It is in these latitudes that most of the human population exists. These climates range from some of the most productive to some of the harshest climates on the planet. Again, vegetation, animal life, and human activities have all had to adapt to the specific climatic conditions. And once again, changes such as global warming and destruction of our ozone shield threaten these climates and the life forms they support.

# The Effects of Altitude on the Human Body

The changes in climate, soils, and vegetation as an individual climbs into higher altitudes are discussed elsewhere in the text. However, it is interesting to note some of the ways in which high altitudes affect the human body. Besides food and water, the metabolism of a human being also depends on the consumption of oxygen. A lessening of the normal amount of oxygen intake can have some profound physiological effects. One way to look at the amount of oxygen available for human respiration is to consider something called the *partial pressure of oxygen gas* ($pO_2$). Partial pressure in this case refers to the portion of total atmospheric pressure attributed to oxygen alone. At sea level, atmospheric pressure is 1013.2 millibars, and the $pO_2$ is about 212 millibars. In other words, 212 of the 1013.2 millibars are attributed to oxygen gas. At an altitude of 10 kilometers (6.2 mi) the $pO_2$ drops to only 55 millibars!

Even at moderate altitudes, the effects of *hypoxia* (oxygen starvation) can cause headaches and nausea. Above 6 kilometers (3.7 mi), this lack of oxygen can seriously affect the brain. Because the body's need for oxygen does not alter with major changes in altitude, any significant drop in $pO_2$ with height can cause severe body stress. Pressure-controlled cabins of high-flying aircraft and the use of oxygen by mountain climbers provide dramatic examples of ways to meet the vital need for oxygen.

Fortunately, the human body can acclimatize to moderate changes in altitude, but it takes time (days to sometimes weeks, depending on the altitude). During this time, however, one may experience some of the following symptoms: sleeplessness, headaches, loss of weight, thyroid deficiency, increased excitability, muscle pain, gastrointestinal disturbances, swelling of the lungs, severe infections, psycho-

logical and mental disturbances, and others. At the very least, when visiting higher altitude locations, most people can expect headaches (perhaps leading to nausea), a drop in physical endurance (climbing stairs will become more difficult), and hampered mental function (solutions to simple problems may be difficult to grasp). There is little need to worry, though; humans do acclimatize and these symptoms will pass.

It is also important to remember that, in addition to hypoxia, higher altitudes may cause increased susceptibility to severe sunburn. If you climb to a higher altitude, this means that more of Earth's atmosphere is below you. Correspondingly, there is less of the protective atmosphere above you to filter harmful ultraviolet radiation. Keeping latitude constant, sunburn of exposed skin is more likely on a high-altitude mountainside than on a sea-level beach.

Mountain climbers carry an oxygen supply to help with high altitudes.

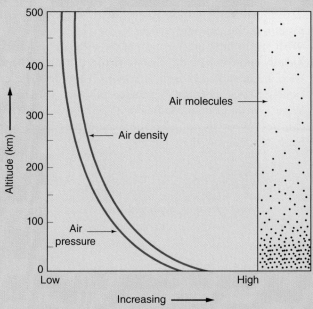

Both air pressure and air density decrease rapidly with increasing altitude.

# ■ DEFINE AND RECALL

Mediterranean climate
sclerophyllous
chaparral
humid subtropical climate
marine west coast climate
humid continental, hot-summer climate

humid continental, mild-summer
climate
subarctic climate
permafrost
patterned ground (frost polygons)
tundra climate

muskeg
ice-sheet climate
exposure
slope aspect
tree line
snow line

# ■ DISCUSS AND REVIEW

1. Summarize the special adaptations of vegetation and soils in Mediterranean regions.

2. Compare the humid subtropical and Mediterranean climates. What are their most obvious similarities and differences?

3. What factors combine to cause a precipitation maximum in late summer in most of the humid subtropical regions?

4. How are temperature, precipitation, and geographic distribution of marine west coast regions linked to the controlling factors for this climate?

5. Explain why the microthermal climates are limited to the Northern Hemisphere.

6. List several features that all humid microthermal climates have in common. How do these features differ from those displayed by the humid mesothermal climates?

7. Describe the relationship between vegetation and climate in the humid continental, mild-summer regions.

8. Refer to Figure 10.22. Using the climographs for Eagle, Alaska, and Verkhoyansk, Russia, describe the temperature patterns of the subarctic regions.

9. What factors limit precipitation in the subarctic regions?

10. Identify and compare the climate factors that strongly influence the tundra and ice-sheet regions. How do these controlling factors affect the distribution of these climates?

11. What kind of plant and animal life can survive in the polar climates? What special adaptations must they make to the harsh conditions of these regions?

12. How do elevation, exposure, and slope aspect affect the microclimates of highland regions? What are the major climatic differences between highland regions and nearby lowlands?

13. Do each of the following for these climates: Mediterranean; humid subtropical; marine west coast; humid continental, hot-summer; humid continental, mild-summer; subarctic; tundra; ice-sheet.

    a. Identify the climate from a set of data or a climograph indicating average monthly temperature and precipitation for a representative station within a region of that climate (use one of the climographs in this chapter).

    b. Match the climate type with a written statement that includes one or more of the following: the statistical parameters of the climate in the modified Köppen classification; the particular climate controls (controlling factors) that produce the climate; the geographic distribution of the climate as stated in terms of physical or political location; the unique climate characteristics or combination of characteristics that distinguishes the climate from others; types of plants, animals, and soils associated with the climate; and the human utilization typical of the climate.

    c. Distinguish between the important subtypes (if any) of each climate by identifying the characteristics that separate them from one another.

# CONSIDER AND RESPOND

1. Based on the classification scheme presented in the "Graph Interpretation" exercise at the end of Chapter 8, classify the following climate stations from the data provided.

| | | J | F | M | A | M | J | J | A | S | O | N | D | Yr |
|---|---|---|---|---|---|---|---|---|---|---|---|---|---|---|
| a. | Temp. (°C) | −42 | −47 | −40 | −31 | −20 | 15 | −11 | −18 | −22 | −36 | −43 | −39 | −30 |
| | Precip. (cm) | 0.3 | 0.3 | 0.5 | 0.3 | 0.5 | 0.8 | 2.0 | 1.8 | 0.8 | 0.3 | 0.8 | 0.5 | 8.6 |
| b. | Temp. (°C) | 3 | 3 | 5 | 8 | 19 | 13 | 15 | 14 | 13 | 10 | 7 | 5 | 9 |
| | Precip. (cm) | 4.8 | 3.6 | 3.3 | 3.3 | 4.8 | 4.6 | 8.9 | 9.1 | 4.8 | 5.1 | 6.1 | 7.4 | 65.8 |
| c. | Temp. (°C) | 23 | 23 | 22 | 19 | 16 | 14 | 13 | 13 | 14 | 17 | 19 | 22 | 18 |
| | Precip. (cm) | 0.8 | 1.0 | 2.0 | 4.3 | 13.0 | 18.0 | 17.0 | 14.5 | 8.6 | 5.6 | 2.0 | 1.3 | 88.1 |
| d. | Temp. (°C) | −27 | −28 | −26 | −18 | −8 | 1 | 4 | 3 | −1 | −8 | −18 | −24 | 12 |
| | Precip. (cm) | 0.5 | 0.5 | 0.3 | 0.3 | 0.3 | 1.0 | 2.0 | 2.3 | 1.5 | 1.3 | 0.5 | 0.5 | 10.9 |
| e. | Temp. (°C) | −4 | −2 | 5 | 14 | 20 | 24 | 26 | 25 | 20 | 13 | 3 | −2 | 12 |
| | Precip. (cm) | 0.5 | 0.5 | 0.8 | 1.8 | 3.6 | 7.9 | 24.4 | 14.2 | 5.8 | 1.5 | 1.0 | 0.3 | 62.2 |
| f. | Temp. (°C) | 9 | 9 | 9 | 10 | 12 | 13 | 14 | 14 | 14 | 12 | 11 | 9 | 11 |
| | Precip. (cm) | 17.0 | 14.8 | 13.3 | 6.8 | 5.5 | 1.9 | 0.3 | 0.3 | 1.6 | 8.1 | 11.7 | 17.0 | 97.6 |
| g. | Temp. (°C) | −3 | −2 | 2 | 9 | 16 | 21 | 24 | 23 | 19 | 13 | 4 | −2 | 11 |
| | Precip. (cm) | 4.8 | 4.1 | 6.9 | 7.6 | 9.4 | 10.4 | 8.6 | 8.1 | 6.9 | 7.1 | 5.6 | 4.8 | 84.8 |
| h. | Temp. (°C) | 0 | 0 | 4 | 9 | 16 | 21 | 24 | 23 | 20 | 14 | 8 | 2 | 12 |
| | Precip. (cm) | 8.1 | 7.4 | 10.7 | 8.9 | 9.4 | 8.6 | 10.2 | 12.7 | 10.7 | 8.1 | 8.9 | 8.1 | 111.5 |

2. The data in the previous table represent the following eight locations, although not necessarily in this order: Beijing, China; Point Barrow, Alaska; Chicago, Illinois; Eismitte, Greenland; Eureka, California; Edinburgh, Scotland; New York, New York; Perth, Australia. Use an atlas and your knowledge of climates to match the climatic data with the locations.

3. Eureka, California; Chicago; and New York City are located within a few degrees latitude of one another, yet they represent three distinctly different climate types. Discuss these differences and identify the primary cause, or source, of the differences.

4. The precipitation recorded at Albuquerque, New Mexico (see Consider and Respond, Chapter 9) is almost twice that recorded at Point Barrow, Alaska, yet Albuquerque is considered a dry climate and Point Barrow is a humid climate. Why?

5. Why is the *Dw* climate type found only in Asia?

6. *Csa* climates are dry during the summer and *Cfa* climates are wet. Why?

# APPLY AND LEARN

Use the data charts in the Geography's Physical Science Perspective: Effective Temperatures box:

1. Determine the heat index values for the following data:
   a. 95°F and 70% relative humidity
   b. 85°F and 90% relative humidity
   c. 80°F and 80% relative humidity
   d. 75°F and 100% relative humidity
   e. 70°F and 50% relative humidity
   f. 100°F and 0% relative humidity

2. Determine the wind chill index values for the following data:
   a. 40°F and 35 mph winds
   b. 20°F and 20 mph winds
   c. 15°F and 35 mph winds
   d. −15°F and 40 mph winds
   e. −30°F and 30 mph winds
   f. −40°F and calm winds

 CourseMate – Make the most of your study time by accessing everything you need to succeed in one place. Read your textbook, take notes, review flashcards, watch videos, complete activities, take practice quizzes, and more online with CourseMate. Log in at **www.cengagebrain.com**.

# 11

# BIOGEOGRAPHY

Fall vegetation patterns in this mountain area of Utah are caused by differences in altitude, temperature, winds, aspect, soil, moisture, and slope steepness. These factors produce varying microenvironmental conditions, which are suitable for different vegetation species. *US Forest Service/Mike Muir*

## ■ OBJECTIVES

WHEN YOU COMPLETE THIS CHAPTER YOU SHOULD BE ABLE TO:

■ Define the four major components of an ecosystem, and explain their interdependence.
■ Recognize that other environmental controls may be more important on a local scale but that climate has the greatest influence over ecosystems on a worldwide basis.
■ Explain how vegetation becomes established on barren or devastated areas, and cite an example of the steps in plant succession.
■ Provide examples of how plants, animals, and the environments in which they live are interdependent, each affecting the others.
■ Outline the climatic factors that have the greatest effect on plants and animals, and summarize the nature of those climatic impacts.
■ Describe Earth's major terrestrial ecosystems (biomes) and the dominant vegetation types that occupy the ecosystems.
■ Cite a few reasons why humans affect ecosystems more than all other life forms, and provide some examples of major impacts.

BIOGEOGRAPHY IS THE STUDY OF how environmental factors affect the locations, distributions, and life processes of plants and animals. Basically, this discipline seeks explanations for the geography of life forms. Biogeographers delineate the spatial boundaries of natural environments and investigate how and why environmental characteristics change spatially and over time. Biogeography is also concerned with human impacts on the organisms living in an ecosystem and their influence on the environmental conditions that support life. Along with ecologists from many other science disciplines, biogeographers often study **ecosystems**—communities of organisms that function together in an interdependent relationship with the environments that they occupy.

It can be said that ecology is an old science. The voyages of discovery that began in the 15th century carried colonists and explorers to uncharted lands with exotic environments. Eventually, scientists accompanied exploratory voyages recording descriptions and illustrations of the flora and fauna they encountered in distant places. It was apparent that certain plants and animals that were found together bore direct relationships to the climate and environment in which they lived. As available information about Earth's environments became more complete and detailed, early biologists studied plant communities and classified vegetation types for much of the world. As the relationships of animals to these plant communities were recognized, naturalists in the early 20th century began dividing Earth's life forms into *biotic associations*. Recently, the functional relationships of plants, animals, and their physical environment have been the primary focus of attention, and the concept of the ecosystem has become widely used. The study of biogeography, ecology, and ecosystems provides an ideal opportunity to demonstrate the environmentally integrative nature of physical geography. Relationships and interactions among the different climate regions, their associated vegetative biomes, and certain soils were introduced in Chapters 9 and 10. In this chapter, we take a closer look at biogeography, ecology, and the nature of certain environments.

# Ecosystems

Our definition of an ecosystem is broad and flexible. The term can be used in reference to the Earth system in its entirety (the ecosphere) or to any group of organisms occupying a given area and functioning together with their nonliving environment. An ecosystem may be large or small, marine or terrestrial (on land), short-lived or long lasting ( ● Fig. 11.1). It may even be an artificial ecosystem, such as an agricultural field. When farmers plant crops, spread fertilizer, practice weed control, and spray insecticides, a new ecosystem is created, but plants and animals still live together in interdependent relationships with the soil, rainfall, temperatures, sunshine, and other factors that constitute the physical environment ( ● Fig. 11.2).

As noted in Chapter 1, ecosystems are *open systems*. There is free movement of both energy and materials into and out of these systems. They are usually so closely related to nearby ecosystems or so integrated with the larger ecosystems of which they are a part that they are not isolated in nature or readily delimited. Nevertheless, the concept of the ecosystem is a valuable model for examining the structure and function of life on Earth.

## Major Components

Despite their great variety on Earth, the typical ecosystem has four basic components. The first of these is the nonliving, or **abiotic,** part of the system. This is the physical environment in which the plants and animals of the system live. In an aquatic ecosystem (a pond, for example), the abiotic component would include such inorganic substances as calcium, mineral salts, oxy-

R. Gabler

● **FIGURE 11.1** This woodland ecosystem in New Hampshire on the slopes of Mount Cardigan demonstrates the close relationships between living organisms and their nonliving environment.

*What environmental characteristics influence the vegetation here?*

gen, carbon dioxide, and water. Some of these would be dissolved in the water, but the majority would lie at the bottom as sediments—a natural reservoir of nutrients for both plants and animals. In a terrestrial ecosystem, the abiotic component provides life-supporting elements and compounds in the soil, groundwater, and atmosphere.

The second component of an ecosystem consists of the basic **producers,** or **autotrophs** (meaning "self-nourished"). Plants are important autotrophs because they can use solar energy to convert water and carbon dioxide into organic molecules through photosynthesis. The sugars, fats, and proteins produced by plants through photosynthesis supply the food that supports other forms of life. Some bacteria are also capable of photosynthesis, and sulfur-dependent organisms that dwell at thermal vents on the sea floor are also classified as autotrophs.

A third component of most ecosystems consists of consumers, or **heterotrophs** (meaning "other-nourished"). These are animals that survive by eating plants or other animals. Heterotrophs are classified on the basis of their feeding habits. **Herbivores** eat only living plant material; **carnivores** eat other

USDA/Peggy Greb

● **FIGURE 11.2** Hybrid seed, fertilizers, insecticides, and irrigation systems may be used to ensure the success of crops in this agricultural ecosystem in the Corn Belt.

*What are some ways that an "artificial" ecosystem, changed nearly completely by human activities, differs from a natural one?*

animals; **omnivores** feed on both plants and animals. Animals make an essential contribution to the Earth ecosystem of which they are a part. They use oxygen in their respiration and return carbon dioxide to the atmosphere, which plants require for photosynthesis. Animals can influence soil development through their digging and trampling activities; two activities that may affect local plant distributions.

We might assume that plants, animals, and a supporting environment are all that are required for a functioning ecosystem, but such is not the case. Without the fourth component of ecosystems, the decomposers, plant growth would soon come to a halt. The **decomposers,** or **detritivores,** feed on dead plant and animal material and waste products. They promote decay and return mineral nutrients to the soil and to bodies of water in a form that plants can use.

## Trophic Structure

The living components of an ecosystem are organized in a sequence that is defined by their eating habits. The components form a sequence in their levels of eating: Herbivores eat plants, carnivores may eat herbivores or other carnivores, and decomposers feed on dead plants and animals and their waste products. The pattern of feeding in an ecosystem is called the **trophic structure,** and the sequence of levels in the feeding pattern is referred to as a **food chain.** The simplest food chain would include only plants and decomposers. However, the chain usually includes at least four steps—for example, grass–field mouse–owl–fungi (plants–herbivore–carnivore–decomposer). More complex food chains may include six or

**TABLE 11.1**
**Trophic Structure of Ecosystems**

| Ecosystem Component | Trophic Level | Examples |
|---|---|---|
| Autotroph | First | Trees, shrubs, grass |
| Heterotroph | Second | Locust, rabbit, field mouse, deer, cow, bear |
| | Third | Praying mantis, owl, hawk, coyote, wolf, bear |
| | Fourth, etc. | Bobcat, wolf, hawk, bear |
| Decomposer | Last | Fungi, bacteria |

more levels as carnivores feed on other carnivores—for example, zooplankton eat plants, small fish eat zooplankton, larger fish eat small fish, bears eat larger fish, and decomposers consume the bear after it dies.

Organisms within a food chain are often identified by their **trophic level,** or the number of steps they are removed from the autotrophs or plants in a food chain (Table 11.1). Plants occupy the first trophic level, herbivores the second, carnivores feeding on herbivores the third, and so forth until the last level—the decomposers—is reached. Omnivores may belong to several trophic levels because they eat both plants and animals.

In reality, most food chains do not operate in a simple linear sequence; they overlap and interact to form a feeding network within an ecosystem called a food web ( ● Fig. 11.3). Food chains and food webs can be used to trace the movement of food and energy from one level to another in an ecosystem.

**Nutrient Cycles** Some biologists and ecologists find it helpful to separate the trophic structure into specific nutrient cycles. There are several such cycles, which at times intertwine and help explain the routing for most of the nutrients through our ecosystems. Cycles have been developed for water, carbon, nitrogen, oxygen, sulfur, and phosphorous. Some of these cycles may be familiar to you. Parts of the oxygen and carbon cycles were discussed in various sections of Chapter 4. The water (hydrologic) cycle was highlighted in Chapter 6. Although each of these cycles can be singled out individually, ● Figure 11.4 shows a summary diagram, which incorporates the major processes involved in a nutrient cycle. Knowledge of chemical nutrient cycles is essential to an understanding of energy flow in ecosystems.

## Energy Flow and Biomass

When physical geographers study ecosystems, they trace the flow of energy through the system just as they do when they study energy flow in other systems, such as streams or glaciers.

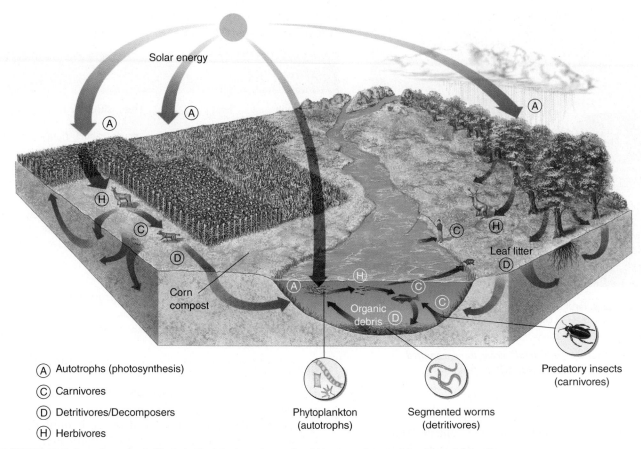

Ⓐ Autotrophs (photosynthesis)

Ⓒ Carnivores

Ⓓ Detritivores/Decomposers

Ⓗ Herbivores

Phytoplankton (autotrophs)

Segmented worms (detritivores)

Predatory insects (carnivores)

● **FIGURE 11.3** Ecosystems clearly illustrate the interdependence of variables in environmental systems. A focus is on the close relationships between living organisms (of the biosphere) and the nonliving (abiotic) components in an environmental system (the atmosphere, hydrosphere, and lithosphere).

*Can you trace a trophic structure through this diagram?*

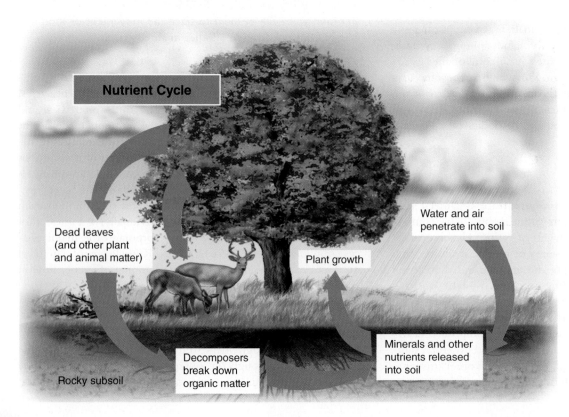

**Nutrient Cycle**

Dead leaves (and other plant and animal matter)

Plant growth

Water and air penetrate into soil

Decomposers break down organic matter

Minerals and other nutrients released into soil

Rocky subsoil

● **FIGURE 11.4** This simplified diagram displays the processes used by the nutrient cycles to travel through the ecosystem.

*How are aspects of the biosphere linked with the atmosphere, lithosphere, and hydrosphere in this scene?*

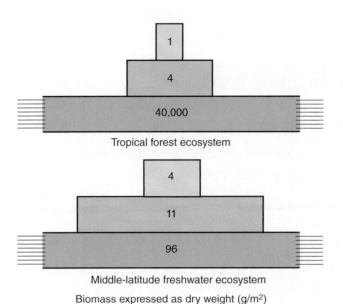

Tropical forest ecosystem

Middle-latitude freshwater ecosystem

Biomass expressed as dry weight (g/m²)

● **FIGURE 11.5** Trophic pyramids showing biomass of organisms at various trophic levels in two contrasting ecosystems. Dry weight is used to measure biomass because the proportion of water to total mass differs from one organism to another.

*Why is there an exceptionally large reduction in biomass between the first and second trophic levels of a tropical forest ecosystem?*

Just as in other systems, the laws of thermodynamics apply to ecosystems. For example, as the first law of thermodynamics states, energy cannot be created or destroyed; it can only be changed from one form to another. Sunlight provides energy to an ecosystem, which is used by plants in photosynthesis, and the energy is stored in the organic materials of plants and animals. Energy flows through the system along food chains and webs from one trophic level to the next. It is finally released from the system when oxygen combines with the chemical compounds of the organic material through the process of oxidation. Respiration, which involves the combination of oxygen with chemical compounds in living cells and can occur at any trophic level, is the major form of oxidation. Fire is yet another form.

The total amount of living material in an ecosystem is referred to as the **biomass.** Because the energy of a system is stored in the biomass, scientists measure the biomass at each trophic level to trace the energy flow through the system. They usually find that the biomass decreases with each successive trophic level ( ● Fig. 11.5). There are a number of explanations for this, each involving a loss of energy. The first instance occurs between trophic levels. The second law of thermodynamics states that whenever energy is transformed from one state to another, there will be a loss of energy through heat. Hence, when an organism at one trophic level feeds on an organism at another, not all of the food energy is used. Some is lost to the system. Additional energy is lost through respiration and movement. At each successive trophic level, the amount of

energy required is greater. A deer may graze in a limited area, but the wolf that preys on the deer must hunt over a much larger territory. Whatever the reason for energy loss, it follows that as the flow of energy decreases with each successive trophic level, the biomass also decreases. This principle also applies to agriculture. A great deal more biomass (and food energy) is available in a field of corn than there is in the cattle that eat the corn.

## Primary and Secondary Productivity

Productivity is defined as the rate at which new organic material is created at a particular trophic level. **Primary productivity** refers to the formation of new organic matter through photosynthesis by autotrophs; **secondary productivity** refers to the rate of formation of new organic material at the heterotroph level.

### Primary Productivity
How efficient are plants at producing new organic matter through photosynthesis? The answer to this question depends on a number of variables. Photosynthesis requires sunlight, the amount of which depends on the length of day and the angle of the sun's rays, which in turn differ widely with latitude. Photosynthesis is also affected by factors such as soil moisture, temperature, the availability of mineral nutrients, the carbon dioxide content of the atmosphere, and the age and species of the individual plants.

Most studies of productivity in ecosystems have been concerned with measuring the net biomass at the autotroph level (Table 11.2). Wherever figures have been compiled on the efficiency of photosynthesis, the efficiency has been surprisingly low. Most studies indicate that less than 5% of the available sunlight is used to produce new biomass in ecosystems. For Earth as a whole, the figure is probably less than 1%. Nonetheless, the net primary productivity of the ecosphere is enormous. It is estimated to be in the range of 170 billion metric tonnes (a metric tonne is about 10% greater than a U.S. ton) of organic matter annually. Although oceans cover approximately 70% of Earth's surface, slightly more than two thirds of net annual productivity is from terrestrial ecosystems and less than one third is from marine ecosystems. Perhaps even more surprising is the fact that humans consume less than 1% of Earth's primary productivity as plant food. However, humans also use biomass in a variety of other ways—for example, using lumber for construction and paper production and using biomass energy for feedstock and as fodder for range animals.

Table 11.2 illustrates the wide range of net primary productivity displayed by various ecosystems. Latitudinal control of insolation on photosynthesis results in a noticeable decrease in terrestrial productivity from tropical ecosystems to those in middle and higher latitudes. Even the tropical savannas, which are dominated by grasses, produce more biomass in a year than the boreal forests, which are found in the colder climates. Today, satellites monitor Earth's biological

## TABLE 11.2
## Net Primary Productivity of Selected Ecosystems

| Net Primary Productivity, gm$^2$ per Year | | |
|---|---|---|
| Type of Ecosystem | Normal Range | Mean |
| Tropical rainforest | 1000–3500 | 2200 |
| Middle-latitude evergreen forest | 600–2500 | 1300 |
| Middle-latitude deciduous forest | 600–2500 | 1200 |
| Boreal forest (taiga) | 400–2000 | 800 |
| Woodland and shrubland | 250–1200 | 700 |
| Savanna | 200–2000 | 900 |
| Middle-latitude grassland | 200–1500 | 600 |
| Tundra and alpine | 10–400 | 140 |
| Desert and semidesert scrub | 10–250 | 90 |
| Extreme desert, rock, sand, and ice | 0–10 | 3 |
| Swamp and marsh | 800–3500 | 2000 |
| Lake and stream | 100–1500 | 250 |
| Open ocean | 2–400 | 125 |
| Upwelling zones | 400–1000 | 500 |
| Algal beds and reefs | 500–4000 | 2500 |

Source: R. H. Whittaker, *Communities and Ecosystems* (2nd ed.). New York: Macmillan, 1975.

productivity and give us a global perspective on our biosphere ( ● Fig. 11.6).

The reasons for differences among aquatic, or water-controlled, ecosystems are not quite as apparent. Swamps and marshes are especially well supplied with plant nutrients and therefore have a relatively large biomass at the first trophic level. In contrast, water depth has the greatest impact on ocean ecosystems. Most nutrients in the open ocean sink to the bottom, beyond the depth where sunlight can penetrate and make photosynthesis possible. Hence, the most productive marine ecosystems are found in the sunlit, shallow waters of estuaries, continental shelves, or coral reefs or in areas where ocean upwelling carries nutrients nearer to the surface ( ● Fig. 11.7).

Some agricultural (artificial) ecosystems can be fairly productive when compared with the natural ecosystems they have replaced. This is especially true in the warmer latitudes where farmers may raise two or more crops in a year or in arid lands where irrigation supplies the water essential to growth. However, Table 11.2 indicates that mean productivity for cultivated land does not approach that of forested land and is just about the same as that of middle-latitude grasslands. Most

quantitative studies have shown that agricultural ecosystems are significantly less productive than natural systems in the same environment.

**Secondary Productivity** Secondary productivity results from the conversion of plant materials to animal substances. We have also noted that the ecological efficiency, or the rate of energy transfer from one trophic level to the next, is low ( ● Fig. 11.8). The efficiency of transfer from autotrophs to heterotrophs varies widely from one ecosystem to another. The amount of net primary productivity actually eaten by herbivores may range from as high as 15% in some grassland areas to as low as 1% or 2% in certain forested regions. In ocean ecosystems, the figure may be much higher, but there is a greater loss during the digestion process. Once the food is eaten, energy loss through respiration or body movement reduces secondary productivity to a small fraction of the biomass available as net primary productivity.

Most authorities consider 10% to be a reasonable estimate of ecological efficiency for both herbivores and carnivores. If both herbivores and carnivores have ecological efficiencies of only 10%, the ratio of biomass at the first trophic level to biomass of carnivores at the third trophic level is several thousand to one. It

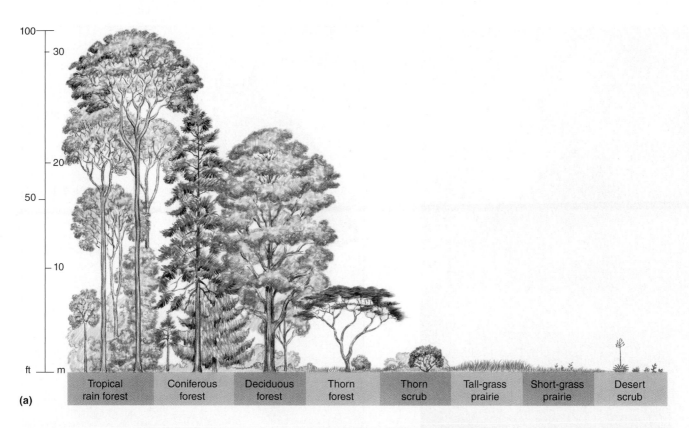

(a)

| Tropical rain forest | Coniferous forest | Deciduous forest | Thorn forest | Thorn scrub | Tall-grass prairie | Short-grass prairie | Desert scrub |

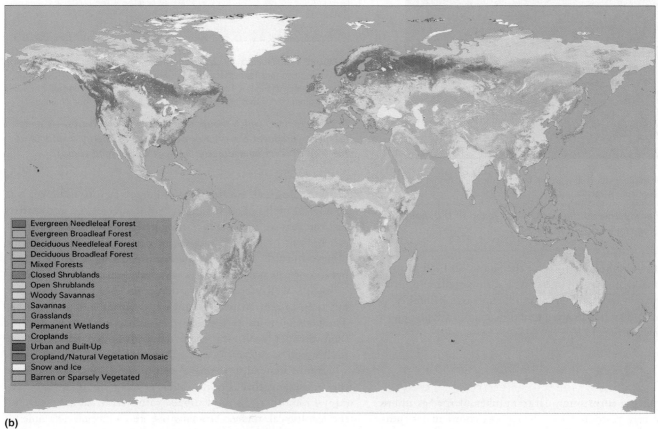

(b)

● **FIGURE 11.6** (a) Examples of global ecosystems and (b) a map of global vegetation patterns based on environmental satellite observations. Compare this image with the world map of natural vegetation in Figure 11.23.

*Does the vegetation shown in the image of environments (Fig. 11.6a) compare well with the figures in Table 11.2?*

1 m² of field for 1 yr

d. Net carnivore productivity:
0.4 g (0.15% of a; 6.7% of c)

c. Net herbivore productivity:
6 g (2.2% of a; 20.7% of b)

b. Every year herbivores eat:
29 g (10.7% of a)

a. Net primary productivity
(plants): 270 g (dry weight)

● **FIGURE 11.8** Productivity at the autotroph, herbivore, and carnivore trophic levels measured in a Tennessee field. The figures represent productivity for 1 square meter of field in 1 year. Note the extremely small proportion of primary productivity that reaches the carnivore level of the food chain.

*In this example, which group—carnivores or herbivores—is more efficient (produces the greater percentage of energy available at the trophic level immediately below it in the food chain)?*

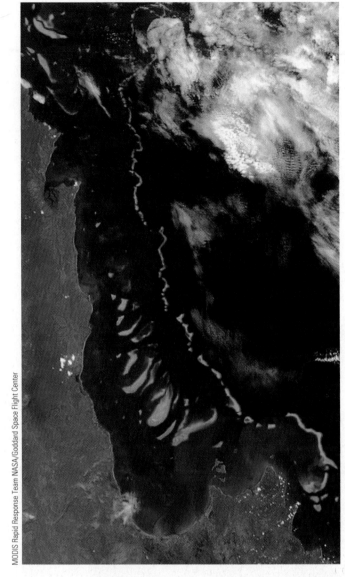

● **FIGURE 11.7** Nutrient-rich waters weave through the Great Barrier Reef off the coast of Queensland, Australia.

*Why are very productive marine ecosystems found in the shallow waters bordering the world's continents?*

obviously requires a huge biomass at the autotroph level to support one animal that eats only meat. As human populations grow at increasing rates and agricultural production lags behind, it is fortunate that human beings are omnivores and can adopt a more vegetarian diet ( ● Fig. 11.9).

## Ecological Niche

Ecosystems have a surprisingly large number of species unless they are severely restricted by adverse environmental conditions. Each organism performs a specific role in the system and lives in a certain location, described as its **habitat.** The combination of role and habitat for a particular species is referred to

as its **ecological niche.** A number of factors influence the ecological niche of an organism. Some species are **generalists** and can survive on a wide variety of food. The North American brown, or grizzly, bear, as an omnivore, will eat berries, honey, and fish. However, the koala of Australia is a specialist because it eats only the leaves of certain eucalyptus trees. Specialists do well when their particular food supply is abundant, but they cannot adapt to changing environmental conditions. The generalists are in the majority in most ecosystems because their broader ecological niche allows survival on alternative food supplies.

Some generalist species occupy an ecological niche in one ecosystem that is quite different from the niche they occupy in another. As food supply varies with habitat, so varies the ecological niche. Humans are the extreme example of the generalist: In some parts of Earth, they are carnivores; in some parts, herbivores; and in some parts, omnivores. It is also true that different species may occupy the same ecological

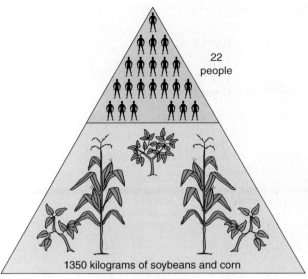

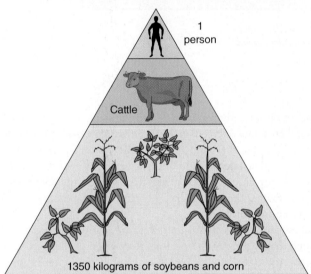

● **FIGURE 11.9** The triangles illustrate the advantages of a vegetarian diet as the human population experiences another century of rapid population growth. People are fortunate that they are omnivores and can choose to eat grain products. The same 1350 kilograms of grain, if converted to meat, supports only one person, but it will support 22 people if cattle or other animals are omitted from the human food chain.

*What does this relationship mean if the population continues to grow unimpeded?*

niche in habitats that are similar but located in separate ecosystems.

# Succession and Climax Communities

Up to this point, ecosystems have mainly been considered in general terms. In the remainder of this chapter we note that it is the species that occupy the ecosystem that give the ecosystem its char-

acter. At least for terrestrial ecosystems, it is the autotrophs—the plant species at the first trophic level—that most easily distinguish one ecosystem from another. All other species in an ecosystem depend on the autotrophs for food, and the association of all living organisms determines the energy flow and the trophic structure of the ecosystem. It should also be noted that the species that occupy the first trophic level are greatly influenced by climate; again we see the interconnections between the major Earth subsystems.

If the plants that comprise the biomass of the first trophic level are allowed to develop naturally without obvious interference from or modification by humans, the resulting association of plants is called **natural vegetation.** These plant associations, or **plant communities,** are compatible because each species within the community has different requirements in relation to major environmental factors such as light, moisture, and mineral nutrients. If two species within a community were to compete, one would eventually eliminate the other. The species forming a community at any specific place and time will be an aggregation of those that together can adapt to the prevailing environmental conditions.

## Succession

Natural vegetation of a particular location develops in a sequence of stages involving different plant communities. This developmental process, known as **succession,** usually begins with a relatively simple plant community. There are two major types of succession, *primary* and *secondary*. In primary succession, a bare substrate is the beginning point. No soil or seedbed exists yet. A *pioneer* community invades the bare substrate (whether it be volcanic lava, glacially deposited sediment, or a bare beach, among others) and begins to alter the environment. As a result, the species structure of the ecosystem does not remain constant. In time, the environmental alteration becomes sufficient to allow a new plant community (that could not have survived under the original conditions) to appear and eventually to dominate the original community. The process continues with each succeeding community rendering further changes to the environment. Because of the initial absence of soil, primary succession can take hundreds or even a few thousand years.

Secondary succession begins when some natural process, such as a forest fire, tornado, or landslide, has destroyed or damaged a great deal of the existing vegetation. Ecologists refer to this process as **gap** creation. Even with such damage, however, soil still typically exists and seeds may be lying dormant in that soil ready to invade the newly opened gap. Secondary succession, therefore, can occur much more quickly than primary succession.

A common form of secondary succession associated with agriculture in the southeastern United States is depicted in ● Figure 11.10. After agriculture has ceased, weeds and grasses are the first vegetative types to adapt to the somewhat adverse conditions associated with bare fields. These low-growing plants stabilize the topsoil, add organic matter, and generally pave the way for the development of hardwood brush such as sassafras, persimmon, and sweet gum. During the brush stage, the soil will become richer in nutrients and organic matter, and its ability to retain water will increase. These conditions encourage the development of pine forests, the next

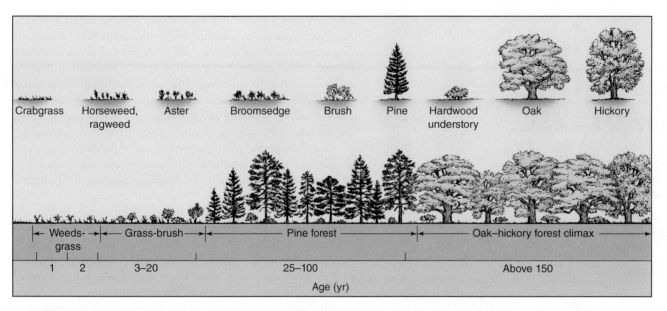

● **FIGURE 11.10** A common plant succession in the southeastern United States. Each succeeding vegetation type alters the environment in such a way that species having more stringent environmental requirements can develop.

*Why would plant succession differ in another region of the United States?*

stage in this vegetative evolutionary process. Pine forests thrive in the newly created environment and will eventually dwarf and dominate the weeds, grasses, and brush that preceded them.

Ironically, the growth and dominance of a pine forest can also lead to its demise. Pine trees require much sunlight for their seeds to germinate. When competing with scattered brush, grasses, and weeds, there is adequate sunlight for germination. Once a pine forest develops, however, shade and litter cover will not allow the pine seeds to germinate. Hardwood trees, such as oak and hickory, whose seeds can germinate in shady conditions, begin to grow as an understory and eventually will replace the pines. Thus, pines eventually will give way to hardwood trees, such as oak and hickory, whose seeds can germinate under those conditions. These seeds may have been present and dormant in the soil, or blown into the forest, or carried into it by animals. The end result is an oak–hickory forest. In this southeastern United States example, a complete succession from field to oak–hickory forest will take about 100 to 200 years if it continues unimpeded. Through succession, the area can return to the natural oak–hickory forest that existed prior to agricultural clearing. In other ecosystems, such as tropical rainforests, succession from deforestation back to a natural forest may take many centuries.

## The Climax Community

The theory of plant succession was introduced early in the 20th century. However, some of the original ideas have undergone considerable modification. Succession was considered to be an orderly process that included various *predictable* steps or phases and ended with a dominant vegetative cover that would remain in balance with the environment until disturbed by human activity—or until there were major changes (for example, a cli-

mate change) in the environment. The final step in the process of succession has been referred to as the **climax community.** It was generally agreed that such a community was self-perpetuating and had reached a state of equilibrium or stability with the environment. In our illustration of plant succession in the southeastern United States, the oak–hickory forest would be considered the climax community.

Succession is still a useful model in the study of ecosystems. Why, then, have some of the original ideas been challenged, and what is the most recent thought on the subject? For one thing, early proponents of succession emphasized the predictable nature of the theory. One plant community would follow another in regular order as the species structure of the ecosystem evolved. But it has been demonstrated on many occasions that changes in ecosystems do not occur in such a rigid fashion. More often than not, the movement of species into an area to form a new community occurs in a random fashion and may be largely a function of chance events.

Many scientists today no longer believe that only one type of climax vegetation is possible for each major climate region of the world. Some suggest that one of several different climax communities might develop within a given area, influenced not only by climate, but also by drainage conditions, nutrients, soil, or topography. The dynamic nature of climate is now more fully understood than it was when the theories of succession and climax were developed, so it is now recognized that by the time the species structure of a plant community has adjusted to new climatic conditions, the climate may change again. Because of the dynamic nature of each habitat and the conditions under which it exists, no climax community will remain, unchanged, within its environment forever.

Today many biogeographers and ecologists view plant communities and their ecosystems as a *landscape* that is an

● **FIGURE 11.11** This patch in the tropical rainforest matrix of Jamaica is the result of land cleared for shifting (slash-and-burn) cultivation.

*What human activities might be responsible for patches in the vegetation matrix of a middle-latitude forest?*

● **FIGURE 11.12** This patch in a middle-latitude forest in British Columbia, Canada, was the result of a 160-kph (100 mph) windstorm.

*Why do you suppose the jagged tree trunks were sawed flat?*

expression of all the various environmental factors functioning together. They view the landscape of an area as a **mosaic** of interlocking parts, much like the tiles in a mosaic artwork. In a pine forest, for example, other plants also exist, and some areas do not support pine trees. The dominant area of the mosaic—that is, the pine forest—is referred to as the **matrix.** Gaps within the matrix, resulting from areas of different soil conditions or gaps created by human-created or natural processes, are referred to as **patches** within the matrix ( ● Fig. 11.11 and ● Fig. 11.12). Relatively linear features cutting across the mosaic, including natural features such as rivers and human-created structures such as roads, fence lines, power lines, and hedgerows, are termed **corridors** ( ● Fig. 11.13 and ● Fig. 11.14). Each particular habitat is unique and constantly changing, and resultant plant and animal

● **FIGURE 11.13** This aerial view of Taskinas Creek, Virginia, shows riparian corridors passing through a forest matrix on the Atlantic coastal plain.

*How does a corridor differ from a patch?*

communities must constantly adjust to these changes. The dominant environmental influence is climate.

As discussed in Chapter 8, climates are also changing. Climate changes can occur over relatively short time periods, on the scales of decades and centuries, and in much longer times measured in millennia. They may be subtle, or they may be sufficiently drastic to create ice ages or warm periods between ice ages. Plant and animal communities must be able to respond to these ongoing changes, or they will not survive. Many modern biogeographers are deeply involved in the reconstruction of the

● **FIGURE 11.14** This view near Franklin, Tennessee shows a human-made power line corridor.

*Why was it important to clear this type of corridor?*

vegetation communities of past climate periods through the examination of evidence such as tree rings, pollen, insect fossils, and plant fossils. By determining how past climate changes affected and induced change in Earth's ecosystems, biogeographers hope to be able to suggest how future changes may develop as climate continues to change.

# Environmental Controls

The plants and animals that exist in a particular ecosystem are those that have been successful in adjusting to their habitat's environmental conditions. Every living organism requires a certain range of environmental conditions to survive. For example, some plants can survive under a wide range of temperature conditions, whereas others have narrow temperature requirements. Biogeographers and ecologists refer to this characteristic as an organism's range of **tolerance** for a particular environmental condition. The ranges of tolerance for a species will determine where on Earth that species may be found, and species with wide ranges of tolerance will be the most widely distributed. The *ecological optimum* refers to the environmental conditions under which a species will thrive ( ● Fig. 11.15). The farther away a species is from its ecological optimum or from the geographic core of its plant or animal community, the more difficult conditions will become for that species or the community to survive. However, those same conditions may be more amenable for another species or community. An **ecotone** is the overlap, or zone of transition, between two plant or animal communities (see Fig. 11.15).

Climate has the greatest influence over natural vegetation when we consider plant communities on a worldwide basis. The major types of terrestrial ecosystems, or **biomes,** are associated with certain temperature ranges and critical annual or seasonal precipitation and evaporation characteristics. Climate influences the sizes and shapes of tree leaves and determines whether or not trees can exist in a region. At the local scale, however, other environmental factors can be just as important. A plant's range of tolerance for acidity, salinity, soil moisture, or other environmental characteristics may be the critical factor in determining whether a plant will grow, flourish, or die. The discussion that follows serves to illustrate how the major environmental factors influence the organization and structure of ecosystems.

## Natural Factors

### Climatic Factors
Sunlight is one of the most critical climatic factors that influence an ecosystem. Sunlight is the energy source for plant photosynthesis, and it also strongly influences the behavior of both plants and animals. Competition for light can make forest trees grow taller, thereby limiting plant growth on the forest floor to shade-tolerant species such as ferns. The sizes, shapes, and colors of leaves may result from variations in light reception, with large

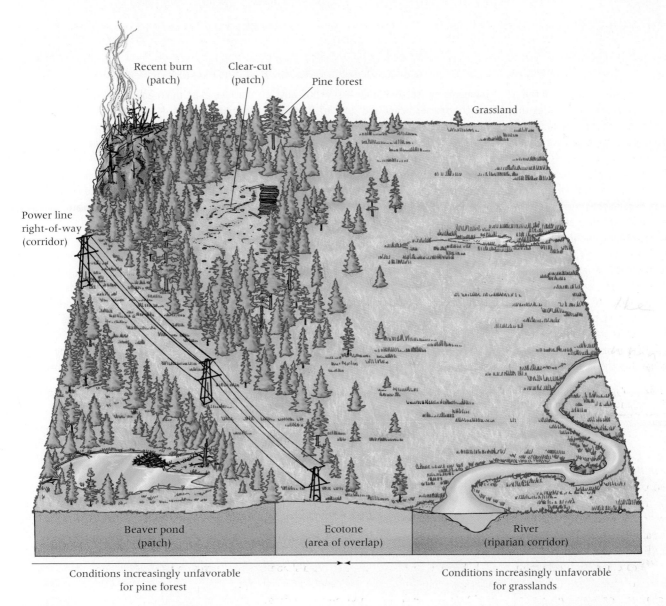

Recent burn (patch)

Clear-cut (patch)

Pine forest

Grassland

Power line right-of-way (corridor)

Beaver pond (patch)

Ecotone (area of overlap)

River (riparian corridor)

Conditions increasingly unfavorable for pine forest

Conditions increasingly unfavorable for grasslands

● **FIGURE 11.15** The concepts of *ecotone, ecological optimum, range of tolerance, mosaic, matrix, patch,* and *corridor* are each illustrated in the diagram.

*What effect would a change to more arid conditions have on the relative sizes of these two ecosystems and the ecotone's position?*

leaves developing in areas of limited light. The *quality* of light is also important, especially in mountain areas, where plant growth may be severely retarded by excess ultraviolet radiation. This radiation does significant damage in the thin air at higher elevations but is effectively screened out by the denser atmosphere at lower elevations. Light *intensity* affects the rate of photosynthesis and hence the rate of primary productivity in an ecosystem. Intense sunlight in the low latitudes produces a greater biomass in the tropical forests compared to the much lower light intensity that reaches the high latitude Arctic regions. *Duration* of daylight, which varies seasonally and with latitude, has a profound effect on the flowering of plants and on animal mating and migration.

A second important climatic control of ecosystems is the availability of water. Virtually all organisms require water to survive. Plants require water for germination, growth, and reproduction, and most plant nutrients are dissolved in soil water before they can be absorbed by plants. Marine plants are adapted to living completely in water (seaweed); some plants, such as mangrove ( ● Fig. 11.16) and bald cypress ( ● Fig. 11.17a and b), rise from coastal marshes and inland swamps; others thrive in the constantly wet rainforest. Certain tropical plants drop their leaves and become dormant during dry seasons; others store water received during periods of rain to survive seasons of drought. Desert plants, such as cacti, are especially adapted to obtaining and

# The Theory of Island Biogeography

Biogeographers are intrigued by the life forms and species diversities that exist on islands that are isolated from larger landmasses by extensive ocean environments. How can land plants and animals be living on an island surrounded by a wide expanse of sea? How did this flora and fauna expand their spatial distribution to become established and flourish on these distant, often geologically recent, and originally barren terrains (volcanic islands, for example)? The farther an island is from the nearest landmass, the more difficult it is for species to migrate to the island and establish a viable population there. Winds, birds, or ocean currents can carry some seeds to islands, where they germinate to develop the vegetative environments on isolated landmasses. But why did the species adapt and survive? Humans have also introduced many species into island environments.

The theory of island biogeography offers an explanation for how natural factors interact to affect either the successful colonization or the extinction of species that come to live on an island. The theory considers an island's locational isolation (the distance from a mainland source of migrating species), the island's size, and the number of species living on an island. Generally, the biological diversity on islands is low compared with mainland areas with similar climates and other environmental characteristics. Low species diversity typically means that the floral and faunal populations exist in an environmentally challenging location. Many extinctions have occurred on islands because of the introduction of some factor that made the habitat nonviable for that species to survive.

Several factors affect the species diversity on islands (as long as other environmental conditions such as climate are comparable):

1. The farther an island is from the area from which species must migrate, the lower the species diversity. Islands nearer to large landmasses tend to have higher diversity than those that are more distant.
2. The larger the island, the greater the species diversity. This is partly because

⊕ In Google Earth fly to: 5.878°N, 162.073°W

A beach on Palmyra Island in the Pacific Ocean, one of the world's most remote islands, illustrates how palm trees become established on tropical islands. Coconuts are the seeds for coconut palm trees. Washed away by surf from their original location, they float and drift hundreds of kilometers from one island to another. Waves deposit and bury coconuts on the beach, and palms spout and grow. Note the coconuts, recently sprouted palms, and fully-grown coconut palms. Very few other plant species grow on this very small island. Greater species diversities tend to exist on larger islands and on islands that are closer to major landmasses.

larger islands tend to offer a wider variety of environments to colonizing organisms than smaller islands do. Larger islands also offer more space for species to occupy.

3. An island's species diversity results from an equilibrium between the rates of extinction of species on the island and the colonization of species. If the island's extinction rate is higher, only a few hardy species will live there; if the extinction rate is lower compared to the colonization rate, more species will thrive and the diversity will be higher.

The theory of island biogeography has also been useful in understanding the ecology and biota of many other kinds of isolated environments, such as high mountain areas that stand, much like islands, above surrounding deserts. In those regions, plants and animals adapted to cool, wet environments live in isolation from similar populations on nearby mountains, separated by inhospitable arid environments.

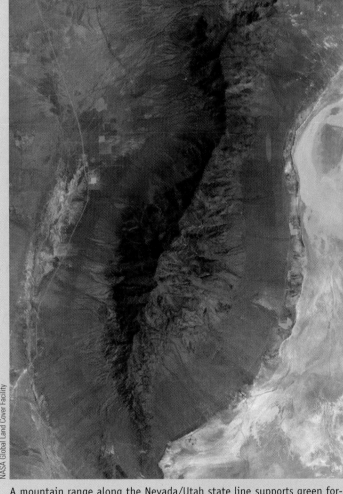

NASA Global Land Cover Facility

A mountain range along the Nevada/Utah state line supports green forests, alpine zones, and animal species associated with those environments. The blue and white areas to the east *(left)* are the edge of the Bonneville Salt Flats. Flora and fauna that could not survive in the surrounding desert environments flourish in cooler, wetter, higher elevation zones. Isolated mountains surrounded by arid environments tend to fit the concept of island biogeography.

Heather Henkel, USGS

● **FIGURE 11.16** Mangrove thicket along the southern Florida coast on the Gulf of Mexico.

*How might this vegetation type have influenced the routes that early Spanish adventurers followed as they first explored Florida?*

storing water when it is available while minimizing their water loss from transpiration.

Animals, too, are severely restricted when water is in short supply. In arid regions, animals must make special adaptations to environmental conditions. Many become inactive during the hottest and driest seasons, and most leave their burrows or the shade of plants and rocks only at night. Others, like the camel, can travel for great distances and live for extended periods without a water supply.

Organisms are affected less by temperature variations than by sunlight and water availability. Many plants can tolerate a wide range of temperatures, although each species has optimum conditions for germination, growth, and reproduction. These functions, however, can be impeded by temperature extremes. Temperatures may also have indirect effects on vegetation. For example, high temperatures will lower the relative humidity, thus increasing transpiration. If a plant's root system cannot extract enough moisture from the soil to meet this increase in transpiration, the plant will wilt and eventually die.

Animals, because of their mobility, are not as dependent as plants are on climatic conditions. Despite the great advantage afforded by mobility, however, animals are nevertheless subject to

R. Gabler

**(a)**

National Park Service Digital Archive

**(b)**

● **FIGURE 11.17** (a) This stand of bald cypress trees is located in extreme southern Illinois at the poleward limit of growth for this type of vegetation. (b) Extensive cypress forests exist in swampy areas of the southeastern United States. These are cypress "knees," root extensions that stick up above the water to supply air to the trees.

*What Köppen climatic type does this site in Illinois represent?*

climatic stress. The geographic distribution of some groups of animals reflects this sensitivity to climate. Cold-blooded animals are, for example, more widespread in warmer climates and more restricted in colder climates. Some warm-blooded animals develop a layer of fat or fur and are able to shiver to protect themselves against the cold. In hot periods, they may sweat, shed fur, or lick their fur in an attempt to stay cool. In extremely cold or arid regions, animals may hibernate. During hibernation, the body temperature of the animal changes roughly in response to outside and ground temperatures. Cold-blooded animals such as the desert rattlesnake move in and out of shade in response to temperature change. Warm-blooded animals may migrate great distances out of environmentally harsh areas.

Some warm-blooded animals exhibit an interesting linkage between body shape and size and variations in average environmental temperature. These adaptations have been described by biologists as *Bergmann's Rule* and *Allen's Rule*. Bergmann's Rule states that, within a warm-blooded species, the body size of the subspecies usually increases with the decreasing mean temperature of its habitat; Allen's Rule notes that, in warm-blooded species, the relative size of exposed portions of the body decreases with the decrease of mean temperature. These rules essentially mean that members of the same species living in colder climates eventually evolve to develop shorter or smaller appendages (ears, noses, arms, legs, and so on) than their relatives in warmer climates (Allen's Rule) and that in cold climates body size will be larger with more mass to provide the body heat needed for survival and for protection of the main trunk of the body where vital organs are located (Bergmann's Rule). In cold climates, small appendages are advantageous because they reduce the amount of exposed area subject to temperature loss, frostbite, and cellular disruption ( ● Fig. 11.18). In warm climates, large body sizes are not necessary for the protection of internal organs, but long limbs, noses, and ears allow for heat dissipation in addition to that provided by panting or fur licking.

As a climatic control on vegetation, the wind is most significant in deserts, polar regions, coastal zones, and highlands. Wind may directly injure vegetation and can have an indirect effect by increasing evapotranspiration. To prevent water loss in areas of severe wind stress, plants twist and grow close to the ground, minimizing their wind exposure ( ● Fig. 11.19). Trees that have taken on this low twisted form because of harsh conditions near the limit of tree growth are referred to as *krummholz*, a German word for crooked wood. During severe winters, they are better off being buried by snow than being ex-

● **FIGURE 11.18** The Arctic polar bear is an excellent example of both Allen's and Bergmann's Rules.

*What warm-climate animal might you suggest as a prime example of Allen's Rule?*

● **FIGURE 11.19** Stunted trees *(krummholz)* at the upper edge of the treeline in the Mosquito Range of the Colorado Rockies. The green vegetation has been covered by snow much of the year and has been protected from the bitterly cold temperatures associated with gale-force winter winds. Note the flag-shaped trees, with no limbs on one side, which indicates the prevailing wind direction.

*What type of biome would be found at elevations higher than the one depicted in this photograph?*

posed to bitterly cold gales. In some windy coastal regions, the shoreline may be devoid of trees or other tall plants. In windy coastal and mountain regions where the trees do grow, they are often misshapen or swept bare of leaves and branches on their windward sides.

## Soil and Topography
In terrestrial ecosystems, soils supply much of the moisture and minerals for plant growth. Soil variations are among the most conspicuous influences on plant distribution and often produce sharp boundaries in vegetation type. Soil variations can strongly influence plant distribution and can produce sharp boundaries between vegetation types. This is partly a consequence of varying chemical requirements of different plant species and partly a reflection of factors such as soil texture. Clay soils may retain too much moisture for certain plants, whereas sandy soils retain too little. Pines generally thrive in sandy soils, grasses in clays, cranberries in acid soils, and chili peppers in alkaline soils. The subject of soils and their influence on vegetation will be explored more thoroughly in Chapter 12.

Topography, particularly in highlands, influences ecosystems by providing diverse microclimates within a relatively small area. Plant communities vary from place to place in highland areas in response to the differing microclimatic conditions. *Slope aspect* has a direct effect on vegetation patterns in areas outside of the equatorial tropics. North-facing slopes in the middle and high latitudes of the Northern Hemisphere have microclimates that are cooler and wetter than those on south-facing exposures. Some plants thrive on the sunny south-facing slopes of highland areas in the Northern Hemisphere; others survive on the colder, shaded, north-facing slopes. The steepness and shape of slopes also affect the amount of time water is present there before draining downslope.

Luxuriant forests tower above the well-watered windward sides of mountain ranges such as the Sierra Nevada and Cascades, but semiarid grasslands and sparse forests cover the leeward sides. Spatial variations in precipitation drainage and resulting vegetation differences would not exist were it not for the presence of the topographic barrier inducing orographic uplift. Each major increase in elevation also produces a different mixture of plant species that can tolerate the lower ranges of temperature found at the higher elevation.

## Natural Catastrophes
The distributions of plants and animals are also affected by a diversity of natural processes that are sometimes referred to as *catastrophes*. It should be noted, however, that this term is applied from a strictly human perspective. What may be catastrophic to a human, such as a hurricane, forest fire, landslide, or tsunami ( ● Fig. 11.20), is simply a natural process operating to produce openings (gaps) in the prevailing vegetative mosaic of a region. The resulting successional processes, whether primary or secondary in nature, produce a variety of patch habitats within the broader regional matrix of vegetation. The study of natural catastrophes and the resulting patch dynamics among the plant and animal residents of an area

USGS/P. Carrara

● **FIGURE 11.20** The vegetation mosaic of this area in Glacier National Park, Montana is coniferous forest, but snow avalanches keep conifer trees from invading the patch of open, low shrubs and grasses.

*Why are there so many broken tree stumps in the foreground?*

is a topic of strong interest and ongoing research in modern landscape biogeography.

## Biotic Factors
Although their influence on a particular species might tend to be overlooked, other plants and animals may also affect whether a given organism exists as part of an ecosystem. Some interactions between organisms are beneficial to both species involved; this is called a **symbiotic relationship.** Other relationships may be directly *competitive* and have an adverse effect on one or both species. Because most ecosystems are suitable to a wide variety of plants and animals, there is always competition between species and among members of a given species to determine which organisms will survive. The greatest competition occurs between species that occupy the same ecological niche, especially during the earliest stages of life cycles when organisms are most vulnerable. Depending on the climate, the greatest competition among plants is for light or water. Those trees that become dominant in the forest are those

● **FIGURE 11.21** Overgrazing is a major cause of desertification at this location in Somalia, Africa. The environment normally would have been a grassy savanna region with scattered trees.

*What are some other causes of desertification?*

that grow the tallest and partially shade the plants growing beneath them. Other competition occurs underground, where the roots compete for soil water and plant nutrients. In sunny arid locales, competition for water may be the greatest determinant for plant survival.

Interactions between animals and plants as well as competition both within and among animal species also can significantly affect an ecosystem. Many animals are helpful to plants during pollination or in the dispersal of seeds, and plants are the basic food supply for many animals. Grazing may also influence the species that make up a plant community. During dry periods, herbivores may be forced to graze an area very closely and the taller plants are grazed out. Plants that are unpalatable, have thorns, or have the strongest root development are the ones that survive. Grazing is a part of the natural selection process, yet serious overgrazing rarely results under natural conditions because wild animal populations increase or decrease with the available food supply. To be more precise, the number of animals of a given species will fluctuate between the maximum number that can be supported when its food supply is greatest and the minimum number required for reproduction of the species. For most animals, predators are also an important control of numbers. Fortunately, when a predator's favorite species is scarce, it will seek an alternative species for its food supply.

## Human Impact on Ecosystems

Throughout history, humans have modified ecosystems and influenced their natural development. Except in regions too remote to be altered significantly by civilization, humans have eliminated much of Earth's natural vegetation. Farming, fire, grazing of domesticated animals, deforestation, road building, urban development, dam building and irrigation, raising and lowering of water tables, mining, and the filling in or draining of wetlands are just a few ways in which humans have modified the plant communities around them. Overgrazing by domesticated animals can seriously harm marginal environments in semiarid climates. Trampling and soil compaction by grazing herbivores may reduce the soil's ability to absorb moisture, leading to increased surface runoff of precipitation. In turn, the decreased absorption and increased runoff may respectively lead to *land degradation* and gully erosion.

As humans alter ecosystems and natural environments, the changes can often produce negative effects on humans themselves. The desertification of large tracts of semiarid portions of East Africa has resulted periodically in widespread famine in countries such as Ethiopia and Somalia ( ● Fig. 11.21). Elsewhere, the continuing destruction of wetlands not only eliminates valuable plant and animal communities but also often seriously threatens the quality and reliability of the water supplies for the people who drained the land.

Human populations have so changed the vegetation in some parts of the world that we can characterize classes of cultivated vegetation cared for by humans—for example, flowers, shrubs, and grasses to decorate our living areas and grains, vegetables, and fruits that we raise for our own food and to feed the animals that we eat. Our focus in the remainder of this chapter is on the major ecosystems of Earth as we assume they would appear without human modification. Human impact on and adaptation to the various vegetation types are described in more detail along with the appropriate climate types in Chapters 9 and 10.

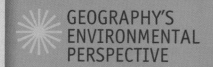

# Invasive Exotic Species— Burmese Pythons

Exotic species are plants or animals that people have introduced to a place that is not their native environment. Some exotics are regarded as beneficial; many landscaping plants were originally native to distant regions. Many problems, however, have occurred with exotics, whether they were introduced to a new area purposefully or inadvertently. A significant ecological problem occurs when the exotics thrive in the new environment to the detriment of native plants or animals. Introduced exotic species that are unable to adapt will die off. Exotics that outcompete native species and reproduce in great numbers are invasive exotic species, which can seriously affect the ecology and biodiversity of a region. Invasive exotics are a major cause of endangered or threatened native flora and fauna and a widespread problem virtually worldwide. Many hundreds of invasive exotic species of both plants and animals are causing problems in the aquatic and terrestrial environments of the United States.

One example is a large snake, a constrictor native to south and Southeast Asia, the Burmese python. These snakes were originally brought to the United States by the exotic pet trade. Acquired as a pet when they are young, they can grow as large 6 meters long (20 ft) and may weigh 90 kilos (200 lbs). Burmese pythons are semiaquatic snakes, very powerful, and excellent swimmers, and they have

NFS/Bob DeGross

A Park Ranger in Everglades National Park holds the tail of a large Burmese python.

# Classification of Terrestrial Ecosystems

The geographic classification of natural vegetation is as difficult as the classification of any other complex phenomena that is influenced by a variety of factors. Plant communities are among the most highly visible natural phenomena, and they can be categorized on the basis of form and structure or major physical characteristics. Of course, the composition of natural vegetation changes from place to place in a transitional manner, just as temperature and rainfall do, and although distinctly different biomes are apparent, there may be broad transition zones (ecotones) between them. Nevertheless, over the world certain recurring plant communities have distinctive similarities, indicating a consistent botanical response to environmental controls that are essentially climatic. It is the dominant vegetation of these plant communities that we recognize when we classify Earth's major terrestrial ecosystems.

Earth's terrestrial ecosystems can be categorized into one of four basic and easily recognized types: forest, grassland, desert, and tundra ( ● Fig. 11.22). Because vegetation adapted to cold climates may occur not only in high-latitude regions, but also at

proliferated in the wetlands of the Florida Everglades. Although it is not exactly known how they were introduced to the Everglades, there have been releases of pets that had become too large, and some snakes escaped their keepers after damage by hurricane Andrew in 1992.

Park rangers in Everglades National Park collected two Burmese pythons in 2000 and more than 1500 in the past five years. Their population is exploding. They reproduce prolifically, sometimes with more than 40 hatchlings from eggs. Today it is estimated that there are tens of thousands of Burmese pythons in the Everglades. These snakes are voracious carnivores that are disrupting the ecology by preying on native wildlife in great numbers, including mammals, birds, and other reptiles. Some of their prey are listed as threatened or endangered species. The American alligator, once an endangered species, may also be threatened. These pythons have been known to consume alligators that are 1.5 to 1.8 meters (5-6 feet) long.

The National Park Service is working to control the population of Burmese pythons in the Everglades. This process is not only important to the ecology in the park, but there is also concern that these pythons may eventually spread to other southern wetland regions along the Gulf and Atlantic coasts.

NPS/Lori Oberhofer

This python is fighting a losing battle with a large alligator. Large alligators can generally fend them off or kill them. The next generations of younger and smaller alligators may be in peril, however, along with many other animal species that are important to the ecology of the Everglades.

high elevations at any latitude, biogeographers often refer to the last of the major types as either arctic or alpine tundra. The forests in equatorial lowlands, however, are entirely different from those of Siberia or of New England, and the original grasslands of Kansas bore little resemblance to those of Kenya or the Sudan. Hence, the four major types of ecosystems can be subdivided into distinctive biomes, each of which is an association of plants and animals of many different species. Pure stands of particular trees, shrubs, or even grasses are extremely rare and are limited to small areas having peculiar soil or drainage conditions.

Earth's major biomes are mapped in ● Figure 11.23 on the basis of the dominant associations of natural vegetation that give each its distinctive character and appearance. The direct influence of climate on the distribution of these biomes is immediately apparent. Temperature (or latitudinal effect on temperature and insolation) and the availability of moisture are the key factors in determining the location of major biomes on a world regional scale ( ● Fig. 11.24). Because climatic elements are those that effect vegetation the most, detailed descriptions of major vegetation types are discussed with their corresponding climate types found in Chapters 9 and 10.

(a)

(b)

(c)

(d)

R. Gabler

Andrea Grosse/NBII

J. Petersen

M. Trapasso

● **FIGURE 11.22** The four major types of Earth biomes: (a) forest biome near Dornbirn, Austria; (b) grassland biome in Paraguay; (c) desert biome in Big Bend National Park, Texas; and (d) tundra biome in Coastal Greenland.

*Which one of these images shows an excellent example of an ecotone?*

# Forest Biomes

Forests are associations of large, woody, perennial tree species, generally several times the height of a human and with a more or less closed canopy of leaves overhead. They vary enormously in density and physical appearance. Forest trees also vary—they may be needle leaf or broadleaf, and either evergreen or deciduous, which means they drop their leaves to reduce moisture loss during seasons that are dry or cold. Forests are found only where the annual moisture balance is positive—where moisture availability exceeds potential evapotranspiration in the growing season. Thus, they occur in the tropics, where either the intertropical convergence zone (ITCZ) or monsoonal circulation brings plentiful rainfall, and in the middle latitudes, where precipitation is associated with polar front cyclones, summer convectional rainfall, or orographic uplift.

Tropical and middle-latitude forests have evolved different characteristics in response to the physical limitations in each area. In general, tropical forests have developed in less climatically restrictive forest environments. Temperatures are always high, although not extreme, in the humid tropics, encouraging rapid and luxuriant growth. Middle-latitude forests, however, must adapt to either seasonal cold (ranging from occasional frosts to subzero temperatures) or seasonal drought (which may occur at the worst possible time for vegetative growth).

## Tropical Forests

The forests of the tropics are far from uniform in appearance and composition. They grade poleward from the equatorial rainforests, which support Earth's greatest biomass, to the last scattering of low trees that overlook seemingly endless expanses of tall grass or desert shrubs on the tropical margins. We have subdivided the tropical forests into three distinct biomes: the tropical rainforest, the monsoon rainforest, and other tropical forest types (primarily thornbush and scrub). Of course, there are gradations (ecotones) between the different types and distinctive variations that are found only in certain localities.

**Tropical Rainforest** In the equatorial lowlands dominated by Köppen's tropical rainforest climate, the only physical limitation for vegetation growth is competition among adjacent species. The competition is for light. Temperatures are high enough to promote constant growth, and water is always sufficient. Thus, we find forests consisting of an amazing number of broadleaf evergreen tree species of similar appearance because special adaptations are not required. A cross-section of the rainforest often reveals concentrations of leaf canopies at several different levels. The trees composing the distinctive individual tiers have similar light requirements—lower than those of the higher tiers but higher than those of the lower tiers ( ● Fig. 11.25). Little or no sunlight reaches the forest floor, which may support ferns but is often rather sparsely vegetated. The forest is literally bound together by vines, **lianas,** which climb the trunks of the forest trees and intertwine in the canopy in their own search for light. Aerial plants may cover the limbs of the forest giants, deriving nutrients from the water and the plant debris that falls from higher levels. Light and variable wind conditions in the rainforest (associated with the tropical belt of the doldrums) preclude wind from being an effective agent of seed and pollen dispersal, so large colorful fruits and flowers, designed to attract animals that will unwittingly carry out seed and pollen dispersal, prevail.

The forest trees commonly depend on widely flared or buttressed bases for support because their root systems are shallow. This is a consequence of the richness of the surface soil and the poverty of its lower levels. The rainforest vegetation and soil are intimately associated. The forest litter quickly decomposes, releasing its nutrients, which are almost immediately reabsorbed by the forest root systems. In this way, a rainforest biomass and the available soil nutrients maintain an almost *closed system*. Tropical soils that maintain the amazing biomass of the rainforest are fertile only as long as the forest remains undisturbed. Clearing the forest interrupts the crucial cycling of nutrients between the vegetation and the soil; the copious amounts of water percolating through the soil leach away its soluble constituents, leaving behind only inert iron and aluminum oxides that cannot support forest growth. The present rate of clearing threatens to wipe out the worldwide tropical rainforests within the foreseeable future. The largest remaining areas of unmodified rainforests are in the upper Amazon Basin, where they cover hundreds of thousands of square kilometers.

Because of the darkness and extensive root systems present on the forest floor, animals of the tropical rainforest are primarily arboreal. A wide variety of species of tree-dwelling monkeys and lemurs, snakes, tree frogs, birds, and insects characterize the rainforest. Even the large herbivorous and carnivorous mammals—such as sloths, ocelots, and jaguars, respectively—are primarily arboreal.

Within the areas of rainforest and extending outward beyond its limits along streams are patches or strips of jungle. Jungle consists of an almost impenetrable tangle of vegetation contrasting strongly with the relatively open nature (at ground level) of the true rainforest. It is often composed of secondary growth that quickly invades the rainforest where a clearing has allowed light to penetrate to the forest floor. Jungle commonly extends into the drier areas beyond the forest margins along the courses of streams. There it forms a gallery of vegetative growth closing over the watercourse and hence has been called **galleria forest.**

**Monsoon Rainforest** In areas of monsoonal circulation, there is an alternation between the dry monsoon season, when the dominant flow of air is from the land to the sea, and the wet monsoon season, when the atmospheric circulation reverses, bringing moist air onshore along tropical coasts. The wet monsoon season rainfall may be very high, even hundreds of centimeters where air is forced upward by topographic barriers. In any case, it is sufficient to produce a forest that, once established, remains despite the dry monsoon season. Monsoon forests may have discernible tiers of vegetation related to the varying light demands of different species, and they are included with the tropical rainforest in Figure 11.25. However, the number of

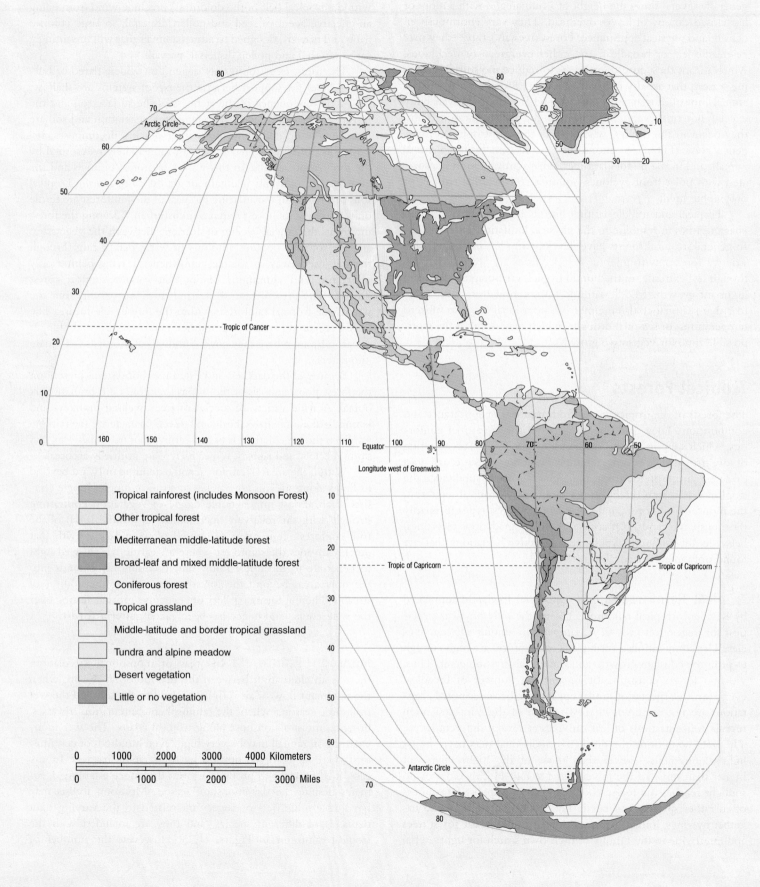

● **FIGURE 11.23** World map of natural vegetation.

A Western Paragraphic Projection
developed at Western Illinois University

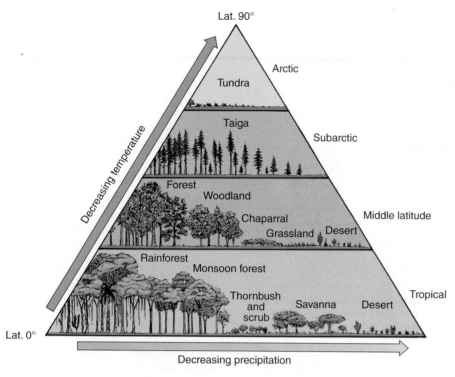

Lat. 90°

Arctic

Tundra

Taiga

Subarctic

Forest

Woodland

Chaparral

Middle latitude

Grassland    Desert

Rainforest

Monsoon forest

Thornbush
and
scrub    Savanna    Desert    Tropical

Lat. 0°

*Decreasing temperature*

*Decreasing precipitation*

Influence of latitude and moisture on distribution of biomes

● **FIGURE 11.24** This schematic diagram shows distribution of major biomes as they relate to temperature (latitude) and to moisture availability.

*What major biome dominates the wetter margins of all latitudes but the Arctic?*

● **FIGURE 11.25** A tropical rainforest on the island of St. Croix in the U.S. Virgin Islands. The dense forest canopy effectively conceals the vast number of different broadleaf evergreen trees and the relatively open forest floor.

*How might this rainforest differ from the rainforests of the Pacific Northwest of North America?*

Sean Linehan/NOAA, NGS, Remote Sensing

species is less than in the true rainforest, and the overall height and density of vegetation are also somewhat less. Some of the species are evergreen, but many are deciduous.

## Other Tropical Forests, Thornbush, and Scrub

Where seasonal drought has precluded the development of true rainforest or where soil characteristics prevent the growth of such vegetation, variant types of tropical forests have developed. These tend to be found on the subtropical margins of the rainforests and on old plateau surfaces where soils are especially poor in nutrients. The vegetation included in this category varies enormously but is generally low growing in comparison to rainforest, without any semblance of a tiered structure, and is denser at ground level. It is commonly thorny, indicating defensive adaptation against browsing animals, and it shows resistance to drought in that it is generally deciduous, dropping its leaves to conserve moisture during the dry winter season. Ordinarily, grass is present beneath the trees and shrubs. As we move away from the equatorial zone, we find the trees more widely spaced and the grassy areas becoming dominant. Along tropical coastlines, a specially adapted plant community, known as mangrove, thrives (see Fig. 11.16). Here trees are able to grow in salt water.

## Middle-Latitude Forests

The forest biomes of the middle latitudes differ from those of the tropics because the dominant trees have evolved mechanisms to withstand periods of water deprivation resulting from low temperatures and annual variations in precipitation. Evergreen and deciduous plants are present and equipped to cope with seasonal extremes not encountered in tropical latitudes.

**Mediterranean Sclerophyllous Woodland** Surrounding the Mediterranean Sea and on the southwest coasts of the continents between approximately 30° and 40°N and S latitude, we have seen that a distinctive climate exists—Köppen's mesothermal hot- and dry-summer type (Mediterranean). Here, annual temperature variations are moderate and freezing temperatures are rare. However, little or no rainfall occurs during the warmest months, and plants must be drought resistant. This requirement has resulted in the evolution of distinctive vegetation that is relatively low growing, with small, hard-surfaced leaves and roots that probe deeply for water. The leaves must be capable of photosynthesis with minimum transpiration of moisture. The general look of the vegetation is a thick scrub plant community, called chaparral in the western United States and *maquis* in the Mediterranean region. Wherever moisture is concentrated in depressions or on the cooler north-facing hill slopes, deciduous and evergreen oaks occur in

● **FIGURE 11.26** A distinctive sclerophyllous evergreen vegetation association grows wherever hot, dry summers alternate with rainy winters. Oaks commonly occupy relatively moist sites such as the valleys and north-facing slopes seen here in southern California.

*What are the general characteristics of sclerophyllous vegetation?*

groves ( ● Fig. 11.26). Drought-resistant needle-leaf trees, especially pines, are also part of the overall vegetation association. Thus, the vegetation is a mosaic related to site characteristics and microclimate. Nevertheless, the similarity of the natural vegetative cover in such widely separated areas as Spain, Turkey, and California is astonishing. (Note the location of Mediterranean middle-latitude forest in Fig. 11.23.)

### Broadleaf Deciduous Forest

The humid regions of the middle latitudes experience a seasonal rhythm dominated by warm tropical air in the summer and invasions of cold polar air in the winter. To avoid frost damage during the colder winters and to survive periods of total moisture deprivation when the ground is frozen, trees whose leaves have large transpiring surfaces drop these leaves and become dormant, coming to life and producing new leaves only when the danger period is past. A large variety of trees have evolved this mechanism; certain oaks, hickory, chestnut, beech, and maples are common examples. The seasonal rhythms produce some beautiful scenes, particularly during the periods of transition between dormancy and activity, with the sprouting of new leaves in the spring and the brilliant coloration of the fall as chemical substances draw back into the plant for winter storage ( ● Fig. 11.27).

The trees of the deciduous forest may be almost as tall as those of the tropical rainforests and, like them, produce a closed canopy of leaves overhead or, in the cold season, an interlaced network of bare branches. However, lacking a multistoried structure and having lower density as a whole, the middle-latitude deciduous forests allow much more light to reach ground level.

Forests of this type are the natural vegetation in much of western Europe, eastern Asia, and eastern North America. To the north and south they merge with mixed forests composed of broadleaf deciduous trees and conifers. (Broadleaf deciduous forests and mixed forests are combined in Figure 11.23.) Both the broadleaf deciduous and mixed forests have been largely logged off or cleared for agricultural land, and the original vegetation of these regions is rarely seen.

### Broadleaf Evergreen Forest

Beyond the tropics, broadleaf trees remain evergreen (active throughout the year) in significant areas only in certain Southern Hemisphere locations. Here the mild maritime influence is strong enough to prevent either dangerous seasonal droughts or severely low winter temperatures. Southeastern Australia and portions of New Zealand, South Africa, and southern Chile are the principal areas of this type. In the Northern Hemisphere, broadleaf evergreen forest may once have been significant in eastern Asia, but it has long since been cleared for cultivation. Limited areas occur in the United States in Florida and along the Gulf Coast as a belt of evergreen oak and magnolia.

### Mixed Forest

Poleward and equatorward, the broadleaf deciduous forests in North America, Europe, and Asia gradually merge into mixed forests, including needle-leaf coniferous trees, normally pines. In general, where conditions permit the growth of broadleaf deciduous trees, coniferous trees cannot compete successfully with them. Thus, in mixed forests the conifers, which are actually more adaptable to soil and moisture deficiencies, are found in the less hospitable sites: in sandy areas, on acid soils, or where the soil itself is thin. The northern mixed forests reflect the transition to colder climates with increasing latitude; eventually, conifers become dominant in this direction. The southern mixed forests are more problematic in origin. In the United States they are transitional to pine forests situated on sandy soils of the coastal plain. In Eurasia they coincide with highlands dominated by conifers during a stage in the plant succession that began with the change of climate environments at the end of the ice ages, some 10,000 years ago.

### Coniferous Forest

The coniferous forests occupy the frontiers of tree growth. They survive where most of the broadleaf species cannot endure the climatic severity and impoverished soils. The hard, narrow needles of coniferous species transpire much less moisture than do broad leaves so that needle-leaf species can tolerate conditions of physiologic drought (unavailability of moisture because of excessive soil permeability, a dry season, or frozen soil water) without defoliation. Pines, in particular, also demand little from the soil in the form of soluble plant nutrients, especially basic elements such as calcium, magnesium, sodium, and potassium. Thus, they grow in sandy places and where the soil is acid in character. As a whole, conifers are particularly well adapted to regions having long, severe winters combined with summers warm enough for vigorous plant growth. Because all but a few exceptions retain their leaves (needles) throughout the

(a) R. Gabler

(b) R. Gabler

(c) R. Gabler

● **FIGURE 11.27** Deciduous hardwood forests in middle-latitude regions with cold winters change dramatically with the seasons. The green leaves of summer (a) will change to reds, golds, and browns in fall (b) and drop to the forest floor by winter (c). Leaf-dropping in cold-winter regions, such as here in western Illinois, minimizes transpiration and moisture loss when trees are unable to access the frozen soil water.

*What length of growing season (frost-free period) is associated with the climate of western Illinois?*

year, they are ready to begin photosynthesis as soon as temperatures permit without having to produce a new set of leaves to do the work.

Thus, we find a great band of coniferous forests (the **boreal forests,** or **taiga)** dominated by spruce and fir species, with pines on sandy soils, sweeping the full breadth of North America and Eurasia northward of the 50th parallel of latitude, approximately occupying the region of Köppen's subarctic climate. Conifers differ from other trees in that their seeds are not enclosed in a case or fruit but are carried naked on cones. All are needle leaf and drought resistant, but a few are not evergreen. Thus, a large portion of eastern Siberia is dominated by larch, a deciduous tree, which produces a deciduous coniferous forest. In this area, January mean temperatures may be −35°C to −51°C (−30°F to −60°F). This is the most severe winter climate in which trees can maintain themselves, and even needle-leaf foliage must be shed for the vegetation to survive. Hardy broadleaf deciduous birch trees share this extreme climate with the robust larches.

Extensive coniferous forests are not confined to high-latitude areas of short summers. Higher elevations in middle-latitude mountains of the Northern Hemisphere have forests of pine, hemlock, and fir ( ● Fig. 11.28), with subalpine larch and specially adapted pine species characterizing the harshest and highest forest sites. The forests along the sandy coastal plain of the eastern United States are there in part because of the sandy soils, but they may also reflect a stage in plant succession that in time will lead to domination by broadleaf types. Similarly, a temporary stage in the postglacial vegetation succession of the Great Lakes area included magnificent forests of white pine and hemlock that were completely logged off during the late 19th century.

A more maritime coniferous forest occupies the west coast of North America extending from southern Alaska to central California. It is made up of sequoias, Douglas fir, cedar, hemlock, and Sitka spruce (farther north). Many of the California sequoias are thousands of years old and more than 100 meters (330 ft) high. The southern regions experience summer drought, and farther north sandy, acidic, or coarse-textured soils dominate.

Aldo Aguilar/USFS

● **FIGURE 11.28** As this scene in the Gifford Pinchot National Forest in Washington State indicates, evergreen coniferous forest is the characteristic vegetation of higher-elevation middle latitude regions. These needle-leaf trees are well adapted to the drought of the winter season, which is longer and more severe at higher elevations. The broadleaf trees are in fall color prior to losing their leaves.

*Why are needle-leaf trees better adapted to physiologic drought than broadleaf trees?*

# Grassland Biomes

Grasses, like conifers, appear in a variety of settings and are part of many diverse plant communities. They are, in fact, an initial form in most plant successions. However, there are enormous, continuous expanses of grasslands on Earth. In general, it is thought that grasses are dominant only where trees and shrubs cannot maintain themselves because of either excessive or deficient moisture in the soil. On the global scale, grassland biomes are located in continental interiors where most, if not all, of the precipitation falls in the summer. Two great geographic realms of grasslands are generally recognized: the tropical and the middle-latitude grasslands. However, it is difficult to define grasslands of either type using any specific climatic parameter, and geographers suspect that human interference with the natural vegetation has caused expansion of grasslands into forests in both the tropical and middle latitudes.

M. Trapasso

● **FIGURE 11.29** The savanna biome of east Africa is a classic landscape of grasses and scattered trees *(background)*. In the foreground, vultures (detritivores) eat the carcass of a zebra (herbivore) left behind by a lion (carnivore).

*What organisms will take over when the vultures are finished?*

## Tropical Savanna Grasslands

The tropical grassland biome differs from grassland biomes of the middle latitudes in that it ordinarily includes a scattering of trees; this is implied in the term **savanna** ( ● Fig. 11.29). In fact, the demarcation between tropical scrub forest and savanna is seldom a clear one. The savanna grasses tend to be tall and coarse with bare ground visible between the individual tufts. The related tree species generally are low growing and wide-crowned forms, having both drought- and fire-resisting qualities, indicating that fires frequently sweep the savannas during the drought season. Savannas occur under a variety of temperature and rainfall conditions but generally fall within the limits of Köppen's tropical savanna type. They commonly occur on red-colored soils, leached of all but iron and aluminum oxides, which become brick-like when dried. They likewise coincide with areas in which the level of the water table (the zone below which all soil and rock pore space is saturated by water) fluctuates dramatically. The up-and-down movement of the water table may in itself inhibit forest development, and it is no doubt a factor in the peculiar chemical nature of savanna soils. Large migratory herds of grazing animals and associated predators, responding to the periodically abundant grasses followed by seasonal drought, characterized the savanna prior to widespread human disruption.

## Middle-Latitude Grasslands

The middle-latitude grasslands occupy the zone of transition between the middle-latitude deserts and forests. On their dry margins, they pass gradually into deserts in Eurasia and are cut off westward by mountains in North America. However, on their humid side, they terminate rather abruptly against the forest margin, again raising questions as to whether their limits are natural or have been created by human activities, particularly the intentional use of fire to drive game animals. The middle-latitude grasslands of North America, like the African savannas, formerly supported enormous herds of grazing animals—in this case, antelope and bison, which were the principal means of support of the American Plains Indians.

Like the savannas, the middle-latitude grasslands were diverse in appearance. They, too, consisted of varying associations of plant species that were never uniform in composition. In North America, the grasses were as much as 3 meters (10 ft) tall in the more humid sections, as in Iowa, Indiana, and Illinois, but only 15 centimeters (6 in.) high on the dry margins from New Mexico to western Canada. Thus, the middle-latitude grassland biomes are usually divided into tall-grass and short-grass prairie, often with a zone of mixture recognized between them. Unlike growth in the tropical savannas, the germination and growth of middle-latitude grasses are attuned to the melting of winter snows followed by summer rainfall. Whether the grasses are annuals that complete their life cycle in one growing season or perennials that grow from year to year, they are dormant in the winter season. Also, unlike in the savannas, the soils beneath these grasslands are extremely rich in organic matter and soluble nutrients. As a consequence, most of the middle-latitude grasslands have been completely transformed by agricultural activity. Their wild grasses have been replaced by domesticated varieties—wheat, corn, and barley—and they have become the "breadbaskets" of the world.

Photo courtesy of Tallgrass Prairie National Preserve

**(a)**

Photo courtesy of Tallgrass Prairie National Preserve

**(b)**

● **FIGURE 11.30** (a) Spring wildflowers bloom in the Tallgrass Prairie National Preserve, Kansas, one of the remaining protected areas of that type of prairie. The grasses will grow taller in later spring and summer. (b) Short-grass prairie, with bison grazing in South Dakota's Wind Cave National Park.

*Both develop under a semiarid climate, but which of these two types of prairie grassland receives a higher annual precipitation?*

**Tall-Grass Prairie** The **tall-grass prairies** ( ● Fig. 11.30a) were an impressive sight; in some better-watered areas, they made up endless seas of grass moving in the breeze, reaching higher than a horse's back. Flowering plants were conspicuous, adding to the effect. Unfortunately, this tall-grass prairie scene, which inspired much vivid description by those first encountering it, is barely visible anywhere today. The tall-grass prairies, which once reached continuously from Alberta to Texas, have been almost completely destroyed. Compared to their once vast extent, today only a few tall-grass prairie areas remain in government-protected preserves. Their tough sod, formed by the dense grass root network, defeated the first wooden plows, which had served well enough in breaking up the forest soils. But the steel plow, invented in the 1830s, subdued the sod and was aided by the introduction of subsurface tile for draining the nearly flat uplands and by the simultaneous appearance of well-digging machinery and barbed wire. These four innovations transformed the tall-grass prairie from grazing land to cropland.

In North America, the tall-grass prairie pushed as far eastward as Lake Michigan. Why trees did not invade the prairie in this relatively humid area remains an unanswered question. Farther west, shallow-rooted grass cover is fully understandable because the lower soil levels, to which tree roots must penetrate for adequate support and sustenance, are bone dry. In such areas, trees can survive only along streams or where depressions collect water.

In Eurasia, tall-grass prairies were found on a large scale in a discontinuous belt from Hungary through Ukraine, Russia, and central Asia to northern China. The grasslands are known in South America as the pampas of Uruguay and Argentina and in South Africa as the veldt. Today, all these areas have been changed by agriculture. The factor that seems to account best for the tall-grass prairie—precipitation that is both moderate and variable in amount from year to year—is the principal hazard in the use of these regions as farmland. However, this hazard becomes much greater in the areas of short-grass prairie.

**Short-Grass Prairie** West of the 100th meridian in the United States and extending across Eurasia from the Black Sea to northern China, roughly coinciding with the areas of Köppen's middle-latitude steppe climate, are vast, nearly level grasslands composed of a mixture of tall and short grass, with short grass becoming dominant in the direction of lower annual precipitation totals (Fig. 11.30b) and **tall-grass prairie** in semiarid regions that have higher annual precipitation (compare to Fig. 11.30a). On the Great Plains between the Rocky Mountains and the tall-grass prairies, the **short-grass prairie** zone more or less coincides with the zone in which moisture rarely penetrates more than 60 centimeters (2 ft) into the soil, so the subsoil is permanently dry. Moving toward the drier areas, the grassland vegetation association dwindles in diversity and, more conspicuously, in height to less than 30 centimeters (1 ft). This is a consequence of reduction in numbers of tall-growing species and greater abundance of shallow-rooted and lower-growing types. The total amount of ground cover also declines toward the drier margins as the deeply rooted, sod-forming grasses of the prairie grassland give way to bunchgrass species (so called because, instead of forming a continuous grass cover, they occur in isolated clumps or bunches). In their natural conditions, the short-grass prairies of North America and Eurasia supported higher densities of grazing animals—bison and antelope in the former, wild horses in the latter—than did the tall-grass prairies. Indeed, it is suspected that the specific plant association of the short-grass regions may have been a consequence of overgrazing under natural conditions. The short-grass prairies cannot be cultivated without the use of irrigation or dry farming methods, so they remain primarily the domain of wide-ranging grazing animals; however, today's animals are domesticated cattle, not the thundering self-sufficient herds of wild species that formerly made these plains one of Earth's marvels.

# Desert Environments

Eventually, lack of precipitation can become too severe even for the hardy grasses. Where evapotranspiration demands greatly exceed available moisture throughout the year, as in Köppen's desert climates, either special forms of plant life have evolved or the surface is bare. Plants that actively combat low precipitation are equipped to probe deeply or widely for moisture, to reduce moisture losses to the minimum, or to store moisture when it is available. Other plants evade drought by merely lying dormant, perhaps for years, until enough moisture is available to ensure successful growth and reproduction. The desert biome is recognized by the presence of plants that are either drought resisting or drought evading ( ● Fig. 11.31).

Plants that have evolved mechanisms to combat drought are known as *xerophytes*. They are perennial shrubs whose root systems below ground are much more extensive than their visible parts or they have evolved tiny leaves with a waxy covering to combat transpiration. They may have leaves that are needle-like or trunks and limbs that photosynthesize like leaves or that have expandable tissues or accordion-like stems to store water

● **FIGURE 11.31** Desert plants employ a variety of strategies for surviving in an arid environment, with little rain and high potential evapotranspiration.

*What aspects of vegetation that help them survive in a desert can you see in this image?*

Anton Foltin/Shutterstock.com

● **FIGURE 11.35** These elephant seals basking in the sun rely on food sources related to the nutrient-rich and cool, upwelled waters along the California coast.

*To which group of marine organisms do these elephant seals belong?*

plankton and then become the main food source for birds, penguins, seals, and whales.

The marine food chain just described is part of a full food cycle in the ocean because, through the excretions of animals and the decomposition of both plants and animals in the ocean, chemical nutrients are returned to the water and are again made available for transformation by phytoplankton into usable foods. Phytoplankton also play an essential role in the production of oxygen for Earth's

atmosphere. In fact, the greatest concern of scientists who study the annual Antarctic "ozone hole" (see Chapter 4) is that excessive ultraviolet radiation will destroy or diminish the phytoplankton in the Antarctic and other oceans.

The uneven distribution of nutrients in the oceans and the fact that sunlight can penetrate only to a depth of about 120 meters (400 ft), depending on the clarity of the water, means that the distribution of marine organisms is also variable. Most organisms are concentrated in the upper layers of the ocean where the most solar energy is available. In deep waters, where the ocean floor lies below the level to which sunlight penetrates, the benthos organisms depend on whatever nutrients and plant and animal detritus filter down to them. For this reason, benthos animals are scarce in deep and dark ocean waters. They are most common in shallow waters near coasts, such as coral reefs and tide pools, where there is sunlight, a rich supply of nutrients, and abundant phytoplankton and zooplankton.

The waters of the continental shelf have the highest concentration of marine life. The supply of chemical nutrients is greater in waters near the continents where nutrients are washed into the sea from rivers. Marine organisms are also concentrated where deeper waters rise (upwelling) to the surface layers where sunlight is available. Such vertical exchanges are sometimes the result of variations in salinity or density. A similar situation occurs where convection causes bottom layers of water to rise and mix with top layers, as is the case in cold polar waters and in middle-latitude waters during the colder winter months. Marine life is also abundant in areas where there is a mixing of cold and warm ocean currents, as there is off the northeastern coast of the United States ( ● Fig. 11.36).

● **FIGURE 11.36** The distribution of chlorophyll-producing marine plankton can be mapped on satellite images. Blue to light blue and green and then to yellow to red indicates increasingly higher levels of chlorophyll concentration.

*What two spatial observations can you make about geographic locations that have high plankton concentrations?*

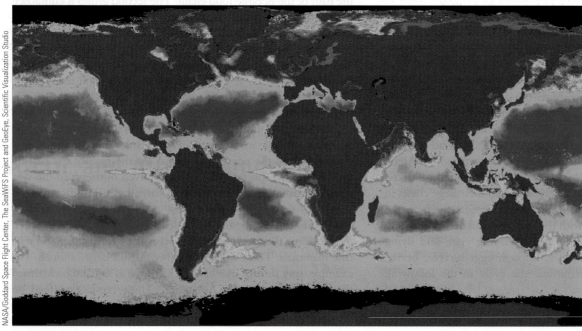

During the 1977 dives of the manned submersible vessel *Alvin* off the East Pacific Rise, scientists for the first time observed abundant sea life on the floor of the ocean at depths of more than 2500 meters (8100 ft). It was previously presumed that these cold (2°C/36°F), dark waters were a virtual biological desert. However, the undersea volcanic mountain range produces vents of warm, mineral-rich waters, which nourish bacteria and large colonies of crabs, clams, mussels, and giant 3-meter (10-ft) red tube worms. The discovery of this deep-ocean ecosystem, which exists without the benefit of sunlight, has caused scientists to rethink old theories about the ocean and its chemistry. This unusual "chemosynthetic" vent community has been observed more recently at several other oceanic ridge sites in the Pacific and Atlantic Oceans.

## The Resilience of Life Forms

Throughout this chapter one overarching theme becomes obvious: Earth's life forms are extremely resilient. First came the *autotrophs* (plant life), followed by the *heterotrophs* (animal life) that reside within and feed upon the plant communities. Temperature, precipitation, sunlight, and winds are among the most important of the climatic factors controlling the geography of life forms, but it is also essential to note that wherever a climatic limit can be breached, or a pocket of favorable climatic elements is present, life can be found. Living things find a way to exist, even in rather inhospitable environments—where one life form fails, another more adaptive life form will survive.

As climates change through time—and they inevitably will—life forms will either adjust to the new climatic conditions or migrate to other areas where favorable conditions exist. Today and in the future mapping the geographic distribution of vegetation and animal migration will be an effective way to document climate change and the environmental shifts related to those climatic conditions.

# CHAPTER 11 ACTIVITIES

## ■ DEFINE AND RECALL

| | | |
|---|---|---|
| ecosystem | habitat | biome |
| abiotic | ecological niche | symbiotic relationship |
| autotroph (producer) | generalist | liana |
| heterotroph | natural vegetation | galleria forest |
| herbivore | plant community | boreal forest (taiga) |
| carnivore | succession | savanna |
| omnivore | gap | tall-grass prairie |
| detritivore (decomposer) | climax community | short-grass prairie |
| trophic structure | mosaic | tundra |
| food chain | matrix | plankton |
| trophic level | patch | phytoplankton |
| biomass | corridor | zooplankton |
| primary productivity | tolerance | nekton |
| secondary productivity | ecotone | benthos |

# DISCUSS AND REVIEW

1. Why is the study of ecosystems important in the world today? Give several examples of natural ecosystems within easy driving distance of your own residence.
2. What are the four basic components of an ecosystem?
3. What factors are most critical in affecting the net primary productivity of a terrestrial ecosystem?
4. In what ways has the original theory of succession been modified?
5. How do the terms *mosaic, matrix, patch, corridor,* and *ecotone* relate to each other and to a vegetation landscape?
6. What are the four major types of Earth biomes? What important climate characteristics are related to each of the major biomes?

7. Describe a true tropical rainforest. How does such a forest differ from jungle?
8. What are the distinctive features of chaparral vegetation? What climate conditions are associated with chaparral?
9. What conditions of climate or soil might be anticipated for each of the following in the middle latitudes: broadleaf deciduous forest; broadleaf evergreen forest; needle-leaf coniferous forest?
10. How have xerophytes adapted to desert climate conditions?
11. What are the chief characteristics of tundra vegetation? What adaptations to climate does tundra vegetation make?
12. The highest concentrations of marine life (see Fig. 11.36) are found in which parts of the oceans? Why?

# CONSIDER AND RESPOND

1. Refer to Table 11.2. Based on the means, which is more important for terrestrial ecosystem productivity between the equator and the polar regions—annual temperatures or the availability of water? Use examples from the table to explain your choice.
2. What broad groups might you use if you were to classify Earth's vegetation on the basis of latitudinal zones? Why do you suppose your textbook authors chose not to do this, electing instead to identify broad groups based on major terrestrial ecosystems (biomes)?

3. Refer to Figure 11.23, which shows considerable variation in the natural vegetation of the United States between 30° and 40° north latitude. Describe the broad changes in vegetation that occur within that latitudinal band as you move from the East Coast to the West Coast of the United States.
4. Examine the climate variation (see Fig. 8.6) within the same latitudinal band described in Question 4. Does there appear to be a relationship between natural vegetation and climate? If so, describe this relationship.

# APPLY AND LEARN

1. Kudzu is a climbing, woody, perennial vine that originates from Japan. From 1935 to 1950, farmers in the southeastern United States were encouraged to plant it to help reduce soil erosion. Without natural enemies, this vine began to grow out of control. Since 1953, it has been identified as a pest weed, and efforts to eradicate this invader continue to this day. Kudzu inhabits about 30,000 km² of the southeastern United States. From 1935 to present, what is the spread of this pest in km²/year?

2. Research has shown that the harvest dates for vineyards in Switzerland vary systematically with the local mean temperatures between April and August. With every 1°C rise in temperature, the harvest date for the grapes moves back by 12 days. Assuming an average harvest date of September 15, what would be the expected harvest date if the April to August mean temperature was the following: 1°C above normal, 1.25°C below normal, 1.5°C above normal, and 1°F above normal?

 CourseMate – Make the most of your study time by accessing everything you need to succeed in one place. Read your textbook, take notes, review flashcards, watch videos, complete activities, take practice quizzes, and more online with CourseMate. Log in at **www.cengagebrain.com**.

# 12

# SOILS AND SOIL DEVELOPMENT

Scientists work to understand and control erosional losses and other problems that threaten our precious natural resource— soil. USDA/ARS/Photo by Jack Dykinga

## ▬ OBJECTIVES

WHEN YOU COMPLETE THIS CHAPTER YOU SHOULD BE ABLE TO:

▪ Appreciate the interaction and integration of subsystems that influence the soil system.
▪ Describe the major components of a soil and how they vary to produce different soil types.
▪ Discuss the role of water in soil processes and how different amounts of available soil water can affect the vegetation growing on a soil.
▪ Understand the factors that determine a soil's formation, development, and fertility, especially the roles of climate and vegetation.
▪ Explain why soils are among the world's most critical resources and the need for effective soil conservation practices.
▪ Cite a few examples of major human impacts on soils.

WHAT ARE THE FOUR MOST important natural constituents that permit life as we know it to exist on Earth? Many people if asked that question would reply "air, water, and sunlight" right away, but they might have to think harder and longer about their fourth answer. Most people give little attention to that fourth natural resource, but it is essential for their life on Earth, and it lies right below their feet. Soil is that critical resource. The soil mantle that covers most land surfaces is indispensable, but fragile, and threatened by erosion, pollution, or being covered over by the human-built environment. Soils provide nutrients that directly or indirectly support much of life on Earth.

**Soil** is a dynamic natural body capable of supporting a vegetative cover. It contains chemical solutions, gases, organic refuse, flora, and fauna. The physical, chemical, and biological processes that take place among the components of a soil are integral parts of its dynamic character. Soil responds to climatic conditions (especially temperature and moisture), to the land surface configuration, to vegetative cover and composition, and to animal activity.

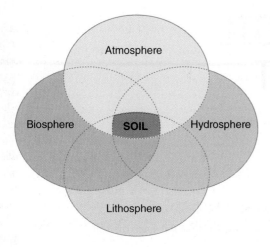

● **FIGURE 12.1** The intertwined links between soil and the major Earth subsystems.

*Why is soil considered to be such an integrator of Earth systems?*

Soil has been called "the skin of the Earth." The condition and nature of a soil reflects both the ancient environments under which it formed and today's environmental conditions. A soil functions as an environmental system, adapting, reflecting, and responding to a great variety of natural and human-influenced processes. Soil is an exceptional example of the integration, interdependence, and overlap among Earth's subsystems because the characteristics of a soil reflect the atmospheric, hydrologic, lithologic, and biotic conditions under which it developed ( ● Fig. 12.1). In fact, because soils integrate these major subsystems so well, they are sometimes considered a separate system called the *pedosphere* (from Greek: *pedon,* ground).

Soil is also home to numerous living organisms forming the environments in which they live, both above and below the ground surface. The life forms that live in or on a soil play significant roles in the development and characteristics of a soil, and through human population growth and expanding civilizations, potentially negative impacts on soils have increased dramatically.

How a soil develops and its resulting characteristics depend on a great number of factors. But when soils are viewed on a world regional scale, a strong and significant influence is climate. The relationships between soils, climate types, and their associated environments were considered in Chapters 9 and 10.

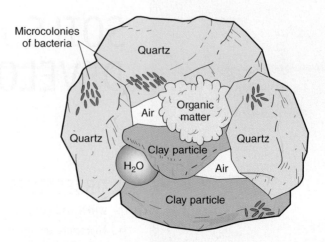

● **FIGURE 12.2** The four major components of soil. Soil contains a complex assemblage of inorganic minerals and rocks, along with water, air, and organic matter. The interaction among these components and the proportion of each are important factors in the development of a soil.
From Purves et al., *Life: The Science of Biology,* Fourth Edition. Used with permission of Sinauer Associates, Inc.

*How do each of these soil components contribute to making a soil suitable to support plant life?*

# Major Soil Components

What is soil actually made of? What does a shovel contain when it scoops up a load of soil? What soil characteristics support and influence variations in Earth's vegetational environments? Soils contain four major components, and many processes act on these components. The four major components of soil are inorganic materials, soil water, soil air, and organic matter ( ● Fig. 12.2).

## Inorganic Materials

Soils contain varying amounts of insoluble materials—rock fragments and minerals that will not readily dissolve in water. Soils also contain soluble minerals, which supply dissolved chemicals held in solution. Most minerals found in soils are combinations of the common elements of Earth's surface rocks: silicon, aluminum, oxygen, and iron. Some of these constituents occur as solid chemical compounds, and others are found in the air and water that are also vital components of a soil. Soils sustain Earth's land ecosystems by providing a great variety of necessary chemical elements and compounds to life forms. Carbon, hydrogen, nitrogen, sodium, potassium, zinc, copper, iodine, and compounds of these elements are important in soils. The chemical constituents of a soil typically come from many sources—the breakdown *(weathering)* of underlying rocks, deposits of loose sediments, and minerals dissolved in water. Organic activities help to disintegrate rocks, create new chemical compounds, and release gases into the soil.

Plants need many chemical substances for growth, and having a knowledge of a soil's mineral and chemical content is necessary for determining its potential productivity. **Soil fertilization** is the process of adding nutrients or other constituents in order to meet the soil conditions that certain plants require ( ● Fig. 12.3).

● **FIGURE 12.3** Fertilizers increase the productivity of soils. This farmer is adding nitrogen fertilizer into the soil on an Iowa farm.

*Why can soil fertilizer be either useful or detrimental when it is introduced into the soil system?*

## Soil Water

The original source of soil water is precipitation. When precipitation falls on the land, the water that is not evaporated away is either absorbed into the ground or by vegetation, or it runs downslope. Soil water is both an ingredient and a catalyst for chemical reactions that sustain life and influence soil development. Water also provides nutrients in a form that can be extracted by vegetation. As water moves through a soil, it washes over and through various soil components, dissolving some of these materials and carrying them through the soil. Soil water is not pure; it is a solution that contains soluble nutrients. Plants need air, water, and minerals to function, live, and grow, and they depend on soil for much of these necessities. A soil functions as an *open system*. Matter and energy flow into and out of a soil, and they are also held in storage ( ● Fig. 12.4). Understanding these flows—inputs and outputs, the components and processes involved, and how they vary from soil to soil—is a key to appreciating the complexities of this natural resource.

The water in a soil is found in several different circumstances (see Fig. 12.4). Soil water adheres to soil particles and soil clumps by surface tension (the property that causes small water droplets to form rounded beads instead of spreading out in a thin film). This soil water, called **capillary water,** serves as a stored water supply for plants. Capillary water can move in all directions through soil be-

cause it migrates from areas with more water to areas with less. Thus, during dry periods, when there is no gravitational water flowing through the soil, capillary water can move upward or horizontally to supply plant roots with moisture and dissolved nutrients.

Capillary water migrates upward and moves minerals from the subsoil toward the surface. If this capillary water evaporates away, the formerly dissolved minerals remain, generally as alkaline or saline deposits in the topsoil. High concentrations of certain mineral deposits, like these, can be detrimental to plants and animals existing in the soil. Lime (calcium carbonate) deposited by evaporating soil water can build up to produce a cement-like layer, called *caliche,* which like a clay **hardpan** prevents the downward percolation of water.

Soil water is also found as a very thin film, invisible to the naked eye, that is bound to the surfaces of soil particles by strong electrical forces. This is **hygroscopic water,** which does not move through the soil, and it does not supply plants with the moisture they need.

Water that percolates down through a soil under the force of gravity is called **gravitational water.** Gravitational water moves downward through voids between soil particles and toward the *water table*—the level below which all available spaces are filled with water. The quantity of gravitational water in a soil is related to several conditions, including the amount of precipitation, the time since it fell, evaporation rates, the space available for water storage, and how easily the water can move through the soil.

Gravitational water performs several functions in a soil ( ● Fig. 12.5). As gravitational water percolates downward, it dissolves soluble minerals and carries them into deeper levels of the soil, perhaps to the zone where all open spaces are saturated. Depleting nutrients in the soil by the through flow of water is called **leaching.** In regions of heavy rainfall, leaching is common and can be intense, robbing a topsoil of all but the insoluble substances.

Gravitational water moving down through a soil also takes with it the finer particles (clay and silt) from the upper soil layers. This downward removal of soil components by water is called **eluviation.** As gravitational water percolates downward, it deposits the fine materials that were removed from the topsoil at a lower level in the soil. Deposition by water in the subsoil is called **illuviation.** Gravitational water also mixes soil particles as it moves them downward. One result of eluviation is that the texture of a topsoil tends to become coarser as the fine particles are removed. Consequently, the topsoil's ability to retain water is reduced. Illuviation may eventually cause the subsoil to become dense and compact, forming a clay hardpan.

Leaching and eluviation both strongly influence the characteristic layered changes with depth, or **stratification.** Fine particles and substances dissolved from the upper soil are deposited in lower levels, which become dense and may be strongly colored by accumulated iron compounds.

## Soil Air

Much of a soil—in some cases, approaching 50%—consists of spaces between soil particles and between *clumps* (aggregates of soil particles). Voids that are not filled with water contain air or certain gases. Compared to the composition of the lower atmosphere, the

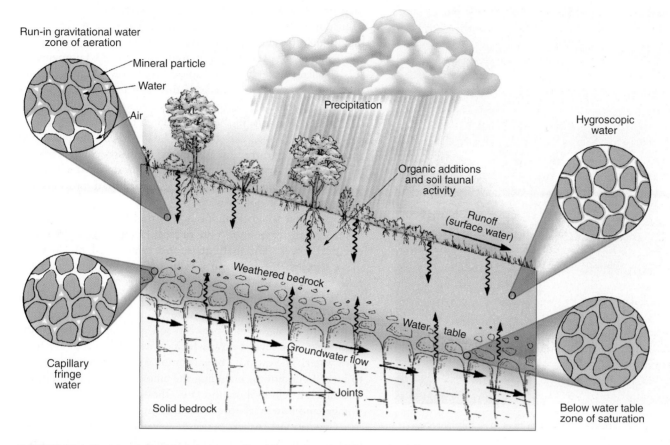

**● FIGURE 12.4** The interrelationships between a soil and the environmental factors that influence soil development. Soil is an example of an open system because it receives inputs of matter and energy, stores part of these inputs, and outputs both matter and energy. Note the inputs and outputs on this diagram.

*What are some examples of energy and matter that flow into and out of the soil system?*

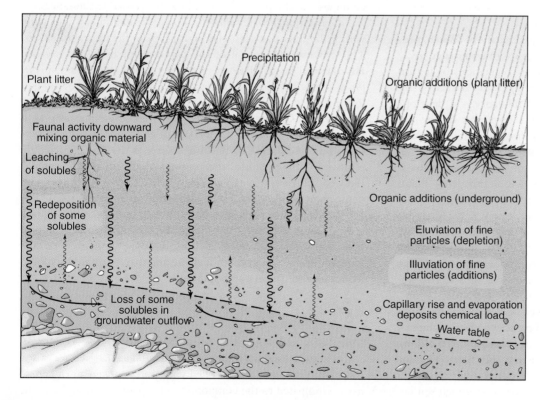

**● FIGURE 12.5** Water plays several important roles in the processes that affect soil development. Water is important in moving nutrients and particles vertically, both up and down, in a soil.

*How does deposition by capillary water differ from deposition (illuviation) by gravitational water?*

air in a soil is likely to have less oxygen, more carbon dioxide, and a fairly high relative humidity because of the presence of capillary and hygroscopic water.

For most microorganisms and plants that live in the ground, soil air supplies oxygen and carbon dioxide necessary for life processes. The problem with a water-saturated soil is not necessarily excess water but, if all pore spaces are filled with water, there is no air supply. The lack of air is why many plants find it difficult to survive in water-saturated soils.

## Organic Matter

Soil contains organic matter in addition to minerals, gases, and water. The decayed remains of plant and animal materials, partially transformed by bacterial action, are collectively called **humus.** Humus is an important catalyst in chemical reactions that help plants to extract soil nutrients. Humus also supplies nutrients and minerals to the soil. Soils that contain humus are workable and have a good capacity to retain water. Humus also provides an abundant food source for microscopic soil organisms.

Most soils are environments that teem with life, ranging from microscopic bacteria and fungi to earthworms, rodents, and other burrowers. Animals contribute to the soil development and enrichment by creating humus from plant litter. They also mix organic material deeper into the soil and move inorganic fragments toward the surface. In addition, the functions of plants and their root systems are integral parts of the soil-forming system.

Soils vary at local, regional, and global scales. Particularly strong relationships exist between a soil and the vegetation and climate at its location. For example, soils in middle-latitude grasslands normally have a very high proportion of organic matter; those in deserts are thin and rich in minerals left behind by evaporating water, like lime and salts; and tropical soils typically have a high content of iron and aluminum oxides. Knowing a soil's water, mineral, and organic components and their proportions can help us determine its productivity and what the best use for that particular soil might be.

# Characteristics of Soil

Several soil properties that can be readily tested or examined are used to describe and differentiate soil types. The most important properties include color, texture, structure, acidity or alkalinity, and capacity to hold and transmit water and air.

## Color

Color is the most visible soil characteristic, but it might not be the most important attribute. Most people are aware of how soils vary in color from place to place. For example, the well-known red clay soils of Georgia are not far from Alabama's belt of black soils. Soils vary in color from black to brown to red, yellow, gray, and near-white. A soil's color is generally related to its physical and chemical characteristics. When describing soils in the field or samples in the laboratory, soil scientists use a book of standardized colors to clearly and precisely identify this coloration ( ● Fig. 12.6).

Courtesy of James P. Shroyer, Kansas State University Research and Extension

● **FIGURE 12.6** Determining soil color. A standardized classification system is used to determine precise color by comparing the soil to the color samples found in Munsell soil color books.

*In general, how would you describe the color of the soils where you live?*

Decomposed organic matter is black or brown, so soils with high humus contents tend to be dark. If the humus content of soil decreases because of either low organic activity or loss of organics through leaching, soil colors typically fade to light brown or gray. Soils rich in humus are usually very fertile. For this reason, dark-brown or black soils are often referred to as *rich*. However, this is not always true because some black or dark brown soils have little or no humus but are dark because of other soil-forming factors.

Soils that are red or yellow typically indicate the presence of iron. In moist climates a light-gray or white soil indicates that iron has been leached out, leaving oxides of silicon and aluminum; in dry climates, the same color typically indicates a high proportion of calcium or salts.

Soil colors provide useful clues to the physical and chemical characteristics of soils and make the job of recognizing different soil types easier. But color alone does not answer all the important questions about a soil's qualities or fertility.

## Texture

**Soil texture** refers to the particle sizes (or distribution of sizes) that make up a soil ( ● Fig. 12.7). In **clayey** soils, the dominant size is **clay** particles, defined as having diameters of less than 0.002 millimeter (soil scientists universally use the metric system). In **silty** soils, the dominant **silt** particles are defined as being between 0.002 and 0.05 millimeter. **Sandy** soils have mostly **sand**-sized particles, with diameters between 0.05 and 2.0 millimeters. Rocks larger than 2.0 millimeters are regarded as pebbles, gravel, or rock fragments and technically are not soil particles.

The proportion of particle sizes determines a soil's texture. For example, a soil composed of 50% silt-sized particles, 45% clay, and

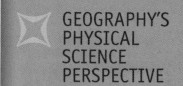
# Basic Soil Analysis

After studying just one chapter on soils, no one would expect you to be able to do a detailed soil analysis. However, there are some basic observations that anyone can perform to better understand a few properties of a local soil. No equipment is required to make analyses; only visual and hands-on examinations are necessary.

## Soil Color

Soil color can hold clues to the composition and/or the formation processes of that soil. Figure 12.6 shows the Munsell color book, a standard guide for matching and recognizing precise colors of soil types. The book includes common soil colors, but each color can also appear in a wide variety of tones.

**Red:** Reddish soil usually indicates that oxidation has been an active process—oxygen has chemically reacted with the soil minerals. Red also indicates that iron is in the soil. Just like rusting iron, many iron-rich minerals turn red when oxidized. The formula for this process is $FeO + O_2 \rightarrow Fe_2O_3$ (ferrous oxide + oxygen becomes hematite, a reddish iron oxide).

**Blue/Silver/Gray:** These tones mean that the soil has likely been reduced; in other words, oxygen has been removed from the soil.

$Fe_2O_3 \rightarrow FeO + O_2$ (the previous formula in reverse)

**White:** Usually denotes that calcium carbonate ($CaCO_3$) or salts (such as NaCl) may be present in the soil.

**Black:** A very dark color may indicate a high amount of organic material present in the soil.

More sophisticated field or laboratory analyses are required for absolute identification, but these examples will allow good working hypotheses for soil characteristics represented by colors.

## Soil Texture

The particle sizes in a soil determine its texture. Soil texture is a property that you can feel, and your fingers can help in the analysis. Sand-sized particles can be easily recognized because they feel gritty to the touch. Wetting the soil and working it with your hands can help in this process. If the sample is not gritty but is smooth to the touch, then the soil contains silt or clay. If the sample feels sticky and you can squeeze a small soil sample into a ribbon (like with modeling clay), then clay-sized particles are abundant. Actual percentages of particle sizes in a soil sample are best established in a laboratory.

## Soil Structure

The shape of clumps that a soil makes when it is broken apart is called structure and can be examined by breaking up a handful of soil. The peds (or small clumps of soil) may take on some distinctive shapes. Although the peds may form a variety of shapes, some of the more common are granular (denoting a presence of sand) and platy (showing a presence of clay). Other soil structures, such as blocky, columnar, or prismatic, are shown in Figure 12.9.

These simple procedures will not yield a complete analysis of a soil sample, but they can certainly be the first steps in the process. It is interesting to note that pedologists (soil scientists) in the field perform many of these same procedures.

© Jeff Vanuga/USDA Natural Resources Conservation Service

Soil analysis in the field is a hands-on process.

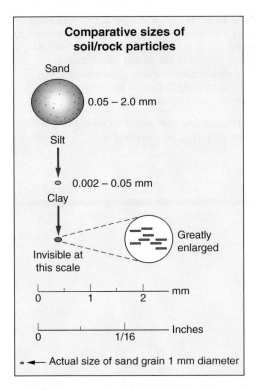

### Comparative sizes of soil/rock particles

Sand
0.05 – 2.0 mm

Silt
0.002 – 0.05 mm

Clay

Invisible at this scale

Greatly enlarged

0      1      2      mm

0          1/16          Inches

Actual size of sand grain 1 mm diameter

● **FIGURE 12.7** Particle sizes in soil. *Sand, silt,* and *clay* are terms that refer to the size of these particles for scientific and engineering purposes. Here, the sizes of each can be compared. Clay particles are tiny, sheetlike particles that cannot be seen.

5% sand would be identified as a silty clay. A triangular graph ( ● Fig. 12.8) is used to discern different classes of soil texture based on the plot of percentages for each **soil grade** (as sand, silt, and clay are called) within each class. Point A within the silty clay class represents the example just given. A second soil sample (B) that is 20% silt, 30% clay, and 50% sand would be referred to as a sandy clay loam. **Loam** soils, which occupy the central areas of the triangular graph, are soils with a mix of the three grades (sizes) of soil particles without any size being greatly dominant. It is interesting to note that loam soils are generally best suited for supporting vegetation growth.

Soil texture helps determine a soil's capacity to retain moisture and air that are necessary for plant growth. Soils with a higher proportion of larger particles tend to be well aerated and allow water to **infiltrate** (seep through) the soil quickly—sometimes so quickly that plants are unable to use the water. Clay soils present the opposite problem because they retard water movement, becoming waterlogged and deficient in air. Aeration of the soil is an important process in cultivation, and plowing a soil opens its structure and increases its air content.

## Structure

In most soils, particles clump into distinctive masses known as **soil peds,** which give a soil a distinctive structure. Soil structure influences a soil's **porosity**—the amount of space that may contain fluids. Soil structure also influences **permeability**—the rate at which fluids such as water can pass through. Permeability is usually greatest in sandy soils, and porosity is usually greatest in clayey soils. Both of these factors control soil drainage and the available moisture in a soil. Soils with similar textures may have different structures and vice versa.

Soil structure can be influenced by outside factors such as moisture regime and the nutrient cycles that plants use to interchange chemicals with the soil, keeping certain ones in the system while others are leached away. We have all seen the structural change in soils that occurs when soils are wet compared to when they are dry. Human activities also influence soil structure through cultivation, irrigation, and fertilization. Fertilizers, and lime or decayed organic debris, encourage clumping of soil particles and the maintenance of clumps. Excess sodium and magnesium have the opposite effect, causing clay soils to become a sticky muck when wet and like concrete when dry. The absence of smaller particles typically hinders the development of a well-defined soil structure.

Scientists classify soil structures according to their form. These range from columns, prisms, and angular blocks to nutlike spheroids, laminated plates, crumbs, and granules ( ● Fig. 12.9). Soils with massive and fine structures tend to be less useful than aggregates of intermediate size and stability, which permit good drainage and aeration.

## Acidity and Alkalinity

An important aspect of soil chemistry is a soil's departure from neutrality toward either acidity or alkalinity (baseness). Levels of acidity or alkalinity are measured on the **pH scale** of 0 to 14. A pH reading indicates the concentration of reactive hydrogen ions present. The pH scale is logarithmic, meaning that each change in a whole pH number represents a tenfold change. It is also an inverted scale—a lower pH means a greater amount of hydrogen ions present (higher acidity). Low pH values indicate an acid soil, and high pH indicates alkaline conditions ( ● Fig. 12.10).

Soil acidity or alkalinity helps determine the available nutrients that affect plant growth. Plants absorb nutrients that are dissolved in liquid. However, soil water that lacks some degree of acidity has little ability to dissolve these nutrients. As a result, even though nutrients are in the soil, plants may not have access to them.

Most complex plants will grow only in soils with levels between pH 4 and pH 10, although the optimum pH for vegetation growth varies with the plant species. Around the world, vegetation has evolved in and adapted to a variety of climates and soil environments, both of which can affect soil pH. Certain species tolerate alkaline soils, and others thrive under more acid conditions.

Leaching caused by high rainfall gradually replaces soil elements such as sodium (Na), potassium (K), magnesium (Mg), and calcium (Ca) with hydrogen. Falling rain picks up atmospheric carbon dioxide and becomes slightly acidic: $H_2O + CO_2 = H_2CO_3$ (carbonic acid), and this is one reason why desert soils tend to be alkaline and soils in humid regions tend to be acidic

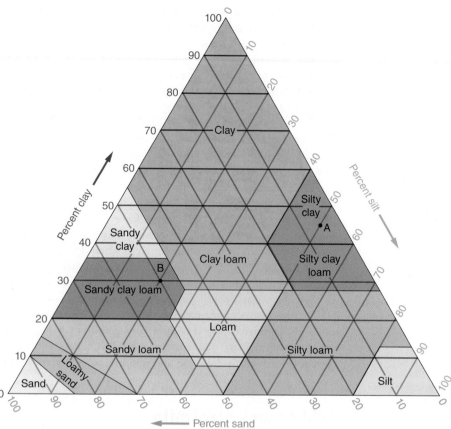

*What would a soil that contains 40% sand, 40% silt, and 20% clay be classified as?*

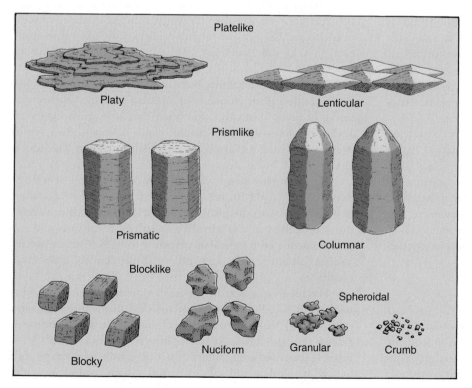

● **FIGURE 12.9** This guide to classifying the structure of a soil on the basis of soil peds can be used to help determine other characteristics of a soil.

*How does soil structure affect a soil's usefulness or suitability for agriculture?*

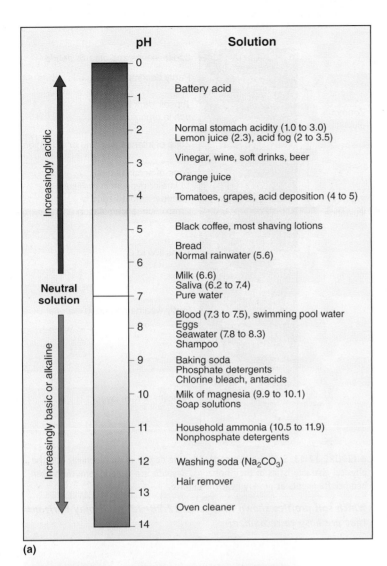

**pH** | **Solution**

- 0
- 1 — Battery acid
- 2 — Normal stomach acidity (1.0 to 3.0) / Lemon juice (2.3), acid fog (2 to 3.5)
- 3 — Vinegar, wine, soft drinks, beer / Orange juice
- 4 — Tomatoes, grapes, acid deposition (4 to 5)
- 5 — Black coffee, most shaving lotions
- 6 — Bread / Normal rainwater (5.6)
- 7 — Milk (6.6) / Saliva (6.2 to 7.4) / Pure water
- 8 — Blood (7.3 to 7.5), swimming pool water / Eggs / Seawater (7.8 to 8.3) / Shampoo
- 9 — Baking soda / Phosphate detergents / Chlorine bleach, antacids
- 10 — Milk of magnesia (9.9 to 10.1) / Soap solutions
- 11 — Household ammonia (10.5 to 11.9) / Nonphosphate detergents
- 12 — Washing soda ($Na_2CO_3$)
- 13 — Hair remover
- 14 — Oven cleaner

Increasingly acidic / Neutral solution / Increasingly basic or alkaline

**(a)**

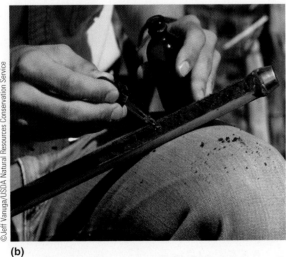

**(b)**

● **FIGURE 12.10** (a) The pH scale of acidity, neutrality, and alkalinity. The degree of acidity or of alkalinity, called pH, can be easily understood when numbers on the scale are linked to common substances. Low pH means acidic, and high pH means alkaline; a reading of 7 is neutral. (b) Alkalinity in a soil can be tested in the field with drops of a dilute acid. If the soil fizzes in the acid solution, alkalinity is typically high.

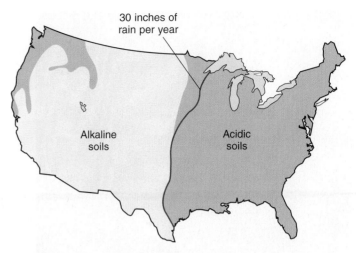

● **FIGURE 12.11** The distribution of alkaline and acidic soils in the United States is generally related to climate. Soils in the East tend to be acidic and those in the West usually are alkaline. The dividing line corresponds fairly well with the 30-inch annual precipitation isohyet.

*Other than climate, what environmental factors might cause this east–west variation, and why are some places west of the 30-inch line acidic?*

( ● Fig. 12.11). The humus content in the soils of humid areas also contributes to higher soil acidity.

To correct soil alkalinity, common in the arid regions, and to make the soil more productive, the soil can be flushed with irrigation water. Strongly acidic soils are also detrimental to plant growth. In acidic soils, soil moisture dissolves nutrients, but they may be leached away before plant roots can absorb them. Soil acidity can generally be corrected by adding lime to the soil. In addition to affecting plant growth, soil acidity or alkalinity also affects microorganisms in the soil. Microorganisms are highly sensitive to a soil's pH, and each species has an optimum environmental setting.

# Development of Soil Horizons

Soil development begins when plants and animals colonize rocks or deposits of rock fragments, the **parent material** on which soil will form. Once organic processes begin among mineral particles or rock fragments, differences begin to develop from the surface down through the parent material.

Initially, vertical differences result from the surface accumulation of organic litter and the removal of fine particles and dissolved minerals from upper layers by percolating water that deposits these materials at a lower level. The vertical cross-section of a soil from the surface down to the parent material is called a **soil profile** ( ● Fig. 12.12). Examining soil profiles and the vertical differences they contain is important to recognizing different soil types and how those soils developed. As climate, vegetation, animal life, and characteristics of the land surface affect soil formation over time, this vertical differentiation becomes more and more apparent.

● FIGURE 12.12 A soil profile is examined by digging a pit with vertical walls to clearly show variations in color, structure, composition, and other characteristics that occur with depth. This soil is in a grassland region of northern Minnesota.

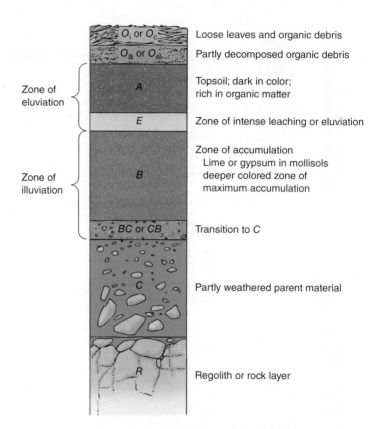

● FIGURE 12.13 Soils are categorized by the degree of development and the physical characteristics of their horizons. *Regolith* is a generic term for broken bedrock fragments at or very near the surface.

*Which soil profiles shown in Figures 12.28–12.31 display horizons that are easy to recognize?*

## Soil Horizons

Within their soil profiles, well-developed soils typically exhibit several distinct layers, called **soil horizons,** that are distinguished by their physical and chemical properties. Soils are classified largely on the differences in their horizons and in the processes responsible for those differences ( ● Fig. 12.13). Soil horizons are designated by a set of letters that refer to their composition, dominant process, and/or position in the soil profile.

At the surface, but only in locations where there is a sufficient cover of decomposed vegetation litter, there will be an O *horizon.* This is a layer of organic debris and humus; the "O" designation refers to this horizon's high organic content. Immediately below is the *A horizon,* commonly referred to as "topsoil." In general, the *A* horizon is dark because it contains decomposed organic matter. Beneath the *A* horizon, certain soils have a lighter-colored *E horizon,* named for the action of strong eluvial processes. Below this is a zone of accumulation, the *B horizon,* where much of the materials removed from the *A* and *E* horizons are deposited. Except in soils with a high organic content that has been mixed vertically, the *B* horizon generally has little humus.

The *C horizon* is the weathered parent material from which the soil has developed—either fragments of the bedrock, or deposits of rock materials that were transported to the site by water, wind, glacial, or other surface process.

The lowest layer, sometimes called the *R horizon,* is unchanged parent material, either bedrock or transported deposits of rock fragments. Certain horizons in some soils may not be as well developed as others, and some horizons may be missing altogether. Because soils and the processes that form them vary widely and can be transitional between horizons, the horizon boundaries may be either sharp or gradual. Variations in color and texture within a horizon are also not unusual.

## Factors Affecting Soil Formation

Because of the great variety among the components of soils and the processes that affected them, no two soils are identical in all of their characteristics. One important factor is rock *weathering,* which refers to the many natural processes that break down rocks into smaller fragments (weathering will be discussed in detail in Chapter 15). Chemical reactions can cause rocks and minerals to decompose, and physical processes also cause the breakup of rocks. Just as statues, monuments, and buildings become "weather-beaten" over time, rocks exposed to the elements eventually break up and decompose.

Hans Jenny, a distinguished soil scientist, observed that soil development was a function of *climate, organic matter, relief, parent material,* and *time*—factors that are easy to remember by their initials arranged in the following order: **Cl, O, R, P, T.** Among these factors, parent material is distinctive because it is the raw material. The other factors influence the type of soil that forms from the parent material.

## Parent Material

All soil contains weathered rock fragments. If these weathered rock particles have accumulated in place—through the physical and chemical breakdown of bedrock directly beneath the soil—we refer to the fragments as **residual parent material.**

If the rock fragments that form a soil have been carried to the site and deposited by streams, waves, winds, gravity, or glaciers, this mass of deposits is called **transported parent material.** The development and action of organic matter through the life cycles of organisms and the climatic conditions are primarily responsible for changing the fragmented rocks or other parent material into a soil.

Parent material influences the characteristics of a soil in varying degrees. Some parent materials, such as a *sandstone* that contains extremely hard and resistant sand-sized fragments, are far less subject to weathering than others. Soils that develop from weathering-resistant rocks tend to have a high level of similarity to their parent materials. If the bedrock is easily weathered, the soils that develop tend to be more similar to soils in other regions that have a similar climate than to those of comparable parent materials, which formed in a different climate.

On a global basis, climate and the associated plant communities produce greater variations in soil characteristics than do parent materials. Soil differences that are related to variations in parent material are most visible on a local level.

As a soil develops, the influence of parent material on its characteristics diminishes. Given the same soil-forming conditions, recently developed soils will show more similarity to its parent material, compared to a soil that has developed over a long time.

Many of the chemicals and nutrients in a soil reflect the composition of its parent material ( ● Fig. 12.14). For example, calcium-deficient parent materials will produce soils that are low in calcium, and its natural fauna and plant cover will consist of the types that require little calcium. Likewise, a parent material with a high aluminum content will produce a soil that is rich in aluminum. In fact, the ore of aluminum is bauxite, found in tropical soils where it has been concentrated by intense leaching away of the other bases.

The particle sizes that result from the breakdown of parent material are a prime determinant of a soil's texture and structure. A rock material such as sandstone, which contains little clay and

● **FIGURE 12.14** Despite strong leaching under a wet tropical climate, Hawaiian soils remain high in nutrients because their parent material is of recent volcanic origin.

*What other parent materials provide the basis for continuously fertile soils in wet tropical climates?*

weathers into relatively coarse fragments, will produce a soil of coarse texture. Parent materials are also an important influence on the availability of air and water to a soil's living population.

## Organic Activity

Plants and animals affect soil development in many ways. The life processes of plants growing in a soil are as important as its microorganisms—the microscopic plants and animals that live in a soil.

Generally, a dense vegetative cover protects a soil from being eroded by running water or wind. Forests form a protective canopy and produce surface litter, which keeps rain from beating directly on the soil and increases the proportion of rainwater entering the soil rather than running off its surface. Variations in vegetation species and density of cover can also affect the evapotranspiration rates. A sparse vegetative cover will allow greater evaporation of soil moisture, and dense vegetation tends to maintain soil moisture.

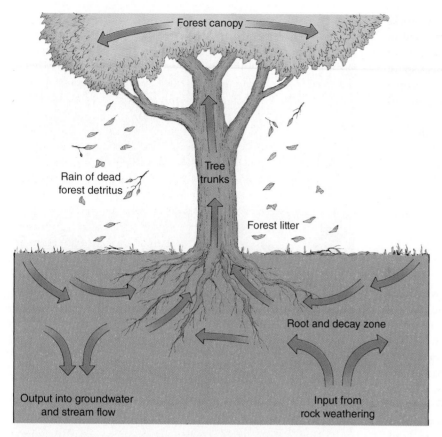

● **FIGURE 12.15** The nutrient cycle in a forest. Trees take up nutrients from the soil through soil water absorbed into their root systems. Nutrients are supplied by the breakdown of rocks and minerals and by leaf and other organic litter.

The characteristics of a plant community affect the nutrient cycles that are involved in soil development ( ● Fig. 12.15). As plants die and decompose or leaves fall to the ground, nutrients are returned to the soil. Soils, however, can become impoverished if soluble nutrients that are not used by plants are lost through leaching. The roots of plants help to break up the soil structure, making it more porous, and roots also absorb water and nutrients from the soil.

Leaves, bark, branches, flowers, and root networks contribute to the organic composition of soil through litter and through the remains of dead plants. The organic content of soil depends on its associated plant life. For example, a grass-covered prairie supplies much more organic matter than the thin vegetative cover of desert regions. There is some question, however, as to whether forests or grasslands (with their thick root networks and annual life cycle) furnish the soil with greater organic content. Many of the world's grassland regions, like the North American prairies, provide some of the world's most fertile soils for cultivation in part because of the high amount of organic matter that a grass cover generates.

In terms of their contribution to soil formation, bacteria are perhaps the most important microorganisms that live in soils. Bacteria break down organic matter, humus, and the debris of living things into organic and inorganic components, allowing the formation of new organic compounds that promote plant growth. It has been suggested that the number of bacteria, fungi, and other microscopic plants and animals living in a soil may be 1 billion per gram (a fifth of a teaspoon) of soil. The activities and remains of these microorganisms, minute though they are individually, add considerably to the organic content of a soil.

Earthworms, nematodes, ants, termites, wood lice, centipedes, burrowing rodents, snails, and slugs also stir up the soil, mixing mineral components from lower levels with organic components from the upper portion. Earthworms contribute greatly to soil development because they take soil in, pass it through their digestive tracts, and excrete it in casts. The process not only helps mix the soil, but also changes the texture, structure, and chemical qualities of the soil. In the late 1800s, Charles Darwin estimated that earthworm casts produced in a year would equal as much as 10 to 15 tons per acre. As for the number of earthworms, a study suggested that the total weight of earthworms beneath a pasture in New Zealand equaled the weight of the sheep grazing above them.

## Climate

Chapters 9 and 10 demonstrated that, on a world regional scale, climate is a major factor in soil formation. Of course, if the climate is the same in a region where the soils vary, other factors must be responsible for the local variation. Soil differences that are apparent at a local level tend to reflect the influence of factors such as parent materials, land surface configurations, vegetation types, and time.

Temperature directly affects soil microorganism activity, which in turn affects the decomposition rates of organic matter. In hot equatorial regions, intense activities by soil microorganisms preclude thick accumulations of organic debris or humus. ● Figure 12.16 shows that the amounts of organic matter and humus in a soil increase toward the middle latitudes and away from polar regions and the tropics. In the mesothermal and microthermal climates *(C and D),* microorganism activity is slow enough to allow decaying organic matter and humus to accumulate in rich layers. Moving poleward into colder regions, retarded microorganism activity and limited plant growth tends to result in thin accumulations of organic matter.

Chemical activity increases and decreases directly with temperature, given equal availability of moisture. As a result, parent materials of soils in hot, humid equatorial regions are altered to a far greater degree by chemical means than are parent materials in colder zones.

Temperature affects soil indirectly through its influence on vegetation associations that are adapted to certain climatic regimes. Soils generally reflect the character of plant cover because of nutrient cycles that tend to keep both vegetation and soil in chemical equilibrium. The combined effects of vegetative cover and the climatic regime tend to produce soil profiles and charac-

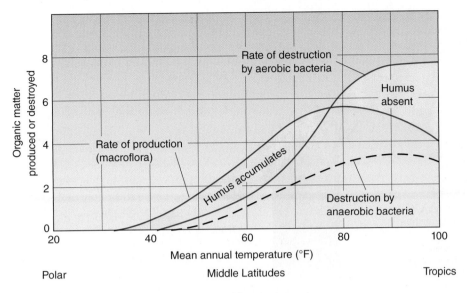

● **FIGURE 12.16** The relationship of temperature to production and destruction of organic matter in the soil.

*What range of mean annual temperatures is most favorable for the accumulation of humus?*

teristics that tend to share certain characteristics among different regions that have similar climates and vegetation associations ( ● Fig. 12.17).

Moisture conditions affect the development and character of soils more directly than any other climatic factor. Without precipitation and the soil water it provides, terrestrial plant life would be impossible. Ample precipitation supports plant growth that can greatly increase the organic content and thereby the fertility of a soil. However, extremely high rainfall will cause leaching of nutrients and a relatively infertile soil.

Gravitational and capillary water have pronounced effects on soil development, structure, texture, and color. Precipitation is the original source of soil water (disregarding the minor contribution of dew), and the amount of precipitation received affects leaching, eluviation, and illuviation and thereby rates of soil formation and horizon development. The evaporation rate is also an important factor. Salt and gypsum deposits from the upward migration of capillary water are more extensive in hot, dry regions—such as the southwestern United States where evaporation rates are high—than in colder, dry regions (see Fig. 12.17).

## Land Surface Configuration

The slope of the land, its relief, and its *aspect* (the direction it faces) all influence soil development. Steep slopes are generally better drained than gentler ones, and they are also subject to rapid runoff of surface water. As a consequence, there is less infiltration of water on steeper slopes, which inhibits soil development, sometimes to the extent that there will be no soil. In addition, rapid runoff on steep slopes can erode surfaces as fast or faster than soil can develop on them. On gentler slopes, where there is less

runoff and higher infiltration, more water is available for soil development and to support vegetation growth, so erosion is not as intense. In fact, erosion rates in areas of gently rolling hills may be just enough to offset the development of soils. Well-developed soils typically form on land that is flat or has a gentle slope.

*Slope aspect* has a direct effect on microclimates in areas outside of the equatorial tropics. North-facing slopes in the middle and high latitudes of the Northern Hemisphere have microclimates that are cooler and wetter than those on south-facing exposures, which receive the sun's rays at a steeper angle and are therefore warmer and drier. Local variations in soil depth, texture, and profile development result directly from microclimate differences.

Topography, through its effects on vegetation, indirectly influences soil development. Steep slopes prevent the formation of a soil that would support abundant vegetation, and a modest plant cover yields less organic debris for the soil.

## Time

Soils have a tendency to develop toward a state of equilibrium with their environment. A soil is sometimes called "mature" when it has reached such a condition of equilibrium. Young soils are still in the process of developing toward being in equilibrium with their environmental conditions. Mature soils have well-developed horizons that indicate the conditions under which they formed. Young or "immature" soils typically have poorly developed horizons or perhaps none at all ( ● Fig. 12.18).

Another effect of time is that, as soils develop, their influence of their parent material decreases and they increasingly reflect their climate and vegetative environments. On a global scale, climate typically has the greatest influence on soils, provided sufficient time has passed for the soils to become well developed.

The importance of time in soil formation is especially clear in soils developed on transported parent materials. Depositional surfaces are in many cases quite recent in geologic terms and have not been exposed to weathering long enough for a mature soil to develop. Deposition occurs in a variety of settings: on river floodplains where the accumulating sediment is known as *alluvium*; downwind from dry areas where dust settles out of the atmosphere to form blankets of wind-deposited silts, called *loess*; and in volcanic regions showered by ash and covered by lava. Ten thousand years ago glaciers withdrew from vast areas, leaving behind jumbled deposits of rocks, sand, silt, and clay.

Because of the great number and variability of materials and processes involved in the formation of soils, there is no fixed amount of time that it takes for a soil to become mature. The Natural Resources Conservation Service, however, estimates that it takes about 500 years to develop 1 inch of soil in the agricultural regions of the United States. Generally, though, it takes thousands of years for a soil to reach maturity.

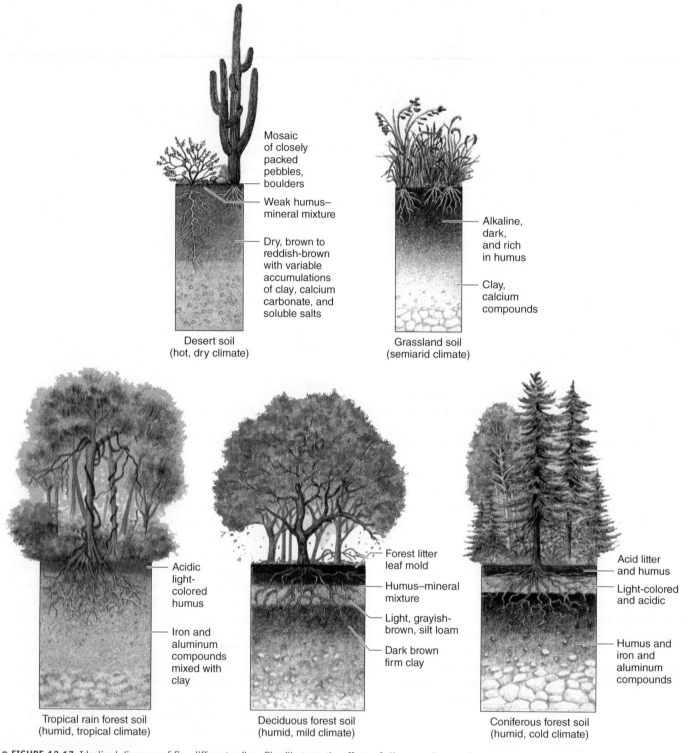

Mosaic of closely packed pebbles, boulders

Weak humus–mineral mixture

Dry, brown to reddish-brown with variable accumulations of clay, calcium carbonate, and soluble salts

Desert soil (hot, dry climate)

Alkaline, dark, and rich in humus

Clay, calcium compounds

Grassland soil (semiarid climate)

Acidic light-colored humus

Iron and aluminum compounds mixed with clay

Tropical rain forest soil (humid, tropical climate)

Forest litter leaf mold

Humus–mineral mixture

Light, grayish-brown, silt loam

Dark brown firm clay

Deciduous forest soil (humid, mild climate)

Acid litter and humus

Light-colored and acidic

Humus and iron and aluminum compounds

Coniferous forest soil (humid, cold climate)

● **FIGURE 12.17** Idealized diagrams of five different soil profiles illustrate the effects of climate and vegetation on the development of soils and their horizons.

*Which two environments produce the most humus, and which two produce the least?*

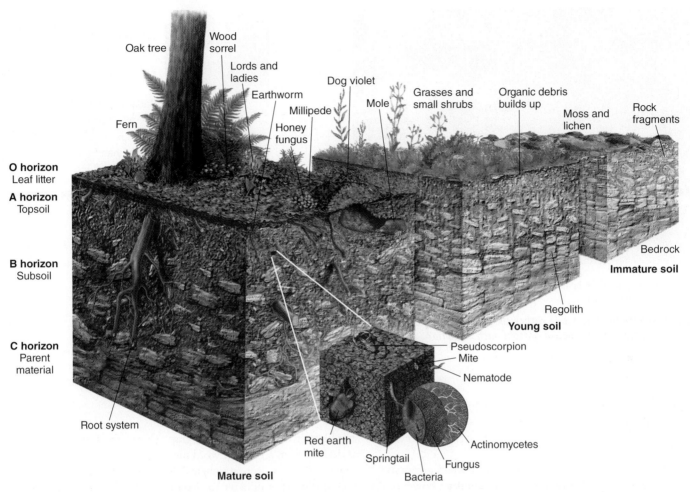

● **FIGURE 12.18** The time that a soil has been developing is important to its composition and physical character. Given enough time and the proper environmental conditions, soils will become more maturely developed with a deeper profile and stronger horizon development.
From Derek Elsom, *Earth*, 1992. Copyright © 1992 by Marshall Editions Developments Limited. New York. Macmillan. Used by permission.

*What major changes occur as the soil illustrated here becomes better developed over time?*

# Soil-Forming Regimes

The characteristics that make major soil types distinctive from one another result from their **soil-forming regimes,** which vary mainly because of differences in climate and vegetation. At the broadest scale of generalization, climate differences produce three primary soil-forming regimes: laterization, podzolization, and calcification.

## Laterization

**Laterization** is a soil-forming regime that occurs in humid tropical and subtropical climates as a result of high temperatures and abundant precipitation. These climatic environments encourage rapid breakdown of rocks and decomposition of nearly all minerals. A soil of this type is known as **laterite,** and these soils are generally reddish in color from iron oxides; the term *laterite* means "brick-like." In tropical areas laterite is quarried for building material ( ● Fig. 12.19).

Despite the dense vegetation that is typical of these climate regions, little humus is incorporated into the soil because the

plant litter decomposes so rapidly. Laterites do not have an *O* horizon, and the *A* horizon loses fine soil particles and most minerals and bases except for iron and aluminum compounds, which are insoluble primarily because of the absence of organic acids ( ● Fig. 12.20). As a result, the topsoil is reddish, is coarse textured, and tends to be porous. In contrast to the *A* horizon, the *B* horizon in a lateritic soil has a heavy concentration of illuviated materials.

In the tropical forests, soluble nutrients released by weathering are quickly absorbed by vegetation, which eventually returns them to the soil where they are reabsorbed by plants. This rapid cycling of nutrients prevents the total leaching away of bases, leaving the soil only moderately acidic. Removal of vegetation permits total leaching of bases, resulting in the formation of crusts of iron and aluminum compounds (laterites), and accelerated erosion of the *A* horizon.

Laterization is a year-round process because of the small seasonal variations in temperature or soil moisture in the humid tropics. This continuous activity and strong weathering of parent material cause some tropical soils to develop to depths of as much as 8 meters (25 ft) or more.

● **FIGURE 12.19** Laterite cut for building stone and stacked along a village road in the state of Orissa, India.

*Why is building with brick or stone rather than wood so important in heavily populated, less developed nations such as India?*

## Podzolization

**Podzolization** occurs mainly in the high middle latitudes where the climate is moist with short, cool summers and long, severe winters. The coniferous forests of these climate regions are an integral part of the podzolization process.

Where temperatures are low much of the year, microorganism activity is reduced enough that humus does accumulate; however, because of the small number of animals living in the soil, there is little mixing of humus below the surface. Leaching and eluviation by acidic solutions remove the soluble bases and aluminum and iron compounds from the *A* horizon ( ● Fig. 12.21). The remain-

● **FIGURE 12.20** Soil profile horizons in laterite, one of three major soil-forming regimes. Laterization is a soil development process that occurs in tropical and equatorial zones that experience warm temperatures year round and wet climates.

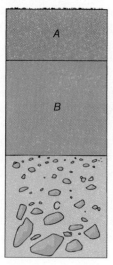

*A* — Little or no organic debris, little silica, much residual iron and aluminum, coarse texture

*B* — Some illuvial bases, much accumulated laterite

*C* — Much of the soluble material lost to drainage

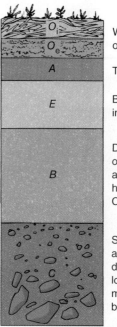

Well-developed organic horizons

*A* — Thin, dark

*E* — Badly leached, light in color, largely Si

*B* — Darker than E; often colorful; accumulations of humus; Fe, Al, N, Ca, Mg, Na, K

*C* — Some Ca, Mg, Na, and K leached down from B is lost to lateral movement of water below water table

● **FIGURE 12.21** Soil profile horizons in a podzol soil, another of the three major soil-forming regimes. Podzolization typically occurs under cool, wet climates in regions of coniferous trees or in boggy environments and is a very acidic soil type.

ing silica gives a distinctive ash-gray color to the *E* horizon (*podzol* is derived from a Russian word meaning "ashy"). The needles that coniferous trees drop are chemically acidic and contribute to the soil acidity. It is difficult to determine whether the soil is acidic because of the vegetative cover or whether the vegetative cover is adapted to the acidic soil.

Podzolization can take place outside the typical cold, moist climate regions if the parent material is highly acidic—for example, on the sandy areas common along the East Coast of the United States. The pine forests that grow in such acidic conditions return acids to the soil, promoting the process of podzolization.

## Calcification

A third distinctive soil-forming regime is called **calcification.** In contrast to both laterization and podzolization, which require humid climates, calcification occurs in regions where evapotranspiration significantly exceeds precipitation. Calcification is important in the climate regions where moisture penetration is shallow. The subsoil is typically too dry to support tree growth, and shallow-rooted grass or shrubs are the primary forms of vegetation. Calcification is enhanced as grasses use calcium, drawing it up from lower soil layers and returning it to the soil when the annual grasses die. Grasses and their dense root networks provide large amounts of organic matter, which is mixed deep into the soil by burrowing animals. Middle-latitude grassland soils are rich in bases and in humus and are the world's most productive agricultural soils. The deserts of the American West generally have no humus, and the

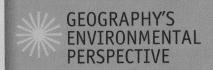

# How Much Good Soil Is There on Earth?

A widely used analogy for understanding the amount of soil that exists on Earth uses an apple to demonstrate that soil, which is adequate for the world's agricultural needs, is a rather limited natural resource. If the apple is cut into quarters, three of those pieces (75%) represent the water bodies on Earth and should be put away. The piece that remains represents the Earth's land area (25%) where soil *could* exist. Half of the remaining piece should be cut off and put away, representing rocky desert, polar, or high mountain regions, where soil does not exist or will not grow much because of harsh climatic conditions. Now there is an eighth (12.5%) of the apple remaining. This eighth of the apple should then be cut into four pieces, and three of them should be put away because they represent areas with local conditions that preclude agriculture (too rocky, steep, wet, and so on) and the areas that we have already covered over with cities, towns, and roads. The remaining apple slice (about 3% of Earth) represents areas with good soils for agriculture to feed the world's population.

## Soil Degradation and Soil Loss

Today even areas with good soils are under continuing threat. It may take 200 to 1000 years or more to develop 2.5 centimeters (1 in.) of soil, but through erosion by water or wind, thousands of years of soil development may be lost in a season or in days. Land degradation by humans is a major cause of soil loss, but impacts from changing climates, particularly desertification, also play a role. Overgrazing, deforestation, overuse of the land, and poor agricultural practices are the major human-related causes of soil loss. The United Nations estimates that, just through erosion, the world loses between 5 and 7 million hectares (12.36–17.30 million acres) of farmland every year. This is an area equal to the size of West Virginia.

In addition to the degradation of soil that is ongoing, farmlands with excellent soils are being taken out of production by urbanization/suburbanization. Many farmlands are attractive to land developers because they are already cleared, the soils are good for lawns and landscaping, and cities and suburbs are expanding into surrounding agricultural lands.

Our planet is losing its arable soil, and the world's population is growing—trends that cannot continue forever. One estimate is that soils are being lost at a rate 17 times faster than the rate at which they form. Many government and private agencies around the world are working to educate people about the critically important problem of soil degradation and to support solutions to minimize human impacts on soils. The problems associated with soil degradation are globally significant, but the solutions are proper conservation and management at local levels.

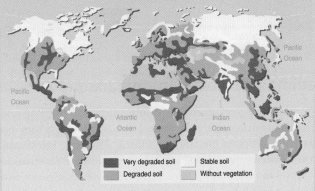

**Soil degradation**

Very degraded soil  Stable soil
Degraded soil  Without vegetation

A world map done by the United Nations shows global patterns—the geography—of soil degradation by various degrees.
Source: UNGP, International Soil Reference and Information Centre (ISRIC), World Atlas of Desertification, 1997. Philippe Rekacewicz, UNEP/GRID-Arendal

*What major geographic factors explain the areas with the highest levels of degradation, and what explains areas with the lowest impact on soils?*

USDA/NRCS/Lynn Betts

These prime farmlands in Iowa are being converted into suburbs as populations grow and towns expand.

*Are there agricultural lands surrounding the place where you live that are currently being converted into housing, commercial, or industrial areas?*

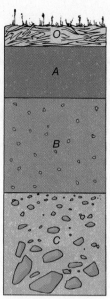

Dark color, granular structure, high content of residual bases

Lighter color, very high content of accumulated bases, caliche nodules

Relatively unaltered, rich in base supply, virtually no loss to drainage water

● **FIGURE 12.22** Soil profile horizons in a calcified soil, formed by the third major soil-forming regime. Calcification is a soil development process that is most prominent in cool to hot subhumid or semiarid climate regions, particularly in grassland regions but also in deserts.

*© State of Victoria, Australia, Department of Primary Industries, 1997*

● **FIGURE 12.23** This shallow-rooted grass is growing in a thin layer of topsoil over a much thicker layer of calcium carbonate called calcrete.

*What precipitation characteristics are associated with the calcification soil-forming process?*

rise of capillary water can leave calcium carbonate ($CaCO_3$) and sodium chloride (NaCl [salt]) at the surface.

In many areas of low precipitation, the air is often loaded with alkali dusts such as calcium carbonate. When calm conditions prevail or when it rains, the dust settles and accumulates in the soil. The rainfall produces an amount of soil water that is just sufficient to translocate these materials to the *B* horizon ( ● Fig. 12.22). Over hundreds to thousands of years, the $CaCO_3$-enriched dust concentrates in the *B* horizon, forming hard layers of *caliche*. Much thicker accumulations called *calcretes* ( ● Fig. 12.23)

*USDA/NRCS/Tim McCabe*

● **FIGURE 12.24** Salinization is indicated by these white deposits on this field in Colorado. Surface salinity has resulted from the upward capillary movement of water and evaporation at the surface causing deposits of salt. The soil cracks also indicate shrinkage caused by evaporative drying of the soil.

*What negative soil effects can result when humans practice irrigated agriculture in regions that experience great evaporation rates?*

form by the upward (capillary) movement of dissolved calcium in groundwater when the water table is near the surface.

## Regimes of Local Importance

Two additional localized soil-forming regimes merit attention. Both characterize areas with poor drainage, although they occur under very different climate conditions. The first, **salinization,** or the concentration of salts in the soil, is often detrimental to plant growth. Salinization occurs in stream valleys, interior basins, and other low-lying areas, particularly in arid regions with high groundwater tables. The high groundwater levels can be the result of water from adjacent mountain ranges, stream flow originating in humid regions, or a wet–dry seasonal precipitation regime ( ● Fig. 12.24). Salinization can also be a consequence of intensive irrigation under arid conditions. Rapid evaporation leaves behind a high concentration of soluble salts and may destroy a soil's agricultural productivity. An extreme example of salinization exists in certain areas of the Middle East, where thousands of years of irrigated agriculture in the desert have made the soils too saline to cultivate today.

Another localized soil regime, **gleization,** occurs in poorly drained areas under cold and wet environmental conditions. Gley soils, as they are called, are typically associated with peat bogs where the soil has an accumulation of humus overlying a blue–gray layer of thick, gummy, water-saturated clay. In poorly drained regions that were formerly glaciated, such as northern Russia, Ireland, Scotland, and Scandinavia, peat has long been harvested and used as a source of energy.

# Soil Classification

Soils, like climates, can be classified by their characteristics and mapped by their spatial distributions. In the United States the Soil Survey Division of the Natural Resources Conservation Service (NRCS), a branch of the Department of Agriculture, is responsible for soil classification (termed **soil taxonomy**). As with any classification system, the methods and categories are continually being updated and refined.

Soil classifications are published in **soil surveys,** books that outline and describe the kinds of soils in a region and include maps that show the distribution of soil types, usually at the county level. These documents, available for most parts of the United States, are useful references for factors such as soil fertility, irrigation, and drainage.

## The NRCS Soil Classification System

The NRCS soil classification system is based on the development and composition of soil horizons. The largest division in the classification of soils is the **soil order,** of which 12 are recognized by the NRCS. To provide greater detail, soil orders can also be subdivided into suborders, great groups, subgroups, families, and series. More than 10,000 soil series have been recognized in the United States.

The NRCS system of soil types uses names derived from root words of classical languages such as Latin, Arabic, and Greek to refer to the different soil categories. The names, like the system, are precise and consistent and were chosen to describe the characteristics that distinguish one soil from another. Some soil orders reflect regional climate conditions; however, other soil orders reflect the recency or type of parent material, so the distribution of these soils does not conform to climate regions.

When examining a soil for classification under the NRCS system, particular attention is paid to characteristic horizons and textures. Some of these horizons are below the surface (**subsurface horizons**); others, called **epipedons,** are surface layers that usually exhibit a dark shading associated with organic material (humus). Examples of some of the more common horizons, illustrating how names were chosen to represent actual soil properties, are found in Table 12.1.

## NRCS Soil Orders

The 12 soil orders are based on a variety of characteristics and processes that can be recognized by examining a soil and its profile. The soil descriptions that follow are based on the sequence shown in ● Figure 12.25, which illustrates the links between climate and soils. ● Figure 12.26 is a map showing the distribution of dominant soil orders in the United States. Global soils based on the NRCS classification of soil orders are shown

**TABLE 12.1**
**Common Soil Horizons (NRCS Soil Classification System)\***

**Oxic horizon (from *oxygen*)**
Subsurface horizon, in low-elevation tropical and subtropical climates, that contains oxides of iron and aluminum.

**Argillic horizon (from Latin: *argilla,* clay)**
Layer formed beneath the *A* horizon by illuviation that contains a high content of accumulated clays.

**Ochric epipedon (from Greek: *ochros,* pale)**
A surface horizon that is light in color and either very low in organic matter or very thin.

**Albic horizon (from Latin: *albus,* white)**
An *A2* horizon, sandy and light-colored as a result of the removal of clay and iron oxides, that is above a spodic horizon.

**Spodic horizon (from Greek: *spodos,* wood ash)**
Beneath an *A2* horizon, this layer is dark-colored from illuviated humus and oxides of aluminum and/or iron.

**Mollic epipedon (from Latin: *mollis,* soft)**
A dark-colored surface layer with a high content of basic substances (calcium, magnesium, potassium).

**Calcic horizon (from *calcium*)**
A subsurface horizon that is rich in accumulated calcium carbonate or magnesium carbonate.

**Salic horizon (from *salt*)**
A soil layer, common in desert basins, that is at least 6 inches thick and contains at least 2% salt.

**Gypsic horizon (from *gypsum*)**
A subsurface soil horizon that is rich in accumulated calcium sulfate (gypsum).

\*This table includes only some of the more common horizons.

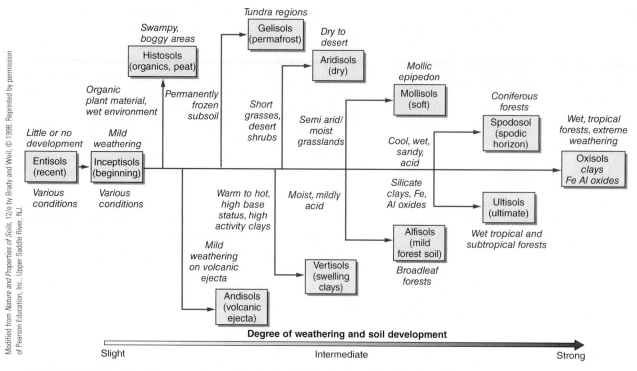

● **FIGURE 12.25** The Natural Resources Conservation Service (NRCS) soil orders. The soil orders of the NRCS can be linked to the parent materials, climate, and vegetation of the region in which they formed. The linkages form a treelike pattern, as seen here.

*How is degree of weathering related to climatic characteristics?*

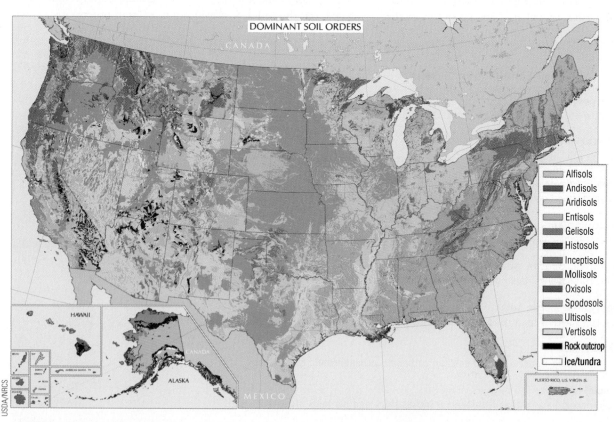

● **FIGURE 12.26** The distribution of soils in the United States according to the National Resource Conservation Service's classification.

*What kind of soil dominates the place where you live, according to this map?*

in ● Figure 12.27. Frequent comparison of Figure 12.27 with the map of world climates in Figure 8.6 will illustrate the relationships between the global distributions of soil and climate.

**Entisols** are soils that have undergone little or no soil development and lack horizons because they have only recently begun to form ( ● Fig. 12.28a). They are often associated with the continuing erosion of sloping land in mountainous regions or with the frequent deposition of alluvium by flooding or in areas of wind-blown sand.

**Inceptisols** are young soils with weak horizon development (Fig. 12.28b). The processes of *A* horizon depletion (eluviation) and *B* horizon deposition (illuviation) are just beginning, usually because of a very cold climate, repeated flood-related deposition, or a high rate of soil erosion. In the United States inceptisols are most common in Alaska, the lower Mississippi River floodplain, and the western Appalachians. Globally, inceptisols are especially important along the lower portions of the great river systems of South Asia, such as the Ganges-Brahmaputra, the Irrawaddy, and the Mekong. In these areas, sediments associated with periodic flooding constantly enrich inceptisols and form the basis for agriculture that supports millions of people.

**Histosols** develop in poorly drained areas, such as swamps, meadows, or bogs, as a product of gleization (Fig. 12.28c). They are largely composed of decomposing plant material. The water-logged soil conditions deprive bacteria of the oxygen necessary to decompose the organic matter. Although histosols may be found in low areas with poor drainage at all latitudes, they are most common in tundra areas or in recently glaciated, high-latitude locations such as Scandinavia, Canada, Ireland, and Scotland. Histosols in the subpolar latitudes are commonly acidic and only suitable for special bog crops such as cranberries. Histosols are the primary source of peat, which is a fuel source in some regions, and also is used in landscaping.

**Andisols** are soils that develop on volcanic parent materials, usually *volcanic ash,* the dust-sized particles emitted by volcanoes ( ● Fig. 12.29a). Because many of these soils are replenished by eruptions, they are often fertile. Intensive agriculture atop andisols supports dense populations in the Philippines, Indonesia, and the West Indies. In the United States, andisols are most common on the slopes of and downwind from volcanoes in the Pacific Northwest and to a lesser extent in Hawaii and Alaska.

**Gelisols** are soils that experience frequent freezing and thawing of the ground, above *permafrost,* permanently frozen subsoil (Fig. 12.29b). When soil freezes, the ice that forms takes up 9% more space than the liquid water that it replaces. To accommodate the increased space taken up by ice, the soil and the particles in it are pushed upward and outward, away from forming ice cores. When the surface soil thaws, gravity pulls the waterlogged ground back downward. Repeated cycles of freezing and thawing mix and churn the upper soil in a process called *cryoturbation*—mixing *(turbation)* related to freezing *(cryo).* Only the upper part of the soil undergoes freeze–thaw cycles. Permafrost does not permit soil water to percolate downward, so gelisol soils are typically water saturated when they are not frozen at the surface.

Gelisols occur in tundra and subarctic climate regions where soil development tends to be slow because chemical processes operate slowly in cold environments. This soil type is found in north and central Alaska and Canada, in Siberia, and in high-altitude tundra areas.

**Aridisols** are soils of desert regions that develop primarily under conditions where precipitation is much less than half of potential evaporation (Fig. 12.29c). Consequently, most aridisols reflect the calcification process. Where groundwater tables are high, evidence of salinization may also be present.

Although aridisols tend to have weak horizon development because of limited water movement in the soil, there is often a subsurface accumulation of calcium carbonate (calcic horizon), salt (salic horizon), or calcium sulfate (gypsic horizon). Soil humus is minimal because vegetation is sparse in deserts; therefore, aridisols are often light in color. Aridisols are usually alkaline, but because few nutrients have been leached, they can support productive agriculture if irrigated to reduce the pH and salinity. Geographically, aridisols are the most common soils on Earth because deserts cover such a large portion of the land surface.

**Vertisols** are typically found in regions of strong seasonality of precipitation such as the tropical wet and dry climates ( ● Fig. 12.30a). In the United States they are most common where the parent materials produce clay-rich soils. The combination of clayey soils in a wet and dry climate leads to the drying of the soil and consequent shrinkage that forms deep cracks during the dry season, followed by expansion of the soil during the wet season. The constant shrink–swell process disrupts horizon formation to the point that soil scientists often describe vertisols as "self-plowing" soils. Vertical soil movement may damage highways, sidewalks, foundations, and basements that are built on shrink–swell soils. Vertisols are dark-colored, are high in bases, and contain considerable organic material derived from the grasslands or savanna vegetation with which they are normally associated. Although they harden when dry and become sticky and difficult to cultivate when swollen with moisture, vertisols can be agriculturally productive.

**Mollisols** are most closely associated with grassland regions and are among the best soils for sustained agriculture (Fig. 12.30b). Because they are located in semiarid climates, mollisols are not heavily leached and they have a generous supply of bases, especially calcium. The characteristic horizon of a mollisol is a thick, dark-colored surface layer rich in organic matter from the decay of abundant root material. Grasslands and associated mollisols served as the grazing lands for countless herds of antelope, bison, and horses. Before the invention of the steel plow, the thick root material of grasses made this soil nearly uncultivable in the United States and thus led to the widespread public image of the Great Plains as a "Great American Desert."

In regions of adequate precipitation, such as the tall-grass prairies of the American Midwest, the combination of soils and climate is unexcelled for agriculture. In areas of lesser precipitation, periodic drought is a constant threat, and the temptation of fertile soils was the downfall of many farmers prior to the advent of center pivot irrigation.

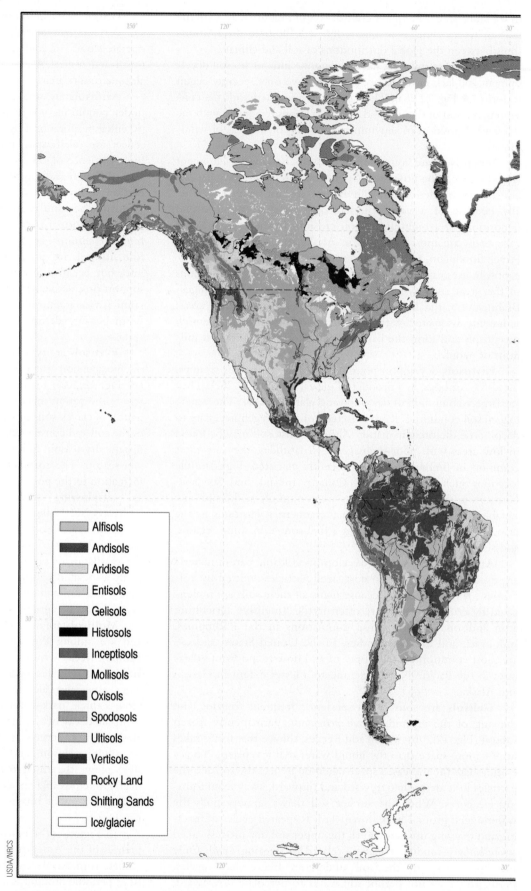

Alfisols
Andisols
Aridisols
Entisols
Gelisols
Histosols
Inceptisols
Mollisols
Oxisols
Spodosols
Ultisols
Vertisols
Rocky Land
Shifting Sands
Ice/glacier

USDA/NRCS

● **FIGURE 12.27** The global distribution of soils by Natural Resources Conservation Service soil orders.

*How do these patterns resemble the spatial distribution of world climates?*

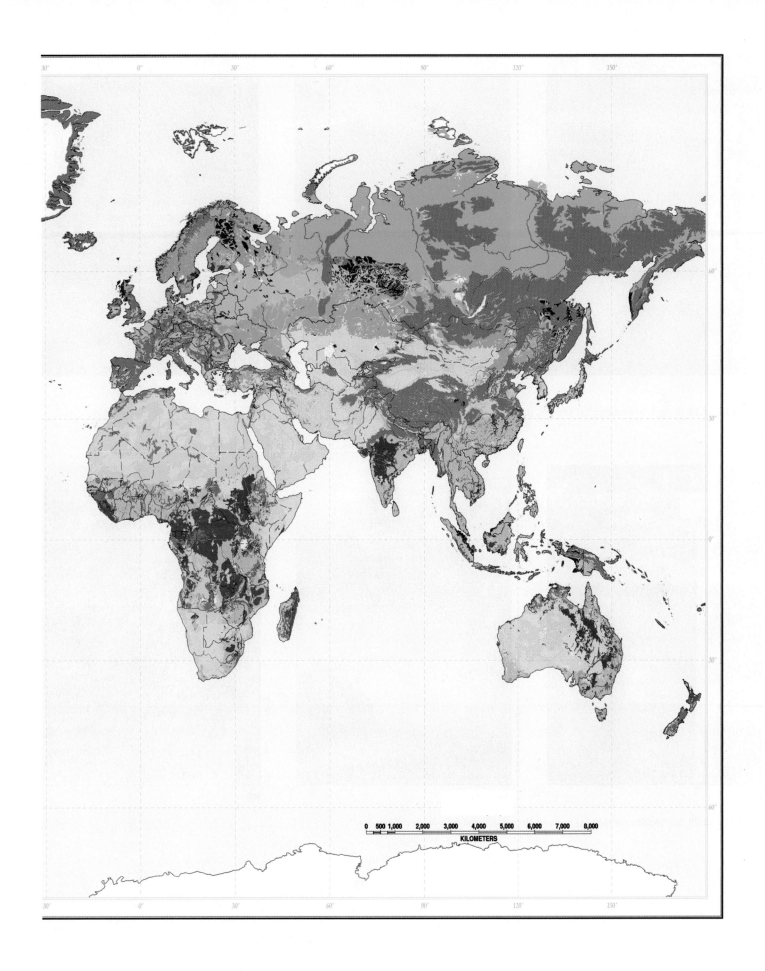

(a)

(b)

(c)

● **FIGURE 12.28** Soil profile examples: (a) entisols, (b) inceptisols, and (c) histosols.

(a)

(b)

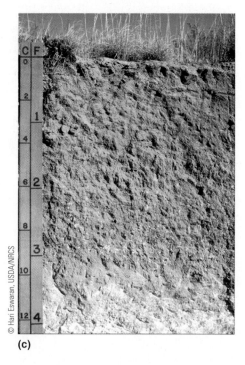
(c)

● **FIGURE 12.29** Soil profile examples: (a) andisols, (b) gelisols, and (c) aridisols.

(a)　　　　　　　　　　　　(b)　　　　　　　　　　　　(c)

● **FIGURE 12.30** Soil profile examples: (a) vertisols, (b) mollisols, and (c) alfisols.

**Alfisols** occur in a wide variety of climate settings. They are characterized by a subsurface clay horizon (argillic *B* horizon), a medium to high base supply, and a light-colored ochric epipedon (Fig. 12.30c). The five suborders of alfisols reflect climate types and exemplify the hierarchical nature of the classification system: *aqualfs* are seasonally wet and can be found in mesothermal areas such as Louisiana, Mississippi, and Florida; *boralfs* are found in moist, microthermal climates such as Montana, Wyoming, and Minnesota; *udalfs* are common in both microthermal and mesothermal climates that are moist enough to support agriculture without irrigation, such as Wisconsin, Ohio, and Tennessee; *ustalfs* are found in mesothermal climates that are intermittently dry, such as Texas and New Mexico; and *xeralfs* are found in California's Mediterranean climate, which is characterized by wet winters and long, dry summers.

Because of their abundant bases, alfisols can be very productive agriculturally if local deficiencies are corrected: irrigation for the dry suborders and properly drained fields for the wet suborders.

**Spodosols** are most closely associated with the podzolization soil-forming process. They are readily identified by their strong horizon development ( ● Fig. 12.31a). There is often a white or light-gray *E* horizon (albic horizon) covered with a thin, black layer of partially decomposed humus and underlain by a colorful *B* horizon enriched in relocated iron and aluminum compounds (spodic horizon).

Spodosols are generally low in bases and form in porous substrates such as glacial drift or beach sands. In New England and Michigan, spodosols are also acidic. In these regions, and in simi-

lar regions in northern Russia, Scandinavia, and Poland, only a few types of agricultural plants, such as cucumbers and potatoes, can tolerate the microthermal climates and sandy, acidic soils. Consequently, the cuisine of these regions directly reflects the spodosols that dominate the areas.

**Ultisols,** like spodosols, are low in bases because they develop in moist or wet regions. Ultisols are characterized by a subsurface clay horizon (argillic horizon) and are often yellow or red because of residual iron and aluminum oxides in the *A* horizon (Fig. 12.31b). In North America the ultisols are most closely associated with the southeastern United States. When first cleared of forests, these soils can be agriculturally productive for several decades, but a combination of high rainfall with the associated runoff and erosion from the fields decreases the natural fertility of the soils. Ultisols remain productive only with the continuous application of fertilizers. Today, forests cover many former cotton and tobacco fields of the southeastern United States because of a reduction in soil fertility and extensive soil erosion.

**Oxisols** have developed over long periods in tropical regions with high temperatures and heavy annual rainfall. They are almost entirely leached of soluble bases and are characterized by a thick development of iron and aluminum oxides (Fig. 12.31c). The soil consists mainly of minerals that resist weathering (for example, quartz, clays, hydrated oxides). Oxisols are most closely associated with the humid tropics, but they also extend into savanna and tropical thorn forest regions. In the United States oxisols are present only in Hawaii. Oxisols are dominated by laterization and retain their natural fertility only

(a)

(b)

(c)

● **FIGURE 12.31** Soil profile examples: (a) spodosols, (b) ultisols, and (c) oxisols.

as long as the soils and forest cover maintain their delicate equilibrium. The bases in the tropical rainforests are stored mainly in the vegetation. When a tree dies, epiphytes and insects must recycle the bases rapidly before the heavy rainfall leaches them from the system.

The burning of vegetation associated with slash-and-burn agriculture in rainforests releases the nutrients necessary for crop growth but quickly results in their loss from the ecosystem. Many tropical oxisols that once supported lush forests are now heavily dissected by erosion and only support a combination of weeds, shrubs, and grasses.

## Soil as a Critical Natural Resource

Regardless of their composition, origin, or state of development, Earth's soils remain one of our most important and vulnerable resources. The word *fertility,* so often associated with soils, has a meaning that takes into consideration the usefulness of a soil to humans. Soils are fertile in respect to their effectiveness in producing specific vegetation types or associations. Some soils may be fertile for corn and others for potatoes. Other soils retain their fertility only as long as they remain in delicate equilibrium with their vegetative cover.

It is clearly the responsibility of all of us who enjoy the agricultural products of farms, ranches, and orchards—and appreciate the natural beauty of Earth's diverse biomes—to help protect our valuable soils. Soil erosion, soil depletion, and the mismanagement of land are problems that we should have great concern about in the world today ( ● Fig. 12.32). We should also be aware that these problems have reasonable solutions ( ● Fig. 12.33). Conserving soils and maintaining soil fertility are critical challenges that are essential to maintaining natural environments and supporting life on Earth today and for the future.

R. Gabler

● **FIGURE 12.32** Gully erosion on farmlands is a significant problem that can often be avoided or overcome by proper agricultural practices. Gullying, if unchecked, can alter the landscape to the point that the original productivity of the land cannot be regained.

*What could have been done to prevent the kind of soil loss shown in this example?*

USDA/NRCS/Lynn Betts

● **FIGURE 12.33** This farm's contour farming techniques, and the use of buffer zones between fields and along the water course, are excellent examples of soil conservation methods.

*What other soil conservation practices are often used to preserve the soil resource?*

# CHAPTER 12 ACTIVITIES

## ■ DEFINE AND RECALL

soil
soil fertilization
capillary water
hardpan
hygroscopic water
gravitational water
leaching
eluviation
illuviation
stratification
humus
soil texture
clay (clayey)
silt (silty)
sand (sandy)
soil grade
loam
infiltrate

soil ped
porosity
permeability
pH scale
parent material
soil profile
soil horizon
Cl, O, R, P, T
residual parent material
transported parent material
soil-forming regime
laterization
laterite
podzolization
calcification
salinization
gleization

soil taxonomy
soil survey
soil order
subsurface horizon
epipedon
entisol
inceptisol
histosol
andisol
gelisol
aridisol
vertisol
mollisol
alfisol
spodosol
ultisol
oxisol

# DISCUSS AND REVIEW

1. Why is soil an outstanding example of the integration and interaction among Earth's subsystems?
2. Describe the different circumstances in which water is found in soil.
3. Under what conditions does leaching take place? What is the effect of leaching on the soil and, consequently, on the vegetation that it supports?
4. How can capillary water contribute to the formation of caliche? What is the effect of caliche on drainage?
5. How is humus formed? What relation does humus have to soil fertility?
6. How is texture used to classify soils? Describe the ways scientists have classified soil structure.
7. What pH range indicates soil suitable for most complex plants?
8. What are the general characteristics of each horizon in a soil profile? How are soil profiles important to scientists?
9. What factors are involved in the formation of soils? Which is most important on a global scale?
10. How does transported parent material differ from residual parent material? List those factors that help determine how much effect the parent material will have on the soil.
11. What are the most important effects of parent material on soil?
12. How does the presence of earthworms and other burrowing animals affect soil?
13. Describe the various ways in which temperature and precipitation are related to soil formation.
14. Describe the three major soil-forming regimes.

# CONSIDER AND RESPOND

1. Refer to Figure 12.13 and associated pages in the text.
   a. What horizons make up the zone of eluviation?
   b. What are two processes that occur in the zone of eluviation?
   c. The various *B* horizons are in what zone?
   d. Weathered parent material is the major constituent of what horizon?
   e. Partly decomposed organic debris makes up which horizon?
2. Refer to Table 12.1.
   a. What materials accumulate in an argillic horizon?
   b. Which would generally be better suited for agriculture—a soil with an ochric epipedon or a mollic epipedon? Why?
   c. What name would be given to a 7-inch-thick horizon that contained at least 2% salt?
3. How important do you consider soils to be among a nation's, and the world's, environmental resources? Explain why.
4. Give your opinion of the overall value of soils in the United States and the extent to which these soils are preserved and protected.

# APPLY AND LEARN

1. Refer to Figure 12.8. Using the texture triangle, determine the textures of the following soil samples.

|        | Sand | Silt | Clay |
|--------|------|------|------|
| Soil a. | 35% | 45% | 20% |
| Soil b. | 75% | 15% | 10% |
| Soil c. | 10% | 60% | 30% |
| Soil d. | 5% | 45% | 50% |

What are the percentages of sand, silt, and clay of the following soil textures? (Note: Answers may vary, but they should total 100%.)
   a. Sandy clay
   b. Silty loam

2. Obtain a small sample of soil (a handful or so) and try to discover its texture by using the following method: Wet the soil a bit and work it in your hand.

I. **First, if you can form a ribbon of soil by kneading it with your fingers:**
   If you can form a ribbon that is long relatively strong and flexible, the soil is a clay.
   If you can form a ribbon that is weak and breaks easily, and the soil can be rolled into a coherent ball, the soil is a clay loam.
   If the clay loam looks powdery when dry, the soil is a silty clay loam.
   If the clay loam has a gritty feel with visible sand, it is a sandy clay loam.

II. **Second, if you cannot form a ribbon because the soil breaks up:**
   If damp soil breaks up easily but is gritty yet still sticky enough to make a ball, the soil is a loam.
   If it feels gritty and sand can be seen and the ball breaks up easily in your hands, the soil is a sandy loam.

III. **Third, if the soil is very loose with visible grains of about the same size:**
   If you can see grains in the soil and the mass in your hand breaks up easily, the soil is a sand.

 CourseMate – Make the most of your study time by accessing everything you need to succeed in one place. Read your textbook, take notes, review flashcards, watch videos, complete activities, take practice quizzes, and more online with CourseMate. Log in at **www.cengagebrain.com**.

# EARTH MATERIALS AND PLATE TECTONICS

## ◼ OBJECTIVES

WHEN YOU COMPLETE THIS CHAPTER YOU SHOULD BE ABLE TO:

- ◼ Compare the relative size and material properties of Earth's core, mantle, and crust.
- ◼ Relate the principal differences between oceanic and continental crust.
- ◼ Contrast the lithosphere with the asthenosphere.
- ◼ Differentiate between minerals and rocks.
- ◼ Recall the definitions of the major categories of igneous, sedimentary, and metamorphic rocks.
- ◼ Explain the meaning of the rock cycle.
- ◼ Understand the relationship between the theories of continental drift and plate tectonics.
- ◼ Discuss the theory of plate tectonics.
- ◼ Provide evidence for the theory of plate tectonics.
- ◼ Describe major Earth features associated with plate convergence, divergence, and transform motion.

**IF WE COULD TRAVEL BACK** in time to view Earth as it was 90 million years ago, in addition to seeing now-extinct life forms, including dinosaurs, we would notice an altered spatial distribution of land and water. The shapes of the oceans and continents, their locations, and their orientations relative to the poles and equator were very different then from what they are today. A vast inland sea cut across what is now the heartland of North America. Dinosaurs left their footprints in large trackways on floodplains of rivers that flowed out of the early Rocky Mountains. Forests grew above the present Arctic Circle, and grasses did not exist yet. The dramatic differences between then and now in the size, shape, and distribution of mountain ranges and water bodies, and the accompanying differences in climate, soils, and organisms, require scientific explanation.

Like the atmosphere, hydrosphere, and biosphere, the large rock portion of the Earth system that lies beneath our feet—the lithosphere—undergoes change resulting from flows of energy and matter. Processes inside Earth help determine the present distribution of various rock types, mineral resources, and natural hazards. Over long intervals of geologic time, the flows of energy and matter within Earth

Relative age of rock on the ocean floor progresses from youngest *(red)* at midoceanic, submarine mountain ranges (spreading centers) to oldest *(blue)* along those continental margins that are tectonically stable. *Elliot Lim, CIRES & NOAA/NDGC*

significantly alter the size, shape, and location of major Earth surface features and environments. In essence, forces inside Earth build the structural foundation that other processes, those originating at Earth's surface, modify. Together, the internal and surface processes create the familiar landscapes in which we live. In this chapter we begin our study of the solid, or rock, portion of Earth—the lithosphere.

# Earth's Planetary Structure

The science of physical geography predominantly focuses on that part of the Earth system that lies at the interface of the atmosphere, hydrosphere, biosphere, and lithosphere, and these come together at Earth's surface. Still, basic knowledge of our planet's internal structure, composition, properties, and processes is needed to understand many aspects of Earth's natural surface characteristics.

From low-density gas molecules in the outermost layer of the atmosphere to high-density iron and nickel at the center of the planet, all of the gas, liquid, and solid matter comprising the Earth system is held within that system by gravitational attraction. Sir Isaac Newton determined that the degree to which particles are drawn to each other by gravity depends on the mass of each particle, which is commonly expressed in units of grams or kilograms. The gravitational force of attraction is greater for objects that have a larger mass than for those with a smaller mass. Scientists commonly use mass per unit volume (for example, grams per cubic centimeter), which is *density,* to compare how equal amounts of various materials differ in mass. Those types of Earth materials that have the greatest density have the greatest gravitational force of attraction, and as a result they have tended to concentrate close together at and near the center of Earth.

Earth's interior is primarily composed of solids, the densest of the three states of matter. Liquids are not as dense as solids, thus most of Earth's liquid water lies at the planet's surface thousands of kilometers above the densest solids located deep inside Earth. Gases, with an even lower density than liquids, have the weakest gravitational attractive force and are held relatively loosely around Earth as the atmospheric envelope, rather than within the planet or on its surface. Traveling outward from the center of the Earth system, there exists a density continuum (spectrum) that extends from the densest materials at Earth's center to the least dense substances at the outer edge of the atmosphere.

Planet Earth has a radius of about 6400 kilometers (4000 mi). Through direct means, by mining and drilling, we have been able to penetrate and examine only an extremely small part of that distance. In South Africa the lure of gold has taken miners to a depth of 3.5 kilometers (2.2 mi), whereas in drilling for oil and gas our machinery has reached a depth of about 12 kilometers (7.5 mi). These explorations have been helpful in providing information about the solid Earth. For example, they have helped establish the notion of the *geothermal gradient*—that temperature increases with increasing depth inside Earth. Our direct explorations into the solid Earth, however, have just barely scratched the planet's surface. Scientists are continually working to learn more about Earth's interior. Extending scientific knowledge about the structure, composition, and processes operating within Earth enhances our understanding of such lithospheric phenomena as earthquakes, volcanic eruptions, the formation and distribution of mineral deposits, and the origins of continents. It can even help us learn more about the origin of the planet itself.

Most of what we know about Earth's internal structure and composition has not been determined directly but has instead been deduced through indirect means by various forms of remote sensing. Thus far, the most important evidence that scientists have used to gain indirect knowledge of Earth's interior is the behavior of various shock waves, known as **seismic waves,** as they move through the planet. Scientists generate some of these shock waves artificially through controlled explosions, but they mainly use evidence derived by tracking the natural seismic waves of earthquakes as they travel through Earth. A sensitive instrument called a **seismograph** can record seismic waves from an earthquake even when the earthquake is centered thousands of kilometers away from the seismograph's location ( ● Fig. 13.1).

Earthquakes produce two major types of seismic waves that travel at different speeds through varying types and densities of Earth material. Of the two types, P (primary) waves travel faster and are the first to arrive at a recording seismograph. S (secondary) waves travel more slowly than P waves, thus they arrive at

● **FIGURE 13.1** This seismograph trace, or seismogram, shows the ground shaking experienced in Athens, Ohio, when seismic waves first started arriving there from the 2010 Chilean earthquake. Data from multiple seismographs are used to determine characteristics of Earth's interior and characteristics of individual earthquakes. Over Earth's surface, Athens lies 8420 kilometers (5230 mi) from the site of the earthquake; the straight-line distance through Earth is 7873 kilometers (4890 mi).

*At what average speed did the seismic waves travel through Earth from Chile to Athens? What maximum vertical motion (vertical deflection) did Athens experience from this earthquake?*

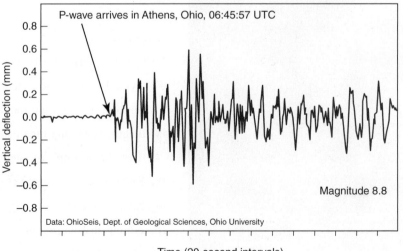

Chilean earthquake occurs 6:34 UTC February 27, 2010

Data: OhioSeis, Dept. of Geological Sciences, Ohio University

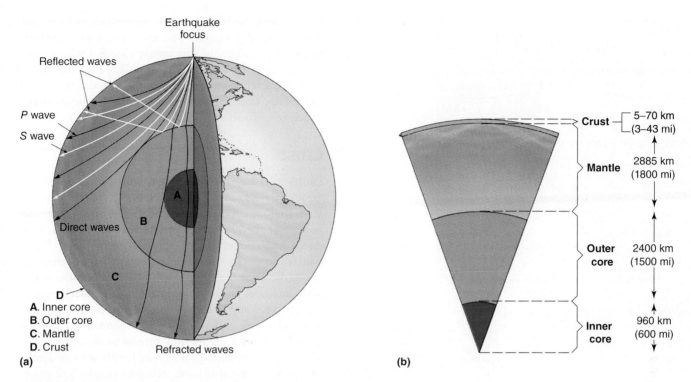

A. Inner core
B. Outer core
C. Mantle
D. Crust

(a)

Crust — 5–70 km (3–43 mi)

Mantle — 2885 km (1800 mi)

Outer core — 2400 km (1500 mi)

Inner core — 960 km (600 mi)

(b)

● **FIGURE 13.2** Earth's internal structure as revealed by seismic waves. (a) Seismic P (primary) waves refract (bend) where material density changes significantly, whereas seismic S (secondary) waves cannot pass through liquids, including Earth's outer core. (b) Cross section through Earth's internal structural zones.

*Which of Earth's internal zones is the largest?*

the seismograph later. Studies of numerous seismograph records show that P and S waves are either refracted (bent) or reflected when they reach the zones of major density change that mark the boundaries between Earth's various interior layers. Both types of waves speed up in denser material and slow down when passing through material that is less dense. P waves pass through all types of matter, including liquids and gases. In fact, P waves traveling through the atmosphere are responsible for the rumbling sound that we hear during an earthquake. S waves, however, can only move through solids; they do not travel through fluids (liquids or air) ( ● Fig. 13.2a). By considering these and other properties of P and S waves and by analyzing global data on travel patterns of earthquake waves collected over decades, scientists have been able to develop a general model of the composition of Earth's interior. This information, supplemented by studies of Earth's magnetism and gravitational pull, reveals a series of layers, or zones, in Earth's internal structure distinguished on the basis of density and composition. From the center of Earth to the surface, these zones are the core, mantle, and crust (Fig. 13.2b). Using a different criterion, how rock material behaves under stress, the outer approximately 350 kilometers (220 mi) of Earth consists of two distinct layers known as the lithosphere and asthenosphere.

## Core

Earth's innermost section, the **core,** contains one third of Earth's mass and has a radius of about 3360 kilometers (2100 mi), which is larger than the planet Mars. Earth's core is under enormous

pressure—several million times atmospheric pressure at sea level. Scientists have deduced that the core is composed primarily of iron and nickel and consists of two distinct sections, the inner core and the outer core.

Earth's **inner core** has a radius of about 960 kilometers (600 mi). The speed of P waves traveling through the inner core shows that it is a solid with a very high material density of about 13 grams per cubic centimeter (0.5 lbs/in.$^3$). The **outer core** forms a 2400-kilometer (1500-mi) thick band around the inner core. Rock matter at the top of the outer core has a density of about 10 grams per cubic centimeter (0.4 lbs/in.$^3$). Because the outer core blocks the passage of seismic S waves, Earth scientists know that the outer core is molten, that is, it consists of liquid (melted) rock matter. The high density of both sections of Earth's core supports the notion that they are composed of iron and nickel.

Temperatures are estimated to be 4800°C (8600°F) at the top of the outer core, increasing to 6900°C (12,400°F) at the very center of Earth. Why, then, is Earth's inner core solid but its cooler outer core molten? The answer involves the facts that the melting temperature of rock matter increases with pressure, and pressure increases with depth beneath Earth's surface. Rock existing under greater pressure needs to achieve a higher temperature to melt than rock at a lower pressure does. The actual temperature of rock material in the outer core exceeds that material's pressure-altered melting temperature, thus rock in the outer core is molten. The extreme high pressure of the inner core, however, has elevated the melting temperature there to a value that lies above the rock's actual temperature, leaving the material of the inner core solid.

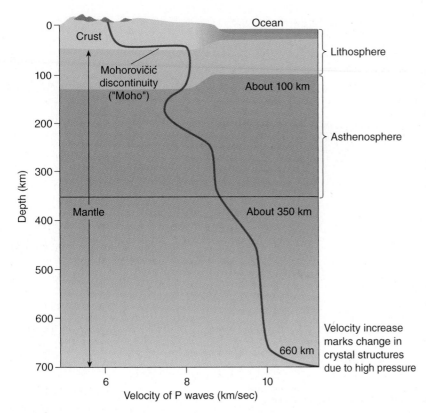

● **FIGURE 13.3** Velocity of seismic P (primary) waves increases with depth as material density increases. An abrupt increase in seismic wave velocity marks the Mohorovičić discontinuity. From Thompson & Turk, *Earth Science and the Environment*, 2007

## Mantle

With a thickness of approximately 2885 kilometers (1800 mi) and representing nearly two thirds of Earth's mass, the **mantle** is the largest of Earth's interior zones. Earthquake waves that pass through the mantle indicate that it is composed of solid rock matter, in contrast to the molten outer core that lies beneath it. Mantle material is also less dense than that of the core, with values ranging from 3.3 to 5.5 grams per cubic centimeter (0.12–0.20 lbs/in.³). Scientists agree that the mantle consists of silicate rocks (high in silicon and oxygen) that also contain significant amounts of iron and magnesium. Heat from decaying radioactive materials inside Earth drives the thermal convection currents that exist in much of the mantle.

The interface between the mantle and the overlying crust is marked by a significant change of density, called a *discontinuity*, which is indicated by an abrupt decrease in the velocity of seismic waves as they travel up through this internal boundary ( ● Fig. 13.3). Scientists call this zone the **Mohorovičić discontinuity,** or **Moho** for short, after the Croatian geophysicist who first detected it in 1909. The Moho does not lie at a constant depth but generally mirrors the surface topography, being deepest under mountain ranges where the crust is thick and rising to within 8 kilometers (5 mi) of the thin ocean floor.

No geologic drilling has yet penetrated through the Moho into the mantle, but an international scientific partnership, called the Integrated Ocean Drilling Program, is working on such a project. Rock samples eventually retrieved from cores drilled through the Moho will add greatly to our understanding of the composition and structure of Earth's interior.

## Crust

Earth's solid exterior is the **crust,** which is composed of a great variety of rock types that respond in diverse ways and at varying rates to surface processes. The crust is the only portion of the lithosphere of which Earth scientists have direct knowledge, yet it represents only about 1% of Earth's planetary mass. As the outermost layer of the solid Earth, the crust comprises the ocean floor and the continents and is of primary importance in understanding surface processes and landforms. Earth's deep interior components, the core and mantle, are of concern to physical geographers primarily for how they affect the crust.

The density of rock matter in Earth's crust is significantly lower than that in the core and mantle, and ranges from 2.7 to 3.0 grams per cubic centimeter (0.10–0.11 lbs/in.³). The crust is also extremely thin in comparison to the size of the planet. Two kinds of Earth crust, oceanic and continental, are distinguished by their location, thickness, and composition ( ● Fig. 13.4). Crustal thickness varies from 3 to 5 kilometers (2–3 mi) in the ocean basins to as much as 70 kilometers (44 mi) under some continental mountain systems. The crust is cold compared with the mantle and behaves in a more rigid and brittle manner, especially in its upper 10 to 15 kilometers (6–9 mi). The crust responds to stress by fracturing, crumpling, or warping.

**Oceanic crust** is composed of *basalt,* a heavy, dark-colored, iron-rich rock that is also high in silicon (Si) and magnesium (Mg). Oceanic crust has a density of 3.0 g/cm³ and is only a few kilometers thick. Forming the vast, deep ocean floors and lava flows on all of the continents, basaltic rocks are the most common rocks on Earth.

● **FIGURE 13.4** Earth has two distinct types of crust: oceanic and continental. The crust and the uppermost part of the mantle, which is rigid like the crust, make up the lithosphere. Below the lithosphere, but still in the upper part of the mantle, is the plastic asthenosphere.

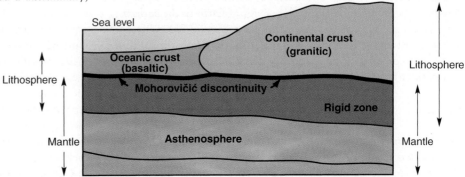

**Continental crust** comprises the major landmasses on Earth that are exposed to the atmosphere. At 2.7 g/cm³, the average density of continental crust is less than that of oceanic crust (3.0 g/cm³). Continental crust, however, is considerably thicker than oceanic crust. It ranges from about 20 to 70 kilometers (12–44 mi) with an average thickness of 32 to 40 kilometers (20–25 mi). At places where continental crust extends to high elevations in mountain ranges it also descends to great depths below the surface. Continental crust contains more light-colored rocks than oceanic crust does and can be regarded as *granitic* in composition. Granite, basalt, and other common rocks are discussed more fully later in this chapter.

## Lithosphere and Asthenosphere

The extreme uppermost part of the mantle, with a thickness of about 100 kilometers (60 mi), has a chemical composition like the rest of the mantle, but it responds to applied stress like the overlying Earth layer, the crust. This response is that of an **elastic solid.** Elastic solids are rigid and brittle. They do not flow, but instead withstand a certain amount of applied stress (force per unit area) with little deformation until a threshold limit of stress is reached. At the threshold value, elastic solids fail by fracturing. Behaviorally, then, this uppermost mantle and overlying crust form a single structural unit, called the **lithosphere.** The term *lithosphere* has traditionally been used to describe the entire solid Earth (as discussed in Chapter 1 and earlier in this chapter). In recent decades, however, the term *lithosphere* has also been used in a separate, structural sense to refer to the brittle outer shell of Earth, including the crust and the rigid, uppermost mantle layer ( ● Fig. 13.5).

● **FIGURE 13.5** A cross section of the lithosphere and asthenosphere and extending lower into the mantle.

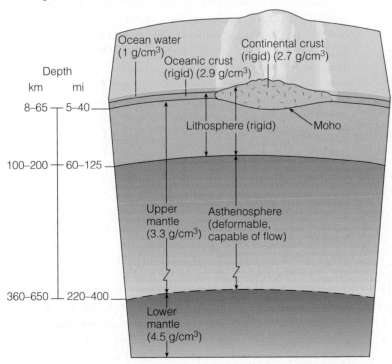

Beneath the lithosphere lies the **asthenosphere** (from Greek: *asthenias,* without strength), a 180-kilometer (110-mi) thick layer of the upper mantle that responds to stress by deforming and flowing slowly rather than by fracturing. In other words, the asthenosphere has the characteristics of a **plastic solid.** Rock in the asthenosphere can flow vertically or horizontally at rates of a few centimeters per year. As material in the asthenosphere flows, it drags segments of the overlying, rigid lithosphere along with it. Earth scientists believe that the energy for **tectonic forces,** large-scale forces that break and deform Earth's crust, sometimes resulting in earthquakes and often responsible for mountain building, comes from movement within the plastic asthenosphere. Movement in the asthenosphere, in turn, is produced by thermal convection currents in the rest of the mantle below the asthenosphere that are driven by heat from decaying radioactive materials in the planet's interior.

# Minerals

Earth's crust is composed of various types of rocks and minerals. A basic understanding of rocks requires us to first consider minerals, which are the building blocks of rocks.

A **mineral** is an inorganic, naturally occurring, crystalline substance represented by a specific chemical formula. A *crystalline* substance displays a specific, repeated, three-dimensional structure at the molecular level. In some cases, the crystalline molecular nature is only visible with a microscope, but minerals sometimes form actual *crystals,* which are geometric forms visible to the unaided eye consisting of smooth faces and sharp edges ( ● Fig. 13.6). The shape of a crystal is an expression of the mineral's molecular structure.

The mineral halite, which is used as table salt, has the specific chemical formula NaCl and as a crystal adopts a cubic form. Quartz, calcite, fluorite, talc, topaz, and diamond are just a few examples of other minerals.

Every mineral has distinctive and recognizable physical characteristics that aid in its identification. Some of these characteristics include hardness, color, luster, cleavage, tendency to fracture, and specific gravity (weight per unit volume), in addition to the shape of the crystal. Luster specifies the shininess of a mineral. Cleavage describes how a mineral tends to break along preferred planes determined by its molecular structure, whereas fracture refers to irregular breaks not along the preferred planes.

Chemical bonds hold together the atoms and molecules that compose a mineral. The strength and nature of those chemical bonds affect the resistance and hardness of minerals and of the rocks that they form. Minerals with weak internal bonds undergo chemical alteration most easily. Charged particles, or *ions,* that form part of a molecule in a mineral may leave or be traded for other substances, generally weakening the mineral structure and forming the chemical basis of *rock weathering.*

Out of more than 100 known elements, Table 13.1 lists the eight most common in Earth's crust and therefore in the minerals and rocks that make up the crust. These eight elements account for almost 99% of Earth's crust by weight.

The most abundant minerals are combinations of these eight elements.

Minerals are categorized into groups based on their chemical composition. Certain elements, particularly silicon, oxygen, and carbon, combine readily with many other elements. As a result, the most common mineral groups are silicates, oxides, and carbonates. Calcite ($CaCO_3$), for example, is a relatively soft but widespread mineral that consists of one atom of calcium (Ca) linked together with a carbonate molecule ($CO_3$), which consists of one atom of carbon (C) plus three atoms of oxygen (O). The silicates, however, are by far the largest and most common mineral group, comprising 92% of Earth's crust.

The two most abundant elements in Earth's crust, oxygen and silicon (Si), frequently combine to form $SiO_2$, which is called *silica*. *Silicate minerals* are compounds of oxygen and silicon that also include one or more metals and/or bases. They are generally created when molten rock matter containing these elements cools and solidifies, causing the crystallization of different minerals at successively lower temperatures and pressures. Dark, heavy, iron-rich silicate minerals crystallize first (at high temperatures), and light-colored, lower density, iron-poor minerals crystallize later at cooler temperatures. The order of mineral crystallization parallels their relative chemical stability in a rock, with silicate minerals that crystallize later tending to be more stable and more resistant to breakdown. In rocks composed of a variety of silicate minerals, the dark, heavy minerals are the first to decompose, whereas the iron-poor minerals decompose later. The mineral quartz is crystalline silica; as a crystal, it adopts a distinctive prismatic shape. Because quartz is one of the last silicate minerals to form from solidifying molten rock matter, it is a relatively hard and resistant mineral.

# Rocks

A **rock** is a consolidated aggregate of various types of minerals or a consolidated aggregate of multiple individual pieces (grains) of the same kind of mineral. In other words, a rock is not one single, uniform crystal. Although a few rock types are composed of many particles of a single mineral, most rocks consist of several different minerals ( ● Fig. 13.7). Each constituent mineral in a rock remains separate and retains its own distinctive characteristics. The properties of the rock as a whole are a composite of those of its various mineral constituents. The number of rock-forming minerals that are common is limited, but they combine through a multitude of processes to produce an enormous variety of rock types (refer to Appendix C for further information and pictures of common rocks mentioned in the text). Rocks are the

● **FIGURE 13.6** Crystals of the mineral calcite, which has the chemical formula $CaCO_3$. The geometric arrangement of atoms determines the crystal form of a mineral.

**TABLE 13.1**
**Most Common Elements in Earth's Crust**

| Element | Percentage of Earth's Crust by Weight |
|---|---|
| Oxygen (O) | 46.60 |
| Silicon (Si) | 27.72 |
| Aluminum (Al) | 8.13 |
| Iron (Fe) | 5.00 |
| Calcium (Ca) | 3.63 |
| Sodium (Na) | 2.83 |
| Potassium (K) | 2.70 |
| Magnesium (Mg) | 2.09 |
| Total | 98.70 |

Source: J. Green, "Geotechnical Table of the Elements for 1953," *Bulletin of the Geological Society of America 64* (1953).

● **FIGURE 13.7** Most rocks consist of several different minerals. This piece of cut and polished granite displays intergrown crystals of differing composition, color, and size that give the rock a distinctive appearance.

*How many different kinds of minerals can you see in this sample of granite?*

● **FIGURE 13.8** Masses of solid rock that are exposed (crop out) at the surface are referred to as outcrops.

*What physical characteristics of this rock outcrop have caused it to protrude above the general land surface?*

fundamental building materials of the lithosphere. They are lifted, pushed down, and deformed by large-scale tectonic forces originating in the lower mantle and asthenosphere. At the surface, rocks are weathered and eroded to be deposited as sediment elsewhere.

A mass of consolidated rock that has not been weathered is termed **bedrock.** Bedrock may be exposed at the surface of Earth or it may be overlain by a cover of broken and decomposed rock fragments, called **regolith.** Soil may or may not have formed on the regolith. On steep slopes, regolith may be absent and the bedrock may be exposed if running water, gravity, or some other surface process removed the weathered rock fragments. A mass of exposed bedrock is typically referred to as an **outcrop** ( ● Fig. 13.8).

Geologists distinguish three major categories of rocks based on mode of formation. These rock types are igneous, sedimentary, and metamorphic.

## Igneous Rocks

When molten rock material cools and solidifies it becomes an **igneous rock.** Molten rock matter below Earth's surface is called **magma,** whereas molten rock material at the surface is known specifically as **lava** ( ● Fig. 13.9). Lava, therefore, is the only form of molten rock matter that we can see. Lava erupts from volcanoes or fissures in the crust at temperatures as high as 1090°C (2000°F). There are two major categories of igneous rocks: extrusive and intrusive.

Molten material that solidifies at Earth's surface creates **extrusive igneous rock,** also called *volcanic rock.* Extrusive igneous rock, therefore, is made from lava. Very explosive eruptions of molten rock material can cause the accumulation of fragments of volcanic rock, dust-sized or larger, termed **pyroclastics** (from

Greek: *pyros,* fire; *clastus,* broken), also known as **tephra,** that settle out of the air ( ● Fig. 13.10a). When molten rock beneath Earth's surface, that is, magma, changes to a solid (freezes), it forms **intrusive igneous rock,** also referred to as *plutonic rock* after Pluto, Roman god of the underworld. Igneous rocks are classified in terms of their mineral composition and the size of constituent minerals, which is referred to as *texture.* Igneous rocks vary in texture, chemical composition, crystalline structure, tendency to fracture, and presence or absence of layering.

Rocks composed of small-sized individual minerals not visible to the unaided eye are described as having a fine-grained texture, whereas those with large minerals that are visible without magnification are referred to as coarse-grained. Molten rock matter extruded onto the surface cools quickly—at Earth surface temperatures—and are fine-grained as a result of the little time available for crystal growth prior to solidification. An extreme example is the extrusive rock *obsidian,* which cools so rapidly that it is essentially a glass (Fig. 13.10b). Large masses of intrusive rock matter solidifying deep inside Earth cool very slowly because, like all rocks, those surrounding the magma are poor conductors and inhibit heat loss. Slow cooling allows more time for crystal formation prior to solidification. In specific circumstances, such as if magma lies in thin stringers

● **FIGURE 13.9** Recently erupted basaltic lava on the island of Hawaii. Molten basalt typically erupts at very high temperatures, flows easily, and cools rapidly into a fine-grained extrusive igneous rock.

*What features besides the red color indicate the part of the flow that was added most recently?*

or close to Earth's surface, it can cool rapidly enough to produce a fine-grained intrusive rock.

The chemical composition of igneous rocks varies from *felsic,* which is rich in light-colored, lighter weight minerals, especially silicon and aluminum ( *fel* for the mineral feldspar; *si* for silica), to *mafic,* which is lower in silica and rich in heavy minerals, such as compounds of magnesium and iron (*ma* for magnesium; *f* for *ferrum,* Latin for iron). Granite, a felsic, coarse-grained, intrusive rock,

has the same chemical and mineral composition as *rhyolite,* a fine-grained, extrusive rock. Likewise, basalt is the dark-colored, mafic, fine-grained extrusive chemical and mineral equivalent of *gabbro,* a coarse-grained intrusive rock that cools at depth ( ● Fig. 13.11).

Igneous rocks also form with an intermediate composition represented by a rough balance between felsic and mafic minerals. The intrusive rock *diorite* and the extrusive rock andesite (named for the Andes where many volcanoes erupt this type of lava) represent this intermediate composition (see Fig. 13.11).

Many igneous rocks are fractured, often by multiple cracks that may be evenly spaced or arranged in regular geometric patterns. In the Earth sciences, simple fractures or cracks in bedrock are called **joints.** Joints caused by regional stresses in the crust are common features in any type of rock, including igneous rocks. Igneous rocks, however, also develop joints because molten rock shrinks in volume as it cools and solidifies, resulting in fractures.

## Sedimentary Rocks

As their name implies, **sedimentary rocks** are derived from sediment, which is a loose accumulation of unconsolidated fragments. After the fragments accumulate, often in horizontal layers, pressure from the addition of more material above compacts the sediment, expelling water and reducing pore space. Cementation occurs when silica, calcium carbonate, or iron oxide bonds the fragments together. The processes of compaction and cementa-

● **FIGURE 13.10** (a) Pyroclastic rocks are made of fragments ejected during a volcanic eruption. (b) Obsidian—volcanic glass—results when molten lava cools too quickly for crystals to form.

**(a)**

**(b)**

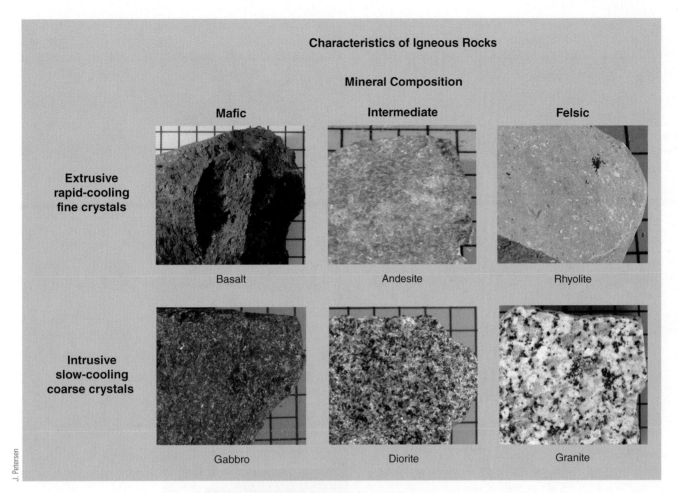

## Characteristics of Igneous Rocks

### Mineral Composition

|  | Mafic | Intermediate | Felsic |
| --- | --- | --- | --- |
| **Extrusive rapid-cooling fine crystals** | Basalt | Andesite | Rhyolite |
| **Intrusive slow-cooling coarse crystals** | Gabbro | Diorite | Granite |

J. Petersen

● **FIGURE 13.11** Igneous rocks are distinguished on the basis of texture (mineral grain size) and whether their mineral composition is mafic, felsic, or intermediate. Rocks that cool rapidly, at and near Earth's surface, have fine (small) crystals. Those that cool slowly, deep beneath the surface, have a coarse crystalline texture (large mineral grains).

*How does granite differ from basalt in composition and texture?*

tion transform, or *lithify,* sediments into solid, coherent layers of rock. There are three major categories of sedimentary rocks: clastic, organic, and chemical precipitate.

Broken fragments of solids are called **clasts** (from Latin: *clastus,* broken). In order of increasing size, clasts range from clay, silt, and sand, to gravel, which is a general category for any fragment larger than sand (larger than 2.0 mm) and includes granules, pebbles, cobbles, and boulders. Clasts consist of fragments of previously existing rocks, shells, or bones that were deposited on a river bed, beach, sand dune, lake bottom, the ocean floor, and other environments where fragments of solids accumulate. Sedimentary rocks that form from fragments of preexisting rocks, shells, or bones are called **clastic sedimentary rocks.**

Examples of clastic sedimentary rocks include *conglomerate, sandstone, siltstone,* and *shale* ( ● Fig. 13.12). Conglomerate is a lithified mass of cemented, roughly rounded pebbles, cobbles, and boulders and may have clay, silt, or sand filling in spaces between those larger particles. A somewhat similar rock composed of lithified fragments that are angular rather than rounded is called *breccia.* Sandstone consists of cemented sand-sized particles, most commonly grains of quartz. Sandstone is usually granular (has vis-

ible grains), porous, and resistant to weathering, but the cementing material influences its strength and hardness. If cemented by silica, sandstone tends to be more resistant to weathering than if it is cemented by calcium carbonate or iron oxide. Unlike sandstone, individual grains in siltstone, which is composed of silt-sized particles, are not easily visible with the unaided eye. Shale is produced from the compaction of very fine-grained sediments, especially clays. Shale is often thinly layered with a smooth surface and a low permeability. It is, however, easily cracked, broken apart, and eroded.

Sedimentary rocks may be further classified by their origin in either a marine or terrestrial (continental) environment. Marine sandstones typically form in nearshore coastal zones; terrestrial sandstones commonly originate in desert or floodplain environments. The nature and arrangement of sediments in a sedimentary rock provide a great deal of evidence for the kind of environment in which they were deposited, for example, whether on a stream bed, a beach, or the deep-ocean floor.

Some clastic sedimentary rocks form in lakes and seas from skeletal remains of organisms. The remains of shellfish, corals, and drifting microscopic organisms called *plankton* sink to the bottom

**Composition of Common Sedimentary Rocks**

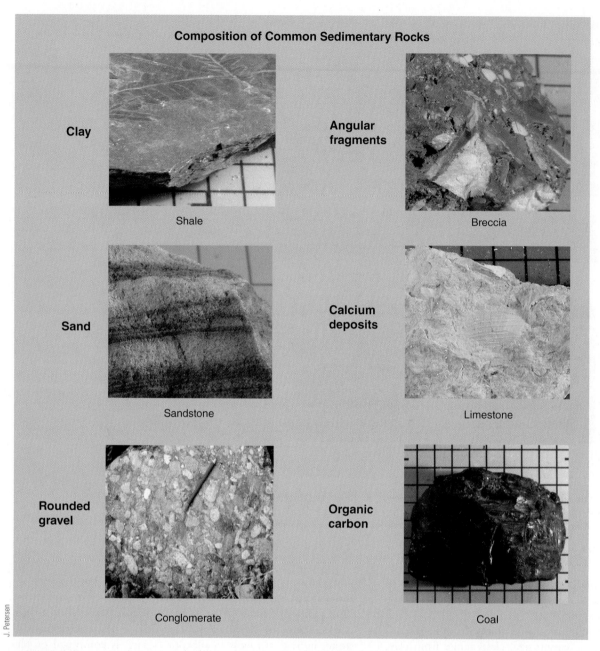

| | | | |
|---|---|---|---|
| **Clay** | Shale | **Angular fragments** | Breccia |
| **Sand** | Sandstone | **Calcium deposits** | Limestone |
| **Rounded gravel** | Conglomerate | **Organic carbon** | Coal |

J. Petersen

● **FIGURE 13.12** Clastic sedimentary rocks are classified by the size and/or shape of the sediment particles they contain.

*Why do the shapes and sizes of the sediments in sedimentary rocks differ?*

of such water bodies, where the hard fragments are compacted and cemented together. Rich in calcium carbonate ($CaCO_3$), they form a type of *limestone* that typically contains fossil shell and coral fragments ( ● Fig. 13.13).

**Organic sedimentary rocks** lithify from the remains of organisms, both plants and animals. *Coal* is created by the accumulation and compaction of partially decayed vegetation in acidic, swampy environments where water-saturated ground prevents oxidation and complete decay of the organic matter. The initial transformation of such organic material produces peat, which, when subjected to deeper burial and further compaction,

is lithified to produce coal. Most of the world's greatest coal deposits originated between 300 and 354 million years ago during an interval of geologic time known as the Mississippian and Pennsylvanian Periods.

When the amounts of dissolved minerals in ocean or lake water reach saturation they begin to precipitate and build up as a deposit on the sea or lake bottom. When lithified, these sediments become **chemical precipitate sedimentary rocks.** Many fine-grained limestone rocks form in this manner from chemical precipitates of calcium carbonate. Limestone, therefore, may vary from a jagged and cemented complex of visible shells or

In Google Earth fly to: 51.102°N, 1.251°E.

● **FIGURE 13.13** The White Cliffs of Dover, England. These striking, steep white cliffs along the English Channel are made of chalky limestone from the skeletal remains of microscopic marine organisms.

fossil skeletal material (a clastic sedimentary rock) to a smooth-textured chemical precipitate sedimentary rock. Where magnesium is a major constituent along with calcium carbonate, the chemical precipitate sedimentary rock is *dolomite.* Because the calcium carbonate in limestone can slowly dissolve in water, limestone tends to be a weak rock in humid environments, but in arid or semiarid climates it tends to be resistant.

Mineral salts that have reached saturation in evaporating seas or lakes precipitate to form a variety of sedimentary deposits that are useful to humans. These include *gypsum* (used in wallboard); *halite* (common salt); and *borates,* which are important in hundreds of products such as fertilizer, fiberglass, detergents, and pharmaceuticals.

Many types of sediment accumulate in distinct layers, or *strata,* that remain visible after lithification. This layering in rocks is known as **stratification. A bedding plane** is the boundary between two sedimentary layers that represent separate depositional events. Bedding planes denote a pause or some change in the nature of deposition, and the strata may differ in grain size, color, or composition ( ● Fig. 13.14). Where markedly mismatching strata meet along an irregular, eroded surface, that contact between the rocks is an **unconformity.** An unconformity indicates a gap in the section caused by removal of sediment from the top of one layer before deposition of the overlying sediment. Another type of stratification, called **cross bedding,** is characterized by a pattern of thin sediment layers that accumulated at an angle to the main strata, often reflecting shifts of direction by waves along a coast, currents in streams, or winds over sand dunes ( ● Fig. 13.15). All types of stratification provide evidence about the environment in which the sediments were deposited, and changes from one sedimentary layer to the next reflect elements of the local geologic history. For example, shale representing an offshore environment may lie directly on top of sandstone strata deposited on an ancient beach, suggesting that first this was a beach that the sea later covered.

Sedimentary rocks become jointed, or fractured, when they are subjected to crustal stresses after they lithify. The impressive "fins" of rock at Arches National Park, Utah, owe their vertical, tabular shape to joints in great beds of sandstone ( ● Fig. 13.16). Structures such as bedding planes and joints are important in the development of physical landscapes because they are weak points in the rock that weathering and erosion can attack with relative ease. Often, water travels preferentially along joints, causing faster rock breakdown and removal along them than elsewhere in the rocks.

## Metamorphic Rocks

*Metamorphic* means "changed form." Enormous heat and pressure deep in Earth's crust can alter (metamorphose) an existing rock into a new rock type that is completely different from the original by recrystallizing the minerals without creating molten rock matter. Compared with the original rocks, the resulting **metamorphic rocks** are typically harder and more compact, have a reoriented crystalline structure, and are more resistant to weathering. There are two major categories of metamorphic rocks, based on the presence (foliated) or absence (nonfoliated) of platy surfaces or wavy alignments of light and dark minerals that form during metamorphism.

Metamorphism occurs most commonly where crustal rocks are subjected to great pressures by tectonic processes or deep burial, or where rising magma generates heat that modifies the nearby rock. Metamorphism causes minerals to recrystallize and, with enough heat and pressure, to re-precipitate perpendicular to the applied stress, forming platy surfaces (cleavage) or wavy bands known as **foliation** ( ● Fig. 13.17). Some shales change to a hard metamorphic rock known as *slate,* which exhibits a tendency to break apart, or cleave, along smooth, flat surfaces that actually represent extremely thin foliation planes ( ● Fig. 13.18a). Where foliation layers are moderately thin, individual minerals have a flattened but wavy, "platy" structure, and the rocks tend to flake apart along these bands. A common metamorphic rock with thin foliation layers is *schist.*

● **FIGURE 13.14** Stratification in sedimentary rocks, Grand Canyon National Park, Arizona. Bedding planes are boundaries between differing strata (layers) and mark a change in the nature of the deposited sediment.

● **FIGURE 13.15** Cross bedding in sandstone composed of sand dune sediments, Valley of Fire, Nevada.

● **FIGURE 13.16** Vertical jointing of sandstone in Arches National Park, Utah, is responsible for creation of these vertical rock walls, called fins. Rock has been preferentially eroded along the joints. Only rock that was far from the locations of the joints remains standing.

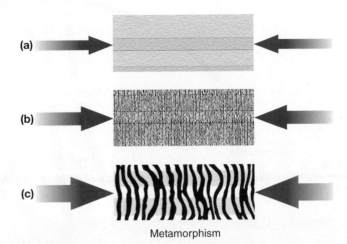

Metamorphism

● **FIGURE 13.17** During metamorphism, applied stress *(arrows)* can lead to an alignment of minerals, known as foliation. (a) Layered rocks under moderate pressure. (b) At greater pressure, metamorphism may realign minerals perpendicular to the applied stress, creating thin foliation layers and a platy structure. (c) With even greater pressure, broader foliation layers may develop as wavy bands of light and dark minerals.

***How does foliation differ from bedding planes?***

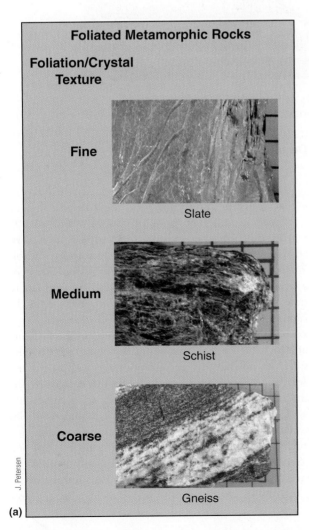

**Foliated Metamorphic Rocks**

**Foliation/Crystal Texture**

**Fine**

Slate

**Medium**

Schist

**Coarse**

Gneiss

(a)

● **FIGURE 13.18** Examples of metamorphic rocks. (a) Slate, schist, and gneiss illustrate an increase in thickness of foliation planes. (b) Marble and quartzite are nonfoliated metamorphic rocks that have a harder, recrystallized composition compared with the limestone and sandstone from which they were made.

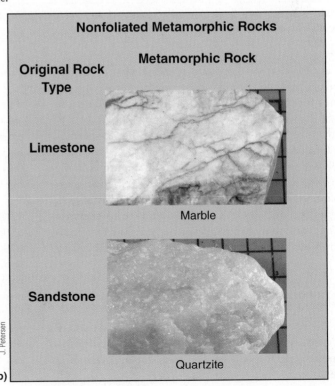

**Nonfoliated Metamorphic Rocks**

**Original Rock Type**    **Metamorphic Rock**

**Limestone**

Marble

**Sandstone**

Quartzite

(b)

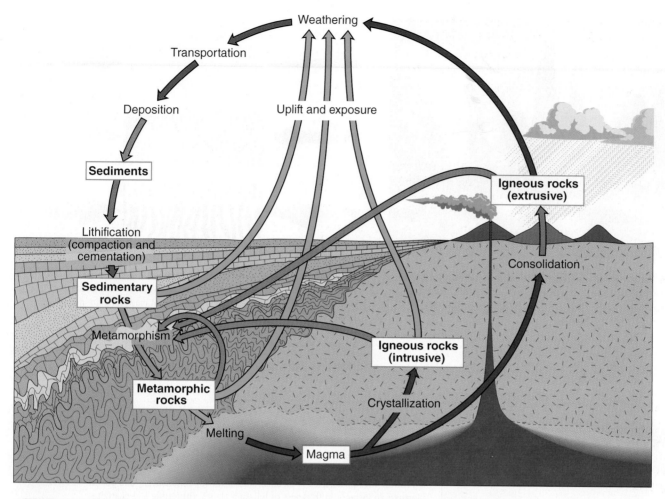

● **FIGURE 13.19** The rock cycle helps illustrate how igneous, sedimentary, and metamorphic rocks are formed and how rock-forming materials can cycle through different rocks over geologic time. Note the links that bypass some parts of the cycle.

*What conditions are necessary to change an igneous rock to a metamorphic rock? Can a metamorphic rock be metamorphosed?*

Where the foliation develops into broad mineral bands, the rock is extremely hard and is known as *gneiss* (pronounced "nice"). Coarse-grained rocks such as granite generally metamorphose into gneiss, whereas fine-grained rocks tend to become schist.

Rocks that originally were composed of one dominant mineral are not foliated by metamorphism (Fig. 13.18b). Limestone is metamorphosed into much denser *marble*, and impurities in the rock can produce a beautiful variety of colors. Silica-rich sandstones fuse into *quartzite*. Quartzite is brittle, harder than steel, and almost inert chemically. It is virtually immune to chemical weathering and commonly forms cliffs or rugged mountain peaks after the surrounding, less resistant rocks have been removed by erosion.

## The Rock Cycle

Many rocks do not remain in their original form indefinitely but instead, over a long time, undergo processes of transformation. The **rock cycle** is a conceptual model for understanding processes that generate, alter, transport, and deposit mineral materials to produce different kinds of rocks ( ● Fig. 13.19). The term *cycle* emphasizes that existing rocks supply the materials to make new and sometimes very different rocks. Whole existing rocks can be "recycled" to form new rocks. The geologic age of a rock is based on the time when it assumed its current state; melting, lithification, or metamorphism reset the age of origin.

A complete cycle is shown in the outer circle of Figure 13.19, but as indicated by the arrows that cut across the diagram, rock matter does not have to go through every step of the full rock cycle. For example, after igneous rocks are created by the cooling and crystallizing of magma or lava, they can weather into fragments that lithify into sedimentary rocks. Igneous rocks, however, could also be remelted and recrystallized to make new igneous rocks or changed into metamorphic rocks by heat and pressure. Sedimentary rocks consist of particles and deposits derived from any of the three basic rock types. Metamorphic rocks can be created by means of heat and pressure changing any preexisting rock—igneous, sedimentary, or metamorphic—into a new rock type. In addition, with sufficient heat, any rock can melt completely into magma that will eventually cool and solidify into an igneous rock. The rock cycle includes all the possible pathways for the recycling of rock matter over time.

# Plate Tectonics

Scientists in all disciplines constantly seek broad explanations that account for the detailed facts, recurring patterns, and interrelated processes that they observe and analyze. From the 17th to the early 19th centuries, many Earth scientists believed that present landscapes were created by a previous period of great and sudden cataclysms. They might have believed, for example, that the Grand Canyon split open one violent day and has remained that way ever since or that the Rocky Mountains appeared overnight. This concept, called **catastrophism,** was largely rejected after about 1830. For almost two centuries, physical geographers, geologists, and other Earth scientists have instead accepted the concept of **uniformitarianism,** which is the idea that internal and external Earth processes operated in the geologic past as they do today, that is, slowly and not by past monumental cataclysms.

Uniformitarianism does not imply that processes have always operated at the same rate or with equal strength everywhere on Earth. In fact, the spatial variability of our planet's surface features is the result of changing intensities of internal and external processes, influenced by geographic position. These processes have varied in intensity and location throughout Earth's history. Furthermore, regular or episodic changes in the Earth system that may seem relatively small to us can dramatically alter a landscape after progressing, even on an irregular basis, for millions of years.

Is there one broad theory compatible with the principle of uniformitarianism that can explain how and why Earth's lithospheric processes work? Such a theory would have to explain many diverse phenomena, including the growth of continents, the movement of solid rock that causes earthquakes, the location of great mountain ranges, differing patterns of temperature in the rocks of the seafloor, and violent volcanic eruptions. One theory explains all of these phenomena; the concept, plate tectonics, involves the continual movement of segments of Earth's surface over millions of years of time.

Sometimes it requires many years to develop, test, and refine a scientific theory to the point where it is more fully understood and broadly acceptable. As data and information are gathered and analyzed, new methods and technologies contribute to the process of testing hypotheses via the scientific method, and bit by bit an acceptable explanatory framework emerges. This was the case in the 20th century with the idea that segments of Earth's outer shell undergo changes in location and orientation over long periods. Over the last century, an initial theoretical framework postulating moving continents was eventually replaced by the theory of plate tectonics, which has been well tested using a great deal of evidence collected from the lithosphere. The theory of plate tectonics revolutionized the Earth sciences and our understanding of Earth's history.

## The Theory of Continental Drift

Most of us have probably observed that on world maps it looks as though the Atlantic coasts of South America and Africa would fit together if we could move them closer to each other. In fact, several widely separated landmasses on Earth appear as though they would fit alongside each other without large gaps or overlaps if we could slide them together ( ● Fig. 13.20). What is the scientific explanation for this phenomenon?

In the early 1900s Alfred Wegener, a German climatologist, proposed the theory of **continental drift,** the idea that continents and other landmasses have shifted their positions during Earth history. Wegener's evidence for continental drift included the close fit of continental coastlines on opposite sides of oceans and the trends of mountain ranges that also appeared to match across oceans. Wegener cited comparable geographic patterns of fossils and rock types found on different continents. To explain the spatial distributions of these features, he reasoned that the now-separate continents must have been joined together at some time in the past. Wegener also noted indications of great climate change, such as evidence of former glacial conditions in the Sahara Desert and tropical fossils in Antarctica, that could be explained best by large landmasses moving from one climate zone to another.

Wegener hypothesized that all the continents had once been part of a single supercontinent, which he called *Pangaea,* that later split into two large landmasses, one in the Southern Hemisphere

● **FIGURE 13.20** The close fit of the continents that face one another across the Atlantic Ocean has been used as geographic evidence supporting the notion of continental drift. The actual fit is even closer if the true continental edges, currently submerged, are used.

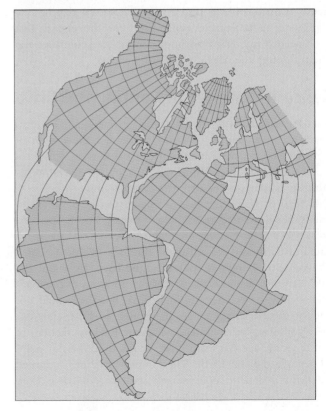

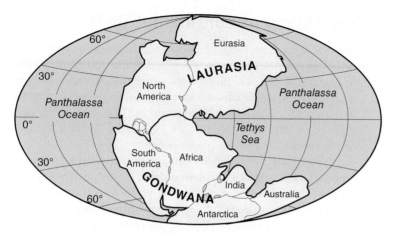

● **FIGURE 13.21** The supercontinent of Pangaea included all of today's major landmasses joined together. Pangaea later split to make Laurasia and Gondwana. Further plate motion left the continents arranged as they are today.

*How would continental movement affect the climates of landmasses?*

(Gondwana) and one in the Northern Hemisphere (Laurasia). Laurasia in the Northern Hemisphere consisted of North America, Europe, and Asia. Gondwana in the Southern Hemisphere was made up of South America, Africa, Australia, Antarctica, and India ( ● Fig. 13.21). He thought these two supercontinents also later broke apart to produce the present continents, which eventually drifted to their current positions by plowing through the ocean.

The reaction of most of the scientific community to Wegener's proposal of drifting continents ranged from skepticism to ridicule. A major objection to his hypothesis was that no one could provide an acceptable explanation for the energy that would be needed to break apart huge continental landmasses and slide them through the ocean over Earth's surface.

## Evidence in Support of Continental Drift

It was almost half a century after Wegener first presented his ideas that Earth scientists began to seriously consider the notion of slowly moving continents. Increasingly, new data were being collected and additional observations were made that seemed to require drifting landmasses for adequate explanation.

One field of study that helped initiate this reassessment of Wegener's theory involves measurement of magnetic-field orientations in rocks. Iron-bearing minerals in rock record the magnetic field of Earth as it existed when the rock solidified, which is a phenomenon known as **paleomagnetism.** By about the mid-20th century, scientists knew that the positions of the magnetic poles shift, or wander, through time, but they could not account for the confusing range of magnetic-field orientations indicated from iron-bearing rocks. Magnetic-field orientations of rocks of the same age but from different locations did not point toward a single spot on Earth. The position that they indicated for the north magnetic pole ranged widely, and some even

pointed toward the present south magnetic pole. The observed variations were more than could be accounted for by the known magnetic polar wandering.

Scientists eventually used the paleomagnetic data in the opposite way—to reconstruct where the sampled rocks would have to have been located relative to a common north magnetic pole. Successful alignment was only possible if the continents had previously occupied different positions than they do today. Using rocks of different age, scientists reconstructed locations of the continents during past periods in geologic history ( ● Fig. 13.22). The paleomagnetic data revealed that the continents were grouped together about 200 million years ago, just as Wegener's hypothesized two supercontinents began to split apart to form the beginnings of the modern Atlantic Ocean. The paleomagnetic data also revealed that the polarity of Earth's magnetic field has reversed many times in the past. A record of these polarity reversals is imprinted within the iron-rich basaltic rocks of the seafloor.

Other supporting evidence for crustal movement eventually came from a variety of sources. The widely separated locations having similar reptile and plant fossils found in Australia, India, South Africa, South America, and Antarctica, previously noted by Wegener, received further study. Fossils within both the reptile and plant groups were considered too similar and specialized to have originated in multiple distant locations. When the positions of the continents were reassembled on a paleomap derived from paleomagnetic data and representing the time when the organisms were living, the fossil locations

● **FIGURE 13.22** Paleomagnetic properties of rocks of the same age indicate the location of magnetic north at that time. Former positions and orientations of the continents are reconstructed by determining where the rocks would have to have originated to point to a common magnetic pole.

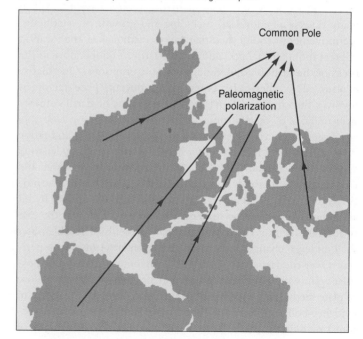

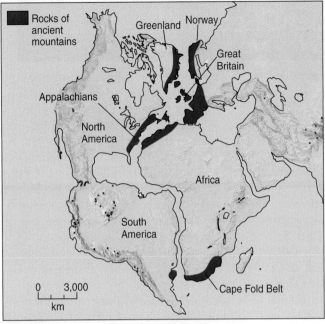

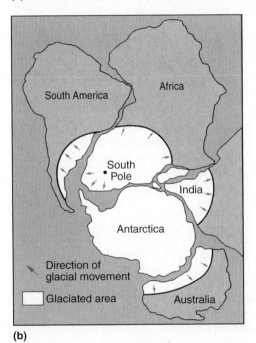

● **FIGURE 13.23** Multiple types of paleogeographic evidence help reveal the previous locations and distributions of Earth's landmasses in the geologic past: (a) rocks of ancient mountain ranges, and (b) evidence of ancient glaciations.

grouped together spatially. Other types of ancient environmental evidence, such as that left by glaciations, also formed logical geographic patterns on reconstructed paleomaps of the continents or the world ( ● Fig. 13.23).

Improved knowledge of the true, submerged edges of continents, which lie a few hundred meters below sea level, supported the notion of continental drift. The precision with which Earth's landmasses fit together when joined on a paleomap was even bet-

ter when using the true continental edges rather than their outlines at sea level. Mountain ranges line up and rock ages and types match well on opposite sides of oceans, as Wegener had suggested when joining landmasses along their true (submerged) continental edges. Refined understanding of the geographic distribution of Earth's environments relative to latitude and climate zones provided additional insight. Evidence that an ancient glaciation occurred simultaneously in India and South Africa while tropical forest climates existed in the northeastern United States and in Great Britain could only be explained by the latitudinal movement of landmasses, and their locations came together well on *paleogeographic* reconstructions.

Despite this evidence in support of continental drift, it remained difficult for many Earth scientists in the mid-20th century to accept Wegener's theory. The processes responsible for moving continents, and how they would slide across the ocean basins, remained unexplained.

## Seafloor Spreading and Subduction

In the years after World War II, studying and mapping of the ocean floor, aided by sonar, radioactive dating of rocks, and improvements in equipment for measuring Earth's magnetism, yielded some surprising results. First, detailed undersea mapping was conducted on a system of midoceanic ridges (also called oceanic ridges or rises) that revealed trends remarkably similar to continental coastlines. Second, it was discovered in the Atlantic and Pacific Oceans that basaltic seafloor displays matching patterns of magnetic properties in rocks of the same age but on opposite sides of midoceanic ridges. Third, scientists made the surprising discovery that although some continental rocks are 3.6 billion years old, rocks on the ocean floor are all geologically young, having been in existence less than 250 million years. Fourth, researchers found that the oldest rocks of the seafloor lie in trenches beneath the deepest ocean waters or close to the continents and that rocks become progressively younger toward the midoceanic ridges where the youngest basaltic rocks exist ( ● Fig. 13.24). Finally, it was determined that temperatures of rocks on the ocean floor vary significantly, being hottest near the midoceanic ridges and becoming progressively cooler farther away.

Only one logical explanation emerged to fit all of this ocean-floor evidence. It became apparent that new oceanic crust forms along the midoceanic ridges, where basaltic rocks are youngest, and that the oceanic crust moves slowly in opposite directions away from the axis of a ridge. This phenomenon is called **seafloor spreading** ( ● Fig. 13.25). Movement of ocean floor in both directions away from a midoceanic ridge is evidenced by the symmetrical pattern of increasing age with distance from the ridge and the symmetrical pattern of paleomagnetism measured in those rocks. As molten basalt cools and crystallizes in the seafloor, the iron minerals record the orientation of Earth's magnetic field at the time. The iron-rich basalts of the seafloor have preserved a historical record of changes in Earth's magnetic field, including **polarity reversals,** when the north and south magnetic poles switch location.

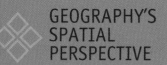
# Paleomagnetism: Evidence of Earth's Ancient Geography

Earth's magnetic field encircles the globe with field lines that converge at two opposite magnetic poles. The geographic North and South Poles do not coincide with their magnetic counterparts, but beyond the polar regions the magnetic poles are useful for navigating by compass. Because the compass needle points toward magnetic north, it is necessary to adjust for the magnetic declination (angle between the geographic and magnetic poles) to determine geographic direction accurately.

Over geologic time, the intensity of Earth's magnetic field has varied and Earth's polarity has reversed many times. Before the last reversal, about 700,000 years ago (to what we call normal polarity), a compass needle would have pointed to the south. *Paleomagnetism* is the study of magnetic fields in mineral crystals within rocks of varying ages. Basaltic rocks, which are iron rich, are most commonly used for paleomagnetic research. When basalt solidifies, iron oxide crystals in the rock record several properties related to Earth's magnetic field at the time. Radiometric dating and paleomagnetic studies help scientists de-

termine a rock's geographic location when it cooled. Paleomagnetic studies have yielded much evidence of plate tectonics, and they help scientists reconstruct the shifting geographic positions of landmasses during Earth history.

Three important characteristics that these rocks record are polarity (normal, like that of today, or reversed), declination, and inclination, (which is measured with a vertically mounted compass needle). Each property provides different evidence about changes in the magnetic field and about how Earth's paleogeography varied as plate tectonics moved the landmasses. Numerous measurements of these three paleomagnetic variables worldwide have given scientists a good picture of Earth's continually changing paleogeography throughout the last several hundred million years.

- **Polarity** Seafloor spreading was confirmed by polarity changes discovered in stripelike patterns of oceanic basalts that match on opposite sides of the Mid-Atlantic Ridge spreading center, where basalts form. Going farther away from a midoceanic ridge, the rocks are

progressively older, and each stripe has a counterpart of the same age and same magnetic polarity on the opposite side of the ridge. The basaltic seafloor has recorded the polarity history of Earth's magnetic field and the widening of the Atlantic Ocean.

- **Declination** Declination shows the direction to the magnetic pole. By studying basalts of the same age but on several continents, it is possible to triangulate directions to the position of the magnetic north pole at the time they formed (see Fig. 13.22). The information provided by these paleodeclinations is the *orientation* of ancient landmasses (whether or not they rotated relative to north as they drifted).

- **Inclination** The magnetic field surrounding Earth causes not only a magnetic compass needle to point north but also to dip downward in a straight-line direction to north. This is called magnetic dip, and a needle's angle off of horizontal approximates its *latitudinal* location. Paleoinclinations recorded from ancient basalts provide the latitude of their location at the time of cooling.

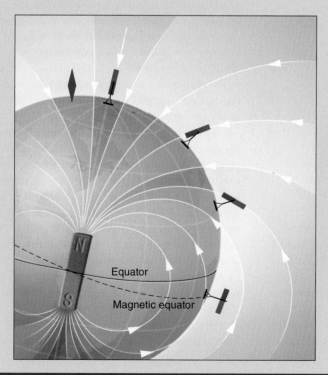

Earth's magnetic field, circling the planet, makes a magnetized dip needle point downward at an angle that equals the latitude of the needle's location. At the equator, the magnetic dip would be 0° (horizontal), and at the north magnetic pole the needle would point straight down (90°).

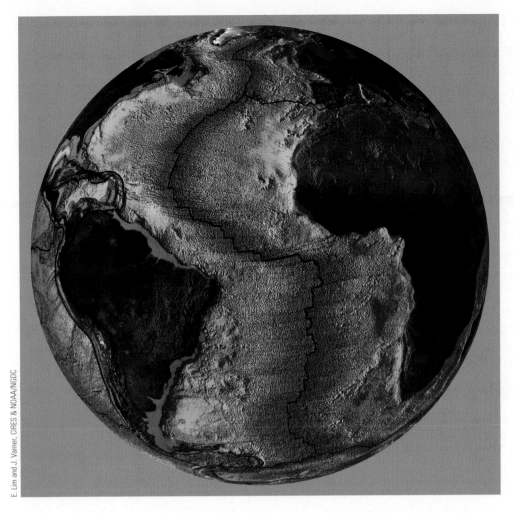

● **FIGURE 13.24** Basaltic rocks forming the floor of the Atlantic Ocean are youngest *(red)* along the extensive midoceanic ridge and become progressively older with increasing distance from the ridge axis.

*What is the relative age of oceanic crust along the East Coast of the United States?*

The young age of oceanic crust as a whole, its increasing age with both distance from the midoceanic ridges and proximity to the oceanic trenches, and the occurrence of deep-centered earthquakes along the trenches indicate that the oldest oceanic crust returns to the subsurface at the trenches near the ocean basin margins. **Subduction** is the process by which Earth material moves down to the subsurface in these zones.

These and other related scientific efforts discovered much new evidence that pointed to the horizontal movement of segments of the entire structural lithosphere, including the uppermost mantle, oceanic crust, and continental crust, rather than just the continents, as Wegener had suggested. Clearly, a new theory more expansive than Wegener's was needed to account for all of these observations.

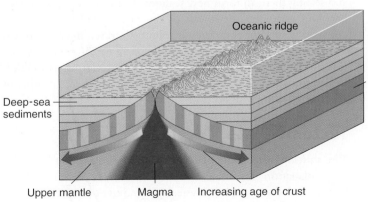

● **FIGURE 13.25** Seafloor spreading at an oceanic ridge produces new oceanic crust.

Oceanic ridge

Oceanic crust

Deep-sea sediments

Upper mantle    Magma    Increasing age of crust

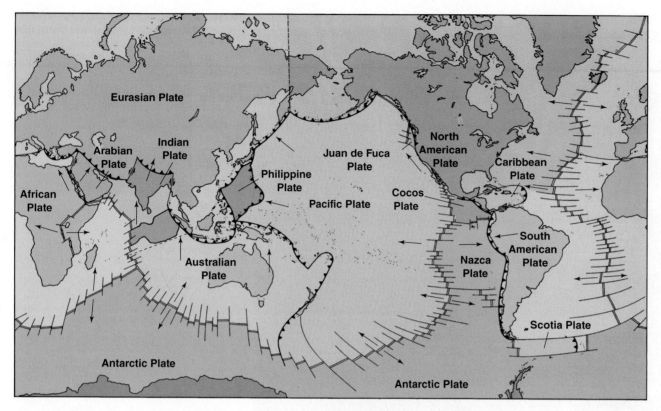

● **FIGURE 13.26** Earth's major lithospheric plates, also called tectonic plates, and their general directions of movement. Most tectonic and volcanic activity occurs along plate boundaries where the large segments separate, collide, or slide past each other. Barbs indicate where the edge of one plate is overriding another.

*Does every lithospheric plate include a continent?*

## The Theory of Plate Tectonics

**Plate tectonics** is the modern theory that explains seafloor spreading, subduction, and the movement of **lithospheric plates,** which are large segments of Earth's lithosphere that include continental crust, oceanic crust, and the uppermost rigid part of the mantle. *Tectonics* involves large-scale forces originating within Earth. According to the theory, thermal convection currents in the mantle cause the deformable, plastic asthenosphere to flow. As material in the asthenosphere flows horizontally, it carries along the overlying rigid and brittle lithosphere causing it to break into several large plates ( ● Fig. 13.26). Seven major plates have proportions as large as or larger than continents or ocean basins. Five other plates are of minor size, although they have maintained their own identity and direction of movement for some time. Several additional plates are even smaller and exist in active zones at the boundaries between major plates. All major plates consist of both continental and oceanic crust, although the largest, the Pacific plate, is primarily oceanic. In some places, plates pull away from each other (diverge), in other places they push together (converge), and elsewhere they slide alongside each other (move laterally).

Unlike Wegener's theory of continental drift, plate tectonics includes a plausible mechanism—convection in the mantle—

for lithospheric plate movement. Hot mantle material travels upward toward Earth's surface and cooler material moves downward as part of huge sublithosphere convection cells ( ● Fig. 13.27). The emergence of new oceanic crust is associated with the movement of lithospheric plates in opposite directions away from the midoceanic ridges. Where rising mantle material reaches the asthenosphere, it spreads out laterally and flows in opposite directions (diverges), pulling apart the rigid overlying lithosphere. Pulling apart the brittle lithosphere in the ocean basins breaks open a midoceanic ridge between two lithospheric plates. Molten basalt wells up into the fractures, cooling and sealing them to form new seafloor. In this process, the ocean becomes wider by the width of the now-sealed fracture. The convective motion continues as solidified crustal material moves away from the ridges. In a time frame of up to 250 million years, older oceanic crust is recycled into Earth's interior by subduction in the trenches located near plate boundaries where sections of the lithospheric plates meet.

## Tectonic Plate Movement

The shifting of tectonic plates relative to one another provides an explanation for many of Earth's surface features. Plate tectonics theory enables physical geographers to better under-

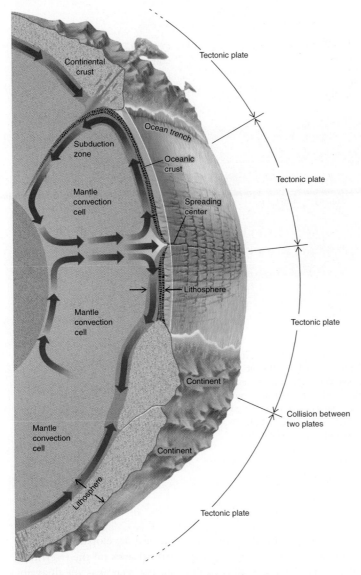

● **FIGURE 13.27** Convection is the mechanism for plate tectonics. Heat causes convection currents of material in the mantle to rise toward the base of the solid lithosphere where the flow becomes more horizontal. As the asthenosphere undergoes its slow, lateral flow, the overlying lithospheric plates are carried along because of friction at the boundary between the asthenosphere and lithosphere.

*Why is plate tectonics a better name than continental drift for the lateral movement of Earth's solid outer shell?*

stand not only our planet's ancient geography, but also the modern global distributions and spatial relationships among such diverse, but often related, phenomena as earthquakes, volcanic activity, zones of crustal movement, and major landform features ( ● Fig. 13.28). We will next examine the three ways in which lithospheric plates relate to one another along their boundaries as a result of tectonic movement: by pulling apart, pushing together, or sliding alongside each other.

**Plate Divergence** The pulling apart of plates, as occurs with seafloor spreading, is tectonic *plate divergence* (see Fig. 13.25). Tectonic forces that act to pull rock masses apart cause the crust to thin and weaken. Shallow earthquakes are often associated with this crustal stretching, and basaltic magma from the mantle wells up along crustal fractures. When oceanic crust is pulled apart the process creates new ocean floor as the plates move away from each other along a spreading center. The diverging plates separate at an average rate of 2 to 5 centimeters (1–2 in.) per year as they are carried along with the flowing plastic asthenosphere in the mantle. The formation of new crust in these spreading centers gives the label *constructive plate margins* to these zones. In some places, volcanoes, like those of Iceland, the Azores, and Tristan da Cunha, mark such boundaries ( ● Fig. 13.29).

Most plate divergence occurs along oceanic ridges, but this process can also break apart continental crust, eventually reducing the size of the continents involved ( ● Fig. 13.30a). The Atlantic Ocean floor formed as the continent that included South America and Africa broke up and moved apart a few centimeters per year over millions of years. The Atlantic Ocean continues to grow today at about the same rate. The best modern example of divergence on a continent is the rift valley system of East Africa, stretching from the Red Sea south to Lake Malawi. Crustal blocks that have moved downward with respect to the land on either side, with lakes occupying many of the depressions, characterize the entire system, including the Sinai Peninsula and the Dead Sea. Measurable widening of the Red Sea suggests that it may be the beginning of a future ocean that is forming between Africa and the Arabian Peninsula, similar to the young Atlantic between Africa and South America about 200 million years ago (Fig. 13.30b).

**Plate Convergence** A wide variety of crustal activity occurs at areas of tectonic *plate convergence*. Despite the relatively slow rates of plate movement in terms of human perception, the incredible energy involved in convergence causes the crust to crumple as one plate overrides another. Zones where plates are converging mark locations of major, and some of the tectonically more active, landforms on our planet. Deep trenches, volcanic activity, and mountain ranges may arise at convergent plate boundaries, depending on the type of crust involved in the plate collision. The distinctive spatial arrangement of these features worldwide can best be understood within the framework of plate tectonics.

If one or both margins of a convergent plate boundary consist of oceanic crust, the margin of one plate—always one composed of oceanic crust—is forced deep below the surface in the process of subduction. Deep ocean trenches, such as the Peru–Chile trench and the Japanese trench, occur where oceanic crust is dragged downward in this way. The subducting plate is heated and rocks are melted as it plunges downward into the mantle. As the subducting plate grinds downward, enormous friction is

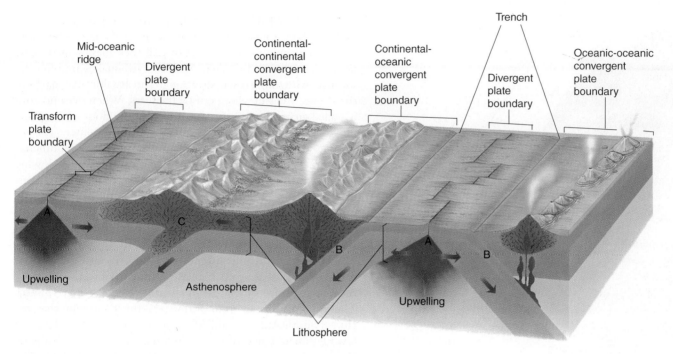

● **FIGURE 13.28** The plate tectonics system derives its energy from Earth's interior. As lithospheric plates move as a result of heat-driven convection cells in the mantle, they interact with adjoining plates, forming different boundary types, each displaying distinct landform features. Spreading centers *(A)* are divergent plate boundaries that have new crustal material emerging along active rift zones, eventually pushing older rock progressively away from the boundary in both directions. Earth's oceanic divergent plate boundaries form the midoceanic ridge system, which extends through all of the major oceans. Subduction zones *(B)* occur where two plates converge, with the margin of at least one of them consisting of oceanic crust. The plate with the denser oceanic crust is subducted beneath the less dense plate, whether it is continental or oceanic crust. Deep ocean trenches and either volcanic mountain ranges (continental crust) or island arcs (oceanic crust) lie along subduction zones. Continental collision zones *(C)* are places where two continental plates collide. Massive nonvolcanic mountains are built in those locations as the crust thickens because of compression.

produced, which explains the occurrence of major earthquakes in these regions.

Where oceanic crust collides with continental crust, the oceanic crust, which is denser, is subducted beneath the less dense continental crust ( ● Fig. 13.31). This is the situation along South America's Pacific coast, where the Nazca plate subducts beneath the South American plate, and in Japan, where the Pacific plate dips under the Eurasian plate. As oceanic crust, and the lithospheric plate of which it forms a part, is subducted, it descends into the asthenosphere to be partially melted and recycled into Earth's interior. Frequently, sediments deposited at continental margins are carried along down into the deep trenches. As these sediments melt, the resulting magma migrates upward into the overriding plate. Where molten rock reaches the surface, it produces a series of volcanic peaks, as in the Cascade Range of the northwestern United States. Rocks can also be squeezed and contorted between colliding plates, becoming uplifted and greatly deformed or metamorphosed. The great mountain ranges, such as the Andes, are produced at convergent plate margins by these processes.

Where oceanic crust lies on either side of a convergent plate boundary, the plate with the denser oceanic crust will subduct

below the other plate. Volcanoes may also develop at this type of boundary, creating major volcanic **island arcs** on the overriding plate. The Aleutians, the Kuriles, and the Marianas are all examples of island arcs lying near oceanic trenches that border the Pacific plate.

Continental crust converging with continental crust is termed **continental collision** and causes two continents or major landmasses to fuse or join together, creating a new larger landmass ( ● Fig. 13.32). This process closes an ocean basin that once separated the colliding landmasses and therefore has also been called *continental suturing*. The crustal thickening that occurs along this type of plate boundary generally produces major mountain ranges resulting from massive folding and crustal block movement rather than volcanic activity. The Himalayas, the Tibetan Plateau, and other high Eurasian ranges formed in this way as the plate containing the Indian subcontinent collided with Eurasia some 40 million years ago. India is still pushing into Asia today to produce the highest mountains in the world. In a similar fashion, the Alps were created as the African plate was thrust against the Eurasian plate.

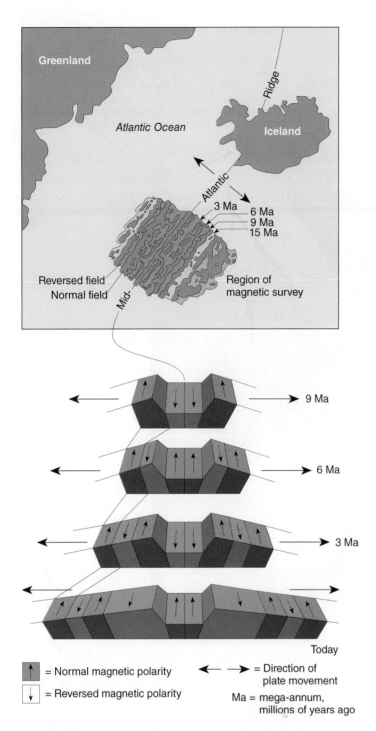

■ = Normal magnetic polarity     ◄──►= Direction of plate movement

↓ = Reversed magnetic polarity     Ma = mega-annum, millions of years ago

● **FIGURE 13.29** Iceland represents a part of the Mid-Atlantic Ridge that reaches above sea level, forming the volcanic island. The striped pattern of polarity reversals documented in the basaltic rocks along the Mid-Atlantic Ridge helped scientists understand the process of seafloor spreading.

**Transform Movement** Lateral sliding along plate boundaries, called **transform movement,** occurs where plates neither pull apart nor converge but instead slide past each other as they move in opposite directions. Such a boundary exists along the San Andreas Fault zone in California ( ● Fig. 13.33). Mexico's Baja Peninsula and Southern California are west of the fault on the Pacific plate. San Francisco and other parts of California east of the fault zone are on the North American plate. In the fault zone, the Pacific plate is moving laterally northwestward in relation to the North American plate at a rate of about 8 centimeters (3 in.) a year (80 km or about 50 mi per million years). If movement continues at this rate, Los Angeles will lie alongside San Francisco (725 km northwest) in about 10 million years and eventually pass that city on its way to finally colliding with the Aleutian Islands at a subduction zone.

Another type of lateral plate movement occurs on ocean floors in areas of plate divergence. As plates pull apart, they usually do so along a series of fracture zones that tend to form at right angles to the major zone of plate contact. These crosshatched plate boundaries along which lateral movement takes place are *transform faults.* Transform faults, or fracture zones, are common along midoceanic ridges, but examples can also be seen elsewhere, as on the seafloor offshore from the Pacific Northwest coast between the Pacific and Juan de Fuca plates (see Fig. 13.33). Transform faults are caused as adjacent plates travel at variable rates, causing lateral movement of one plate relative to the other. The most rapid plate motion is on the East Pacific rise where the rate of movement is more than 15 centimeters (6 in.) per year.

## Hot Spots in the Mantle

The Hawaiian Islands, like many major landform features, owe their existence to processes associated with plate tectonics. As the Pacific plate in that region moves toward the northwest, it passes over a mass of molten rock in the mantle that does not move with the lithospheric plate. Called **hot spots,** these almost stationary molten masses occur in a few other places in both continental and oceanic locations. Melting of the upper mantle and oceanic crust causes undersea eruptions and the outpouring of basaltic lava on the seafloor, eventually constructing a volcanic island. This process is responsible for building the Hawaiian Islands and the chain of islands and undersea volcanoes that extends for thousands of kilometers northwest of Hawaii. Today the hot spot causes active volcanic eruptions on the island of Hawaii. The other islands in the Hawaiian chain came from a similar origin, also having formed over the hot spot, but these volcanoes have now drifted along with the Pacific plate away from their magmatic source. Evidence of the plate motion is indicated by the fact that the youngest islands of the Hawaiian chain, Hawaii and Maui, are to the southeast and the older islands, such as Kauai and Oahu, are located to the northwest ( ● Fig. 13.34). A newly forming undersea volcano, named Loihi, is now developing southeast of the island of Hawaii and will someday be the next member of the Hawaiian chain.

**(a)**

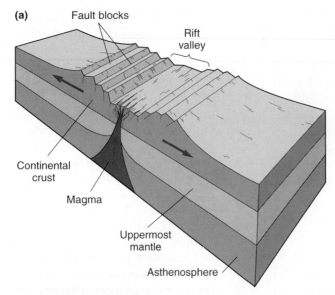

Fault blocks

Rift valley

Continental crust

Magma

Uppermost mantle

Asthenosphere

● **FIGURE 13.30** (a) A continental divergent plate boundary breaks continents into smaller landmasses. (b) The roughly triangular-shaped Sinai Peninsula, flanked by the Red Sea *(lower left)* to the south, Gulf of Suez *(photo center)* on the west, and Gulf of Aqaba *(lower right)* toward the east, illustrates the breakup of a continental landmass. The Red Sea rift and the narrow Gulf of Aqaba are both zones of spreading.

NASA, Johnson Space Center

**(b)**

● **FIGURE 13.31** An oceanic–continental convergent plate boundary where continent and seafloor collide. An example is the west coast of South America, where collision has formed the Andes and an offshore deep ocean trench.

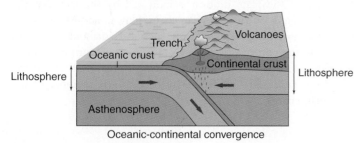

Oceanic-continental convergence

● **FIGURE 13.32** Continental collision along a convergent plate boundary fuses two landmasses together. The Himalayas, the world's highest mountains, were formed when India drifted northward to collide with Asia.

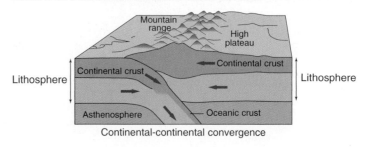

Continental-continental convergence

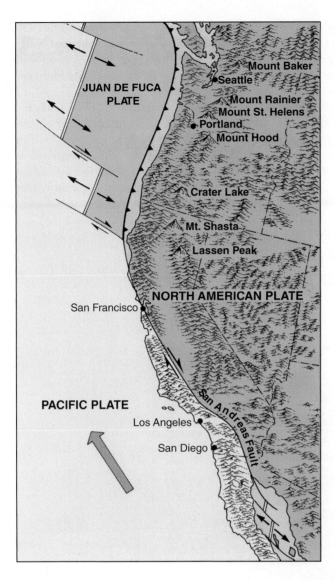

● **FIGURE 13.33** Along this lateral plate boundary, marked by the San Andreas Fault in western North America, the Pacific plate moves northwestward relative to the North American plate. Note that north of San Francisco the boundary type changes.

*What boundary type is found north of San Francisco, and what types of surface features indicate this change?*

● **FIGURE 13.34** Over the last few million years, a stationary zone of molten material in the mantle—a hot spot—has created each of the volcanic Hawaiian Islands in succession. Because they move to the northwest with the Pacific plate, the islands are progressively older to the northwest (ages are in millions of years). The hot spot is currently located at the island of Hawaii.

*Approximately how long did it take the Pacific plate to move Oahu to its current position?*

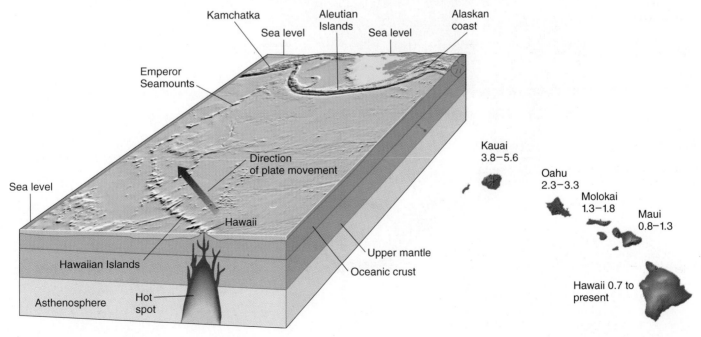

# Growth of Continents

The origin of continents is still being debated. It is clear that the continents tend to have a core area of very old igneous and metamorphic rocks that are interpreted as representing the deeply eroded roots of ancient mountains. These core regions have been worn down by hundreds of millions of years of erosion to create areas of relatively low relief that are located far from active plate boundaries. As a result, they have a history of tectonic stability over an immense period of time. These ancient crystalline rock areas are known as **continental shields** ( ● Fig. 13.35). The

Canadian, Scandinavian, and Siberian shields are outstanding examples. Around the peripheries of the exposed shields, flat-lying, younger sedimentary rocks at the surface indicate the presence of a stable and rigid rock mass below, as in the American Midwest, western Siberia, and much of Africa.

Most Earth scientists consider continents to grow by **accretion,** that is, by adding numerous chunks of crust to the main continent by collision. Western North America grew in this manner over the past 200 million years by adding segments of crust, known as *microplate terranes,* as it moved westward over the Pacific and former oceanic plates. Paleomagnetic data show that

● **FIGURE 13.35** Map of North America showing the continental shield (Canadian shield) and the general ages of rocks.

*Going outward from the shield toward the coast, what generally happens to the age of the rocks?*

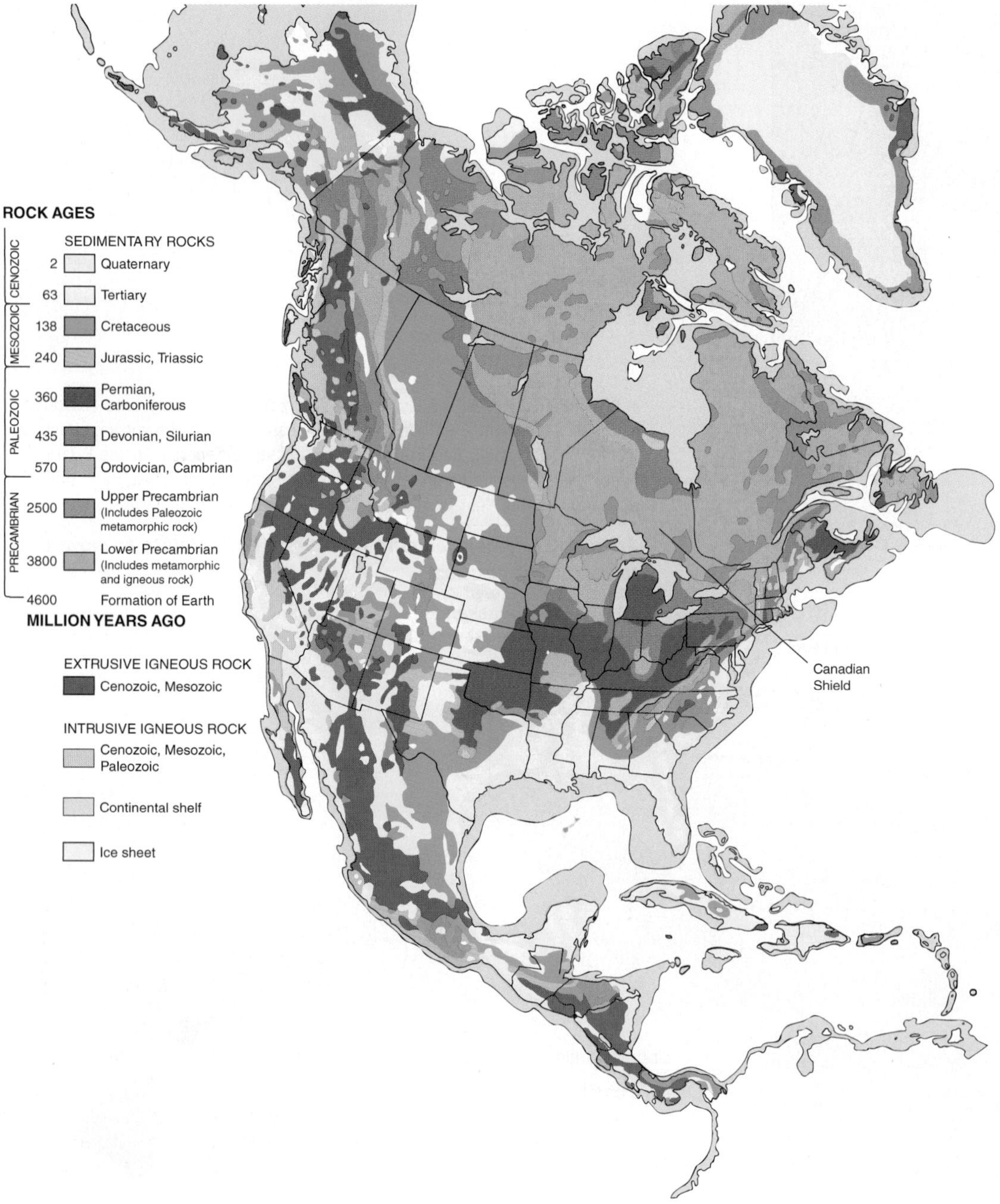

ROCK AGES

SEDIMENTARY ROCKS

| CENOZOIC | 2 | Quaternary |
| | 63 | Tertiary |
| MESOZOIC | 138 | Cretaceous |
| | 240 | Jurassic, Triassic |
| PALEOZOIC | 360 | Permian, Carboniferous |
| | 435 | Devonian, Silurian |
| | 570 | Ordovician, Cambrian |
| PRECAMBRIAN | 2500 | Upper Precambrian (Includes Paleozoic metamorphic rock) |
| | 3800 | Lower Precambrian (Includes metamorphic and igneous rock) |
| | 4600 | Formation of Earth |

**MILLION YEARS AGO**

EXTRUSIVE IGNEOUS ROCK
- Cenozoic, Mesozoic

INTRUSIVE IGNEOUS ROCK
- Cenozoic, Mesozoic, Paleozoic

- Continental shelf

- Ice sheet

Canadian Shield

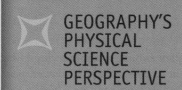
# Isostasy: Balancing Earth's Lithosphere

Structurally, the solid uppermost mantle, oceanic crust, and continental crust constitute the rigid and brittle lithosphere, which rests on top of the plastic, deformable asthenosphere. Mantle material in the asthenosphere flows like a very thick fluid at a rate of about 2 to 5 centimeters (1–2 in.) per year. The lithosphere is broken into several plates (segments) that behave like rafts moving along with currents in the flowing asthenosphere. The plates float because material in the lithosphere is less dense than material in the asthenosphere.

Because of the principle of buoyancy an object will float in a fluid as long as its weight per unit volume (specific weight) is less than that of the fluid. The volume of water displaced by a floating object is the amount that has the same total weight as the object. The difference in weight per unit volume between the object and the fluid is represented by the proportion of the object that floats above the surface. An iceberg having 90% of the specific weight of ocean water floats with 10% of the iceberg extending above the water surface. As long as the specific weight of a cargo ship is less than that of the water, a balance (equilibrium) will be maintained and the ship will float. If the ship takes on so much cargo that its weight per unit volume exceeds that of the water, the ship will sink because it will displace a volume of water that exceeds the volume of the ship.

Isostasy is the term for a similar concept regarding the equalization of hydrostatic pressure (fluid balance) between the lithosphere and asthenosphere. Isostasy suggests that a column of lithosphere (and the overlying hydrosphere) anywhere on Earth weighs about the same as a column of equal diameter from anywhere else, regardless of vertical thickness. The lithosphere is thicker (taller and deeper) where it contains a high percentage of low-density materials and thinner where it contains more high-density materials. Oceanic crust is thinner than continental crust because oceanic crust has a higher density than continental crust.

If an additional load is placed on an area of Earth's surface by a massive accumulation of glacial ice, lake water, or sediments, the lithosphere there will subside in a process called isostatic depression until it attains a new equilibrium level. If the surface accumulation is later removed, the region will tend to rise in a process called isostatic rebound. Neither subsidence nor uplift of the lithosphere will be instantaneous because flow in the asthenosphere is only a few centimeters per year.

Isostasy suggests that mountains are made of relatively low-density crustal materials and thus exist in areas of very thick crust, whereas regions of low elevation have thin, high-density crust. Similarly, a tall iceberg requires a massive amount of ice below the surface in order to expose ice so high above sea level, and as ice above the surface melts, ice from below will rise above sea level to replace it until the iceberg has completely melted.

Isostatic balance helps to explain many aspects of Earth's surface, including the following:

- Why most of the continental crust lies above sea level.
- Why wide areas of the seafloor are at a uniform depth.
- Why many mountain ranges continue to rise even though erosion removes material from them.
- Why some regions where rivers are depositing great amounts of sediments are subsiding.
- Why the crust subsided in areas that were covered by thick accumulations of ice during the last glacial age and now continues to rebound after deglaciation.

The density of ice is 90% that of water, thus icebergs (and ice cubes) float with 90% of their volume below the water surface and 10% above it.

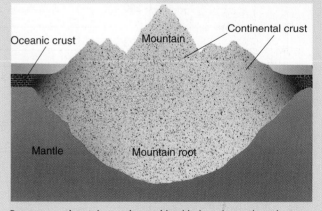

Because continental crust is considerably less dense than the material in the asthenosphere, where continental crust reaches high elevations it also extends far below the surface. Oceanic crust is also less dense than mantle material, but because it is denser than continental crust, it is thinner than continental crust.

parts of western North America from Alaska to California originated south of the equator and moved to join the continent. Terranes, which have their own distinct geology from that of the continent to which they are now joined, may have originally been offshore island arcs, undersea volcanoes, or islands made of continental fragments, such as New Zealand or Madagascar are today.

## Geologic Time and Paleogeography

The study of past geographic environments is known as **paleogeography.** The goal of paleogeography is to try to reconstruct the past environment of a geographic region based on geologic and climatic evidence. For students of physical geography, it may seem that the present is complex enough without trying to know what the geography of ancient times was like. However, peering into the past helps us forecast and prepare for changes in the future.

The immensity of geologic time over which major events or processes (such as plate tectonics, ice ages, or the formation and erosion of mountain ranges) have taken place is difficult to picture in a human time frame of days, months, and years. The geologic timescale is a calendar of Earth history (Table 13.2). It is divided into *eras,* which are typically long units of time, such as the Mesozoic Era (*meso,* middle; *zoic,* life), and eras are divided into *periods,* such as the Cretaceous Period. *Epochs*—for example, the Pleistocene Epoch (recent ice ages)—are shorter time units and are used to subdivide the periods of the Cenozoic

**TABLE 13.2**
**Geologic Timescale**

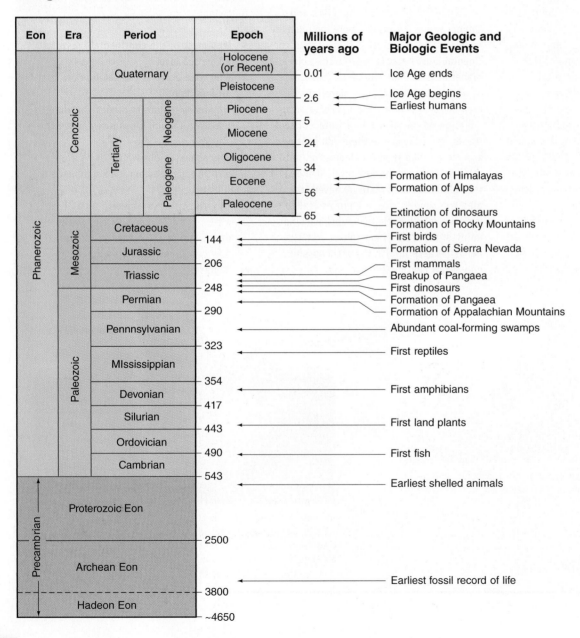

| Eon | Era | Period | | Epoch | Millions of years ago | Major Geologic and Biologic Events |
|---|---|---|---|---|---|---|
| Phanerozoic | Cenozoic | Quaternary | | Holocene (or Recent) | 0.01 | Ice Age ends |
| | | | | Pleistocene | 2.6 | Ice Age begins / Earliest humans |
| | | Tertiary | Neogene | Pliocene | 5 | |
| | | | | Miocene | 24 | |
| | | | Paleogene | Oligocene | 34 | |
| | | | | Eocene | 56 | Formation of Himalayas / Formation of Alps |
| | | | | Paleocene | 65 | Extinction of dinosaurs |
| | Mesozoic | Cretaceous | | | 144 | Formation of Rocky Mountains / First birds / Formation of Sierra Nevada |
| | | Jurassic | | | 206 | First mammals |
| | | Triassic | | | 248 | Breakup of Pangaea / First dinosaurs / Formation of Pangaea / Formation of Appalachian Mountains |
| | Paleozoic | Permian | | | 290 | |
| | | Pennnsylvanian | | | 323 | Abundant coal-forming swamps |
| | | MIssissippian | | | 354 | First reptiles |
| | | Devonian | | | 417 | First amphibians |
| | | Silurian | | | 443 | First land plants |
| | | Ordovician | | | 490 | First fish |
| | | Cambrian | | | 543 | Earliest shelled animals |
| Precambrian | | Proterozoic Eon | | | 2500 | |
| | | Archean Eon | | | 3800 | Earliest fossil record of life |
| | | Hadeon Eon | | | ~4650 | |

Era (*ceno,* recent; *zoic,* life), for which geologic evidence is more abundant. Today we are in the Holocene Epoch (last 10,000 years) of the Quaternary Period (last 2.6 million years) of the Cenozoic Era (last 65 million years). In a sense, these divisions are used like we would use days, months, and years to record time.

If we took a 24 hour day to represent the approximately 4.6 billion year history of Earth, the Precambrian, an era of which we know very little, would consume the first 21 hours. The current period, the Quaternary, which has lasted about 2.6 million years, would take less than 30 seconds and human beginnings, over about the last 4 million years, would take about 1 minute.

Each era, period, and epoch in Earth's geologic history had a unique paleogeography with its own distribution of land and sea, climate regions, plants, and animal life. If we look at evidence for the paleogeography of the Mesozoic Era (248 million to 65 million years ago), for instance, we would find a much different physical geography than exists now. This was a time when the supercontinents, Gondwana and Laurasia, each gradually split apart as new ocean floors widened, creating the continents that are familiar to us today. Global and local Mesozoic climates were very different from those of today but were changing as North America drifted to the northwest. During the Cretaceous Period, much of the present United States experienced warmer climates than now. Ferns and conifer forests were common. The Mesozoic was the "age of the dinosaurs," a class of large animals that ruled the land and the sea. Other life also thrived, including marine plants and invertebrates, insects, mammals, and the earliest birds.

The Mesozoic Era ended with an episode of great extinctions, including the end of the dinosaurs. Geologists, paleontologists, and paleogeographers are not in agreement as to what caused these great extinctions. Some of the strongest evidence points to a large meteorite striking Earth 65 million years ago, disrupting global climate and causing global environmental change. Other evidence points to plate tectonic changes in the distribution of oceans and continents or increased volcanic activity, either of which could cause rapid climate changes that might possibly trigger mass extinctions.

Available maps depicting Earth in early geologic times show only approximate and generalized patterns of mountains, plains, coasts, and oceans, with the addition of some environmental characteristics. These maps portray a general picture of how global geography has changed through geologic time ( ● Fig. 13.36). Much of the evidence and the rocks that bear this information have

● **FIGURE 13.36** Paleomaps showing Earth's tectonic history over the last 250 million years.

*How has the environment at the location where you live changed through geologic time?*

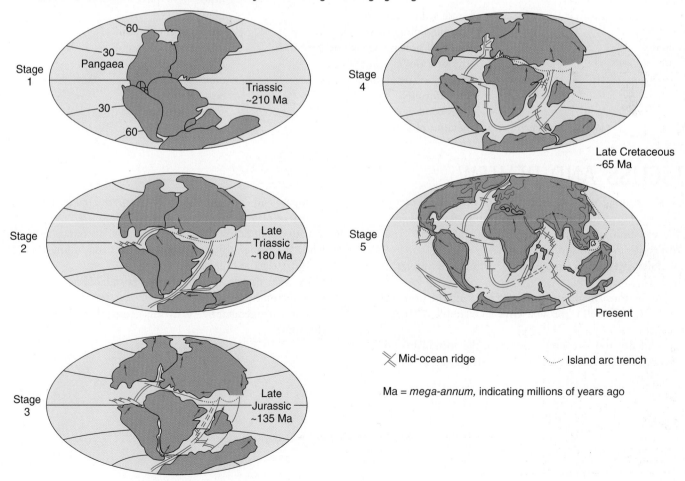

been lost through metamorphism or erosion, buried under younger sediments or lava flows, or recycled into Earth's interior. The further back in time, the sketchier is the paleoenvironmental information presented on the map. Paleomaps, like other maps, are simplified models of the regions and times they represent.

As time passes and additional evidence is collected, paleogeographers may be able to fill in more of the empty spaces on those maps of the past that are so unfamiliar to us. These paleo-geographic studies aim not only at understanding the past, but also at understanding today's environments and physical landscapes, how they have developed, and how processes act to change them. By applying the concept of uniformitarianism and the theory of plate tectonics to our knowledge of how the Earth system and its subsystems function, we can gain a better understanding of our planet's geologic past, as well as its present, and this will facilitate better forecasts of its potential future.

# CHAPTER 13 ACTIVITIES

## ■ DEFINE AND RECALL

seismic wave
seismograph
core
inner core
outer core
mantle
Mohorovičić discontinuity (Moho)
crust
oceanic crust
continental crust
elastic solid
lithosphere (an element of planetary structure)
asthenosphere
plastic solid
tectonic force
mineral
rock
bedrock

regolith
outcrop
igneous rock
magma
lava
extrusive igneous rock
pyroclastics (tephra)
intrusive igneous rock
joint
sedimentary rock
clast
clastic sedimentary rock
organic sedimentary rock
chemical precipitate sedimentary rock
stratification
bedding plane
unconformity
cross bedding
metamorphic rock

foliation
rock cycle
catastrophism
uniformitarianism
continental drift
paleomagnetism
seafloor spreading
polarity reversal
subduction
plate tectonics
lithospheric plate
island arc
continental collision
transform movement
hot spot
continental shield
accretion
paleogeography

## ■ DISCUSS AND REVIEW

1. Name the major zones of Earth's interior from the center to the surface. How do these zones differ from one another?
2. Define and distinguish (a) continental crust and oceanic crust, and (b) the lithosphere and asthenosphere.
3. How do minerals differ from rocks? Provide examples of each.
4. Describe the three major categories of rock and the principal means by which each is formed. Give an example of each.
5. What is the rock cycle?
6. What evidence did Wegener rely on in the formulation of his theory of continental drift? What evidence did he lack?

7. Compare and contrast the theory of continental drift and the theory of plate tectonics.
8. What type of lithospheric plate boundary is found (a) paralleling the Andes, (b) at the San Andreas Fault, (c) in Iceland, and (d) near the Himalayas?
9. How does the formation of the Hawaiian Islands support plate tectonic theory?
10. Define paleogeography. Why are geographers interested in this topic?

## ■ CONSIDER AND RESPOND

1. What evidence has been found to support the theory that lithospheric plates move around on Earth's surface?
2. How is the concept of uniformitarianism related to plate tectonics theory and the arrangement of the continents today?

3. Explain why the eastern portion of the United States has relatively little tectonic activity compared with the western portion of the United States.

## ■ APPLY AND LEARN

1. Two plates are diverging and both are moving at a rate of 3 centimeters per year. How long will it take for them to move 100 kilometers apart?
2. An area of oceanic crust has a density of 3.0 grams per cubic centimeter and a thickness of 4 kilometers. An area of con-

tinental crust has a density of 2.7 grams per cubic centimeter. How thick in kilometers would the continental crust have to be for its total mass to equal that of the oceanic crust?

 CourseMate – Make the most of your study time by accessing everything you need to succeed in one place. Read your textbook, take notes, review flashcards, watch videos, complete activities, take practice quizzes, and more online with CourseMate. Log in at **www.cengagebrain.com**.

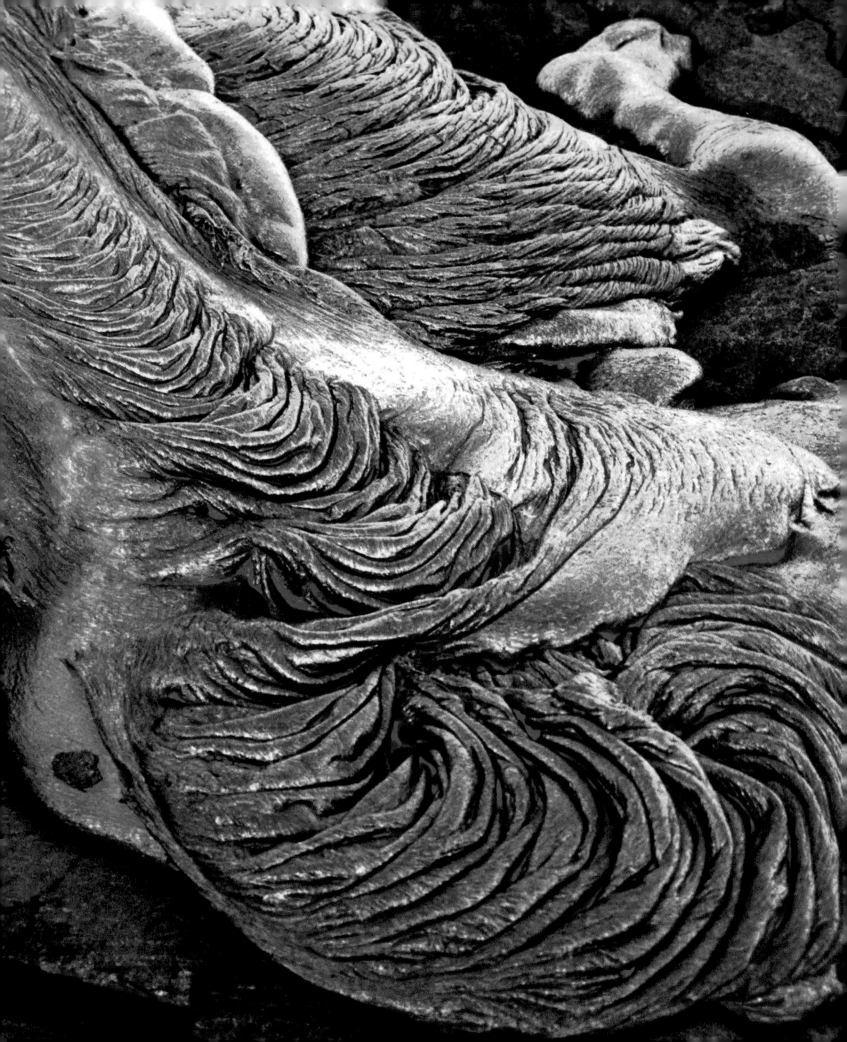

# 14

# TECTONIC AND VOLCANIC PROCESSES AND LANDFORMS

## ■ OBJECTIVES

WHEN YOU COMPLETE THIS CHAPTER YOU SHOULD BE ABLE TO:

■ Distinguish the endogenic from the exogenic landforming processes.
■ Demonstrate how compressional, tensional, and shearing forces each stress rock matter.
■ Sketch examples of the different types of faults, indicating direction of motion.
■ Associate the different types of faults with the type of tectonic force responsible for them.
■ Draw a cross section of a fault that shows the relationship between an earthquake's focus and epicenter.
■ Discuss the two ways in which the severity of an earthquake is measured.
■ List various factors that help determine the devastation caused by an earthquake.
■ Explain why some volcanic eruptions are explosive and others are effusive.
■ Describe the principal differences among the six major types of volcanic landforms.
■ Differentiate the various types of igneous intrusions.

OUR PLANET'S SURFACE **TOPOGRAPHY,** THE distribution of landscape highs and lows, is complex and intriguing. Landscapes consist of rugged mountains, gently sloping plains, rolling hills and valleys, or elevated plateaus cut by steep canyons. These are just a few examples of the types of surface terrain features, referred to as **landforms,** that contribute to the beauty and diversity of Earth's environments. Landforms are one of the most appealing and impressive elements of Earth's surface. Each year, local, state, and national parks attract millions of visitors seeking to observe and experience firsthand spectacular examples of landforms. Landforms owe their existence and development to processes and materials that originate in Earth's interior, at its surface, or, most typically, some combination of both.

Understanding landforms and landscapes—how they originate, why they vary, and their significance in a local, regional, or global context—is the primary goal of **geomorphology,** a major subfield of physical geography devoted to the scientific study of landforms. Geomorphologists seek explanations for the origin, shape, and spatial distribution of terrain features of all kinds and for the processes that modify

Basaltic lava flows from an erupting volcano on the island of Hawaii. Lava that has solidified and turned dark in color at the surface is carried along by the hot, red molten lava that is still flowing. *Copyright and photograph by Dr. Parvinder S. Sethi*

and destroy them. Landforms represent the interplay of one set of processes that elevate, depress, or disrupt Earth's surface, creating topographic inequalities, and another set of processes that wear down, fill in, and work to level the landscape.

Landforming *tectonic processes* (from Greek: *tekton,* carpenter, builder), which are movements of parts of the crust and upper mantle, and *igneous processes* (from Latin: *ignis,* fire), which are related to the eruption and solidification of molten rock matter, constitute the primary geomorphic mechanisms that increase the topographic irregularities on Earth's surface. These tectonic and igneous forces originate in the planet's interior, yet they contribute significantly to the nature of Earth's exterior topography, affecting the composition, shape, size, and appearance of surface terrain features in many regions. Areas of the crust can be uplifted or downdropped by tectonism, or built up by the ejection or intrusion of molten rock matter from Earth's interior through igneous processes. Tectonic and igneous processes build extensive mountain systems and a great variety of other landforms. The geographic distribution of these terrain features, moreover, is not random. The greatest tectonic forces, most dangerous earthquake zones, and largest mountain ranges lie along lithospheric plate boundaries, and plate margins are the most common locations for extensive volcanic landforms.

Tectonic and igneous processes have produced many impressively scenic landscapes, but they also present serious natural hazards to people and their property. This chapter and those that follow focus on understanding how various landforms and landscapes develop and on the potential hazards that are related to them. It is extremely important for us to understand how geomorphic processes work to shape Earth's surface landforms because they are active, ongoing, and often powerful processes that affect human welfare. Landforms are a fundamental, dynamic, beautiful, diverse, and sometimes dangerous aspect of the human habitat.

# Landforms and Geomorphology

A fundamental characteristic of all landforms and landscapes is their relative amount of **relief,** which is the difference in elevation between the highest and lowest points within a specified area or on a particular surface feature ( ● Fig. 14.1). With no variations in relief our planet would be a smooth, featureless sphere and certainly much less interesting. It is hard to imagine Earth without dramatic terrain as seen in the high-relief mountainous regions of the Himalayas, Alps, Andes, Rockies, and Appalachians or in the huge chasm that we call the Grand Canyon. Interspersed with high-relief features, large expanses of low-relief features, like the Great Plains, are equally impressive and inspiring.

Earth's landforms result from mechanisms that act to increase relief by raising or lowering the land surface and from mechanisms that work to reduce relief by removing rock matter from high places and using it to fill in depressions. Geomorphic processes that originate within Earth, called **endogenic processes** (*endo,* within; *genic,* originating), tend to increase the amount of surface relief, whereas the **exogenic processes** (*exo,* external), those that originate at Earth's surface, work to decrease relief. Tectonic and igneous processes constitute the endogenic geomorphic processes. Exogenic processes consist of various means of rock breakdown, collectively termed *weathering,* and the removal, movement, and relocation of those weathered rock products in the continuum of processes known as **erosion, transportation,** and **deposition.** Erosion, transportation, and deposition act through the force of gravity alone, as in the fall of a weathered clast from a cliff to the ground below, or operate with the help of a **geomorphic agent,** a medium that picks up, moves, and eventually lays down pieces of broken rock matter. The most common geomorphic agents are flowing water, wind, moving ice, and waves, but organisms, including people, also accomplish some erosion, transportation, and deposition of weathered pieces of Earth material.

Mountains and other landforms of high relief exist despite the leveling tendencies of exogenic processes because of various factors, either alone or in combination. In many cases high-relief features endure because they are made predominantly of resistant rocks, on which weathering and erosion proceed slowly. Some highland areas have originated so recently that exogenic processes have had insufficient time to make a substantial impact on them. High-relief features also persist where endogenic processes operate, or have operated, at faster rates than weathering and erosion ( ● Fig. 14.2). In contrast, areas of low relief tend to represent terrain dominated by exogenic processes. In those regions, sediment eroded and transported from higher elevations has been deposited in low areas, thereby reducing topographic irregularities.

Natural landforming processes proceed at a wide variety of rates, usually expressed in millimeters or centimeters of change per year, century, or millennium, averaged over the time span of available data. However, it is important to remember that these are average rates and that many natural processes operate in an episodic, rather than in a steady and continuous, manner. In episodic processes, a period of relative calm and little change continues until a certain threshold condition is reached, resulting in a short burst of rapid, often intense activity that causes significant environmental

**(a)**                                                                    **(b)**

● **FIGURE 14.1** An area displaying (a) low relief in western Utah lies about 100 kilometers (60 mi) east of (b) an area of high relief in Great Basin National Park of eastern Nevada.

● **FIGURE 14.2** The Teton range of Wyoming stands high above the valley floor because rates of uplift resulting from endogenic processes exceed the rates of the exogenic processes of weathering and erosion.

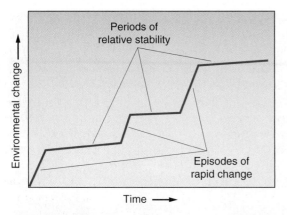

● **FIGURE 14.3** Punctuated equilibrium describes Earth processes that operate slowly most of the time but that are occasionally interrupted by events that cause relatively major change in a landscape or natural system in a short time. A great variety of processes that affect the Earth system operate in this episodic manner.

*What Earth processes can you think of that operate this way?*

modification. The exact timing of these periods of increased activity is difficult or impossible to predict with certainty. Many physical processes in the Earth system operate according to this *punctuated equilibrium* ( ● Fig. 14.3). Earthquakes and volcanic eruptions provide good examples of this type of Earth change. Long periods of limited activity are occasionally interrupted by large earthquakes or major eruptions that are short-lived but cause great change in the landscape. After an occurrence of major change, the landscape adjusts to the newly created conditions and stays in relative equilibrium, experiencing comparatively limited modification by exogenic processes until the next major event again disrupts the system ( ● Fig. 14.4). Like endogenic processes, exogenic processes may

also operate episodically with extended periods of slow or moderate activity interrupted by short periods of rapid change, such as occurs during a stream flood or a landslide.

In this chapter we study the processes related to the buildup of relief through tectonic and igneous activity and we examine the rock structures and landforms associated with these endogenic processes. Subsequent chapters focus on the exogenic processes and the landforms and landscapes formed by the various geomorphic agents.

# Tectonic Forces, Rock Structure, and Landforms

Tectonic forces, which at the largest scale move the lithospheric plates, also cause bending, warping, folding, and fracturing of Earth's crust at continental, regional, and even local scales. Such deformation is documented by **rock structure**, the nature, orientation, inclination, and arrangement of affected rock layers. For example, rock layers that have undergone significant tectonic forces may be tilted, folded, or fractured or, relative to adjacent rocks masses, offset, uplifted, or downdropped. Sedimentary rocks are particularly useful for identifying tectonic deformation because most are originally horizontal, with each successive, overlying rock layer younger than the one beneath it. If strata are bent, fractured, offset, or otherwise out of sequence, some kind of structural deformation has occurred.

Earth scientists describe the orientations of inclined rock layers by measuring their strike and dip. **Strike** is the compass direction of the line that forms at the intersection of a tilted rock layer and a horizontal plane ( ● Fig. 14.5). A rock layer, for example, might strike northeast. Note that it would be equally correct to

● **FIGURE 14.4** (a) Mount Vesuvius, near Naples, Italy, destroyed the ancient city of Pompeii in an eruption in A.D. 79. The eruption is an example of an episodic process. (b) A plaster cast shows a victim who attempted to cover his face from hot gases and the volcanic ash that buried Pompeii.

(a)

(b)

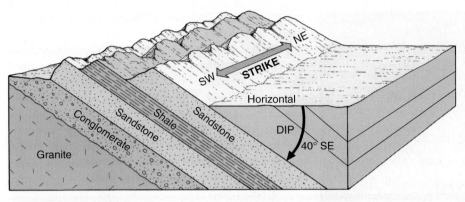

**● FIGURE 14.5** Geoscientists use strike and dip to describe the orientation of tilted rock layers.

*Does a strike to the southwest differ from a strike to the northeast?*

express that orientation as a southwest strike. The inclination of the rock layer, the **dip,** is always measured at right angles to the strike and in degrees of angle from the horizontal (0° dip = horizontal). The direction toward which the rock dips down is provided with the general compass direction. For example, a rock layer that strikes northeast and is inclined 11° from the horizontal down to the southeast would be described as having a dip of 11° to the southeast (see Fig. 14.5).

Earth's crust has been subjected to tectonic forces throughout its history, although the forces have been greater during some geologic periods than others and have varied widely over Earth's surface. Most of the resulting changes in the crust have occurred slowly over hundreds of thousands or millions of years, but others have been rapid and cataclysmic. The response of crustal rocks to tectonic forces yields a variety of configurations in rock structure, depending on the nature of the rocks and the nature of the applied forces.

Tectonic forces are divided into three principal types that differ in the direction of the applied forces (● Fig. 14.6). **Compressional tectonic forces** push crustal rocks together. **Tensional tectonic forces** pull parts of the crust away from each other. **Shearing tectonic forces** slide parts of Earth's crust past each other.

## Compressional Tectonic Forces

Tectonic forces that push two areas of crustal rocks together tend to shorten and thicken the crust. How the affected rocks respond to compressional forces depends on how brittle (breakable) the rocks are and the speed with which the forces are applied. **Folding,** which is a bending or crumpling of rock layers, occurs when compressional forces are applied to rocks that are *ductile* (bendable), as opposed to

brittle. Rocks that lie deep within the crust and that are therefore under high pressure are generally ductile and particularly susceptible to behaving plastically, that is, to deforming without breaking. As a result, rocks deep within the crust typically fold rather than break in response to compressional forces (● Fig. 14.7). Folding is also more likely than fracturing when the compressional forces are applied slowly. Eventually, however, if the force per unit area, the *stress,* is great enough, the rocks may still break with one section pushed over another.

As elements of rock structure, upfolds are called **anticlines** and downfolds are called **synclines** (● Fig. 14.8). The rock layers that form the flanks of anticlinal crests and synclinal troughs are the *fold limbs*. Folds in some rock layers are very small, covering a few centimeters, whereas others are enormous with vertical distances between the upfolds and downfolds measured in kilometers. Folds can be tight or broad, symmetrical or asymmetrical. Folds are symmetrical—that is, each limb has about the same dip angle—if they were formed by compressional forces that were relatively equal from both sides. If compressional forces were stronger from one side, a fold may be asymmetrical, with the dip of one limb being much steeper than that of the other. Eventually, asymmetrically folded rocks may become overturned and perhaps so compressed that the fold lies horizontally; these are known as **recumbent folds** (see Fig. 14.8).

**● FIGURE 14.6** Three types (directions) of tectonic force cause deformation of rock layers. (a) Compressional forces push rocks together and can bend (fold) rocks or cause rocks to break and slide along the breakage zone (fault). (b) Tensional forces pull rocks apart and may also lead to the breaking and shifting of rock masses along faults. (c) Shearing forces slide rocks past each other horizontally, and this likewise can cause movement along a fault.

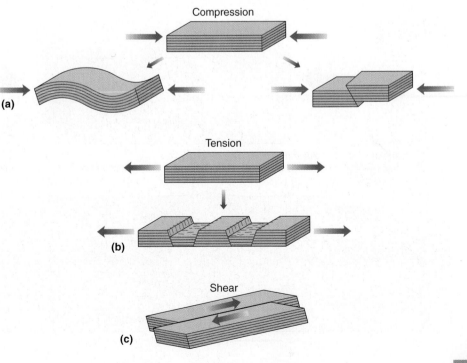

● **FIGURE 14.7** Compressional forces have made complex folds in these layers of sedimentary rock.

*How can solid rock be folded without breaking?*

Almost all mountain systems exhibit some degree of folding. Much of the Appalachian Mountain system is an example of folding on a large scale. Spectacular folds exist in the Rocky Mountains of Colorado, Wyoming, and Montana and in the Canadian Rockies. Highly complex folding created the European Alps, where folds are overturned, sheared off, and piled on top of one another.

Rock layers that are near Earth's surface and not under high confining pressures are too rigid to bend into folds when experiencing compressional forces. If the tectonic force is large enough, or if it is applied rapidly, these rocks will break rather than bend and the rock masses will move relative to each other along the fracture. **Faulting** is the slippage or displacement of rocks along a fracture surface, and the fracture along which movement has occurred is a **fault.**

When compressional forces cause faulting, either one mass of rock is pushed up relative to the other mass of rock along a steep-angled fault or one mass of rock slides along a shallow, low-angle fault over the other rock mass. The steep, high-angle fault resulting from compressional forces is termed a **reverse fault** ( ● Fig. 14.9a). Where compression pushes a mass of rock along a low-angle fault so that it overrides rocks on the other side of the fault, the fracture surface is called a **thrust fault,** and the shallow displacement is an **overthrust** (Fig. 14.9b). Major overthrusts exist in the northern Rocky Mountains and in the southern Appalachians, and movement along a thrust fault related to a subduction zone caused the 2011 earthquake in Japan.

In both reverse and thrust faults, one block of crustal rocks is wedged up relative to the other. Direction of motion along faults is always given in relative terms because, even though it may seem obvious that one block was pushed up along the fault, the other block may have slid down some distance also, and it is not always possible to determine with certainty if one or both blocks moved.

● **FIGURE 14.8** Folded rock structures become increasingly complex as the applied compressional forces become more unequal from the two directions.

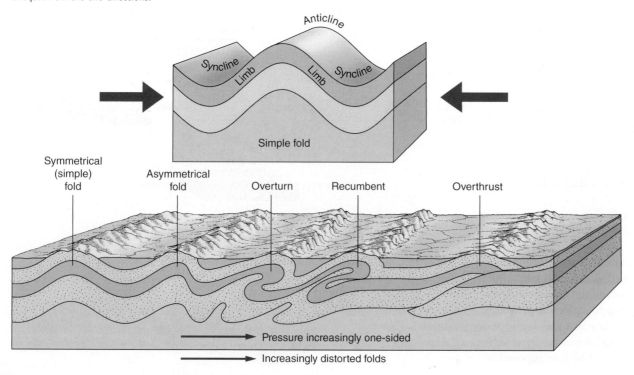

In some cases, reverse or thrust faulting resulting from the rapid application of compressional forces affects rocks that have already responded to the force by folding. Where folded rocks undergo thrust faulting, the upper part of a fold breaks, sliding over the lower rock layers along the low-angle fault forming an overthrust. Together, recumbent folds and overthrusts are important rock structures that have formed in complex mountain ranges such as the Andes, Alps, and Himalayas.

## Tensional Tectonic Forces

Tensional tectonic forces pull in opposite directions in a way that stretches and thins the impacted part of the crust. Rocks, however, typically respond to tensional forces by faulting, rather than bending or stretching plastically. Tensional forces commonly cause the crust to break into discrete blocks, called **fault blocks,** that are separated from each other by **normal faults** (Fig. 14.9c). To accommodate the extension of the crust, one crustal fault block slides downward along the normal fault relative to the adjacent fault block. Notice that the direction of motion along a normal fault is opposite to that along a reverse or thrust fault (see Fig. 14.9a).

In map view, tensional forces affecting a large region produce a repeated pattern of roughly parallel normal faults, creating a series of alternating downdropped and upthrown fault blocks. Each block that slid downward between two normal faults, or that remained in place while blocks on either side slid upward along the faults, is called a **graben** ( ● Fig. 14.10). A fault block that moved relatively upward between two normal faults—that is, it actually moved up or it remained in place while adjacent blocks slid downward—is a **horst.** The great Ruwenzori Range of East Africa is a horst, as is the Sinai Peninsula between the fault troughs in the Gulfs of Suez and Aqaba (see again Fig. 13.30). Horsts and grabens are rock structural features that are identified by the nature of the offset of rock units along normal faults; topographically, horsts form mountain ranges and grabens form basins. The large Basin and Range region of the western United States, which extends eastward from California to Utah and southward from Oregon to New Mexico, is undergoing tensional tectonic forces that are pulling the region apart to the west and east. A transect from west to east across that region, for example from Reno, Nevada, to Salt Lake City, Utah, encounters an extensive series of alternating downdropped and upthrown fault blocks comprising the basins and ranges for which the region is named. Some of the ranges and basins are simple horsts and grabens, but others are tilted fault blocks that resulted from the uplift of one side of a fault block while the other end of the same block rotated downward ( ● Fig. 14.11). Death Valley, California is a classic example of the downtilted side of a tilted fault block ( ● Fig. 14.12).

In some cases, large-scale tensional tectonic forces create **rift valleys,** which are long but relatively narrow zones of crust downdropped between normal faults. Examples of rift

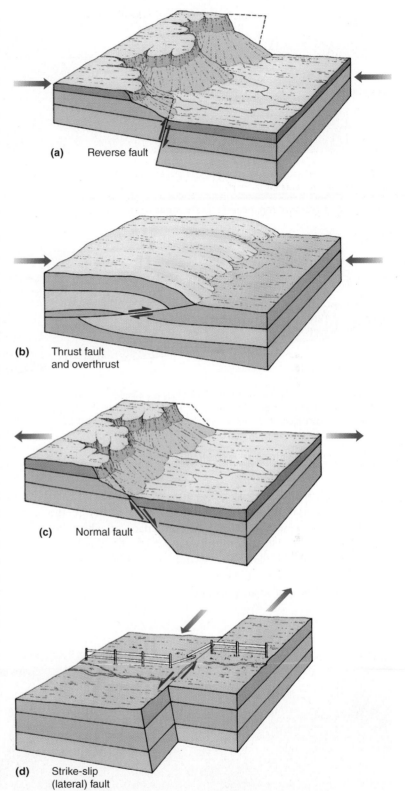

**(a)** Reverse fault

**(b)** Thrust fault and overthrust

**(c)** Normal fault

**(d)** Strike-slip (lateral) fault

● **FIGURE 14.9** The major types of faults and the tectonic forces that cause them (indicated by large arrows). Compressional forces form (a) reverse or (b) thrust faults. (c) Tensional tectonic forces break rocks along normal faults. (d) Shearing forces move rocks horizontally past each other along strike-slip (lateral) faults.

*How does motion along a normal fault differ from that along a reverse fault?*

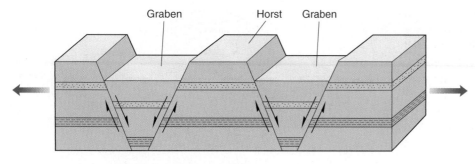

● **FIGURE 14.10** Horsts (upthrown blocks) and grabens (downdropped blocks) are bounded by normal faults.

*What kind of tectonic force causes these kinds of fault blocks?*

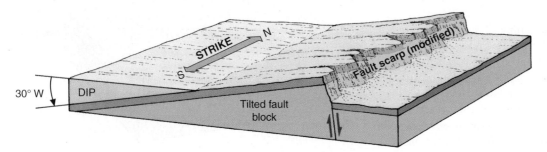

● **FIGURE 14.11** This diagram of a tilted fault block is a simplified version of the kind of faulting that produced Death Valley, which occupies the downtilted part of a tilted fault block. Here, the east-facing cliff is a fault scarp that has been worn back to lower slope by erosion.

*Is erosion an endogenic or an exogenic process?*

valleys include the Rio Grande rift of New Mexico and Colorado and the Great Rift Valley of East Africa. A rift valley in the Middle East contains the Dead Sea, the surface of which lies about 390 meters (1280 ft) below the level of the Mediterranean Sea, which is only 64 kilometers (40 mi) away. Other rift valleys extend along the crests of the oceanic ridges where diverging lithospheric plates pull apart oceanic crust.

An **escarpment,** often shortened to *scarp,* is a steep cliff, which may be tall or short. Scarps form on Earth surface terrain for many reasons and in many different settings. A cliff that results from movement along a fault is specifically a **fault scarp.** Fault scarps are commonly visible in the landscape along normal fault zones, where they may consist of impressive rock faces on fault blocks that have undergone extensive uplift over long periods. Much lower *piedmont fault scarps* instead offset unconsolidated sediments that exogenic processes have deposited in the normal fault zone along the base of the upthrown block ( ● Fig. 14.13).

Some fault scarps account for spectacular mountain walls, especially in regions like much of the western United States that have a history of recent tectonic activity. In southern Utah and northern Arizona, the northwestern part of the elevated Colorado Plateau steps down to the Great Basin in an impressive series of west-facing fault scarps. In California, the 645-kilometer-long (400 mi) Sierra Nevada Range is a great tilted fault block with the east side faulted upward and the west side tilted down (see Fig. 14.11). The east face of the Sierra Nevada essentially represents a fault scarp that rises abruptly a steep 3350 meters (11,000 ft) above the desert ( ● Fig. 14.14). In contrast, the west side of the Sierra (the "back slope") dips down gradually over a distance of 100 kilometers (60 mi), forming terrain that includes a band of foothills. Like the Sierra Nevada, the Grand Tetons of Wyoming rise dramatically along an east-facing fault scarp. Another excellent example of a fault scarp lies along the fault block across which the Rio Grande has carved Santa Elena Canyon in Big Bend National Park, Texas. Other than that 500-meter-deep (1640 ft) canyon, the fault

● **FIGURE 14.12** Death Valley, California, is a fine example of a topographic basin created by a tilted fault block. The lowest elevation in North America, 86 meters (282 ft) below sea level, is found on the valley floor.

● **FIGURE 14.13** This piedmont fault scarp in Nevada is the topographic expression of a normal fault.

*On which side of the fault does the horst lie?*

block has been so little modified by erosion in the arid climate that it retains much of its blocklike shape ( ● Fig. 14.15).

Major uplift of faulted mountain ranges has a strong impact on other aspects of the physical geographic system, as exemplified by the uplift of the Sierra Nevada, which began several million years ago. As the mountain range rose in response to endogenic processes, precipitation increased on its windward (western) side because of orographic lifting of air masses, and the greater elevation brought cooler temperatures. Along the steep, lee (eastern) side of the tilted Sierra fault block, the rain shadow effect led to increased aridity. These climate changes influenced the vegetation, soils, and animal life. Exogenic processes attacked the rocks as uplift progressed. Weathering, stream erosion, soil erosion, and the downslope gravitational

movement of rock material intensified as a result of the greater runoff, elevation, and slope. As in many high mountain ranges, glaciers formed, altering and etching the terrain further. Thus, over time exogenic processes carved and shaped valleys in the Sierra fault block leaving spectacular canyons and mountain peaks. High relief persists because the Sierra Nevada continues to rise rapidly, in a geologic sense, on average about a centimeter per year.

## Shearing Tectonic Forces

Vertical displacement along a fault occurs when the rocks on one side move up or drop down in relation to rocks on the other side. Faults with this kind of movement, up or down along the dip of the fault plane extending into Earth, are known as **dip-slip faults.** Normal and reverse faults have dip-slip motion. There also exists, however, a completely different category of fault along which displacement of rock units is horizontal rather than vertical. In this case, the direction of slippage is parallel to the surface trace, or strike, of the fault; thus it is called a **strike-slip fault** or, because of the horizontal motion, a **lateral fault** (see Fig. 14.9d). Offset along strike-slip faults is most easily seen in map view (from above), rather than in cross-sectional view. Active strike-slip faults cause horizontal displacement of roads, railroad tracks, fences, streambeds, and other features that extend across the fault. The direction of motion along a strike-slip fault is described as left lateral or right lateral. Direction of motion is determined by imagining yourself standing on one block looking across the strike-slip fault to assess if the other block moved to your left or right. Movement along a left-lateral fault caused the 2010 earthquake in Haiti. The San Andreas Fault, which runs through much of California, has right lateral strike-slip movement. A long and narrow, rather linear valley composed of rocks that have been

● **FIGURE 14.14** The east front of the Sierra Nevada in California is essentially the steep scarp side of a tilted fault block.

● **FIGURE 14.15** The steep fault scarp at Santa Elena Canyon, along the Texas–Mexico border, has undergone limited modification by weathering and erosion. The canyon has been carved by the Rio Grande during uplift of the fault block.

⊕ In Google Earth fly to: 29.165°N, 103.612°W (Santa Elena Canyon, Alpine, TX).

crushed and weakened by faulting marks the trace of the San Andreas Fault zone ( ● Fig. 14.16).

The amount that Earth's surface is offset during instantaneous movement along a fault varies from fractions of a centimeter to several meters. Faulting moves rocks laterally, vertically, or both.

● **FIGURE 14.16** The San Andreas fault extends from upper left to lower right across this false-color image of an area west of San Mateo, California. Crust and upper mantle to the west *(left)* of the fault is moving to the northwest relative to that on the east side of the fault.

*Is the San Andreas a left or right lateral fault?*

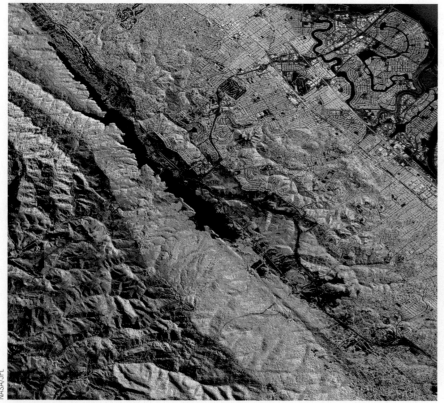

The maximum horizontal displacement along the San Andreas Fault in California during the 1906 San Francisco earthquake was more than 6 meters (21 ft). Vertical offset of more than 10 meters (33 ft) occurred during the Alaskan earthquake of 1964. Over millions of years, the cumulative displacement along a major fault may be tens of kilometers vertically or hundreds of kilometers horizontally, although the majority of faults have offsets that are much smaller than these values.

## Relationships between Rock Structure and Topography

Tectonic activity produces different types of structural features that range from microscopic fractures to major folds and fault blocks. At Earth's surface, these structural features comprise various landforms and are subject to modification by weathering, erosion, transportation, and deposition. It is important to distinguish between elements of rock structure and topographic features because rock structure reflects endogenic factors but landforms reflect the balance between endogenic and exogenic factors. As a result, a specific type of structural element can assume a variety of topographic expressions ( ● Fig. 14.17). For instance, a structural upfold is an anticline even though geomorphically it may comprise a ridge, a valley, or a plain, depending on surface erosion processes and the nature of broken or weak rocks. Nashville, Tennessee, occupies a topographic valley, yet it is sited in the remains of a structural dome (a circular domal anticline). Likewise, even though synclines are structural downfolds, topographically a syncline may comprise a valley or a ridge. Some mountaintops in the Alps are the erosional remnants of synclines. Words like *mountain,*

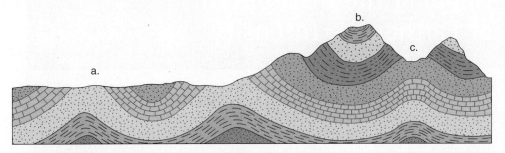

● FIGURE 14.17 Rock structure versus surface topography in a region of folded rocks. Structure, the rock response to applied tectonic forces, may or may not be represented directly in the surface topography, which depends on the nature and rate of both exogenic and endogenic geomorphic processes. (a) The structure is an anticline, but the surface landform is a plain. (b) The erosionally resistant center of a downfold (a syncline) supports a mountain peak. (c) A valley has been eroded into the crest of an anticline.

**Why is it that not all anticlines form mountains?**

*ridge, valley, basin,* and *fault scarp* are geomorphic terms that describe the surface topography, whereas *anticline, syncline, horst, graben,* and *normal fault* are structural terms that describe the arrangement of rock layers. Elements of rock structure may or may not be directly reflected in the surface topography. It is important to remember that the topographic variation on Earth's surface results from the interaction of three major factors: endogenic processes that create relief, exogenic processes that shape landforms and reduce relief, and the relative strength or resistance of different rock types to weathering and erosion.

# Earthquakes

**Earthquakes,** evidence of present–day tectonic activity, are ground motions of Earth caused when accumulating tectonic stress is relieved by the sudden displacement of rocks along a fault. The abrupt, lurching movement of crustal blocks past one another to new positions releases energy that moves through Earth as traveling *seismic waves*. We saw in Chapter 13 how seismic waves help us understand the characteristics of Earth's interior, but they also have a great impact on Earth's surface. Seismic waves passing along the crustal exterior or emerging at Earth's surface from below cause the damage, injury, and loss of life that we associate with major tremors. The vast majority of earthquakes are so slight that we cannot feel them and they produce no injuries or damage.

The subsurface location where rock displacement and the resulting earthquake originate is the earthquake **focus,** which may be located anywhere from near the surface to a depth of 700 kilometers (435 mi). The earthquake **epicenter** is the point on Earth's surface that lies directly above the focus, and it is where the strongest shock is normally felt ( ● Fig. 14.18). Most earthquakes occur at a focus deep enough that no displacement is visible at Earth's surface. Others cause mild shaking that rattles a few dishes. A few earthquakes are so strong that they topple buildings and break power lines, gas mains, and water pipes. In addition to their impact on the part of our environment built by people, ground shaking or surface offset during an earthquake can

trigger rockfalls, landslides, and avalanches. Earthquakes that occur beneath or near the ocean, moreover, are one of the main causes of *tsunamis,* which are sets of large ocean waves generated by the sudden displacement of water. Aftershocks commonly follow a large earthquake as crustal adjustments continue to occur, and foreshocks sometimes precede larger quakes.

## Measuring Earthquake Size

Scientists express the severity of an earthquake in two distinct ways: (1) the size of the event as a physical Earth process, and (2) the degree of its impact on humans. These two characteristics are sometimes related because, all other factors being equal, powerful earthquakes should have a greater effect on humans than smaller earthquakes. However, large tremors that strike in sparsely inhabited places will have limited human impact, whereas smaller earthquakes that strike densely populated areas can cause considerable damage and suffering. Many factors other than earthquake size affect the damage and loss of human life resulting from a tremor. Measuring the physical size of earthquakes and, separately, their effects on people help scientists and planners understand local and regional earthquake hazard potential.

In 1935 Charles F. Richter developed a numeric scale of **earthquake magnitude** to express the size of earthquakes as recorded by *seismographs,* sensitive instruments that measure ground shaking. Earthquake magnitude is reported as a number, generally with one decimal place, that represents the amount of energy released in an earthquake. As established by Richter, any whole number increase in earthquake magnitude (for example,

● FIGURE 14.18 The relationship between an earthquake's focus, the location where movement along the fault began, and its epicenter, the point on Earth's surface directly above the focus.

**Why is the epicenter in this example not located where the fault crosses Earth's surface?**

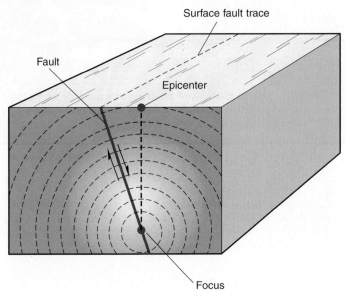

# GEOGRAPHY'S ENVIRONMENTAL PERSPECTIVE

# Mapping the Distribution of Earthquake Intensity

When an earthquake strikes a populated area, one of the first pieces of scientific information released is the magnitude of the tremor. Magnitude is a numeric expression of an earthquake's size at its focus in terms of energy released. In this sense, earthquakes can be compared with explosions. For example, a magnitude 4.0 earthquake releases energy equivalent to exploding 1000 tons of TNT. Because the scale is logarithmic rather than linear, a 5.0-magnitude tremor is the equivalent of 32,000 tons of TNT, whereas energy released in a 6.0-magnitude quake approximates 1 million tons of TNT.

Because of their greater energy, earthquakes of larger magnitude have the potential to cause much more damage and human suffering than those of smaller magnitude, but the reality is much more complex than that. A moderate earthquake in a densely populated area may cause great injury and damage, whereas a very large earthquake in an isolated region may not affect humans at all. Many factors relating to physical geography influence an earthquake's impact on people and their built environment. In general, the farther a location is from the earthquake epicenter, the less the effect of shaking, but this

generalization does not apply in every case. An earthquake in 1985 caused great damage in Mexico City, including the complete collapse of buildings, even though the epicenter was 385 kilometers (240 mi) away.

The Mercalli Scale of earthquake intensity (I–XII) was devised to measure the impact of a tremor on people and their built environment. Although every earthquake has only one magnitude, intensity often varies greatly from place to place, and a single tremor typically generates a range of intensity values. The spatial variation in Mercalli intensity for a given

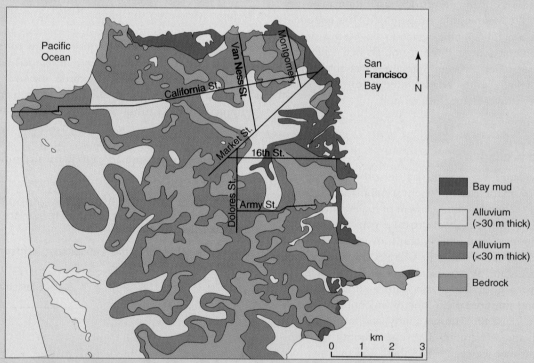

The locations of different Earth materials during the 1906 San Francisco earthquake.
Copyright © 2011 Brooks/Cole, Cengage Learning

earthquake is portrayed on maps by *isoseismals,* lines connecting points of equal shaking and earthquake damage expressed in Mercalli intensity levels. Patterns of isoseismals help scientists assess the local conditions that contributed to spatial variations in shaking and impact. Earthquake intensity factors vary with the nature of the substrate, construction type and quality, topography, and population density. Areas with unconsolidated Earth materials, poor construction, or high population densities generally suffer more from ground shaking and experience greater damage.

The 1906 San Francisco earthquake and ensuing fire caused the destruction of many buildings, numerous injuries, and an estimated 3000 deaths. The fire resulted from earthquake-damaged electrical and gas lines. Neither the magnitude nor the intensity scales existed in 1906. Subsequent studies suggest that the earthquake magnitude was about 8.3. Its Mercalli intensities have been reconstructed and mapped, showing the great variations in ground shaking and damage that the earthquake caused. In some cases, buildings on one side of a street were destroyed, whereas those on the other side of the street suffered little damage. Geographic patterns of the isoseismals have been analyzed and compared with the distribution of other factors to determine why some areas suffered more than others. These analyses reveal that areas of bedrock experienced lower intensities (less damage) than areas of unconsolidated Earth materials. Much of the worst damage occurred where bayfront lands and stream valleys had been artificially filled with sediment to allow construction.

Analyzing the nature of sites where intensities in an earthquake were higher or lower than expected helps us understand the reasons for spatial variations in earthquake hazards. The geographic patterns of Mercalli intensity that are generated even by small tremors help in planning for larger earthquakes in the same area. The overall patterns of ground shaking and isoseismals should be similar for a larger earthquake having the same epicenter as a smaller one, but the amount of ground shaking, Mercalli intensity, and size of the area affected would be greater for the larger magnitude tremor.

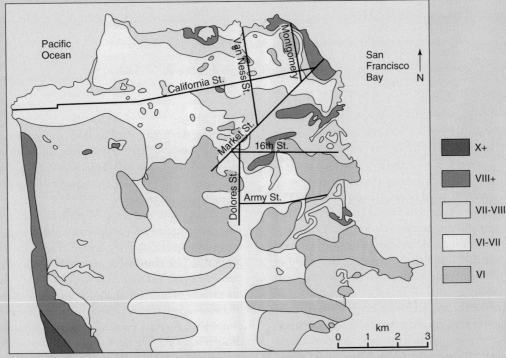

Geographic patterns of Mercalli intensity caused by the 1906 earthquake. The areas of intensity VIII+ in the northeast quarter of the city are on artificial landfill.

**TABLE 14.1**

**Modified Mercalli Intensity Scale**

| | |
|---|---|
| I. | Not felt except by a very few under especially favorable conditions. |
| II. | Felt only by a few persons at rest, especially on upper floors of buildings. |
| III. | Felt quite noticeably by persons indoors, especially on upper floors of buildings. Many people do not recognize it as an earthquake. Standing motor cars may rock slightly. Vibrations similar to the passing of a truck. Duration estimated. |
| IV. | Felt indoors by many, outdoors by few during the day. At night, some awakened. Dishes, windows, doors disturbed; walls make cracking sound. Sensation like heavy truck striking building. Standing motor cars rocked noticeably. |
| V. | Felt by nearly everyone; many awakened. Some dishes, windows broken. Unstable objects overturned. Pendulum clocks may stop. |
| VI. | Felt by all, many frightened. Some heavy furniture moved; a few instances of fallen plaster. Damage slight. |
| VII. | Damage negligible in buildings of good design and construction; slight to moderate in well-built ordinary structures; considerable damage in poorly built or badly designed structures; some chimneys broken. |
| VIII. | Damage slight in specially designed structures; considerable damage in ordinary substantial buildings with partial collapse. Damage great in poorly built structures. Fall of chimneys, factory stacks, columns, monuments, walls. Heavy furniture overturned. |
| IX. | Damage considerable in specially designed structures; well-designed frame structures thrown out of plumb. Damage great in substantial buildings, with partial collapse. Buildings shifted off foundations. |
| X. | Some well-built wooden structures destroyed; most masonry and frame structures destroyed with foundations. Rails bent. |
| XI. | Few, if any (masonry) structures remain standing. Bridges destroyed. Rails bent greatly. |
| XII. | Damage total. Lines of sight and level are distorted. Objects thrown into the air. |

Source: Abridged from *The Severity of an Earthquake: A U.S. Geological Survey General Interest Publication*. U.S. Government Printing Office: 1989, 288–913.

from 5.0 to 6.0) represents a 10 times increase in amplitude on the seismograph trace (see Fig. 13.1), but the actual energy released is about 32 times greater. The extremely destructive 1906 San Francisco earthquake occurred before the magnitude scale was devised but is estimated to have been a magnitude 8.3. The 1989 Loma Prieta earthquake near San Francisco had a Richter magnitude of 7.1 but was still responsible for dozens of fatalities. The strongest earthquake in North America to date, a magnitude 8.6, occurred in Alaska in 1964, and the strongest in the world in recent history was a 9.5 magnitude earthquake that occurred off the coast of Chile in 1960.

Over the last several decades, scientists have derived alternative ways for calculating earthquake magnitude from seismograph data. Today the energy released in most moderate and major earthquakes is determined using formulas for the more accurate **moment magnitude** instead of using Richter's original procedure (Richter magnitude). Regardless of how it is calculated, earthquake magnitude is reported using the same numeric scheme devised by Richter, where a whole number increase in magnitude represents 10 times greater amplitude and approximately 32 times greater energy released. Furthermore, as with Richter magnitude, each earthquake has only one moment magnitude representing the earthquake's size in terms of energy released. Like most recent earthquakes, the 9.1 magnitude reported for the December 2004 tsunami-generating earthquake off the west coast of northern Sumatra, Indonesia (called the Sumatra–Andaman earthquake), the 7.0 magnitude determined for the January 2010 earthquake in Haiti, the 8.8-magnitude quake off the coast of Maule, Chile, in February 2010, and the 9.0 magnitude Tohoku earthquake in Japan in March 2011 are all moment magnitudes.

A very different type of scale is that used to record and investigate patterns of **earthquake intensity,** the damage caused by an earthquake and the degree of its impact on people and their property. The **modified Mercalli scale** of earthquake intensity uses categories numbered from I to XII (Table 14.1) to describe the effects of an earthquake on humans and the spatial variation of those impacts. The categories are represented with Roman numerals to avoid confusion with earthquake magnitude. Although a maximum intensity is determined for each earthquake, one earthquake typically produces a variety of intensity levels, depending on varying local conditions. These conditions include distance from the epicenter, how long shaking lasted, the severity of shaking as influenced by local surface materials, population density, and building construction in the affected area. After an earthquake, observers gather information about Mercalli intensity levels by noting the damage and talking to residents about their experiences. The variation in intensity levels is then mapped so that geographic patterns of damage and shaking intensity can be analyzed. Understanding the spatial patterns of damage and ground response helps us plan and prepare so that we can reduce the hazards of future earthquakes.

## Earthquake Hazards

Unfortunately, there is plenty of evidence concerning the deadliness of earthquakes as illustrated by just a few recent examples. The Tohoku earthquake and tsunami in Japan in 2011 left almost 28,000 people missing or confirmed dead ( ● Fig. 14.19a). In 2010 the 7.0-magnitude earthquake in Haiti killed approximately 225,000, injured 300,000, and displaced 1.3 million people (Fig. 14.19b), whereas the 8.8-magnitude quake in Chile killed hundreds, injured thousands, and displaced 800,000 people. In 2008 a magnitude 7.9 earthquake devastated extensive areas of eastern Sichuan, China, causing 68,800 fatalities. A magnitude 7.6 tremor in Pakistan in 2005 resulted in 86,000 deaths. Together with the ensuing tsunami, the 2004 Sumatra–Andaman earthquake led to almost 300,000 fatalities ( ● Fig. 14.20). In 2003, 31,000 people perished from a magnitude 6.6 earthquake in southeastern Iran.

The impact of an earthquake on humans and their property depends on many factors. Some of the most important pertain to the likelihood of structural collapse and number of people potentially affected by such failures. Factors influencing structural collapse include the location of an earthquake epicenter relative to population density, construction materials and methods, and the stability of Earth materials on which buildings and other structures were erected. A strong tremor of magnitude 7.5 that struck the Mojave Desert of Southern California in 1992 caused little loss of life because of sparse settlement. Structures made of brick, unreinforced masonry, or other inflexible materials or without adequate structural support do not withstand ground shaking well, and this contributed greatly to the high number of injuries and fatalities in the 2010 earthquake in Haiti. Freeway spans collapsed in the San Francisco Bay area in the Loma Prieta earthquake of 1989 and in the 1994 6.7-magnitude Northridge earthquake in the San Fernando Valley area of Los Angeles ( ● Fig. 14.21). Wood frame houses generally fare better than brick, adobe, or block structures because the wood frame has greater flexibility during ground shaking. Buildings located on loosely deposited Earth materials (unconsolidated sediment) tend to shake violently compared with areas of solid bedrock. The effects of an 8.1-magnitude earthquake with an epicenter 385 kilometers (240 mi) from Mexico City in 1985 caused a death toll of more than 9000 and extensive damage to many buildings, including high-rise sections of the city. Mexico's densely populated capital city is built on an ancient lake bed made of soft sediments that shook greatly, causing much destruction even though the earthquake was centered in a distant mountain region.

In addition to structural collapse, loss of life in an earthquake is influenced by several other factors including time of day, time of year, weather conditions at the time of the quake, resulting fires from downed power lines and broken gas mains, and whether ground shaking triggers rockfalls, landslides, avalanches, or tsunamis. Freeway spans that collapse from an earthquake will cause much greater loss of life if the tremor occurs at rush hour instead of in the middle of the night. Contributing to the large death toll from a

U.S. Air Force photo by Master Sgt. Jeremy Lock

**(a)**

Department of Defense

**(b)**

● **FIGURE 14.19** (a) Some damage in Ofunato, Japan, from the March 2011 earthquake. (b) Left-lateral, strike-slip movement along the boundary between the North American and Caribbean plates caused tremendous damage and loss of life in the 2010 earthquake in Haiti. The 7.0-magnitude quake destroyed a majority of the homes in this southern Haitian city of Jacmel.

*What factors contributed to the large death toll for the quake in Haiti?*

● FIGURE 14.20 The 9.1-magnitude Sumatra–Andaman earthquake, and the tsunami that it generated, devastated the city of Banda Aceh on the Indonesian island of Sumatra. Extensive parts of the city were flattened by the catastrophe. Here, only a lone piece of a building is left standing. The dark scratches above the lowest set of windows were made by debris carried in the tsunami.

● FIGURE 14.21 Elevated freeways collapsed in both the 1994 Northridge earthquake in the Los Angeles area (pictured here) and the 1989 Loma Prieta earthquake in the San Francisco Bay area. Each tremor killed close to 60 people, but lives were spared in 1994 because the Northridge earthquake occurred at 4:30 a.m. on a holiday.

7.8-magnitude earthquake in Peru in 1970 were enormous earthquake-induced avalanches that obliterated entire mountain villages. Even small slope failures that block roads can increase the death toll by hampering access for rescue operations.

Although most major earthquakes are related to known faults and have epicenters in or near mountainous regions, the one most widely felt in North America was not. It was one of a series of tremors that occurred during 1811 and 1812, centered near New Madrid, Missouri, and was felt from Canada to the Gulf of Mexico and from the Rocky Mountains to the Atlantic Ocean. Fortunately, the region was not densely settled at that time in history. Because of that huge earthquake, St. Louis is considered to be at risk from a major earthquake, as shown on the seismic-risk map for the conterminous United States ( ● Fig. 14.22). Although not common in these regions, recent earthquakes have occurred in New England, New York, and the Mississippi Valley. Probably no area on Earth could be called entirely "earthquake safe."

# Igneous Processes and Landforms

Landforms resulting from igneous processes are related to eruptions of extrusive igneous rock material or emplacements of intrusive igneous rock. **Volcanism** refers to the extrusion of rock matter from Earth's subsurface to the exterior and the creation of surface terrain features as a result. **Volcanoes** are mountains or hills that form in this way. **Plutonism** refers to igneous processes that occur below Earth's surface, including the cooling of magma to form intrusive igneous rocks and rock masses. Some masses of intrusive igneous rock are eventually exposed at Earth's surface where they comprise landforms of distinctive shapes and properties.

## Volcanic Eruptions

Few spectacles in nature are as awesome as a volcanic eruption ( ● Fig. 14.23). Although large violent eruptions tend to be infrequent events, they can devastate the surrounding environment and completely change the nearby terrain. Yet volcanic eruptions are natural processes and should not be unexpected by people who live or travel in the vicinity of active volcanoes.

Eruptions vary greatly in their size and character, and the volcanic landforms that result are extremely diverse. *Explosive eruptions* violently blast pieces of molten and solid rock into the air, whereas molten rock pours less violently onto the surface as flowing streams of lava in *effusive eruptions*. Variations in eruptive style and in the landforms produced by volcanism stem mainly from differences in the **viscosity,** that is, resistance to flowing, of the magma feeding the eruption, and in its gas content. Viscosity and gas content are determined especially by the chemical properties of the magma, but temperature is also involved.

Like honey compared with water at the same room temperature, some molten rock matter is naturally more resistant to

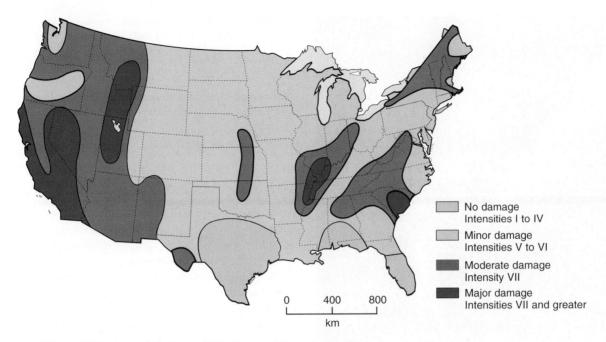

No damage
Intensities I to IV

Minor damage
Intensities V to VI

Moderate damage
Intensity VII

Major damage
Intensities VII and greater

0    400    800
km

● **FIGURE 14.22** Earthquake hazard potential in the conterminous United States.

*What is the earthquake hazard potential where you live?*

flowing (more viscous) than other molten rock matter at the same temperature because of differences in chemical composition. The primary material component that contributes to a higher viscosity in molten rock is silica ($SiO_2$) content. Everything else being equal, because they are rich in silica, *felsic* magmas have a higher viscosity than *mafic* magmas, which contain less silica. Temperature also affects the viscosity of a liquid—consider the viscosity difference between cooler and warmer honey. As it does with honey, an increase in temperature re-

● **FIGURE 14.23** Few natural events are as spectacular as a volcanic eruption. This eruption of Italy's Stromboli volcano, on an island off Sicily, lit up the night sky.

USGS/VHP/B. Chouet

duces to at least some extent the viscosity of any type of magma (makes it flow more easily).

Magma viscosity influences the explosiveness of a volcanic eruption through its effect on the gas content of the magma. Magmas contain large amounts of gases that remain dissolved when under high pressure at great depths. As molten rock rises closer to the surface, the decrease in pressure from above releases expanding gases from the magma. The gases move with comparative ease through low viscosity, mafic magma into adjacent rocks and through cracks into the atmosphere relieving the eruptive pressure of the expanding gases. The relief in gas pressure delays eruption, often allowing the magma to become hotter, which further reduces its viscosity. As a result, mafic magmas, such as those with a basaltic composition, typically erupt in the less violent effusive fashion, with streams of flowing lava, rather than by exploding. If, however, the magma has a felsic composition and therefore a higher viscosity—for example, if it has a rhyolitic composition—expanding gases move more slowly through the molten rock and do not vent easily to the atmosphere. Gas pressure builds quickly in this type of magma and can soon exceed the strength of the overlying rock. When the overlying rock breaks, the gases surge with great speed and force from the high pressure in the magma toward the much lower pressure at Earth's surface taking molten rock out with it in an explosive eruption.

Explosive eruptions hurl into the air fragments of solidified lava, clots of molten lava that solidify in flight, or molten clots that solidify once they land. All of these represent *pyroclastic materials* (fire fragments), also referred to as *tephra*. These rock fragments erupt

● **FIGURE 14.24** Volcanic ash streaming from Mount Etna on the Italian island of Sicily was captured on this photograph taken from the International Space Station in 2001. The ash cloud reportedly reached a height of about 5200 meters (17,000 ft) on that day.

*What do you think conditions were like at the time of this eruption for settlements located under the ash cloud?*

Appendix C), but basalt is by far the most common because its low viscosity and hot eruptive temperature allow gases to escape, greatly reducing the potential for an explosive eruption.

Solidified lava flows tend to have many fractures, known as *joints*. When basaltic lava cools and solidifies, it shrinks, and this contraction can produce a network of vertical fractures that break the rock into numerous hexagonal columns. This creates columnar-jointed basalt flows ( ● Fig. 14.25).

Lava flows display distinctive surface characteristics. Extremely fluid lavas tend to flow rapidly and for long distances before solidifying. In this case, a thin surface layer of lava in contact with the atmosphere solidifies, while the molten lava beneath continues to move, carrying the thin, hardened crust along and wrinkling it into a ropy surface form called **pahoehoe.** Lavas of slightly greater viscosity flow more slowly, allowing a thicker surface layer to harden while the still-molten interior lava keeps on flowing. This causes the thick layer of hardened crust to break up into sharp-edged, jagged blocks, making a surface known as **aa.** The terms pahoehoe and aa both originated in Hawaii, where

in a range of sizes from **volcanic ash,** which is sand-sized or smaller (<2 mm), to *cinders* (2–4 mm), *lapilli* (4–64 mm), and *blocks* (>64 mm). Pyroclastic materials include volcanic "bombs," large spindle-shaped clasts that solidified while flying through the air. In the most explosive eruptions, clay and silt-sized volcanic ash is hurled into the atmosphere to an altitude of 10,000 meters (32,800 ft) or more ( ● Fig. 14.24). The 1991 eruptions of Mount Pinatubo in the Philippines ejected a volcanic aerosol cloud that circled the globe. The suspended material caused spectacular reddish orange sunsets from increased scattering and lowered global temperatures slightly for 3 years by increasing reflection of solar energy back to space.

## Volcanic Landforms

The type of landform that results from a volcanic eruption depends mostly on the explosiveness of that eruption. We will consider six major kinds of volcanic landforms, beginning with those associated with the most effusive (least explosive) eruptions. Four of the six major landforms are types of volcanoes.

**Lava Flows** Layers of erupted rock matter that poured or oozed over the landscape when they were molten are **lava flows.** After cooling and solidifying, the rock retains the appearance of having flowed. Lava flows can be made from any composition of lava (see

● **FIGURE 14.25** Columnar-jointed lava at Devil's Postpile National Monument, California.

*Why are the cliffs shown in this photograph so steep?*

effusive eruptions of basalt, and these surface characteristics, are common ( ● Fig. 14.26).

Lava flows do not have to emanate directly from volcanoes but may pour out of deep fractures in the crust, called **fissures,** that can be independent of mountains or hills of volcanic origin. Very fluid basaltic lava erupting from fissures has traveled up to 150 kilometers (93 mi) before solidifying. In some regions, multiple layers of basalt flows have constructed relatively flat-topped, but elevated, tablelands known as **basaltic plateaus.** The Columbia Plateau in Washington, Oregon, and Idaho, covering 520,000 square kilometers (200,000 sq mi), is a major example of a basaltic plateau ( ● Fig. 14.27), as is the Deccan Plateau in India. In the geologic past, huge amounts of basalt issued from some fissures, eventually burying existing landscapes under thousands of meters of lava flows. These very extensive flows are often referred to as *flood basalts.*

● **FIGURE 14.26** Lava flow surfaces often consist of (a) the ropy-textured pahoehoe, where the molten lava was extremely fluid, and (b) the blocky and angular aa, where viscosity of the molten rock matter was slightly higher (less fluid).

Copyright and photograph by Dr. Parvinder S. Sethi

(a)

D. Sack

(b)

**Shield Volcanoes** When numerous successive basaltic lava flows occur in a given region, they may eventually pile up into the shape of a large mountain, called a **shield volcano,** which resembles a giant knight's shield resting on Earth's surface ( ● Fig. 14.28a). The gently sloping, dome-shaped cones of Hawaii best illustrate this largest type of volcano ( ● Fig. 14.29). Shield volcanoes erupt extremely hot, mafic lava at temperatures near 1090°C (2000°F). Escape of gases and steam occasionally hurls fountains of molten lava a few hundred meters into the air ( ● Fig. 14.30), causing some accumulations of solidified pyroclastic materials, but the major characteristic of shield volcanoes is the outpouring of fluid basaltic lava flows. Compared with other volcano types, these eruptions are not very explosive, although they are still potentially dangerous and damaging. The extremely hot and fluid basalt can flow long distances before solidifying, and the accumulation of flow layers develops broad, dome-shaped volcanoes with very gentle slopes.

On the island of Hawaii, active shield volcanoes also erupt lava from fissures on their flanks so that living on the island's edges, away from the summit craters, does not guarantee safety from volcanic hazards. Neighborhoods in Hawaii have been destroyed or threatened by lava flows. The Hawaiian shield volcanoes form the largest volcanoes on Earth in terms of both their height—beginning at the ocean floor—and diameter.

**Cinder Cones** The smallest type of volcano, typically only a couple of hundred meters high, is a **cinder cone.** Cinder cones consist largely of gravel-sized pyroclastics. Gas-charged eruptions throw molten lava and solid pyroclastic fragments into the air. Falling under the influence of gravity, these particles accumulate around the almost pipelike conduit for the eruption, the *vent,* as a large pile of tephra (Fig. 14.28b). Each eruptive burst ejects more pyroclastics that fall and cascade down the sides to build an internally layered volcanic cone. Cinder cone volcanoes typically have a rhyolitic composition but can be made of basalt if conditions of temperature and viscosity keep gases from escaping easily. The form of a cinder cone is distinctive, with steep straight sides and a crater (depression) at the top of the hill ( ● Fig. 14.31). The steep slopes of accumulated pyroclastics lie at or near the **angle of repose,** the steepest angle that a pile of loose material can maintain without rocks sliding or rolling downslope. The angle of repose for unconsolidated rock matter generally ranges between 30° and 34°. Cinder cone examples include several in the Craters of the Moon area in Idaho, Capulin Mountain in New Mexico, and Sunset Crater, Arizona. In 1943 a remarkable cinder cone called Paricutín grew from a fissure in a Mexican cornfield to a height of 92 meters (300 ft) in 5 days and to more than 336 meters (1100 ft) in a year. Eventually, the volcano began erupting basaltic lava flows, which completely buried a nearby village except for the top of a church steeple.

● **FIGURE 14.27** East of Portland, Oregon, the Columbia River has cut an impressive gorge exposing layers of Columbia Plateau flood basalts.

Ocean/Corbis

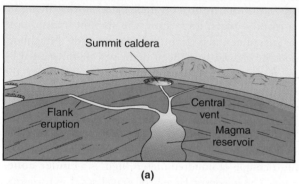

(a)

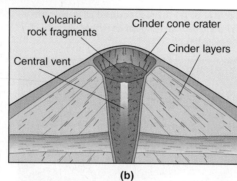

(b)

● **FIGURE 14.28** The four basic types of volcanoes are (a) shield volcano, (b) cinder cone, (c) composite cone, also known as stratovolcano, and (d) plug dome.

*What are the key differences in the shapes of these volcanoes? What properties are alike or different in their internal structure?*

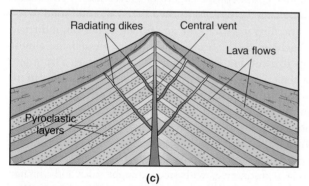

(c)

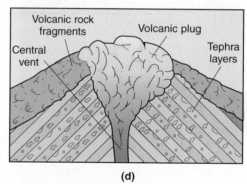

(d)

● **FIGURE 14.29** Mauna Loa, on the island of Hawaii, is the largest volcano on Earth and clearly displays the dome, or convex, shape of a classic shield volcano. Mauna Loa reaches to 4170 meters (13,681 ft) above sea level, but its base lies far beneath sea level, creating almost 17 kilometers (56,000 ft) of relief from base to summit.

*Why do Hawaiian volcanoes erupt less explosively than volcanoes of the Cascades or Andes?*

D. Sack

● **FIGURE 14.30** This fountain of lava in Hawaii reached a height of 300 meters (1000 ft).

● **FIGURE 14.31** Cinder cones grow as volcanic fragments (pyroclastics) ejected during gas-charged eruptions pile up around the eruptive vent. Here, a cinder cone stands among lava flows in Lassen Volcanic National Park, California.

*Why is the crater so prominent on this volcano?*

● **FIGURE 14.32** Composite cones are composed of both lava flows and pyroclastic material and have distinctive concave side slopes. Oregon's Mount Hood is a composite cone in the Cascade Range.

*Along what type of lithospheric plate boundary is this volcano located?*

**Composite Cones** A third kind of volcano, a **composite cone,** results when formative eruptions are sometimes effusive and sometimes explosive. Composite cones are therefore composed of a combination—that is, they represent a composite—of lava flows and pyroclastic materials (Fig. 14.28c). They are also known as **stratovolcanoes** because they are constructed of layers (strata) of pyroclastics and lava. The topographic profile of a composite cone represents what many people might consider the classic volcano shape, with concave slopes that are gentle near the base and steep near the top ( ● Fig. 14.32). Composite volcanoes form from andesite, which is a volcanic rock intermediate in silica content and explosiveness between basalt and rhyolite. Although andesite is only intermediate in terms of these characteristics, composite cones are dangerous. As a composite cone grows larger, the vent eventually becomes plugged with unerupted andesitic rock. When this happens, the pressure driving an eruption builds to the point where either the plug is explosively forced out or the mountain side is pushed outward until it fails, allowing the great accumulation of pressure to be relieved in a lateral explosion. Such explosive eruptions are commonly accompanied by **pyroclastic flows,** fast-moving density currents of airborne volcanic ash, hot gases, and steam that flow downslope close to the ground like avalanches. The speed of a pyroclastic flow can reach 100 kilometers per hour (62 mi/hr) or more.

Most of the world's best-known volcanoes are composite cones. Some examples include Mounts Fujiyama in Japan, Cotopaxi in Ecuador, Vesuvius and Etna in Italy, Rainier in Washington, and Shasta in California. The highest volcano on Earth, Nevados Ojos del Salado, is an andesitic composite cone that reaches an elevation of 6887 meters (22,595 ft) on the border between Chile and Argentina in the Andes, the mountain range for which andesite was named.

On May 18, 1980, residents of the American Pacific Northwest were stunned by the eruption of Mount St. Helens in the Cascade Mountain Range. Mount St. Helens, a composite cone of southwestern Washington that had been venting steam and ash for several weeks, exploded with incredible force on that day. A menacing bulge had been growing on the side of Mount St. Helens, and Earth scientists warned of a possible major eruption, but no one could forecast the magnitude or the exact timing of the blast. Within minutes, nearly 400 meters (1300 ft) of the mountain's north summit had disappeared by being blasted into the sky and down the mountainside ( ● Fig. 14.33). Unlike most volcanic eruptions, in which the eruptive force is directed vertically, much of the explosion blew pyroclastic debris laterally outward from the site of the bulge. An eruptive blast composed of an intensely hot cloud of steam, noxious gases, and volcanic ash burst outward at more than 320 kilometers per hour (200 mi/hr), obliterating forests, lakes, streams, and campsites for nearly 32 kilometers (20 mi). Volcanic ash and water from melted snow and ice formed huge mudflows that choked streams, buried valleys, and engulfed everything in their paths. More than 500 square kilometers (200 sq mi) of forests and recreational lands were destroyed. Hundreds of homes were buried or badly damaged. Choking ash several centimeters thick covered nearby cities, untold numbers of wildlife were killed, and more than 57 people lost their lives in the eruption. It was a minor event in Earth's history but a sharp reminder to the region's residents of the awesome power of natural forces.

Some of the worst natural disasters in history have occurred in the shadows of composite cones. Mount Vesuvius, in Italy, killed more than 20,000 people in the cities of Pompeii and Herculaneum in A.D. 79. Mount Etna, on the Italian island of Sicily, destroyed 14 cities in 1669, killing more than 20,000 people, and is still active much of the time. The greatest volcanic eruption in recent history was the 1883 explosion of Krakatoa in what is now Indonesia. The explosive eruption killed more than 36,000 people, many as a result of the tsunamis generated by the eruption that swept the coasts of Java and Sumatra. In 1985 the Andean composite cone Nevado del Ruiz, in the center of Colombia's coffee-growing region, erupted and melted its snowcap sending torrents of mud and debris down its slopes, burying cities and villages, and resulting in a death toll in excess of 23,000. The 1991 eruption of Mount Pinatubo in the Philippines killed more than 300 people, and airborne ash caused climatic effects for 3 years following the eruption. In 1997 a series of violent eruptions from the Soufriere Hills volcano destroyed more than half of the Caribbean island of Montserrat with volcanic ash and pyroclastic flows ( ● Fig. 14.34). Mexico City, one of the world's most populous urban areas, continues to be threatened by ash falls from eruptions of a large, active composite cone 70 kilometers (45 mi) away. At this distance, ash fall from a major eruption is the most severe hazard expected. Volcanic ash is much like tiny slivers of glass. It is abrasive and causes respiratory problems in people and animals, clogging of air intakes for engines, destructive

(a)

(b)

(c)

● **FIGURE 14.33** (a) Prior to the 1980 eruption, Mount St. Helens towered majestically over Spirit Lake in the foreground. (b) On May 18, 1980, the violent eruption removed more than 3.0 cubic kilometers (1 cu mi) of material from the mountain's north slope, leaving a crater more than 400 meters (1300 ft) deep. The blast and mudflows decimated forests and took 57 human lives. (c) Two years after the eruption, the volcano continued to spew small amounts of gas, steam, and ash.

*Could other volcanoes in the Cascade Range, such as Oregon's Mount Hood, erupt with the kind of violence that Mount St. Helens displayed in 1980?*

scratching and pitting of metal and glass, and the collapse of building roofs under the heavy weight of significant ash accumulations. The damaging effects of volcanic ash on airplane engines, fuel lines, windshields, and the multitude of other sensitive parts on airplanes makes flying through volcanic ash extremely hazardous. Ash erupting from an Icelandic composite cone in 2010 shut down air travel over most of Europe for 6 days, affecting millions of travelers and having major economic impacts.

**Plug Domes** Where extremely viscous silica–rich magma pushes up into the vent of a volcanic cone but does not flow farther, it forms a **plug dome** (Fig. 14.28d). Solidified outer portions of this blockage create the dome-shaped summit, and jagged blocks that break away from the plug, or preexisting parts of the cone, form the steep side slopes of the volcano. Great pressures build up causing more blocks to break off from the plug and producing the potential for extremely violent explosive eruptions, including pyroclastic flows. In 1903 Mount Pelée, a plug dome on the French West Indies island of Martinique, caused the deaths in a single blast

● **FIGURE 14.34** Beginning in 1995 the Caribbean island of Montserrat was struck by a series of volcanic eruptions, including pyroclastic flows, that devastated much of the island. The town of Plymouth, shown here, has been completely abandoned because of the amount of destruction and threat of future eruptions. Prior to the 1995 disaster, the volcano had not erupted for 400 years.

USGS/Cascades Volcano Observatory

● FIGURE 14.35 Plug dome volcanoes extrude stiff silica-rich lava and have steep slopes. Lassen Peak, located in northern California, is a plug dome and the southernmost volcano in the Cascade Range. The lava plugs are the darker areas protruding from the volcanic peak. Lassen was last active between 1914 and 1921.

*Why are plug dome volcanoes considered dangerous?*

of all but a few people from a town of 30,000. Lassen Peak in California is a large plug dome that erupted with great violence less than 100 years ago (● Fig. 14.35). Other plug domes exist in Japan, Guatemala, the Caribbean, and the Aleutian Islands.

**Calderas** Occasionally, the eruption of a volcano expels so much material and relieves so much pressure within the magma chamber that only a large and deep depression remains in the area that previously contained the volcano's summit. A large depression made in this way is termed a **caldera.** The best-known caldera in North America is the basin in south-central Oregon that contains Crater Lake, a circular body of water 10 kilometers (6 mi) across, almost 595 meters (1950 ft) deep, and surrounded by near-vertical cliffs as much as 610 meters (2000 ft) high. The caldera that contains Crater Lake was formed by the prehistoric eruption and collapse of a composite volcano. A cinder cone, Wizard Island, subsequently built up from the floor of the caldera to a height almost 235 meters (770 ft) above the lake's surface (● Fig. 14.36). The area of Yellowstone National Park is the site of three ancient calderas, and the Valles Caldera in New Mexico is another excellent example. Krakatoa in Indonesia and Santorini in Greece have left island remnants of their calderas. Calderas are

● FIGURE 14.36 (a) Crater Lake, Oregon, formed about 7,700 years ago when a violent eruption of Mount Mazama blasted out solid and molten rock matter, leaving behind a deep crater, the caldera, which later accumulated water. (b) Wizard Island is a later, secondary volcano that has risen within the caldera.

*Could other Cascade volcanoes erupt to the point of destroying the volcano summit and leaving a caldera?*

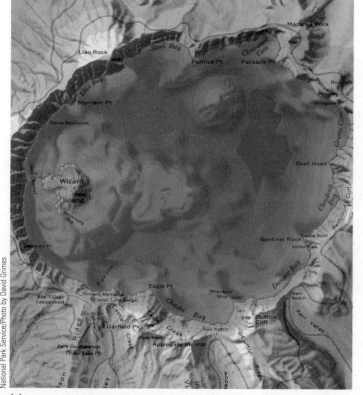

National Park Service/Photo by David Grimes

(a)

National Park Service/Photo by David Grimes

(b)

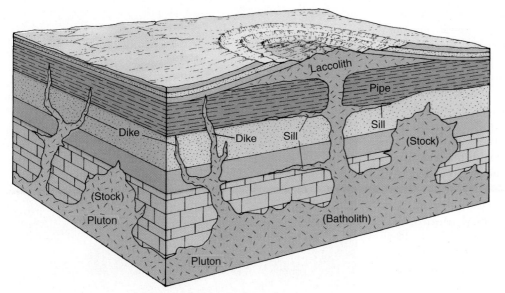

also found in the Philippines, the Azores, Japan, Nicaragua, Tanzania, and Italy, many of them occupied by deep lakes.

## Plutonism and Intrusions

Bodies of magma that exist beneath the surface of Earth or masses of intrusive igneous rock that cooled and solidified beneath the surface are **igneous intrusions,** or *plutons.* A great variety of shapes and sizes of magma bodies result from intrusive igneous activity, also known as plutonism. When they are first formed, smaller plutons have little or no effect on the surface terrain. During their formation, larger plutons may be associated with uplift of the land surface under which they are intruded.

The many different kinds of intrusions are classified by their size, shape, and relationship to the surrounding rocks ( ● Fig. 14.37). After millions of years of uplift and erosion of overlying rocks, even small intrusions may be exposed at the surface to become part of the landscape. Uplifted plutons that crop out at the surface tend to stand higher than the surrounding landscape because intrusive igneous rocks that constitute the plutons, such as granite, have a resistance to weathering and erosion exceeding that of many other kinds of rocks.

When exposed at Earth's surface, an irregularly shaped intrusion is a **stock** if it covers an area less than 100 square kilometers (40 sq mi); if larger, it is known as a **batholith.** Batholiths are complex masses of solidified magma, usually granite. Batholiths represent large plutons that, during their development kilometers beneath Earth's surface, melted, metamorphosed, or pushed aside other rocks. Some batholiths are as much as several hundred kilometers across and thousands of meters thick. Because of their size and the resistance of intrusive igneous rocks to weathering and erosion, batholiths form the core of many major mountain ranges. The Sierra Nevada batholith, Idaho batholith, and the Peninsular Ranges batholith of Southern and Baja California comprise hundreds of thousands of square kilometers of granite landscapes in western North America.

Magma creates other kinds of igneous intrusions by forcing its way into fractures and between rock layers without melting the surrounding rock. A **laccolith** is a mushroom-shaped intrusion that develops where molten magma pushes its way between preexisting horizontal layers of other rock, causing the overlying strata to bulge upward as the intrusion grows. Laccoliths have a mushroom-like shape because the upper dome part is usually connected to the magma source by a narrower pipe or stem-like component. As the laccolith is forming and causing the overlying rock strata to bulge, Earth's surface arches upward like a giant blister, with magma beneath the overlying layers comparable to the fluid beneath the skin of a blister. Although laccoliths are smaller, like batholiths they form mountains or hills after erosion has worn away the overlying less resistant rocks. The La Sal, Abajo, and Henry Mountains in southern Utah are excellent examples of exposed laccoliths ( ● Fig. 14.38).

● **FIGURE 14.38** The La Sal Mountains in southern Utah, near Moab, are composed of a laccolith that was exposed at the surface by uplift and subsequent erosion of the overlying sedimentary rocks.

*How do laccoliths deform the rocks they are intruded into?*

Landforms smaller than stocks, batholiths, and laccoliths also result when erosion of overlying rocks exposes igneous intrusions at Earth's surface. Magma sometimes intrudes between rock layers without bulging them upward, solidifying into a horizontal sheet of intrusive igneous rock called a **sill.** The Palisades, along New York's Hudson River, provide an example of a sill made of gabbro, the intrusive compositional equivalent of basalt ( ● Fig. 14.39). Molten rock under pressure may also intrude into a nonhorizontal fracture that cuts across the surrounding rocks. The solidified magma in this case has a wall-like shape and is known as a **dike.** When exposed by erosion of surrounding, less resistant rocks, dikes appear as vertical or near-vertical walls of rock rising above the surrounding topography ( ● Fig. 14.40). At Shiprock, in New Mexico, resistant dikes many kilometers long rise vertically more than 90 meters (300 ft) above the surrounding plateau ( ● Fig. 14.41). Shiprock is a **volcanic neck,** a tall rock spire made of the now-exposed, formerly subsurface pipe that about 30 million years ago fed an overlying volcano. Erosion has removed the volcanic cone, exposing the resistant dikes and neck that were once internal features of the volcano.

# Distribution of Endogenic Processes

The geographic distributions of earthquake and volcanic activity are quite similar, and plate tectonics determines where much of that activity occurs. The greatest tectonic forces, most extensive faults, most dangerous earthquakes, largest mountain ranges, and most of the world's volcanic activity are concentrated in relatively narrow zones along the boundaries of lithospheric plates. Recent major earthquakes in Japan, Haiti, Chile, and Indonesia, for example, occurred along faults associated with plate boundaries. Likewise, recent volcanic eruptions in Iceland, Indonesia, Central America, and the Andes of South America highlight the geographic relationship between volcanic forces and plate margins. Because plate margins bordering the Pacific Ocean are particularly prone to both types of endogenic processes, the boundary of that ocean is often described as the Pacific Ring of Fire.

Many regions with earthquake or volcanic activity are incredibly scenic and offer attractive environments in which to live and vacation, so it is not surprising that some of these hazard-prone regions are densely populated. To minimize the impacts of tectonic, volcanic, and other related hazards, such as landslides and tsunamis, it is essential that people in these regions have a basic understanding of the nature of the hazards with which they live, and for residents and governmental agencies to make detailed preparations for coping with disasters before they occur.

Anthony G. Taranto Jr., Palisades Interstate Park – NJ Section

● **FIGURE 14.39** The Palisades of the Hudson River, the impressive cliffs along the river near New York City, are made from a thick sill of igneous rock that was intruded between layers of sedimentary rocks.

*Why does the sill at the Palisades form a cliff?*

J. Petersen

● **FIGURE 14.40** When erosion of overlying rocks exposes them at Earth's surface, dikes, like this one in Big Bend National Park, Texas, often stand somewhat higher than the rock into which they were intruded.

*How does a dike differ from a sill?*

Copyright and photograph by Dr. Parvinder S. Sethi

● **FIGURE 14.41** Shiprock, New Mexico, is a volcanic neck exposed by erosion of surrounding rock. Volcanic necks are resistant remnants of the intrusive pipe of a volcano.

# Plate Boundaries, Earthquakes, and Volcanoes

Tectonic and volcanic activity are concentrated along the boundaries of lithospheric plates. Although the locations of earthquake and volcanic activity correlate fairly well, there are exceptions, and their nature and severity differ from place to place. In general, the frequency and severity of earthquakes or volcanic eruptions vary according to their proximity to a specific type of lithospheric plate boundary or specific site in the central part of a plate.

Regardless of whether it breaks a continent or the seafloor, the tensional tectonic forces associated with plate divergence cause faulting and create fractures that provide avenues for molten rock to reach the surface. The divergent midoceanic ridges experience small to moderate earthquakes that originate at a shallow depth and rather mild volcanic eruptions. People are affected when these tectonic and volcanic activities occur on islands associated with midocean ridges, such as Iceland and the Azores. Where continental crust is diverging and breaking, as in rift valleys, earthquakes tend to be small to moderate, but continental crust mixed with mafic magma produces a wider variety of volcanic eruptions than along diverging oceanic crust, and some of the continental eruptions are potentially quite violent. Examples of resulting volcanoes in the East African rift valleys include Mount Kilimanjaro and Mount Kenya.

The potential severity of earthquakes and volcanic eruptions is much greater where plates are converging rather than diverging. Along the oceanic trenches where crustal rock material is subducted, earthquakes are common and volcanoes typically develop along the edge of the overriding plate. The largest region where this occurs is the Pacific Ring of Fire, the earthquake-prone and volcanically active margin around the Pacific Ocean. Although most are small to moderate, the largest earthquakes ever recorded have been related to subduction in this region. Points where earthquakes originate along an oceanic trench become deeper toward the overriding plate, indicating the subducting plate's progress downward toward where it is recycled into the mantle. Where oceanic crust subducts beneath continental crust along an oceanic trench, some of it melts into magma that moves upward under the continental crust. Subduction along the Pacific Ocean is associated with extensive volcanoes in the Andes, the Cascades, and the Aleutians; the Kuril Islands and the Kamchatka Peninsula in Russia; and Japan, the Philippines, New Guinea, Tonga, and New Zealand. Many of these volcanoes erupt rock and lava of andesitic composition and can be dangerously explosive.

Another volcanic and seismic belt occupies the continental collision zone between northward-moving Southern Hemisphere lithospheric plates and the Eurasian plate. The volcanoes of the Mediterranean region, Turkey, Iran, and Indonesia are located along this collision zone. Seismic activity is common in that zone and has included some major, deadly earthquakes.

Transform plate boundaries, where lateral sliding occurs, also experience many earthquakes. The potential for major earthquakes mainly exists in places such as the Caribbean and along the San Andreas Fault zone in California where thick continental crust is resistant to sliding easily. Volcanic activity along transform plate boundaries ranges from moderate on the seafloor to slight in continental locations.

Areas far from active plate boundaries are not necessarily immune from earthquakes and volcanism. The Hawaiian Islands, the Galapagos Islands, and the Yellowstone National Park area are examples of intraplate "hot spots" located away from plate margins and associated with a plume of magma rising from the mantle. Oceanic crustal areas that lie over hot spots, like the Hawaiian Islands, have strong volcanic activity and moderate earthquake activity. In midcontinental areas large earthquakes occur in suture zones where continents are colliding, such as in the Himalayas, or where broken edges of ancient landmasses shift even though they are today situated in midcontinent and are deeply buried by more recent rocks.

Earthquake and volcanic activities that are located away from active plate margins are intriguing and show that we still have much to learn about Earth's internal processes and their impact on the surface. Still, plate tectonics has contributed greatly to our understanding of the variations in volcanism, earthquake activity, and the landforms associated with these processes.

Smithsonian Institution Hologlobe Project

The spatial correspondence among plate margins, active volcanoes, earthquake activity, and hot spots is not coincidental but is strongly related to lithospheric plate boundaries. This map shows plate boundaries and the global distribution of active volcanoes (1960–1994).

Smithsonian Institution Hologlobe Project

This map shows plate boundaries and the global distribution of earthquake activity (magnitude 4.5+, 1990–1995).

# CHAPTER 14 ACTIVITIES

## ■ DEFINE AND RECALL

topography
landform
geomorphology
relief
endogenic process
exogenic process
erosion
transportation
deposition
geomorphic agent
rock structure
strike
dip
compressional tectonic force
tensional tectonic force
shearing tectonic force
folding
anticline
syncline
recumbent fold
faulting
fault

reverse fault
thrust fault
overthrust
fault block
normal fault
graben
horst
rift valley
escarpment (scarp)
fault scarp
dip-slip fault
strike-slip fault (lateral fault)
earthquake
focus
epicenter
earthquake magnitude
moment magnitude
earthquake intensity
modified Mercalli scale
volcanism
volcano
plutonism

viscosity
volcanic ash
lava flow
pahoehoe
aa
fissure
basaltic plateau
shield volcano
cinder cone
angle of repose
composite cone (stratovolcano)
pyroclastic flow
plug dome
caldera
igneous intrusion (pluton)
stock
batholith
laccolith
sill
dike
volcanic neck

## ■ DISCUSS AND REVIEW

1. Explain how endogenic processes differ from exogenic processes.
2. Distinguish among compressional, tensional, and shearing forces.
3. Draw a diagram to illustrate folding, showing anticlines and synclines.
4. How does a reverse fault differ from a normal fault?
5. How is a fault different from a joint?
6. What causes an earthquake?
7. What is the relationship between the focus and the epicenter of an earthquake?
8. How do earthquake magnitude and earthquake intensity differ? Why are there two systems for evaluating earthquakes?
9. How do plutonism and volcanism differ from each other? How are they alike?
10. What are the four basic types of volcanoes, and what are the distinguishing features of each?
11. Which is the most dangerous type of volcano? Explain why.
12. How is a sill different from a dike?
13. Explain the distribution of most of the world's earthquakes and volcanoes.

## ■ CONSIDER AND RESPOND

1. Name the type of tectonic or volcanic activity that is primarily responsible for the following:
   a. Sierra Nevada
   b. Cascade Mountain Range
   c. Basin and Range region
   d. Appalachian Mountains
2. Name several areas in the United States that are highly susceptible to natural hazards from earthquakes or volcanic activity.
3. List five countries bordering the Pacific Ocean that exhibit evidence of major volcanic activity.
4. Name a few countries not in the Pacific region that face tectonic hazards.
5. Recall the locations of some recent earthquakes or volcanic eruptions reported in the news. Can you explain why they occurred where they did?
6. Assume that you are a regional planner for an urban area in the western United States. What hazards must you plan for if the region has active fault zones and volcanoes? What recommendations for land use and settlement patterns might reduce the danger from earthquake and volcanic hazards?

# ■ APPLY AND LEARN

1. If the first seismic wave to arrive at the epicenter of an earthquake traveled 6 kilometers per second and took 4.3 seconds to reach the surface, how deep was the earthquake focus?

2. A 6.5-magnitude earthquake rocked a city on the California coast. In an older community along a bay everyone felt it, and those who were asleep were awakened by it. People were frightened and stood under doorways for protection. Ultimately, there were no fatalities and only minor injuries. Almost every house experienced at least broken dishes, and there were some reports of fallen plaster, broken windows, and damaged chimneys. Three homes experienced collapse of overhangs that had been added to cover the porch after the homes were built. What intensity was the earthquake for this coastal community?

# ■ LOCATE AND EXPLORE

1. Using Google Earth and the USGS Earthquake Layer (download from CourseMate at www.cengagebrain.com), describe the distribution of recent earthquakes in the United States. In what areas do they tend to cluster and why? What was the largest recent earthquake in Utah within the last 7 days? Was the event large enough to reach the national media?

2. The Hawaiian hot spot is a mantle plume that has persisted for millions of years. As the Pacific plate moved to the northwest, a progression of volcanoes was created and then died as their source of magma was shut off. As shown in Figure 13.34 of your text, formation of the island of Kauai (22.05°N, 159.54°W) was over by about 3.8 million years before present, whereas the island of Hawaii (19.60°N, 155.50°W) started to form approximately 0.7 million years before present. Using the ruler tool in Google Earth, calculate the distance between the two islands and use the distance to calculate the speed of the Pacific Plate in meters per century.

3. Using Google Earth, identify the landforms or structural features at the following locations (latitude, longitude) and provide a brief explanation of why the feature is found there. Use the zoom, tilt, rotate, and elevation exaggeration tools to help view and interpret the feature and the area in which it is found.
   a. 34.5449°N, 115.7907°W
   b. 27.45°N, 54.44°E
   c. 38.876°N, 113.010°W
   d. 52.822°N, 169.947°W
   e. 27.682°N, 16.622°E
   f. 35.2705°N, 119.8264°W
   g. 19.47°N, 155.59°W
   h. 36.688°N, 108.837°W

 CourseMate - Make the most of your study time by accessing everything you need to succeed in one place. Read your textbook, take notes, review flashcards, watch videos, complete activities, take practice quizzes, and more online with CourseMate. Log in at **www.cengagebrain.com**.

# ■ VOLCANIC LANDFORMS

## The Map

The Folsom, New Mexico, map area is located in the far northeastern part of New Mexico, about 300 kilometers (190 mi) northeast of Albuquerque and 80 kilometers (50 mi) west of the Oklahoma panhandle. This region is found within the high, western strip of the Great Plains, adjacent to the southern Rocky Mountains. The Basin and Range province, a large region undergoing tensional forces, lies to the southwest and extends south into Mexico and Texas and west and then north into several western states. Elevation on the entire Folsom map, only a part of which is shown here, ranges from about 1868 meters (6130 ft) to almost 22,500 meters (8200 ft) above sea level. Vegetation is transitional between the grassland of the plains and the for-

ests of the Rocky Mountains; grasses dominate the lower map areas, whereas vegetated higher terrain exhibits juniper and pinyon trees.

Topographically, the map area is dominated by Capulin Mountain, one of almost 100 volcanic landforms in the region's 21,000 square kilometer (8000 mi$^2$) Raton-Clayton volcanic field. Capulin Mountain formed about 62,000 years ago near the end of an 8 million year period of activity for the Raton-Clayton volcanic field. Activity was marked by both effusive and explosive eruptions, occurring in some cases at different times from approximately the same eruptive source. The beautiful Mount Capulin has been protected since 1891. It acquired National Monument status in 1916.

## Interpreting the Map

1. What type of volcanic landform is Capulin Mountain? On what landform characteristics is your decision based?
2. What is the maximum local relief of Capulin Mountain from its base to its peak? What is the drop in elevation from the peak to the bottom of the mountain's crater (hachured contours indicate a depression)?
3. What is Capulin Mountain's maximum diameter? What geometric shape do you think the map view shape of the mountain most closely resembles?
4. Describe how the terrain adjacent to and slightly beyond the base of the mountain's east and southeast side differs from the terrain at the base of the mountain on its northwest side. What accounts for the difference?
5. Between the northwest base of Capulin Mountain and the nearby main highway lies the narrow, southern end of a terrain feature that widens substantially as it extends to the north edge of the map. Use the aerial image and the topographic map to help you determine what type of landform that is. What relationship, if any, do you think it has to Capulin Mountain?
6. A roughly north-trending secondary road runs just west of the feature you identified in question 5. Consider the landform occupying the map area west of that road. In what major way does the surface terrain of that feature differ from the surface terrain of the landform to the east of the road? What might account for the difference in surface terrain?
7. Provide a reasonable explanation for the landform labeled Mud Hill. What observations support your explanation?
8. Volcanic activity from the last several million years is not typical of either the Great Plains or the Southern Rockies. What large-scale processes associated with another nearby region might have contributed to the volcanism in this map area?
9. How helpful is the aerial photograph in analyzing the terrain displayed on the map?

USDA/NAIP

Black and white orthophotograph of Capulin Mountain and vicinity.

⊕ In Google Earth fly to: 36.784°N, 103.968°W (Capulin Mountain).

Opposite:
Folsom, New Mexico
Scale 1:24,000
Contour interval = 20 ft
USGS

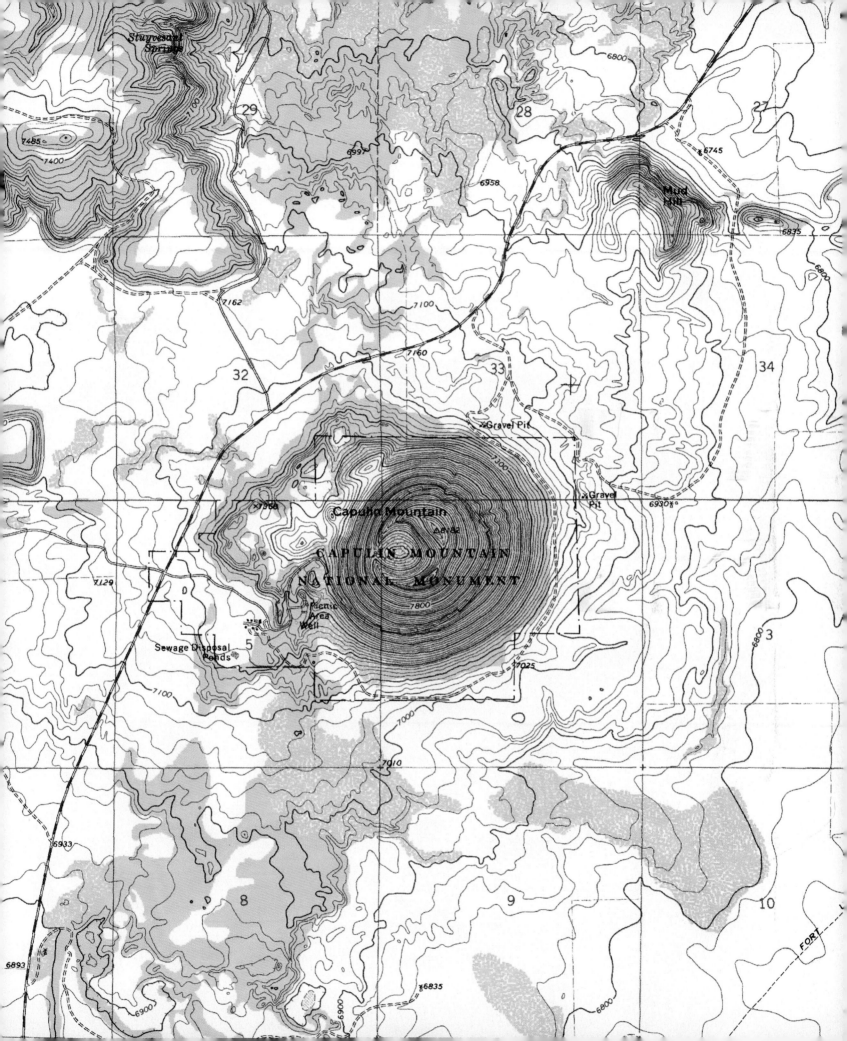

# 15

# WEATHERING AND MASS WASTING

## ▧ OBJECTIVES

WHEN YOU COMPLETE THIS CHAPTER YOU SHOULD BE ABLE TO:

▧ Appreciate that by disintegrating and decomposing rock, weathering prepares rock fragments for erosion, transportation, and deposition.

▧ Explain the principal differences among the various physical and chemical weathering processes.

▧ Discuss how variations in climate, rock type, and rock structure influence weathering rates.

▧ Provide examples of the topographic effects of differential weathering and erosion.

▧ Understand the role of gravity in mass wasting.

▧ Use appropriate terms to describe the types of material involved in slope failures.

▧ Categorize various mass wasting events as slow or fast.

▧ Describe the various ways by which materials can move downslope.

▧ Recognize some of the landforms and landscape features created by mass wasting.

▧ Identify ways in which people influence weathering and mass wasting and ways in which weathering and mass wasting influence people.

IN THE PREVIOUS TWO CHAPTERS we examined the materials, processes, and structures associated with the construction of topographic relief at the surface of Earth. We also noted that the relief-building endogenic geomorphic processes of tectonism and volcanism, which arise from within Earth, are opposed by the relief-reducing exogenic geomorphic processes, which originate at the surface. These exogenic processes break down rocks and erode rock fragments from higher energy sites, transporting them to locations of lower energy. The relocation of rock fragments can be accomplished by the force of gravity alone or with the help of one of the geomorphic agents—flowing water, wind, moving ice, or waves. This chapter focuses on the exogenic processes that cause rocks to decay and on the ways that erosion, transportation, and deposition of surficial Earth materials are accomplished when gravity, rather than a geomorphic agent, is the dominant factor in transporting these materials.

Gravity constantly pulls downward on all Earth surface materials. Weakened rock and broken rock fragments are especially susceptible to downslope movement by gravity, which may be slow and barely noticeable or rapid and catastrophic. Slope

Pressure release and expansion have caused an outer shell of this granite outcrop to break parallel to the main mass and then into smaller sections. Many broken pieces of the outer shell have already moved downslope off the outcrop.  J. Petersen

instability causes costly damage to buildings, roadways, pipelines, and other types of construction and is also responsible for injury and loss of life. Some gravity-induced slope movements are entirely natural in origin, but human actions contribute to the occurrence of others. Understanding the processes involved and circumstances that lead to slope movements can help people avoid these costly and often hazardous events.

# Nature of Exogenic Processes

Earth's surface provides a harsh environment for rocks. Most rocks originate under much higher temperatures and pressures and in very different chemical settings than those found at Earth's surface. Thus, surface and near-surface conditions of comparatively low temperature, low pressure, and extensive contact with water cause rocks to undergo varying amounts of disintegration and decomposition ( ● Fig. 15.1). This breakdown of rock material at and near Earth's surface is known as **weathering.** Rocks weakened and broken by weathering become susceptible to the other exogenic processes—erosion, transportation, and deposition. A rock fragment broken (weathered) from a larger mass will be removed from that mass (eroded), moved (transported), and set down (deposited) in a new location. Together, weathering, erosion, transportation, and deposition actually represent a chain or continuum of processes that begins with the breakdown of rock.

Erosion, transportation, and deposition of weathered rock material often occur with the assistance of a geomorphic agent, such as stream flow, wind, moving ice, and waves. Sometimes, however, the only factor involved is gravity. Gravity-induced

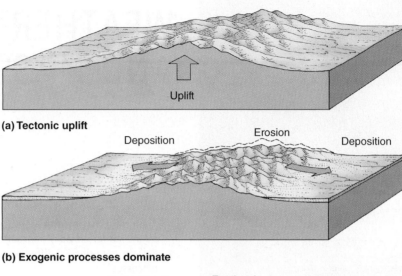

**(a) Tectonic uplift**

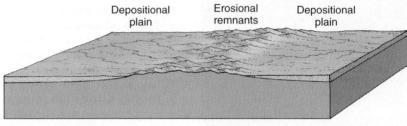

**(b) Exogenic processes dominate**

**(c) Reduced relief**

● **FIGURE 15.2** (a) Tectonic uplift, an endogenic process, is opposed by (b) the exogenic processes of weathering, mass wasting, erosion, transportation, and deposition that (c) eventually decrease relief at Earth's surface significantly if no additional uplift occurs.

downslope movement of rock material that occurs without the assistance of a geomorphic agent, as in the case of a rock falling from a cliff, is known as **mass wasting.** Although gravity also plays a role in the redistribution of rock material by geomorphic agents, the term mass wasting is reserved for movement caused by gravity alone. Whether it is mass wasting or a geomorphic agent doing the work, fragments and ions of weathered rock are removed from high-energy locations and transported to positions of low energy, where they are deposited.

The variations in elevation at Earth's surface, and the shape of individual landforms, reflect the opposing tendencies of endogenic and exogenic processes. The relief created by tectonism and volcanism will decrease over time if endogenic processes cease or operate at a slow rate compared with the exogenic processes ( ● Fig. 15.2). Rates of the exogenic processes depend on such factors as rock resistance to weathering and erosion, the amount of relief, and climate.

● **FIGURE 15.1** This boulder, which was once hard and solid, has undergone disintegration and decomposition because of conditions at Earth's surface.

*Why are some exposed parts of the boulder darker than others?*

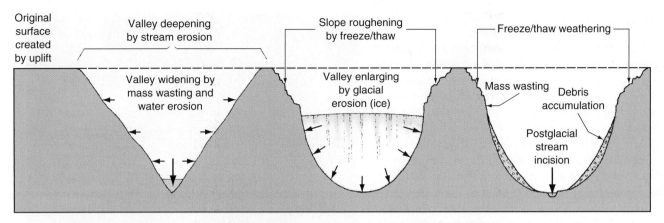

● **FIGURE 15.3** The landscape of most high mountain regions has been produced by at least three different phases of landforming processes: stream cutting and valley formation during tectonic uplift, glacial enlargement of former stream valleys with intense freeze–thaw weathering above the level of the ice, and postglacial weathering and mass wasting with recent stream cutting.

*How does the cross-sectional profile of the valley change at each phase?*

Different exogenic geomorphic processes impart visually distinctive features to a landform or landscape. Typically, weathering, mass wasting, or one of the various geomorphic agents do not work alone in shaping and developing a landform. More often they work together to modify the landscape, and evidence of multiple processes can be discerned in the appearance of the resulting land-form ( ● Fig. 15.3). For example, both the northern Rockies and the Sierra Nevada were produced by tectonic uplift (an endogenic process), but much of the spectacular terrain seen in both mountain ranges today is the result of weathering, mass wasting, running water, and glacial activity (exogenic processes) that have sculpted the mountains and valleys in distinctive ways ( ● Fig. 15.4).

● **FIGURE 15.4** In the foreground of this aerial view of mountain summits in the northern Rockies are the White Cloud Peaks in the Sawtooth Range of Idaho.

*Can you identify evidence of the three phases shown in Figure 15.3?*

USGS/Austin Post

# Weathering

Environmental conditions at and near Earth's surface subject rocks to temperatures, pressures, and substances, especially water, that contribute to physical and chemical breakdown of exposed rock. Broken fragments of rock, called *clasts,* that detach from the original rock mass can be large or small. These detached pieces continue to weather into smaller particles. Fragments may accumulate close to their source or be widely dispersed by mass wasting and the geomorphic agents. Many weathered rock fragments become sediments deposited in such landforms as floodplains, beaches, or sand dunes, whereas others blanket hillslopes as *regolith,* the inorganic portion of soils. Weathering is the principal source of the inorganic constituents of soil, without which most vegetation could not grow. Likewise, ions chemically removed from rocks during weathering are transported in surface or subsurface water to other locations. Ions represent a major source of nutrients in terrestrial as well as aquatic ecosystems, including rivers, ponds, lakes, and the ocean.

The several types of rock weathering fall into two basic categories. **Physical weathering,** also known as mechanical weathering, disintegrates rocks, breaking smaller fragments from a larger block or outcrop of rock. **Chemical weathering** decomposes rock through chemical reactions that remove ions from the original rock-forming minerals. Many different physical and chemical processes lead to rock weathering, and water plays an important role in almost all of them.

## Physical Weathering

The mechanical disintegration of rocks by physical weathering is especially important to landscape modification in two ways. First, the resulting smaller clasts are more easily eroded and transported than the initial larger ones. Second, the breakup of a large rock into smaller ones encourages additional weathering because it increases the surface area exposed to weathering processes. Rocks can be physically weathered in several ways; even a person breaking a rock with a hammer is carrying out physical weathering. Although people, plants, and animals are responsible for breaking some rocks ( ● Fig. 15.5), most physical weathering is not accomplished through the direct action of organisms. Five principal types of physical weathering are discussed here.

**Unloading** Most rocks form under much higher pressures (weight per unit area) than the 1013.2 millibars (29.92 in. Hg; 15 lbs/in.²) of average atmospheric pressure that exists at Earth's surface. Intrusive igneous rocks solidify slowly deep beneath the surface under great pressure from the weight of the overlying rocks. Sedimentary rocks solidify partly or wholly as a result of compaction from the weight of overlying sediments. Metamorphic rocks form when high pressure and temperature substantially change preexisting rocks. Many rocks that originated under conditions of high pressure from deep burial have been uplifted through mountain-building tectonic processes and eventually exposed at Earth's surface. For example, a large pluton of an intrusive igneous rock, such as granite, can be uplifted in a fault

● **FIGURE 15.5** A growing tree has broken a concrete sidewalk and a retaining wall, just as trees and other plants can break up natural rock into smaller fragments, contributing to physical weathering.

*How might an animal cause physical weathering?*

block during tectonism. High elevation helps drive erosional stripping of the overlying rocks, and ultimately through this removal of overlying weight—the **unloading** process—the upper part of the granite is exposed at the surface, where it experiences the low pressure of the atmosphere. As a result of the pressure differential between the higher pressure on the rock at depth and the lower pressure on it at Earth's surface, the outer few centimeters to meters of the rock mass expand outward toward the low pressure of the atmosphere ( ● Fig. 15.6). This expansion causes an extensive crack to form a few centimeters or meters below, and roughly parallel to, the exposed rock surface. Additional cracks may develop parallel to the initial one. These expansion

● **FIGURE 15.6** Outward expansion as a result of unloading has caused this granite in the Sierra Nevada, California, to break, forming joints and sheets of rock that parallel the surface.

● **FIGURE 15.7** Aerial view of Enchanted Rock, a huge granite exfoliation dome in central Texas.

*Why is granite so susceptible to unloading and exfoliation?*

● **FIGURE 15.8** This split rock may be the result of physical weathering by thermal expansion and contraction.

cracks are joints, not faults, and they break the rock into concentric sheets. The breakage is mechanical weathering that resulted from erosional removal of all the rock originally overlying the pluton. Weathering by unloading is especially common on granite but can affect other rocks that, like granite, generally lack internal bedding or other planes of weakness.

As the outer sheet of an unloaded rock continues to weather, segments of it may slide off and slough away, further reducing the load on the underlying rock and allowing additional concentric joints to form. The successive removal of these outer rock sheets is known as **exfoliation,** and each concentric broken layer of rock is an *exfoliation sheet.* The term **exfoliation dome** designates an unloaded, exfoliating outcrop of rock with a dome-like surface form. Well-known examples of granite exfoliation domes are Stone Mountain in Georgia; Half Dome in Yosemite National Park; Sugar Loaf Mountain, which overlooks Rio de Janeiro, Brazil; and Enchanted Rock in central Texas ( ● Fig. 15.7).

**Thermal Expansion and Contraction** Many early Earth scientists believed that the extreme diurnal temperature changes common in deserts physically weather rocks by expansion and contraction as they heat and cool. Those scientists cited widespread existence of split rocks in arid regions as evidence of the effectiveness of this **thermal expansion and contraction** weathering ( ● Fig. 15.8). Laboratory studies in the early 20th century seemed to refute the idea of rock weathering by this process, yet the notion remained popular. Recent field studies in the desert of the American Southwest now lend scientific support to the notion that alternating heating and cooling can lead to the mechanical splitting of rocks. Less controversial has been the notion that differential thermal expansion and contraction of individual mineral grains in coarse crystalline rocks contributes to **granular disintegra-**

**tion,** the breaking free of individual mineral grains from a rock ( ● Fig. 15.9). The broken rocks found after a forest or brush fire provide additional evidence for the effectiveness of thermal expansion and contraction weathering but under much greater extremes of temperature.

● **FIGURE 15.9** Weathering of this intrusive igneous boulder has taken the form of granular disintegration, as evidenced by the numerous, individual mineral grains that have collected around the base of the boulder.

*What other evidence exists on the boulder to suggest that it has been subjected to considerable weathering?*

● **FIGURE 15.10** Within rock fractures, the force of expansion of water freezing into ice is great enough to cause some rock weathering.

*How important is freeze–thaw weathering where you live?*

**Freeze–Thaw Weathering** In areas subject to numerous diurnal cycles of freeze/thaw, water repeatedly freezing in fractures and small cracks in rocks contributes significantly to rock breakage by **freeze–thaw weathering,** sometimes referred to as *frost weathering* or *ice wedging*. When water freezes, it expands in volume up to 9%, and this can cause large pressures to be exerted on the walls and bottom of the crack, widening it and eventually leading to a piece of rock breaking off ( ● Fig. 15.10). The damaging effects of freezing water expanding is why vehicles driven in freezing temperatures must have antifreeze instead of water in their radiator systems. Likewise, water pipes to and within buildings will burst if they are not sufficiently insulated to protect the water inside them from freezing. Freeze–thaw weathering is particularly effective in the upper-middle and lower-high latitudes, and results of freeze–thaw weathering are especially noticeable in mountainous regions near the tree line where angular blocks of rock attributed to freeze–thaw weathering are common ( ● Fig. 15.11). Because of warmer temperatures, freeze–thaw weathering is not significant at lower latitudes except in areas of high elevation.

**Salt Crystal Growth** The development of salt crystals in cracks, fractures, and other void spaces in rocks causes physical disintegration in a way that is similar to freeze–thaw weathering. With **salt crystal growth,** water containing dissolved salts accumulates in these spaces. When the water evaporates, the salts stay behind growing into crystals that are capable of wedging pieces of rock apart. This physical weathering process is most common in arid regions and in rocky coastal locations where salts are abundant, but people also contribute to salt weathering when they use salt to melt ice from roads and sidewalks in winter. Salt crystal growth leads to granular disintegration in coarse crystalline (intrusive igneous) rocks or to the removal of clastic particles from sedimentary rocks, especially sandstones. Continued salt weathering, accompanied by removal of the weathered fragments,

● **FIGURE 15.11** Angular blocks of rock attributed to freeze–thaw weathering, like these near the tree line in Great Basin National Park, Nevada, are common in mountainous areas that experience numerous freeze–thaw cycles every year.

*Why are these rocks angular in shape rather than rounded?*

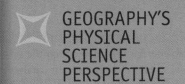
# Expanding and Contracting Soils

Expansion and contraction of rocks, sediments, or materials in fractures and other void spaces in rocks and sediments is a common theme in weathering and mass wasting. Physical weathering by thermal expansion and contraction, freeze/thaw of water, and hydration/dehydration of salts and clay minerals all involve cycles of expansion and contraction, as does downslope movement of soil by creep.

Soils consist of inorganic and organic materials, soil water, and soil air. The inorganic fraction of soils—rock fragments and minerals—derives from rock weathering. These fragments are of sand, silt, and clay size. As weathering continues, clasts become smaller and smaller. In addition, the chemical weathering process of hydrolysis produces very small substances called clay minerals. Clay minerals are colloidal in size, meaning they are smaller than 0.0001 millimeter. In addition to rock fragments and clay minerals, some decaying organic matter (humus) in the soil is colloidal in size.

Having colloidal-sized materials in the soil benefits soil fertility because colloids increase the cation exchange capacity of the soil, which is essentially the ability of the soil to hold plant nutrients. Colloids are so small that their negative electrical charge becomes an important characteristic. Plant nutrients, such as potassium, calcium, magnesium, phosphorous, copper, zinc, and several others, are positively charged ions, or cations. The negatively charged colloids, therefore, attract nutrient cations. When a plant takes up soil water, nutrient cations are also drawn up. A high cation exchange capacity means that the soil has plenty of colloids and therefore the potential to maintain a large number of nutrient cations in the soil for plants to use. The specific capacity of a colloid to hold cations is a function of its composition and structure. The cation exchange capacity of a clay mineral depends on the type of clay mineral, which, in turn, depends on the kind of parent material and the extent of weathering. Illite and vermiculite are two clay minerals associated with slightly weathered soils; montmorillonite is an important component of moderately weathered soils; and kaolinite is characteristic of highly weathered soils.

Nutrient cations are not the only substances that join and leave some colloids. Water molecules come into and leave the crystal structure of colloidal-sized clay minerals during hydration and dehydration. Clay minerals have layered, "sandwichlike" structures. Water molecules enter and exit their layered crystalline structure and cause the clay mineral to expand when wet and contract when dry. Of the four basic clay minerals, montmorillonite expands and contracts the most upon hydration and dehydration. The expansion and contraction, or swelling and shrinking, of clay minerals applies physical pressure on surrounding soil particles and rock surfaces and contributes to physical weathering. According to some sources, the expansion in clay mineral volume can vary from a very small percentage to more than 100%. In most cases, however, the expansion in volume is less than 50%.

A high content of clay minerals advantageous for soil fertility sometimes interferes with the human-built environment. Soils with a significant amount of expansive clays swell and shrink considerably as they become wet and dry. The expansion and contraction in soil volume can shift and crack roads, sidewalks, and building foundations. In regions with typically fertile, expansive clay-rich soils, this causes an estimated $7 billion damage a year to structures in the United States alone. Special construction techniques can be used to mitigate the negative effects of shrinking and swelling on the built environment.

The concrete driveway slab here has been heaved upward by expansive soils, also causing damage to the adjacent railing and retaining wall.

Foundation cracks resulting from expansive soils can cause serious damage to buildings that is expensive to repair.

● **FIGURE 15.12** Physical weathering by salt crystal growth helps break apart rocks, especially along bedding planes exposed in cliffs and on rocky seacoasts. Weathered fragments are removed by gravity, wind, and water, leaving behind hollows.

*Once a small hollow is formed, how might it affect further weathering at that site?*

contributes to the creation of the numerous small hollows that occur along many sandstone cliffs in arid and coastal locations ( ● Fig. 15.12).

**Hydration**  In weathering by **hydration,** water molecules attach to the crystalline structure of a mineral without causing a permanent change in that mineral's composition. The water molecules are able to join and leave the "host" mineral during hydration and dehydration, respectively. A mineral will expand when hydrated and shrink when dehydrated. As in freeze–thaw and salt crystal growth weathering, when hydrating materials expand in cracks or voids within rocks, pieces of the rock can wedge apart. Clasts, mineral grains, and thin flakes are broken from rock masses because of hydration weathering. Salts and **clay minerals,** which are clay-sized materials formed during chemical weathering, commonly occupy cracks and voids in rocks and are subject to hydration and dehydration.

## Chemical Weathering

Chemical reactions between rock-forming minerals and other matter at Earth's surface also work to break down rocks. In chemical weathering, ions from a rock are either released into water or recombine with other substances to form new materials, such as clay minerals. New materials made by chemical weathering are more stable at Earth's surface than the original rocks. The most important catalysts and reactive agents performing chemical weathering are water, oxygen, and carbon dioxide, all of which are common in soil, precipitation, surface water, groundwater, and air.

**Oxidation**  Water that has regular contact with the atmosphere contains plenty of oxygen. When oxygen in water comes in contact with certain elements in rock-forming minerals, a chemical reaction can occur. In this reaction, the element releases its bond with the mineral, leaving behind a substance with an altered chemical formula, and establishes a new bond with the oxygen. This chemical union of oxygen atoms with another substance to create a new product is **oxidation.** Metals, particularly iron and aluminum, are commonly oxidized in rock weathering and form iron and aluminum oxides as the new products. Compared with the original rock, these oxides, including $Fe_2O_3$ and $Al_2O_3$, are chemically more stable, are not as hard, are larger in volume, and have a distinctive color. Iron oxides produced in this way often have a red, orange, or yellow color, whereas oxidized aluminum from rock weathering frequently appears yellow ( ● Fig. 15.13). Oxidation of iron is common, and when it affects iron and steel objects we call it rust.

● **FIGURE 15.13** The reddish-orange coloration on this boulder reveals that the rock contains iron-bearing minerals and that it has been weathered by oxidation.

*What is a likely chemical formula for the reddish-orange substance?*

**FIGURE 15.14** The light-green patches of lichens growing on this reddish rock contribute to solution weathering with the acids that they secrete.

*How does the fact that lichens retain moisture also contribute to weathering?*

**Solution and Carbonation** Under certain circumstances, some rock-forming minerals dissolve in water. In this process of **solution,** a chemical reaction causes mineral-forming ions to dissociate and the separated ions are carried away in the water. Rock salt, which contains the mineral halite (NaCl), is susceptible to solution in water. Most minerals that are insoluble or only slightly soluble in pure water will dissolve more readily if the water is acidic. Water at or near Earth's surface increases in acidity by picking up organic acids from the soil or carbon dioxide from the air or soil. Lichens and mosses growing on rocks contribute to the weathering of those rocks by secreting acidic substances which mix with water from precipitation ( ● Fig. 15.14).

The chemical weathering process of **carbonation** is a common type of solution that involves carbon dioxide and water molecules reacting with, and thereby decomposing, rock material. Carbonation weathering is most effective on carbonate rocks (those containing $CO_3$), particularly limestone, which is an abundant sedimentary rock composed of calcium carbonate ($CaCO_3$). When water with sufficient carbon dioxide comes into contact with limestone, the chemical reaction decomposes some of the calcium carbonate into separate, detached ions of calcium ($Ca^{2+}$) and bicarbonate ($HCO_3^-$) that are carried away in the water. The weathering event occurs as follows: $H_2O + CO_2 + CaCO_3 = Ca^{2+} + 2HCO_3^-$. Similar reactions take place when other types of carbonate rocks undergo carbonation weathering. Because of the role of water in carbonation limestone in particular, but carbonate rocks overall, tend to weather extensively in humid regions but are resistant and often form cliffs in arid climates ( ● Fig. 15.15). Because

water can obtain carbon dioxide by moving through the soil, carbonation operates on rocks affected by soil water or groundwater in the subsurface as well as on rocks exposed at the surface.

**Hydrolysis** In the weathering process of **hydrolysis,** water molecules alone, rather than oxygen or carbon dioxide in water, react with chemical components of rock-forming minerals to create new compounds, of which the $H^+$ and $OH^-$ ions of water are a part. Many common minerals are susceptible to hydrolysis, particularly the silicate minerals (those containing oxygen and silica) that comprise igneous rocks. For example, hydrolysis of the feldspar mineral $KAlSi_3O_8$ can proceed as follows: $KAlSi_3O_8 + H_2O = HAlSi_3O_8 + KOH$.

Hydrolysis of silicate minerals often produces clay minerals. The specific type of clay mineral produced by hydrolysis depends on the composition of the preexisting mineral and other substances that may be present and active in the water. Because water is the weathering agent, hydrolysis is not limited to rocks that are exposed at the surface but can occur in the subsurface through the action of soil and groundwater; in tropical humid climates, it can decay rock to a depth of 30 meters (100 ft) or more.

● **FIGURE 15.15** Limestone weathered by carbonation often takes on a fluted, pitted, or even honeycombed appearance.

*Why does the rock near the bottom of the outcrop seem to be more weathered than that at the top?*

The chemical weathering process of hydrolysis differs from the physical weathering process of hydration, in which water molecules join and leave a substance, causing it to swell and shrink without changing its inherent chemical formula. Both weathering processes, however, may involve clay minerals because those clay-sized substances are often produced by hydrolysis, and many clay minerals swell and shrink substantially during hydration and dehydration.

# Variability in Weathering

How and why types and rates of weathering vary, at both the regional and local spatial scales, are of particular interest to physical geographers. Climate, the type of rock (lithology), and the nature and amount of fractures or other weaknesses in the rock are major influences on the effectiveness of the various weathering processes. Understanding how specific rocks weather in natural settings is important for explaining the characteristics of regolith, soil, and relief, but it is also important in our cultural environment because weathering affects building stone in addition to rock in its natural setting.

## Climate Factors

In almost all environments, physical and chemical weathering processes operate together, even though one of these categories usually dominates. Water plays a role in all but two of the physical weathering processes, but it is essential for all types of chemical weathering. Also, chemical weathering increases as more water comes into contact with rocks. Chemical weathering, then, is particularly effective and rapid in humid climates ( • Fig. 15.16). Most arid regions have enough moisture to allow some chemical weathering, but it is much more restricted than in humid climates. Arid regions typically receive sufficient moisture for physical weathering by salt crystal growth and the hydration of salts. Abundant salts, high humidity, and contact with seawater make salt weathering of rocks very effective in marine coastal locations.

The other principal climatic variable, temperature, also influences dominant types and rates of weathering. Most chemical reactions proceed faster at higher temperatures. Low-latitude regions with humid climates consequently experience the most intense chemical weathering. In the tropical rainforest, savanna, and monsoon climates, chemical weathering is more significant than physical weathering, soils are deep, and landforms appear rounded. Although chemical weathering is somewhat less ex-

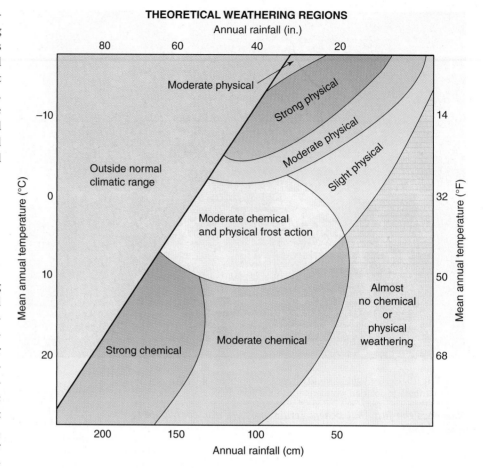

**THEORETICAL WEATHERING REGIONS**

● **FIGURE 15.16** This generalized, theoretical diagram indicates that, although chemical and physical weathering both occur in any climate, chemical weathering is more intense in regions of high temperature and rainfall. Physical weathering effects tend to be more pronounced where temperature and rainfall are both low.

treme in the midlatitude humid climates, its influence is apparent in the moderate soil depth and rounded forms of most landscapes in those regions. In contrast, the landforms and rocks of both arid and cold regions, where physical weathering dominates, tend to be sharper, angular, and jagged, but this depends to some extent on rock type, and rounded features may remain in an arid landscape as relics from prehistoric times when the climate was wetter ( • Fig. 15.17). Comparatively low rates of chemical weathering are reflected in the thin soils found in arid, subarctic, and polar climate regimes. Daily temperature ranges are the critical factor for thermal expansion and contraction weathering in arid climates and freeze–thaw weathering in areas with cold winters.

Air pollution that contributes to the acidity of atmospheric moisture accelerates weathering rates. Extensive damage has already occurred in some regions to historic cultural artifacts made of limestone and marble (metamorphosed limestone), both containing calcium carbonate ( • Fig. 15.18). Because many of the world's great monuments and sculptures are made of limestone or marble, there is a growing concern about

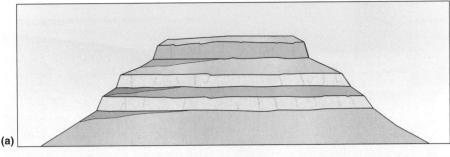

(a)

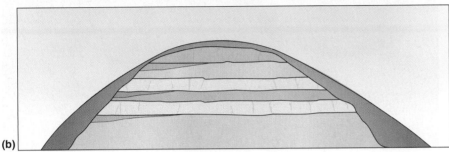

(b)

● **FIGURE 15.17** (a) Because of the dominance of physical weathering, which is comparatively slow, and the sparseness of vegetation, slopes in arid and semiarid environments tend to be bare and angular. Slope angles reflect differences in rock resistance to weathering and erosion. (b) Because chemical weathering dominates in humid regions and abundant vegetation holds the regolith in place longer, slopes in wetter climates have a deep weathered mantle and a rounded appearance.

weathering damage to these treasures. The Parthenon in Greece, the Taj Mahal in India, and the Great Sphinx in Egypt are just a few examples of cultural artifacts undergoing pollution-induced solution, salt crystal growth, hydration of salts, and other destructive weathering processes.

● **FIGURE 15.18** These limestone tombstones have undergone extensive chemical weathering, accelerated by air pollution.

***What kind of chemical weathering has affected the iron fence?***

J. Petersen

## Rock Type

Some rock types are more resistant to weathering than other rock types. Commonly, a rock that is strong under certain environmental conditions may be easily weathered and eroded in a different environmental setting. Rocks that are resistant in a climate dominated by chemical weathering may be weak where physical weathering processes dominate and vice versa. Quartzite is a good example. It is chemically nearly inert and harder than steel, but it is brittle and can be fractured by physical weathering. Shale is chemically inert but mechanically weak. In humid regions, limestone is highly susceptible to solution and carbonation, but under arid conditions, limestone is much more resistant. Granitic outcrops in an arid or semiarid region resist weathering. However, the minerals in granite are susceptible to alteration by oxidation, hydration, and hydrolysis, particularly in regions with warm, humid conditions. Accordingly, granitic areas are often covered by a deeply weathered regolith where they have been exposed to a tropical humid environment.

## Structural Weaknesses

In addition to rock type, the relative resistance of a rock to weathering depends on other characteristics, such as the presence of joints, faults, folds, and bedding planes that make rocks susceptible to enhanced weathering. Some sedimentary rocks, such as sandstone, are only as strong as their cement, which varies from soluble calcium carbonate to inert and resistant silica. In general, the more *massive* the rock, that is, the fewer the joints and bedding planes it has, the more resistant it is to weathering.

The processes of volcanism, tectonism, and rock formation produce fractures in rocks that can be exploited by exogenic processes, including weathering. Joints can be found in any solid

(a)

(b)

⊕ **In Google Earth fly to: 37.290°N, 113.049°W (Zion National Park).**

● **FIGURE 15.19** (a) Extensive jointing influences the location and shape of canyons in and around Zion National Park, Utah. To get an idea of the scale of this satellite image, note the smoke from the wildfire at upper left. (b) Farther north, in Bryce Canyon National Park, Utah, numerous closely spaced vertical joints in sedimentary rocks are sites of preferential weathering and erosion, leaving narrow rock walls between joints.

rock that has been subjected to crustal stresses, and some rocks are intensely jointed ( ● Fig. 15.19). Joints and other fractures that commonly develop in igneous, sedimentary, and metamorphic rocks represent zones of weakness that expose more surface area of rock, provide space for flow or accumulation of water, collect salts and clay minerals, and offer a foothold for plants ( ● Fig. 15.20). Rock surfaces along fractures tend to experience pronounced weathering compared with parts of the rock farther from a fracture. Chemical and physical weathering both proceed faster along any kind of gap, crack, or fracture than in places without such voids.

Because joints are sites of concentrated weathering, the spatial pattern of joints strongly influences the landforms and the appearance of the landscapes that develop. Multiple joints that parallel each other form a **joint set,** and two joint sets will cross each other at an angle ( ● Fig. 15.21). Over time, preferential weathering and erosion in crossing joint sets leave rock in the central area between the fractures only slightly weathered, whereas the rock near the fractures acquires a more rounded appearance. This distinctive, rounded weathered form, known as **spheroidal weathering,** develops especially well on jointed crystalline rocks, such as granite ( ● Fig. 15.22). Spheroidal weathering is not a specific weathering process; it is a form that some rocks acquire as a result of weathering. Once a rock becomes rounded, weathering rates of spheroidal outcrops and boulders decrease because there are

no more sharp, narrow corners or edges for weathering to attack, and a sphere exposes the least amount of surface area for a given volume of rock.

## Differential Weathering and Erosion

Wherever several different rock types occupy a given landscape, some will naturally be more resistant and others will be less resistant to the weathering processes operating there. Because erosion removes small, weathered rock fragments more easily than large, intact rock masses, areas of diverse rock types undergo **differential weathering and erosion.**

Variation in rock resistance to weathering exerts a strong and often highly visible influence on the appearance of landforms and landscapes. Given sufficient time, rocks that are resistant to weathering and erosion tend to stand higher than less resistant rocks. Resistant rocks stand out in the topography as cliffs, ridges, or mountains, whereas weaker rocks undergo greater weathering and erosion to create gentler slopes, valleys, or subdued hills.

An outstanding example of how differential weathering and erosion can expose rock structure and enhance its expression in the landscape is the scenery at Arizona's Grand Canyon ( ● Fig. 15.23). In the arid climate of this region, limestone is resistant, as are sandstones and conglomerates, but shale is relatively weak. Strong and resistant rocks are necessary to maintain steep or vertical cliffs. The stair-stepped walls of the Grand Canyon have cliffs composed of limestone, sandstone, or conglomerate, separated by gentler slopes of shale. At the canyon base, ancient resistant metamorphic rocks have produced a steep-walled inner gorge. The topographic effects of differential weathering and erosion tend to be more prominent and obvious in landscapes of arid and semiarid climates. In dry environments, chemical weathering is minimized, so slopes and varying rock units are not generally covered under a significant mantle of soil or weathered rocks. In addition, sparse vegetation makes details of topography easy to see in arid regions.

Another outstanding example of differential weathering and erosion is the Appalachian Ridge and Valley region of the eastern United States ( ● Fig. 15.24). The rock structure here consists of sandstone, conglomerate, shale, and limestone folded into anticlines and synclines. These folds have been eroded so that the edges of steeply dipping rock layers are exposed as prominent ridges. In this humid climate region, forested ridges composed of resistant sandstones and conglomerates stand up to 700 meters (2000 ft) above agricultural lowlands that have been excavated by weathering and erosion out of weaker shales and soluble limestones.

● **FIGURE 15.20** High up at a narrow spot along this vertical joint, enough soil has accumulated to allow a cactus to grow, while enhanced weathering and erosion lower down has noticeably widened the joint.

● **FIGURE 15.21** Multiple cross-cutting joint sets are visible in this aerial view of part of the Colorado Plateau.

*With north at the top of this photo, what directions do the two most apparent joint sets trend?*

● **FIGURE 15.22** Spheroidally weathered blocks of rock in cross-jointed granite east of the Sierra Nevada in California.

● **FIGURE 15.23** Rock layers of varying thickness and resistance to weathering and erosion along the Grand Canyon create a distinctive array of cliffs composed of strong rocks and gentler slopes formed by less resistant rocks.

● **FIGURE 15.24** A radar image of Pennsylvania's Ridge and Valley section of the Appalachians clearly shows the effects of weathering and erosion on folded rock layers of different resistance. Resistant rocks form ridges, and weaker rocks form valleys.

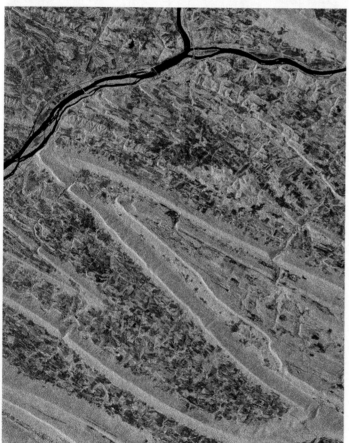

# Mass Wasting

Mass wasting, also called *mass movement,* is a collective term for the downslope transport of surface materials in direct response to gravity. Everywhere on the planet's surface, gravity pulls objects toward Earth's center. This gravitational force is represented by the weight of each object. Heavier objects have a greater downward pull from gravity than lighter objects. The force of gravity encourages rock, sediment, and soil to move downhill on sloping surfaces.

Mass wasting operates in a wide variety of ways and at many scales. A single rock rolling and tumbling downhill is a form of this gravitational transfer of materials ( ● Fig. 15.25), as is an entire hillside sliding hundreds or thousands of meters downslope, burying homes, cars, and trees. Some mass movements act so slowly that they are imperceptible by direct observation and their effects appear gradually over long periods of time. In these cases, tilted telephone poles, gravestones, fence posts, retaining walls, trees, or cracks in buildings can reveal how mass wasting processes are affecting the ground beneath those objects. Other types of mass wasting occur rapidly, sometimes with great violence and disastrous consequences.

● **FIGURE 15.25** A massive boulder, which was loosened by heavy rains and pulled downhill by gravity, blocks a road in Southern California. The boulder weighed an estimated 270,000 kilograms (300 tons) and had to be dynamited to clear the highway.

*What other kinds of problems on roads are related to mass wasting?*

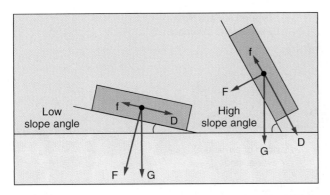

● **FIGURE 15.26** Besides slope angle and material strength, vegetation cover and soil moisture can influence the occurrence of mass wasting. G = total force of gravity (weight of the block); F = component of the block's weight resisting motion; f = frictional forces resisting motion; D = downslope component of gravity.

*How might vegetative cover or moisture content affect the potential for downslope movement of soil?*

The cumulative impact of all forms of mass wasting rivals the work of running water as a modifier of physical landscapes because gravitational force is always present. Wherever loose rock, regolith, or soil lies on a slope, some particles will sooner or later move downslope because of gravity. Friction and rock strength are factors that resist this downslope movement of materials. Friction increases with the roughness and angularity of a rock fragment and the roughness of the surface on which it rests. Rock strength depends on physical and chemical properties of the rock and is reduced by any kind of break or gap in the rock. Fractures, joints, faults, bedding planes, and spaces between mineral grains or clasts all weaken the rock. Furthermore, because all of these gaps invite the accumulation of water, their bonds to the outcrop continue to weaken over time through weathering.

Slope angle also helps determine whether mass wasting will occur. Gravitational forces act to pull objects straight downward, toward the center of Earth. The closer a slope is to being parallel to that downward direction—in other words, the steeper the slope angle—the easier it is for the gravitational forces to overcome the resistance provided by friction and rock strength. Gravity is more effective at pulling rock particles downslope on steep hillsides and cliffs than on gently sloping or level surfaces. The steeper the slope, the stronger the friction or rock strength must be to resist downslope motion ( ● Fig. 15.26). Any surface materials on a slope that do not have the strength or stability to resist the force of gravity will respond by creeping, falling, sliding, or flowing downslope until stopping at the bottom of the slope or wherever there is enough friction to resist further movement. As a result, soil and regolith are thinner on steep slopes and thicker on gentle slopes, and intense mass wasting is one reason why bedrock tends to be exposed in areas of steep terrain.

Gravity is the principal force responsible for mass wasting, but water commonly plays a role and can do so in several ways. Water (1) contributes to weathering, which prepares rock matter for mass movement, (2) adds weight to porous materials on a slope, (3) decreases the strength of unconsolidated slope sedi-

ments, and (4) can increase the slope angle. We have seen that water is involved in many weathering processes that break and weaken rocks, making them more susceptible to mass movement. Unconsolidated soil and regolith have a considerable volume of voids, or pore spaces, between particles. Usually some of these voids contain air and some contain water, but storms, wet seasons, broken pipelines, irrigation, and other situations can cause the voids to fill with water. Saturated conditions encourage mass wasting because water adds weight to the sediments, and an object's weight represents the amount of gravitational force pulling on it. As water replaces air in the voids, unconsolidated sediments and soil also experience decreasing strength as the rock fragments come into greater contact with the liquid, which tends to flow downslope. Finally, streams, and in coastal locations waves, undercut the base of slopes by erosion, thereby increasing the slope angle and facilitating mass movement.

In some cases, factors besides water contribute to the occurrence of mass wasting. When people remove rock material from the base of a slope for construction projects, they steepen the slope angle, making mass movement more likely. Ground shaking, especially during earthquakes, often triggers mass wasting by briefly separating particles supported by other particles. This separation causes a decrease in material strength and friction, both of which resist downslope movement.

Understanding the conditions and processes that affect mass wasting is important because gravity-induced movements of Earth materials are common and frequently affect people and the built environments in which we live. Although this natural hazard cannot be eradicated, we can avoid actions that aggravate the hazard potential and pay close attention to evidence of impending failure in susceptible terrain.

## Materials and Motion

Physical geographers categorize mass wasting events according to the kinds of Earth materials involved and the ways in which they move. Mass wasting events are described with a specific term, for example *rockfall,* that summarizes the type of material and the type of motion. The various kinds of motion are also divided into two general groups according to the speed with which they occur. The types of motion are distinguished and common material and motion combinations are described after a brief overview of the material categories and speed.

Anything on Earth's surface that exists in or on an unstable, or potentially unstable, land surface is susceptible to gravity-induced movement and can therefore be transported downslope as a result of mass wasting. Mass wasting involves almost all kinds of surface materials. Rock, snow and ice, soil, earth, debris, and mud commonly experience downslope movements. In a mass wasting sense, **soil** means a relatively thin unit of predominantly fine-grained, unconsolidated surface material. A thicker unit of the same type of material is referred to as **earth. Debris** specifies a given mass of sediment that contains a wide range of grain sizes, at least 20% of which is gravel. **Mud** indicates saturated sediment composed mainly of clay and silt, which are the smallest particle sizes.

**TABLE 15.1**
**Common Types of Mass Wasting**

| Motion | Common Material | Typical Speed | Effect |
|---|---|---|---|
| Creep | Soil | Slow | |
| Solifluction | Soil | Slow | |
| Fall | Rock | Fast | |
| Avalanche | Ice and snow or debris or rock | Fast | |
| Slump (rotational slide) | Earth | Fast | |
| Slide (linear) | Rock or debris | Fast | |
| Flow | Debris or mud | Fast | |

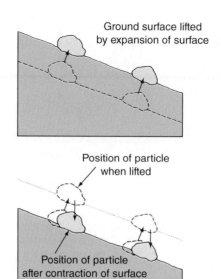

● **FIGURE 15.27** Repeated cycles of expansion and contraction cause soil particles to be lifted at right angles to the surface slope but to fall straight downward by the force of gravity, resulting in soil creep.

*Are there places near where you live that show evidence of soil creep?*

Surface materials move in response to gravity in many different ways, depending on the material, its water content, and characteristics of the setting. Some types of mass movement happen so slowly that no one can watch the motion occurring. With these **slow mass wasting** types we can only measure the movement and observe its effects over long periods. The motion of **fast mass wasting** can be witnessed by people. The speed of downslope movement of material varies greatly according to details of the slope, the rock matter, and if a triggering factor is involved. In addition, slow and fast mass movements often work in combination. Mass movement that is initially slow may be a precursor to more destructive rapid motion, and most materials that have undergone rapid mass wasting continue to shift with slow movements. Specific types of motion, and common Earth materials associated with them, are discussed in the following sections on slow and fast mass wasting. These are summarized in Table 15.1.

## Slow Mass Wasting

Slow mass wasting has a significant, cumulative effect on Earth's surface. Landscapes dominated by slow mass wasting tend toward rounded hillcrests and an absence of sharp angular features.

**Creep** Most hillslopes covered with weathered rock or soils undergo **creep**, the slow migration of particles to successively lower elevations. This gradual downslope motion often occurs as *soil creep*, primarily affecting a relatively thin surface layer of weathered rock particles. Creep is so gradual that it is visually imperceptible; the rate of movement is usually less than a few centimeters per year, yet creep is the most widespread and persistent form of mass wasting because it affects nearly all slopes that have weathered rock fragments at the surface.

Creep typically results from some kind of **heaving** process, which causes individual soil particles or rock fragments to be first pushed upward perpendicular to the slope, and then eventually fall straight downward because of gravity. Freezing and thawing of soil water, and wetting and drying of soils or clays, can lead to soil heaving. For example, when soil water freezes, it expands, pushing overlying soil particles upward relative to the surface of the slope. When that ice thaws, the soil particles move back but not to their original positions because the force of gravity pulls downward on them, as shown in ● Figure 15.27. Soil creep results from repeated cycles of expansion and contraction related to freezing and thawing, or wetting and drying, which cause lifting followed by the downslope movement of soil and rock particles.

Organisms contribute to soil creep and also to other kinds of mass wasting. The traversing of slopes by people and animals tends to push surface material downhill. Every step up, down, or across a steep slope shifts some surface material downhill to a slightly lower position. Even small burrowing animals, such as ground squirrels and chipmunks, are effective soil movers. When they dig their tunnels on a sloping surface, excavated material tends to fall downslope. Plant roots can also move soils slowly outward and in a downward direction on a slope.

Rates of soil creep are greater near the surface and diminish into the subsurface of a slope because the factors instigating it are more frequent near the surface and because frictional resistance to movement increases with depth. As a result, telephone poles, fence posts, gravestones, retaining walls, other structures built by people, and even trees ( ● Fig. 15.28) become tilted when affected by the downward movement of creep.

For the most part, creep does not produce discrete distinctive landforms, but it contributes to rounded, rather than angular, topography in hilly terrain. After gravitational forces deposit these surface sediments at the base of the slope, they may eventually be carried away by one of the geomorphic agents, usually running water.

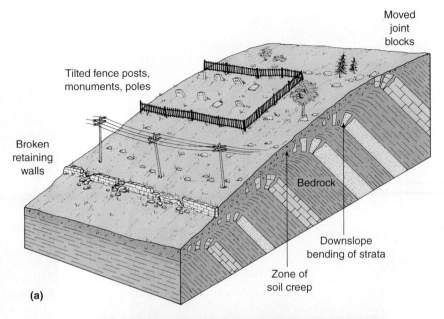

(a)

**Solifluction** The word **solifluction,** which literally means "soil flow," refers to the slow downslope movement of water-saturated soil and/or regolith. Solifluction is most common in high-latitude or high-elevation tundra regions that have **permafrost,** a subsurface layer of permanently frozen ground. Above the permafrost layer lies the **active layer,** which freezes during winter but thaws during summer. During the summer thaw of the centimeters-to meters-thick active layer, the permanently frozen ground beneath it prevents downward percolation of melted soil water. As a result, the active layer becomes a heavy, water-saturated soil mass that, even on a gentle incline, sags slowly downslope by the force of gravity until the next surface freeze arrives. Movement rates by solifluction are typically only centimeters per year.

Evidence of solifluction exists in many tundra landscapes. It consists of relatively thin, irregular, tongue-shaped lobes of soil that may partially override each other to produce hummocky terrain or low mounds ( ● Fig. 15.29).

## Fast Mass Wasting

Four major kinds of mass wasting usually occur so quickly—from seconds to days—that people can watch the material move. The actual speed of movement varies with the situation and depends on the quantity and composition of the material, the steepness of slope, the amount of water involved, the vegetative cover, and the triggering factor. The effects of fast mass wasting events on the land surface are more dramatic than those of slow mass wasting. Rapid mass movements usually leave a visible upslope scar on the landscape, revealing where material has been removed, and a definite deposit where transported Earth material has come to rest at a lower elevation.

**Falls** Mass wasting events that consist of Earth materials plummeting downward freely through the air are **falls. Rockfalls** are the most common type of fall. Rocks fall from steep bedrock

(b)

● **FIGURE 15.28** (a) Effects of soil creep are visible in natural and cultural landscapes. (b) Trees attempt to grow vertically, but their trunks become bent if surface creep is occurring.

*What other constructed features might be damaged by creep?*

cliffs, either (1) one by one as weathering weakens the bonds between individual clasts and the rest of the cliff, or (2) as large rock masses that fall from a cliff face or an overhanging ledge ( ● Fig. 15.30). Large slabs that fall typically break into angular individual clasts when they hit the ground at the base of the cliff.

● **FIGURE 15.29** Solifluction has formed these tongue-shaped masses of soil on a slope near Suslositna, Alaska.

*How does solifluction differ from soil creep?*

In steep mountainous areas, rockfalls are particularly common during the spring when snowmelt, rains, and alternating freezing and thawing loosen and disturb rocks positioned on cliffs or steep slopes. Ground shaking caused by earthquakes is another common trigger for rockfalls.

Over time, a sloping accumulation of angular, broken clasts piles up at the base of a cliff that is subject to rockfall. This slope is known as a **talus,** sometimes referred to as a *talus slope* or, where cone-shaped, a *talus cone* ( ● Fig. 15.31). The presence of talus is good evidence that the cliff above is undergoing rockfall. Like other accumulations of loose, unconsolidated sediment, such as the slopes of cinder cones or piles of gravel in a gravel pit, talus slopes typically lie at or near the steepest angle they can maintain without instability, their *angle of repose* ( ● Fig. 15.32). The angle of repose commonly lies between 30° and 34°, varying with the size and angularity of the clasts. Large angular clasts have a steeper angle of repose than small, more rounded rock fragments.

Falling rocks create hazardous conditions wherever bedrock cliffs are exposed, including mountainous regions and areas with steep roadcuts. Rockfall hazard mitigation remains a high priority along transportation routes in these locations. Massive rockfalls have contributed to the scenic beauty of the towering and steep granite cliffs in Yosemite National Park, California, but frequent

● **FIGURE 15.30** Eventually this overhanging sandstone ledge will fall in a rockfall. As more individual rocks fall from the damper, shaded zone beneath the ledge, the size of the overhang will increase until its weight exceeds the strength of the bonds holding the ledge in place.

*What weathering processes might be acting on the sandstone cliff at the base of the ledge when it becomes wet?*

rockfalls also represent serious hazards there ( ● Fig. 15.33). In 2010 a young girl was rescued after being pinned beneath a 4100 kilogram (9000 lb) boulder that had fallen 15 meters (50 ft) from the nearby steep slope. A larger, fatal rockfall in the park in 1996 moved downslope at an estimated 250 kilometers per hour (160 mph). In that case, the huge mass of moving rock also generated a destructive blast of compressed air that destroyed trees hundreds of meters from the cliff as the rock crashed to the valley floor.

**Avalanches** An **avalanche** is a type of mass movement in which much of the involved material is pulverized—that is, broken into small, powdery fragments—and then flows rapidly as an airborne density current along Earth's surface. Although the word

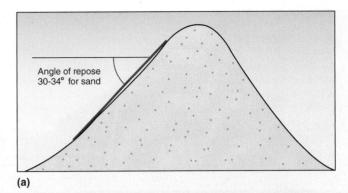
(a)

● **FIGURE 15.31** Rockfall deposits in Glacier National Park, Montana, have accumulated into a talus cone, the slope of angular clasts.

(b)

● **FIGURE 15.32** (a) The angle of repose is the steepest natural slope angle that loose material can maintain. Particles are held at this angle by friction between clasts. (b) Nearly identical piles of sediment containing the same size and shape of particles lie at the same angle of repose.

*How would the angle of repose of rounded particles differ from that of angular particles of the same size?*

avalanche may bring to mind billowing torrents of snow and ice roaring down a steep mountainside, *snow avalanches* are not the only kind. Avalanches of pulverized bedrock, called *rock avalanches,* and those of a very poorly sorted mixture of gravel, sands, silts, and clays, called *debris avalanches,* are also common and have caused considerable loss of life and destruction in mountain communities. Many avalanches are triggered by falls, when the falling snow, ice, rock, or debris pulverizes upon impact with a lower surface. Snow avalanches, the type of avalanche best known to the public, present serious hazards in areas with steep slopes and deep accumulations of snow ( ● Fig. 15.34). In some regions, public safety groups try to limit snow avalanche hazards by purposefully triggering small events or restricting public access when avalanches seem likely. Regardless of the specific type of Earth material involved, avalanches are powerful and dangerous, traveling up to 100 kilometers per hour (60 mph). They easily knock down trees and demolish buildings and have destroyed entire towns.

Slides   In **slides,** a cohesive or semicohesive unit of Earth material slips downslope in continuous contact with the land surface. Water plays a somewhat greater role in most slides than it does in falls. Slides of all kinds threaten the lives and property of people who live in regions with considerable sloping terrain along with such characteristics as tilted layers of alternating strong and weak rocks.

Slides of large sections of bedrock are called *rockslides,* and they are frequent in mountainous terrain where originally horizontal sedimentary rock layers have been tilted by tectonism. The importance of water in reducing the resisting forces of rock strength and friction is seen in the fact that rockslides are most common in wet years, or after a rainstorm or snowmelt. As weathering and erosion by water weaken contacts between successive rock layers, the force of gravity can exceed the strength of the bonds between two of the rock layers. When this happens, a unit of rock, often of massive size, detaches and slides along the tilted planar surface of the contact ( ● Fig. 15.35). Rockslides sometimes end as rockfalls if the topography is such that free fall is needed to transport the rock to a stable position on more level ground. Deposits from rockslides tend to consist of larger blocks of rock than comprise rockfall deposits.

National Park Service

● **FIGURE 15.33** In 2009, rocks from this large rockfall at Ahwiyah Point in Yosemite National Park, California, fell about 550 meters (1800 ft) onto a hiking trail and the adjacent forest.

Some rockslides are enormous, with volumes measured in cubic kilometers. Anything in their path is obliterated. Rockslide deposits in canyons and river valleys may act as a temporary dam that blocks drainage and causes a lake to form. When the lake becomes deep enough, it may wash out the rockslide dam, pro-ducing a sudden and disastrous downstream flood. Thus, immediately after this kind of major rockslide occurs, engineers work to stabilize the resulting dam and control the overflow outlet of the newly formed lake. This was done success-fully after the Hebgen Lake slide in southwestern Montana in 1959 ( ● Fig. 15.36). Triggered by an earthquake, this rockslide, one of the largest in North American history, killed 28 people camped along the Madison River.

Huge rockslides may result from instability related to rock structure and from the undercut-ting of slopes by streams, glaciers, or waves. To-day there are many locations in mountain regions where enormous slabs of rock supported by weak materials are poised on the brink of detachment, waiting only for an unusually wet year or a jarring earthquake to set them in motion.

Rock is not the only Earth material prone to mass wasting by sliding. *Debris slides*, which contain a poorly sorted mixture of gravel and fines, and *mudslides,* which are dominated by wet silts and clays, are also common ( ● Fig. 15.37).

**Slumps** are rotational slides where a thick block of soil, called earth, moves along a concave, curved surface. Because of this curved surface of failure, slump blocks undergo a backward rotation as they slide ( ● Fig. 15.38), causing what was initially the level ground surface at the top of the slump to tilt backward. Slumps are most common in wet years and during wet seasons in many regions with substantial relief, including the Appalachians, New England, and mountainous parts of the western United

● **FIGURE 15.34** (a) A small snow avalanche in the Colorado Rockies. (b) Many people erroneously believe that snow avalanche deposits are light and powdery. Note the heavy, packed nature of the snow deposited by this avalanche, which destroyed homes in Alaska.

Courtesy Lynn French, National Snow and Ice Data Center

(a)

FEMA/Dave Saville

(b)

● **FIGURE 15.35** Rock units that dip in the same direction as the topographic slope of the land are especially susceptible to rockslide, as recognized along this stretch of highway in Wyoming.

D. Sack

States. During exceptionally wet winters in Mediterranean climate regions, like California, slumps frequently damage hillside homes. Slumps, debris slides, and mudslides, like rockslides, can be triggered by earthquakes, which greatly reduce friction and material strength. People contribute to slumping by purposefully

● **FIGURE 15.36** The 1959 earthquake-induced rockslide on the Madison River in Montana completely blocked the river valley and created a new body of water, Earthquake Lake, seen in the background. The massive slide killed 28 people in a valley campground.

**Why can earthquakes trigger landslides?**

© Lloyd Cluff/CORBIS

or accidentally adding water to hillslope sediments and by increasing slope angles by excavating for construction purposes.

**Landslide** has become a general term popularly used to refer to any form of rapid mass movement. However, in some cases, Earth scientists apply the term to large, rapid mass wasting events that are difficult to classify because they contain elements of more than one category of motion or because multiple types of materials—rock, debris, earth, soil, and mud—are involved in a single massive slide. Such large slides are relatively rare but are often newsworthy because of their destructive qualities ( ● Fig. 15.39).

**Flows** Mass wasting **flows** are masses of water-saturated unconsolidated sediments that move downslope by the force of gravity. Flows carry water in moving sediments, whereas rivers carry sediments in moving water. Compared with slides, which tend to move as cohesive units, flows involve considerable churning and mixing of the materials as they move.

When a relatively thick unit of predominantly fine-grained, unconsolidated hillside sediment or shale becomes saturated and mixes and tumbles as it moves, the mass movement is an *earthflow*. Earthflows occur as independent gravity-induced events or in association with slumps in a compound feature called a *slump-earthflow* (see Fig. 15.38). A slump-earthflow moves as a cohesive unit along a concave surface in the middle and upper reaches of the failure, the slump part. Downslope of the failure plane, the mass continues to move but in the more fluid-like, less cohesive manner of an earthflow.

**Debris flows** and **mudflows** differ from each other primarily in grain size and sediment attributes. Both flow faster than earthflows, often move down gullies or canyon stream channels for at least part of their travel, create raised channel rims called *flow levees,* and leave lobate

● **FIGURE 15.37** Mudslide damage in California. Mudslides can be generated by heavy rains or rapid snowmelt on steep hillslopes that have predominantly fine-grained sediments at and near the surface. Like rockslides, mudslides and debris slides travel along an approximately linear, inclined path.

(tongue-shaped) deposits where they spill out of the channel. They result from torrential rainfall or rapid snowmelt on steep, poorly vegetated slopes and are the most fluid of all mass movements. Debris flows transport more coarse-grained sediment than mudflows do.

Debris flows often originate on steep slopes, especially in arid or seasonally dry regions. In humid regions they may occur on steep slopes that have been deforested by human activity

● **FIGURE 15.38** Slump is the common name for a rotational earthslide. Many slumps transition into the more fluid motion of an earthflow in their lower reaches.

*How does the earthflow component differ from the slump component?*

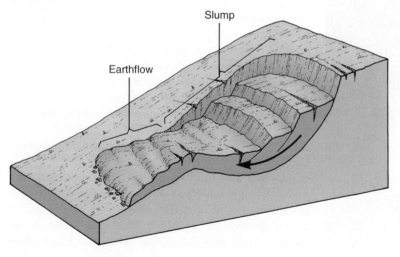

Slump

Earthflow

or wildfire. In each climate setting, rain or meltwater flush weathered rock material from the steep slopes into canyons where it acquires additional water from surface runoff. The result is a chaotic, saturated mixture of fine and coarse sediment, ranging in size from tiny clays to large boulders. As it flows down stream channels, some of the debris is piled along the sides as levees. Where a flow spills out of the channel, the unconfined mass spreads out and velocity decreases, resulting in deposition of a lobe of sediment ( ● Fig. 15.40). Debris flows are powerful mass wasting events that can destroy bridges, buildings, and roads ( ● Fig. 15.41). With dry summers and wet winters, Mediterranean climate regions are susceptible to mudflows and debris flows, particularly when rainy winters follow dry seasons in which considerable vegetation was destroyed by fires.

Serious mudflow hazards also exist in many active volcanic regions. Here, steep slopes may be covered with hundreds of meters of volcanic ash. During eruptions, emitted steam, cooling and falling as rain, saturates the ash, sending down dangerous and fast-moving volcanic mudflows known as **lahars.** Of particular concern are high volcanic peaks capped with glaciers and snowfields. Should an eruption melt the ice and snow, rapid and catastrophic lahars could rush down the mountains with little warning and bury entire valleys and towns. In the United States there is concern that some of the high Cascade volcanoes in the Pacific Northwest may pose a risk of eruptions and associated lahars. Lahars accompanied the 1980 eruption of Mount St. Helens, and Mounts Rainier, Baker, Hood, and Shasta all have the conditions in place, including nearby populated areas, for potentially disastrous lahars to occur ( ● Fig. 15.42).

● **FIGURE 15.39** Very large slides that move a variety of materials are referred to as landslides. Landslides modify the landscape but can also cause much destruction when buildings are constructed in areas susceptible to these large mass movements.

● **FIGURE 15.40** Although small, this recent debris flow in western Utah left well-developed levees on either side of the fresh channel and deposited a tongue-shaped mass (lobe) of sediment where the flow spread out at the base of the slope.

***What evidence is there to indicate this is a site of repeated debris flows?***

● **FIGURE 15.41** A 1995 debris flow in La Conchita, California, destroyed several homes. Steep slopes consisting of weak, unstable sediments of a variety of grain sizes failed during a period of heavy rainfall. A similar precipitation event triggered massive movement there again 10 years later, damaging 36 homes and killing 10 people.

*Why might a specific site experience repeated slope failures over time?*

● **FIGURE 15.42** The violent 1980 eruption of Mount St. Helens in Washington generated lahars—mudflows consisting of volcanic ash. This house was half buried in lahar deposits associated with that volcanic eruption.

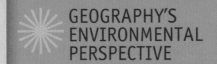
# The Frank Slide

**M**any mass wasting events are compound, having elements of more than one type of motion. Avalanches and flows often begin as falls or slides. Slumps commonly grade into earthflows in their lower reaches. The massive and deadly slope failure that occurred more than a century ago at Turtle Mountain in the Canadian Rockies appears to have comprised two of the most catastrophic types of mass movement: rockfall and rock avalanche. This 1903 Turtle Mountain failure is known as the Frank Slide after the town of Frank, Alberta, a portion of which was obliterated by the very rapidly moving 30 million cubic meters (82 million tons) of rock, resulting in the loss of an estimated 70 lives.

Frank was situated at the base of Turtle Mountain along the Canadian Pacific Railroad line. Many of the town's 600 citizens worked as underground coal miners within the steep mountain. Most of the townsfolk didn't know what hit them as they lay sleeping at 4:10 a.m. when the mountain gave way. Others were at work in the mine inside the mountain when the force of gravity overcame the strength of the 1-kilometer (3280-ft) wide, 425-meter (1395-ft) high, and 150-meter (490-ft) thick mass of limestone. The resistance of the rock mass was weakened by underground mining activities, including blasting, and weathering and erosion along fractures near the mountain summit. Severe weather conditions may have

also played a role. In less than 2 minutes, the rockfall that became an avalanche on impact destroyed homes, buildings, roads, and the railroad line in its path and left a huge expanse of broken rock that extends across to the far side of the valley. Amazingly, 17 miners survived and dug their way out of the mountain through the rubble.

Despite having occurred more than a century ago, tremendous evidence of the huge rock failure still exists in the landscape today. The scar on the flank of Turtle Mountain and the rock rubble strewn over more than 3 square kilometers (1.2 sq mi) across the valley floor serve as a reminder of the incredible and deadly power that can be unleashed by the force of gravity.

The massive 1903 rockfall–avalanche known as the Frank Slide left a huge scar on Turtle Mountain, Alberta, that remains very obvious in the landscape more than a century later.

Rubble from the rapidly moving rockfall–avalanche was strewn across the valley floor, well beyond the partly buried town of Frank.

# Weathering, Mass Wasting, and the Landscape

In this chapter we have concentrated on the exogenic processes of weathering and mass movement. Although neither weathering nor the slower forms of mass movement usually attract much attention from the general public, they are critical to soil formation and, like faster forms of gravity-induced motion, they are significant factors in shaping the landscape. Weathering and mass wasting processes also significantly affect people and our built environment. The

impact, however, is reciprocal; our actions accelerate some weathering processes and help induce mass movements.

Every slope reflects the local weathering and mass wasting processes that have acted on it. These, in turn, are largely determined by the properties of the rocks and the local climate factors. Slow weathering of resistant rocks leaves steep hillslopes, whereas rapid weathering of weak rocks produces gentle hillslopes that are typically blanketed by a thick mantle of soil or regolith. Differential weathering and erosion in areas of multiple rock types or variations in structural weakness produce complex landscapes of variable slopes.

Weathering proceeds rapidly in warm, humid climates where heat and moisture accelerate the reactions that cause minerals and rocks to decompose chemically. Deep mantles of soil and regolith, and rounded forms subject to creep, dominate these regions. In contrast, rocks in arid and cold climates weather more slowly, mainly by physical processes. Arid-region slopes typically have thin, discontinuous, sparsely vegetated accumulations of predominantly coarse-grained regolith that are easily mobilized during intense precipitation events. A tendency toward fast mass wasting is reflected in the angular slopes that are common in many arid regions of mountainous terrain.

In reading the following chapters it will be important to remember the key role that weathering plays in preparing Earth materials for erosion, transportation, and deposition by the geomorphic agents. Because weathered rock fragments are often delivered to a geomorphic agent by gravity-induced movement from adjacent slopes, mass wasting is also important in preparing sediment for redistribution by streams, wind, ice, and waves.

# CHAPTER 15 ACTIVITIES

## ■ DEFINE AND RECALL

| | | |
|---|---|---|
| weathering | solution | solifluction |
| mass wasting | carbonation | permafrost |
| physical (mechanical) weathering | hydrolysis | active layer |
| chemical weathering | joint set | fall |
| unloading | spheroidal weathering | rockfall |
| exfoliation | differential weathering and erosion | talus (talus slope, talus cone) |
| exfoliation dome | soil (as a mass wasting material) | avalanche |
| thermal expansion and contraction | earth (as a mass wasting material) | slide |
| granular disintegration | debris | slump |
| freeze–thaw weathering | mud | landslide |
| salt crystal growth | slow mass wasting | flow |
| hydration | fast mass wasting | debris flow |
| clay mineral | creep | mudflow |
| oxidation | heaving | lahar |

## ■ DISCUSS AND REVIEW

1. In what ways is mass wasting similar to, yet different from, the action of the geomorphic agents?
2. How does physical weathering encourage chemical weathering in rock?
3. What are several processes involving expansion and contraction that can contribute to rock weathering?
4. Distinguish between hydration and hydrolysis.
5. Why is chemical weathering more rapid in humid climates than in arid climates?
6. How are joints, fractures, and other voids in a rock related to the rate at which weathering takes place?
7. What are the impacts of differential weathering and erosion on shaping landforms?
8. Distinguish among the principal types of Earth materials moved in mass wasting.
9. What factors facilitate creep?
10. Describe the principal differences between (a) a rockslide and a slump, and (b) a debris flow and a mudflow.

## ■ CONSIDER AND RESPOND

1. What would you recommend as a solution to prevent the loss of valuable historical monuments to weathering processes?
2. What are some ways in which climate influences mass wasting?
3. If you were an urban planner in a city with numerous steep slopes, what major hazards would you have to plan for? What recommendations would you make to reduce these dangers to the community?

# APPLY AND LEARN

1. A mass of wet sediment, including numerous large boulders, traveled rapidly down a desert canyon. The mass stopped moving when it spread out downslope of the canyon mouth, leaving a deposit in the form of a lobe. Field and laboratory analyses determined that the sediment consisted of 18% clay, 29% silt, 27% sand, and 26% gravel. Based on this information, what specific type of mass wasting was it?

2. Find the climate data for the place where you live, or for the nearest climate data station with a similar climate to that of where you live. Using that data, determine which theoretical weathering region you live in from the graph in Figure 15.16. Based on your observational evidence of weathering in the local environment (perhaps of natural rocks, tombstones, or building materials like stone, asphalt, and so on), write a short explanation and cite examples of why the rock weathering in your area fits or does not fit the theoretical weathering region from Figure 15.16.

# LOCATE AND EXPLORE

1. Using Google Earth, identify the landforms at the following locations (latitude, longitude) and provide a brief discussion of how the landform developed and why the landform is found in that general area.

   a. 33.805°N, 84.145°W
   b. 51.56°N, 116.36°W
   c. 49.59°N, 114.40°W
   d. 37.740°N, 119.609°W
   e. 30.506°N, 98.818°W
   f. 49.305°N, 121.241°W

 CourseMate – Make the most of your study time by accessing everything you need to succeed in one place. Read your textbook, take notes, review flashcards, watch videos, complete activities, take practice quizzes, and more online with CourseMate. Log in at **www.cengagebrain.com**.

# 16

# UNDERGROUND WATER AND KARST LANDFORMS

## ■ OBJECTIVES

WHEN YOU COMPLETE THIS CHAPTER YOU SHOULD BE ABLE TO:

■ Recall the general distribution of Earth's freshwater resources.
■ Outline the principal differences among the subsurface water zone of aeration, the intermediate zone, and the zone of saturation.
■ Explain how water enters the subsurface and what facilitates its ability to move through the rock matter found there.
■ Describe the principal factors responsible for variations in groundwater supplies over space and time.
■ Draw a cross-sectional sketch showing a combined topographic and subsurface setting that would lead to an artesian spring.
■ Recount why underground water is susceptible to pollution.
■ Discuss why karst landforms are more common in warm and humid climates than in cold or dry climates.
■ Distinguish among the major landforms produced by solution of rock.
■ Understand what caverns are and how they are formed.
■ Explain how water dripping in caverns leads to the development of stalactites and stalagmites.

AS PART OF OUR UNDERSTANDING of the distribution and effects of flowing water on Earth, we must consider the part of the hydrologic cycle that operates beneath the surface as well as surface flow. Like water flowing at the surface, water beneath Earth's surface moves, carries other substances, influences the form and appearance of the landscape, and represents an important source of freshwater for human use.

Freshwater is a precious, but limited, natural resource. There is much concern today throughout the world about the quantity and quality of our freshwater resources. Many populated regions have limited supplies of freshwater, whereas, ironically, in some sparsely populated or uninhabited areas, such as tundra and tropical rainforest regions, potable water is plentiful. Ice and snow in the polar regions represent 70% of the freshwater on Earth, but it is generally unavailable for human use. When most people think of freshwater, they typically envision rivers and lakes, both of which are important sources, but together they represent less than 1% of this crucial resource. The remaining freshwater, nearly 30%, lies close to, but beneath, Earth's surface. These underground resources make up an impressive

Underground water first dissolves limestone to form caverns. Later, slow deposition of calcium carbonate creates the beautiful features inside, as these at Carlsbad Caverns National Park, New Mexico. Natalia Bratslavsky/ Shutterstock.com

Gulf Coast from Louisiana to Florida. Where the landscape is one of hills and narrow valleys, the lowest points of the water table are often indicated by the location of stream channels on the valley floors. The base of most humid region stream channels lies below the elevation of the water table. In these cases, water from the zone of saturation seeps directly into the channel beneath the stream's water surface, thereby supplying the stream with *base flow*. This **effluent** condition, where groundwater is seeping into a stream, keeps the stream flowing between rains or during dry seasons ( ● Fig. 16.3a).

Many streams in semiarid and arid regions flow only seasonally or immediately following a significant rainfall. In semiarid regions, the water table typically lies some distance below the streambed during dry periods and rises to intersect the streambed during wet periods. These streams receive groundwater only during wet seasons; during dry, **influent** periods the streams lose water by seepage into the ground below (Fig. 16.3b). In most truly arid regions, surface streams flow just during and immediately after rains. With

the water table deep beneath the channel throughout the year, only influent flow occurs; these streams lose water by infiltration into the channel bed whenever they flow (Fig. 16.3c).

## Groundwater Supply

The amount and availability of groundwater in an area depend on a variety of factors. Most fundamental is the amount of precipitation that falls in the area and the amount that falls in adjacent regions that drain into it. A second important factor is the rate of evaporation. Third is the ability of the ground surface to allow water to infiltrate into the underground water system. A fourth factor is the amount and type of vegetation cover. Although dense vegetation transpires great amounts of moisture back to the atmosphere, it also inhibits rapid surface runoff of rainfall, encourages infiltration of water into the ground, and lowers evaporation rates by providing shade. Thus, the overall effect of vegetation in humid regions is to increase the supply of groundwater.

Two additional factors that affect the amount and availability of groundwater are the porosity and permeability of the sediments and rocks ( ● Fig. 16.4). *Porosity* refers to the amount of

● **FIGURE 16.3** Water may exit or enter the subsurface system at surface streams. (a) In humid regions, groundwater seeps out (effluent condition) into major stream channels all year, providing them with a source of continuous flow. (b) In semiarid or seasonally dry regions, the water table may fall below the stream bed, causing the stream to dry up until the next wet period (seasonally influent condition). (c) In arid regions, significant depth to the water table means that streams flow only immediately after a rainfall event, and water seeps from the streambed into the subsurface (influent condition).

● **FIGURE 16.4** The relationship between porosity and permeability. (a) Sandy sediment and unjointed sandstones tend to have low porosity and high permeability, but fractures (joints) make this sandstone porous and highly permeable. (b) Unjointed shale (clay-sized clasts) has a high porosity but the pores are small and very poorly connected, making for a low permeability. (c) Jointed rock that is nonporous gains porosity and permeability through its fractures. Modified from *U.S. News and World Report* (18 March 1991): 72–73. Copyright 1991 *U.S. News & World Report,* L. P. Reprinted with permission.

*Which of these three rock types would be best for obtaining water?*

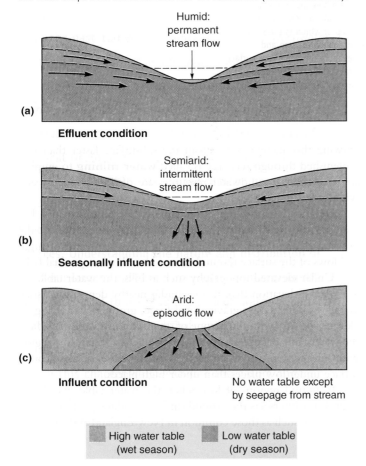

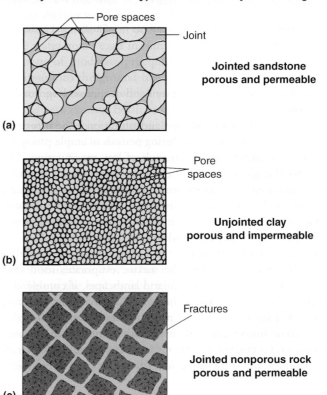

space between the particles that make up the sediments or rocks. Porosity is expressed as the volume of voids compared to the total volume of the material (including voids). Sediments and rocks consisting of clay-sized clasts, perhaps surprisingly, have a relatively high porosity and therefore can contain considerable amounts of water in the large number of very tiny pores. Sand and gravel have comparatively low porosity, but the actual value depends on the packing of grains and uniformity of the grain sizes present. **Permeability** expresses the relative ease with which water flows through void spaces in Earth material. Despite the high porosity of clay sediments and shale, the pore spaces are poorly interconnected, giving them a low permeability. It is therefore very difficult to obtain the water held within unjointed clays and shales. Permeability will increase significantly if these materials have joints or fractures that provide interconnections to facilitate flow. The inherently high permeability of sands and gravels often offsets their low porosity to make them good sources from which to obtain water.

Rocks that are composed of interlocking crystals, such as granite, have virtually no pore space and can hold little water within the rock itself. These crystalline rocks, however, may contain water within joints, which allow the passage of groundwater rather freely. Thus, jointed granite can be described as permeable, even though the rock itself is not porous. In contrast, basaltic lavas often contain considerable pore space formed by gas bubbles frozen into the rock, but typically these holes are not interconnected, so the rock itself is porous but not permeable. The presence of fractures is critical for providing permeability in many areas of volcanic rocks.

It is important to understand the distinction between porosity and permeability. Porosity controls the potential amount of groundwater storage by providing available spaces for the water to be held. Permeability affects the rates and volumes of groundwater movement and is facilitated by the presence of large pore spaces, bedding planes, joints, faults, and even caverns. Porosity and permeability both influence the availability of groundwater resources.

An **aquifer** (from Latin: *aqua,* water; *ferre,* to carry) is a sequence of porous and permeable layers of sediment or rock that acts as a storage medium and transmitter of water ( ● Fig. 16.5). Although any rock material that is sufficiently porous and permeable can serve as an aquifer, most aquifers that supply water for human use are sandstones, limestones, or deposits of loose, coarse sediments (sand and gravel). A rock layer that is relatively impermeable, such as slate or shale, restricts the passage of water and therefore is called an **aquiclude** (from Latin: *aqua,* water; *claudere,* to close off).

Sometimes an aquifer will exist between two aquicludes. In this case, water flows in the aquifer much as it would in a water pipe or hose; water moves within the aquifer but does not escape into the enclosing aquicludes. An aquiclude can also prevent downward percolating soil water from reaching the zone of saturation. An accumulation of groundwater above an aquiclude is called a **perched water table** (see Fig. 16.5). Careless drilling can puncture the aquiclude supporting a perched water table so that the water drains farther down into the subsurface, requiring much deeper drilling to obtain it.

**Springs** are natural outflows of groundwater to the surface. They are related to many causes—landform configuration, bedrock structure, level of the water table, and the relative position of aquifers and aquicludes. Springs may occur along a valley wall where a stream or river has cut through the land to a level lower than a perched water table. The impermeable aquiclude below the

● **FIGURE 16.5** An aquifer is a natural underground storage medium for groundwater. A perched water table can develop where impermeable rock lies between a permeable layer and the regional water table. Beneath the water table water in the regional zone of saturation flows toward the nearby river.

*Is a perched water table a reliable source of groundwater?*

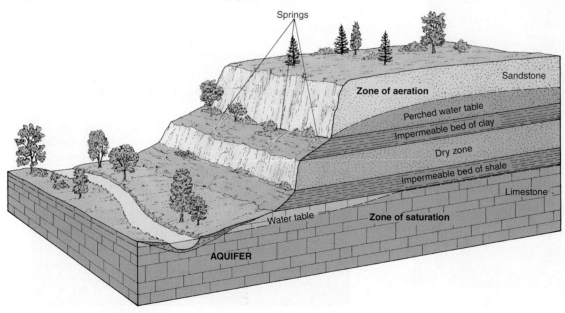

perched water table prevents further downward percolation of water, forcing the water above to move horizontally until it reaches an outlet on the land surface. A spring flows continuously if the water table always remains at a level above the spring's outlet; otherwise the spring flow is intermittent, flowing only when the water table is at a high enough level to feed water to the outlet.

# Groundwater Utilization

Groundwater is a vital resource to most of the world. Half of the population of the United States derives its drinking water from groundwater, and some states, such as Florida, draw almost all of their drinking water from this source. Irrigation, however, consumes the bulk of the groundwater—more than two thirds—used in the United States today. One of the largest aquifers supplying groundwater for irrigation is the Ogallala Aquifer, also known as the High Plains aquifer, which underlies the Great Plains from west Texas northward to South Dakota ( ● Fig. 16.6). The

Ogallala Aquifer alone supplies more than 30% of the groundwater used for irrigation in the United States. Considerable concern exists about the future of this aquifer. Much of the water withdrawn from it accumulated thousands of years ago, and with the now-semiarid climate of the region, recharge is limited.

Groundwater also plays a major role in supporting many wetlands and in forming shallow lakes and ponds, all of which are ecologically invaluable resources. Water bodies and wetlands, fed by groundwater, are critical habitats for thousands of resident and migratory birds. Adequate groundwater flow is vital for the survival of the Everglades in southern Florida. This "river of grass," its great variety of birds, and many other animals are totally dependent on the continued southward movement of groundwater flow through the region.

## Wells

**Wells** are artificial openings dug or drilled below the water table to extract water. Water is drawn from wells by lifting devices ranging from simple rope-drawn water buckets to pumps powered by gasoline, electricity, or wind. In many shallow wells, the supply of water varies with fluctuations of the water table. Deeper wells that penetrate into aquifers beneath the zone of water table fluctuation provide more reliable sources of water and are less affected by seasonal periods of drought.

In areas where there are many wells or a limited groundwater supply, the rate of groundwater removal may exceed the rate of its natural replenishment through groundwater recharge. Many areas that are irrigated from wells have had their water table fall below the depth of the original wells ( ● Fig. 16.7). Progressively deeper wells must be dug, or the old ones extended, to reach the supply of water. In the Ganges River Valley of northern India, the excavation of deep, modern wells to replace shallow hand-dug wells has increased the amount of groundwater being

● **FIGURE 16.6** The Ogallala Aquifer supplies water to a large, semiarid area of the High Plains. Much of the water in the Ogallala Aquifer, the largest freshwater aquifer in the United States, accumulated during wetter times thousands of years ago, and the water table is now falling.

*Why do you think the drop in water supply has been greatest in the southern part of the aquifer?*

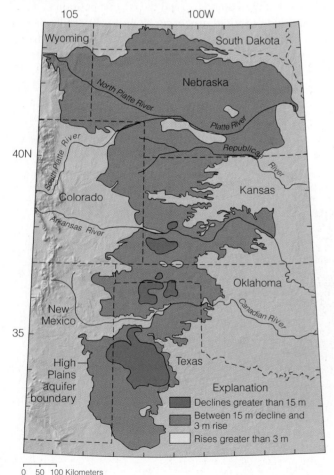

● **FIGURE 16.7** Cones of depression may develop in the water table as a result of the pumping of water from wells. In areas with many wells, adjacent cones of depression intersect, lowering the regional water table and causing shallow wells to go dry.

*What impact might this scenario have on some of the local natural vegetation?*

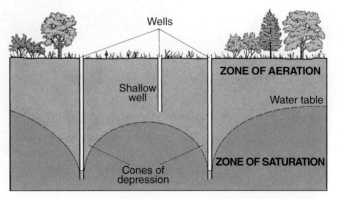

● **FIGURE 16.8** A high-capacity well irrigating crops with water from the Ogallala Aquifer.

brought to the surface, but this greater usage has lowered the water table significantly. The previously mentioned Ogallala Aquifer has been experiencing alarming water level declines, a condition that can be attributed to wells used mainly for agricultural irrigation ( ● Fig. 16.8).

In certain environments, particularly where high groundwater demand has led to extensive pumping, a sinking of the land, called *subsidence,* can occur as a result of compaction related to the water withdrawal. Mexico City, Venice (Italy), and the Central Valley of California, among many other places, have subsidence problems related to groundwater withdrawal. In parts of Southern California, withdrawn groundwater has been replaced artificially by diverting streams so that they flow over permeable deposits. This process is known as **artificial recharge.**

## Artesian Systems

In some cases, groundwater exists in *artesian* conditions, meaning the water is under so much pressure that if it finds an outlet, it will flow upward to a level above the local water table. Where this water achieves outflow to the surface, it creates an **artesian spring** if natural; where the outlet to the surface is artificial, the result is a **flowing artesian well.** In flowing artesian wells, water tapped by the well rises to the surface and flows out under its own pressure, without pumping. In fact, the word artesian is derived from the Artois region of France, where the first known free-flowing well of this kind was dug in the Middle Ages. If water in a well rises above the local water table, but not to the point of flowing out of the well and onto the surface, it is a **nonflowing artesian well.**

Certain conditions are required for artesian water flow ( ● Fig. 16.9). First, a permeable aquifer, often sandstone or limestone, must be exposed at the surface in an area of high recharge by precipitation. The aquifer must receive water from the surface by infiltration, incline downward often hundreds of meters below the surface, and be confined between impermeable layers that prevent escape of the water except to the artesian springs or wells. These conditions cause the aquifer to act as a pipe that conducts water through the subsurface. Water at a given point in the "pipe" is under pressure from the water in the aquifer that lies upslope from that point, closer to the area of intake at the surface. As a consequence of this pressure, water flows toward any available outlet. If that outlet happens to be a well drilled through the impervious layer and into the aquifer, the water will rise in the well. The height to which the water rises in a well is determined by the amount of pressure exerted

● **FIGURE 16.9** Special conditions produce an artesian system. As part of the Northern High Plains Aquifer, the Dakota Sandstone, which averages 30 meters (100 ft) in thickness, transmits water from the Black Hills to locations more than 320 kilometers (200 mi) eastward beneath South Dakota.

*What is unique about artesian wells?*

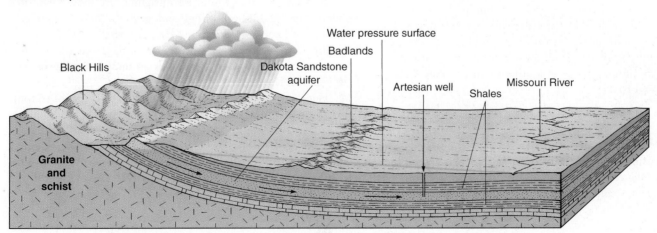

at that outlet from upslope in the aquifer. That pressure depends on the quantity of water in the aquifer, the angle of incline, and the number of other outlets, usually wells, available to the water. The pressure, and the water height in the wells, increase with greater amounts of water, steeper inclines, and fewer outlets.

An excellent example of an artesian groundwater system occurs in South Dakota, where sandstone exposed at the surface along the flanks of the Black Hills in the western part of the state transmits artesian water eastward across the state to wells as far as 320 kilometers (200 mi) away (see Fig. 16.9). Other well-known artesian systems are found in Olympia, Washington; the western Sahara; and eastern Australia's Great Artesian Basin, which is the largest artesian system in the world.

## Groundwater Quality

Because most subsurface water percolates down through a considerable amount of soil and rock, by the time it reaches the zone of saturation it is mostly free of clastic sediment. However, it often carries a large amount of minerals and ions dissolved from the materials through which it passed. As a result of this large mineral content, groundwater is often described as "hard water," in comparison with "softer" (less mineralized) rainwater. Moreover, just as increases in population, urbanization, and industrialization have resulted in the pollution of some of our surface waters, they have also resulted in the pollution of some of our groundwater supplies. For example, subsurface water moving through underground mines in certain types of coal deposits that are widespread in the eastern United States becomes highly acidic. If this **acid mine drainage** reaches the surface, it can have detrimental effects on the local aquatic organisms.

Other dangers to groundwater quality stem from the percolation of toxic substances or salt water into the zone of saturation. Excessive applications of pesticides and incompletely sealed surface or subsurface storage facilities for toxic substances, including gasoline and oil, are situations that can lead to groundwater pollution through downward percolation. In coastal regions with excessive groundwater pumping, denser salt water from the ocean seeps inland to replace freshwater withdrawn from the zone of saturation. This problem of saltwater replacing freshwater in the groundwater zone has occurred in many coastal localities, notably in southern Florida, New York's Long Island, and Israel.

## Geothermal Water

Springs are classified as nonthermal or thermal, depending on the temperature of the emerging groundwater. Water temperature of a *nonthermal spring* (cold spring) is approximately equal to the

George Marler/National Park Service

● **FIGURE 16.10** One of the world's most famous geysers is Old Faithful in Yellowstone National Park, Wyoming.

*How do geysers differ from hot springs?*

mean annual temperature of the atmosphere in the area of the spring. *Thermal springs* have water temperatures that are higher than the surrounding mean annual atmospheric temperature and are further separated into *warm springs* and **hot springs** by the temperature of 36.7°C (98°F). Because water issuing from hot springs is heated by contact with hot rocks in the subsurface, it is referred to as **geothermal water.**

In hot springs, geothermal water flows out to the surface fairly continuously. Where geothermal water flow is intermittent and somewhat eruptive it produces a **geyser,** an impressive phenomenon with sporadic bursts of steam and hot water expelled from a fissure or vent. The Old Faithful geyser in Yellowstone National Park, Wyoming, is a well-known example ( ● Fig. 16.10). Geysers erupt when temperature and pressure of the water at depth reach a critical level, forcing a column of superheated water and steam out of the fissure in an explosive manner. The word geyser is an Icelandic term for these intermittent eruptions that are so common in that volcanic terrain.

Hot springs and geysers are intriguing groundwater-related phenomena. Most hot springs and geysers contain significant amounts of minerals in solution. These minerals, typically rich in calcium carbonate or silica, precipitate out of the water, often forming colorful terraces or mounds around the spring or vent ( ● Fig. 16.11).

Geothermal water is usually associated with areas of tectonic and volcanic activity, especially along lithospheric plate boundaries and over hot spots. In several of these locations geothermal water has been used to produce electricity. The best geothermal water for energy uses is not only very hot, but also "clean"—that is, relatively free of dissolved minerals that can clog pipes and generating equipment.

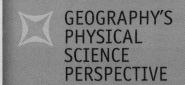

# Acid Mine Drainage

Acid mine drainage (AMD) typically occurs when subsurface water flowing through mines or mine tailings undergoes chemical reactions that leave the water highly acidic (low pH) and metal-rich. AMD is a serious environmental concern in some coal- and metal-mining regions, including much of the coal-mining area of the eastern United States and parts of Australia, South America, and South Africa. When sufficient quantities of the affected water flow onto the surface as springs or seeps into streams, lakes, or ponds, the mineral content and low pH endanger aquatic organisms and make the water unsuitable for human consumption. Fish have completely disappeared from some streams with very low pH and high metal content because of AMD.

The chemical reactions involved in the formation of AMD occasionally happen under natural conditions but at much slower rates and in much smaller amounts than on disturbed lands. Mining greatly increases the permeability of susceptible rocks, allowing much more water to undergo the chemical reactions. Susceptible rocks contain pyrite ($FeS_2$), a common substance in the extensive coals of Pennsylvanian age in the eastern United States. When water containing oxygen comes into contact with pyrite in an abandoned underground coal mine, the pyrite oxidizes readily. Iron, sulfate ($SO_4$), and hydrogen ions are released into the water as a result of the chemical weathering; it is the presence of hydrogen ions that increases the water's acidity. Additional chemical reactions lead to hydrolysis of the iron ions in the water, which releases more hydrogen ions, and a further increase in acidity. Rate of hydrogen production is greatly accelerated if certain microorganisms that thrive in conditions of low pH are present.

Mines that lie in the zone of the fluctuating water table between the zone of aeration and the zone of saturation, like many mines in the Appalachian coal fields, are particularly susceptible to AMD because of the frequent introduction of new, moving, oxygenated water. Pumping removes water from these mines while they are operational. When they are abandoned, however, flowing underground water returns to the now highly permeable and chemically reactive environment and produces AMD.

Controlling the flow of underground water represents a principal way to limit production of AMD. Approaches include locking susceptible mines out of subsurface water circulation by sealing water out or flooding them with very slow moving, stagnant groundwater that has little opportunity to obtain new oxygen. AMD prevention and reduction remains an active field of research, and understanding the chemistry and circulation of underground water is critical to the ongoing investigation.

D. Sack

Orange-colored stream water and orange coating on channel rocks result from acid mine drainage in this Appalachian coal-mining region. Acid mine drainage lowers the pH of local streams, which encourages precipitation of these orange, oxidized iron compounds.

● **FIGURE 16.11** Because of the water's high mineral content, this hot spring in Yellowstone National Park, Wyoming, has left extensive calcareous deposits, known as travertine.

# Landform Development by Subsurface Water and Solution

In areas where the bedrock is soluble in water, subsurface water is an important agent in shaping landform features at the surface and underground. Underground water is a vital ingredient in subsurface chemical weathering processes, and, like surface water, subsurface water dissolves, removes, transports, and deposits rock-forming materials.

The principal mechanical role of subsurface water in landform development is to encourage mass movement by adding weight and reducing the strength of soil and sediments, thereby contributing to slumps, mud or debris slides, mud or debris flows, and landslides. Chemically, subsurface water also contributes to processes that shape landforms. Through the chemical breakdown of rock materials by carbonation and other forms of solution, and the deposition of those dissolved substances elsewhere, underground water is an effective land-shaping agent, especially in areas where limestone is present. Like surface water, subsurface water can dissolve limestone through carbonation or simple solution in acidic water. In many of these areas, surface outcrops of limestone are pitted and pockmarked by chemical solution, especially along joints, sometimes forming large, flat, furrowed limestone platforms ( ● Fig. 16.12). Wherever water acts on any rock type that is significantly soluble in water, a distinctive landscape will develop.

## Karst Landforms

The most common soluble rock is limestone, the chemical precipitate sedimentary rock composed of calcium carbonate ($CaCO_3$). Landform features created by the solution and reprecipitation (redeposition) of calcium carbonate by surface or sub-

● **FIGURE 16.12** Solution of limestone is most intense in fracture zones where the dissolved minerals from the rock are removed by surface water infiltrating to the subsurface. This landscape shows a limestone "platform" where intersecting joints have been widened by solution.

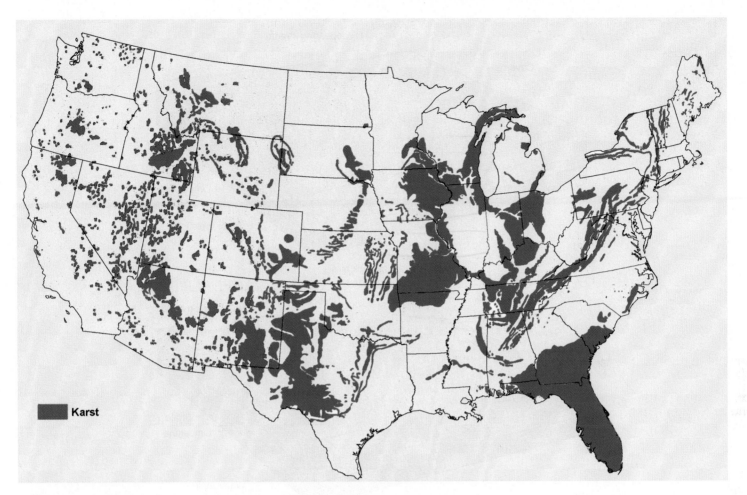

● **FIGURE 16.13** The distribution of limestone in the conterminous United States indicates where varying degrees of karst landform development exists, depending on climate and local bedrock conditions.

*Where is the nearest karst area to where you live?*

surface water are found in many parts of the world. The north-central Mediterranean region, in particular, exhibits large-scale limestone solution features. These are most clearly developed on the Karst Plateau along Croatia's scenic Dalmatian Coast. Landforms developed by solution are called **karst** landforms after this impressive locality. Other extensive karst regions are located in Mexico's Yucatán Peninsula, the larger Caribbean islands, central France, southern China, Laos, Vietnam, and many areas of the United States ( ● Fig. 16.13).

The development of a classic karst landscape, in which solution has been the dominant process in creating and modifying landforms, requires several special circumstances. A warm, humid climate with ample precipitation is most conducive to karst development. In arid climates, karst features are typically absent or are not well developed, but some arid regions have karst features that originated during previous periods of geologic time when the climate was much wetter than it is today. Compared with colder humid climates, warmer humid climates have greater amounts of vegetation, which supplies carbon dioxide to subsurface water. Carbon dioxide is necessary for carbonation of limestone and it increases the acidity of water, which encourages solution in general.

Another important factor in the development of karst landforms is movement of subsurface water. This movement allows water that has become saturated with dissolved calcium carbonate to flow away and be replaced by water unsaturated with calcium carbonate and therefore capable of dissolving additional limestone. Groundwater moves vigorously where it flows toward a low outlet, as might be provided along a deeply cut stream valley or a tectonic depression. In addition, everything else being equal, faster groundwater flow occurs in Earth material that has a greater permeability.

Because infiltration of water into the subsurface tends to be concentrated where cross-cutting sets of joints intersect, in limestone these intersections are subject to intense solution. Such concentrated solution may produce roughly circular surface depressions, called **sinkholes** or **dolines,** that are prominent features of many karst landscapes ( ● Figs. 16.14a and b). Two types of sinkholes are distinguished by differences in their mode of formation ( ● Fig. 16.15). If the depressions result primarily from the surface or near-surface solution of rock and the removal of dissolved materials by water infiltrating downward into the subsurface, the depressions are **solution sinkholes**

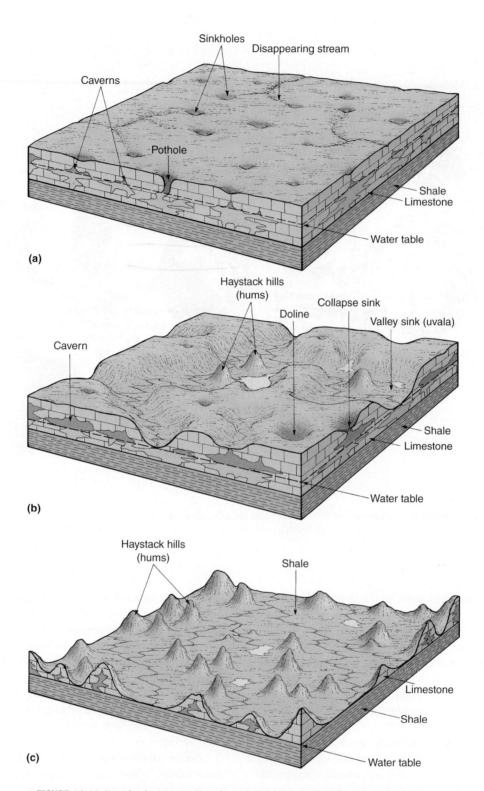

● **FIGURE 16.14** Karst landscapes can be quite varied. (a) Solution at joint intersections in limestone encourages sinkhole (doline) development. Caverns form by groundwater solution along fracture patterns and between bedding planes in the limestone. Cavern ceilings may collapse, causing larger and deeper sinkholes. Surface streams may disappear into sinkholes to join the groundwater flow. (b) Some limestone landscapes have more relief, such as this irregular terrain with merged sinkholes that make karst valleys, called uvalas, and some conical haystack hills (hums). (c) Areas that have experienced more intense solution may be dominated by limestone remnants known as haystack hills that are isolated above an exposed surface of insoluble rock, such as shale.

*Why are there no major depositional landforms created at the surface in areas of karst terrain?*

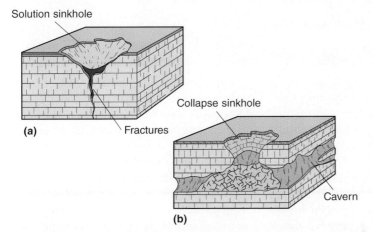

● **FIGURE 16.15** The two major types of sinkholes (dolines) based on principal mode of formation. (a) Solution sinkholes develop gradually where surface water funneling into the subsurface dissolves bedrock to create a closed surface depression. (b) Collapse sinkholes form when either the bedrock or the regolith above a large subsurface void fails, falling into the void.

● **FIGURE 16.17** This large collapse sinkhole appeared in Winter Park, Florida, when a drought-induced lowering of the water table caused the surface to collapse into an underground cavern system.

*What human activities might contribute to the occurrence of such hazards?*

● **FIGURE 16.16** A solution sinkhole in southern Indiana.

( ● Fig. 16.16). If the depressions are caused by the caving-in of the land surface above voids created by subsurface solution in bedrock below, they are referred to as **collapse sinkholes** ( ● Fig. 16.17).

The two processes, solution and collapse, actually cooperate to create most sinkholes in soluble rocks. Whether the depressions are termed solution or collapse sinkholes depends on which of these two processes was dominant in their formation. Solution and collapse sinkholes often occur together in a region. Sinkhole form varies greatly, but solution sinkholes tend to be funnel-shaped whereas collapse sinkholes tend to be steep-walled.

Sudden collapse of sinkholes is a significant natural hazard that every year causes severe property damage and human injury. The rapid formation of sinkholes may result from excessive

groundwater withdrawal for human use or from significant drought periods. Either of these conditions lowers the water table, causing a loss of buoyant support for the ground above, followed by collapse. Rapid sinkhole collapse has damaged roads and railroads and has even swallowed buildings.

Sinkholes may grow in area and merge over time to form larger karst depressions called **uvalas** (from Croatian: *uvala,* hollow or depression) or **valley sinks** (see Fig. 16.14b). In many areas, uvalas are linearly arranged along former routes of subsurface water flow.

Despite their humid climates, many karst regions have few continually flowing surface streams. Surface water seeping into fractures in limestone widens the fractures by solution. These widened avenues for water flow increase the downward permeability, accelerate infiltration, and direct water rapidly toward the zone of saturation. In some cases, surface streams flowing on bedrock of low permeability upstream will encounter a highly permeable rock downstream, where it rapidly loses its surface flow by infiltration (see Fig. 16.12). These are called **disappearing streams** because they "vanish" from the surface as the water flows into the subsurface ( ● Fig. 16.18a).

Water moving along joints and bedding planes below the surface can also dissolve limestone, sometimes creating a system of connected passageways within the soluble bedrock (see Figs. 16.14a and b). If the water table falls, leaving these passageways above the zone of saturation, they are called **caverns** or **caves.**

In many well-developed karst landscapes with caverns, a complex underground drainage system all but replaces surface water flow. In these cases, the surface terrain includes many valleys that contain no streams. Surface streams originally existed and excavated the valleys, but the water flow was eventually diverted to underground conduits within the cavern system. The site where a surface

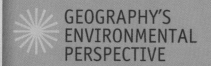
# Sudden Sinkhole Formation

Although many sinkholes develop slowly through solution and infiltration of water, regolith, and soil along joint intersections in soluble rocks, others appear almost instantaneously as the ground unexpectedly collapses into subsurface voids, including caverns. The sudden collapse of the ground caused by the creation of a sinkhole is a problem in many areas of the United States that have soluble rocks at or near the surface. An example is shown in the accompanying photographs of a road collapse near Bowling Green, Kentucky.

Earth scientists recognize two types of collapse processes that form sinkholes: regolith collapse (or cover collapse) and bedrock collapse. In both cases, subsurface solution begins long before the surface sinkhole appears.

Sinkhole formation by sudden regolith collapse occurs where vertical gaps in the subsurface rock exist as a result of solution along joints or other forms of weakness in the rock. The gaps often open to larger passageways at depth. Regolith and soil lie on top of the bedrock and also cross over the gaps as regolith bridges or arches. Initially, the water table lies above the rock within the regolith layer. Sudden regolith collapse is most common after the water table has fallen below the subsurface gaps, such as during drought periods or by excessive pumping from wells. The drop in the water table leaves the regolith in a zone in which some pore spaces fill with air and through which water percolates downward toward the air-filled gap or cavern below. If the force of gravity pulling downward on the soil and regolith exceeds the frictional and cohesive strength holding up the regolith bridge, the regolith collapses into the gap, creating a sinkhole at the ground surface. Adding additional weight to the regolith and soil layer by constructing roads and buildings or by impounding water at the surface increases the force of gravity and therefore the likelihood that the regolith bridges will suddenly collapse.

The formation of a sinkhole by bedrock collapse is associated with a growing underground cavern relatively near the ground surface. Caverns can grow vertically as well as horizontally. If continued solution makes the rock roof of the cavern thinner and thinner, eventually the rock can become so thin that it fails, crashing into the cavern void below. A sinkhole is formed when collapse of the bedrock cavern roof causes overlying Earth material up to the ground surface to lose its support and also slump into the hole.

Physical geographers who have examined the Dishman Lane sinkhole in Bowling Green report evidence of cave roof collapse in this case. It is likely that much of the roof collapsed thousands of years ago, but the fact that limestone blocks were found in the rubble indicated that part of the cave roof collapsed in this event. The resulting sinkhole affected over 0.4 hectares (1 acre) of surface area and caused considerable property damage.

Sinkhole collapse near Bowling Green, Kentucky, caused severe road damage in an area larger than a football field. Fixing the problem required completely filling in the sinkhole with rock and repaving the road at a cost of more than $1 million.

*Courtesy of the Center for Cave and Karst Studies at Western Kentucky University*

Several vehicles were damaged, but luckily no one was injured by the sudden collapse of the sinkhole during rush hour.

*Courtesy of the Center for Cave and Karst Studies at Western Kentucky University*

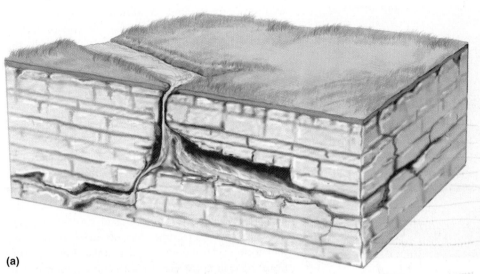

**(a)**

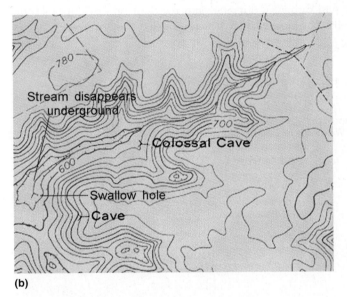

**(b)**

● **FIGURE 16.18** Disappearing streams and swallow holes are common in some karst areas. (a) A surface stream disappears through a solution-widened hole in the ground, the swallow hole, to flow instead through a subsurface cavern system. The stream may emerge back onto the surface at a major spring. (b) This topographic map shows a disappearing stream and a swallow hole. The hachured contours indicate the closed depression into which the stream is flowing.

stream disappears into the cavern system is called a **swallow hole** (Fig. 16.18b). In some cases these underground-flowing "lost rivers" reemerge at the surface as springs where they encounter impervious beds below the limestone.

After intense and long-term karst development, especially in wet tropical conditions, only limestone remnants are left standing above insoluble rock below. These remnants usually take the form of small, steep-sided, and cave-riddled karst hills called **haystack hills, conical hills,** or **hums** (see Figs. 16.14b and c). Multiple terms exist because similar features of slightly different form were independently named in different parts of the world. Areas dominated by these features have been described as "egg box" landscapes because an aerial view of the numerous haystack

hills and sinkholes resembles the shape of an egg carton ( ● Fig. 16.19). Excellent examples of these steep, karstic hills are found in Puerto Rico, Cuba, and Jamaica. If the limestone hills are particularly high and steep-sided, the landscape is called **tower karst.** Spectacular examples of tower karst landscapes are found in southern China and Southeast Asia ( ● Fig. 16.20).

## Limestone Caverns and Cave Features

Of the various types of landforms created by limestone solution, caverns are the best known and most spectacular. Globally, nine cavern systems have been designated as, or are part of, UNESCO World Heritage sites because of their outstanding natural features. Many other beautiful caverns exist in a variety of countries, attracting large numbers of visitors each year. Examples of famous limestone caverns in the United States are Carlsbad Caverns in New Mexico, Mammoth and Colossal Caves in Kentucky, and Luray and Shenandoah Caverns in Virginia. In fact, 34 states have caverns that are open to the public. Some are quite extensive with rooms more than 30 meters (100 ft) high and with kilometers of connecting passageways.

Limestone caverns originate when subsurface water, sometimes flowing much like an underground stream, dissolves rock, leaving networks of passageways. If the water table should drop to the floor of the cavern or lower, typically as a result of climate change or tectonic uplift, the caves will become filled with air. Interaction between the cave air and mineral-saturated water percolating down to the cavern from above will then begin to precipitate minerals, especially calcium carbonate, on the cave ceiling, walls, and floor, decorating them with often-intricate depositional forms.

The nature of any fracturing that exists in soluble bedrock exerts a strong influence on cavern development in karst regions. Groundwater solution widens joints, faults, and bedding planes to produce passageways. The relationship between caverns and fracture distributions is evident on cave maps that show linear and parallel patterns of cave passageways ( ● Fig. 16.21).

Limestone caverns vary greatly in size, shape, and interior character. All caves formed by solution, however, show some evidence of previous water flow, such as deposits of clay and silt on the cavern floor; in some caves water still actively flows. Continued subterranean flow of water further deepens some caverns, and ceiling collapse enlarges them upward. Many cavern systems have several levels and are almost spongelike in the pattern of their passageways. Some caves contain elaborate deposits of chemical precipitates. The variations in cavern size and form may indicate differences in mode of origin. Some small caves might have formed above the water table by water percolating downward through the zone of aeration. The majority of caverns, however, developed just below the water table, where the rate of

● **FIGURE 16.19** Intense solution in wet tropical environments can produce a karst landscape consisting of a maze of haystack hills and intergrown sinkholes.

solution is most rapid. A subsequent decline in the water table level, caused by the incision of surface streams, climate change, or tectonic uplift, fills the cavern with air, allowing calcium carbonate deposition to begin.

**Speleothem** is the generic term for any chemical precipitate feature deposited in caves. Speleothems develop in a great variety of textures and shapes, and many are delicate and ornate. Speleothems originate when previously dissolved substances, particularly calcium carbonate, precipitate out of subsurface water to produce some of the most beautiful and intricate forms found in nature. The dripping water leaves behind a deposit of calcium carbonate called *travertine,* or *dripstone.* As these travertine deposits grow downward, they form icicle-like spikes called **stalactites** that hang from the ceiling ( ● Fig. 16.22). Water saturated with calcium carbonate dripping onto the floor of a cavern builds up similar but more massive structures called **stalagmites.** Stalactites

● **FIGURE 16.20** Guilin, a limestone region in southern China, is famous for its beautiful tower karst.

*At one time in the geologic past, could this region have looked much like the landscape in Figure 16.19?*

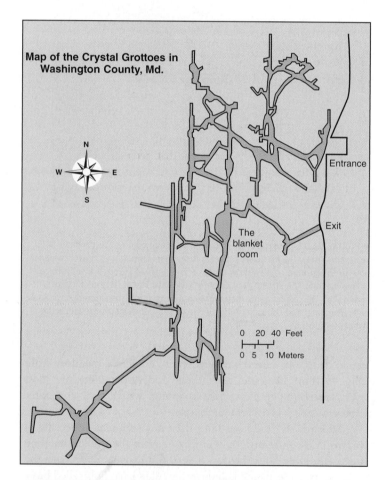

● **FIGURE 16.21** This map of the Crystal Grottoes in Washington County, Maryland, illustrates the influence of fractures on the growth of the cavern system, as evidenced by the geometric arrangement and spacing of passageways. Groundwater flow widens fractures and other zones of weakness to develop a cavern system.

● **FIGURE 16.22** Stalactites form where mineralized water that has percolated down to the cave from above reaches the cave ceiling and deposits some of its dissolved minerals, typically calcium carbonate, as hanging features.

and stalagmites often meet and continue growing to form pillar-like **columns** ( ● Fig. 16.23).

People who have not studied caves often think that evaporating water that leaves behind deposits of calcium carbonate is the dominant process that produces the spectacular speleothems seen in caverns, but that is not the case. Caves that foster active development of speleothems typically have air that is fully saturated with water, having a relative humidity near 100%, so evaporation is minimal. Instead, much of the water that percolates into the cave from above first acquired carbon dioxide from the soil, then used that to dissolve calcium carbonate (by carbonation) on its way to the cave. Cave air, in contrast, contains a comparatively low amount of carbon dioxide. When the dripping water contacts the cave air, it therefore releases carbon dioxide gas to the air. This degassing of carbon dioxide from the water essentially reverses the carbonation process, causing the water to precipitate calcium carbonate. Eventually, enough calcium carbonate is deposited in this way to form a stalactite, stalagmite, or other depositional cave feature.

Cavern development is a complex process, involving such variables as rock structure, groundwater chemistry, hydrology, and the regional tectonic and erosional history. As a result, the scientific study of caverns, **speleology,** is particularly challenging. Adding to the challenge are the field conditions. Much of our knowledge of cavern systems has come from explorations as deep as hundreds of meters underground by individuals making scientific observations while crawling through mud, water, and even bat droppings in dark, narrow passages ( ● Fig. 16.24). Exploration and mapping of water-filled caves involve scuba diving. Cave diving in fully submerged, totally dark, and confined passageways, which may contain dangerous currents, is an extremely risky undertaking.

Water beneath Earth's surface performs geomorphic work by carving out and decorating caverns and creating other karst landforms in soluble rock at and near Earth's surface, and like most landforms, karst features can be hazardous as well as beautiful. Regardless of the rock type present, any underground water tends to acquire ions from the rock-forming minerals with which it comes in contact, and a host of human actions lead to pollution of underground water. Understanding the basic nature, distribution, and availability of subsurface water, including its importance to humans as a source of freshwater, is crucial to whether it will continue to be available in adequate quality and quantity.

WASHINGTON

*Pacific Ocean*

OREGON   IDAHO

Oregon Caves

NEVADA

● **FIGURE 16.23** Speleothems in Oregon Caves illustrate how stalactites and stalagmites join to create a column.

● **FIGURE 16.24** Cave exploration, called caving or spelunking, can be exciting but also hazardous.

## ■ DEFINE AND RECALL

| | | |
|---|---|---|
| subsurface water | aquifer | sinkhole (doline) |
| infiltration | aquiclude | solution sinkhole |
| recharge | perched water table | collapse sinkhole |
| zone of aeration | spring | uvala (valley sink) |
| percolation | well | disappearing stream |
| soil water | artificial recharge | cavern (cave) |
| zone of saturation | artesian spring | swallow hole |
| groundwater | flowing artesian well | haystack hill (conical hill or hum) |
| water table | nonflowing artesian well | tower karst |
| intermediate zone | acid mine drainage | speleothem |
| water mining | hot spring | stalactite |
| effluent | geothermal water | stalagmite |
| influent | geyser | column |
| permeability | karst | speleology |

## ■ DISCUSS AND REVIEW

1. Why is it important to understand the basic nature and properties of underground water?
2. What is the difference between infiltration and percolation?
3. When water soaks into the subsurface, what two zones does it pass through on its way to the zone of saturation? How is the zone of saturation related to groundwater and the water table?
4. Define porosity and permeability. How are these two properties related to groundwater supply?
5. What is the difference between a spring and an artesian spring? What is the difference between a well and a flowing artesian well?

6. What is an aquifer? Describe the conditions necessary for an aquifer.
7. Define karst. What conditions encourage the development of karst landscapes?
8. Describe a sinkhole. How are sinkholes formed?
9. Describe a cavern. How do caverns form?
10. Explain how calcium carbonate is deposited by subsurface water to form a stalactite in a humid cave.

## ■ CONSIDER AND RESPOND

1. If you were a water resources geographer and had to plan for the development of groundwater resources for a community, what would be your major considerations?
2. What are some human impacts and environmental problems related to groundwater resources?

3. Describe the major landform features in a region of karst topography. Over time, what changes might be expected in the landscape?

# APPLY AND LEARN

1. Drilling at seven sites spaced every 5 kilometers along a straight-line transect heading from west to east has provided the following data on depth to the water table and the rock type present just below the water table: (a) 15 meters in sandstone, (b) 14.5 meters in sandstone, (c) 14 meters in sandstone, (d) 5 meters in limestone, (e) 13 meters in sandstone, (f) 12.5 meters in sandstone, and (g) 12 meters in sandstone. What is the regional slope of the water table? Provide a reasonable explanation for the data obtained from site d.

2. Scientists studying an air-filled cavern located in an arid region determined that the speleothems present began forming 20,000 years ago and stopped growing 12,000 years ago. Suggest a reasonable history of climatic and hydrologic events related to the formation of the cavern evidence.

# LOCATE AND EXPLORE

1. Using Google Earth and Figure 16.13 from this chapter, examine the spatial distribution of karst regions in the United States. The development of karst landscapes depends on geology, precipitation, topography, tectonic activity, vegetation cover, and climate change history. Provide an explanation for karst development in Kentucky, Texas, and Florida.

2. Using Google Earth, identify the landforms at the following locations (latitude, longitude) and provide a brief discussion of how the landform developed. Include a brief discussion of why the landform is found in that general area of the United States or the world.
   a. 18.40°N, 66.53°W
   b. 25.05°N, 110.37°E
   c. 29.658°N, 81.861°W
   d. 38.207°N, 90.175°W
   e. 20.87°N, 107.19°E
   f. 39.4032°N, 113.8625°W

 CourseMate – Make the most of your study time by accessing everything you need to succeed in one place. Read your textbook, take notes, review flashcards, watch videos, complete activities, take practice quizzes, and more online with CourseMate. Log in at **www.cengagebrain.com**.

# Map Interpretation

## ■ KARST TOPOGRAPHY

### The Map

The Interlachen area is in northeastern Florida. Florida's peninsula is the emerged portion of a gentle anticline called the Peninsular Arch. The region is underlain by thousands of meters of marine limestones and shales. This great thickness of marine sediments originated in the Mesozoic Era when Florida was a marine basin. As the arch rose, Florida became a shallow shelf and was eventually elevated above sea level.

Although outsiders think of Florida mainly as a state with magnificent beaches and warm winter weather because of its humid subtropical climate, it is also a state with hundreds of lakes dotting its center. The lake region is formed on the Ocala Uplift, a gentle arch of limestone that reaches to 46 meters (150 ft) above sea level. Lake Okeechobee is the largest of these lakes and has an average depth of less than 4.5 meters (15 ft). Most of the lakes, such as those in the Interlachen map area, are much smaller.

Florida's lake region is an ideal area for studying karst topography. Both the surface and the subsurface features express the geomorphic effects of subsurface water. Extensive cavern systems exist beneath the surface. Much of the state's runoff is channeled through huge aquifers, and springs are quite common.

### Interpreting the Map

1. This area would be classified as what major type of landform (for example, mountain)? What is its average elevation?
2. What contour interval is used on this map? Why do you think the cartographers chose this interval?
3. On what type of bedrock is the map area situated? Do you think the climate has any influence on the landforms in the area? Explain.
4. What landform features on the map indicate that this is a karst region?
5. What are the round, steep depressions called? Why do lakes occupy some of the depressions?
6. Locate Clubhouse Lake on the full-page map (scale 1:62,500) and the smaller map (1:24,000). What is the elevation of Clubhouse Lake? What is its maximum width?
7. What type of feature is the area north of Lake Grandin?
8. What is the approximate elevation of the water table? (Note: You can determine this from the elevation of the lakes' water surface.)
9. Underground, the water flows through an aquifer. Define an aquifer and list the characteristics an aquifer must have. What is the general direction of groundwater flow in the aquifer underlying the Interlachen area?
10. Because much of central Florida is rapidly urbanizing, what problems and hazards do you anticipate in this karst area?

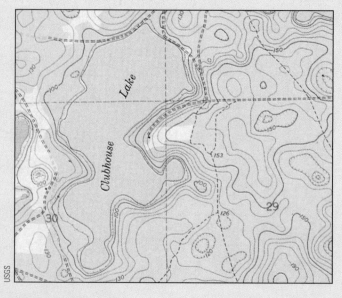

USGS

Putnam Hall, Florida

Scale 1:24,000

Contour interval = 10 ft

◆ In Google Earth fly to: 29.641°N, 81.936°W.

Opposite:

Interlachen, Florida

Scale 1:62,500

Contour interval = 10 ft

USGS

# 17

# FLUVIAL PROCESSES AND LANDFORMS

Streams are the dominant geomorphic agent on Earth's surface, but many are also beautiful landscape features in their own right, such as this small stream meandering through a meadow in Grand Tetons National Park.  Lee Prince/Shutterstock

## ■ OBJECTIVES

WHEN YOU COMPLETE THIS CHAPTER YOU SHOULD BE ABLE TO:

■ Describe how surface runoff is generated and how channels are initiated.
■ Understand how stream systems are organized within drainage basins and how streams and drainage basins are ordered.
■ Discuss how the amount of water flowing in a stream channel is determined.
■ Identify the variables that affect stream energy.
■ Appreciate the geomorphic work that can be accomplished by stream energy.
■ Recognize the major landforms associated with meandering stream systems.
■ Recall the distinctions among the upper, middle, and lower course of an ideal stream system.
■ Explain how tectonism and climate change can affect stream systems.
■ Understand flood hazards.
■ Appreciate the value of quantitative methods for stream system analysis.

FLOWING WATER IS MORE INFLUENTIAL in shaping the surface form of our planet than any other exogenic geomorphic process, primarily because of the sheer number of streams on Earth. Through erosion and deposition, water flowing downslope over the land surface, particularly when concentrated in channels, modifies existing landforms and creates others. Nearly every region of Earth's land surface in arid as well as humid climates exhibits at least some topography that has been shaped by the power of flowing water, and many regions exhibit considerable evidence of stream action. Polar landscapes buried under thick accumulations of perennial ice are the major exception to Earth's extensive areas of stream-dominated topography.

The study of flowing water as a land-shaping process, together with the study of the resulting landforms, is **fluvial geomorphology** (from Latin: *fluvius,* river). Fluvial geomorphology includes the action of channelized and unchannelized flow moving downslope by the force of gravity.

Stream is the general term for natural, channelized flow. In the Earth sciences, the word stream pertains to water flowing in a channel of any size, even though in general usage we refer to large streams as rivers and use local terms, such as creek, brook, run, draw, and bayou, for smaller streams. The land between adjacent channels in a stream-dominated landscape is the **interfluve** (from Latin: *inter,* between; *fluvius,* river).

Understanding stream processes, landforms, and hazards is fundamental for maintaining human safety and quality of life. Because of the common and widespread occurrence of stream systems and their key role in providing freshwater for people and our agricultural, industrial, and commercial activities, a substantial portion of the world's population lives in close proximity to streams. Most streams, however, occasionally expand out of the confines of their channel. Although these floods typically last only a few days at most, they reveal the tremendous—and often very dangerous—geomorphic power potential of flowing water.

This chapter begins by considering how water accumulates into organized channel systems at Earth's surface. We then investigate factors that influence the amount and variability of water flow at any given channel reach and how those flow conditions affect a stream's ability to erode, transport, and deposit rock material. As streams perform that erosion, transportation, and deposition, they create distinctive landforms and landscape features that over long periods may also be influenced by tectonic and climate-change factors. Like short-term floods, the long-term effects of stream flow, whether dominated by erosion or deposition, may also be dramatic. Two impressive examples in the United States that illustrate the effectiveness of flowing water in creating landforms are the Black Canyon of the Gunnison River, carved by long-term river erosion into the Rocky Mountains, and the Mississippi River delta, where fluvial deposition is building new land into the Gulf of Mexico ( ● Fig. 17.1).

● **FIGURE 17.1** (a) Long-term fluvial erosion by the Gunnison River, with the aid of weathering and mass wasting, has carved the Black Canyon of the Gunnison through a part of the Rocky Mountains in Colorado. The narrow canyon is 829 meters (2722 ft) deep, with resistant rocks comprising the steep walls. (b) The Mississippi River has been building its delta outward into the Gulf of Mexico for thousands of years. Where the river enters the gulf, the slowing current deposits large amounts of fine-grained sediment that came from the river's drainage basin. On this false-color image, muddy water appears light blue; clearer, deeper water is dark blue.

(a)

(b)

# Surface Runoff

Liquid water flowing over the surface of Earth, that is, **surface runoff,** can originate as ice and snow melt or as outflow from springs, but most runoff originates in direct response to precipitation arriving as rain. When rain strikes the ground, several factors interact to determine whether surface runoff will occur. Runoff is generated when the amount of precipitation exceeds the ability of the ground to soak up the moisture. Because the process of water soaking into the ground is infiltration, the amount of water that voids and pore spaces in the soil and surface sediments can hold is known as the *infiltration capacity*. A portion of infiltrated water will seep (percolate) down to lower subsurface positions and reach the zone of saturation beneath the water table, while much of the rest will eventually return to the atmosphere by evaporation from the soil or by transpiration from plants. When more precipitation falls than can be infiltrated into the ground, the excess water flows downslope by the force of gravity as surface runoff.

Various factors act individually or together to either enhance or inhibit the generation of surface runoff. Greater infiltration to the subsurface, and therefore less runoff, occurs under conditions of permeable surface materials, deeply weathered sediments and soils, gentle slopes, dry initial soil conditions, and a dense cover of vegetation. **Interception** of precipitation by vegetation allows greater infiltration by slowing down the rate of delivery of precipitation to the ground. Vegetation makes space in soil pores for additional infiltration when it takes up soil water and returns it to the atmosphere through transpiration. Given the same precipitation event, surface materials of low permeability and limited weathering, thin soils, steep slopes, preexisting soil moisture, and sparse vegetation each contribute to increased runoff by decreasing infiltration ( ● Fig. 17.2). Human activities can affect many of these variables, and in some places the generation of runoff has been greatly modified by urbanization, mining, logging, or agriculture.

Once surface runoff forms, it first starts to flow downslope as a thin sheet of unchannelized water, known as **sheet wash,** or unconcentrated flow. Because of gravity, after a short distance the sheet wash will begin to move preferentially into any preexisting swales or depressions in the terrain. This concentration of flow leads to the formation of tiny channels, called **rills,** or somewhat larger channels, called **gullies.** Rills are on the order of a couple of centimeters deep and a couple of centimeters wide ( ● Fig. 17.3), whereas gully depth and width may approach as much as a couple of meters.

Water does not flow in rills and gullies all the time but only during and shortly after a precipitation (or snowmelt) event. Channels that are empty of water much of the time are described as having **ephemeral flow.** As these small, ephemeral channels continue downslope, rills join to form slightly larger rills, which may join to make gullies. In humid climates, following these successively larger ephemeral channels downslope will eventually lead us to a point where we first encounter **perennial flow.** Perennial streams flow all year but

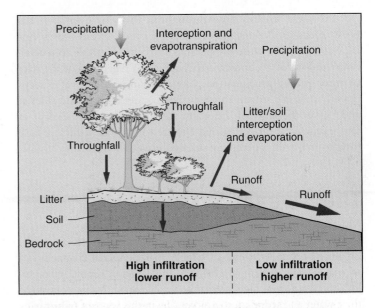

● **FIGURE 17.2** The amount of runoff that occurs is a function of several factors, including the intensity and duration of a rainstorm. Surface features that increase infiltration and evapotranspiration will reduce the amount of water available to run off and vice versa. Deep soil, dense vegetation, fractured bedrock, and gentle slopes reduce runoff. Thin or absent soils, sparse vegetation, and steep slopes increase runoff.

not always with the same volume or at the same velocity. Most arid region streams flow on an ephemeral basis, although some may instead have **intermittent flow,** which lasts for a couple of months in response to an annual rainy season or spring snowmelt. Because of this contrast in flow duration, and other differences between arid and humid region streams, a full discussion of arid region stream systems appears in the separate chapter on arid region landforms, Chapter 18.

● **FIGURE 17.3** Rills are the smallest type of channel. They are commonly visible on devegetated surfaces, like these mine tailings.

*Is there a gully in this photo?*

Perennial streams flow throughout the year even if it has been several weeks since a precipitation event. In most cases, this is possible only because the perennial streams continue to receive direct inflow of groundwater (Chapter 16) regardless of the date of most recent precipitation. Slow-moving groundwater seeps directly into the stream through the channel bottom and sides at and below the level of the water surface as *base flow*. Except in rare instances, it takes a humid climate to generate sufficient base flow to maintain a perennial stream between rainstorms.

# The Stream System

Most flowing water becomes quickly channelized into streams as it is pulled downhill by gravity. Continuing downslope, streams form organized channel systems in which small perennial channels join to make larger perennial channels and larger perennial channels join to create even bigger streams. Smaller streams that contribute their water and sediment to a larger one in this way are **tributaries** of the larger channel, which is called the *trunk stream* ( ● Fig. 17.4).

## Drainage Basins

Each individual stream occupies its own **drainage basin** (also known as *watershed* or *catchment*), the expanse of land from which it receives runoff. *Drainage area* refers to the measured extent of a drainage basin and is typically expressed in square kilometers or square miles. Because the runoff from a tributary's drainage basin is delivered by the tributary to the trunk stream, the tributary's drainage basin also constitutes part of the drainage basin of the trunk stream. In this way, small tributary basins are nested within, or are *subbasins* of, a succession of larger and larger trunk stream drainage basins. Large river systems drain extensive watersheds that consist of numerous inset subbasins.

Drainage basins are open systems that involve inputs and outputs of water, sediment, and energy. Knowing the boundaries of a drainage basin and its component subbasins is critical to properly managing the water resources of a watershed. For example, pollution discovered in a river almost always comes from a source within the river's drainage basin, entering the stream system either at the point where the pollutant was first detected or at a

● **FIGURE 17.4** The stream system represents the surface runoff subsystem of the hydrologic cycle. Its major water input is from precipitation, but groundwater also contributes to the stream system, particularly in humid regions. The major water output for most stream systems is flow delivered to the ocean. Output or loss of water from a stream channel also occurs by evaporation back to the atmosphere and by infiltration into the underground water system. Materials transported by streams, known as load, enter the stream system by erosion and mass movement, particularly in the headwaters of a drainage basin. As the number and size of tributaries increase downstream, the amount of load carried by the stream increases dramatically. Load leaves the stream system when deposited in the sea at the river mouth. Streams also deposit sediment adjacent to their channels as they overflow their banks during floods.

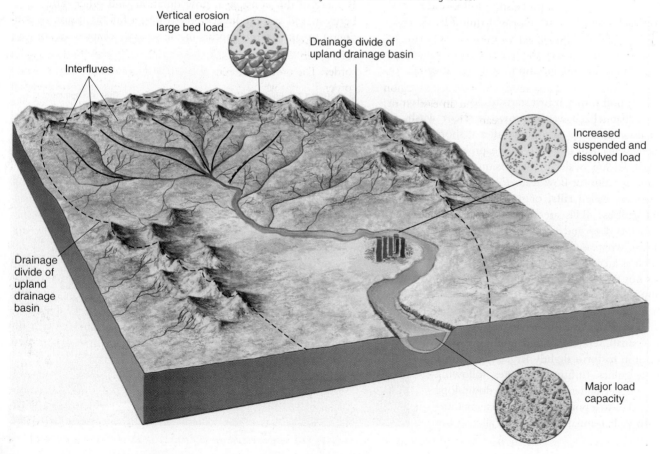

Vertical erosion
large bed load

Drainage divide of
upland drainage basin

Interfluves

Increased
suspended and
dissolved load

Drainage
divide of
upland
drainage
basin

Major load
capacity

● **FIGURE 17.5** An aerial photograph of a small stream channel network, basin, and divide. Every tributary of the stream system occupies its own drainage basin (bounded by a drainage divide). These small drainage basins combine to form the trunk stream's drainage basin.

*Can you mentally trace the outline of the main drainage basin shown here?*

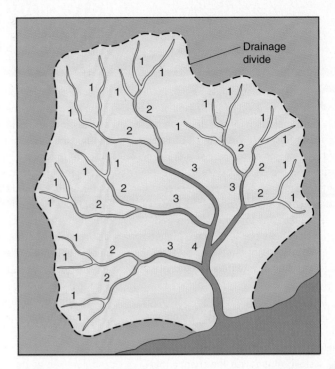

Drainage divide

● **FIGURE 17.6** The concept of stream order is illustrated by the channels of this fourth-order drainage basin. Stream order changes only when two streams of equal order join, thereby creating a stream of the next higher order.

*What is the highest-order stream in a selected drainage basin called?*

location upstream from that site. This knowledge helps us track, detect, and correct sources of pollution.

The **drainage divide** represents the outside perimeter of a drainage basin and thus also the boundary between it and adjacent basins ( ● Fig. 17.5). The drainage divide follows the crest of the interfluve between two adjacent drainage basins. In some places, this crest is a definite ridge, but the higher land that constitutes the divide is not always ridge-shaped, nor is it necessarily much higher than the rest of the interfluve. Surface runoff generated on one side of the divide flows toward the channel in one drainage basin, while runoff on the other side travels in a very different direction toward the channel in the adjacent drainage basin. The *Continental Divide* separates North America into a western region where most runoff flows to the Pacific Ocean and an eastern region where runoff flows to the Atlantic Ocean. The Continental Divide generally follows the crest line of high ridges in the Rocky Mountains, but in some locations the highest point between the two huge basins lies along the subtle crest of gently sloping high plains.

A stream, like the Mississippi River, that has a very large number of tributaries and encompasses several levels of nested subbasins, will differ in some major ways from a small creek that has no perennial tributaries and lies high up in the drainage basin near the divide. Knowing where each stream lies in the hierarchical order of tributaries helps Earth scientists make more meaningful comparisons among different streams. **Stream ordering** is the technique used to describe quantitatively the position of a stream and its drainage basin in the nested hierarchy of tributaries. *First-order streams* have no perennial tributaries. Even though they are the smallest perennial channels in the drainage basin, first-order channels are evident on large-scale topographic maps. Most first-order streams lie high up in the drainage basin near the drainage divide, the *source* area of the stream system. Two first-order streams must meet to form a *second-order stream*, which is larger than each of the first-order streams. It takes the intersection of two second-order

channels to make a *third-order stream* regardless of how many first-order streams might independently join the second-order channels. The ordering system continues in this way, requiring two streams of a given order to combine to create a stream of the next higher order. The order of a drainage basin derives from the largest stream order found within it ( ● Fig. 17.6). For example, the Mississippi is a tenth-order drainage basin because the Mississippi River is a tenth-order stream. Stream ordering allows us to compare various attributes of streams quantitatively by relative size, which helps us better understand how stream systems work. Among other things, comparing streams on the basis of order has shown that as stream order increases, basin area, channel length, channel size, and amount of flow also increase.

Water in a stream system proceeds downslope through a succession of channels of ever-increasing order toward the downstream end, or *mouth,* of the stream. The mouths of most humid region, perennial streams lie at sea level where the channel system finally ends and the stream water is delivered to the ocean. Drainage basins with channel systems that convey water to the ocean have **exterior drainage.** Many arid region streams have **interior drainage** because they do not have enough flow to reach the ocean but terminate instead in local or regional areas of low elevation.

Every stream has a **base level,** the elevation below which it cannot flow. Sea level is the *ultimate base level* for virtually all stream action. Streams with exterior drainage reach ultimate base level; the low point of flow for a stream with interior drainage is referred to as a *regional base level*. In some drainage basins, a resistant rock layer

# Watersheds as Critical Natural Regions

Perhaps the most environmentally logical way to divide Earth's land surface into regions is by the drainage basins for stream systems. Virtually all of Earth's land surface comprises part of a drainage basin. Like the channel systems that they contain, drainage basins are hierarchical and are conveniently subdivided into smaller subbasins for local studies and management. At the same time, higher-order watersheds of large river systems are subject to broad, integrated, regional-scale analyses.

The various components of a drainage basin are strongly interrelated. Problems in one part of a drainage basin system are likely to cause problems in other parts of the system. Along with the channel network that occupies each watershed, drainage basins consist of water, soil, rock, terrain, vegetation, people, wildlife, and domesticated animals that together form complex natural biotic habitats. Monitoring and managing these complex watershed systems requires an interdisciplinary effort that considers aspects of all four of the world's major spheres.

Surface water from watershed sources provides much of the potable water resources for the world's population. Increasing human populations and land-use intensities in previously natural drainage basins place pressures on these habitats and their water quality. Maintaining water quality within local subbasins contributes to maintaining water quality of larger river systems at the regional scale. Because streams comprise a network of flow in one direction (downstream), identifying sources of pollution involves upstream tracing of pollutants to the source and downstream spreading of pollution problems. Humans have a vital responsibility to monitor, maintain, and protect the quality of freshwater resources, and doing so at the local watershed scale has far-reaching and local benefits.

In recent years many governmental agencies have designated watersheds as critical regions for environmental management. Some of these efforts involve the establishment of locally administered river conservation districts representing relatively small-order drainage basins in which community involvement plays a vital role.

There are several sound reasons, including the connectivity and hierarchical nature of watersheds, for this spatial strategy.

Watersheds are clearly defined, well-integrated, natural regions of critical importance to life on Earth, and they make a logical spatial division for environmental management. However, a stream system may flow through many cities, counties, states, and countries, and management problems can arise where regional political and administrative boundaries do not coincide with the divide that defines the edge of a watershed. Each of these political jurisdictions may have different needs, goals, and strategies for using and managing their part of a watershed, and often a variety of strategies results in conflict. Still, most administrative units recognize that cooperative management of the watershed system as a whole is the best approach. Many cooperative river basin authorities have been established to encourage a united effort to protect a shared watershed. This watershed-oriented management strategy, based on a fundamental natural region, is an important step in protecting our freshwater resources.

The Mississippi River drainage area covers an extensive region of the United States.

D. Sack

As a tributary of the Ohio River, the fourth-order Hocking River watershed in southeastern Ohio is a subbasin within the Mississippi River system. Community involvement in improving and maintaining the health of the Hocking River watershed, and other drainage basins, has regional as well as local benefits.

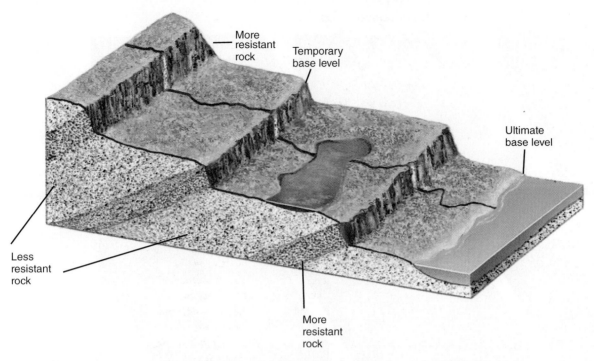

● **FIGURE 17.7** The lowest point to which a stream can flow is its base level. Sea level represents ultimate base level for all of Earth's streams, and most humid region streams have sufficient flow to reach the ocean. Many arid region streams lose so much water by evaporation to the air and infiltration into the channel bed that they cannot flow to the sea, but instead reach a regional base level in a basin on the continent. A temporary base level is formed where a rock unit lying in the pathway of a stream is significantly more resistant than the rock upstream from it. The stream will not be able to cut into the less resistant rock any faster than it can cut into the resistant rock of the temporary base level.

located somewhere upstream from the river mouth can act as *temporary base level*, temporarily controlling the lowest elevation of the flow upstream from it until the stream is finally able to cut down through the resistant layer ( ● Fig. 17.7).

## Drainage Density and Drainage Patterns

Each system of stream tributaries exhibits spatial characteristics that provide important information about the nature of the drainage basin. The extent of channelization can be represented by measuring **drainage density** $(D_d)$, where $D_d = L/A_d$, the total length $(L)$ of all channels in the drainage basin, divided by the area of the drainage basin $(A_d)$. Because drainage density indicates how dissected the landscape is by channels, it reflects both the tendency of the drainage basin to generate surface runoff and the erodibility of the surface materials ( ● Fig. 17.8). Regions with high drainage densities have limited infiltration, considerable runoff, and at least moderately erodible surface materials.

In addition to the factors noted previously that reduce infiltration and promote runoff—impermeable sediments, thin soils, steep slopes, sparse vegetation—the ideal climate for high drainage densities is one that is semiarid. Humid climates encourage extensive vegetation cover, which promotes infiltration because of interception and reduces channel formation by holding soils

● **FIGURE 17.8** Drainage density—the length of channels per unit drainage area—varies according to several environmental factors. Everything else being equal, highly erodible and impermeable rocks tend to have higher drainage density than areas dominated by resistant or permeable rocks. Slope and vegetation cover can also affect drainage density.

*What kind of drainage density would you expect in an area of steep slopes and sparse vegetation cover?*

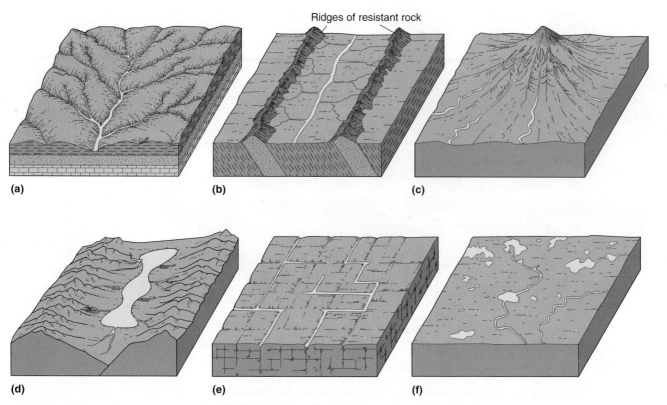

Ridges of resistant rock

(a)   (b)   (c)

(d)   (e)   (f)

● **FIGURE 17.9** Drainage patterns often reflect bedrock structure. (a) The dendritic pattern is found where rocks have uniform resistance to weathering and erosion. (b) A trellis pattern indicates parallel valleys of weak rock between ridges of resistant rock. (c) Multiple channels trending away from the top of a domed upland or volcano form the radial pattern. (d) The centripetal pattern shows multiple channels flowing inward toward the center of a structural lowland. (e) Rectangular patterns indicate sets of linear joints in the bedrock structure. (f) A deranged pattern typically results following the retreat of continental ice sheets; it is characterized by a chaotic arrangement of channels connecting small lakes and marshes.

and surface sediment in place.][In arid climates, although the vegetation cover is sparse, there is insufficient precipitation to create enough runoff to carve many channels.]Semiarid climates have enough precipitation input to produce overland flow but not enough to support an extensive vegetative cover. For example, the easily eroded Dakota Badlands, which are located in a steppe climate, have an extremely high drainage density of more than 125 kilometers per square kilometer (125 km of channel per 1 sq km of land), whereas very resistant granite hills in a humid climate may have a drainage density of only 5 kilometers per square kilometer. Another way to understand the concept of drainage density is to think about what would happen on a hillside in a humid climate if the natural vegetation were burned off in a fire. Erosion would rapidly cut gullies, creating more channels than previously existed there. In other words, the drainage density would increase. We could use the quantitative measure of $D_d = L/A_d$ to determine the change in channelization precisely and to monitor it over time.

When viewed by looking down at Earth's surface from the air or on maps, networks of stream tributaries display distinct spatial arrangements called **drainage patterns.**[Two primary factors that influence drainage pattern are bedrock structure and surface topography.]A *dendritic* (from Greek: *dendros,* tree) drainage pattern ( ● Fig. 17.9a) resembles a branching tree with tribu-

taries joining larger streams at acute angles (less than 90°). Dendritic patterns are common and develop where the rocks have a roughly equal resistance to weathering and erosion and are not intensely jointed. A *trellis* pattern consists of long, parallel streams with short tributaries entering at right angles (Fig. 17.9b). Trellis drainage indicates folded terrain where short streams flow down the sides of resistant rock ridges into a larger stream that occupies the adjacent valley of erodible rock, as in the Ridge and Valley region of the Appalachians. A *radial* pattern develops where streams flow away from a common high point on cone- or dome-shaped geologic structures, such as volcanoes (Fig. 17.9c). The opposite pattern is *centripetal,* with the streams converging on a central lowland as in an arid region basin of interior drainage (Fig. 17.9d). *Rectangular* patterns occur where streams follow sets of intersecting fractures to produce a blocky network of straight channels with right-angle bends (Fig. 17.9e). In some regions that were recently covered by extensive glacier ice, streams flow on low-gradient terrain left by the receding glaciers, and wander between marshes and small lakes in a chaotic pattern called *deranged* drainage (Fig. 17.9f).

[Many streams follow the "grain" of the topography or the bedrock structure, eroding valleys in weaker rocks and flowing away from divides formed on resistant rocks.] Many examples also exist of streams that flow across—that is, transverse to—the struc-

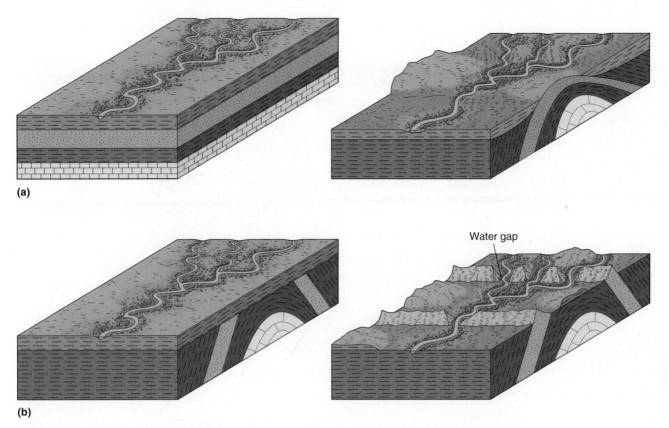

(a)

(b)

Water gap

● **FIGURE 17.10** Transverse streams form valleys and canyons across mountains. (a) An antecedent stream maintains its course by cutting a valley through a mountain that is being uplifted gradually over much time. (b) A superimposed stream has uncovered and excavated ancient structures that were buried beneath the surface. As the stream erodes the landscape downward, it cuts across and through the ancient structures.

ture, cutting a gorge or canyon through mountains or ridges. These **transverse streams** can be puzzling, giving rise to questions as to how a stream can cut a gorge through a mountain range or otherwise cross from one side of a mountain range to the other. Such streams are probably either *antecedent* or *superimposed*.

Antecedent streams existed before the formation of the mountains that they flow across, maintaining their courses by cutting an ever-deepening canyon as gradual mountain building takes place across their paths. The Columbia River Gorge through the Cascade Range in Washington and Oregon and many of the great canyons in the Rocky Mountain region, such as Royal Gorge and the Black Canyon of the Gunnison, both in Colorado, probably originated as antecedent streams ( ● Fig. 17.10a). Other rivers, including some in the central Appalachians, have cut gaps through mountains in a very different manner. These streams originated on earlier rock strata, since stripped away by erosion, so that the streams have been superimposed onto the rocks beneath (Fig. 17.10b). This sequence would explain why, in many instances, the rivers flow across folded rock structure, creating water gaps through mountain ridges. Examples include the Cumberland Gap and the gap formed by the Susquehanna River in Pennsylvania, both of which were important travel routes for the first European settlers crossing the Appalachians.

# Flow Properties
## Stream Discharge

The amount of water flowing in a stream depends not only on the impact of recent weather patterns, but also on such drainage basin factors as its size, relief, climate, vegetation, rock types, and land-use history. Stream flow varies considerably from time to time and place to place. Most streams experience occasional brief periods when the amount of flow exceeds the ability of the channel to contain it, resulting in the flooding of channel-adjacent land areas.

Just as it was important to develop the technique of stream ordering to indicate numerically a channel's place in the hierarchy of tributaries, it is also crucial to be able to describe quantitatively the amount of flow being conveyed through a stream channel. **Stream discharge** $(Q)$ is the volume of water $(V)$ flowing past a given channel cross section per unit time $(t)$: $Q = V/t$. Discharge is most commonly expressed in units of cubic meters per second $(m^3/s)$ or cubic feet per second $(ft^3/s)$. A cross section is essentially a thin slice extending from one stream bank straight across the channel to the other stream bank and oriented perpendicular to the channel. If a drainage basin experiences a rainfall event that produces significant runoff, the volume of water $(V)$ reaching the channel will increase. Notice from the discharge equation that

**TABLE 17.1**
**Ten Largest Rivers of the World**

| | Length | | Area of Drainage Basin | | Discharge | |
|---|---|---|---|---|---|---|
| | km | mi | sq km (×1000) | sq mi | 1000 m³/s | 1000 ft³/s |
| Amazon | 6276 | 3900 | 6133 | 2368 | 112–140 | 4000–5000 |
| Congo (Zaire) | 4666 | 2900 | 4014 | 1550 | 39.2 | 1400 |
| Chang Jiang (Yangtze) | 5793 | 3600 | 1942 | 750 | 21.5 | 770 |
| Mississippi–Missouri | 6260 | 3890 | 3222 | 1244 | 17.4 | 620 |
| Yenisei | 4506 | 2800 | 2590 | 1000 | 17.2 | 615 |
| Lena | 4280 | 2660 | 2424 | 936 | 15.3 | 547 |
| Paraná | 2414 | 1500 | 2305 | 890 | 14.7 | 526 |
| Ob | 5150 | 3200 | 2484 | 959 | 12.3 | 441 |
| Amur | 4666 | 2900 | 1844 | 712 | 9.5 | 338 |
| Nile | 6695 | 4160 | 2978 | 1150 | 2.8 | 100 |

*Source:* Adapted from Morisawa, *Streams: Their Dynamics and Morphology.* New York: McGraw-Hill Book Company, 1968.

this increase in volume *(V)* will cause an increase in stream discharge *(Q)*.

It is important to collect and analyze discharge data for several reasons. These data can be used to compare the amount of flow carried in different streams, at different sites along one stream, or at different times at a single cross section. Discharge data indicate the size of a stream and, in times of excessive flow, provide an index of flood severity. In general, streams in larger drainage basins have greater discharge and are longer than streams in smaller drainage basins. Among the major rivers of the world, the Amazon has by far the largest drainage basin and the greatest discharge; the Mississippi River system is ranked fourth in terms of discharge (Table 17.1).

The volume of water rushing through a cross section of a stream per second is extremely difficult to measure directly. In reality, discharge is determined not by measuring $Q = V/t$ directly but by using the fact that discharge *(Q)* is also equal to the area of the cross section *(A)* times the average stream velocity *(v)*. This equation, $Q = Av$, can also be expressed as $Q = wdv$ because the cross-sectional area *(A)* is approximately equivalent to channel width *(w)* times channel depth *(d)*, two factors that are relatively easy to measure in the field ( ● Fig. 17.11). Notice that, because cross-sectional area *(A)* is measured in square meters or square feet and average velocity is measured in meters per second or feet per second, solving the equation $Q = Av$ yields discharge values in units of volume per unit time (cubic meters per second or cubic feet per second). This analysis of the measurement units should convince you that volume per unit time is indeed equivalent to cross-sectional area times average velocity—that stream discharge is both volume per unit time and cross-sectional area times average velocity, $Q = V/t = Av$.

As is true for any equation, a change on one side of the discharge equation must be accompanied by a change on the other. If discharge increases because a rainstorm delivers a large volume of water to the stream, that increase in volume *(V)* will occupy a larger cross-sectional area *(A)* and flow through the cross section at a faster rate *(v)*. In other words, a large rainstorm will cause the level of a stream to rise and the water to flow faster. As the water level (flow depth, *d*) rises, streams also experience an increase in width *(w)* because most channels flare out a bit as they rise up toward their banks. Stream channels continually adjust their

● **FIGURE 17.11** These university students are measuring a stream's velocity, depth, and width to find discharge, the volume of water flowing through a given cross section in a stream per unit time.

J. Petersen

cross-sectional area and flow velocity in response to changes in flow volume. Understanding the relationships among the factors involved in discharge is very important for understanding how streams work.

## Stream Energy

When people have more energy, we can accomplish more, or do more work, than when we have less energy. The same is true for streams and the other geomorphic agents. The ability of a stream to erode and transport sediment—that is, to perform geomorphic work—depends on its available energy. Picking up pieces of rock and moving them require the stream to have *kinetic energy,* the energy of motion. When a stream has more kinetic energy, it can pick up and move more clasts (rock particles) and heavier clasts than when it has less energy. The kinetic energy $(E_k)$ equation, $E_k = 1/2mv^2$, shows that the amount of energy a stream has depends on its mass *(m)* but especially on its velocity of flow *(v)* because kinetic energy varies with velocity squared $(v^2)$. Thus, stream velocity is a critical factor in determining the amount of geomorphic work that can be accomplished by a stream. As we saw in the previous section, when stream discharge *(Q)* varies because of changes in runoff *(V)*, so does stream velocity *(v)* because $Q = V/t = wdv$.

Besides discharge variations, another major way to alter stream velocity, and thereby stream energy, is through a change in the slope, or gradient, of the stream. *Stream gradient* is the drop in stream elevation over a given stream distance and is typically expressed in meters per kilometer (m/km) or feet per mile (ft/mi). Everything else being equal, water flows faster down channels with steeper slopes and slows down over gentler slopes. Steeper channels typically occupy locations that are farther upstream and higher in the drainage basin and places where a stream flows over rock types that are more resistant to erosion. Gentler stream gradients tend to occur closer to the mouth of a stream and where the stream crosses easily eroded rock types. As water moves through a channel system, its ability to erode and transport sediment—that is, its energy—is continuously changing as variations in stream gradient and discharge cause changes in flow velocity.

Different factors work to decrease the energy of a stream. Friction along the bottom and sides of a channel and even between the stream surface and the atmosphere slows down the stream velocity and therefore contributes to a decrease in stream energy. A channel with numerous substantial irregularities, such as those from large rocks and vegetation sticking up into the flow, has a great amount of **channel roughness,** which causes a considerable decrease in stream energy because of the resulting frictional effects. Smooth, bedrock channels without these irregularities have lower channel roughness and therefore a smaller loss in stream energy caused by frictional effects. In all cases, however, friction along the bottom and sides of a channel slows down the flow and slows it down the most at the channel boundaries so that the maximum velocity usually occurs somewhat below the stream surface at the deepest part of the cross section. In addition to the external friction at the channel boundary, streams lose energy because of internal friction in the flow related to eddies, currents,

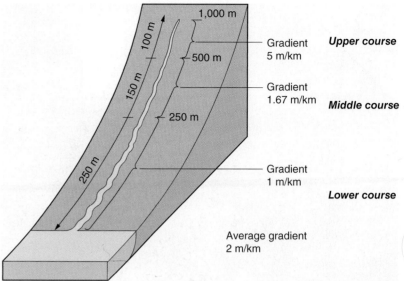

● **FIGURE 17.12** Stream gradient decreases from source to mouth for the ideal stream.
Copyright © 2011 Brooks/Cole, Cengage Learning

and the interaction among water molecules. About 95% of a stream's energy is consumed in overcoming all types of frictional effects. Only the remaining energy, probably less than 5%, is available for eroding and transporting sediment.

Stream gradients are usually steepest at the headwaters and in new tributaries and diminish in the downstream direction ( ● Fig. 17.12). Discharge, however, increases downstream in humid regions as the size of the contributing drainage basin increases. As the water originally conveyed in numerous small channels high in the drainage basin is collected in the downstream direction into a shrinking number of larger and larger channels, flow efficiency improves and frictional resistance decreases. As a result, flow velocity tends to be higher in downstream parts of a channel system than over the steep gradients in the headwaters. The ability of the stream to carry sediment, therefore, may be equivalent or even greater downstream than upstream.

Sediment being transported by a stream is called the **stream load.** Carrying sediment is a major part of the geomorphic work accomplished by a stream. In order from smallest to largest, the size of sediment that a stream may transport includes clay, silt, sand, granules, pebbles, cobbles, and boulders. Sand marks the boundary between fine-grained (small) clasts and coarse-grained (large) clasts. Gravel is a general term for any sediment larger in size than sand. The maximum size of rock particles that a stream is able to transport, referred to as the **stream competence** (measured as particle diameter in centimeters or inches), and the total amount of load being moved by a stream, termed **stream capacity** (measured as weight per unit time), depend on available stream energy and thus on the flow velocity. As a stream flows along, its velocity is always changing, reflecting the constant variations in stream discharge, stream gradient, and frictional resistance factors. As a result, the size and amount of load that the stream can carry is also changing constantly. If the material is available when flow velocity increases, the stream will pick up from its own channel bed larger clasts and more load. When its velocity decreases and the stream can no longer transport such large clasts and so much material, it drops the larger

(a)

Streams have no influence over the amount of water entering the channel system, nor can they change the type of rocks over which they flow. Streams do, however, have some control over channel size, shape, and gradient. For example, when a stream undergoes a decrease in energy so that it deposits some of its sediment, the deposit raises the channel bottom and locally creates a steeper gradient downchannel from the deposit. With continued deposition, the location will eventually attain a slope steep enough to cause a sufficient increase in velocity so that the flow will again entrain and carry that deposited sediment. Streams tend to adjust their channel properties so they can move the sediment supplied to them. *Dynamic equilibrium* is maintained by adjustments among channel slope, shape, and roughness; amount of load eroded, carried, or deposited; and the velocity and discharge of stream flow. A stream is described as **graded** if it has just the velocity and discharge necessary to transport the load eroded from the drainage basin.

(b)

● **FIGURE 17.13** It is not just the power of flowing water that causes damage in a flood. (a) A flooding stream can ruin structures and belongings by moving heavy objects, carrying floating debris, depositing sediment, and causing mold and mildew. (b) Even the fine-grained sediment typical of many floodwaters destroys people's possessions.

# Fluvial Processes

## Stream Erosion

Fluvial erosion is the removal of rock material by flowing water. Fluvial erosion may take the form of the chemical removal of ions from rocks or the physical removal of rock fragments (clasts). Physical removal of rock fragments includes breaking off new pieces of bedrock from the channel bed or sides and moving them as well as picking up and removing preexisting clasts that were temporarily resting on the channel bottom. Breaking off new pieces of bedrock proceeds very slowly where highly resistant rock types are found.

The removal of rock material by erosion does not necessarily mean that the landscape is undergoing long-term lowering. If sediments eroded from the bed of a stream channel are replaced by the deposition of other fragments transported in from upstream, there will be no net drop in the position of the channel bottom. Such lowering, known as channel incision, occurs only when there is net erosion compared with deposition. Net erosion results in the lowering of the affected part of the landscape

clasts onto its bed, depositing them there until a new increase in velocity provides the energy to entrain and transport them. Thus, the particular geomorphic process being carried out by a stream at any moment, whether it is eroding or depositing, is determined by the complex of factors that control its energy.

Stream capacity and stream competence both increase in response to even a relatively small increase in velocity. During a flood, a stream that doubles its velocity may increase its amount of sediment load six to eight times. The boulders seen in many mountain streams arrived there during some past flood that greatly increased stream competence; they will be moved again when a flow of similar magnitude occurs. Rivers do most of their heavy clast-moving work during short periods of flood ( ● Fig. 17.13).

and is termed **degradation.** Net deposition of sediments results in a building up, or **aggradation,** of the landscape.

One way that streams erode occurs when stream water chemically dissolves rock material and then transports the ions away in the flow. This fluvial erosion process, called **corrosion** (or *solution*, or *dissolution*), has a limited effect on many rocks but can be significant in certain rock types, such as limestone.

**Hydraulic action** refers to the physical, as opposed to chemical, process of stream water alone removing pieces of rock. As stream water flows downslope by the force of gravity, it exerts stress on the streambed. Whether this stress results in entrainment and removal of a preexisting clast currently resting on the channel bottom, or even the breaking off of a new piece of bedrock from

National Park Service – Nay Aug Park

● **FIGURE 17.14** A plunge pool has been eroded at the base of Nay Aug Falls by Roaring Brook in Scranton, Pennsylvania.

*Why do deep plunge pools form at the base of most waterfalls?*

Courtesy Sheila Brazier

● **FIGURE 17.15** These potholes were cut into a streambed of solid rock during times of high flow. Potholes result from stream abrasion and the swirling currents in a fast-flowing river.

the channel, depends on several factors including the volume of water, flow velocity, flow depth, stream gradient, friction with the streambed, the strength and size of the rocks over which the stream flows, and the degree of stream turbulence. **Turbulence** is chaotic flow that mixes and churns the water, often with a significant upward component, that greatly increases the rate of erosion and the load-carrying capacity of the stream. Turbulence is influenced by channel roughness and the gradient over which the stream is flowing. A rough channel bottom increases the intensity of turbulent flow. Likewise, even a small increase in velocity caused by a steeper gradient can result in a significant increase in turbulence. Turbulent currents contribute to erosion by hydraulic action when they wedge under or pound away at rock slabs and loose fragments on the channel bed and sides, dislodging clasts that are then carried away in the current. *Plunge pools* at the base of waterfalls and in rapids reveal the power of turbulence-enhanced hydraulic action where it is directed toward a localized point ( ● Fig. 17.14).

As soon as a stream begins carrying rock fragments as load, it can start to erode by **abrasion,** a process even more powerful than hydraulic action. As rock particles bounce, scrape, and drag along the bottom and sides of a stream channel, they break off additional rock fragments. Because solid rock particles are denser than water, the impact of having clastic load thrown against the channel bottom and sides by the current is much more effective than the impact of water alone. Under certain conditions, stream abrasion makes distinctive round depressions called *potholes* in the rock of a bedrock streambed ( ● Fig. 17.15). Potholes generally originate in special circumstances, such as below waterfalls or swirling rapids, or at points of structural weakness, which include joint intersections in the streambed. Potholes range in diameter and depth from a few centimeters to many meters. If you peer into a pothole, you can often see one or more round stones at the bottom. These are the abraders, or *grinders*. Swirling whirlpool movements of the stream water cause such stones to grind the bedrock and enlarge the pothole by abrasion, whereas finer sediments are carried away in the current.

In a process related to abrasion, as rock fragments moving as load are transported downstream, they are gradually reduced in size and their shape changes from angular to rounded. This wear and tear experienced by sediments as they tumble and bounce against one another and against the stream channel is called **attrition.** Attrition explains why gravels found in streambeds are rounded and why the load carried in the lower reaches of most large rivers is composed primarily of fine-grained sediments and dissolved minerals.

Stream erosion widens and lengthens stream channels and the valleys they occupy. Lengthening occurs primarily at the source through **headward erosion,** accomplished partly by surface runoff flowing into a stream and partly by springs undermining the slope. The lengthening of a river's course in an upstream direction is particularly important where erosional gullies are rapidly dissecting agricultural land. Such gullying may be counteracted by soil conservation practices to reduce erosional soil loss. Channel lengthening also occurs if the path of the stream channel becomes more winding, or *sinuous,* and this will cause a decrease in stream gradient.

## Stream Transportation

A stream directly erodes some of the sediment that it transports, and most chemical sediments are delivered to the channel in base flow, but a far greater proportion of its load is delivered to the stream channel by surface runoff and mass movement. Regardless of the sediment source, streams transport their load in several ways ( ● Fig. 17.16). Some minerals are dissolved in the water and are thus carried in the transportation process of **solution.** The finest solid particles are carried in **suspension,** buoyed by vertical turbulence. Such small grains can remain suspended in the water column for long periods, as long as the force of upward turbulence is stronger than the downward settling tendency of the particles. Some grains too large and heavy to be carried in suspension bounce along the channel bottom in a process known as

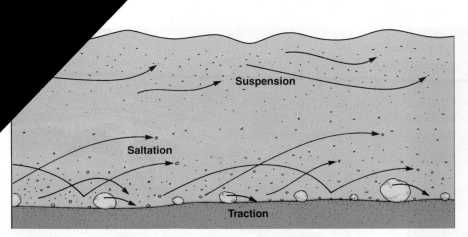

● **FIGURE 17.16** Transportation of solid load in a stream. Clay and silt particles are carried in suspension. Sand typically travels by suspension and saltation. The largest (heaviest) particles move by traction.

*What is the difference between traction and saltation?*

**saltation** (from French: *sauter,* to jump). Particles that are too large and heavy to move by saltation may slide and roll along the channel bottom in the transportation process of **traction.**

There are three main types of stream load. Ions of rock material held in solution constitute the **dissolved load. Suspended load** consists of the small clastic particles being moved in suspension. Larger particles that saltate or move in traction along the streambed comprise the **bed load.** The total amount of load that a stream carries is expressed in terms of the weight of the transported material per unit time.

The relative proportion of each load type present in a given stream varies with such drainage basin characteristics as climate, vegetative cover, slope, rock type, and the infiltration capacities and permeabilities of the rock and soil types. Dissolved loads will be larger than average in basins with high amounts of infiltration and base flow because slow-moving groundwater that feeds base flow acquires ions from the rocks through which it moves. The intense weathering in humid regions produces much fine-grained sediment that is incorporated into streams as suspended load. Large amounts of suspended load give streams a characteristically muddy appearance ( ● Fig. 17.17). The Huang He in northern China, known as the "Yellow River" because of the color of its silty suspended load, carries a huge amount of sediment in suspension, with more than 1 million tons of suspended load per year. Compared with the "muddy" Mississippi River, the Huang He transports five times the suspended sediment load with only 20% of the discharge. Streams dominated by bed load tend to occur in arid regions because of the limited weathering rate in arid climates. Limited weathering leaves considerable coarse-grained sedi-

ment in the landscape available for transportation by the stream system.

## Stream Deposition

Because the capacity and competence of a stream to carry material depend on flow velocity, a decrease in velocity will cause a stream to reduce its load through deposition. Velocity decreases over time when flow subsides—for example, after the impact of a storm—but it also varies from place to place along the stream. A distinctly shallower part of a channel cross section that lies far from the deepest and fastest flow typically experiences low flow velocity and becomes the site of recurring deposition. The resulting accumulation of sediment, like what forms on the inside of a channel bend, is referred to as a **bar.** Sediment also collects in locations where velocity falls because of a reduction in stream gradient, where the river current meets the standing body of water at its mouth, and on the land adjacent to the stream channel during floods.

**Alluvium** is the general name given to fluvial deposits, regardless of the type or size of material. Alluvium is recognized by the characteristic sorting and/or rounding of sediments that streams perform. A stream sorts particles by size, transporting the sizes that it can and depositing larger ones. As velocity fluctuates from changes in discharge, channel gradient, and roughness, particle sizes that can be picked up, transported, and deposited vary accordingly ( ● Fig. 17.18). The alluvium deposited by a stream with fluctuating velocity will exhibit alternating layers of coarser and finer sediment.

● **FIGURE 17.17** Some rivers carry a tremendous load of sediment in suspension and show it in their muddy appearance, as in this aerial view of the Mississippi River in Louisiana. Much of the load carried by the Mississippi River is delivered to it by its major tributaries.

*What are some of the major tributaries that enter the Mississippi River?*

Ron Sherman/Getty Images

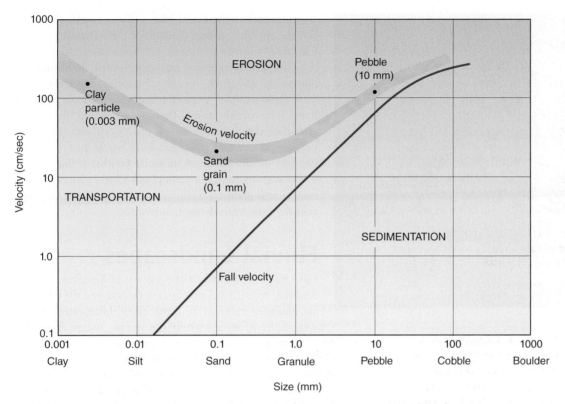

● **FIGURE 17.18** The relationship between stream flow velocity and the ability to erode or transport material of varying sizes. When a stream loses the ability to transport particles of a particular size, those particles will be deposited. Small pebbles (particles with a diameter of 10 mm, for example) need a high stream flow velocity to be moved because of their size and weight. Silts and clays (smaller than 0.05 mm) also need high velocities for erosion because they stick together cohesively. Sand-sized particles (between 0.05 and 2.0 mm) are easier to erode (are picked up at a lower flow velocity) than clasts from smaller or larger size categories.

When streams leave the confines of their channels during floods, the channel cross-sectional width is suddenly enlarged so much that the velocity of flow must slow down to counterbalance it $(Q = wdv)$. The resulting decrease in stream competence and capacity causes deposition of sediment on the flooded land adjacent to the channel. This sedimentation is greatest right next to the channel where aggradation constructs channel-bounding ridges known as **natural levees,** but some alluvium will be left behind wherever load settled out of the receding flood waters.

**Floodplains** constitute the often extensive, low-gradient land areas composed of alluvium that lie adjacent to many stream channels ( ● Fig. 17.19). Floodplains are aptly named because they are inundated during floods and because they are at least partially composed of *vertical accretion deposits,* the sediment that settles out of slowing and standing floodwater. Most floodplains also contain *lateral accretion deposits*. These are generally channel bar deposits that get left behind as a channel gradually shifts its position in a sideways fashion (laterally) across the floodplain ( ● Fig. 17.20).

● **FIGURE 17.19** During floods, sediment-laden water that flows over the stream's banks inundates the adjacent, low floodplain areas and deposits alluvium, mainly silts and clays. This photo shows a flooded farm in Wisconsin in 2008.

*What would the floodwaters leave behind on the ground and in flooded homes and businesses after the water recedes?*

FEMA/Barry Bahler

● **FIGURE 17.20** Sediment deposited in a bar on the inside of a channel bend becomes part of the floodplain alluvium if the stream migrates laterally away leaving the bar deposits behind. Here, winter ice fills swales between ridges that mark the crests of successive bar deposits.

## Channel Patterns

Three principal types of stream channel have traditionally been recognized when considering the form of a given channel segment in map view. Although *straight channels* may exist for short distances under natural circumstances, especially along fault zones, joints, or steep gradients, most channels with parallel, linear banks are artificial features that were totally or partially constructed by people.

If a stream has a high proportion of bed load in relation to its discharge, it deposits much of its load as sand and gravel bars in the streambed. The stream flows in multiple interweaving strands that split and rejoin around the bars to give a braided appearance to the channel, and indeed such a pattern is called **braided** ( ● Fig. 17.21). This channel pattern develops in settings that provide considerable amounts of loose sand and gravel bed load to the stream system and are therefore common in arid regions and in glacial meltwater zones. Braided streams

are common on the Great Plains (for example, the Platte River), in the desert Southwest, in Alaska, and in Canada's Yukon.

The most common channel pattern in humid climates displays broad, sweeping bends in map view. Over time these sinuous, **meandering channels** also wander from side to side across their low-gradient floodplains widening the valley by lateral erosion on the outside of meander bends and leaving behind lateral accretion (bar) deposits on the inside of meander bends ( ● Fig. 17.22). These streams and their floodplains have a higher proportion of fine-grained sediment and thus greater bank cohesion than the typical braided stream.

## Fluvial Landscapes

One way to understand the variety of landform features resulting from fluvial processes is to examine the course of an idealized river as it flows from headwaters in the mountains to its mouth at the ocean, where it reaches base level. Ions and sediment picked up by and delivered to the stream system are transported downstream and deposited in temporary or permanent storage in the channel, adjacent to the channel, or at the mouth of the stream. The gradient of this river would diminish continually downstream as it flows from its source toward base level. In nature exceptions exist to this idealized profile because some streams flow entirely over a low gradient ( ● Fig. 17.23), whereas other streams, particularly small ones on mountainous coasts, flow down a steep gradient all the way to the standing body of water at their mouth. Rather than having a smoothly decreasing slope from headwaters to mouth, we would expect real streams to have some irregularities in this *longitudinal profile,* the stream gradient from source to mouth (see Fig. 17.12).

The following discussion subdivides the ideal perennial stream course into upper, middle, and lower river sections flowing over steep, moderate, and low gradients, respectively, and ending where it reaches base level at the ocean. Fluvial erosion processes dominate in the steep upper course, whereas deposition predominates in the lower course. The middle course displays important elements of both fluvial erosion and deposition. Over long periods these patterns of erosion and deposition can

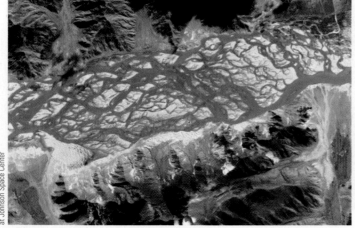

● **FIGURE 17.21** The braided channel of the Brahmaputra River in Tibet, viewed from the International Space Station.

*What does the common occurrence of braided channels just downstream from glaciers tell you about the sediment transported and deposited by moving ice?*

● **FIGURE 17.23** Not all streams fit the generalized pattern of characteristics for upper, middle, and lower stream segments. The Mississippi River, here a relatively small stream, meanders on a low gradient near its headwaters in Minnesota, far upstream from its mouth.

● **FIGURE 17.22** Over time, a meandering (sinuous) stream channel, like the Missouri River shown on this colorized radar image, may swing back and forth across its valley. Where the outside of a meander bend on the floodplain *(purple and blue tones)* impinges on the edge of higher terrain *(orange areas)*, stream erosion undercuts the valley side wall and, with the assistance of mass wasting, contributes to floodplain widening.

be altered by changes in the stream gradient or base level caused by such factors as tectonism and climate change.

## Features of the Upper Course

At the headwaters in the upper course of a river, the stream primarily flows in contact with bedrock. Over the steep gradient high above its base level, the stream works to erode vertically downward by hydraulic action and abrasion. Erosion in the upper course creates a steep-sided valley, gorge, or ravine as the stream channel in the bottom of the valley cuts deeply into the land. Little if any floodplain is present, and the valley walls typically slope directly to the edge of the stream channel. Steep valley sides encourage mass movement of rock material directly into the flowing stream. Valleys of this type, dominated by the downcutting activity of the stream, are often called V-shaped valleys because with their steep slopes they attain the form of the letter *V* ( ● Fig. 17.24).

The effects of differential erosion can be significant in the upper course if rivers cut through rock layers of varying resistance. Rivers flowing over resistant rock have a steeper gradient than where they encounter weaker rock. A steep gradient gives the stream flow more energy, which the stream needs to erode the resistant rock. Rapids and waterfalls may mark the location of resistant materials in a stream's upper course. Where rocks are particularly resistant to weathering and erosion, valleys will be narrow, steep-sided gorges or canyons; where rocks are less resistant, valleys tend to be more spacious.

Many streams spill from lake to lake in their upper courses, either over open land (like the Niagara River at Niagara Falls, between Lake Erie and Lake Ontario) or through gorges. In either case, the lakes will eventually be eliminated if the outflowing stream lowers the lake basin outlets enough or if fluvial sediment deposited at the inflow points fills in the lake basins.

● **FIGURE 17.24** Where the upper course of a stream lies in a mountainous region, its valley typically has a characteristic "V" shape near the headwaters. Such a stream flows in a steep-walled valley, with rapids and waterfalls, as shown here in Yellowstone Canyon, Wyoming.

*How does the gradient of the Yellowstone River compare with that of the stretch of the Mississippi River shown in Fig. 17.23?*

## Features of the Middle Course

In the middle section of the ideal longitudinal profile, the stream flows over a moderate gradient and on a moderately smooth channel bed. Here the river valley includes a floodplain, but remaining ridges beyond the floodplain still form definite valley walls. The stream lies closer to its base level, flows over a gentler gradient, and thus directs less energy toward vertical erosion than in its upper course. The stream still has considerable energy, however, because of the downstream increase in flow volume and reduction in bed friction. The river now uses much of its available energy for transporting the considerable load that it has accumulated and toward

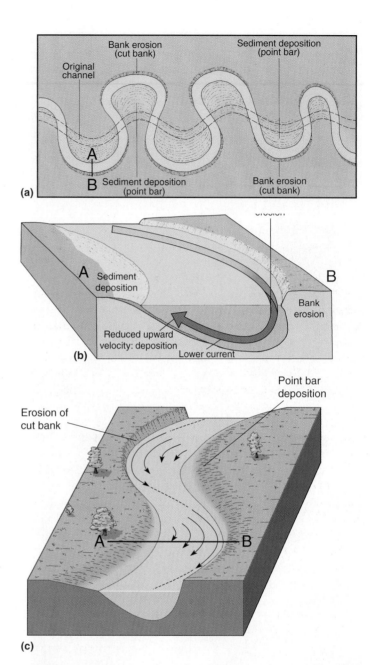

● **FIGURE 17.25** Characteristics of a meandering river channel. Note that water flowing in a channel has a tendency to flow downstream in a helical, or "corkscrew," fashion, which moves water against one side of the channel and then to the opposite side. The up-and-down motion of the water contributes to the processes of erosion, transportation, and deposition.

Copyright © 2011 Brooks/Cole, Cengage Learning

lateral erosion of the channel sides. The stream displays a definite meandering channel pattern with its sinuous bends that wander over time across the valley floor. The stream erodes a **cut bank** on the outside of meander loops, where the channel is deep and centrifugal force accelerates stream velocity. The cut bank is a steep slope, and slumping may occur there, particularly when there is a rapid fall in water level. Slumping on the outside of meander bends contributes to the effect of lateral erosion by the stream and adds load to the stream. In the low velocity and shallow flow on the inside of the meander bends, the stream deposits a **point bar** ( ● Fig. 17.25).

Erosion on the outside and deposition on the inside of river meander bends result in the sideways displacement, or **lateral migration,** of meanders. When the laterally shifting channel reaches and undercuts the confining valley walls it helps increase the area of the gently sloping floodplain. Tributaries flowing into a larger stream also aid in widening the valley through which the trunk stream flows. Although flooding of the valley floor is always a potential hazard, flood sediments bring in nutrients and contribute to the agricultural richness of floodplain soils.

## Features of the Lower Course

The minimal gradient and close proximity to base level along the ideal lower river course make downcutting virtually impossible. Stream energy, now derived almost exclusively from the higher discharge rather than the downslope pull of gravity, leads to considerable lateral shifting of the river channel. The river meanders around helping to create a large depositional plain (see Map Interpretation: Fluvial Landforms). The lower floodplain of a major river is much wider than the width of its meander belt and shows evidence of many changes in course ( ● Fig. 17.26). The stream migrates laterally through its own previously deposited sediment in a channel composed exclusively of alluvium. During floods, these extensive floodplains, called **alluvial plains,** become inundated with sediment-laden water that contributes vertical accretion deposits to the large natural levees and to the already-thick alluvial valley fill of the floodplain in general. Natural levees along the Mississippi River rise up to 5 meters (16 ft) above the rest of the floodplain.

A common landform in this deposition-dominated environment provides evidence of the meandering of a river over time. Especially during floods, **meander cut-offs** occur when a stream seeks a shorter, steeper, and straighter path; breaches through the levees; and leaves a former meander bend isolated from the new channel position. If the cut-off meander remains filled with water, which is common, it forms an **oxbow lake** ( ● Fig. 17.27).

Sometimes people attempt to control streams by building up levees artificially to keep the river in its channel. During times of reduced discharge, however, when a river has less energy, deposi-

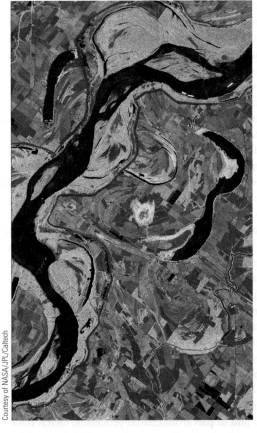

● **FIGURE 17.26** This colorized radar image shows part of the Mississippi River floodplain along the Arkansas–Louisiana–Mississippi state lines. The colors are used to enhance landscape features such as water bodies *(dark)*, field patterns, and forested areas *(green)*. Abandoned portions of former meanders visible on the floodplain show that the river has changed its channel position many times.

tion occurs in the channel. Thus, in an artificially constrained channel, a river may raise the level of its channel bed. In some instances, as in China's Huang He and the Yuba River in northern California, deposition has raised the streambed above the surrounding floodplains. Flooding presents an especially serious danger in this

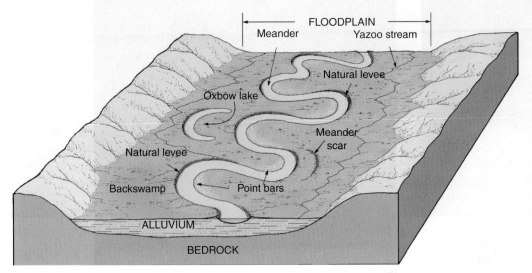

● **FIGURE 17.27** Features of a large floodplain common in the lower courses of major rivers. Backswamps are low, marshy, or swampy parts of the floodplain, generally at the water table.
Copyright © 2011 Brooks/Cole, Cengage Learning

***What is the origin of an oxbow lake?***

● **FIGURE 17.36** The multipurpose Lookout Point Dam on the Middle Fork of the Willamette River, Oregon.

*What are some of the multiple purposes that dams serve?*

( ● Fig. 17.36). Many rivers have been transformed by dam construction into a string of reservoirs. Unfortunately, the life span of these reservoirs, like that of almost any lake, is only a few centuries at most because they will gradually fill with the sediment delivered to them by their inflowing streams. To protect the natural flow and environments of the few remaining undeveloped streams and rivers in the United States, Congress enacted the Wild and Scenic Rivers Act in 1968. In recent years, increasing efforts have been directed toward assessing the impacts of dams and dam removal on stream geomorphology and ecology. This is just one part of a larger trend in the natural sciences aimed at studying and mitigating the effects of human disturbance on natural stream and other systems.

# Quantitative Fluvial Geomorphology

Quantitative methods are important in studying virtually all aspects of the Earth system, and they are used by climatologists, meteorologists, biogeographers, soil scientists, and hydrologists, as well as geomorphologists. The value of quantitative methods to the objective analysis of fluvial systems in particular may be inferred from the material presented in this chapter. Streams are complicated, dynamic systems with input, throughput, and output of energy and matter that depend on numerous, often-interrelated variables. Geomorphologists routinely measure stream channel, drainage basin, and flow properties and analyze them using statistical methods and the principles of fluid mechanics so that they can describe, compare, monitor, predict, and learn more about streams and the geomorphic work that they perform. Drainage area, stream order, drainage density, stream discharge, stream velocity, channel width, and channel depth are just some of the numeric data collected in the field, on topographic maps, from digital elevation models, or from remotely sensed imagery to facilitate the study of stream systems. Extensive efforts are made to gather and analyze numeric stream data because of the widespread occurrence of streams and their great importance to human existence. Quantitative studies not only help us better understand the origins and formational processes of landforms and landscapes, but they will also help us better predict water supplies and flood hazards, estimate soil erosion, and trace sources of pollution.

## ■ DEFINE AND RECALL

fluvial geomorphology
interfluve
surface runoff
interception
sheet wash
rill
gully
ephemeral flow
perennial flow
intermittent flow
tributary
drainage basin (watershed, catchment)
drainage divide
stream order
exterior drainage
interior drainage
base level
drainage density
drainage pattern
transverse stream

stream discharge
channel roughness
stream load
stream competence
stream capacity
graded stream
degradation
aggradation
corrosion
hydraulic action
turbulence
abrasion
attrition
headward erosion
solution
suspension
saltation
traction
dissolved load
suspended load

bed load
bar
alluvium
natural levee
floodplain
braided channel
meandering channel
cut bank
point bar
lateral migration
alluvial plain
meander cut-off
oxbow lake
yazoo stream
delta
distributary
stream terrace
stream hydrograph
recurrence interval
reservoir

# ■ DISCUSS AND REVIEW

1. On a worldwide basis, what is the most important geomorphic agent operating to shape the landscape of the continents? Why is that agent so important?
2. What is the relationship between infiltration and surface runoff? What are some of the factors that enhance infiltration? What are some of the factors that enhance surface runoff?
3. Define the concept of stream and drainage basin order. How would a typical first-order channel differ from a typical fifth-order channel in the same drainage basin in terms of tributaries, length, gradient, cross-sectional area, discharge, and velocity?
4. What factors affect the discharge of a stream? What are the two very different equations that can be used to represent stream discharge?
5. Why do drainage basins in semiarid climates tend to have greater drainage density than those in humid climates?
6. Which is the most effective fluvial erosion process? Why is it so effective?
7. What are the differences among the fluvial transportation processes? Which moves the largest particles?
8. How does a braided channel differ from a meandering channel?
9. Describe the principal landforms of a meandering stream system and the development of natural levees.
10. Explain how tectonism and climate change can each contribute to changes in a stream's base level.
11. Why is it important to collect and study stream flow data from gaging stations?
12. What is the relationship between rivers and lakes, and how do the two types of features interact?

# ■ CONSIDER AND RESPOND

1. How do streams represent the concept of dynamic equilibrium? What are some examples of negative feedback in a stream system?
2. In the study of a drainage basin, what types of geographic observations and quantitative data might prove useful in planning for flood control and water supply?
3. How does urbanization affect runoff, discharge, and flood potential in a drainage basin?

# APPLY AND LEARN

1. A measured cross section of a stream has a width of 6.5 meters, a depth of 2.5 meters, and a flow velocity of 2.0 meters per second. What is the area of the cross section? What is the stream discharge at the cross section?
2. A stream channel lies at an elevation of 1200 meters above sea level at one location and at 1075 meters above sea level at a site 11 kilometers downstream. What is the channel gradient between those two locations?
3. One hundred years of maximum annual flow data show that the stream under study had a maximum annual discharge of 3700 cubic meters per second or greater five times. What is the recurrence interval of this flow? What is the probability that a discharge of 3700 cubic meters per second will occur this year?

# LOCATE AND EXPLORE

1. Using Google Earth, identify the landforms at the following locations (latitude, longitude) and provide a brief discussion of how the landform developed. Include a brief discussion of why the landform is found in that general area of the United States or the world.
   a. 43.73°S, 172.02°E
   b. 52.110°N, 28.571°E
   c. 30.410°N, 85.015°W
   d. 45.120°N, 111.672°W
   e. 37.20°N, 110.00°W
   f. 45.36°N, 111.71°W
   g. 39.27°N, 119.55°W
   h. 30.57°N, 87.19°W
2. Using Google Earth identify the features listed below. Rank the features from youngest to oldest with an explanation for your order.
   a. 34.543999°N, 90.568936°W
   b. 34.659545°N, 90.480867°W
   c. 34.930455°N, 90.333536°W
   d. 34.642015°N, 90.411898°W
   e. 34.520741°N, 90.254868°W

 CourseMate – Make the most of your study time by accessing everything you need to succeed in one place. Read your textbook, take notes, review flashcards, watch videos, complete activities, take practice quizzes, and more online with CourseMate. Log in at **www.cengagebrain.com**.

# Map Interpretation

## ■ FLUVIAL LANDFORMS

### The Map

Campti is in northwestern Louisiana on the Gulf Coastal Plain. The coastal plain region stretches westward from northern Florida to the Texas–Mexico border and extends inland in some places for more than 320 kilometers (200 mi). Elevations on the Gulf Coastal Plain gradually increase from sea level at the shoreline to a few hundred meters far inland. The region is underlain by gently dipping sedimentary rock layers. The surface material includes marine sediments and alluvial deposits from rivers that cross the coastal plain, especially those of the Mississippi drainage system. This is a landscape of meanders, natural levees, and bayous.

The Red River's headwaters are located in the semiarid plains of the Texas Panhandle, but it flows eastward toward an increasingly more humid climate. About midcourse, the Red River enters a humid subtropical climate region, which supports rich farmland and dense forests. The river flows into the Mississippi in southern Louisiana, about 160 kilometers (100 mi) downstream of the map area. The Red River is the southernmost major tributary of the Mississippi.

Louisiana has mild winters with hot, humid summers. Annual rainfall totals for Campti average about 127 centimeters (50 in.). The warm waters of the Gulf of Mexico supply vast amounts of atmospheric energy and moisture, producing a high frequency of thunderstorms, tornadoes, and, on occasion, hurricanes that strike the Gulf Coast.

### Interpreting the Map

1. How would you describe the general topography of the Campti map area? What is the local relief? What is the elevation of the banks of the Red River at the town of Campti?

2. What kind of landform is the low-relief surface on which the river is flowing? Is this mainly an erosional or depositional landform?

3. In what general direction does the Red River flow? Is the direction of flow easy or hard to determine just from the map area? Why?

4. Does the Red River have a gentle or steep gradient? Why is it difficult to determine the river gradient from the map area?

5. What is the origin of Smith Island? Adjacent to Smith Island is Old River; what is this type of feature called?

6. Explain the stippled brown areas in the meanders south of Campti. Are these areas on the inside or the outside of the meander bend?

7. How would you describe the features labeled as "bayou"?

8. Is this map area more typical of the upper, middle, or lower course of a river?

9. Although this is not a tectonically active region, how would the river change if it experienced tectonic uplift?

U.S. Department of Agriculture

Healthy green vegetation appears red on color infrared images, here showing the Red River and Smith Island, Louisiana.

In Google Earth fly to: 31.890°N, 93.182°W.

Opposite:
Campti, Louisiana
Scale 1:62,500
Contour interval = 20 ft
USGS

# 18 ARID REGION LANDFORMS AND EOLIAN PROCESSES

## ■ OBJECTIVES

WHEN YOU COMPLETE THIS CHAPTER YOU SHOULD BE ABLE TO:

- Appreciate the relative role of water and wind in creating landforms within the world's arid regions.
- Apply drainage basin, base level, and stream channel concepts appropriately to arid region landscapes.
- Draw sketches that illustrate distinguishing features among principal arid region landforms of fluvial erosion.
- Compare and contrast pediments and alluvial fans and the specific settings in which each occurs.
- Describe a playa, and distinguish between the two major types of playa.
- Discuss the ways in which wind transports and erodes sediment.
- Provide examples of landforms made by eolian erosion.
- Explain why and how some sand dunes are stabilized and what their stabilization can mean.
- Discuss the major types of sand dunes and the situations under which each forms.
- Understand the origin and importance of the world's large loess deposits.

BECAUSE OF THEIR LOW AMOUNTS of precipitation, arid region landscapes look quite different from those of other climatic environments in many ways. The limited water supply restricts rock weathering and the amount of vegetation present. Without extensive vegetation to hold weathered rock matter (regolith) in place, the weathered rock particles that are produced are often stripped away when storms do occur. As a result, whereas hillslopes in humid regions tend to be rounded and mantled by soil, mountains and hillslopes in arid regions are generally angular, with extensive, barren exposures of bedrock. Desert lowlands may be filled in with sediments eroded from uplands, or they may consist of just a thin cover of sediments overlying rock strata.

Many desert landscapes have a majestic beauty that derives from the stark display of the colors, characteristics, and structure of the rocks that make up the area. The desert's barrenness reveals evidence about landforms and geomorphic processes that is much more difficult to observe in humid environments, with their cover of soil and vegetation. Much of our understanding of how landforms and landscapes develop has come from scientific studies conducted in desert regions.

Zabriskie Point in Death Valley National Park, California and Nevada, illustrates the dramatic beauty of desert landscapes. J. Petersen

503

Although the wind plays an important role in arid region geomorphology, running water does more geomorphic work than the wind does in arid regions. Wind erosion is mainly confined to picking up fine, dust-sized (silt and clay) particles from desert regions and to dislodging loose rock fragments of sand-sized materials. Still, we tend to associate arid environments with wind-derived (eolian) geomorphic processes because of the notable accumulations of wind-deposited sediment found in some desert areas, usually in the form of sand dunes. With its sparse vegetation and other environmental characteristics, the geomorphic work of the wind reaches its optimum in arid environments. However, because air has a much lower density than water, even in deserts the wind is outmatched by fluvial geomorphic processes. We should understand, too, that landforms fashioned by the wind are not confined to arid regions; they are also conspicuous in many coastal areas and in any area where loose sediments—including snow—are frequently exposed to winds strong enough to move them.

## Surface Runoff in the Desert

Landforms, rather than vegetation, typically dominate desert scenery. The precipitation and evaporation regimes of an arid climate result in a sparse cover of vegetation and, because many weathering processes require water, relatively low rates of weathering. With low weathering rates, insufficient vegetation to break the force of raindrop impacts, and a lack of extensive plant root networks to help hold rock fragments in place, a blanket of moisture-retentive soil cannot accumulate on slopes. Soils tend to be thin, rocky, and discontinuous. This absence of a continuous vegetative and soil cover gives desert landforms their unique character. Under these surface conditions of very limited interception and low permeability, much of the rain that falls in the desert quickly becomes surface runoff available to perform fluvial geomorphic work. With little to hold them in place, any grains of rock that have been loosened by weathering may be swept away in surface runoff produced by the next storm. Ironically, although desert landscapes strongly reflect a deficiency of water, the effects of running water are widely evident on slopes and in valley bottoms ( ● Fig. 18.1). Where vegetation is sparse, running water, when it is available, is extremely effective in shaping the land.

J. Petersen

● **FIGURE 18.1** In almost all deserts the effects of running water are prominent in the landscape.

***Why do you think the drainage density is so high on the terrain in this photo?***

Desert climates characteristically receive small amounts of precipitation and are subjected to high rates of potential evapotranspiration. In exceptional circumstances, years may pass without any rain in some desert areas. Most desert locations, however, receive precipitation each year, but the frequency and amount are highly unpredictable, and rains that do fall are often quite intense. The most important impact of rain on landform development in deserts is that when rainfall does occur, much of it falls on impermeable surfaces, producing intense runoff, often generating flash floods, and operating as a powerful agent of erosion.

The visible evidence of water as a geomorphic agent in arid regions stems not only from the climate of those areas today, but also from past climates. Paleogeographic studies reveal that many deserts have not always had the arid climate that exists there today. Geomorphologists studying arid regions have found landforms in some locations that are incompatible with the present arid climate and that can be attributed to the work of water under earlier, wetter climates. For midlatitude and subtropical deserts, the most recent major wet period was during the Pleistocene Epoch. While glaciers were advancing in high latitudes and in high-elevation mountain regions during the Pleistocene, precipitation was also greater than it is today in the now-arid basins, valleys, and plains found in parts of the middle and subtropical latitudes. At the same time, cooler temperatures for these regions meant that they also experienced lower evaporation rates. In many deserts, evidence of past wet periods includes deposits and wave-cut shorelines of now-extinct lakes ( ● Fig. 18.2) and large canyons occupied by streams that are now too small to have eroded such large valleys.

Running water is a highly effective geomorphic agent in deserts even though it operates only occasionally. In most desert areas, running water flows just during and shortly after rainstorms. Desert streams, therefore, typically have *ephemeral flow,* containing water only for brief intervals and remaining dry the rest of the time. In contrast to *perennial streams,* which flow all year and are typical of humid environments, ephemeral streams do not receive seepage from groundwater to sustain them between episodes of surface runoff. Ephemeral streams instead lose water to the groundwater system through infiltration into the channel bed. Because of the low weathering rates in desert environments, most arid region streams receive an abundance of coarse sediment that they must transport as bed load. As a result, *braided channels,* in which multiple threads of flow split and rejoin around temporary deposits of coarse-grained sediments, are common in deserts ( ● Fig. 18.3).

Unlike the typical situation for humid region streams, many desert streams undergo a downstream decrease, rather than increase, in discharge (volume of flow per unit time). A discharge decrease occurs for two major reasons: (1) infiltration into the gravelly streambed continues downstream, removing more and more water from the channel, and (2) evaporation losses increase downstream because of warmer temperatures at lower elevations. As a result of the diminishing discharge, many desert streams terminate before reaching the ocean. The same mountains that contribute to aridity through the rain-shadow effect can effectively block desert streams from flowing to the sea. Without sufficient discharge to reach that ultimate

J. Petersen

● **FIGURE 18.2** Many desert basins have remnant shorelines that were created by wave action from lakes they contained during the Pleistocene. The linear feature extending across this mountain front, below which is a smoother slope with smaller channels, was the highest shoreline of one of these ancient lakes.

*Why does the terrain look so much smoother below than above the highest shoreline?*

● **FIGURE 18.3** This braided stream in Canyon de Chelly National Monument near Chinle, Arizona, splits and rejoins multiple times as it works to carry its extensive bed load of coarse sand.

*Why do the number and position of the multiple channels sometimes change rapidly?*

D. Sack

base level, desert streams terminate in depressions in the continental interior where they commonly form shallow, ephemeral lakes. Ephemeral lakes evaporate and disappear only to reappear when rain provides another episode of adequate inflow. During the cooler and wetter times of the Pleistocene Epoch, many of these topographically blocked, closed basins in now-arid regions were filled with considerable amounts of water that in some cases accumulated into large perennial freshwater lakes instead of the shallow ephemeral lakes that they contain today.

Where surface runoff drains into closed desert basins, sea level does not govern erosional base level as it does for streams that flow into the ocean and thereby attain *exterior drainage*. Desert drainage basins characterized by streams that terminate in closed interior depressions are known as basins of *interior drainage* ( ● Fig. 18.4);

● **FIGURE 18.4** Death Valley, here depicted on a radar image, is a basin of interior drainage. Water accumulates on the basin floor from surface inflow, groundwater sources, and rarely by direct precipitation. No streams flow out of the basin.

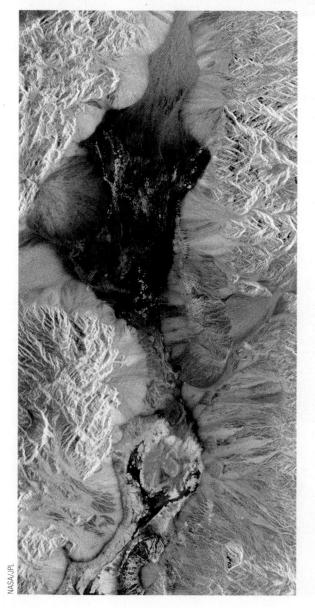

such streams are controlled by a **regional base level** instead of ultimate base level. When sedimentation raises the elevation of the desert basin floor located at the stream's terminus, the stream's base level rises, which causes a decrease in the stream's slope, velocity, and energy. If tectonic activity lowers the basin floor, the regional base level falls, which may lead to rejuvenation of the desert stream. Tectonism has even created some desert basins of interior drainage with floors below sea level, as in Death Valley, California; the Dead Sea Basin in the Middle East; the Turfan Basin in western China; and Australia's Lake Eyre.

Many streams found in deserts originate in nearby humid regions or in cooler, wetter mountain areas adjacent to the desert ( ● Fig. 18.5). Even these, however, rarely have sufficient discharge to sustain flow across a large desert. With few tributaries and virtually no inflow from groundwater, stream water lost to evaporation and underground seepage is not replenished. In most cases, the flow dwindles and finally disappears. The Humboldt River in Nevada is an outstanding example; after rising in the mountains of central Nevada and flowing 465 kilometers (290 mi), the river disappears into the Humboldt Basin, a closed depression. Only a few large rivers that originate in humid uplands have enough discharge to survive a long journey across hundreds of kilometers of desert to reach the sea ( ● Fig. 18.6). These **exotic streams,** like the Nile River in Sudan and Egypt, provide the desert with exterior drainage. Under natural conditions, the Colorado River of the United States and Mexico would reach ultimate base level in the Gulf of California, but it no longer does so because of the huge volume of water that people are withdrawing from the river.

● **FIGURE 18.5** This stream flows through a deep gorge in the Atlas Mountains of Morocco, on the arid, rain-shadow side of the mountains, facing the Sahara. This stream loses water by infiltration and evaporation and disappears into the Sahara.

● **FIGURE 18.6** A false-color satellite image of the exotic Nile River meandering across the Sahara in Egypt from its headwaters in the wetter climates of the Ethiopian Highlands and lakes in the East African Rift Zone. The irrigated croplands that appear dark red contrast with the lighter tones of the surrounding desert terrain.

# Water as a Geomorphic Agent in Arid Lands

When rain falls in the desert, sheets of water run down unprotected slopes, picking up and moving sediment. Dry channels quickly change to flooding streams. The material removed by sheet wash and surface streams is transported, just as in humid lands, until flow velocity decreases sufficiently for deposition to occur. Eventually these streams disappear when seepage and evaporation losses deplete their discharge. Huge amounts of sediment can be deposited along the way as a stream loses volume and velocity. The processes of erosion, transportation, and deposition by running water are essentially the same in both arid and humid lands. However, the resulting landforms differ because of the sporadic nature of desert runoff, the lack of vegetation to protect surface materials against rapid erosion, and the common occurrence of streams that do not reach the sea.

## Arid Region Landforms of Fluvial Erosion

Among the most common desert landforms created by surface runoff and erosion are the channels of ephemeral streams. Known as **washes** or **arroyos** in the southwestern United States and **wadis** in North Africa, these channels usually form where rushing surface waters cut into unconsolidated alluvium ( ● Fig. 18.7). These typically gravelly and braided channels are prone to flash floods, which makes them potentially dangerous sites. Although it may sound strange, many people have drowned in the desert—during flash floods.

In areas of weak, easily eroded clays or shales, rapid erosion from surface runoff can produce barren slopes and ridges dissected by a dense maze of steep gullies and ravines. This type of rugged, barren, and highly dissected terrain is termed **badlands** topography ( ● Fig. 18.8). Badlands topography has an extremely high *drainage density,* defined as the length of stream channels per unit area of the drainage basin. Extensive badlands exist in the Dakotas; Death Valley National Park in California; Big Bend National Park in Texas; and southern Alberta, Canada. Badlands do not generally form under natural conditions in humid climates because the vegetation in those regions inhibits runoff and erosion, leading to lower drainage densities. Removing vegetation from clay or shale areas by overgrazing, mining, or logging, however, can cause badlands topography to develop even in humid environments.

A **plateau** is an extensive, elevated region with a fairly flat top surface. Most plateaus consist of tectonically uplifted, horizontal rock layers. Many striking plateaus exist in the deserts and semiarid regions of the world, including the extensive tectonically uplifted Colorado Plateau, centered on the Four Corners area of Arizona, Colorado, New Mexico, and Utah. In such desert plateau regions, streams have responded to the uplift by cutting narrow, steep-sided canyons ( ● Fig. 18.9). Where the canyon walls consist of horizontal layers of alternating resistant and erodible rocks, differential weathering and erosion have created stair-stepped walls, with near-vertical cliffs made from resistant rock layers (ordinarily sandstone, limestone, or basalt) and weaker strata (often shales) forming gentler slopes. The cliff-forming rock strata at the top (rim) of the sequence is a **caprock,** a term that refers to a resistant horizontal layer that forms (caps) the top of a landform. Such horizontal rock layers of varying thickness and resistance provide the distinctive appearance to much of the landscape of the Colorado Plateau, including the Grand Canyon.

Caprocks top plateaus and constitute canyon rims, but they also form the summits of other, smaller kinds of flat-topped landforms that, although they are found in many climate regions, are most characteristic of deserts. Weathering and erosion will eventually reduce the extent of a caprock until only flat-topped, steep-sided **mesas** remain (in Spanish: *mesa,* table). A mesa has a smaller surface area than a plateau and is roughly as broad across as it is tall. Mesas are relatively common landscape features in the Colorado Plateau region. Through additional erosion of the caprock from all sides, a mesa is reduced to a **butte,** which is a similar, flat-topped erosional remnant but with a smaller surface area

● **FIGURE 18.7** This dry streambed, or wash, has a channel bed of coarse alluvium and conveys water only during and slightly after a rainstorm.

*Why would this desert stream channel have a high risk for flash floods?*

( ● Fig. 18.10). Mesas and buttes in a landscape are generally evidence that uplift occurred in the past and that weathering and erosion have been extensive since that time. Variations in the form of the slope extending down the sides of buttes, mesas, and plateaus are related to the height of the cliff at the top, which is controlled by the thickness of the caprock in comparison to the overall size of landform. Monument Valley, in the Navajo Tribal Reservation on the Utah–Arizona state line, is an exquisite example of such a landscape formed with a caprock that is particularly thick, contributing to the distinctive scenery ( ● Fig. 18.11). Many classic western movies were filmed in Monument Valley and in nearby sections of the Colorado Plateau because of the striking, colorful, and photogenic desert landscape.

Over time, sheet wash and gully development can accomplish extensive erosion of mountainsides and hillslopes fringing a desert basin or plain. Particularly in desert regions with exterior drainage or a sizable trunk stream

● **FIGURE 18.8** In badlands, such as these in South Dakota, impermeable clays that lack a soil cover produce rapid runoff, leading to intensive gully erosion and a high drainage density.

*What makes it difficult to cross this terrain on foot or horseback?*

● **FIGURE 18.9** The earliest European American explorers of the Grand Canyon took along an artist to record the geomorphology of the canyon, here beautifully shown in great detail. Aridity creates an environment where the bare rocks are exposed to our view; differential weathering and erosion give the stair-stepped quality to the walls of the Grand Canyon.

USGS

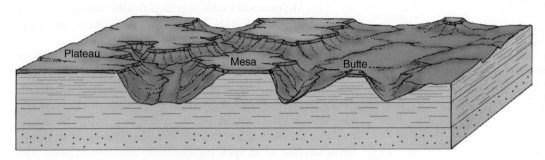

Plateau     Mesa     Butte

● **FIGURE 18.10** Plateaus, mesas, and buttes are developed through weathering and erosion in areas of horizontal rock layers with a resistant caprock. An excellent example of this terrain is the Colorado Plateau of Arizona, Colorado, New Mexico, and Utah.

● **FIGURE 18.11** The caprock in Monument Valley, Arizona, is particularly thick and represents a rock layer that once covered the entire region. The prominent buttes are erosional remnants of that layer.

Copyright and photograph by Dr. Parvinder S. Sethi

on the basin or plain, this fluvial action, aided by weathering, may lead to the gradual, erosional retreat of the position of the mountain front. As the bedrock slope of the mountain front migrates slowly away from the basin floor or plain, it leaves behind a gently sloping surface of the eroded bedrock, known as a **pediment** ( ● Fig. 18.12). Characteristically in desert areas there tends to be a sharp break in slope between the base of steep hills or mountains (the mountain front), which rise at angles of 20° to 30° or steeper, and the gentle slope of the bedrock pediment, which is usually 2° to 7°. Resistant knobs of the bedrock comprising the pediment

project up above the surface on some pediments. These resistant knobs are referred to as **inselbergs** (from German: *insel,* island; *berg,* mountain).

Multiple explanations of pediment formation have been suggested, and different processes may be responsible for their formation in different regions. There is general agreement that most pediments are erosional surfaces created or partially created by the action of running water. In some areas, weathering, perhaps when the climate was wetter in the past, seems to also have played a strong role in the development of pediments.

J. Petersen

● **FIGURE 18.18** Salt deposits accumulate on playas that evaporate considerable groundwater. In the foreground a mixture of salt and mud has created the playa microtopography known as puffy ground.

contrast, **salt-crust playas,** also known as **salt flats,** receive much of their water from groundwater, are damp most of the time, and are encrusted with salt mineral deposits crystallizing out of the evaporating groundwater ( ● Fig. 18.18). Some playas, such as Utah's famous Bonneville Salt Flats, are the floors of desiccated, ancient lakes that occupied now-desert basins during the Pleistocene Epoch. In those cases, salts also accumulated on the basin floor by precipitating out of the evaporating water body.

Playas are useful in several ways. Mineral companies mine the rich deposits of evaporite minerals, including such important industrial chemicals as potassium chloride, sodium chloride, sodium nitrate, and borates, that have been deposited in some playa beds. Also, the extensive, flat surfaces of some playas make them suitable as racetracks and airstrips. Utah's famous Bonneville Salt Flats represent the bed of an extinct Pleistocene lake, called Lake Bonneville, of which Great Salt Lake is a small remnant. The western portion of the Salt Flats, where world land-speed records are set, often floods to a depth of 30 to 60 centimeters (1–2 ft) in the winter. Hard, flat playa surfaces at Edwards Air Force Base, in California's Mojave Desert, have served for many decades as landing sites for military aircraft and the space shuttle. These landings have occasionally been disrupted because of flooding of the playa.

# Wind as a Geomorphic Agent

Overall, wind is less effective than running water, waves, groundwater, moving ice, or mass movement in accomplishing geomorphic work. Under certain circumstances, however, wind can be a significant agent in the modification of topography. Landforms—whether in the desert or elsewhere—that are created by wind are called **eolian** (or aeolian) landforms, after Aeolus, the god of winds in classical Greek mythology ( ● Fig. 18.19). The three principal conditions necessary for wind to be effective as a geomorphic agent are a sparse vegetative cover; the presence of dry, loose materials at the surface; and a wind velocity that is high enough to pick up and move those surface materials. These three conditions occur most commonly in arid regions and in coastal locations, but areas of recent fluvial or glacial deposits, snow, newly plowed fields, and overgrazed lands are also subject to significant wind action ( ● Fig. 18.20).

A dense cover of vegetation reduces wind velocity near the surface by providing frictional resistance. It also prevents wind from being directed against the land surface and holds materials in place with its root network. Without such a protective cover, fine-grained and sufficiently dry surface materials are subject to removal by strong gusts of wind. If surface particles are damp, they tend to adhere together in wind-resistant aggregates as a result of increased cohesion provided by the water. The arid conditions of deserts, therefore, make these regions very susceptible to wind erosion.

Eolian processes have many things in common with fluvial processes because air and water are both fluids. Some important contrasts also exist, however, as a result of fundamental differences between gases and liquids. For example, rock-forming materials cannot dissolve in air, as some can in water; thus air does not erode by corrosion or move load in solution. Otherwise, the wind detaches and transports rock fragments in ways comparable to flowing water, but it does so with less overall effectiveness because air has a much lower density than water. Another difference is that, compared to streams, the wind has fewer lateral or vertical limitations on movement. As a result, the dissemination of material by the wind can be more widespread and unpredictable than that by streams.

A principal similarity between the geomorphic properties of wind and running water is that flow velocity controls their competence—that is, the size of particles each can pick up and carry. However, because of its low density, the competence of moving air is generally limited to rock fragments that are sand sized or smaller. Wind erosion selectively entrains small particles, leaving behind the coarser and heavier particles that it is not able to lift. Like water-laid sediments, wind deposits are stratified according to changes in its velocity although within a much narrower range of grain sizes than occurs with most alluvium.

## Wind Erosion and Transportation

Strong winds blow frequently in arid regions, whipping up loose surface materials and transporting them within turbulent air currents. The finest particles transported by winds, the clays and silts, are

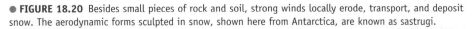

● **FIGURE 18.19** Wind-deposited sand dunes in Death Valley illustrate the stunningly beautiful landscapes that can be created by eolian processes.

*In what other environments besides arid lands might sand dunes be found?*

● **FIGURE 18.20** Besides small pieces of rock and soil, strong winds locally erode, transport, and deposit snow. The aerodynamic forms sculpted in snow, shown here from Antarctica, are known as sastrugi.

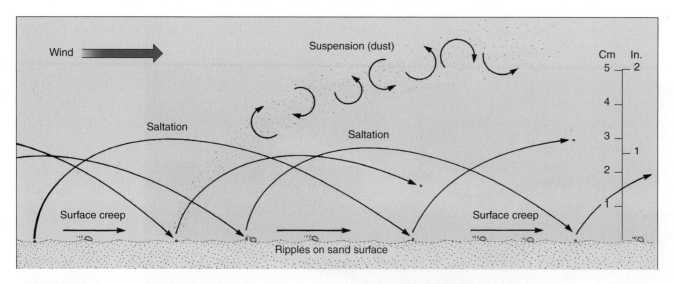

● **FIGURE 18.21** Wind moves sediment in the transportation processes of suspension, saltation, and surface creep. Some of the bed load forms ripples, which can be seen moving forward when the wind is strong.

*Why are grains larger than sand not generally moved by the wind?*

moved in *suspension,* buoyed up by vertical currents ( ● Fig. 18.21). Such particles essentially comprise a fine dust that will remain in suspension as long as the strength of upward air currents exceeds the tendency of the particles to settle out onto the ground because of gravity. Sediments carried in suspension by the wind make up its suspended load. If the wind velocity surpasses 16 kilometers (10 mi) per hour, surface sand grains can be put into motion. As with fluvial transportation, particles that are too large to be carried in eolian suspension bounce along the ground as part of the bed load in the transportation process of *saltation.* When particles moving in eolian saltation, which are typically sand sized, bounce on the ground, they generally dislodge other particles that are then added to the wind's suspended or saltating load. Larger sand grains too heavy to be lifted into the air are pushed forward along the ground surface by the impact of the saltating grains in a process called **surface creep.** Grains moving along the ground in surface creep often become organized into small wave forms termed **ripples.**

Wind erodes surface materials by two main processes. **Deflation,** which is similar to the hydraulic force of running water, occurs when wind blowing fast enough or with enough turbulence over an area of loose sediment is able to pick up and remove on its own small fragments of rock. Like all of the geomorphic agents, once the wind obtains some load, it also erodes by *abrasion.* In eolian abrasion particles already being transported by the wind strike rocks or sediment and break off or dislodge additional rock fragments. Wind-driven solid particles are more effective than the wind alone in dislodging and entraining other grains and in breaking off new fragments of rock. Most eolian abrasion is quite literally sandblasting, and quartz sand, which is common in many desert areas, can be a very effective abrasive agent in eolian processes. Yet sand grains are typically the largest size of clast that the wind

can move, and they rarely are lifted higher than 1 meter (3 ft) above the surface. Consequently, the effect of this natural sandblast is limited to a zone close to ground level.

Where loose fine-grained particles exist on the land surface, during strong winds they will be picked up primarily by deflation and carried off in suspension. Sometimes the result is a thick, dark, swiftly moving dust cloud that swirls over the land and greatly reduces visibility. These *dust storms* are sometimes so severe that visibility drops to nearly zero and almost all sunlight is blocked ( ● Fig. 18.22). Dust storms are often highly destructive,

● **FIGURE 18.22** Dust storms occur when strong winds entrain and suspend a large amount of silt and clay particles. This dust storm occurred in Texas during the Dust Bowl era of the 1930s.

*Can you suggest a continent that might be a source of major dust storms today?*

NOAA/George E. Marsh Album

removing layers of surface materials in one area and depositing them elsewhere over the course of a few hours. The infamous Dust Bowl era of the 1930s particularly affected the southern Great Plains of the United States in this way when devastating dust storms were brought about by years of drought and poor agricultural practices. *Sandstorms* can occur in areas where sand is abundant at the surface. Because sand grains are larger and heavier than silts and clays, most sandstorms are restricted to a zone near the ground, where objects are severely abraded by the abundant saltating sand. Evidence of the restricted height of desert sandstorms is visible on vehicles that have traveled through the desert and on fence posts, utility poles, and other structures, all of which are more intensely pitted and abraded in their lower than in their higher sections.

Erosion by deflation produces shallow depressions in a barren surface of unconsolidated materials. These depressions, which vary in diameter from a few centimeters to a few kilometers, are called **deflation hollows.** Deflation hollows are particularly common in nonmountainous arid regions. They tend to collect rainwater and may hold water for a time, depending on permeability and evaporation rates. Thus deflation hollows, like bolsons, frequently contain playas. Many thousands of deflation hollows that contain playas occur in the semiarid southern High Plains of west Texas and eastern New Mexico in the United States. Often deflation hollows form at sites that were already exhibiting a slight depression or where vegetation cover has been breached by overgrazing, fire, or other means.

Deflation has traditionally been considered one of several possible factors that help to produce **desert pavement** (or reg), a close-fitting mosaic of gravel-sized stones that overlies a deposit of mostly finer-grained sediments. Desert pavement is common in many arid regions, particularly in parts of the Sahara, interior Australia, the Gobi in central Asia, and the American Southwest. If deflation selectively removes the smaller particles from a desert surface of mixed particle sizes, the gravel-sized clasts left behind form a concentration of stones at the surface overlying the mixed grain sizes below ( ● Fig. 18.23). Sheet wash (unchannelized running water) also appears to contribute to the formation of some desert pavements by selectively eroding only the fine-grained clasts from an area of mixed grain sizes. Research has shown that other occurrences of pavement form by eolian deposition, that is, by accumulation of wind-deposited fine-grained sediments that subsequently fell, washed, and sifted down beneath the stony surface layer. Regardless of its origin, once formed, desert pavement helps to stabilize desert surfaces by protecting the fine-grained material below the pavement from erosion. Desert pavement is fragile, however, and easily disturbed by vehicles. Use of off-road vehicles in the desert severely disrupts the desert pavement and contributes greatly to accelerated erosion, the amount of wind-blown dust in the atmosphere, the occurrence of dust storms, and damage to desert ecological systems.

Like deflation under natural conditions, eolian abrasion is also responsible for creating interesting desert landform features. Where the land surface is exposed bedrock, wind abrasion can polish, groove, or pit the rock surface and in some cases produces **ventifacts,** which are individual wind-fashioned rocks. A venti-

● **FIGURE 18.23** The surface of desert pavement is a mosaic of gravel. Immediately underneath the gravel lies a zone of fine-grained sediment with little gravel.

*What problem might result if the surface stones are removed?*

fact is a rock that has been trimmed back to a smooth slope on one or more sides by sandblast. Because of frictional effects at the surface, the ability of the wind to erode by abrasion increases with increasing distance from the ground surface, at least up to a certain height. Thus, abrasion carves the windward side of a rock into a smooth, sloping surface, or face. Ventifacts subjected to multiple sand-transporting wind directions have multiple faces, called *facets,* which meet along sharp edges ( ● Fig. 18.24).

A distinctive feature sometimes attributed incorrectly to wind abrasion is the pedestaled, or balanced, rock, in which a larger top part of the rock sits on a narrower lower part, the pedestal, below. It is tempting to think that eolian abrasion wore away the lower part faster than the upper part. Actually, such forms result from various physical and chemical weathering pro-

● **FIGURE 18.24** A ventifact on Mars. With high wind speeds, plenty of loose surface materials, and no vegetation, rocks on the Martian surface display facets, pits, and other strong evidence of eolian abrasion.

● **FIGURE 18.25** Rocks with pedestals are also known as rock mushrooms. Eolian deflation and abrasion may contribute to the narrowing of some pedestals, but weathering processes and sheet wash are considered to be more important for their formation.

*How might the rock on top of the pedestal differ from the rock forming the pedestal?*

cesses in the damper environment at the base of an exposed rock and are not typically related to eolian abrasion ( ● Fig. 18.25).

Rates of eolian erosion in arid regions often reflect the strength and kinds of the exposed surface materials. Where eolian abrasion affects rocks of varying resistance, differential erosion etches away softer rocks faster than the more resistant rocks. Even in desert locations of extensive soft rock, such as shale, or semiconsolidated sediments, like ancient lake deposits, abrasion may not act in a uniform fashion over the entire exposure. A **yardang** is a wind-sculpted remnant ridge, often of easily eroded rock or semilithified sediments, left behind after the surrounding material has been eroded by abrasion ( ● Fig. 18.26), perhaps with deflation assisting in removal of fine-grained fragments. Everything else being equal, abrasion and deflation are most effective where rocks are soft or weak.

● **FIGURE 18.26** Eolian erosion carves yardangs, aerodynamically shaped ridges like this one in the Kharga Depression of Egypt.

## Wind Deposition

All materials transported by the wind are eventually deposited whenever and wherever the wind energy falls below the minimum amount needed to keep the grains moving. The nature of the resulting eolian depositional landform depends primarily on the characteristics of the wind, which is commonly affected by topography and vegetation, and the grain size of the sediments. Fine-grained particles, such as silts, are often transported in suspension long distances from their source area before being deposited, sometimes quite thickly, as a blanket covering the pre-existing topography. In comparison, the coarser, sand-sized particles, transported in saltation and surface creep, accumulate much closer to their source area.

## Sand Dunes

In some circumstances, eolian sand deposits form a **sand sheet,** a rather uniform cover of sand-sized particles displaying little surface relief. More often, eolian sand deposits accumulate in the shape of hills, mounds, or ridges, called **sand dunes.** In fact, to many people, the word desert evokes the image of endless sand dunes, perhaps accompanied by blinding sandstorms, a blazing sun, mirages, and an occasional palm oasis. Although these features do exist, particularly in parts of North Africa and the Arabian Peninsula, most areas of the world's deserts have rocky or gravelly surfaces, scrubby vegetation, and few or no sand dunes. Nevertheless, sand dunes are the most spectacular landforms of wind deposition, whether they occur as seemingly endless dune regions called **sand seas** (or ergs), as small dune fields, or as sandy ridges inland from beaches ( ● Fig. 18.27).

Dune topography is highly variable. For instance, dunes in the great sand seas of the Sahara and Arabian Deserts look like

● **FIGURE 18.27** Coastal dunes lie just landward of many sandy beaches. Grass has been planted on this dune ridge along the Maryland coast to help stabilize the sand.

*Why are coastlines such good locations for dune formation?*

(a)

(b)

● **FIGURE 18.28** Active and stabilized dunes. (a) Sand moves freely in active dunes, which typically have little or no vegetation. (b) Vegetation reduces wind speed over the dune and helps increase material cohesion, leading to dune stabilization.

*What effect might the trails that are visible on the stabilized dunes have on the dune system?*

rolling ocean waves. Others have aerodynamic crescent forms. The specific type of dune that forms depends on the amount of sand available, the strength and direction of sand-transporting winds, and the amount of vegetative cover. As wind that is carrying sand encounters surface obstacles or topographic obstructions that decrease its velocity, the sand is deposited and piles up in drifts. Once initiated, these sand piles themselves interfere with wind velocity and the sand-transporting capabilities of the wind. The dunes grow larger until equilibrium is reached between dune size and the ability of the wind to supply the dunes with sand.

Sand dunes are either *active* or *stabilized* ( ● Fig. 18.28). Active dunes change their shape or advance downwind as a result of wind action. Active dunes change their shape with variations in wind direction and/or wind strength. Many active dunes travel forward as the wind erodes sand from their upwind (windward) slope, depositing it on their downwind (leeward) side. Sand entrained on the upwind slope moves by saltation and surface creep up to and over the dune crest onto the steep leeward slope, which is the **slip face.** The slip face of a dune lies at the *angle of repose* for dry, loose sand. The angle of repose—the steepest slope that a pile of dry, loose material can maintain without experiencing slipping, sliding, or flow down the slope—is about 34° for sand. When wind direction and velocity are relatively constant, a dune can move forward while maintaining its general form by this downwind transfer of sediment from the windward slope to the slip face ( ● Fig. 18.29). The speed at which active dunes move downwind varies greatly, but as with many processes, the movement is episodic; dunes advance only when the wind is strong enough to move sand from the upwind to the downwind side. Because of their greater height and especially their greater volume of sand, large dunes travel

more slowly than smaller dunes, which migrate up to 40 meters (130 ft) a year. During sandstorms, a dune may migrate more than 1 meter (3 ft) in a single day. In places where plants, including trees, lie in the path of an advancing dune, the sand cover can move onto and smother the vegetation. Some active dunes are affected by seasonal wind reversals so that they do not advance, but the crest at the top moves back and forth annually under the influence of seasonally opposing winds.

Stabilized dunes maintain their shape and position over time. Vegetation normally stabilizes dunes. Vegetation can stabilize an active sand dune if plants gain a foothold and send roots down to moisture deep within the dune. This task is difficult for most plants because sand offers limited nutrients and has high permeability. If the vegetation cover becomes breached on a stabilized dune, per-

● **FIGURE 18.29** Active dunes migrate downwind as sediment from the upwind side is transported over the dune and deposited on the slip face.

*Why does the inside of the migrating dune consist of former slip faces?*

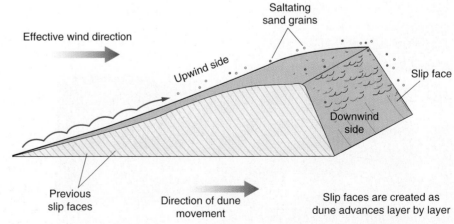

● **FIGURE 18.30** A sand dune advances into a vegetated area on the Oregon coast. In coastal dune regions landscape change is common as dunes move inland and are subsequently invaded by plants.

***Explain how plants can stabilize dunes.***

haps from damage by range animals or off-road vehicles, the wind can then remove some of the sand, creating a local **blowout.** Where invading dunes and blowing sands are a problem, attempts are frequently made to plant grasses or other vegetation to stabilize the dunes, halting their advance ( ● Fig. 18.30).

One extensive area of stabilized dunes in North America is the Sand Hills of Nebraska. This region features large dunes that formed during a drier period between glacial advances in the Pleistocene Epoch. Subsequent climate change affected sand supply, wind patterns, and moisture availability so that these impressive dunes are now stabilized by grasses ( ● Fig. 18.31). Similar stabilized dunes are found along the southern edge of the Sahara, where sand dunes that were active during drier conditions in the recent geologic past are now stabilized by vegetation.

● **FIGURE 18.31** The rolling grazing lands of the Sand Hills in central Nebraska were once a major region of active sand dunes.

***Why are these dunes no longer active today?***

## Types of Sand Dunes

Different types of sand dunes are distinguished on the basis of their shape and on their orientation relative to the wind direction. The different types are also related to the amount of available sand, which affects not only the size, but also the shape of sand dunes.

**Barchans** are a kind of crescent-shaped dune ( ● Fig. 18.32a). The two *arms* of the crescent, also called the dune's *horns,* point downwind ( ● Fig. 18.33). The main body of the crescent lies on the upwind side of the dune. From the desert floor at its upwind edge, the dune rises as a gentle slope up which the sand moves until it reaches the highest point, or crest, of the dune and, just beyond that, the slip face at the angle of repose. The slip face is oriented perpendicular to the barchan's arms. The arms extend downwind beyond the location of the slip face. Barchans form in areas of minimal sand supply where winds are strong enough to move sand downwind in a single prevailing direction. They may be most common in smaller desert basins surrounded by highlands where they tend to form near the downslope, sandy edge (toe) of alluvial fans, and adjacent to small playas. Although they may form as isolated dunes, barchans often appear in small groups, called barchan fields.

Another type of crescent-shaped dune is the **parabolic dune,** but the orientation of the crescent in this case is reversed from that of a barchan (Fig. 18.32b). With parabolic dunes, the arms of the crescent are stabilized by vegetation, long, and point upwind, trailing behind the unvegetated main body and crest of the dune rather than extending downwind from it. The main body of a parabolic dune points downwind, and the slip face along its downwind edge has a convex shape when viewed from above. Parabolic dunes commonly occur just inland of beaches and along the margin of active dune areas in deserts.

**Transverse dunes** are created where sand-transporting winds blow from a constant direction and the supply of sand is abundant (Fig. 18.32c). As with barchans, the upwind slope of a transverse dune ridge is gentle, whereas the steeper downwind slip face is at the angle of repose. In the downwind direction, transverse dunes form ridge after ridge separated from each other by low swales in a repeating wavelike fashion. Each dune ridge is laterally extensive perpendicular to the sand-transporting wind direction. The dune ridges, slip faces, and interdune swales trend perpendicular to the direction of prevailing winds, hence the name *transverse.* Abundant sand supply derives from such sources as easily eroded sandstone bedrock, from sandy alluvium deposited by exotic streams or during wetter climates in the Pleistocene Epoch, or from sandy deltas and beaches left in the landscape after the desiccation of ancient lakes.

Long dunes aligned parallel to the average wind direction are **longitudinal dunes** (Fig. 18.32d). There is no consistent distinction between the back slopes and slip faces of these dunes, and their summits may be either rounded or sharp. Strong winds are important to the formation of most longitudinal dunes, which do not migrate but instead elongate in the downwind direction. Small longitudinal dunes, such as those found in North America, may simply represent the long trailing ridges of breached parabolic dunes or a sand streak extending from a somewhat isolated source of sand. Much higher and considerably longer than these are the impressive longitudinal dunes that cross vast

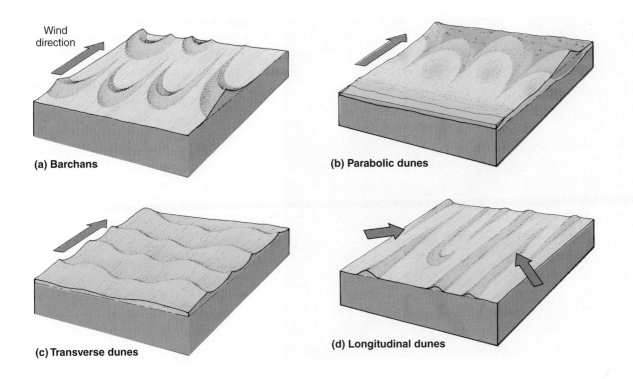

(a) Barchans

Wind direction

(b) Parabolic dunes

(c) Transverse dunes

(d) Longitudinal dunes

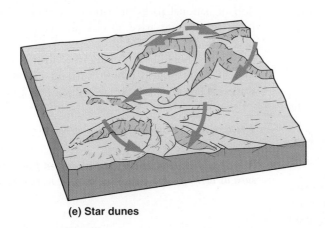

(e) Star dunes

● **FIGURE 18.32** Five principal types of sand dunes. Arrows indicate the dominant wind direction for each.
Copyright © 2011 Brooks/Cole, Cengage Learning

*What factors play a role in which type of dune will be found in a region?*

● **FIGURE 18.33** A small barchan in southern Utah.

*On this photo, does the right or left side of the dune face into the wind (face upwind)?*

D. Sack

NASA

● **FIGURE 18.34** A satellite view of longitudinal dunes in the Sahara. The width of this image represents approximately 160 kilometers (100 mi) of terrain, from left to right.

*What is the approximate ground length of the dunes in this satellite image?*

areas of the flatter, more open desert topography of North Africa, the Arabian peninsula, and interior Australia (● Fig. 18.34). These dunes develop under bidirectional wind regimes, where the two major sand-transporting wind directions come from approximately the same quadrant, such as a northwesterly and southwesterly wind. A type of large longitudinal dune called a **seif** (pronounced *safe;* from the Arabic for sword) is found in the deserts of Arabia and North Africa. Seif dunes are huge, sinuous (rather than linear), sharp-crested dunes, sometimes hundreds of kilometers long, whose troughs are almost free of sand. They attain heights up to 180 meters (600 ft).

The large, widely spaced **star dunes** have a pyramidal shape in which ridges of sand radiate out from a peaklike center to resemble a star in map view (Fig. 18.32e). These dunes are most common in areas where there is a great quantity of sand, changing wind directions, and an extremely hot and dry climate. Star dunes are stationary, but ridges and slip faces shift orientation with wind variations.

## Dune Protection

Many people consider sand dunes one of nature's most beautiful landforms, and they are attractive sites for recreation. They seem particularly inviting for off-road vehicle enthusiasts. Although dunes appear to be indestructible and rapidly changing environments that do not damage easily, this is far from the truth. Dunes are fragile environments with delicate, easily affected ecologies. Because dune regions are the result of an environmental balance between moving dunes and the plants trying to stabilize them, the environmental equilibrium is easily upset. Some of the most spectacular dune areas in the United States, including White Sands (New Mexico), Great Sand Dunes (Colorado), Indiana Dunes (Indiana), and Cape Cod (Massachusetts), lie within the protec-

tion of national parks or national seashores. Many dune areas, however, do not have special protection, and environmental degradation is a constant threat.

Dunes should be left undisturbed for many reasons. Near the shoreline, dunes help protect adjacent inland areas from storm waves (● Fig. 18.35). In this regard, they are particularly important

● **FIGURE 18.35** A sign directs visitors on how to help protect this dune area along the South Carolina coast.

*Why should some dune areas be protected from human activities such as driving dune buggies and other recreational vehicles?*

Copyright and photograph by Dr. Parvinder S. Sethi

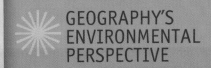

## GEOGRAPHY'S ENVIRONMENTAL PERSPECTIVE

# Off-Road Vehicle Impacts on Desert Landscapes

Whether you prefer to call them off-road vehicles (ORVs) or all-terrain vehicles (ATVs), driving motorized vehicles of any kind over the desert landscape off established roadways has tremendous negative impacts on desert flora and fauna, the habitat of those organisms, the stability of landforms and landforming processes, and the aesthetic beauty of the natural desert environment. Deserts are particularly fragile environments, primarily because of the low amount and high variability of precipitation. The climatic regime leads to slow rates of weathering, slow rates of soil formation, low density of vegetation, unusual species of plants and animals specially adapted to the hostile conditions of severe moisture stress, and rapid generation of surface runoff when precipitation does occur.

ORVs damage desert biota in several ways. ORVs kill and injure plants, animals (for example, birds, badgers, foxes, snakes, lizards, and tortoises, to name just a few), and insects when they hit or ride over them. Even careful ORV riders who drive slowly cannot avoid crushing small animals and insects that lie hidden under a loose cover of sand or shallow roots that extend out far from a plant. Driving ORVs at night is particularly deadly for desert animals, which tend to be nocturnal. ORVs do not have to hit or ride over organisms to cause populations to decrease. Hearing loss experienced by animals in areas frequented by ORVs puts them at a major disadvantage for feeding, defending themselves, and mating. ORVs crush and destroy burrows and other animal and insect homes and nesting sites, including plants. Oil and gas leaking from poorly maintained vehicles are another danger for desert dwellers, as are grass and range fires inadvertently started from such vehicles. No desert subenvironment, not even sand dunes, is immune from these negative impacts.

In addition to the direct impacts on organisms, physical properties of the landscape are also negatively affected by ORVs. One major problem is that ORVs compact the desert surface sediments and soil. Compaction causes a decrease in permeability, which greatly restricts the ability of water to infiltrate into the subsurface. More surface runoff translates to greater erosion of the already thin desert soils. In addition, even long after ORV use ceases in a disturbed area, soil compaction makes it very difficult for desert plants, animals, and insects to become reestablished there. Denuded, eroded swaths made by ORVs often remain visible as ugly scars in the landscape for decades and act as sources of sediment, adding to the severity of dust storms. Where ORV trails traverse steep slopes, they have contributed to the occurrence of debris flows, mudflows, and other forms of mass wasting.

The attributes of desert environments that give them such stark beauty also make them vulnerable to disturbance. With limited but often torrential precipitation, sparse and slow-growing vegetation, and slow rates of weathering and soil formation, deserts require long periods to recover from environmental disturbance. Recovery, moreover, will likely not return the landscape to the condition it would have been in if the disturbance had not occurred.

Motorcycles ridden on a desert hillslope in Utah have stripped the vegetation, compacted the soil, instigated accelerated erosion, and seriously marred the aesthetic beauty of the landscape.

Riding off-road vehicles on active sand dunes harms the sensitive organisms that live there and interferes with the natural dune processes by compacting the sand.

## DEFINE AND RECALL

| | | |
|---|---|---|
| regional base level | piedmont alluvial plain | yardang |
| exotic stream | bolson | sand sheet |
| wash (arroyo, wadi) | playa | sand dune |
| badlands | playa lake | sand sea (erg) |
| plateau | clay pan | slip face |
| caprock | salt-crust playa (salt flat) | blowout |
| mesa | eolian (aeolian) | barchan |
| butte | surface creep | parabolic dune |
| pediment | ripples | transverse dune |
| inselberg | deflation | longitudinal dune |
| alluvial fan | deflation hollow | seif |
| fan apex | desert pavement (reg) | star dune |
| debris flow fan | ventifact | loess |
| bajada | | |

## DISCUSS AND REVIEW

1. What are some ways in which desert streams differ from humid region streams?
2. How do basins of interior drainage differ from basins of exterior drainage, and why are both found in arid lands?
3. What distinguishes a plateau from a mesa and a butte? How are they related?
4. How does an alluvial fan differ from a pediment?
5. What are playas, and why are they commonly found in desert bolsons?
6. Why is wind a weaker geomorphic agent than running water?
7. How does formation of a deflation hollow differ from formation of a ventifact?
8. Explain the main differences between a barchan and a parabolic dune.
9. What is loess? What is an important economic activity related to loess in many regions?
10. What are some ways in which climate and vegetation affect landforms in the desert?

## CONSIDER AND RESPOND

1. How are eolian erosion and transportation processes similar to fluvial erosion and transportation processes, and how do they differ? What are the main reasons for these similarities and differences? Is eolian deposition similar to fluvial deposition? If so, what are some of the similarities?
2. What physical geographic factors contribute to create so many basins of interior drainage in arid regions? What are some ways in which base-level changes can occur in basins of interior drainage?
3. Name several types of arid region landforms commonly found in tectonically active desert regions, such as the American Great Basin. Name several types of arid region landforms commonly found in deserts, such as interior Australia, that have been tectonically stable for a very long time.
4. Deserts are considered fragile environments. What are some ways in which desert landforms fit or contribute to this designation?
5. Why are there so many national parks and monuments in the arid parts of the American Southwest? What landform characteristics and other factors of desert landscapes draw people to these locations?

# APPLY AND LEARN

1. A stream flowing out of a canyon mouth onto the apex of an alluvial fan has a discharge of 12 cubic meters per second, flow depth of 2 meters, and flow width of 3 meters. What is the flow velocity at the fan apex? If the velocity decreases down-fan by 0.1 meter per second, at what distance from the apex will it reach a velocity of zero?

2. A barchan in Arizona migrated the following total annual amounts (in meters) in each of the last 13 years, listed in order from the oldest to the most recent year: 9, 12, 14, 15, 15, 19, 22, 23, 26, 28, 29, 29, and 30. What was the total distance that the dune migrated during the 13 years, and what was its average annual migration rate for the study period? How would you describe the overall trend in dune migration for this period? What factor or factors might be responsible for the observed trend?

# LOCATE AND EXPLORE

1. Using Google Earth, identify the landforms at the following locations (latitude, longitude) and provide a brief discussion of how the landform developed. Include a brief discussion of why the landform is found in that general area of the United States or the world.
   a. 36.415°N, 116.810°W
   b. 37.66°N, 117.625°W
   c. 36.25°N, 116.82°W
   d. 23.42°S, 14.77°E
   e. 36.96°N, 110.11°W
   f. 25.35°S, 131.03°E
   g. 28.07°N, 114.06°W
   h. 36.25°N, 116.82°W
   i. 36.32°N, 116.94°W
   j. 32.77°N, 106.21°W

2. The following are latitude and longitude points for the center of major world deserts. Using Google Earth to view their global position and setting, provide an explanation of why deserts are located in these areas. Consider such characteristics as mountain barriers, continentality, cold ocean currents, and global-scale pressure and wind belts.
   a. Sahara Desert (22.4°N, 12.6°E)
   b. Atacama Desert (23.7°S, 68.8°W)
   c. Simpson Desert (25.3°S, 133.8°E)
   d. Mojave Desert (35.7°N, 115.1°W)
   e. Gobi Desert (38.9°N, 82.6°E)
   f. Namib Desert (24.3°S, 15.1°E)

 CourseMate - Make the most of your study time by accessing everything you need to succeed in one place. Read your textbook, take notes, review flashcards, watch videos, complete activities, take practice quizzes, and more online with CourseMate. Log in at **www.cengagebrain.com**.

# Map Interpretation

## ■ EOLIAN LANDFORMS

### The Map

Eolian processes formed the Sand Hills region, the largest expanse of sand in North America. The region covers more than 52,000 square kilometers (20,000 sq mi) of central and western Nebraska.

The Sand Hills region was part of an extensive North American desert some 5000 years ago. The sand dunes here reached more than 120 meters (400 ft) high and inundated postglacial peat bogs and rivers. As the climate became wetter, vegetation invaded the dunes, greatly reducing eolian erosion and transportation. The vegetation anchored the sand, and the stabilized dunes developed a more rounded form. Underlying the Sand Hills is the Ogallala aquifer. The high water table of the aquifer supports the many lakes nestled between the dunes.

The Sand Hills area has a midlatitude steppe climate (BSk) and receives about 50 centimeters (20 in.) of precipitation annually.

Temperatures have a great annual range, from freezing winters to very hot summers. During summer months, the region is often pelted by thunderstorms and hail; during the winter months, the area is subject to blinding blizzards.

Vegetation is mainly bunch grasses that can survive on the dry, sandy, and hilly slopes. Some species of bunch grasses have extensive root systems that may extend more than 1 meter (3 ft) into the sandy soil. The lakes and marshes in the interdunal valleys support a marsh plant community that in turn supports thousands of migratory and local birds. Currently, the main land use in the region is cattle grazing. Some scientists are predicting that the Sand Hills area will lose its protective grass cover if global warming continues and will revert to an active area of migrating desert sand dunes.

### Interpreting the Map

1. What is the approximate relief between the dune crests and the interdunal valleys?

2. What is the general linear direction in which the dunes and valleys trend?

3. To which direction do the steepest sides of the dunes generally face?

4. Knowing that the slip face is the steepest slope of the dunes, determine what the prevailing wind direction was when the dunes were active.

5. Based on your answers to the previous three questions, determine what type of sand dune formed the Sand Hills.

6. Use the elevation of the lakes to determine the water table elevation at several points. Use that information to identify the general direction of groundwater flow in the aquifer beneath the Sand Hills.

7. Sketch a north–south profile across the middle of the map from School No. 94 to the eastern end of School Section Lake (in section 16). Label the following landform features: dune crests, dune slip faces, interdunal valleys, and lakes.

8. What cultural features on the map indicate the dominant land use for the region?

Location map of the Nebraska Sand Hills, which cover almost one third of the state.

⊕ In Google Earth fly to: 42.369°N, 101.627°W.

Opposite:
Steverson Lake, Nebraska
Scale 1:62,500
Contour interval = 20 ft
USGS

# 19

# GLACIAL SYSTEMS AND LANDFORMS

Ruth Glacier, Denali National Park, Alaska. National Park Service

## ■ OBJECTIVES

WHEN YOU COMPLETE THIS CHAPTER YOU SHOULD BE ABLE TO:

■ Appreciate the role of glaciers in Earth's hydrologic budget.
■ Explain what a glacier is and how glacial ice forms.
■ Differentiate the different types of alpine and continental glaciers.
■ Discuss the ways in which glaciers flow.
■ Draw a cross section of an alpine glacier and identify its major surface and subsurface zones.
■ Understand the processes of glacial erosion.
■ Define each of the major glacial erosional landforms.
■ Relate the principal characteristics of glacial sediments.
■ Distinguish among the various glacial depositional landforms.
■ Discuss what Earth's glacial environment was like during the Pleistocene Epoch and how that environment influences people today.

LARGE MASSES OF FLOWING ICE, which are termed **glaciers,** play several important roles in the Earth system. They are excellent climate indicators because certain environmental conditions are required for glaciers to exist and they respond visibly to climate variation. Glaciers become established, expand, contract, and disappear in response to changes in climate. Their long-term storage of freshwater as ice has a tremendous impact on the hydrologic cycle and the oceans, and the accumulation of ice by glaciers provides a record of past climates that can be studied from ice cores. Where glaciers once existed or where they were once much larger than they are today, much evidence can be found concerning past climatic conditions.

The processes of erosion, transportation, and deposition by glaciers, whether ongoing or in the past, leave a distinctive stamp on a landscape. The term **glaciation** refers to the existence and actions of moving ice and also its effects on the landscape. Some of the most beautiful and rugged terrain in the world exists in mountainous and other highland regions that have been sculpted by glaciers. Virtually every high-mountain region in the world displays glacial landscapes, including the Alps, the Rocky Mountains, the Himalayas, and the Andes. Glaciers have also carved impressive steep-sided coastal valleys in Norway, Chile, New Zealand, and Alaska. Rugged mountain peaks rising high above lake-filled valleys or narrow and deep oceanic embayments create the ultimate in scenic appeal for many people.

Masses of moving ice have transformed the appearance of high mountains and large portions of continental plains into distinctive glacial landscapes. The flowing ice of glaciers is an effective and spectacular geomorphic agent on major portions of Earth's surface.

# Glacier Formation and the Hydrologic Cycle

Glaciers are masses of flowing ice that have accumulated on land in areas where the annual input of frozen precipitation, especially snowfall, has exceeded its yearly loss by melting and other processes. Snow falls as hexagonal ice crystals that form flakes of intricate beauty and variety. Snowflakes have a low density (mass per unit volume) of about 0.1 grams per cubic centimeter (0.06 oz/in.³). Once snow accumulates on the land, it becomes transformed by compaction along with melting and refreezing at pressure points into a mass of smaller, rounded grains ( ● Fig. 19.1). Density in-

● **FIGURE 19.1** The transformation of frozen water from snow to glacial ice changes the size and shape of the crystals and greatly increases the material density.

*How does firn differ from snow?*

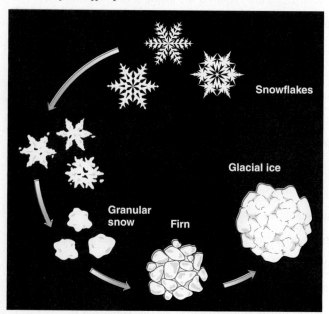

creases as the air space around this more granular snow continues to decrease by compaction and melting and refreezing. Through melting, refreezing, and pressure caused by the increasing weight from burial under newer snowfalls, the granular snow compacts further into a denser, crystalline granular stage, known as **firn,** with a density of about 0.5 grams per cubic centimeter (0.29 oz/in.³). Over time, the small firn granules grow together into larger interlocked ice crystals through pressure, partial melting, and re-freezing. When the ice is deep enough and has a density up to about 0.9 grams per cubic centimeter (0.52 oz/in.³), it becomes glacial ice. Pressure from burial under many layers of snow, firn, and ice causes the glacial ice below to become plastic and flow outward or downward away from the area of greatest snow and ice accumulation.

Glaciers are open systems with input, storage, and output of material. Any addition of frozen water to a glacier is termed **ac-cumulation.** Most accumulation occurs in winter and consists of snowfall onto a glacier, but there are many ways that frozen water can accumulate, including other forms of precipitation onto the ice, atmospheric water vapor freezing directly onto the ice (gas to solid), and transportation of snow and ice to a glacier surface from surrounding terrain by the wind as *snowdrift* and by avalanches. Snowdrift has been an important source of frozen water to gla-ciers in such highlands as the Colorado Rockies and the Ural Mountains in Russia.

**Ablation** is the opposite of accumulation, representing any removal of frozen water mass from a glacier. Although most abla-tion is accomplished in summer through melting, glaciers also lose mass by direct change from ice to water vapor (solid to gas) and by other processes that may be at work. **Calving,** for ex-ample, refers to the loss of large chunks of ice from a glacier to an adjacent lake or ocean, where the chucks float away as icebergs ( ● Fig. 19.2).

A glacial system is controlled by two basic climatic conditions: frozen precipitation and freezing temperatures. First, to establish a glacier, there must be sufficient mass input (accumulation) to exceed the annual loss through ablation. Mountains along mid-latitude coastlines and high mountains near the equator can sup-port glaciers because of heavy orographic snowfalls, despite in-tense sunshine and warm climates in the surrounding lowlands. Some very cold polar regions in subarctic Alaska and Siberia and a few valleys in Antarctica have no glaciers because the climate is too dry.

The second climatic condition that affects a glacier is tem-perature. Summer temperatures must not be high for too long, or all of the accumulation (primarily snowfall) from the previ-ous winter will be lost through ablation (primarily melting). Surplus snowfall is essential for glacial formation because it al-lows the pressure from years of accumulated snow layers to transform older buried snow first into firn and then into glacial ice. When the ice reaches a depth of about 30 meters (100 ft), a pressure threshold is reached that enables the solid, glacial ice to flow.

Glaciers are an important part of Earth's hydrologic cycle and are second only to the ocean in the amount of water they contain. Approximately 2.25% of Earth's total water is

(a)

(b)

● **FIGURE 19.2** Icebergs originate as large chunks of ice that calve (break off) from the end of a glacier into the ocean or a large lake. (a) Ellesmere Island, Canada. (b) Melville and Mapple Glaciers, Antarctica.

currently frozen in glaciers. This frozen water, however, makes up about 70% of the world's freshwater, with the vast majority stored on Greenland and Antarctica. The total amount of ice is even more impressive if we estimate the water that would be released if all of the world's glaciers were to melt. Sea level would rise about 65 meters (215 ft). This would change the geography of the planet considerably. In contrast, if another ice age were to occur, sea level would drop drastically. During the last major glacial expansion, sea level fell about 120 meters (400 ft).

Unlike the water in a stream system, much of which returns rapidly to the sea or atmosphere, water that becomes glacial ice is stored for a long time in a much more slowly flowing system. Glaciers may store water as ice for hundreds or even hundreds of thousands of years before it is released as meltwater. Yet, glacial ice is not stagnant. It moves slowly but with tremendous energy across the land. Glaciers reshape the landscape by engulfing, eroding, pushing, dragging, carrying, and finally depositing rock debris, often in places far from its original location. Long after glaciers recede from a landscape, glacial landforms remain as a

● **FIGURE 19.24** These braided stream channels carry meltwater and large amounts of outwash derived from glaciers in Denali National Park, Alaska.

*Why are braided channels often associated with the deposition of glacial outwash?*

# Continental Glaciers

In terms of their size and shape, continental glaciers, which consist of ice sheets or the somewhat smaller ice caps, are very different from alpine glaciers. However, all glaciers share certain characteristics and processes, and much of what we have discussed about alpine glaciers also applies to continental glaciers. The geomorphic work of the two categories of glaciers differs primarily in scale, attributable to the enormous disparity in size between continental and alpine glaciers.

Ice sheets and ice caps are shaped somewhat like a convex lens in cross section, thicker in the center and thinning toward the edges. They flow radially outward in all directions from where the pressure is greatest, in the thick, central zone of accumulation, to the surrounding zone of ablation (● Fig. 19.25). Like all glaciers, ice sheets and ice caps advance and retreat by responding to changes in temperature and snowfall. As with alpine glaciers flowing down pre-existing stream valleys, movement of advancing continental glaciers takes advantage of paths of least resistance found in pre-existing valleys and belts of softer rock.

● **FIGURE 19.25** Ice in ice sheets and ice caps flows outward and downward from the center where the glacier is thickest.

*How is this manner of ice flow different from and similar to that of an alpine glacier?*

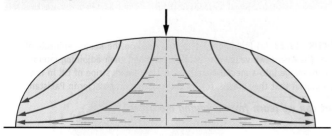

## Existing Continental Glaciers

Glaciers of all categories currently cover about 10% of Earth's land area. In area and ice mass, alpine glaciers are almost insignificant compared to continental glaciers. The huge ice sheets of Greenland and Antarctica alone account for 96% of the area occupied by glaciers today. Ice caps currently exist in Iceland, on the arctic islands of Canada and Russia, in Alaska, and in the Canadian Rockies.

The Greenland ice sheet covers the world's largest island with a glacier that is more than 3 kilometers (2 mi) thick at its center. The only land exposed in Greenland is a narrow, mountainous coastal zone (● Fig. 19.26). Where the ice reaches the ocean, it usually does so through fjords. These flows of ice to the sea resemble alpine glaciers and are called **outlet glaciers.** The action of waves and tides on the outlet glaciers breaks off huge ice masses that float away as icebergs. These large, irregular chunks of ice are a hazard to vessels in the North Atlantic shipping lanes south of Greenland. Tragic maritime disasters, such as the sinking of the *Titanic,* have been caused by collisions with icebergs, which are nine-tenths submerged and thus mostly invisible to ships. Today, with iceberg tracking by radar, satellites, and the ships and aircraft of the International Ice Patrol, deadly encounters with icebergs are much less likely.

The Antarctic ice sheet covers some 13 million square kilometers (5 million sq mi), almost 7.5 times the area of the Greenland ice sheet (● Fig. 19.27). As in Greenland, only a small amount of land is exposed in Antarctica, and the weight of the 4.5-kilometer (nearly 3 mi) thick ice in some interior areas has depressed the land well below sea level. Where the ice reaches the sea, it contributes to the **ice shelves,** enormous flat-topped plates of ice attached to land along at least one side (● Fig. 19.28). The flat-topped ice shelves provide the region's icebergs, thus the icebergs in Antarctic waters do not have the irregular surface form of Greenland's icebergs. Because they do not float into heavily used shipping lanes, these extensive tabular-shaped Antarctic icebergs are not as much of a hazard to navigation as the Greenland icebergs (● Fig. 19.29). They do, however, add to Antarctica's accessibility problem. The huge wall of the ice shelf, massive pieces of sea ice (frozen sea water), and the extreme climate combine to make Antarctica inaccessible to all but the hardiest people and equipment. Scientists from many countries, however, travel to Antarctica to study its glaciology, climatology, geology, and ecology.

## Pleistocene Glaciation

The Pleistocene Epoch of geologic time was an interval of great climate change that began about 2.6 million years ago and ended around 10,000 years before the present. A great number of glacial fluctuations occurred during the Pleistocene, marked by numerous major advances and retreats of ice over large portions of the world's landmasses. It is a common misconception that the continental ice sheets originated, and remain thickest, at the poles. When the Pleistocene glaciers advanced, ice spread outward from centers in Canada, Scandinavia, eastern Siberia, Greenland, and Antarctica,

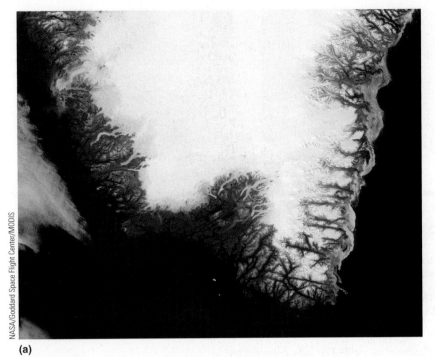

**(a)**

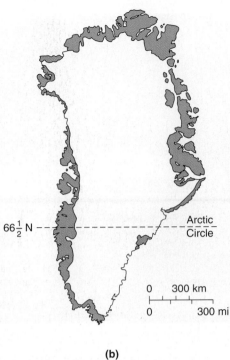

66½ N - - - - - - - - - - - - - - Arctic
Circle

0        300 km

0                300 mi

**(b)**

● **FIGURE 19.26** (a) Only a narrow coastal zone of Greenland is not covered by the ice sheet, which reaches a maximum thickness of more than 3 km (2 mi). (b) Extent of the Greenland ice sheet.

● **FIGURE 19.27** The Antarctic ice sheet covers an area larger than the United States and Mexico combined. White and blue shades on this satellite image represent ice. The only rocky areas (darker tones on the image) are the Antarctic Peninsula, the Ellsworth Mountains, and the Transantarctic Mountains.

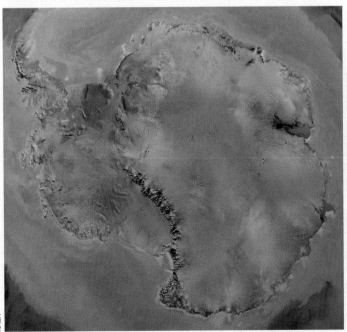

while alpine glaciers spread to lower elevations. At their maximum extent in the Pleistocene, glaciers covered nearly a third of Earth's land surface ( ● Fig. 19.30). At the same time, the extent of sea ice expanded equatorward. In the Northern Hemisphere, sea ice was present along coasts as far south as Delaware in North America and Spain in Europe. Between each period of maximum ice advance, known as a **glacial** (or glacial period), a warmer time called an **interglacial** occurred, during which the enormous continental ice sheets, ice caps, and sea ice retreated and almost completely disappeared. Studies of glacial deposits and landforms have revealed that within each major glacial advance, many minor retreats and advances occurred, reflecting smaller changes in global temperature and precipitation. During each advance of the ice sheets, alpine glaciers were much more numerous, extensive, and massive in highland areas than they are today, but their total extent was still dwarfed by that of the continental glaciers.

North America and Eurasia experienced major glacial expansion during the Pleistocene. In North America, ice sheets extended as far south as the Missouri and Ohio Rivers and covered nearly all of Canada and much of the northern Great Plains, the Midwest, and the northeastern United States. In New England, the ice was thick enough to overrun the highest mountains, including Mount Washington, which has an elevation of 1917 meters (6288 ft). The ice was more than 2000 meters (6500 ft) thick in the Great Lakes region. In Europe, glaciers spread over most of what is now Great Britain, Ireland, Scandinavia, northern Germany, Poland, and western Russia. The great weight of this ice depressed the land surface several hundred meters. As the ice receded and its weight on the land was removed, the land rose by the process of *isostatic rebound*. Measurable amounts of glacial iso-

(a)

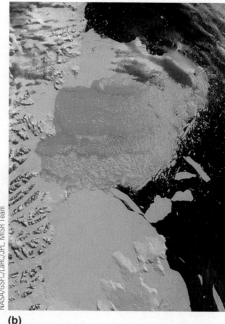

(b)

● **FIGURE 19.28** (a) Antarctica's flat-topped ice shelves form a coastal buffer of ice between the open ocean and the ice-covered continent. (b) The ice shelves have experienced significant recent destruction as in this collapse of Antarctica's Larsen B Ice Shelf in 2002.

static rebound are still raising elevations in parts of Sweden, Canada, and eastern Siberia by up to 2 centimeters (1 in.) per year, and it may cause Hudson Bay and the Baltic Sea to emerge someday above sea level. Should Greenland and Antarctica lose their ice sheets, their depressed central land areas would also rise to reach isostatic balance.

The geomorphic effects of the last major glacial advance, known in North America as the Wisconsinan stage, are the most

● **FIGURE 19.29** Antarctica's flat-topped (tabular) icebergs, here with penguin passengers, look quite different from the irregularly shaped icebergs of the Northern Hemisphere.

***What portion of an iceberg is hidden below the ocean surface?***

visible in the landscape today. The glacial landforms created during the Wisconsinan stage, which ended about 10,000 years ago, are relatively recent and have not been destroyed to any great extent by subsequent geomorphic processes. Consequently, we can derive a fairly clear picture of the extent and actions of the ice sheets and ice caps, as well as alpine glaciers, at that time.

Where did the water locked up in all the ice and snow come from? Its original source was the ocean. During the periods of glacial advance, sea level fell, exposing large portions of the **continental shelves,** which are the edges of the continents and which currently lie below sea level. That lowering of sea level likewise exposed land bridges across the present-day North, Bering, and Java Seas. At the end of the Pleistocene, melting of the glaciers raised sea level by a similar amount—about 120 meters (400 ft). Evidence for this rise in sea level can be seen along many coastlines around the world.

## Continental Glaciers and Erosional Landforms

Ice sheets and ice caps erode the land through plucking and abrasion, carving landscape features that have many similarities to those of alpine glaciers but on a much larger scale. Erosional landforms created by ice sheets stretch over millions of square kilometers of North America, Scandinavia, and Russia, and are far more extensive than those formed by alpine glaciation. As ice sheets flowed over the land, they gouged Earth's surface with striations, enlarged valleys that already existed, scoured out rock basins, and smoothed off existing hills. The ice sheets

The Driftless Area, a region in the upper midwestern United States, mainly covers parts of Wisconsin but also sections of the adjacent states of Iowa, Minnesota, and Illinois. This region was not glaciated by the last Pleistocene ice sheets that extended over the rest of the northern Midwest, so it is free of glacial deposits from that time. Drift is a general term for all deposits of glacial ice and its meltwater, thus the regional name is linked to its geomorphic history. There is some debate about whether earlier Pleistocene glaciations covered the region, but the most recent glacial advance did not override this locality.

The landscape of the Driftless Area is very different from the muted terrain and low rounded hills of adjoining glaciated terrain. Narrow stream valleys, steep bluffs, caves, sinkholes, and loess-covered hills produce a scenic landscape with

many landforms that would not have survived erosion by a massive glacier.

In some parts of the United States, lobes of the ice sheet extended as much as 485 kilometers (300 mi) farther south than the latitude of the Driftless Area. Why this region remained glacier-free is a result of the topography directly to the north. The Superior Upland, a highland of resistant rock, caused the front of the glacier to split into two masses, called lobes, diverting the southward-flowing ice around this topographic obstruction.

The diverging lobes flowed into two valleys that today hold Lake Superior and Lake Michigan. These two lowlands were oriented in directions that channeled the glacial lobes away from the Driftless Area. Although not overridden by these glaciers, the Driftless Area did receive some outwash deposits and a cover of wind-deposited

loess, both derived from the surrounding glaciers and their sediments.

The Driftless Area has a unique landscape because of its isolation, an unglaciated "island" or "peninsula" almost completely surrounded by extensive glaciated regions. The rocks, soils, terrain, relief, vegetation, and habitats contrast strongly with the nearby drift-covered terrains and provide a glimpse of what the preglacial landscapes of this midwestern region may have been like. The Driftless Area also provides distinctive habitats for flora and fauna that do not exist in the adjacent glaciated terrain. Unique and unusual aspects of its topography and ecology attract ecotourists and offer countless opportunities for scientific study. For these reasons, there are many protected natural lands in the Driftless Area, an excellent example of a natural region defined by its physical geographic characteristics.

The Driftless Area was virtually surrounded by the advancing Pleistocene ice sheets, yet it was not glaciated.

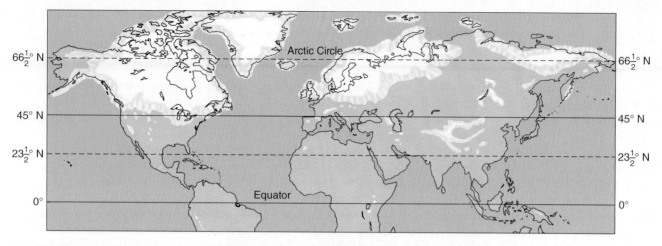

● **FIGURE 19.30** Glacial ice coverage in the Northern Hemisphere was extensive during the Pleistocene. Glaciers up to several thousand meters thick covered much of North America and Eurasia.

*What might be a reason for some areas that were very cold during this time, such as portions of interior Alaska and Siberia, being ice-free?*

removed most of the soil and then eroded the bedrock below. Today these *ice-scoured plains* are areas of low, rounded hills, lake-filled depressions, and wide exposures of bedrock ( ● Fig. 19.31).

When ice sheets expand, they cover and totally disrupt the former stream patterns. Because the last glaciation was so recent in terms of landscape development, some of the new drainage systems have not had time to form well-integrated systems of

● **FIGURE 19.31** Seen from the space shuttle, Lake Manicouagan on the Canadian Shield in Quebec occupies a circular depression that continental glaciers erosionally enhanced in rocks surrounding an asteroid impact structure. The annular (ring-shaped) lake has a diameter of approximately 70 kilometers (43 mi).

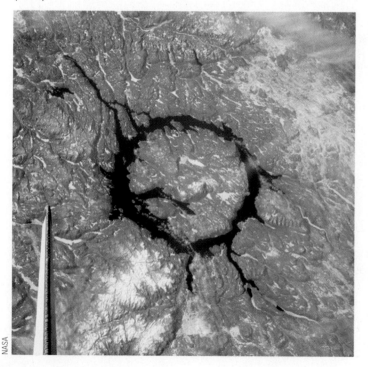

NASA

stream channels. In addition to large expanses of exposed gouged bedrock, ice-scoured plains are characterized by extensive areas of standing water, including lakes, marshes, and muskeg (poorly drained areas grown over with vegetation that form in cold climates).

## Continental Glaciers and Depositional Landforms

The sheer disparity in scale causes depositional landforms of ice sheets and ice caps to differ from those of alpine glaciers. Although terminal and recessional moraines, ground moraines, and glaciofluvial deposits are produced by both categories of glaciers, retreating continental glaciers leave significantly more extensive versions of these features than alpine glaciers do ( ● Fig. 19.32). Because of their size, continental glaciers deposit some ridge and hill forms that alpine glaciers generally cannot.

**Moraines and Plains** Ranging up to about 60 meters (200 ft) in height, terminal and recessional end moraines deposited by Pleistocene ice sheets form substantial belts of low hills and ridges in areas affected by continental glaciation ( ● Fig. 19.33). The last major Pleistocene glacial advance through New England left its terminal moraine running the length of New York's Long Island and created the offshore islands of Martha's Vineyard and Nantucket, Massachusetts. Glacial retreat left recessional moraines forming both Cape Cod and the rounded southern end of Lake Michigan. Both terminal and recessional end moraines are usually arc-shaped and convex toward the direction of ice flow. Their pattern and placement indicate that the ice sheets did not maintain an even front but spread out in tongue-shaped lobes channeled by the underlying terrain ( ● Fig. 19.34). End moraines provide more evidence than just the direction of ice flow. Characteristics of the material deposited in the end

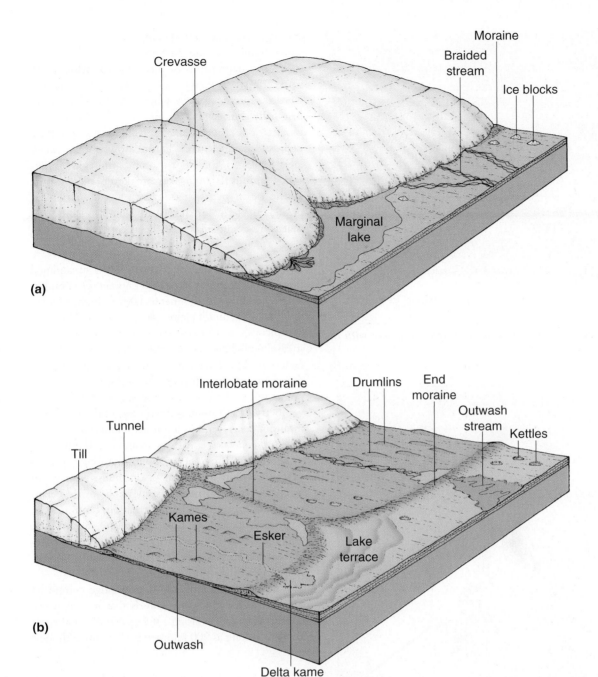

(a)

Crevasse

Moraine
Braided
stream
Ice blocks

Marginal
lake

(b)

Interlobate moraine

Drumlins

End
moraine

Tunnel

Outwash
stream

Kettles

Till

Kames

Esker

Lake
terrace

Outwash

Delta kame

● **FIGURE 19.32** (a) Features associated with stagnation at the edge of a continental glacier. (b) Landforms resulting from further modification of the ice-marginal terrain as the glacier retreats.

*How important is liquid water in creating the landforms shown here?*

moraines also help us detect the sequence of advances and retreats of each successive ice sheet.

Continental glaciers carry massive amounts of poorly sorted sediment, much of it near the base of the ice. As a glacier ablates away at the end of a glacial interval, it abandons its load essentially in place, leaving till strewn over the landscape as a **till plain.** This till often accumulates to depths of 30 meters (100 ft) or more. Because of the uneven nature of deposition from the wasting ice, the topographic configuration of the till-covered land varies from place to place. In some areas, the till is too thin to hide the original con-

tours of the land, whereas in other regions, thick deposits of till make broad, rolling plains of low relief. Small hills and slight depressions, some filled with water, characterize most till plains, reflecting the uneven glacial deposition. The gently rolling till plains of Illinois and Iowa are excellent examples of this glacial landform.

Beyond the belts of hills that represent terminal and recessional moraines lie **outwash plains** composed of meltwater deposits. These extensive areas of relatively low relief consist of glaciofluvial deposits that are sorted as they are transported by meltwater from the ice sheets. Outwash plains, which may cover

● **FIGURE 19.33** Hilly topography of an end moraine deposited by a continental ice sheet in eastern Washington.

*What makes the terrain at the left of the photo appear bumpier compared with the smoother surface of the plain at the right?*

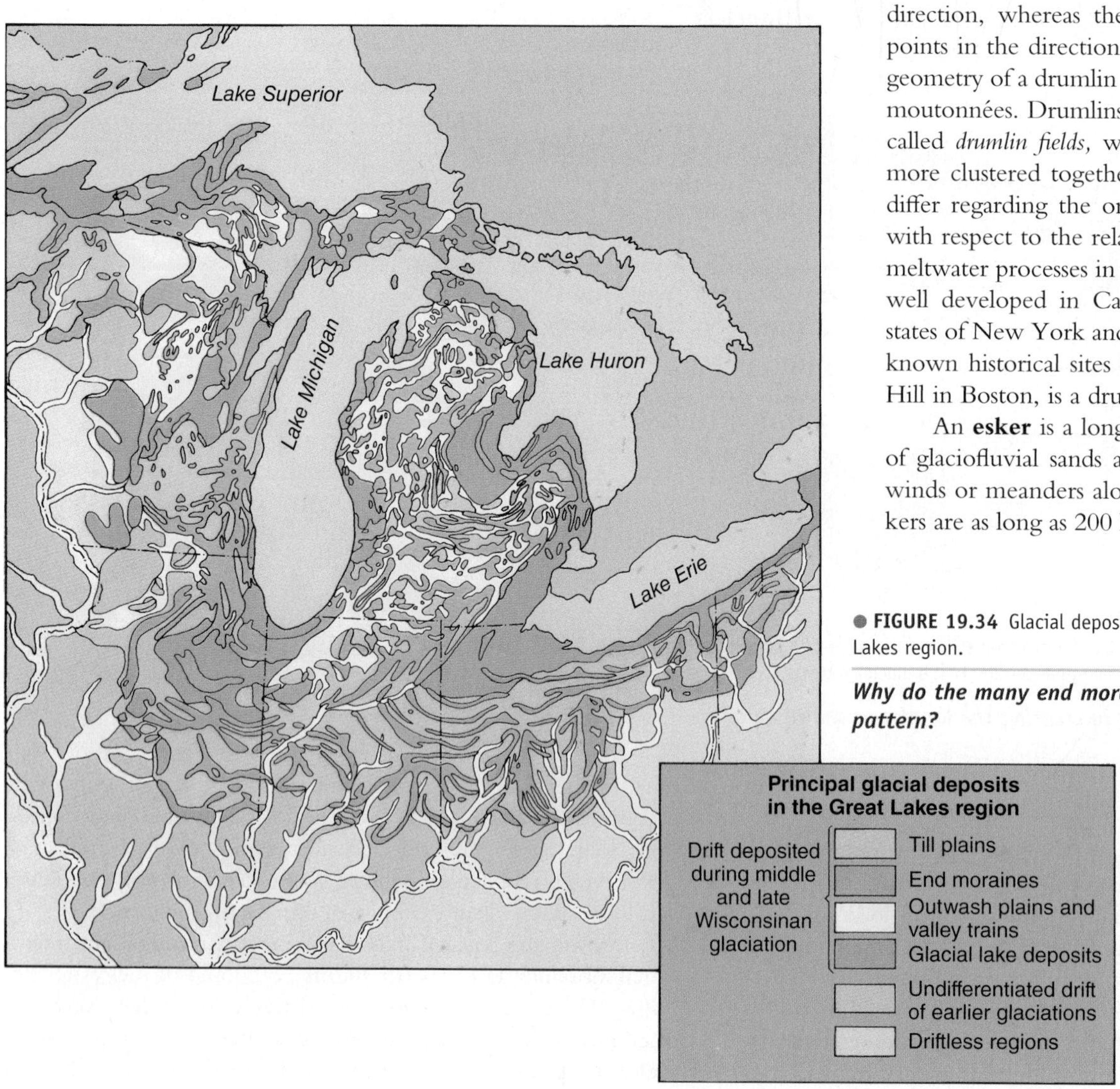

hundreds of square kilometers, are analogous to the valley trains of alpine glaciers.

Small depressions or pits, called **kettles,** mark some outwash plains, till plains, and moraines. Kettles represent places where blocks of ice were originally buried in glacial deposits. When the blocks of ice eventually melt, they leave surface depressions, and many kettles now contain *kettle lakes* ( ● Fig. 19.35). For example, most of Minnesota's famous 10,000 lakes are kettle lakes. Some kettles occur in association with alpine glacial deposits, but the vast majority are found in landscapes that were occupied by ice sheets or ice caps.

**Ridges and Hills** A **drumlin** is a streamlined ridge or hill, typically about 0.5 kilometer (0.3 mi) in length and less than 50 meters (160 ft) high, molded in glacial drift on till plains ( ● Fig. 19.36). The most conspicuous feature of drumlins is the elongated, streamlined shape that resembles the convex side of a teaspoon. The broad, steep end faces in the up-ice direction, whereas the gently sloping tapered end points in the direction that the ice flowed; thus the geometry of a drumlin is the reverse of that of roches moutonnées. Drumlins are usually found in swarms, called *drumlin fields,* with as many as a hundred or more clustered together. Opinions among scientists differ regarding the origin of drumlins, particularly with respect to the relative importance of ice versus meltwater processes in their formation. Drumlins are well developed in Canada, in Ireland, and in the states of New York and Wisconsin. One of the best-known historical sites in the United States, Bunker Hill in Boston, is a drumlin.

An **esker** is a long and narrow ridge composed of glaciofluvial sands and gravels that in map view winds or meanders along ( ● Fig. 19.37). Some eskers are as long as 200 kilometers (125 mi), although

● **FIGURE 19.34** Glacial deposits are widespread in the Great Lakes region.

*Why do the many end moraines have such a curved pattern?*

Lake Superior

Lake Michigan

Lake Huron

Lake Erie

**Principal glacial deposits in the Great Lakes region**

Drift deposited during middle and late Wisconsinan glaciation

Till plains

End moraines

Outwash plains and valley trains

Glacial lake deposits

Undifferentiated drift of earlier glaciations

Driftless regions

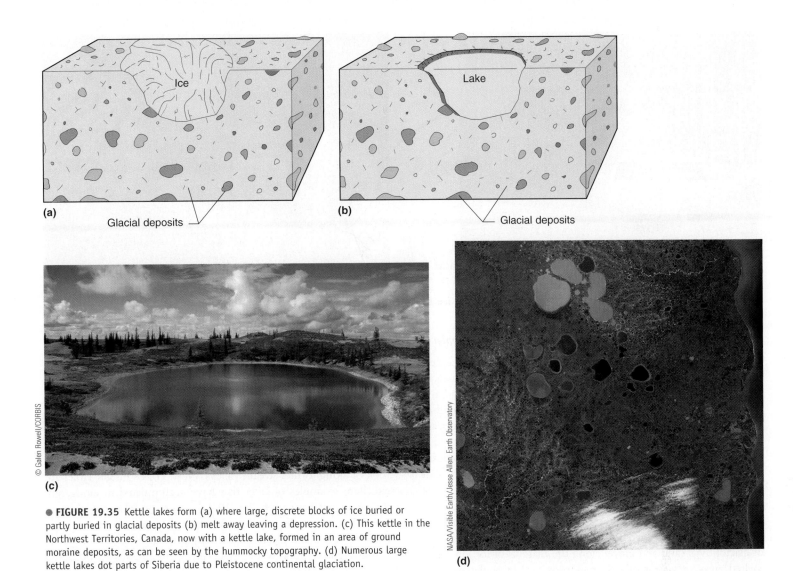

(a)

Ice

(b)

Lake

Glacial deposits

Glacial deposits

© Galen Rowell/CORBIS

(c)

NASA/Visible Earth/Jesse Allen, Earth Observatory

(d)

● **FIGURE 19.35** Kettle lakes form (a) where large, discrete blocks of ice buried or partly buried in glacial deposits (b) melt away leaving a depression. (c) This kettle in the Northwest Territories, Canada, now with a kettle lake, formed in an area of ground moraine deposits, as can be seen by the hummocky topography. (d) Numerous large kettle lakes dot parts of Siberia due to Pleistocene continental glaciation.

● **FIGURE 19.36** Drumlins, such as this one in Montana, are streamlined hills elongated in the direction of ice flow.

J. Petersen

● **FIGURE 19.37** An esker near Albert Lea, Minnesota.

*What economic importance do eskers have?*

© Henry Kyllingstad/Photo Researchers

● **FIGURE 19.38** This huge boulder in Yellowstone National Park is a glacial erratic, transported far from its point of origin by a continental glacier.

*What does this erratic illustrate about the ability of flowing ice to modify the terrain?*

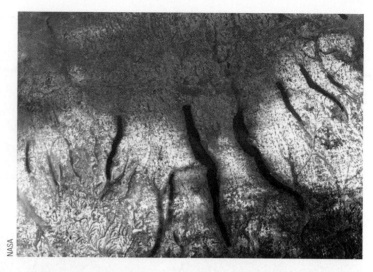

● **FIGURE 19.39** A satellite image of the Finger Lakes of New York. The lakes occupy linear, glacially eroded basins excavated during the Pleistocene.

*What characteristics of the bedrock caused ice to form these narrow lake basins?*

several kilometers is more typical of esker length. Most eskers probably formed by meltwater streams flowing in ice tunnels at the base of ice sheets. Many eskers are mined for their sand and gravel. Being natural embankments, they are frequently used in marshy, glaciated landscapes as places to build roads. Eskers are especially well developed in Sweden, Finland, and Russia.

Roughly conical hills composed of sorted glaciofluvial deposits are known as **kames.** Kames may develop from sediments that accumulated in glacial ice pits, in crevasses, and among jumbles of detached ice blocks. Like eskers, kames are excellent sources for mining sand and gravel and they are especially common in New England. *Kame terraces* are landforms resulting from accumulations of glaciofluvial sand and gravel along the margins of ice lobes that melted away in valleys of hilly regions. Good examples of kame terraces can be seen in New England and New York.

Large boulders scattered in and on the surface of glacial deposits or on glacially scoured bedrock are called **erratics** if the rock they consist of differs from the local bedrock ( ● Fig. 19.38). Moving ice is capable of transporting large rocks very far from their source. The source regions of erratics can be identified by rock type, and this provides evidence of the direction of ice flow. Erratics are known to be of glacial origin because they are marked by glacial striations and are found only in glaciated terrain. Erratics can occur in association with alpine glaciers, but they are best known and more impressive when deposited by ice sheets, which have moved boulders weighing hundreds of tons over hundreds of kilometers. In Illinois, for example, glacially deposited erratics have come from source regions as far away as Canada.

## Glacial Lakes

Perhaps you have noticed how frequently lakes have been mentioned in this chapter. Lakes are commonly found in several types of glaciated terrain. Many, many thousands of glacially created

lakes exist in depressions within deposits of the continental glaciers that once covered much of North America and Eurasia. The Pleistocene ice sheets created numerous other lake basins by erosion, scooping out deep elongated basins along zones of weak rock or along former stream valleys. New York's Finger Lakes are excellent examples of lakes that have accumulated in elongated, ice-deepened basins ( ● Fig. 19.39). Lakes are also common in areas that were affected by erosion and deposition from alpine glaciers. Once the glaciers have disappeared, cirques and glacial troughs tend to pond water, especially if those basins are bordered by an end moraine, as in the case of Lakes Maggiore, Como, and Garda in the Italian Alps ( ● Fig. 19.40). Evidence for many lakes that no longer exist is found in the widespread occurrence of

● **FIGURE 19.40** After withdrawal of the ice, end moraines often contribute to the formation of closed depressions in which glacial lakes accumulate.

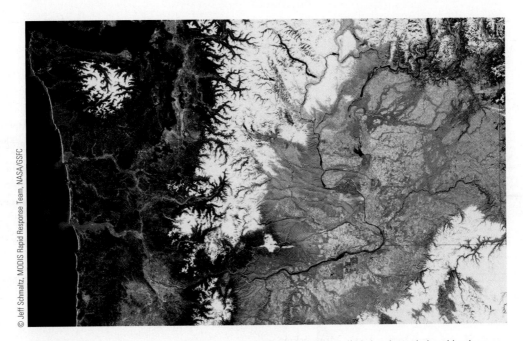

● **FIGURE 19.41** Much of eastern Washington State *(tan colored)* is called the channeled scablands because of the gigantic, largely abandoned river channels that cross the region *(dark tan)*. These channels represent huge outpourings of water that occurred when glacial ice damming large lakes in the Pleistocene failed, releasing an amount of water perhaps ten times the flow of all rivers of the world today.

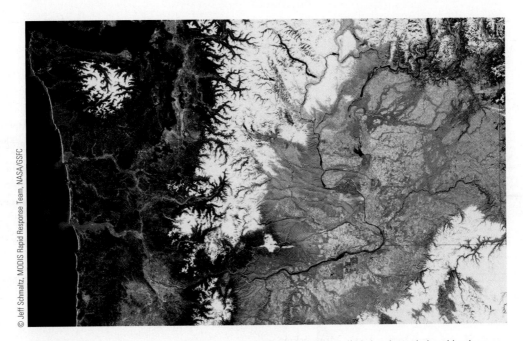

*glaciolacustrine* (from *glacial*, ice; *lacustrine*, lake) deposits that allow us to reconstruct the extent and determine the age of those lakes.

Some lakes formed along the front of the Pleistocene ice sheets in areas where glacial deposits disrupted the surface drainage or where a glacier prevented depressions from being drained of meltwater. These lakes usually accumulated where water became trapped between a large end moraine and the ice front or where the land sloped toward, instead of away from, the ice front. In both situations, *ice-marginal lakes* filled with meltwater. They drained and ceased to exist when the retreat of the ice front uncovered an outlet route for the water body.

During their existence, fine-grained sediment accumulated on the floors of these ice-marginal lakes, filling in topographic irregularities. As a result of this sedimentation, extremely flat surfaces characterize glacial plains where they consist of glaciolacustrine deposits. An outstanding example of such a plain is the valley of the Red River in North Dakota, Minnesota, and Manitoba. This plain, one of the flattest landscapes in the world, is of great agricultural significance. The plain was created by deposition in a vast Pleistocene lake held between the front of the receding continental ice sheet on the north and moraine dams and higher topography to the south. This ancient body of water is named Lake Agassiz for the Swiss scientist who early on championed the theory of an ice age. The Red River flows northward eventually into the last remnant of Lake Agassiz, Lake Winnipeg, which occupies the deepest part of an ice-scoured and sediment-filled lowland.

Another ice-marginal lake in North America produced much more spectacular landscape features but not in the area of the lake itself. In northern Idaho, a glacial lobe moving southward from Canada blocked the valley of a major tributary of the Columbia River, creating an enormous ice-dammed lake known as Lake Missoula. This lake covered almost 7800 square kilometers (3000 sq mi) and was 610 meters (2000 ft) deep at the ice dam. On occasions when the ice dam failed, Lake Missoula emptied in tremendous floods that engulfed much of eastern Washington. The racing floodwaters scoured the basaltic terrain, producing Washington's channeled scablands consisting of intertwining steep-sided troughs (coulees), dry waterfalls, scoured-out basins, and other features quite unlike those associated with normal stream erosion, particularly because of their gigantic size ( ● Fig. 19.41).

The Great Lakes of the eastern United States and Canada make up the world's largest lake system. Lakes Superior, Michigan, Huron, Erie, and Ontario occupy former river valleys that were vastly enlarged and deepened by glacial erosion. All the lake basins except that of Lake Erie have been gouged out to depths below sea level and have irregular bedrock floors lying beneath thick blankets of glacial till. The history of the Great Lakes is exceedingly complex, resulting from the back-and-forth movement of the ice front that produced many changes of lake levels and overflow in varying directions at different times ( ● Fig. 19.42).

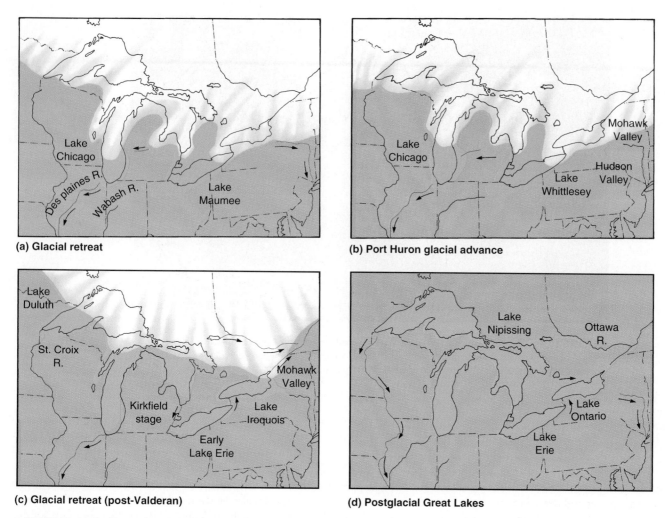

**(a) Glacial retreat**

**(b) Port Huron glacial advance**

**(c) Glacial retreat (post-Valderan)**

**(d) Postglacial Great Lakes**

● **FIGURE 19.42** The Great Lakes of North America formed as the ice sheet receded at the close of the Pleistocene Epoch.

# Periglacial Landscapes

Not all cold regions have sufficient precipitation to lead to permanent accumulation of thick masses of ice. In much of Siberia and interior Alaska, it was too cold and dry during the Pleistocene to generate the massive snow and ice accumulations that occurred elsewhere in North America and northwestern Europe. Instead of glacial processes, these **periglacial** environments (*peri*, near) lacking year-round snow or ice cover undergo intense frost action and are frequently regions of *permafrost* (permanently frozen ground). Large areas of periglacial terrain exist today in Alaska, Canada, Russia, and some areas of high elevation including in mountain and plateau regions of China. During the Pleistocene ice advances, periglacial environments migrated to lower latitudes and lower elevations, leaving relict periglacial features in places, including parts of the Appalachians, where it is no longer actively forming today.

The intense frost action of periglacial landscapes includes freezing of soil moisture and produces angular, shattered rocks. Frost action also causes heaving, thrusting, and size-sorting of stones in the soil that lead to the formation of fascinating repeating designs in *patterned ground* features ( ● Fig. 19.43).

● **FIGURE 19.43** Intensive frost action in periglacial regions leads to the formation of intricate, repetitive patterned ground features.

© Emma Pike

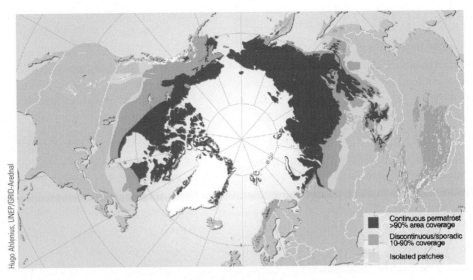

Hugo Ahlenius, UNEP/GRID-Arednal

Continuous permafrost >90% area coverage

Discontinuous/sporadic 10-90% coverage

Isolated patches

● **FIGURE 19.44** Many periglacial regions contain permafrost. This map shows the distribution of permafrost in the Northern Hemisphere.

Where mean annual temperatures are cold enough, ice-free landscapes develop permafrost ( ● Fig. 19.44). As we saw in Chapter 15, extensive areas of permafrost result in considerable solifluction (slowly flowing soil) on slopes when the upper, active layer of the permanently frozen ground thaws in the summer and becomes saturated. Permafrost areas are also prone to the formation of fissures that accumulate ice, leading to the formation of large *ice wedges* in the ground.

Understanding permafrost is important for human activity in periglacial regions. Unless proper construction techniques are used, erecting buildings, roads, pipelines, and other structures on permafrost disrupts the natural thermal environment, often leading to permafrost melting. Saturated ground cannot support the weight of structures resting on top of them. As the ground deforms and slowly flows, the structures are destroyed. To avoid these problems, construction techniques have been developed that keep the permafrost frozen or allow it to experience its natural temperature fluctuations. The latter is generally accomplished by building above the surface so that air can circulate to the permafrost.

# CHAPTER 19 ACTIVITIES

## ■ DEFINE AND RECALL

glacier
glaciation
firn
accumulation
ablation
calving
alpine glacier
valley glacier
piedmont glacier
cirque glacier
cirque
continental glacier
ice sheet
ice cap
crevasse
plucking
till
zone of accumulation

zone of ablation
equilibrium line
glacier advance
glacier retreat
striation
roche moutonnée
tarn
arête
horn
col
glacial trough
paternoster lake
hanging valley
fjord
drift
lateral moraine
medial moraine

end moraine
ground moraine
glacial outwash
valley train
outlet glacier
ice shelf
glacial (glacial period)
interglacial
continental shelf
till plain
outwash plain
kettle
drumlin
esker
kame
erratic
periglacial

# DISCUSS AND REVIEW

1. How does glacial ice differ from snow?
2. What are the three main types of alpine glacier, and what distinguishes them from each other?
3. Explain how glaciers move.
4. Diagram and label the characteristic parts of an alpine glacier.
5. Define accumulation and ablation, and explain how they are related to glacial advance and retreat.
6. How do glaciers accumulate load? Provide some examples of evidence of glacial erosion and movement.
7. What are some major similarities and some major differences between continental and alpine glaciers?

8. Distinguish till from outwash and drift.
9. Where are the two major existing ice sheets located? How do they compare in area and thickness?
10. How does the present extent of continental ice-sheet coverage compare with the maximum extent of the Pleistocene ice-sheet coverage?
11. How has ice-sheet erosion altered the landscape? What kinds of landscape features are produced by continental glacier deposition and recession?
12. What landscape features would indicate that a region is undergoing periglacial processes?

# CONSIDER AND RESPOND

1. How do alpine glaciers differ from streams in terms of flow processes, erosion processes, load characteristics, valley shaping, and nature of tributary valleys?
2. Name four coastal areas of the world where fjords can be found. What is the dominant climate type in coastal areas with fjords? Why are fjords associated with that climate? In what way are fjords related to both alpine and continental glaciers?

3. Describe the ice budget of a glacial system. What two major factors control this budget? How is the ice budget related to the movement of glaciers? Explain how a glacier that is calving can also be undergoing advance.
4. Glaciation is truly an interdisciplinary topic. In addition to physical geographers, what other scientists do you think are involved in the study of glaciers and why?

# APPLY AND LEARN

1. These tables show the position (elevation) of the toe of a Northern Hemisphere valley glacier as measured yearly on September 30 during the 1970s and then again during the 1990s. Use decadal averages and annual rates of change of position to help you describe how the behavior of the glacier in the 1990s differed from that of the 1970s.

| Year | 1970 | 1971 | 1972 | 1973 | 1974 | 1975 | 1976 | 1977 | 1978 | 1979 |
|---|---|---|---|---|---|---|---|---|---|---|
| Elevation (meters) | 2634 | 2631 | 2632 | 2629 | 2630 | 2629 | 2627 | 2625 | 2624 | 2623 |

| Year | 1990 | 1991 | 1992 | 1993 | 1994 | 1995 | 1996 | 1997 | 1998 | 1999 |
|---|---|---|---|---|---|---|---|---|---|---|
| Elevation (meters) | 2642 | 2648 | 2654 | 2661 | 2668 | 2676 | 2684 | 2692 | 2701 | 2710 |

2. Diamonds were recently found at three separate sites in Canada in till deposited by the Pleistocene ice sheet. At Site A, nearby ice-scoured troughs now occupied by lakes are elongated from NNE to SSW. Site B lies 60 kilometers west of Site A, and striations in adjacent bedrock show a NE to SW orientation. Lying 60 kilometers east of Site A, Site C has nearby drumlins with their tail end pointed SSE and their blunt end NNW. Using this information, how might you proceed in narrowing down a search area for the bedrock from which the diamonds originated?

# LOCATE AND EXPLORE

1. Using Google Earth, identify the landforms at the following locations (latitude, longitude). Briefly explain how each landform developed and why it is found in that general area.
   a. 59.75°N, 140.59°W
   b. 42.73°N, 76.73°W
   c. 54.06°N, 98.71°W
   d. 61.34°N, 25.47°E
   e. 44.52°S, 168.84°E
   f. 42.35°N, 71.92°W
   g. 58.44°N, 107.58°W

2. Using Google Earth and the Topographic Map Layer for Gilkey Glacier from the USGS (downloadable from the book's CourseMate at www.cengagebrain.com), look at the change in the glacier (58.76°N, 134.57°W) between 1948 and today and answer the following questions:
   a. Use the measuring tool to determine how far the glacier has retreated.
   b. What is the average annual rate of retreat between 1948 and today?
   c. Battle Glacier was formerly joined to Gilkey Glacier, but it has also retreated. What river now occupies the valley formerly filled by Battle and Thiel Glaciers? Why is this type of river commonly associated with glaciers?
   d. What glacial landform is found along the edge of the valley of Battle Glacier (58.743911°N, 134.582156°W)?

 CourseMate – Make the most of your study time by accessing everything you need to succeed in one place. Read your textbook, take notes, review flashcards, watch videos, complete activities, take practice quizzes, and more online with CourseMate. Log in at **www.cengagebrain.com**.

# Map Interpretation

## ALPINE GLACIATION

### The Map

The Chief Mountain, Montana, map area is located in Glacier National Park, which adjoins Canada's Waterton Lakes National Park located across the international border. Alpine glaciation has created much of the spectacular scenery of this region of the northern Rocky Mountains.

Most of the glaciated landscape in this region was produced during the Pleistocene. Today glaciers still exist in Glacier National Park, but because of rapid melting in recent decades, it is estimated that they may be gone within 70 years. Prior to glaciation, this region was severely faulted and folded during the formation of the northern Rocky Mountains.

The photograph and map clearly show the rugged nature of the terrain of this map area. Steep slopes, horns, U-shaped valleys, lakes, arêtes, and glaciers are obvious landform characteristics of alpine glaciation. Temperature is a primary control of the highland climate of this region. Elevation, in turn, influences temperature and precipitation amounts.

As you would expect, the rapid decrease in temperature with increasing elevation results in a variety of microclimates within alpine regions. Exposure is also an important highland climate control. West-facing slopes receive the warm afternoon sun, whereas east-facing slopes are sunlit only in the cool of the early morning.

### Interpreting the Map

1. What is the approximate amount of local relief depicted on this map? Is the scale of this topographic map larger or smaller than that of the topographic maps in other map exercises in the book?
2. From your examination of the topographic map and photograph of this mountain region, does the landscape seem dominated by glacial erosion or deposition?
3. Locate Grinnell, Swiftcurrent, and Sperry Glaciers. What specific type of glaciers are they?
4. What evidence indicates that the glaciers were once larger and extended farther down the valleys?
5. Note that most of the existing glaciers are located to the northeast of the mountain summits. Explain this orientation. At what elevation are the glaciers found?

6. What types of glacial landform are the following features?
   a. The features occupied by Kennedy, Iceberg, and Ipasha Lakes
   b. The feature occupied by McDonald Lake (in the southwest corner of the map) and Lake Josephine
   c. Mount Gould and Mount Wilbur
7. Along the high ridges runs a dashed line labeled Continental Divide. What is its significance?
8. If you were to hike southeast from Auto Camp on McDonald Creek up Avalanche Creek to the base of Sperry Glacier, how far would you travel and how much elevation change would you encounter?

D. Sack

Glaciated terrain in Glacier National Park.

⊕ In Google Earth, fly to: 48.722°N, 113.737°W.

Opposite:
Chief Mountain, Montana
Scale 1:125,000
Contour interval = 100 ft
USGS

# Map Interpretation

## ■ CONTINENTAL GLACIATION

### The Map

The Jackson, Michigan, map area is located in the Great Lakes section of the Central Lowlands. Massive continental ice sheets covered this region during the Pleistocene. As the glaciers melted and the ice sheets retreated, a totally new terrain, vastly different from that of preglacial times, was exposed. Today evidence of glaciation is found throughout the region. The most obvious glacially produced landforms are the thousands of lakes (including the Great Lakes), the knobby terrain, and moraine ridges. Moraines left by advancing and retreating tongue-shaped ice lobes that extended generally southward from the main continental ice sheet also influenced the shapes of the Great Lakes.

The ice sheet, its deposits of sediment, and its meltwaters also created other, smaller-scale features. The glaciers left a jumbled mosaic of deposits from boulders through smaller gravel to sand, silt, and clay that has produced a hilly and hummocky terrain. The overall relief is low, in part because of glacial erosion. Many landform features of continental glaciation are well illustrated on the Jackson, Michigan, 7.5' quadrangle map.

This region has a humid continental climate. The summers are mild and pleasant. Excessively warm and humid air seldom reaches this area for more than a few days at a time. Instead, cool but pleasant evening temperatures tend to be the rule in summer. Winters, however, are long and often harsh. Snow can be abundant and on the ground continuously for many weeks or even months at a time. The annual temperature range is quite large; precipitation occurs year-round provided primarily by midlatitude cyclonic storms.

### Interpreting the Map

1. Describe the general topography of the map area.
2. What is the local relief? Why is it so difficult to find the exact highest and lowest points on this map?
3. Does the topography of this region indicate glacial erosion or glacial deposition?
4. Does the area appear to be well drained? What are the three main hydrographic features that indicate the drainage conditions?
5. What type of glacial landform is Blue Ridge? What are its dimensions (length and average height)?

6. How is a feature such as Blue Ridge formed? What economic value might it have?
7. What is the majority of the surface material in the map area? What is the term for this type of surface cover?
8. What probably caused the many small depressions and rounded lakes? What are these depressions called?
9. Describe how the topography of the Jackson, Michigan, area appears on the aerial image.
10. What is the advantage of having both aerial image and topographic map coverage of an area of study?

Aerial image of the region near Jackson, Michigan

⊕ In Google Earth, fly to: 42.078°N, 84.422°W.

Opposite:
Jackson, Michigan
Scale 1:62,500
Contour interval = 10 ft
USGS

# 20

# COASTAL PROCESSES AND LANDFORMS

## ■ OBJECTIVES

WHEN YOU COMPLETE THIS CHAPTER YOU SHOULD BE ABLE TO:

- Sketch the coastal zone and label its principal subdivisions.
- Explain how waves are formed and characterized.
- Discuss how wind waves transfer and deliver energy to the coastal zone.
- Explain why coastlines tend to straighten over time.
- Recount how sediment moves along a beach.
- Identify coastal erosion processes and the primary landforms found on erosion-dominated coastlines.
- Describe the types and sources of sediment found in coastal depositional landforms.
- Distinguish among the principal coastal depositional landforms.
- Summarize a global-scale and a regional-scale classification of coasts.
- Describe the different types of coral reef.

◁ The shorelines of the world are extensive, complex, and often spectacular environments. USGS NBII/Randolph Femmer

THE WORLD OCEAN COVERS 71% of Earth's surface, and the boundary between the ocean and dry land is of enormous length. A large percentage of the world's population lives near the coast, with cities and towns established there for reasons of transportation, resources, industry, commerce, defense, or tourism and spectacular scenery. Earth's coastlines have tremendous resources and are biologically and geomorphically diverse. They draw more tourists than any other natural environment and continue to attract new residents.

Coastal zones are popular, but they are also subject to an array of natural hazards and human-induced environmental problems. Coastal communities must cope with powerful storms, the influence of tides, waves, currents, and moving sediment. Low-lying coasts are subject to flooding, storm surges, and, in some places, tsunamis. Coasts of high relief are susceptible to rockfalls, landslides, and other forms of mass wasting. Environmental problems stem from rapid urban development, high population densities, and economic and industrial activities ranging from tourism to port operations, offshore oil production, and agricultural runoff. Some of our most polluted waters are found in coastal locations. If global warming reduces the extent of the continental ice sheets, the ensuing rise in sea level will have a profound impact on the human-built infrastructure in coastal regions and on coastal geography and geomorphology. Understanding the natural processes that operate in the coastal zone is fundamental to solving the present and future problems in this dynamic part of Earth's landscape.

## The Coastal Zone

Most of the processes and landforms of the marine coastal zone are also found along the coastlines of large lakes. All are considered *standing bodies of water* because the water in each occupies a basin and has an approximately uniform still-water level around the basin. This contrasts with the sloping, channelized flow toward lower elevations that constitutes streams.

The **shoreline** of a standing body of water is the exact and constantly changing contact between the ocean or lake surface and dry land. The position of this boundary fluctuates with incoming waves, with storms, and, in the case of the ocean, with the tides. Over the long term, the position of the shoreline is also affected by tectonic movements and by the amount of water held in the ocean or lake basin. **Sea level** is a complexly determined average position of the ocean shoreline and the vertical position (the reference, or *datum)* above and below which other elevations are measured. The *coastal zone* consists of the general region of interaction between the land and the ocean or lake. It ranges from the inland limit of coastal influence through the present shoreline to the lowest submerged elevation to which the shoreline fluctuates.

As waves approach the mainland from an open body of water, they eventually become unstable and break, sending a rush of water toward land. The *nearshore zone* extends from the seaward or lakeward edge of breakers to the landward limit reached by the broken wave water ( ● Fig. 20.1). The nearshore zone contains the *breaker zone* where waves break, the *surf zone* through which a bore of broken wave water moves, and, most landward, the *swash zone* over which a thin sheet of water rushes up to the inland limit of water and then back toward the surf zone. This thin sheet of water rushing toward the land is known

● **FIGURE 20.1** Principal divisions of the coastal zone.
From Komar, P. D., *Beach Processes and Sedimentation*, © 1979, p. 13. Reprinted by permission of Pearson Education, Inc., Upper Saddle River, NJ.

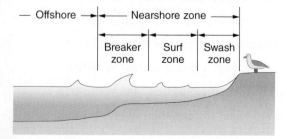

as **swash,** and the return flow is **backwash.** The *offshore zone* accounts for the remainder of the standing body of water, that part lying seaward or lakeward of the outer edge of the breaker zone.

## Origin and Nature of Waves

Waves are traveling, repeating forms that consist of alternating highs and lows, called *wave crests* and *wave troughs,* respectively ( ● Fig. 20.2). The vertical distance between a trough and the adjacent crest is *wave height. Wavelength* is the horizontal distance between successive wave crests. Other important attributes are **wave steepness,** or the ratio of wave height to wavelength, and **wave period,** the time it takes for one wavelength to pass a fixed point.

Waves that have traveled across the surface of a water body are the principal geomorphic agent responsible for coastal landforms. Like streams, glaciers, and the wind, waves erode, transport, and deposit Earth materials, continually reworking the narrow strip of coastal land with which they come in contact. Most of the waves that affect the coastal zone originate in one of three ways. The tides consist of two very long wavelength waves caused by interactions between Earth and the moon and sun. **Tsunamis** result from the sudden displacement of water by movement along faults, landslides, volcanic eruptions, or other impulsive events. Most of the waves that have an impact on the coastal zone, however, are **wind waves,** created when air currents push along the water surface.

### Tides

The two long-wavelength waves that comprise the tides always exist on Earth. There are two wave crests (high tides), each followed by a trough (low tide). As a crest then a trough move slowly through an area of the ocean, they cause a gradual rising and subsequent falling of the ocean surface. Along the marine coastline, the change in water level caused by the tides brings the influence of coastal processes to a range of elevations. Tides are so small on lakes that they have virtually no effect on coastal processes, even in large lakes.

The gravitational pull of the moon, and to a lesser extent the sun, and the force produced by motion of the combined Earth–moon system are the major causes of the tides ( ● Fig. 20.3). The moon is much smaller than the sun, but because it is significantly closer to Earth, its gravitational influence on Earth exceeds that of the sun. The moon completes one revolution around Earth every 29.5 days, but it does not revolve around the center of Earth. Instead, the moon and Earth are a combined system that as a unit moves around the system's center of gravity. Because Earth's mass is much larger than the mass of the moon, the combined system's center of gravity occupies a point within Earth on the side that is facing the moon. Being closest to the moon and more easily deformed than land, ocean at Earth's surface above the center of gravity is pulled toward the moon, making the first tidal bulge (high tide). At the same time, ocean water on the opposite side of Earth experiences the outward–flying, or *centrifugal,* force of inertia and forms the other tidal bulge. Troughs (low

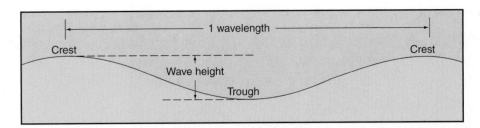

tides) occupy the sides of Earth midway between the two tidal bulges. As Earth rotates on its axis each day, these bulges and troughs sweep across Earth's surface.

The sun has a secondary tidal influence on Earth's ocean waters, but because it is so much farther away, its tidal effect is less than half that of the moon. When the sun, moon, and Earth are aligned, as they are during new and full moons, the added influence of the sun on ocean waters causes higher than average high tides and lower than average low tides. The difference in sea level between high tide and low tide is the *tidal range*. The increased tidal range caused by the alignment of Earth, the moon, and the sun, known as **spring tide,** occurs every 2 weeks. A week after a spring tide, when the moon has revolved a quarter of the way around Earth, its gravitational pull on Earth is exerted at a 90° angle to that of the sun. In this position, the forces of the sun and moon detract from one another. At the time of the first-quarter and last-quarter moon, the counteracting force of the sun's gravitational pull diminishes the moon's attraction. Consequently, the high tides are not as high and the low tides are not as low at those times. This moderated situation, which like spring tides occurs every 2 weeks, is **neap tide** ( ● Fig. 20.4).

The moon completes its 360° orbit around Earth in a month, traveling about 12° per day in the same direction that Earth rotates daily around its axis. Thus, by the time Earth completes one rotation in 24 hours, the moon has moved on 12° in its orbit around Earth (see Fig. 20.3). To return to a given position with respect to the moon takes the Earth an additional 50 minutes. As a result, the moon rises 50 minutes later every day at any given spot on Earth, the tidal day is 24 hours and 50 minutes, and two successive high tides are ideally 12 hours and 25 minutes apart.

The most common tidal pattern approaches the ideal of two high tides and two lows in a tidal day. This *semidiurnal* tidal regime is characteristic along the Atlantic coast of the United States, for example, but it does not occur everywhere, as seen in ● Figure 20.5. In a few seas that have restricted access to the open ocean, such as the Gulf of Mexico, tidal patterns of only one high and one low tide occur during a tidal day. This type of tide, called *diurnal,* is not very common. A third type of tidal pattern consists of two high tides of unequal height and two low tides, one lower than the other. The waters of the Pacific coast of the United States exhibit this *mixed tide* pattern.

● **FIGURE 20.3** Forces responsible for Earth's two tidal bulges and two troughs. A "tidal day" is 24 hours and 50 minutes long because the moon continues in its orbit around Earth while Earth is rotating.

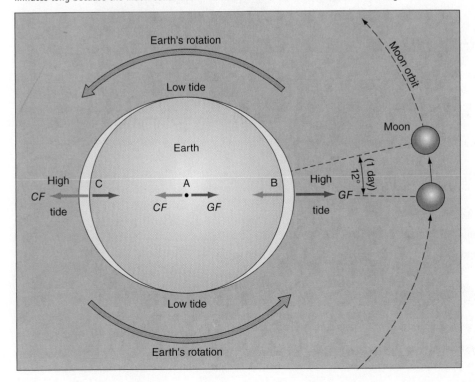

A. Gravitational force (*GF*) and centrifugal force (*CF*) are equal. Thus separation between Earth and the moon remains constant.

B. Gravitational force exceeds centrifugal force, causing ocean water to be pulled toward the moon.

C. Centrifugal force exceeds gravitational force, causing ocean water to be forced outward away from the moon.

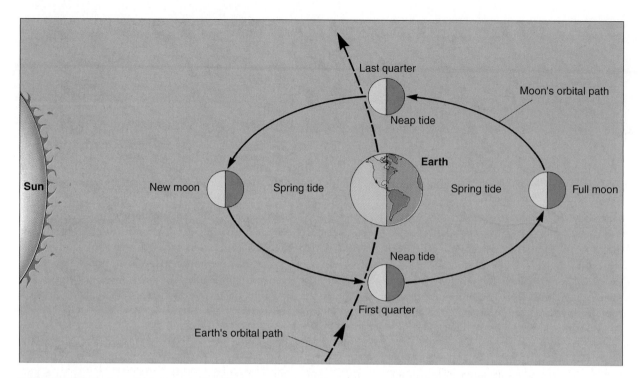

● **FIGURE 20.4** Maximum tidal ranges (spring tides) occur at new and full moon, when the moon and sun are aligned either on the same or opposite sides of Earth, respectively. Minimum tidal ranges (neap tides) occur when the moon and sun act at right angles to each other.

*How many spring tides and neap tides occur each month?*

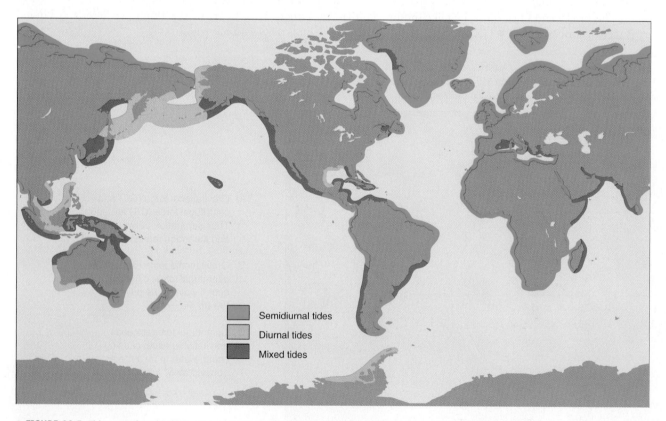

● **FIGURE 20.5** This map of world tidal patterns shows the geographic distribution of diurnal, semidiurnal, and mixed tides.

*What is the tidal pattern on the coastal area nearest where you live?*

● **FIGURE 20.6** The extreme tidal range of the Bay of Fundy in Nova Scotia, Canada, can be seen in the difference between (a) high tide, and (b) low tide at the same point along the coast.

*Why does the Bay of Fundy have such a great tidal range?*

Tidal range varies from place to place in response to the shape of the coastline, water depth, access to the open ocean, submarine topography, and other factors. The tidal range along open-ocean coastlines, like the Pacific coast of the United States, averages between 2 and 5 meters (6–15 ft). In restricted or partially enclosed seas, like the Baltic or Mediterranean Sea, the tidal range is usually 0.7 meters (2 ft) or less. Funnel-shaped bays off major oceans, especially the Bay of Fundy on Canada's east coast, produce extremely high tidal ranges. The Bay of Fundy is famous for its enormous tidal range, which averages 15 meters (50 ft) and may have reached a maximum of 21 meters (70 ft) ( ● Fig. 20.6). Other narrow, elongated coastal inlets that exhibit great tidal ranges are Cook Inlet in Alaska, Washington's Puget Sound, and the Gulf of California in Mexico.

## Tsunamis

Tsunamis are long-wavelength waves that form when a large mass of water displaced upward or downward by an earthquake, volcanic eruption, landslide, or other sudden event works to regain its equilibrium condition. Resulting oscillations of the water surface travel outward from the origin as one wave or a series of waves. In deep water, the displacement may cause wave heights of a meter or more that can travel at speeds of up to 725 kilometers (450 mi) per hour yet pass beneath a ship unnoticed. As the long-wavelength waves approach the shallow water of a coastline, their height can grow substantially. As these very large and extremely dangerous waves surge into low-lying land areas, they acquire huge amounts of debris and can cause tremendous damage, injury, and loss of life in addition to erosion, transportation, and deposition of Earth materials.

In 1946 an earthquake in Alaska caused a tsunami that reached Hilo, Hawaii, where it attained a maximum height greater than 10 meters (33 ft) and killed more than 150 people. When the Krakatoa volcano erupted in 1883, it generated a powerful 40-meter-high (130 ft) tsunami that killed more than 37,000 people in the nearby Indonesian islands. In December 2004 the devastating earthquake-generated Indian Ocean tsunami struck the shorelines of Indonesia, Thailand, Myanmar, Sri Lanka, India, Somalia, and other countries, causing an almost unimaginable 230,000 fatalities from the tsunami alone. This tragic event reinforced the importance of tsunami early-warning systems. An early-warning system had been established for the Pacific Ocean, but none was operating in the Indian Ocean in 2004. The United Nations Educational, Scientific, and Cultural Organization (UNESCO) has since been helping member nations in the region develop a tsunami warning system. The 2011 tsunami resulting from the 9.0 magnitude Tohoku earthquake near the east coast of the Japanese island of Honshu achieved a height of almost 38 m (124 ft) and extended onto the island by as much as 10 km (6 mi) causing tremendous destruction ( ● Fig. 20.7). Although the dead

● **FIGURE 20.7** Example of the horrendous destruction caused by the 2011 tsunami on the east coast of Japan.

● **FIGURE 20.12** Embayments along this coastline have been building seaward by filling with sediment, whereas wave energy focused on the headlands has been eroding them landward.

*What happens to sediment eroded from the headlands?*

● **FIGURE 20.13** Incomplete wave refraction causes waves to break at an angle with respect to the orientation of the shoreline rather than parallel to the shoreline.

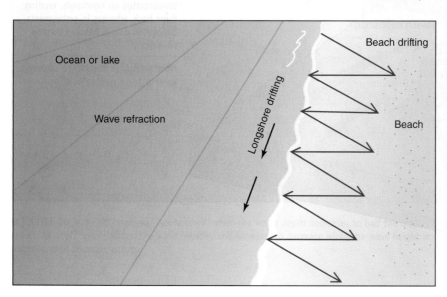

● **FIGURE 20.14** Incomplete wave refraction leads to sediment transport along the coast by beach drifting and longshore drifting.

*Why is the backwash perpendicular to the trend of the shoreline when the swash is at an angle to it?*

it is closest to shore ( ● Fig. 20.14). This velocity decrease spreads progressively along the crestline as more of the wave enters shallower water. With insufficient time for complete refraction before breaking begins, the crest lies at an angle to the beach, not parallel to it, when it breaks. As a result, the broken wave water, and sediment it has entrained, rushes up the beach face diagonally to the shoreline, rather than directly up its slope. Backwash, however, which also moves sediment, flows straight back down the beach face toward the water body by the force of gravity. In this way, as one incompletely refracted wave after another breaks, sediment zigzags along the beach in the swash zone. **Beach drifting** is this zigzag-like transportation of sediment in the swash zone caused by incomplete wave refraction. Over time, beach drifting transports tons of sediment along the shore.

Another outcome of an incompletely refracted oblique wave is that when the crest arrives at the break point at one location, farther along the beach in the direction the waves are traveling, that same position relative to the shoreline is occupied by a wave trough. This difference in water level initiates a current of water, called the *longshore current,* flowing parallel to the shoreline near the breaker zone. Considerable amounts of sediment suspended when incompletely refracted waves break are transported along the shore in this process of **longshore drifting.**

## Coastal Erosion

Because waves and streams both consist of liquid water, similarities exist in how these two geomorphic agents erode rock matter. Like water in streams, water that has accumulated in basins erodes some rock material chemically through *corrosion*. Corrosion is the removal of the ions that have been separated from rock-forming minerals by solution and other chemical weathering processes. Likewise, the power of *hydraulic action* from the sheer physical force of the water alone pounds against and removes coastal rock material, sometimes compressing air or water into cracks to help in the process. The power of storm waves, combined with the buoyancy of water, enables them at times to dislodge and move even large boulders. Once clastic particles are in motion, waves have solid tools to use to perform even more work through the grinding erosive process of *abrasion*. Abrasion is the most effective form of erosion by each of the geomorphic agents, including waves.

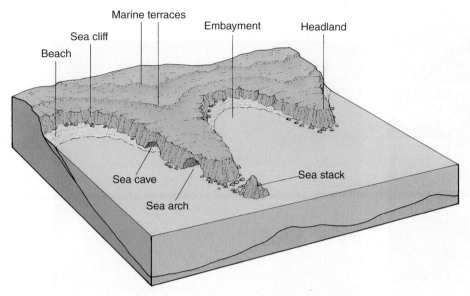

Beach  Sea cliff  Marine terraces  Embayment  Headland

Sea cave  Sea arch  Sea stack

● **FIGURE 20.15** Some of the major landforms found along erosion-dominated coastlines.

Weathering is an important factor in the breakdown of rocks in the coastal zone, as in other environments, preparing pieces for removal by wave erosion. Water is a key element in most weathering processes, and in addition to normal precipitation, rocks near the shoreline are subjected to spray from breaking waves, high relative humidities, and condensation. Salt weathering is particularly significant in preparing rocks for removal through chemical and physical weathering along the marine coast and coasts of salt lakes.

Coasts of high relief are dominated by erosion ( ● Fig. 20.15). *Sea cliffs (or lake cliffs)* are carved where waves pound directly against steep land. If a steep coastal slope continues deep beneath the water, it may reflect much of the incoming wave energy until corrosion and hydraulic action eventually take their toll on the rock. The tides present along marine coasts allow these processes to attack a range of shoreline elevations. Once a recess, or **notch,** has been carved out along the base of a cliff ( ● Fig. 20.16a), weathering and rockfall within the shaded overhang supply clasts that can collect on the notch floor and be used by the water as tools for more efficient erosion by abrasion. Abrasion extending the notch landward leaves the cliff above subject to rockfall and other forms of mass wasting. Stones used as tools in abrasion quickly become rounded and may accumulate at the base of the cliff as a **cobble beach.** Where the cliffs are well jointed but cohesive, wave erosion can create *sea caves* along the lines of weakness (Fig. 20.16b). *Sea arches* result where two caves meet from each side of a headland (Fig. 20.16c). When the top of an arch collapses or a sea cliff retreats and a resistant pillar is left standing, the remnant is called a **sea stack** (Fig. 20.16d).

Landward recession of a sea cliff leaves behind a wave-cut bench of rock, an **abrasion platform,** that is sometimes visible at lower water levels, such as at low tide ( ● Fig. 20.17a). Abrasion platforms record the amount of cliff recession. In some cases deposits accumulate as wave-built terraces just sea-

ward of an abrasion platform. If tectonic activity uplifts these wave-cut benches and wave-built terraces above sea level out of the reach of wave action, they become **marine terraces** (Fig. 20.17b). Successive periods of uplift can create a coastal topography of marine terraces that resembles a series of steps. Each step represents a period of time that a terrace was at sea level. The Palos Verdes peninsula just south of Los Angeles has perhaps as many as ten marine terraces, each representing a period of platform formation separated by episodes of uplift.

Rates of coastal erosion are controlled by the interaction between wave energy and rock type. Coastal erosion is greatly accelerated during high-energy events, such as severe storms and tsunamis. Human actions can also accelerate coastal erosion rates. We commonly do so by interfering with coastal sediment and vegetation systems that would naturally protect some coastal segments from excessive erosion rates. We explore the nature of coastal depositional systems next.

## Coastal Deposition

Significant amounts of sediment accumulate along coasts where wave energy is low relative to the amount or size of sediment supplied. Embayments and settings where waves break at a distance from the shoreline, such as areas with gently shelving underwater topography, tend to sap wave energy and encourage deposition. Amount and size of sediment supplied to the coastal zone vary with rock type, weathering rates, and other elements of the climatic, biological, and geomorphic environment.

Sediment within coastal deposits comes from three principal sources. Most of it is delivered to the standing body of water by streams. At its mouth, the load of a stream may be deposited for the long term in a delta or within an **estuary,** a biologically very productive embayment that forms at some river mouths where saltwater and freshwater meet. Elsewhere, stream load may instead be delivered to the ocean or lake for continued transportation. Once in the standing body of water, fine-grained sediments that stay in suspension for long periods may be carried out to deep water where they eventually settle out onto the basin floor. Other clasts are transported by waves and currents in the coastal zone, being deposited when energy decreases and, if accessible, re-entrained when wave energy increases. The same is true of the second major source of coastal sediment, coastal cliff erosion. Of less importance is sediment brought to the coast from offshore sources. Although we may tend to think of sand-sized sediment when we think of coastal deposits, coastal depositional landforms may be composed of silt, sand, or any size classes of gravel from granules and pebbles through cobbles and boulders.

The most common landform of coastal deposition is the **beach,** a wave-deposited feature that is contiguous with the

D. Sack

(a)

J. Petersen

(b)

NOAA Image Library

(c)

● **FIGURE 20.16** Examples of landforms created by wave erosion. (a) A notch exposed near the base of basaltic sea cliffs in Hawaii. (b) A sea cave carved in limestone along Italy's Mediterranean Sea coast. (c) A sea arch in Alaska. (d) Sea stacks along a cliffed coast.

Richard Price/Getty Images

(d)

● **FIGURE 20.17** (a) An abrasion platform carved on strongly dipping sedimentary rocks in central California. (b) A marine terrace, the wide, beveled plain atop this eroding coastline.

*What other coastal erosional landforms do you see in photo (b)?*

mainland throughout its length ( ● Fig. 20.18). Many beaches are sandy, but beaches of other grain sizes are also common— for example, the cobble beach discussed earlier in the section on coastal erosion. In settings with high wave energy, particles tend to be larger and beaches tend to be steeper than where wave energy is low. Beach sediments come in a variety of colors depending on the rock and mineral types represented. Tan quartz, black basalt, white coral, and even green olivine beaches exist on Earth.

Any given stretch of beach may be a permanent feature, but much of the visible sediment deposited in it is not. Individual grains come and go with swash and backwash; wear away through abrasion; are washed offshore in storms; or move into, along, and out of the stretch of beach by littoral drifting. Because waves generated by closer storms tend to be higher than waves generated in distant storms, some beaches undergo seasonal changes in the amount and size of sediment present.

In the middle latitudes, beaches are generally narrower, steeper, and composed of coarser material in winter than they are in summer. The larger winter storm waves are more erosive and destructive, whereas the smaller summer waves, which often travel from the other hemisphere, are depositional and constructive. On the Pacific coast of the United States, summer beaches are generally temporary accumulations of sand deposited over

coarser winter beach materials ( ● Fig. 20.19). Sand-sized sediment eroded from the beach in winter forms a deposit called a **longshore bar** that lies submerged parallel to shore, with the sediment returning to the beach in summer. On the Atlantic and Gulf coasts of the United States, the late summer to early fall hurricane season is also a time when beach erosion can be severe.

Whereas beaches are attached to the mainland along their entire length, **spits** are coastal depositional landforms connected to the mainland at just one end ( ● Fig. 20.20). Spits project out into the water like peninsulas of sediment. They form where the mainland curves significantly inland while the trend of the longshore current remains at the original orientation. Sediments accumulate into a spit in the direction of the longshore current ( ● Fig. 20.21a). Where similar processes form a strip of sediment connecting the mainland to an island, the landform is a **tombolo** (Fig. 20.21b).

Another category of coastal landforms are **barrier beaches,** elongated depositional features constructed parallel to the mainland. Barrier beaches act to protect the mainland from direct wave attack. All barrier beaches have restricted waterways, called **lagoons,** that lie between them and the mainland. Salinity in the lagoon varies from that of the open water body, depending on freshwater inflow and evaporation, and affects organisms living in

(a)

(b)

(c)

(d)

● **FIGURE 20.18** Beaches are the most common evidence of wave deposition and may be made of any material deposited by waves. (a) A sandy beach on the eastern shore of Lake Michigan. (b) A boulder beach in Arcadia National Park, Maine (note the person for scale). (c) A tropical, white sand beach made of broken coral pieces. (d) A sea turtle enjoying a beach of black basalt on the island of Hawaii.

(a)

(b)

● **FIGURE 20.19** Because of seasonal variations in wave energy, the differences in a beach from summer to winter can be striking, particularly in the midlatitudes. (a) Waves in summer are generally mild and deposit sand on the beach. (b) Winter waves from storms nearer to and at the beach remove the sand, leaving boulders and bare bedrock in the beach area.

*What attribute of waves represents the amount of energy they have?*

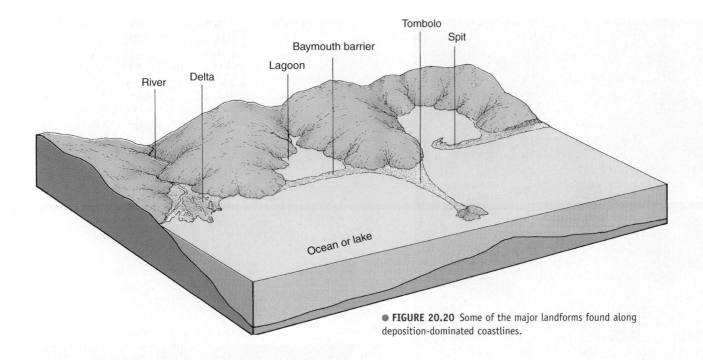

● **FIGURE 20.20** Some of the major landforms found along deposition-dominated coastlines.

(a)

(b)

(c)

USGS Coastal & Marine Geology Program

NOAA/Captain Albert E. Theberge

USGS Coastal & Marine Geology Program

● **FIGURE 20.21** (a) A spit is attached to the mainland at one end. (b) A tombolo connects a nearby island with the mainland at Point Sur, California. (c) A baymouth barrier crosses the mouth of an embayment connecting to land at both ends.

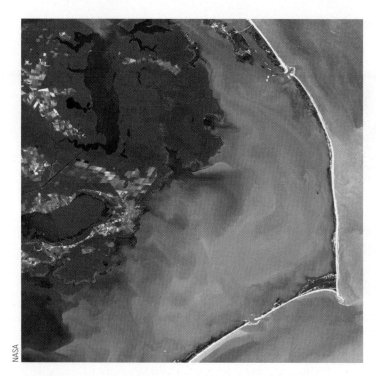

● **FIGURE 20.22** Barrier islands lie parallel to the mainland but are not attached to it. This barrier island is located near Pamlico Sound on the North Carolina coast.

*What feature separates a barrier island from the mainland?*

the lagoon. Like beaches and spits, barrier beaches have a submerged part and a portion that is always above water, except in extreme storm conditions or extremely high tide. This contrasts with bars, like the longshore bars discussed previously, which are submerged except in extreme conditions.

There are three kinds of barrier beaches. A **barrier spit** originated as a spit and thus is attached to the mainland at one end but has extended almost completely across the mouth of an embayment to restrict the circulation of water between it and the ocean or lake. If the barrier spit crosses the mouth of the embayment to connect with the mainland at both ends, it becomes a **baymouth barrier** (Fig. 20.21c). Limited connection is maintained between the lagoon and the main water body through a breach or *inlet* cut across the barrier somewhere along its length. The position of inlets can change during storms. **Barrier islands** are likewise elongated parallel to the mainland and separate lagoons and the mainland from the open water body, but they are not attached to the mainland at all ( ● Fig. 20.22).

Barrier islands are common features of low-relief coastlines. They dominate the Atlantic and Gulf coasts of the United States from New York to Texas. Some excellent examples of long barrier islands are Fire Island (New York), Cape Hatteras (North Carolina), Cape Canaveral and Miami Beach (Florida), and Padre Island (Texas).

Rising sea level since the Pleistocene appears to have played a major role in the formation of barrier islands. They migrate landward over long periods and may change drastically during severe storms, especially hurricanes ( ● Fig. 20.23).

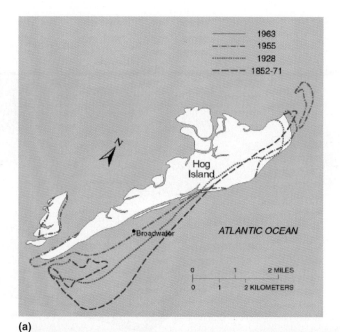

(a)

(b)

(c)

● **FIGURE 20.23** (a) Historic maps show us that barrier islands, like this one off the coast of Virginia, shift their shapes and positions over time but especially during storm events. (b) Homes along the shore of a barrier island before a hurricane. (c) The same homes after a hurricane.

*How can this type of damage be prevented in the future?*

● **FIGURE 20.24** A series of groins on the Atlantic shoreline at Norfolk, Virginia, captures sand to maintain the beach along one stretch of the coastline.

*How do you think the stretch of coast beyond the last groin would be affected by these structures?*

Beaches not only serve recreational needs, but also help protect coastal settlements from erosion and flooding by storm waves. Beach systems are in equilibrium when input and output of sediment are in balance. People interrupt natural coastal depositional processes for various reasons, such as to keep harbors free of sediment or to encourage growth of recreational beaches. To increase the size of a beach, people build artificial obstructions to the principal direction of littoral drifting. A **groin** is an obstruction, usually a concrete or rock wall, built perpendicular to a beach to inhibit sediment removal by littoral drifting while the input of sediment from littoral drifting remains the same. This obstruction, however, starves the adjacent, downcoast beach area of material input while the rate of sediment removal from that adjacent beach does not change ( ● Fig. 20.24). Human actions deplete the natural sediment supply to the coast by damming rivers. With decreased sediment supply, beaches become narrower and lose some of their ability to protect the coastal region against storms. In Florida, New Jersey, and California, hundreds of millions of dollars have been spent to artificially replenish sandy beaches that have lost significant amounts of sediment, and thus their recreational and protective abilities, because of human actions.

## Types of Coasts

Coasts are spectacular, dynamic, and complex systems that are influenced by tectonics, global sea-level change, storms, and marine and continental geomorphic processes. Because of this complexity, there is no single, classification system for coasts. Each major way of classifying coasts focuses on a different characteristic of coastal systems; all aid our understanding of these natural, complex systems.

On a global scale, coastal classification is based on plate tectonic relationships. This system has two major coastal types: passive-margin coasts and active-margin coasts. **Passive-margin coasts** are tectonically quiescent, with little mountain-building or volcanic activity. Like the eastern seaboard of the United States, which is an excellent example, these coasts generally have low relief with broad coastal plains and wide continental shelves ( ● Fig. 20.25), but passive-margin coasts that are relatively young, such as those of the Red Sea and Gulf of California, may have somewhat greater relief. Passive-margin coasts

● **FIGURE 20.25** The general nature of the boundary between the continents (continental crust) and ocean basins (oceanic crust). The gently sloping continental shelf varies in width along different coasts depending on plate tectonic history and proximity to plate margins.

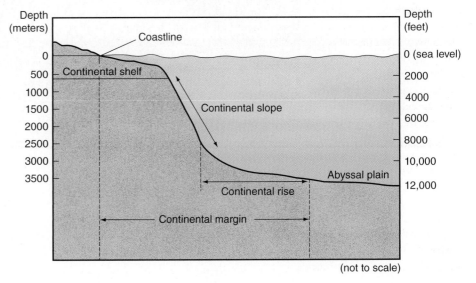

# Beach Protection

Beaches act as buffers to help protect the land behind them from wave erosion. Under natural conditions, sediment eroded from the beach in a storm will eventually be replaced through the action of gentle waves and by littoral drifting bringing in new sediment. People interfere with the natural sediment budget when they dam rivers and build on the normally shifting sediment of beach and dune systems in the coastal zone. Intense human development, like that found at many popular tourist beaches, typically interrupts the supply of replacement sediment and leads to more permanent erosion of beaches. Beach erosion leaves the coastal buildings and infrastructure susceptible to damage from storm waves while eliminating the primary attraction, the beach itself.

Beach protection and restoration strategies include the construction of *groins* built perpendicular to the trend of the coastline to slow down the rate of littoral drifting of sediment out of the area. *Jetties* are also built perpendicular to the coastline, but always in pairs, and act to keep sediment from blocking an inlet, such as a river mouth or a channel for boats. *Breakwaters* are walls built parallel to the shoreline in the breaker zone. Large waves expend the bulk of their energy breaking on the structure, thus limiting the amount of beach erosion. Another strategy has been to add sediment to the beach artificially by dredging harbors or reservoirs and trucking or pumping that sediment through pipes to the beach. Unless the factors that are limiting the

natural sediment supply to the beach are addressed, this artificial form of *beach nourishment* will likely have to continue, at least periodically.

Miami Beach, Florida, has long been a popular destination for millions of annual visitors, who are drawn to the magnificent beaches. After decades of development, by 1970 the beaches had mostly disappeared because of erosion. With the beaches gone, shoreline condominiums and hotels were threatened with serious damage from storm waves, and the hotels were half empty as tourists traveled to other destinations.

To help solve the problem, the U.S. Army Corps of Engineers (the federal agency responsible for the development of inland and coastal waterways) chose Miami Beach to

AP Photo

Visitors and residents of Miami Beach benefit from the artificial replacement of sand to the beach.

experiment with the *beach building* method of shoreline protection. The technology involves transferring millions of cubic meters of sand from offshore to replenish existing beaches or build new ones where beaches have eroded away. Huge barge-mounted dredges dig sand from the sea bottom near the shore, and the sand is pumped as a slurry through massive movable tubes to be deposited on a beach.

The initial project cost $72 million, but it was so successful that within 2 years Miami Beach again had a sandy beach 90 meters (300 ft) wide and 16 kilometers (10 mi) long. With the return of the beach, the number of visitors soon grew to three times the number prior to beach building.

Today beach building has become the accepted way to respond to beach erosion. Nearly $1.7 billion was spent on beach restoration during the last decade of the 20th century. Since 1950, the Corps of Engineers has undertaken beach nourish-ment projects on more than 565 kilometers (350 mi) of coastline. According to the Corps, these projects are meant primarily to protect buildings on or near the beach, rather than to provide beaches for recreational purposes. In the view of the Corps of Engineers, the true value of beach building can best be measured in savings from storm damage. For example, it has been estimated that beach replenishment along the coast of Miami-Dade County prevented property losses of $24 million during Hurricane Andrew in 1992.

When beaches are rebuilt or widened to protect against storms, continued erosion is inevitable. In deciding how much sand must be pumped in, it is important to determine the beach width necessary to protect shoreline property from storms. This is usually about 30 meters (100 ft). Then an amount of additional "sacrificial" sand is added to produce a beach twice the width of the desired permanent beach. In about 7 years, the sacrificial sand will be lost to wave erosion. If beach rebuilding is repeated each time the sacrificial sand has been removed, the permanent beach will remain in place and the coastal zone will be protected. An estimated $5.5 billion has already been committed to the continuous rebuilding of existing projects over the next 50 years.

Is the battle with the environment at the beachfront worth the price we are paying? Many people are not so sure, but most environmental scientists respond to the question with a resounding no because they are concerned with a price that is not measured solely in appropriated dollars. This price is paid in destroyed natural beach environments, reduced offshore water quality, eliminated or displaced species, and repeated damage to food chains for coastal wildlife each time a beach is rebuilt. Clearly, beach building is a mixed blessing.

Seagulls enjoy the swash on this sand-replenished beach in Alameda, California, on San Francisco Bay.

● **FIGURE 20.26** A sandy passive-margin coast on the Atlantic Ocean at Marconi Beach, Cape Cod National Seashore, Massachusetts.

are typically dominated by depositional landforms and are well represented by the coastal regions of continents along the Atlantic Ocean ( ● Fig. 20.26). Most major tectonic activity within the Atlantic occurs in the center of the ocean along the Mid-Atlantic Ridge, rather than along its coastlines.

**Active-margin coasts** are best represented by coastal regions along the Pacific Ocean ( ● Fig. 20.27). There, most tectonic activity occurs around the ocean margins because of active subduction and transform plate boundaries along the "Pacific Ring of Fire." Active-margin coasts are usually characterized by high relief with narrow coastal plains, narrow continental shelves, earthquake activity, and volcanism. These coasts tend to be erosional, having less time within Earth history for the development of marine or continental depositional features. The coast of Japan and the west coast of the United States are excellent examples of active-margin coasts.

On a regional scale, coasts can be classified as coastlines of emergence or coastlines of submergence. **Coastlines of emergence** occur where the water level has fallen or the land has risen in the coastal zone. In either case, land that was once below sea level has emerged above the water. Evidence for emergence includes marine terraces and relict sea cliffs, sea stacks, and beaches found above the reach of present wave action. Many coastlines of emergence would have existed during the glacial phases of the Pleistocene Epoch when sea level was about 120 meters (400 ft) lower than it is today. Features of emergence are common along active-margin coasts like those of California, Oregon, and Washington, where tectonic uplift has raised coastal landforms as much as 365 meters (1200 ft) above sea level ( ● Fig. 20.28). Other emergent coastlines, such as around the Baltic Sea and Hudson Bay, are located where isostatic rebound has elevated the land following the retreat of the continental ice sheets.

● **FIGURE 20.27** The rugged coast of Point Lobos, California, exemplifies an active-margin coastline, having experienced much tectonic unrest that includes general uplift in addition to great lateral movement along the San Andreas Fault.

● **FIGURE 20.28** Cape Blanco, Oregon, represents an emergent coastline. The flat surface on which the lighthouse is built is a marine terrace formed by wave erosion and deposition prior to tectonic uplift. The elevation of the marine terrace is about 61 meters (200 ft) above sea level.

Along **coastlines of submergence** many features of the former shore lie underwater and the present shoreline crosses land areas that are not fully adjusted to coastal processes. Coastlines of submergence were created as global sea level rose in response to the retreat of the Pleistocene ice sheets. Coastlines of submergence also occur where tectonic forces have lowered the level of the land, as in San Francisco Bay. Great thickness of river deposits and compaction of alluvial sediments, as along the Louisiana coast, may also cause coastal submergence. The specific features present along a coastline of submergence are related to the character of coastal lands prior to submergence. Plains, for instance, will produce a far more regular shoreline than will a mountainous region or an area of hills and valleys. When areas of low relief with soft sedimentary rocks are submerged, barrier islands form with shallow bays and lagoons behind them. The classic examples of this type of submerging coastline are the Atlantic and Gulf coasts of the United States.

Two special types of submerged coastlines are ria and fjord coasts. **Rias** are created where river valleys are drowned by a rise in sea level or a sinking of the coastal area ( ● Fig. 20.29). These irregular coastlines result when valleys become narrow bays and the ridges form peninsulas. The Aegean coast of Greece and Turkey is an outstanding example of a ria coastline. *Fjords,* which are drowned glacial valleys, form spectacular scenic shorelines ( ● Fig. 20.30). A coastline with fjords is highly irregular, with deep, steep-sided embayments penetrating far inland in U-shaped valleys originally deepened by glaciers. Tributary streams cascade down fjord sidewalls, which may be a few thousand meters high. Fjord coastlines are found in Norway (where the term originated), Chile, New Zealand, Greenland, and Alaska. Canada, however, has more fjords than any other nation. In many fjords, the glaciers have retreated far inland, but in others, especially in Greenland and Alaska, glaciers still reach the fjords calving icebergs into their cold ocean water.

Some coastlines, such as those composed of river deltas, cannot be classified as either submerging or emerging. Actually, most shorelines show evidence of more than one type of development largely because the land elevation and the level of the ocean have changed many times during geologic history. For this reason, features of both submerged and emerged shorelines characterize some coastal areas.

● **FIGURE 20.29** Submergent coasts like the Chesapeake Bay region are characterized by drowned river valleys, known as rias, that developed as sea level rose at the end of the Pleistocene.

● **FIGURE 20.30** Fjords, like this one in Greenland, are glaciated valleys that were drowned by the sea following the Pleistocene Epoch as the glaciers receded and sea level rose.

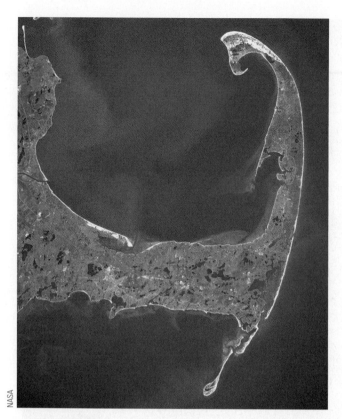

NASA

● **FIGURE 20.31** Cape Cod is an example of a primary shoreline of glacial deposits being modified by waves and currents.

*How has wave action modified the moraine that originally formed Cape Cod?*

Because both continental and marine geomorphic processes shape coastlines, another regional classification system recognizes two types of coasts: primary and secondary coastlines. Erosional and depositional processes of the land dominate **primary coastlines.** Primary coastlines result from such rapid changes in the position of the shoreline that coastal processes are not able to have a significant landforming effect. The major types of primary coastlines (with examples) are drowned river valleys (Delaware and Chesapeake Bays), glacial erosion coasts (southeastern Alaska, British Columbia, Puget Sound in Washington, and from Maine to Newfoundland), glacial deposition coasts (Cape Cod, Massachusetts [ ● Fig. 20.31], and the north shore of New York's Long Island), river deltas (Mississippi Delta, Louisiana), volcanic coasts (Hawaii), and faulted coasts (California).

**Secondary coastlines** are those formed mainly by coastal geomorphic agents, especially waves, and by aquatic organisms. Sea cliffs, arches, sea stacks, and sea caves generally dominate erosional secondary coasts, like that of Oregon. Depositional secondary coasts typically display barrier beaches and spits, such as the coastline of North Carolina. An example of coasts built by aquatic organisms is the coral reef ( ● Fig. 20.32). The Florida

● **FIGURE 20.32** Corals building the coastline at Vieques National Wildlife Refuge, Puerto Rico, provide an example of a secondary shoreline.

Keys were constructed by coral growth. Mangrove trees and salt-marsh grasses also trap sediment to build new land areas in shallow coastal waters.

# Islands and Coral Reefs

The perimeters of islands are subject to the same coastal processes as the boundary between a standing body of water and the mainland. Within the ocean there exist three basic types of islands: continental, oceanic, and atolls. **Continental islands** are geologically part of the continent. Continental islands are usually found on the part of a continent submerged by the ocean, the continental shelf. Most large continental islands became separated from the continent because of global sea-level change or regional tectonic activity. A few large continental islands, such as New Zealand and Madagascar, are isolated "continental fragments" that separated from continents millions of years ago. The world's largest islands—Greenland, New Guinea, Borneo, and Great Britain—are continental. Smaller continental islands include the barrier islands off the Atlantic and Gulf Coasts of the United States, New York's Long Island, California's Channel Islands, and Vancouver Island off the west coast of Canada.

**Oceanic islands** are volcanoes that rise from the deep-ocean floor and are geologically related to oceanic crust, not the continents. Most oceanic islands, such as the Aleutians, Tonga, and the Marianas, occur in island arcs along the edges of the trenches. Others, like Iceland and the Azores, are peaks of oceanic ridges rising above sea level. Many oceanic islands, like the Hawaiian Islands, occur in chains. The oceanic crust sliding over a stationary "hot spot" in the mantle causes these island chains. The Hawaiian Islands are moving northwestward with the Pacific plate and will slowly submerge. A new volcanic Hawaiian island, named Loihi, is forming to the southeast ( ● Fig. 20.33). Evidence of the plate motion is indicated by the fact that the youngest islands of the Hawaiian chain, Hawaii and Maui, are to the southeast, whereas the older islands, such as Kauai and Midway, are located to the northwest.

An **atoll** is an island consisting of a ring of coral reefs that have grown up from a subsiding volcanic island and that encircle a central lagoon ( ● Fig. 20.34a). To understand atolls, we must first consider coral reefs.

**Coral reefs** are shallow, wave-resistant structures made by the accumulation of remains of tiny sea animals that secrete a skeleton of calcium carbonate. Many other organisms, including algae, sponges, and mollusks, add material to the reef structure. Reef corals need special conditions to grow—clear and well-aerated water, water temperatures above 20°C (68°F), plenty of sunlight, and normal marine salinity. These conditions are found in the shallow waters of tropical regions, including Hawaii, the West Indies, Indonesia, the Red Sea, and the coast of Queensland in Australia. Today coastal water pollution, dredging, souvenir coral collecting, and climate change threaten the survival of many coral reefs.

A **fringing reef** is a coral reef attached to the coast (Fig. 20.34b). Fringing reefs tend to be wider where there is more wave action that brings a continuous supply of well-aerated water and additional nutrients for increased coral growth. They are usually absent near river mouths because the coral cannot grow where the waters are laden with sediment or where river water lowers the salinity of the marine environment.

● **FIGURE 20.33** The oceanic island of Hawaii was formed over a "hot spot" like the other Hawaiian Islands before it. Two large shield volcanoes, Mauna Loa and Mauna Kea, dominate the island. To the south, Loihi, an active submarine volcano, may reach sea level to become the newest Hawaiian Island in 50,000 years.

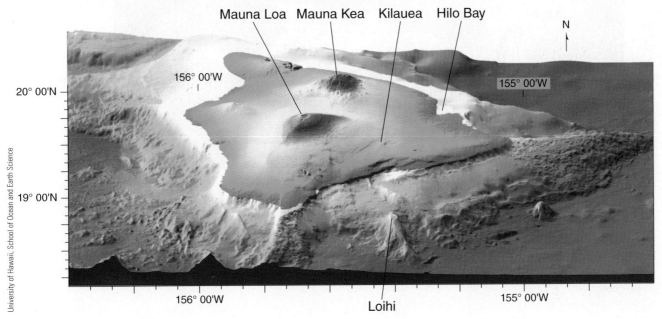

(a)

(b)

(c)

● **FIGURE 20.34** The major types of coral reefs are evident in the Society Island chain of French Polynesia. (a) Atolls are islands consisting of a ring of coral with no surface evidence remaining of its former volcanic core. (b) Fringing reefs are attached to mainland or island coasts. (c) Like coastal barriers in general, barrier reefs are separated from dry land by a lagoon.

Sometimes coral forms a **barrier reef,** which lies offshore, separated from the land by a shallow lagoon (Fig. 20.34c). Most barrier reefs occur in association with slowly subsiding oceanic islands, growing at a pace that keeps them above sea level. Other barrier reefs, including Australia's Great Barrier Reef, the Florida Keys, and the Bahamas, were formed on continental shelves and grew upward as sea level rose after the Pleistocene ice age waned. At more than 2000 kilometers (1250 mi) long, the world's largest organic structure is the Great Barrier Reef of Australia.

● Figure 20.35 illustrates the manner in which atolls develop. As a volcanic island subsides, the fringing reef grows upward, keeping pace with the seafloor subsidence, becoming a barrier reef when a lagoon develops and finally an atoll when the only solid material is the outer ring of coral. Charles Darwin proposed this explanation of atoll formation in the 1830s. Evidence from drilling indicates that there have been as much as 1200 meters (4000 ft) of subsidence and an equal amount of reef development in the past 60 million years.

Atolls pose severe challenges as environments for human habitation. First, they have a low elevation above sea level and provide no defense against huge storm waves and tsunamis that can inundate the entire atoll, drowning all its inhabitants. Potential future sea-level rise from global warming would also threaten these low islands. Second, there is little freshwater available on the porous coral limestone surface. Third, little vegetation can survive in the lime-rich rock and soil of the atoll islands. The coconut palm is an exception, and coconuts were vital to the survival of early inhabitants of the atoll islands in Polynesia and Micronesia.

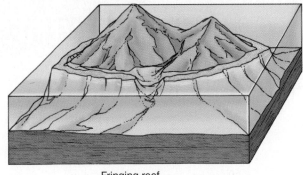

Fringing reef

(a)

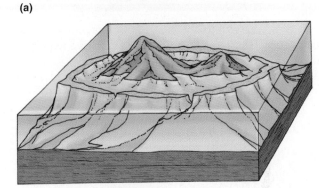

Barrier reef

(b)

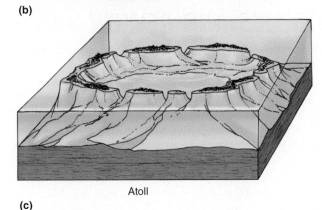

Atoll

(c)

● **FIGURE 20.35** Charles Darwin proposed the three-part theory of how coral reefs develop around oceanic islands. (a) First, a fringing reef grows along the shore. (b) Later, as the island erodes and subsides, a barrier reef develops. (c) Further subsidence causes the coral to build upward while the volcanic core of the island is completely submerged below a central lagoon, forming an atoll.

## ■ DEFINE AND RECALL

| | | |
|---|---|---|
| shoreline | littoral drifting | baymouth barrier |
| sea level | beach drifting | barrier island |
| swash | longshore drifting | groin |
| backwash | notch | passive-margin coast |
| wave steepness | cobble beach | active-margin coast |
| wave period | sea stack | coastlines of emergence |
| tsunami | abrasion platform | coastlines of submergence |
| wind wave | marine terrace | ria |
| spring tide | estuary | primary coastline |
| neap tide | beach | secondary coastline |
| sea | longshore bar | continental island |
| swell | spit | oceanic island |
| fetch | tombolo | atoll |
| wave base | barrier beach | coral reef |
| rip current | lagoon | fringing reef |
| wave refraction | barrier spit | barrier reef |

## ■ DISCUSS AND REVIEW

1. What is the difference between a shoreline and the coastal zone?
2. Describe the major factors that produce the tides. What are some variations in tidal patterns?
3. How do waves change when they enter shallow coastal waters? What is the main factor causing this change in the waves?
4. What is wave refraction? How is wave refraction related to the shape of the coastline?
5. Describe how sea cliffs form. Name three other coastal erosional landforms typically found in areas with sea cliffs.

6. What are the differences between beach drifting and littoral drifting? What causes both?
7. Describe the similarities and differences between beaches and barrier beaches.
8. How does a bar differ from a barrier?
9. What are the major differences between active-margin coastlines and passive-margin coastlines?
10. What is a coral reef, and how does it form?

## ■ CONSIDER AND RESPOND

1. What major changes would you expect in the world's coastal zones if sea level were to rise as a result of global warming? How would the landforms change?
2. Explain some of the various ways that people can influence coastal geomorphology.

3. How are systems of coastal classification useful in studying coastal environments?
4. If you were asked to plan for construction of a major tourist resort on a beautiful tropical atoll, what limitations and concerns would you need to evaluate before construction?

# APPLY AND LEARN

1. A wind wave with a wavelength of 75 meters and a height of 0.8 meter takes 7 seconds to travel past a given point. What is the wave's steepness? What is its period?

2. If a 60-meter-long wave travels through water that is 25 meters deep, is the wave feeling bottom? As the wave comes in toward land, how high can it get before it breaks?

# LOCATE AND EXPLORE

1. Using Google Earth and the Tide Layer (download from the book's CourseMate at www.cengagebrain.com), examine the hourly change in tidal elevation over the last week for the coastal sites listed below. Identify the tides as either diurnal, semidiurnal, or mixed and rank the sites by the tidal range. Compare the predicted and observed tidal variations. What, other than tides, causes the water level to vary?
   a. Portland, Maine
   b. Fort Pulaski, Georgia
   c. Key West, Florida
   d. Dauphin Island, Alabama
   e. Rockport, Texas
   f. Los Angeles, California
   g. South Beach, Oregon
   h. Seattle, Washington

2. Using Google Earth, examine the barrier islands along the Maryland coast north and south of 38.325°N, 75.09°W. What is the main direction of sediment transport along shore, and what evidence do you see to support your answer? Why do you think the island to the south is displaced landward of the more developed island to the north?

3. Using Google Earth and Hurricane Ivan LIDAR Layer for Santa Rosa Island in northwest Florida (download from the book's CourseMate at www.cengagebrain.com), view the topographic data from before and after Hurricane Ivan (September 2004). The topography of the island was created using Light Detection and Ranging (LIDAR), an increasingly used mapping technology that provides spatially dense and accurate topographic data. The data are collected with aircraft-mounted lasers capable of recording elevation measurements at a rate of 2000 to 5000 pulses per second and with a vertical precision of 0.15 meter. In the prestorm data you will see that the dunes along the beach are highly variable in height and extent, whereas in the poststorm data you will see that the dunes are gone and replaced by overwash fans of different sizes.

   a. Change the transparency of the pre-Ivan LIDAR image and the ruler tool to measure the change in the position of the shoreline as a result of Hurricane Ivan. Now make the post-Ivan image transparent and measure the change in shoreline after Hurricane Ivan.

   b. Using both LIDAR images, describe how the overwash fan development is related to prestorm dune height. Can you develop a simple model to predict the impact of a hurricane on a barrier island using the height of the dunes relative to the elevation of the hurricane's storm surge?

 CourseMate – Make the most of your study time by accessing everything you need to succeed in one place. Read your textbook, take notes, review flashcards, watch videos, complete activities, take practice quizzes, and more online with CourseMate. Log in at **www.cengagebrain.com**.

# Map Interpretation

## ACTIVE-MARGIN COASTLINES

### The Map

Point Reyes National Seashore lies north of San Francisco on the rugged California coast. Point Reyes consists of resistant bedrock that is being eroded by coastal processes. Marine life is abundant, including many birds and marine mammals (note Sea Lion Cove below Point Reyes). A hilly area, Punta de Los Reyes, separates Drake's Bay and Point Reyes from Tomales Bay (barely showing in the northeastern corner of the map).

The oblique aerial photo of Point Reyes was taken from a high-altitude NASA aircraft. Healthy vegetation appears red on color infrared images. The view is looking northeast. Trending across the upper portion of the photograph is the San Andreas Fault, which forms the linear Tomales Bay. The San Andreas Fault also separates the Pacific plate, on which Point Reyes is located, from the North American plate, which underlies the area east of the fault. Point Reyes is moving to the northwest along this fault. Emergent coastal features and recent tectonic activity characterize active-margin coastlines.

Point Reyes has a Mediterranean climate, influenced by the cool offshore California current. Here, the ocean is not hospitable for swimming because of the uncomfortably cool seawater, high surf, and dangerous rip currents. The coastal location and onshore westerly winds create a truly temperate climate. Although very hot and very cold temperatures are rarely experienced, this is one of the windiest and foggiest coastlines in the United States. It is an area of rugged natural beauty with wind-sculpted trees, grasslands, rocky sea cliffs, and long sandy beaches.

### Interpreting the Map

1. Which area of the coast is most exposed to wave erosion? What features indicate this type of high-energy activity?
2. Which area of the map is under the influence of strong longshore currents? What is the general direction of flow, and what coastal feature would indicate this flow?
3. Looking into the future, what may happen to Drakes Estero? What would Limantour Spit become?
4. Which area of the map appears to have strong wind activity? What would indicate this? What do you think the prevailing wind direction is?

5. Locate examples of the following coastal erosional landforms:
   a. Headland
   b. Sea stack
   c. Sea cliffs
6. Note that the bays and *esteros* (Spanish for estuary) have mud bottoms. Because the creeks and streams are very small in the region, what is the probable source and cause for movement of the mud?
7. Note the offshore bathymetric contours *(blue isolines)*. Which has a steeper gradient, the Pacific coast *(west side)* or Drakes Bay *(east side)*? Why do you think there is such a great difference?

NASA/JSC-ES&IA

High altitude oblique aerial photo of Point Reyes, California.

In Google Earth fly to: 38.001°N, 122.997°W.

Opposite:
Point Reyes, California
Scale 1:62,500
Contour interval = 80 ft
USGS

## ■ PASSIVE-MARGIN COASTLINES

### The Map

Eastport, New York, is located on the south shore of Long Island, 115 kilometers (70 mi) east of New York City. Long Island is part of the Atlantic Coastal Plain, which extends from Cape Cod, Massachusetts, to Florida. Water bodies such as Delaware Bay, Chesapeake Bay, and Long Island Sound embay much of the Atlantic Coastal Plain in the eastern United States. The south shore of Long Island has low relief, and its coastal location moderates its humid continental climate.

Although this is a coastal region, its recent glacial history influences the landforms that exist today. Two east–west trending glacial terminal moraines deposited during the Pleistocene glacial advances form Long Island. Between the coast and the south moraine is a sandy glacial outwash plain that forms the higher eleva-

tions at the northern part of the map area. As the glaciers melted, sea level rose, submerging the lowland now occupied by Long Island Sound. This water body separates Long Island from the mainland of the Atlantic Coastal Plain.

The Atlantic and Gulf coasts of the United States have nearly 300 barrier islands with a combined length of more than 2500 kilometers (1600 mi). New York, especially the coastal zone of Long Island's south shore, has 15 barrier islands with a total length of more than 240 kilometers (150 mi). Coastal barrier islands protect the mainland from storm waves, and they contribute to the formation of coastal wetlands on their landward side, which are a critical habitat for fish, shellfish, and birds.

### Interpreting the Map

1. What type of landform is the entire long, narrow feature on which lies the straight shoreline labeled Westhampton Beach?

2. What is the highest elevation on the linear coastal feature? What do you think comprises the highest portions of this feature?

3. How much evidence of human use exists on the landform in Question 1? What problems might these cultural features be subjected to?

4. Behind the linear feature is a water body. Is it deep or shallow? Is it a high-energy or low-energy environment? What is a water body such as this called?

5. Is the coastline shown on the Eastport map predominantly erosional or depositional? What features support your answer? Is this a coastline of submergence or emergence?

6. Note Beaverdam Creek in the upper middle of the map area. Does it have a steep or gentle gradient?

7. Based on its gradient, does Beaverdam Creek always flow seaward? If not, what might influence its flow?

8. What kind of topographic feature exists as indicated by the map symbols in the area surrounding Oneck?

9. How do you think geomorphic processes will likely modify this map area in the future? Which area will probably be modified the most? Explain.

10. What type of natural hazard is this coastal area of Long Island most susceptible to? Why?

NASA/GSFC/SVS

Satellite image of Long Island, New York.

⊕ In Google Earth fly to: 40.784°N, 72.680°W.

Opposite:
Eastport, New York
Scale 1:24,000
Contour interval = 10 ft
USGS

# Appendix A

## International System of Units (SI), Abbreviations, and Conversions

| Symbol | Multiply | By | To Find | Symbol |
|---|---|---|---|---|
| **Area** | | | | |
| in.² | square inches | 645.2 | square millimeters | mm² |
| ft² | square feet | 0.093 | square meters | m² |
| yd² | square yards | 0.836 | square meters | m² |
| ac | acres | 0.405 | hectares | ha |
| mi² | square miles | 2.59 | square kilometers | km |
| ac | acres | 43,560 | square feet | ft² |
| mm² | square millimeters | 0.0016 | square inches | in.² |
| m² | square meters | 10.764 | square feet | ft² |
| m² | square meters | 1.195 | square yards | yd² |
| ha | hectares | 2.47 | acres | ac |
| km² | square kilometers | 0.386 | square miles | mi² |
| **Mass** | | | | |
| oz | ounces | 28.35 | grams | g |
| lb | pounds | 0.454 | kilograms | kg |
| g | grams | 0.035 | ounces | oz |
| kg | kilograms | 2.202 | pounds | lb |
| **Length** | | | | |
| in. | inches | 25.4 | millimeters | mm |
| ft | feet | 0.305 | meters | m |
| yd | yards | 0.914 | meters | m |
| mi | miles | 1.61 | kilometers | km |
| ft | feet | 5280 | mile | mi |
| mm | millimeters | 0.039 | inches | in. |
| m | meters | 3.28 | feet | ft |
| m | meters | 1.09 | yards | yd |
| km | kilometers | 0.62 | miles | mi |
| **Volume** | | | | |
| gal | U.S. gallons | 3.785 | liters | l |
| ft³ | cubic feet | 0.028 | cubic meters | m³ |
| yd³ | cubic yards | 0.765 | cubic meters | m³ |
| l | liters | 0.264 | U.S. gallons | gal |
| m³ | cubic meters | 35.30 | cubic feet | ft³ |
| m³ | cubic meters | 1.307 | cubic yards | yd³ |
| **Velocity** | | | | |
| mph | miles/hour | 1.61 | kilometers/hour | km/h |
| knot | nautical miles/hour | 1.85 | kilometers/hour | km/h |
| km/h | kilometers/hour | 0.62 | miles/hour | mph |
| km/h | kilometers/hour | 0.54 | nautical miles/hour | knots |
| **Pressure or Stress** | | | | |
| mb | millibars | 0.75 | millimeters of mercury | mm Hg |
| mb | millibars | 0.02953 | inches of mercury | in. Hg |
| mb | millibars | 0.01450 | pounds per square inch | (lb/in.² or psi) |
| lbs/in.² | pounds per square inch | 6.89 | kilopascals | kPa |
| in. Hg | inches of mercury | 33.865 | millibars | mb |
| kPa | kilopascals | 0.145 | pounds per square inch | (lb/in.² or psi) |

| Standard Sea-Level Pressure |
|---|
| 29.92 in. Hg |
| 14.7 lb/in.$^2$ |
| 1013.2 mb |
| 760 mm Hg |

| Temperature | | | | |
|---|---|---|---|---|
| °F | Fahrenheit | (°F − 32)/1.8 | Celsius | °C |
| °C | Celsius | 1.8°C + 32 | Fahrenheit | °F |
| K | Kelvins | K = °C + 273 | Celsius | °C |

| Powers of Ten | | | | |
|---|---|---|---|---|
| nano | one billionth | = $10^{-9}$ | = 0.000000001 |
| micro | one millionth | = $10^{-6}$ | = 0.000001 |
| milli | one thousandth | = $10^{-3}$ | = 0.001 |
| centi | one hundredth | = $10^{-2}$ | = 0.01 |
| deci | one tenth | = $10^{-1}$ | = 0.1 |
| hecto | one hundred | = $10^2$ | = 100 |
| kilo | one thousand | = $10^3$ | = 1000 |
| mega | one million | = $10^6$ | = 1,000,000 |
| giga | one billion | = $10^9$ | = 1,000,000,000 |

# Appendix B

## Topographic Maps

Mapping has changed considerably in recent years, and with the ever-increasing capabilities of computers to store, retrieve, and display graphics, this trend will continue well into the future. In the United States the U.S. Geological Survey produces the vast majority of topographic maps available. These maps have long been the tried and true tools of geographers and scientists in many other disciplines who study various aspects of the environment. Today virtually all USGS topographic maps are accessible in digital format, for downloading and printing, on computer disks, or for examining on a computer screen. Computers and the Internet have made maps much more available and accessible than they were just a few years ago. This availability makes it easy to print maps or map segments at home, school, or work. A logical question then would be, will paper maps become obsolete, given computer displays?

There are several reasons why paper maps will still be popular and useful, whether they are purchased from the source or downloaded and printed. Topographic maps are particularly important in fieldwork. Maps are highly portable, require no batteries or electrical power that could fail, and do not suffer from technology glitches. They are reliable and easy to use and they also provide a good base for making field notes, and marking routes.

Today the USGS is working to make and maintain a seamless database of the United States, with maps, imagery, and spatial data in digital form, so areas that once were split on adjacent maps can be printed on a single sheet. This is a great change from dividing the country into quadrangles (the roughly rectangular area a map displays). Topographic quadrangles (quads), however, will continue to be in use for a long time. There are several standard quadrangles, each with a specific scale, and many other special-purpose topographic maps.

**7.5-minute quads**—these are printed at a scale of 1:24,000 and cover 7.5 minutes of longitude and 7.5 minutes of latitude.

**15-minute quads**—these are printed at a scale of 1:62,500 and cover 15 minutes of longitude and 15 minutes of latitude.

**1° × 2° quads**—these are printed at a scale of 1:250,000 and cover 1 degree of latitude and 2 degrees of longitude.

**1:100,000 metric quads**—these are printed at a scale of 1:100,000, cover 30 minutes of longitude and 60 minutes of latitude, and use metric measurements for distance and elevation.

Brown topographic contours are used to show elevation differences and the terrain. Some rules for interpreting contours are given in Chapter 2.

### Determining Distances on a Map, Distances from a Map, or an RF Scale

It is important to understand representative fractions, like 1:24,000, which means that any measurement on the map will represent 24,000 of the same measurements on the ground. This knowledge is particularly significant because reproduced maps may not be printed at the original size. On a map that is reduced or enlarged the bar scale will still be accurate, but the printed RF scale will not. Enlarging or reducing a map changes the RF scale from the original.

### How to Find the RF of a Map (or Air Photo, or Satellite Image) of Unknown Scale

Here is the formula:

$$1/\text{RFD} = \text{MD}/\text{GD}$$

The numerator in an RF scale is always the number 1. The RFD is the denominator of the RF (such as 24,000). MD is map distance measured in any particular units on the map (cm, in.). GD is the true ground distance that the map distance represents (expressed in the same units used to measure the map distance). Never mix units in this calculation, but convert values into desired units afterward.

Example: How long in inches is a mile on a 1:24,000 scale map?

***Important information:*** There are 63,360 inches in a mile.
To find MD, for a known distance (mile) on a map of known scale, use this formula:

$$1/24{,}000 = \text{MD}/63{,}360 \text{ in.}$$

$$1 \text{ mile} = 2.64 \text{ inches at } 1{:}24{,}000$$

This statement has now been converted into a **stated scale** so units may be mixed.

### 1:24,000 Bar Scale

Here is a bar scale that can be used to make distance measurements directly from the 1:24,000 maps printed in the Map Interpretation sections. Note: Check the listed RF, because not all are at a 1:24,000 scale.

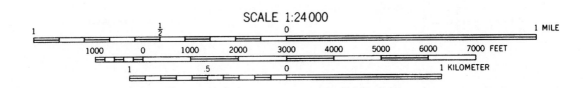

SCALE 1:24 000

## BOUNDARIES

National  
State or territorial  
County or equivalent  
Civil township or equivalent  
Incorporated-city or equivalent  
Park, reservation, or monument  
Small park  

## LAND SURVEY SYSTEMS

U.S. Public Land Survey System:  
  Township or range line  
    Location doubtful  
  Section line  
    Location doubtful  
  Found section corner; found closing corner  
  Witness corner; meander corner  

Other land surveys:  
  Township or range line  
  Section line  
Land grant or mining claim; monument  
Fence line  

## ROADS AND RELATED FEATURES

Primary highway  
Secondary highway  
Light duty road  
Unimproved road  
Trail  
Dual highway  
Dual highway with median strip  
Road under construction  
Underpass; overpass  
Bridge  
Drawbridge  
Tunnel  

## BUILDINGS AND RELATED FEATURES

Dwelling or place of employment: small; large  
School; church  
Barn, warehouse, etc.: small; large  
House omission tint  
Racetrack  
Airport  
Landing strip  
Well (other than water); windmill  
Water tank: small; large  
Other tank: small; large  
Covered reservoir  
Gaging station  
Landmark object  
Campground; picnic area  
Cemetery: small; large  

## RAILROADS AND RELATED FEATURES

Standard gauge single track; station  
Standard gauge multiple track  
Abandoned  
Under construction  
Narrow gauge single track  
Narrow gauge multiple track  
Railroad in street  
Juxtaposition  
Roundhouse and turntable  

## TRANSMISSION LINES AND PIPELINES

Power transmission line; pole; tower  
Telephone or telegraph line  
Aboveground oil or gas pipeline  
Underground oil or gas pipeline  

## CONTOURS

Topographic:  
  Intermediate  
  Index  
  Supplementary  
  Depression  
  Cut; fill  

Bathymetric:  
  Intermediate  
  Index  
  Primary  
  Index Primary  
  Supplementary  

## MINES AND CAVES

Quarry or open pit mine  
Gravel, sand, clay, or borrow pit  
Mine tunnel or cave entrance  
Prospect; mine shaft  
Mine dump  
Tailings  

## SURFACE FEATURES

Levee  
Sand or mud area, dunes, or shifting sand  
Intricate surface area  
Gravel beach or glacial moraine  
Tailings pond  

## VEGETATION

Woods  
Scrub  
Orchard  
Vineyard  
Mangrove  

## COASTAL FEATURES

Foreshore flat  
Rock or coral reef  
Rock bare or awash  
Group of rocks bare or awash  
Exposed wreck  
Depth curve; sounding  
Breakwater, pier, jetty, or wharf  
Seawall  

## BATHYMETRIC FEATURES

Area exposed at mean low tide; sounding datum  
Channel  
Offshore oil or gas: well; platform  
Sunken rock  

## RIVERS, LAKES, AND CANALS

Intermittent stream  
Intermittent river  
Disappearing stream  
Perennial stream  
Perennial river  
Small falls; small rapids  
Large falls; large rapids  
Masonry dam  
Dam with lock  
Dam carrying road  
Intermittent lake or pond  
Dry lake  
Narrow wash  
Wide wash  
Canal, flume, or aqueduct with lock  
Elevated aqueduct, flume, or conduit  
Aqueduct tunnel  
Water well; spring or seep  

## GLACIERS AND PERMANENT SNOWFIELDS

Contours and limits  
Form lines  

## SUBMERGED AREAS AND BOGS

Marsh or swamp  
Submerged marsh or swamp  
Wooded marsh or swamp  
Submerged wooded marsh or swamp  
Rice field  
Land subject to inundation

# Appendix C

# Understanding and Recognizing Some Common Rocks

Rocks are aggregates of minerals, and although there are thousands of kinds of rocks on our planet, they can be classified into three fundamental kinds based on their origin: igneous, sedimentary, and metamorphic. The formation of rocks was outlined in Chapter 13. Having a solid knowledge of how the Rock Cycle operates (see Fig. 13.19) as well as its components and its processes is essential to understanding the solid Earth. Specific rock types are mentioned in several chapters of this book. Although making a

positive identification of a rock type requires examining several physical properties, having a mental image of what different rocks look like will be an aid to understanding Earth processes and landforms. The following is an illustrated guide to a few common rocks to help in their identification. Intrusive igneous rocks form from a molten state by cooling and crystallizing underground; they generally cool very slowly, which allows coarse crystals to form that are easy to see with the naked eye.

## Igneous Rocks

Igneous rocks are subdivided into intrusive and extrusive, depending on whether they cooled within Earth or on its exterior.

### Intrusive Igneous

Copyright and photograph by Dr. Parvinder S. Sethi

**Granite**

Granite forms deep in the crust, and has easily visible intergrown crystals of light and dark minerals, but is dominated by light-colored silicate minerals. Granitic rocks are typically gray or pink, and their mineral composition is similar to that of the continental crust.

Copyright and photograph by Dr. Parvinder S. Sethi

**Diorite**

Diorite is an intermediate intrusive rock, meaning that it has a roughly equal mix of easily visible light and dark minerals, which gives it a spotted appearance. Generally diorite is dark gray in color.

Copyright and photograph by Dr. Parvinder S. Sethi

**Gabbro**

Gabbro is a dark intrusive rock dominated by heavy, iron-rich silicate minerals. The crystals are coarse enough to be easily visible, but because of the overall dark tone, they tend to blend together. Gabbro is black and may contain some very dark green minerals.

# Extrusive Igneous

Extrusive igneous rocks cool at or near the surface, and include lavas, as well as rocks made of tephra (pyroclastics), fragments blown out of a volcano. Extrusive rocks sometimes preserve gas bubble holes, and may contain visible crystals, but typically the grains are small. Cooling relatively rapidly at the surface produces fine-grained lavas. Clastic means made up of cemented rock fragments, such as clay, silt, sand, pebbles, cobbles, or even boulders. The sizes and shapes of clasts within a sedimentary rock provide clues about the environments under which the fragments were deposited (fluvial, eolian, glacial, coastal).

### Rhyolite

Rhyolite is a very thick lava when molten (much like melted glass), light in color and high in silica. Colors of rhyolite vary widely, but light gray, very light brown, and pink or reddish are common. Rhyolite is the extrusive equivalent of granite in terms of mineral composition.

### Andesite

Andesite, named for the Andes, is an intermediate lava in terms of both mineral content and color. Associated with composite cone volcanoes, it is relatively thick when molten. Often mineral crystals are visible in a matrix of finer grains, and the color is gray to brown. Andesite is the extrusive equivalent of diorite.

### Basalt

Basalt is dark, typically black, and heavier than other lavas. Associated with fissure flows and shield volcanoes, basalt is relatively low in silica, so it has a lower viscosity than other lavas. Basalt tends to be hotter than other lavas, and is relatively thin when molten, thus it can flow for many miles before cooling enough to stop. Basalt is the rock of the oceanic crust and is the extrusive equivalent of gabbro.

### Tuff

Tuff is a rock made of fine tephra—volcanic ash—that was blown into the atmosphere by a volcanic eruption, and settled out in layers that blanketed the surface. Tephra (loose fragments) was converted into tuff by burial and compaction, or by being welded together from intense heat. Tuff is gray to tan.

# Sedimentary Rocks

Sedimentary rocks can also be divided into two major categories, clastic and nonclastic.

## Clastic Sedimentary

Clastic means made up of cemented rock fragments, such as clay, silt, sand, granules, pebbles, cobbles, or even boulders. The sizes and shapes of clasts within a sedimentary rock provide clues about the environments under which the fragments were deposited (fluvial, eolian, glacial, coastal).

Copyright and photograph by Dr. Parvinder S. Sethi

### Shale
Shale is a fine-grained clastic rock that contains lithified clays, generally deposited in very thin layers. Shales represent a calm water environment, such as a sea or lake bottom. Shales vary widely in color, but most are gray or black, and they break up into smooth, flat surfaces.

Copyright and photograph by Dr. Parvinder S. Sethi

### Sandstone
Sandstone is made of cemented fragments of sand size, typically made of quartz or other relatively hard mineral. Sandstones can be virtually any color, feel gritty, and may be banded or layered. Sandstone may represent ancient beaches, dunes, or fluvial deposits.

Copyright and photograph by Dr. Parvinder S. Sethi

### Conglomerate
Conglomerate contains rounded pebbles cemented together by finer sediments. Conglomerate may represent deposits from a river's bed load, or a pebble beach.

Copyright and photograph by Dr. Parvinder S. Sethi

### Breccia
Breccia is similar to conglomerate, but the cemented fragments are angular. Breccia is associated with mudflows, pyroclastic flows, and fragments deposited by mass wasting.

# Nonclastic Sedimentary

Nonclastic sedimentary rocks consist of materials that are not rock fragments. Examples include chemical precipitates, such as limestone, evaporites such as rock salt, and deposits of organic materials, for example coal, or limestones made of shells and coral fragments. Nonclastic rocks also represent the ancient environment under which they were deposited.

### Rock Salt
Rock salt consists of sodium chloride, table salt, often with a mix of other salts. Rock salt represents the deposits left behind by the evaporation of a saline inland lake, or an arm of the sea that was cut off from the ocean by sea-level change or tectonic activity. Common color is white.

### Limestone
Limestone is made of calcium carbonate deposits (lime, $CaCO_3$). Limestone can represent a variety of environments and varies widely in color and appearance. Typical colors are white or gray, and the most common depositional environment was in shallow tropical seas, which were rich in lime. Many cave and spring deposits are also varieties of limestone.

### Coal
Coal is a rock made of the carbonized remains of ancient plants. Coal deposits typically represent a swampy lowland environment that was invaded by sea-level rise, which killed and buried dense vegetation.

# Metamorphic Rocks

Metamorphic rocks can also be divided into two general categories, foliated rocks and nonfoliated rocks.

## Foliated

Foliated metamorphic rocks have either wavy, roughly parallel plates, or bands of light and dark minerals that formed under intense heat and pressure. The nature of these foliations indicate the degree of metamorphism or change from the rock's original state.

Copyright and photograph by Dr. Parvinder S. Sethi

**Slate**
Slate is metamorphosed shale, and looks much like shale, except it is harder and has very thin platy foliations. The most common color is black.

Copyright and photograph by Dr. Parvinder S. Sethi

**Schist**
Schist has very prominent, wavy, and platy foliations generally covered with mineral crystals that formed during metamorphism. Schist represents a high degree of metamorphism and could originally have been any of a wide variety of rocks, so colors and appearance vary greatly.

Copyright and photograph by Dr. Parvinder S. Sethi

**Gneiss**
Gneiss (pronounced "nice") is a banded metamorphic rock, with alternating bands of light and dark minerals. Gneiss represents extreme heat and pressure during metamorphism and also may originally have been any of a variety of rocks. Metamorphism of granites commonly produces gneiss.

# Nonfoliated

Nonfoliated metamorphic rocks do not display regular patterns of banding or platy foliation. In general, nonfoliated metamorphics represent a rock that has been changed by the fusing and recrystallization of minerals in the original, often identifiable, rock.

Copyright and photograph by Dr. Parvinder S. Sethi

Copyright and photograph by Dr. Parvinder S. Sethi

**Quartzite**
Quartzite is metamorphosed quartz sandstone in which the former sand grains have fused together to produce an extremely hard, resistant rock.

**Marble**
Marble is metamorphosed limestone that has been recrystallized. Many colors and patterns of marble exist, and its relative softness compared to other rocks makes it easy to cut and polish.

# Glossary

**aa** a blocky, angular surface of a lava flow.

**abiotic** natural, nonliving component of an ecosystem.

**ablation** any removal of frozen water from the mass of a glacier.

**abrasion** erosion process in which particles already being carried by a geomorphic agent are used as tools to aid in eroding more Earth material.

**abrasion platform** wave-cut bench of rock just below the water level; indicates the landward extent of coastal cliff erosion.

**absolute humidity** mass of water vapor present per unit volume of air, expressed as grams per cubic meter, or grains per cubic foot.

**absolute location** location of an object on the basis of mathematical coordinates on an Earth grid.

**absolute zero** zero on the Kelvin temperature scale, indicating absence of molecular motion (no temperature).

**accretion** growth of a continent by adding large pieces of crust along its border by plate tectonic collision.

**accumulation** any addition of frozen water to the mass of a glacier.

**acid mine drainage** seepage to the surface of subsurface water that has become highly acidic by flowing through underground coal mines.

**acid rain** rain with a pH value of less than 5.6, the pH of natural rain; often linked to the pollution associated with the burning of fossil fuels.

**active layer** the upper soil zone that thaws in summer in regions underlain by permafrost.

**active-margin coast** coastal region characterized by active volcanism and tectonism.

**actual evapotranspiration (actual ET)** the actual amount of moisture loss through evapotranspiration measured from a surface.

**adiabatic heating and cooling** changes of temperature within a gas because of compression (resulting in heating) or expansion (resulting in cooling); no heat is added or subtracted from outside.

**advection** horizontal heat transfer within the atmosphere; air masses moved horizontally, usually by wind.

**advection fog** fog produced by the movement of warm, moist air across a cold sea or land surface.

**aerial photograph** photographs of the terrain taken with a camera and film from an aircraft

**aerosols** Tiny solid particles or droplets of liquid suspended in the atmosphere.

**aggradation** building up of an area or landscape that results from more deposition than erosion over time.

**air mass** large portion of the atmosphere, sometimes subcontinental in size, that may move over Earth's surface as a distinct, relatively homogeneous entity.

**albedo** proportion of solar radiation reflected back from a surface, expressed as a percentage of radiation received on that surface.

**Aleutian low** center of low atmospheric pressure in the area of the Aleutian Islands, especially persistent in winter.

**Alfisol** soil that has a subsurface clay horizon, is medium to high in bases and is light colored.

**alluvial fan** fan-shaped depositional landform, particularly common in arid regions, occurring where a stream emerges from a mountain canyon and deposits sediment on a plain.

**alluvial plain** extensive floodplain of very low relief.

**alluvium** general term for clastic particles deposited by a stream.

**alpine glacier** a mass of flowing ice that exists in a mountainous region due to climatic conditions resulting from high elevation.

**Altithermal** an interval of time about 7000 years ago when the climate was hotter than it is today.

**altitude** heights of points above Earth's surface.

**alto** signifies a middle-level cloud (i.e., from 2000 to 6000 m in elevation).

**analemma** a diagram that shows the declination of the sun throughout the year.

**Andisol** soil that develops on volcanic parent material.

**aneroid barometer** a device that uses the expansion and contraction of a sealed, accordion-like cell to measure atmospheric pressure.

**angle of inclination** tilt of Earth's polar axis at an angle of 23½° from the vertical to the plane of the ecliptic.

**angle of repose** maximum angle at which a slope of loose sediment can stand without particles tumbling or sliding downslope.

**annual march of temperature** changes in monthly temperatures throughout the months of the year.

**annual temperature lag** time lag between the maximum (minimum) of insolation during the solstices and the warmest (coldest) temperatures of the year.

**annual temperature range** difference between the mean daily temperatures for the warmest and coolest months of the year.

**Antarctic circle** latitude at 66½°S; the northern limit of the zone in the Southern Hemisphere that experiences a 24-hour period of sunlight and a 24-hour period of darkness at least once a year.

**anticline** the upfolded element of folded rock structure.

**anticyclone (high)** an area of high atmospheric pressure, also known as a high.

**aphelion** position of Earth's orbit at farthest distance from the sun during each Earth revolution.

**aquiclude** rock layer that restricts flow and storage of groundwater; it is impermeable and nonporous.

**aquifer** rock layer that is a container and transmitter of groundwater; it is both porous and permeable.

**Arctic Circle** parallel of latitude at 66½°N; the southern limit of the zone in the Northern Hemisphere that experiences a 24-hour period of sunlight and a 24-hour period of darkness at least once a year.

**arête** jagged, sawtooth spine or wall of rock separating two glacial valleys or cirques.

**arid climate** climate region or consition where annual potential evapotranspiration greatly exceeds annual precipitation.

**Aridisol** soil that develops in deserts where precipitation is less than half the potential evaporation.

**artesian** condition in which groundwater is under sufficient pressure to flow to an outlet if one becomes available.

**artesian spring** natural flow of groundwater to the surface from below due to pressure.

**artesian well** *See* flowing artesian well and nonflowing artesian well.

**artificial recharge** diverting surface water to permeable terrain for the purpose of replenishing groundwater supplies.

**asteroid** sometimes called a minor planet; any solar system body composed of rock and/or metal not exceeding 800 kilometers (500 mi) in diameter.

**asthenosphere** thick, plastic layer within Earth's mantle that flows in response to convection, instigating plate tectonic motion.

**atmosphere** blanket of air, composed of various gases, that envelops Earth.

**atmospheric air pressure (barometric pressure)** force per unit area that the atmosphere exerts on any surface at a particular elevation.

**atmospheric disturbance** refers to variation in the secondary circulation of the atmosphere that cannot correctly be classified as a storm; for example, front, air mass.

**atoll** ring of coral reefs and islands encircling a lagoon, with no inner island.

**attrition** the reduction of size in sediment as it is transported downstream.

**aurora** colorful interaction of solar radiation with ions in Earth's upper atmosphere; more commonly seen in higher latitudes. Called aurora borealis in the Northern Hemisphere (also known as the northern lights), and the aurora australis (southern lights) in the Southern Hemisphere.

**Aurora Australis** The southern equivalent of the Aurora Borealis, (the northern lights).

**Aurora Borealis** Also called the Northern Lights—illuminations in the night sky, most common in the high latitudes, caused by charged particles interacting with the Earth's magnetic field.

**autotroph (producer)** organism that, because it is capable of photosynthesis, is at the foundation of a food web and is considered a basic producer.

**autumnal equinox** The day on which the subsolar point is at the equator and that marks the first day of autumn; occurs on the September equinox for the Northern Hemisphere and the March equinox for the Southern Hemisphere.

**avalanche** density current of pulverized (powdered) Earth material traveling rapidly downslope by the pull of gravity.

**axis** an imaginary line between the geographic North Pole and South Pole, around which the planet rotates.

**azimuth** an angular direction of a line, point, or route measured clockwise from north 0–360°.

**azimuthal map** map created with a projection that preserves the true direction from the map center to any other point on the map.

**Azores High** *see* Bermuda High.

**backing wind shift** change in wind direction counterclockwise around the compass; for example, from east to northeast, to north, to northwest.

**backwash** thin sheet of broken wave water that rushes back down the beach face in the swash zone.

**badlands** barren region of soft rock material intensely eroded into ridges and ravines by numerous gullies and washes.

**bajada** an extensive intermediate slope of adjacent, coalescing alluvial fans connecting a steep mountain front with a basin or plain.

**bar** a shallow, submerged, mounded accumulation of sediment in near-shore locations or in streams.

**barchan** crescent-shaped sand dune with arms pointing downwind.

**barometer** instrument for measuring atmospheric pressure.

**barrier beach** a category of elongate depositional coastal landforms that lie parallel to the mainland but separated from it by a lagoon.

**barrier island** a barrier beach constructed parallel to the mainland, but not attached to it; separated from the mainland by a lagoon.

**barrier reef** coral reef parallel to the coast and separated from it by a lagoon.

**barrier spit** a barrier beach attached to the mainland at one end like a spit, but enclosing a lagoon.

**basalt** a dark-colored, fine-grained extrusive igneous rock generally associated with the oceanic crust and oceanic volcanoes.

**basaltic plateau** elevated area of low surface relief consisting of horizontal layers of basaltic lava.

**base flow** groundwater that seeps into stream channels below the water surface; sustains perennial streams between storms.

**base level** elevation below which a stream cannot flow; most humid region streams flow to sea level (ultimate base level).

**base line** east-west lines of division in the U.S. Public Lands Survey System.

**batholith** a large, irregular mass of intrusive igneous rock (pluton).

**baymouth barrier** a barrier beach that extends across the mouth of an embayment, attached to the mainland at both ends, to form a lagoon.

**beach** coastal landform of wave-deposited sediment attached to dry land along its entire length.

**beach drifting** littoral drifting in which waves breaking at an angle to the shoreline move sediment along the beach in a zigzag fashion in the swash zone.

**bearing** an angular direction of a line, point, or route measured from north or from a current location to a desired location (often in 90° compass quadrants).

**bed load** solid particles moved by wind or water by bouncing, rolling, or sliding along the ground or streambed.

**bedding plane** boundary between different sedimentary layers.

**bedrock** solid rock of Earth's crust that underlies soil and other unconsolidated materials

**benthos** classification of marine organisms that live on the ocean floor.

**Bergeron (ice crystal) process** rain-forming process where cloud droplets begin as ice crystals and melt into rain as they fall toward the surface.

**bergschrund** the large crevasse at the head of a valley glacier, beneath the cirque headwall.

**Bermuda High (Azores High)** persistent, high atmospheric pressure center located in the subtropics of the north Atlantic Ocean, also called the Azores High.

**biogeography** The study of the distribution of organisms and their existing and changing relationships.

**biomass** amount of living material or standing crop in an ecosystem or at a particular trophic level within an ecosystem.

**biome** one of Earth's major terrestrial ecosystems, classified by the vegetation types that dominate the plant communities within the ecosystem.

**biosphere** the life forms—human, animal, or plant—of Earth that form one of the major Earth subsystems.

**blizzard** heavy snowstorm accompanied by strong winds (35 mph or greater) that reduces visibility to less than one-quarter mile.

**blowout** local, wind-eroded surface depression in an area dominated by wind-deposited sand.

**bolson** desert basin, surrounded by mountains, with no drainage outlet.

**boreal forest (taiga)** coniferous forest dominated by spruce, fir, and pine found growing in subarctic conditions around the world north of the 50th parallel of north latitude.

**boulder** a rock fragment greater than 256 millimeters (10 in) in diameter.

**braided channel** stream channel composed of multiple subchannels of simultaneous flow that split and rejoin and frequently shift position.

**breaker zone** the part of the nearshore area in which waves break.

**butte** isolated erosional remnant of a tableland with a flat summit, often bordered by steep-sided escarpments. Buttes are usually found in arid regions of flat-lying sediments and are somewhat taller than they are wide.

**calcification** soil-forming process of subhumid and semiarid climates. Soil types in the mollisol order, the typical end products of the process, are characterized by little leaching or eluviation and by the accumulation of both humus and mineral bases (especially calcium carbonate, $CaCO_3$).

**caldera** a large depression formed by a volcanic eruption.

**caliche** hardened layers of lime ($CaCO_3$) deposited at the surface of a soil by evaporating capillary water.

**calorie** amount of heat necessary to raise the temperature of 1 gram of water 1°C.

**calving** breaking off a mass of ice from the toe of a glacier at its junction with the ocean or a lake.

**Canadian High** high atmospheric pressure area that tends to develop over the central North American continent in winter.

**capacity (moisture)** the maximum amount of water vapor that can be contained in a given quantity of air at a given temperature.

**capillary action** the upward movement of water through tiny cracks and pore spaces.

**capillary water** soil water that clings to soil peds and individual soil particles as a result of surface tension. Capillary water moves in all directions through the soil from areas of surplus water to areas of deficit.

**caprock** a resistant horizontal layer of rock that forms the flat top of a landform, such as a butte or a mesa.

**carbonate** substance that contains one or more $CO_3$ ions as part of its chemical formula.

**carbonation** carbon dioxide in water chemically combining with other substances to create new compounds.

**carnivore** animal that eats only other animals.

**cartography** the science of mapmaking.

**catastrophism** once-popular theory that Earth's landscapes developed in a relatively short time by cataclysmic events.

**cavern (cave)** natural void in rock created by solution that is large enough for people to enter.

**Celsius (centigrade) scale** temperature scale in which 0° is the freezing point of water and 100° its boiling point at standard sea-level pressure.

**centrifugal force** force that pulls a rotating object away from the center of rotation.

**channel roughness** an expression of the frictional resistance to stream flow due to irregularities in a stream channel bed and sides.

**chaparral** sclerophyllous woodland vegetation found growing in the Mediterranean climate of the western United States; these seasonal, drought-resistant plants are low-growing, with small, hard-surfaced leaves and deep, water-probing roots.

**chemical precipitate sedimentary rock** rock created from dissolved minerals that have precipitated out of water.

**chemical weathering** breakdown of rock material by chemical reactions that change the rock's mineral composition (decomposition).

**Chinook** dry warm wind on the eastern slopes of the Rocky Mountains (*see* foehn wind).

**cinder cone** hill composed of fragments of volcanic rock (pyroclastics) erupted from a central vent.

**circle of illumination** line dividing the sunlit (day) hemisphere from the shaded (night) hemisphere; experienced by individuals on Earth's surface as sunrise and/or sunset.

**cirque** deep, sometimes steep-sided amphitheater formed at the head of an alpine valley by glacial ice erosion.

**cirque glacier** a generally small alpine glacier restricted to a high-elevation basin (cirque).

**cirro** signifies a high-level cloud (i.e., above 6000 m in elevation).

**cirrus** high, detached clouds consisting of ice particles. Cirrus clouds are white and feathery or fibrous in appearance.

**Cl, O, R, P, T** Hans Jenny's list of soil formation factors: Climate, Organics, Relief, Parent material, and Time.

**clast** solid broken piece of rock, bone, or shell.

**clastic sedimentary rock** rock formed by the compaction and cementation of preexisting rock debris.

**clay (clayey)** a very fine grained mineral particle with a size less than 0.004 millimeter, often the product of weathering.

**clay mineral** A small (clay-sized) hydrous aluminum silicate material typically created by chemical weathering, especially hydrolysis.

**clay pan** an ephemeral lake bed composed mostly of fine-grained clastic particles.

**climate** accumulated and averaged weather patterns of a locality or region; the full description is based on long-term statistics and includes extremes or deviations from the norm.

**climatology** scientific study of climates of Earth and their distribution.

**climax community** the final step in the succession of plant communities that occupy a specific location.

**climograph** graphic means of giving information on mean monthly temperature and rainfall for a select location or station.

**closed system** system in which no substantial amount of materials can cross its boundaries.

**cloud forest** rainforest which is produced by nearly-constant light rain on the windward slopes of mountains.

**coastal zone** general region of interaction between the land and a lake or the ocean.

**coastlines of emergence** coast with formerly submerged land that is now above water, either due to uplift of the land or a drop in sea level.

**coastline of submergence** a coastal area that has undergone sinking or subsidence relative to sea level.

**cobble** rock fragments ranging in diameter from 64 to 256 millimeters (2.5–10 in).

**cobble beach** cobble-sized sediment deposited by waves along the shoreline, often found along the base of sea cliffs or lake cliffs.

**col** a glacially eroded pass between two mountain valleys.

**cold current** a flow of sea water that moves like a river through the ocean and is relatively colder than water in the ocean area that it flows through.

**cold front** leading edge of a relatively cooler, denser air mass that advances upon a warmer, less dense air mass.

**collapse sinkhole** topographic depression formed mainly by the cave-in of the land above a cavern.

**collision-coalescence process** rain-forming process where raindrops form by collision between cloud droplets.

**color composite image** a digital image that combines several wavelength bands—a common color composite that is used resembles a near-infrared color photo.

**column** pillar-shaped speleothem resulting from the joining of a stalactite and a stalagmite.

**columnar joints** vertical, polygonal fractures caused by lava shrinking as it cools.

**comet** a small body of icy and dusty matter that revolves about the sun. When a comet comes near the sun, some of its material vaporizes, forming a large head and often a tail.

**composite cone (stratovolcano)** volcano formed from alternating layers of lava and pyroclastic materials; generally known for violent eruptions.

**compressional tectonic force** pushing together from opposite sides (convergence).

**compromise projection** maps that compromise true shape and true area in order to display both fairly well.

**conceptual model** image in the mind of an Earth feature or landscape as derived from personal experiences.

**condensation (release of heat)** process by which a vapor is converted to a liquid during which energy is released in the form of latent heat.

**condensation nuclei** minute particles in the atmosphere (e.g., dust, smoke, pollen, sea salt) on which condensation can take place.

**conduction** transfer of heat within a body or between adjacent matter by means of internal molecular movement.

**conformal map** a map using a projection that maintains the true shape of small areas on Earth's surface.

**conic projection** map projection achieved by transferring a spherical area (typically the geographic grid) onto a cone.

**continental** pertaining to, or having characteristics derived from, large land masses.

**Continental Arctic (cA)** very cold, very dry air mass originating from the arctic region.

**continental collision** the fusing together of landmasses as tectonic plates converge.

**continental crust** the less dense (ave. 2.7 g/cm³), thicker portion of Earth's crust; underlies the continents.

**Continental Divide** line of separation dividing runoff between the Pacific and Atlantic Oceans. In North America it generally follows the crest of the Rocky Mountains.

**continental drift** theory proposed by Alfred Wegener stating that the continents joined, broke apart, and moved on Earth's surface; it was later replaced by the theory of plate tectonics.

**continental glacier** a very large and thick mass of flowing ice that exists due to climatic conditions resulting from high latitude.

**continental island** island that is geologically part of a continent and usually located on the continental shelf.

**Continental Polar (cP)** cold, dry air mass originating from landmasses approximately 40° to 60° N or S latitude.

**continental shelf** the gently sloping margin of a continent overlain by ocean water.

**continental shield** the ancient part of a continent that consists of crystalline rock.

**Continental Tropical (cT)** warm, dry air mass originating from subtropical landmasses.

**continentality** the distance a particular place is located in respect to a large body of water; the greater the distance, the greater the continentality.

**continuous data** numerical or locational representations of phenomena that are present everywhere—such as air pressure, temperature, elevation.

**contour interval** vertical distance represented by two adjacent contour lines on a topographic map.

**contour map (topographic map)** map that uses contour lines to show differences in elevation (topography).

**controls of temperature** factors which control temperatures around the world, such as: latitude, proximity to water bodies, ocean currents, altitude, and landform barriers and human activities.

**convection** process by which a circulation is produced within an air mass or fluid body (heated material rises, cooled material sinks); also, in tectonic plate theory, the method whereby heat is transferred to Earth's surface from deep within the mantle.

**convectional precipitation** precipitation resulting from condensation of water vapor in an air mass that is rising convectionally as it is heated from below.

**convective thunderstorm** a thunderstorm produced by the convective uplift mechanism.

**convergent wind circulation** pressure-and-wind system where the airflow is inward toward the center, where pressure is lowest.

**coordinate system** a precise system of grid lines used to describe locations.

**coral reef** shallow, wave-resistant structure made by the accumulation of skeletal remains of tiny sea animals.

**core** extremely hot and dense innermost portion of Earth's interior; the molten outer core is 2400 kilometers (1500 mi) thick; the solid inner core is 1120 kilometers (700 mi) thick.

**Coriolis effect** apparent effect of Earth's rotation on horizontally moving bodies, such as wind and ocean currents; such bodies tend to be deflected to the right in the Northern Hemisphere and to the left in the Southern Hemisphere.

**corridor** in biogeographic terms, any linear feature that crosses the general vegetation cover in an area (for example, rivers, power line clearings, roads).

**corrosion** chemical erosion of rock matter by water; the removal of ions from rock-forming minerals in water.

**creep** slow downslope movement of Earth material involving the lifting and falling action of sediment particles.

**crevasse** stress crack commonly found along the margins and at the toe of a glacier.

**cross bedding** thin layers within sedimentary rocks that were deposited at an angle to the dominant rock layering.

**crust** relatively thin, approximately 8–64 kilometers (5–40 mi) deep, low-density surface layer of Earth.

**crustal warping** gentle bending and folding of crustal rocks.

**cumulus** globular clouds, usually with a horizontal base and strong vertical development.

**cut bank** the steep slope found on the outside of a bend in a meandering stream channel.

**cyclone (low)** center of low atmospheric pressure.

**cyclonic (convergence) precipitation** precipitation formed by cyclonic uplift.

**cylindrical projection** map projection achieved by transferring a spherical area (typically the geographic grid) onto a cylinder.

**daily (diurnal) temperature range** difference between the highest and lowest temperatures of the day (usually recorded hourly).

**daily march of temperature** changes in daily temperatures as we go from an overnight low to a daytime high and back to the overnight low.

**daily temperature lag** The time of offset between solar noon, when solar radiation is most effective in heating, and the warmest time of day (usually 2-4 hours after 12 noon).

**debris** unconsolidated slope material with a wide range of grain sizes including at least 20% gravel (>2 mm).

**debris flow** rapid, gravity-induced downslope movement of wet, poorly sorted Earth material.

**debris flow fan** fan-shaped depositional landform, particularly common in arid regions, created where debris flows emerge onto a plain from a mountain canyon.

**deciduous** a plant, usually a tree or shrub, that drops its leaves (usually seasonally).

**decimal degrees** angular degrees with fractions expressed as tenths, hundredths, etc., rather than as the sixty-based traditional subdivisions of minutes and seconds.

**detritivore (decomposer)** organism that promotes decay by feeding on dead plant and animal material and returns mineral nutrients to the soil or water in a form that plants can utilize.

**deep–water wave** wave traveling in depth of water greater than or equal to half the wavelength.

**deflation** entrainment and removal of loose surface sediment by the wind.

**deflation hollow** a wind-eroded depression in an area not dominated by wind-deposited sand.

**degradation** landscape lowering that results from more erosion than deposition over time.

**delta** depositional landform constructed where a stream flows into a standing body of water (a lake or the ocean).

**delta plain** the portion of a delta that lies above the level of the lake or ocean.

**dendritic** term used to describe a drainage pattern that is treelike with tributaries joining the main stream at acute angles.

**dendrochronology** method of determining past climatic conditions using tree rings.

**deposition** accumulation of Earth materials at a new site after being moved by gravity, water, wind, or glacial ice.

**desert climate** climate where the amount of precipitation received is less than one half of the potential ET.

**desert pavement (reg)** desert surface mosaic of close-fitting stones that overlies a deposit of mostly fine-grained sediment.

**dew** tiny droplets of water on ground surfaces, grass blades, or solid objects. Dew is formed by condensation when air at the surface reaches the dew point.

**dew point** the temperature at which an air mass becomes saturated; any further cooling will cause condensation of water vapor in the air.

**differential weathering and erosion** rock types vary in resistance to weathering and erosion, causing the processes to occur at different rates and often producing distinctive landform features.

**digital elevation model (DEM)** three-dimensional views of topography.

**digital image** an image made from computer data displayed like a mosaic of tiny squares, called pixels.

**digital map layers** mapped information and data that are stored in digital form and can be overlaid with other mapped information in a geographic information system.

**digital terrain model** a computer-generated graphic representation of topography.

**dike** igneous intrusion with a wall-like shape.

**dip** inclination of a rock layer from the horizontal; always measured at right angles to the strike.

**dip–slip fault** a vertical fault where the movement is up and down the dip of the fault surface.

**disappearing stream** stream that has its flow diverted entirely to the subsurface.

**discharge** *see* stream discharge.

**discrete data** numerical or locational representations of phenomena that are present only at certain locations—such as earthquake epicenters, sinkholes, tornado paths.

**dissolved load** soluble minerals or other chemical constituents carried in water as a solution.

**distributary** a smaller stream that conducts flow away from the larger main channel, especially on deltas; the opposite of a tributary.

**diurnal** synonym for daily, particularly in reference to a day to night pattern of change.

**divergent wind circulation** pressure-and-wind system where the airflow is outward away from the center, where pressure is highest.

**doldrums** zone of low pressure and calms along the equator.

**doline** *see* sinkhole.

**Doppler radar** advanced type of radar that can detect motion in storms, specifically motion toward and away from the radar signal.

**drainage area** the size of a drainage basin, typically expressed in $km^2$ or $mi^2$.

**drainage basin (watershed, catchment)** the region that provides runoff to a stream.

**drainage density** the summed length of all stream channels per unit area in a drainage basin.

**drainage divide** the outer boundary of a drainage basin.

**drainage pattern** the form of the arrangement of channels in a stream system in map view.

**drainage wind (katabatic wind)** downslope flow of cold, dense air that has accumulated in a high mountain valley or over an elevated plateau or ice cap.

**draping** overlaying a digital image over a digital elevation model of a landscape.

**drift** sediment deposited in association with glacial ice or its meltwater.

**drizzle** fine mist or haze of very small water droplets with a barely perceptible falling motion.

**drumlin** streamlined, elongated hill composed of glacial drift with a tapered end indicating direction of continental ice flow.

**dry adiabatic lapse rate** rate at which a rising mass of air is cooled by expansion when no condensation is occurring ($10°C/1000$ m or $5.6°F/1000$ ft).

**dust storm** a moving cloud of wind-blown dust (typically silt).

**dynamic equilibrium** constantly changing relationship among the variables of a system, which produces a balance between the amounts of energy and/or materials that enter a system and the amounts that leave.

**earth (as a mass wasting material)** thick unit of unconsolidated, predominantly fine-grained slope material.

**Earth system** set of interrelated components, variables, processes, and subsystems (e.g., atmosphere, lithosphere, biosphere, hydrosphere), which interact and function together to make up Earth.

**Earth's energy budget** relationship between solar energy input, storage, and output within the Earth system.

**earthquake** series of shock waves set in motion by sudden movement along a fault.

**earthquake intensity** a measure of the impact of an earthquake on humans and their built environment.

**earthquake magnitude** measurement representing an earthquake's size in terms of energy released.

**easterly wave** trough-shaped, weak, low pressure cell that progresses slowly from east to west in the trade wind belt of the tropics; this type of disturbance sometimes develops into a tropical hurricane.

**eccentricity cycle** the change in Earth's orbit from slightly elliptical to more circular, and back to its earlier shape every 100,000 years.

**ecological niche** combination of role and habitat as represented by a particular species in an ecosystem.

**ecology** science that studies the interactions between organisms and their environment.

**ecosystem** community of organisms functioning together in an interdependent relationship with the environment that they occupy.

**ecotone** transition zone of varied natural vegetation occupying the boundary between two adjacent and differing plant communities.

**effective precipitation** actual precipitation available to supply plants and soil with usable moisture; does not take into consideration storm runoff or evaporation.

**effluent** the condition of groundwater seeping into a stream.

**effusive eruption** streaming flows of molten rock matter pouring onto Earth's surface from the subsurface.

**El Niño** warm countercurrent that influences the central and eastern Pacific.

**elastic solid** a solid that withstands stress with little deformation until a maximum value is reached, whereupon it breaks.

**electromagnetic energy** all forms of energy that share the property of moving through space (or any medium) in a wavelike pattern of electric and magnetic fields; also called radiation.

**electromagnetic spectrum** The full range of wavelengths of electromagnetic energy, portions of which constitute such bands as visible light, ultraviolet radiation, infrared radiation, and others.

**element (atmosphere)** the major elements include solar energy, temperature, pressure, winds, and precipitation.

**elevation** vertical distance from mean sea level to a point or object on Earth's surface.

**eluviation** downward removal of soil components by water.

**empirical classification** classification process based on statistical, physical, or observable characteristics of phenomena; it ignores the causes or theory behind their occurrence.

**end moraine** a ridge of till deposited at the toe or terminus of a glacier.

**endogenic process** landforming process originating within Earth.

**energy balance** the concept that the energy Earth receives from the sun is equaled (balanced) by return of that energy back to space, thus Earth has not continually increased or decreased in temperature over its history of billions of years.

**Enhanced Fujita Scale** enhancement made to the tornado intensity scale which concentrates more specifically on the types of damages that occur.

**enhanced greenhouse effect** refers to the increased global warming impact by human contributions to the atmosphere, that is intended to differentiate it from the natural greenhouse effect.

**Entisol** soil with little or no development.

**entrenched stream** a stream that has eroded downward so that it flows in a relatively deep and steep-sided (trench-like) valley or canyon.

**environment** surroundings, whether of man or of any other living organism; includes physical, social, and cultural conditions that affect the development of that organism.

**eolian (aeolian)** pertaining to the landforming work of the wind.

**ephemeral flow** describes streams that conduct flow only occasionally, during to shortly after precipitation events or due to ice or snowmelt.

**epicenter** point on Earth's surface directly above the focus of an earthquake.

**epiphyte** a plant that grows on another plant, but does not take its nutrients from the host plant.

**epipedon** surface soil layer that possesses specific characteristics essential to the identification of soils in the National Resources Conservation Service System.

**equal-area map** a map created with a projection that preserves correct proportional sizes for any given areas of Earth's surface.

**equator** great circle of Earth midway between the poles; the zero degree parallel of latitude that divides Earth into the Northern and Southern Hemispheres.

**equatorial low (equatorial trough)** zone of low atmospheric pressure centered more or less over the equator where heated air is rising; *see also* doldrums.

**equidistance** a property of some maps that depicts distances equally without scale variation.

**equilibrium** state of balance between the interconnected components of a system, an organized whole.

**equilibrium line** balance position on a glacier that separates the zone of accumulation from the zone of ablation.

**equinox** one of two times each year (approximately March 21 and September 22) when the position of the noon sun is overhead (and its vertical rays strike) at the equator; all over Earth, day and night are of equal length.

**erosion** removal of Earth materials from a site by gravity, water, wind, or glacial ice.

**erratic** large glacially transported boulder deposited on top of bedrock of different composition.

**escarpment (scarp)** a steep slope, or cliff, of any height regardless of the geomorphic or geologic process by which it formed.

**esker** narrow, winding ridge of coarse sediment probably deposited in association with a meltwater tunnel at the base of a continental glacier.

**estuary** coastal embayment where salt and fresh water mix.

**evaporation** process by which a liquid is converted to the gaseous (vapor) state by the addition of latent heat.

**evaporative cooling** the lowered temperature experienced when water evaporates.

**evaporite** mineral salts that are soluble in water and accumulate when water evaporates.

**evapotranspiration** combined water loss to the atmosphere from ground and water surfaces by evaporation and, from plants, by transpiration.

**evergreen** a plant, usually a tree or shrub, does not seasonally drop its leaves.

**exfoliation** successive removal of outer rock sheets or slabs broken from the main rock mass by weathering.

**exfoliation dome** large, smooth, convex (dome-shaped) mass of exposed rock undergoing exfoliation due to weathering by unloading.

**exfoliation sheet** relatively thin outer layer of rock broken from the main rock mass by weathering.

**exogenic process** landforming process originating at or very near Earth's surface.

**exotic stream** stream that originates in a humid region and has sufficient water volume to flow across a desert region.

**explosive eruption** violent blast of molten and solid rock matter into the air.

**exposure** direction of mountain slopes with respect to prevailing wind direction.

**exterior drainage** streams and stream systems that flow to the ocean.

**extratropical disturbance** *see* middle-latitude disturbance.

**extrusive igneous rock** rock solidified at Earth's surface from lava; also called volcanic rock.

**Fahrenheit scale** temperature scale in which 32° is the freezing point of water, and 212° its boiling point, at standard sea-level pressure.

**fall** type of fast mass wasting characterized by Earth material plummeting downward freely through air.

**fan apex** the most upflow point on an alluvial fan; where the fan-forming stream emerges from the mountain canyon.

**fast mass wasting** gravity-induced downslope movement of Earth material that people can witness directly.

**fault** breakage zone along which rock masses have slid past each other.

**fault block** discrete block-like region of crustal rocks bordered on two opposite sides by normal faults.

**fault scarp (escarpment)** the steep cliff or exposed face of a fault where one crustal block has been displaced vertically relative to another.

**faulting** the movement of rock masses past each other along either side of a fault.

**feedback** sequence of changes in the elements of a system, which ultimately affects the element that was initially altered to begin the sequence.

**feedback loop** path of change as its effects move through the variables of a system until the effects impact the variable originally experiencing change.

**fertilization** *see* soil fertilization.

**fetch** distance over open water that winds blow without interruption.

**firn** compact granular snow formed by partial melting and refreezing due to overlying layers of snow.

**fissure** an extensive crack or break in rocks along which lava may be extruded.

**fissure flow** lava flow that emanates from a crack (fissure) in the surface rather than from the vent or crater of a volcano.

**fjord** deep, glacial trough along the coast invaded by the sea after the removal of a glacier.

**flood** stream water exceeding the amount that can be contained within its channel.

**flood basalt** massive outpouring of basaltic lava.

**floodplain** the low-gradient area adjacent to many stream channels that is subject to flooding and primarily composed of alluvium.

**flow** rapid downslope movement of wet unconsolidated Earth material that experiences considerable mixing.

**flowing artesian well** Artificially dug outlet that allows naturally pressurized groundwater to flow to Earth's surface.

**fluvial** term used to describe landforms and processes associated with the work of streams and rivers.

**fluvial geomorphology** the study of streams as landforming agents.

**focus** point within Earth's crust or upper mantle where an earthquake originated.

**foehn** warm, dry, downslope wind on lee of mountain range, caused by adiabatic heating of descending air.

**fog** mass of suspended water droplets within the atmosphere that is in contact with the ground.

**folding** the bending or wrinkling of Earth's crust due to compressional tectonic forces.

**foliation** the occurrence of banding or platy structure in metamorphic rocks.

**food chain** sequence of levels in the feeding pattern of an ecosystem.

**food web** feeding mosaic formed by the interrelated and overlapping food chains of an ecosystem.

**freeze–thaw weathering** breaking apart of rock by the expansive force of water freezing in cracks; also called frost weathering or ice wedging.

**freezing rain (glaze)** rainfall that freezes into ice upon coming in contact with a surface or object that is colder than 0°C (32°F).

**friction** force that acts opposite to the direction of movement or flow; for example, turbulent resistance of Earth's surface on the flow of the atmosphere.

**fringing reef** coral reef attached to the coast.

**front** sloping boundary or contact surface between air masses with different properties of temperature, moisture content, density, and atmospheric pressure.

**frontal precipitation** precipitation resulting from condensation of water vapor by the lifting or rising of warmer, lighter air above cooler, denser air along a frontal boundary.

**frontal thunderstorm** a thunderstorm produced by the frontal uplift mechanism.

**frost** frozen condensation that occurs when air at ground level is cooled to a dew point of 0°C (32°F) or below; also any temperature near or below freezing that threatens sensitive plants.

**fusion (thermonuclear) reaction** the fusing together of two hydrogen atoms to create one helium atom. This process releases tremendous amounts of energy.

**galactic movement** movement of the solar system within the Milky Way Galaxy.

**galaxy** a large assemblage of stars; a typical galaxy contains millions to hundreds of billions of stars.

**galleria forest** junglelike vegetation extending along and over streams in tropical forest regions.

**gap** an area within the territory occupied by a plant community when the climax vegetation has been destroyed or damaged by some natural process, such as a hurricane, forest fire, or landslide.

**Gelisol** soil that experiences frequent freezing and thawing.

**generalist** species that can survive on a wide range of food supplies.

**genetic classification** classification process based on the causes, theory, or origins of phenomena; generally ignoring their statistical, physical, or observable characteristics.

**geographic grid** lines of latitude and longitude form the geographic grid.

**geographic information system (GIS)** versatile computer software that combines the features of cartography and database management to produce new maps of data for solving spatial problems.

**geography** study of Earth phenomena; includes an analysis of distributional patterns and interrelationships among these phenomena.

**geomorphic agent** a medium that erodes, transports, and deposits Earth materials; includes water, wind, and glacial ice.

**geomorphology** the study of the origin and development of landforms.

**geostrophic winds** upper-level winds in which the Coriolis effect and pressure gradient are balanced, resulting in a wind flowing parallel to the isobars.

**geothermal water** water heated by contact with hot rocks in the subsurface.

**geyser** natural eruptive outflow of water that alternates between hot water and steam.

**giant planet (gas planet)** the four largest planets—Jupiter, Saturn, Uranus, and Neptune.

**glacial (glacial period)** a time characterized by the extensive expansion of glaciers in high latitude and high elevation areas, as well as into the middle latitudes. *See also* ice age.

**glacial outwash** the fluvial deposits derived from glacial meltwater streams.

**glacial trough** a U-shaped valley carved by glacial erosion.

**glaciation** an interval of glacial activity.

**glacier** a large mass of ice that flows as a plastic solid.

**glacier advance** expansion of the toe or terminus of a glacier to a lower elevation or lower latitude due to an increase in size.

**glacier head** farthest upslope part of an alpine glacier.

**glacier retreat** withdrawal of the toe or terminus of a glacier to a higher elevation or higher latitude due to a decrease in size.

**glacier toe (terminus)** farthest downslope part of an alpine glacier.

**glaciofluvial deposit** sorted glacial drift deposited by meltwater.

**glaciolacustrine deposit** sorted glacial drift deposited by meltwater in lakes associated with the margins of glaciers.

**gleization** soil-forming process of poorly drained areas in cold, wet climates. The resulting soils have a heavy surface layer of humus with a water-saturated clay horizon directly beneath.

**Global Positioning System (GPS)** GPS uses satellites and computers to compute positions and travel routes anywhere on Earth.

**global warming** climate change that would cause Earth's temperatures to rise.

**gnomonic projection** planar projection with greatly distorted land and water areas; valuable for navigation because all great circles on the projection appear as straight lines.

**graben** block of crustal rocks between two parallel normal faults that has slid downward relative to adjacent blocks.

**graded stream** stream where slope and channel size provide velocity just sufficient to transport the load supplied by the drainage basin; a theoretical balanced state averaged over a period of many years.

**gradient** a term for slope often used to describe the angle of a streambed.

**granite** a coarse-grained intrusive igneous rock generally associated with continental crust.

**granular disintegration** weathering feature of coarse crystalline rocks in which visible individual mineral grains fall away from the main rock mass.

**graphic (bar) scale** a ruler-like device placed on maps for making direct measurements in ground distances.

**gravel** a general term for sediment sizes larger than sand (larger than 2.0 mm)

**gravitational water** meteoric water that passes through the soil under the influence of gravitation.

**gravity** the mutual attraction of bodies or particles.

**great circle** any circle formed by a full circumference of the globe; the plane of a great circle passes through the center of the globe.

**greenhouse effect** warming of the atmosphere that occurs because shortwave solar radiation heats the planet's surface, but the loss of long-wave heat radiation is hindered by the release of gases associated with human activity (e.g., $CO_2$).

**greenhouse gases** atmospheric gases which hinder the escape of Earth's heat energy.

**Greenwich mean time (GMT)** time at zero degrees longitude used as the base time for Earth's 24 time zones; also called Universal Time or Zulu Time.

**groin** artificial structure extending out into the water from a beach built to inhibit loss of beach sediment.

**ground moraine** irregular, hummocky landscape of till deposited on Earth's surface by a wasting glacier.

**ground-inversion fog** see radiation fog.

**groundwater** water in the saturated zone below the water table.

**gully** steep-sided stream channel somewhat larger than rills that even in humid climates flow only in direct response to precipitation events.

**gyre** broad circular patterns of major surface ocean currents produced by large subtropical high pressure systems.

**habitat** location within an ecosystem occupied by a particular organism.

**hail** form of precipitation consisting of pellets or balls of ice with a concentric layered structure usually associated with the strong convection of cumulonimbus clouds.

**hanging valley** tributary glacial trough that enters a main glaciated valley at a level high above the valley floor.

**hardpan** dense, compacted, clay-rich layer occasionally found in the subsoil (B horizon) that is an end product of excessive illuviation.

**haystack hill (conical hill or hum)** remnant hills of soluble rock remaining after adjacent rock has been dissolved away in karst areas.

**headward erosion** gullying and valley cutting that extends a stream channel in an upstream direction.

**heat** energy being transferred from one substance or medium to another because of temperature differences.

**heat island** mass of warmer air overlying urban areas.

**heaving** various means by which particles are lifted perpendicular to a sloping surface, then fall straight down by gravity.

**hemisphere** half of a sphere; for example, the northern or southern half of Earth divided by the equator or the eastern and western half divided by two meridians, the 0° and 180° meridians.

**herbivore** an animal that eats only living plant material.

**heterosphere** layer of the atmosphere which lies between 80 km (50 miles) and the outer limits of the atmosphere; here, atmospheric gases separate into individual ionized gases.

**heterotroph (consumer)** an organism that consumes organic material from other life forms, including all animals, and parasitic plants.

**high** see anticyclone.

**highland climate** a general climate classification for regions of high, yet varying, elevations.

**Histosol** soil that develops in poorly drained areas.

**holistic approach** considering and examining all phenomena relevant to a problem.

**Holocene** the most recent time interval of warm, relatively stable climate that began with the retreat of major glaciers about 10,000 years ago.

**homosphere** layer of the atmosphere where all the atmospheric gases are mixed together in the same proportions; this layer lies between Earth's surface and 80 km (50 miles) aloft.

**horn** pyramid-like peak created where three or more expanding cirques meet at a mountain summit.

**horse latitudes** A historic term for the latitudinal belt influenced by the subtropical high pressure cells.

**horst** crustal block between two parallel normal faults that has slid upward relative to adjacent blocks.

**hot spot** a mass of hot molten rock material at a fixed location beneath a lithospheric plate.

**hot spring** natural outflow of geothermal groundwater to the surface.

**human geography** specialization in the systematic study of geography that focuses on the location, distribution, and spatial interaction of human (cultural) phenomena.

**humid continental, hot-summer climate** climate type characterized by hot, humid summers and mild, moist winters.

**humid continental, mild-summer climate** climate type characterized by mild, humid summers and cold, moist winters.

**humid subtropical climate** climate type characterized by hot humid summers, found in the subtropical regions of the world.

**humidity** amount of water vapor in an air mass at a given time.

**humus** organic matter found in the surface soil layers that is in various stages of decomposition as a result of bacterial action.

**hurricane** severe tropical cyclone of great size with nearly concentric isobars. Its torrential rains and high-velocity winds create unusually high seas and extensive coastal flooding; also called willy willies, tropical cyclones, baguios, and typhoons.

**hydration** rock weathering due to substances in cracks swelling and shrinking with the addition and removal of water molecules.

**hydraulic action** erosion resulting from the force of moving water.

**hydrologic cycle** circulation of water within the Earth system, from evaporation to condensation, precipitation, runoff, storage, and reevaporation back into the atmosphere.

**hydrolysis** water molecules chemically recombining with other substances to form new compounds.

**hydrosphere** major Earth subsystem consisting of the waters of Earth, including oceans, ice, freshwater bodies, groundwater, and water within the atmosphere and biomass.

**hygroscopic water** water in the soil that adheres to mineral particles.

**ice age** period of Earth history when much of Earth's surface was covered with massive continental glaciers. The most recent ice age occurred during the Pleistocene Epoch.

**ice cap** a continental glacier of regional size, less than 50,000 km².

**ice fall** portion of a glacier moving over and down a steep slope, creating a rigid white cascade, criss-crossed with deep crevasses.

**ice sheet** the largest type of glacier; a continental glacier larger than 50,000 km².

**ice shelf** large flat-topped plate of ice overlying the ocean but attached to land; a source of icebergs.

**Icelandic Low** center of low atmospheric pressure located in the north Atlantic, especially persistent in winter.

**ice-sheet climate** climate type where the average temperature of every month of the year is below freezing.

**ice wedge** wedge-shaped ice mass that has accumulated in a ground fissure in a periglacial environment.

**iceberg** free-floating mass of ice broken off from a glacier where it flows into the ocean or a lake.

**ice-marginal lake** temporary lake formed by the disruption of meltwater drainage by deposition along a glacial margin, usually in the area of an end moraine.

**ice-scoured plain** a broad area of low relief and bedrock exposures eroded by a continental glacier.

**igneous intrusion (pluton)** a mass of igneous rock that cooled and solidified beneath the surface.

**igneous processes** processes related to the solidification and eruption of molten rock matter.

**igneous rock** one of the three major categories of rock; formed from the cooling and solidification of molten rock matter.

**illuviation** deposition of fine soil components in the subsoil (B horizon) by gravitational water.

**imaging radar** radar systems designed to sense the ground and convert reflections into a map-like image.

**Inceptisol** young soil with weak horizon development.

**infiltrate** water seeping downward into the soil or other surface materials.

**infiltration** the process of water seeping downward into the soil or other surface materials.

**infiltration capacity** the greatest amount of infiltrated water that a surface material can hold.

**influent** the condition of stream water seeping into the channel bed and adding to groundwater.

**inner core** the solid, innermost portion of Earth's core, probably of iron and nickel, that forms the center of Earth.

**input** energy and material entering an Earth system.

**inselberg** remnant bedrock hill rising above a stream-eroded plain or pediment in an arid or semiarid region.

**insolation** incoming solar radiation, that is, energy received from the sun.

**instability** condition of air when it is warmer than the surrounding atmosphere and is buoyant with a tendency to rise; the lapse rate of the surrounding atmosphere is greater than that of unstable air.

**interception** the delay in arriving at the ground surface experienced by precipitation that strikes vegetation.

**interfluve** the land between two stream channels.

**interglacial** warmer period between glacial advances during which continental ice sheets and many valley glaciers retreat and disappear or are greatly reduced in size.

**interior drainage** streams and stream systems that flow within a closed basin and thus do not reach the ocean.

**intermediate zone** subsurface water layer between the zone of aeration above and the zone of saturation below; saturated only during times of ample precipitation.

**intermittent flow** describes streams that conduct flow seasonally.

**International Date Line** line roughly along the 180° meridian, where each day begins and ends; it is always a day later west of the line than east of the line.

**Intertropical Convergence Zone (ITCZ)** zone of low pressure and calms along the equator, where air carried by the trade winds from both sides of the equator converges and is forced to rise.

**intrusive igneous rock** rock that solidified within Earth from magma; also called plutonic rock.

**inversion** see temperature inversion.

**inversion layer** the altitudinal boundary, which can be sharp, between cold air near the surface and warmer air aloft—which are the conditions of a temperature inversion.

**ionosphere** layer of ionized gasses concentrated between 80 km (50 miles) and outer limits of the atmosphere.

**isarithm** line on a map that connects all points of the same numerical value, such as isotherms, isobars, and isobaths.

**island arc** a chain of volcanic islands along a deep oceanic trench; found near tectonic plate boundaries where subduction is occurring.

**isobar** line drawn on a map to connect all points with the same atmospheric pressure.

**isoline** a line on a map that represents equal values of some numerical measurement such as lines of equal temperature or elevation contours.

**isostasy** theory that holds that Earth's crust floats in hydrostatic equilibrium on the denser plastic layer of the mantle.

**isotherm** line drawn on a map connecting points of equal temperature.

**jet stream** high-velocity upper-air current with speeds of 120–640 kilometers per hour (75–250 mph).

**jetty** artificial structure, typically built in pairs, extending into a body of water; built to protect a harbor, inlet, or river mouth from excessive coastal sedimentation.

**joint** fracture or crack in rock.

**joint set** system of multiple parallel cracks (joints) in rock.

**jungle** dense tangle of trees and vines in areas where sunlight reaches the ground surface (not a true rainforest).

**kame** conical hill composed of sorted glaciofluvial deposits; presumed to have formed in contact with glacial ice when sediment accumulated in ice pits, crevasses, and among jumbles of detached ice blocks.

**kame terraces** landform resulting from accumulation of glaciofluvial sand and gravel along the margin of a glacier occupying a valley in an area of hilly relief.

**karst** unique landforms and landscapes derived by the solution of soluble rocks, particularly limestone.

**Kelvin scale** temperature scale developed by Lord Kelvin, equal to Celsius scale plus 273; no temperature can drop below absolute zero, or 0 degrees Kelvin.

**kettle** depression formed by the melting of an ice block buried in glacial deposits left by a retreating glacier.

**kettle lake** a small lake or pond occupying a kettle.

**kinetic energy** energy of motion; one-half the mass *(m)* times velocity *(v)* squared, $E_k = \frac{1}{2}mv^2$.

**Köppen system** climate classification based on monthly and annual averages of temperature and precipitation; boundaries between climate classes are designed so that climate types coincide with vegetation regions.

**La Niña** cold sea-surface temperature anomaly in the Equatorial Pacific (opposite of El Niño).

**laccolith** massive igneous rock intrusion that bows overlying rock layers upwards in a domal fashion.

**lagoon** coastal embayment of water that is blocked from free circulation with the ocean or lake by a landform barrier.

**lahar** rapid, gravity-driven downslope movement of wet, fine-grained volcanic sediment.

**lake effect snow** Enhanced, often heavy, snowfall in an area downwind from a lake, caused when a storm picks up moisture and rises as it passes over the lake.

**land breeze** air flow at night from the land toward the sea, caused by the movement of air from a zone of higher pressure associated with cooler nighttime temperatures over the land. *See also* sea breeze.

**landform** a terrain feature, such as a mountain, valley, plateau, etc.

**Landsat** a family of U.S. satellites that have been returning digital images since the 1970s.

**landslide** layperson's term for any fast mass wasting; used by some earth scientists for massive slides that involve a variety of Earth materials.

**lapse rate** *see* normal lapse rate.

**latent heat exchange** conversion between sensible heat, which we can feel, and latent heat, which is stored heat energy that we cannot sense, through changes in the state of water—for example evaporation, condensation, and freezing.

**latent heat of condensation** energy release in the form of heat, as water is converted from the gaseous (vapor) to the liquid state.

**latent heat of evaporation** amount of heat absorbed by water to evaporate from a surface (i.e., 590 calories/g of water).

**latent heat of fusion** amount of heat transferred when liquid turns to ice and vice versa; this amounts to 80 calories/gram.

**latent heat of sublimation** amount of heat that is leased when ice turns to vapor without first going through the liquid phase; this amounts to 670 calories/gram.

**lateral migration** the sideways shift in the position of a stream channel over time.

**lateral moraine** a ridge of till deposited along the side margin of a glacier.

**laterite** iron, aluminum, and manganese rich layer in the subsoil (B horizon) that can be an end product of laterization in the wet-dry tropics (tropical savanna climate).

**laterization** soil-forming process of hot, wet climates. Oxisols, the typical end product of the process, are characterized by the presence of little or no humus, the removal of soluble and most fine soil components, and the heavy accumulation of iron and aluminum compounds.

**latitude** angular distance (distance measured in degrees) north or south of the equator.

**lava** molten (melted) rock matter erupted onto Earth's surface; solidifies into extrusive igneous (volcanic) rocks.

**lava flow** erupted molten rock matter that oozed over the landscape and solidified.

**leaching** removal by gravitational water of soluble inorganic soil components from the surface layers of the soil.

**leeward** located on the side facing away from the wind.

**legend** key to symbols used on a map.

**levee** raised bank along margins of a stream or debris flow channel.

**liana** woody vine found in tropical forests that roots in the forest floor but uses trees for support as it grows upward toward available sunshine.

**lidar** an abbreviation of light detection and ranging, which uses lasers to scan and image landscapes—a remote sensing method.

**life-support system** interacting and interdependent units (e.g., oxygen cycle, nitrogen cycle) that together provide an environment within which life can exist.

**light-year** the distance light travels in 1 year—9.5 trillion kilometers (6 trillion mi).

**lightning** visible electrical discharge produced within a thunderstorm.

**lithification** the combined processes of compaction and cementation that transform clastic sediments into sedimentary rocks.

**lithosphere** a general term for the solid (rock) part of the Earth system, in contrast to the atmosphere, hydrosphere, and biosphere.

**lithosphere (planetary structure)** rigid and brittle outer layer of Earth consisting of the crust and uppermost mantle.

**lithospheric plate** Earth's exterior is broken into several of these large regions (plates) of rigid and brittle crust and upper mantle (lithosphere).

**Little Ice Age** a period of generally cooler conditions and growth of glaciers from the mid-16th to the mid-19th centuries.

**littoral drifting** general term for sediment transport parallel to shore in the nearshore zone due to incomplete wave refraction.

**loam** a soil with a texture in which none of the three soil grades (sand, silt, or clay) predominate over the others.

**loess** wind-deposited silt; usually transported in dust storms and derived from arid or glaciated regions.

**longitude** angular distance (distance measured in degrees) east or west of the prime meridian.

**longitudinal dune** a linear ridge-like sand dune that is oriented parallel to the prevailing wind direction.

**longitudinal profile** the change in stream channel elevation with distance downstream from source to mouth.

**longshore bar** submerged feature of wave- and current-deposited sediment lying close to and parallel with the shore.

**longshore current** flow of water parallel to the shoreline just inside the breaker zone; caused by incomplete wave refraction.

**longshore drifting** transport of sediment parallel to shore by the longshore current.

**longwave radiation** electromagnetic radiation emitted by Earth in the form of waves more than 4.0 micrometers in amplitude, which includes heat reradiated by Earth's surface.

**low** *see* cyclone.

**magma** molten (melted) rock matter located beneath Earth's surface from which intrusive igneous rocks are formed.

**magnetic declination** horizontal angle between geographic north and magnetic north.

**mantle** moderately dense, relatively thick (2885 km/1800 mi) middle layer of Earth's interior that separates the crust from the outer core.

**map projection** any presentation of the spherical Earth on a flat surface.

**maquis** sclerophyllous woodland and plant community, similar to North American chaparral; can be found growing throughout the Mediterranean region.

**marine terrace** abrasion platform that has been elevated above sea level and thus abandoned from wave action.

**marine west coast climate** climate type characterized by cool wet conditions much of the year.

**maritime** relating to weather, climate, or atmospheric conditions in coastal or oceanic areas.

**Maritime Equatorial (mE)** hot and humid air mass which originates from the ocean region straddling the equator.

**Maritime Polar (mP)** cold, moist air mass originating from the oceans around 40 to 60 degrees N or S latitude.

**Maritime Tropical (mT)** warm moist air mass which originates from the tropical ocean regions.

**mass** a measure of the total amount of matter in a body.

**mass wasting** gravity-induced downslope movement of Earth material.

**mathematical/statistical model** computer-generated representation of an area or Earth system using statistical data.

**matrix** the dominant area of a mosaic (ecosystem supporting a particular plant community) where the major plant in the community is concentrated.

**meander** a broad, sweeping bend in a river or stream.

**meander cut-off** bend of a meandering stream that has become isolated from the active channel.

**meandering channel** stream channel with broadly sinuous banks that curve back and forth in sweeping bends.

**medial moraine** a central moraine in a large valley glacier formed where the interior lateral moraines of two tributary glaciers merge.

**Mediterranean climate** climate type characterized by warm dry summers and cool moist winters.

**mental map** conceptual model of special significance in geography because it consists of spatial information.

**Mercator projection** mathematically produced, conformal map projection showing true compass bearings as straight lines.

**mercury barometer** instrument measuring atmospheric pressure by balancing it against a column of mercury.

**meridian** a line of longitude that forms half of a great circle on the globe and connects all points of equal longitude; all meridians connect the North and South Poles.

**mesa** flat-topped, steep-sided erosional remnant of a tableland, roughly as broad as tall, characteristic of arid regions with flat-lying sedimentary rocks.

**mesosphere** layer of atmosphere above the stratosphere; characterized by temperatures that decrease regularly with altitude.

**mesothermal climate** climate region or condition with hot, warm, or mild summers that do not have any months that average below freezing.

**metamorphic rock** one of the three major categories of rock; formed by heat and pressure changing a pre-existing rock.

**meteor** mass of rock pulled toward Earth by gravity; most meteors burn up before hitting Earth's surface, and form a streak of light across the night sky ("shooting star").

**meteorite** any fragment of a meteor that reaches Earth's surface.

**meteoroid** stone or iron mass that enters our atmosphere from outer space, becoming a meteor as it burns up in the atmosphere.

**meteorology** study of the patterns and causes associated with short-term changes in the elements of the atmosphere.

**microclimate** climate associated with a small area at or near Earth's surface; the area may range from a few inches to 1 mile in size.

**microplate terrane** segment of crust of distinct geology added to a continent during tectonic plate collision.

**microthermal climates** climate regions or conditions with warm or mild summers that have winter months with temperatures averaging below freezing.

**middle-latitude cyclone (extratropical cyclone)** storm system that travels with the westerly winds of the middle latitudes and the polar front, typically consisting of a cold front, a warm front and three sectors (cold, warm and cool).

**middle-latitude disturbance** convergence of cold polar and warm subtropical air masses over the middle latitudes.

**middle-latitude zone** the latitudinal area poleward of 23½° (N or S), and 66½° (N or S).

**midoceanic ridge** linear seismic mountain range that interconnects through all the major oceans; it is where new molten crustal material rises through the oceanic crust.

**millibar** unit of measurement for atmospheric pressure; 1 millibar equals a force of 1000 dynes per square centimeter; 1013.2 millibars is standard sea-level pressure.

**mineral** naturally occurring inorganic substance with a specific chemical composition and crystalline structure.

**model** simplified representation of a more complex reality that is useful for analysis and prediction.

**modified Mercalli scale** an earthquake intensity scale with Roman numerals from I to XII used to assess spatial variations in the degree of impact on people and structures that a tremor generates.

**Mohorovičić discontinuity (Moho)** zone marking the transition between Earth's crust and the denser mantle.

**Mollisol** soil that develops in grassland regions.

**moment magnitude** the measure of an earthquake's size in terms of energy released.

**monsoon** seasonal wind that reverses direction during the year in response to a reversal of pressure over a large landmass. The classic monsoons of Southeast Asia blow onshore in response to low pressure over Eurasia in summer and offshore in response to high pressure in winter.

**moraine** a category of glacial landform, with multiple subtypes, deposited beneath or along the margins of an ice mass.

**mosaic** a plant community and the ecosystem on which it is based, viewed as a landscape of interlocking parts by ecologists.

**mountain breeze** air flow downslope from mountains toward valleys during the night. *See also* valley breeze.

**mouth** downflow terminus of a stream.

**mud** wet, fine-grained sediment, particularly clay and silt sizes.

**mudflow** rapid mass wasting of wet, fine-grained sediment; may deposit levees and lobate (tongue-shaped) masses.

**multispectral remote sensing** using a number of energy wavelength bands to create images.

**muskeg** poorly drained vegetation-rich marshes or swamps usually overlying permafrost areas of polar climatic regions.

**natural hazards** any natural process that has a detrimental impact on humans and the built environment.

**natural levee** bank of a stream channel (or margin of a mass wasting flow channel) raised by deposition from flood (or flow) deposits; artificial levees are sometimes built along stream banks for flood control.

**natural resource** any element, material, or organism existing in nature that may be useful to humans.

**natural vegetation** vegetation that has been allowed to develop naturally without obvious interference from or modification by humans.

**navigation** the science of location and finding one's way, position, or direction.

**neap tide** the smaller than average tidal range that occurs during the first and third quarter moon.

**Near Earth Objects (NEO)** large celestial bodies like comets, and asteroids, which may come close enough to collide with Earth.

**near-infrared (NIR) image** image created by measuring reflected light in the part of the spectrum just longer than red wavelengths, and not visible to people.

**nearshore zone** area from the seaward or lakeward edge of breaking waves to the landward limit of broken wave water.

**negative feedback** reaction to initial change in a system that counteracts the initial change and leads to dynamic equilibrium in the system.

**nekton** classification of marine organisms that swim in the oceans.

**nimbus (nimbo)** term used in cloud description to indicate precipitation; thus cumulonimbus is a cumulus cloud from which rain is falling.

**nonflowing artesian well** artificially dug outlet into which groundwater flows under its own pressure, but not reaching Earth's surface.

**nor'easters** a New England term for a storm that stalls over the north Atlantic along the east coast of North America, delivering strong winds from the poleward side of the storm that blow from the northeast; nor'easters typically bring heavy rain or snow and strong waves on the coast as the storm winds blow onshore.

**normal fault** breakage zone with rocks on one side sliding down relative to rocks on the other side because of tensional forces; footwall up, hanging wall down.

**normal lapse rate (environmental lapse rate)** decrease in temperature with altitude under normal atmospheric conditions; approximately 6.5°C/1000 meters (3.6°F/1000 ft).

**North Atlantic Oscillation (NAO)** oscillating (see-saw) pressure tendencies between the Azores High and the Icelandic Low.

**North Pole** maximum north latitude (90°N), at the point marking the axis of rotation.

**North Star (Polaris)** the star that is very near to being directly above Earth's north pole of rotation (90°N). Polaris is useful for celestial navigation in the northern hemisphere.

**northeast trades** *see* trade winds.

**notch** a recess, relatively small in height, eroded by wave action along the base of a coastal cliff.

**oblate spheroid** Earth's shape—a slightly flattened sphere.

**obliquity cycle** the change in the tilt of the Earth's axis relative to the plane of the ecliptic over a 41,000-year period.

**occluded front** boundary between a rapidly advancing cold air mass and an uplifted warm air mass cut off from Earth's surface; denotes the last stage of a middle-latitude cyclone.

**ocean current** horizontal movement of ocean water, usually in response to major patterns of atmospheric circulation.

**oceanic crust** the denser (ave. 3.0 g/cm³), thinner, basaltic portion of Earth's crust; underlies the ocean basins.

**oceanic island** volcanic island that rises from the deep ocean floor.

**oceanic trench (trench)** long, narrow depression on the seafloor usually associated with an island arc. Trenches mark the deepest portions of the oceans and are associated with subduction of oceanic crust.

**offshore zone** the expanse of open water lying seaward or lakeward of the breaker zone.

**omnivore** animal that can feed on both plants and other animals.

**open system** system in which energy and/or materials can freely cross its boundaries.

**organic sedimentary rock** rock created from deposits of organic material, such as carbon from plants (coal).

**orographic precipitation** precipitation resulting from condensation of water vapor in an air mass that is forced to rise over a mountain range or other raised landform.

**orographic thunderstorm** a thunderstorm produced by the orographic uplift mechanism.

**outcrop** a mass of bedrock exposed at Earth's surface that is not concealed by regolith or soil.

**outer core** the upper portion of Earth's core; considered to be composed of molten iron liquefied by Earth's internal heat.

**outlet glacier** a valley glacier that flows outward from the main mass of a continental glacier.

**output** energy and material leaving an Earth system.

**outwash** glacial drift deposited beyond the terminus of a glacier by glacial meltwater.

**outwash plain** extensive, relatively smooth plain covered with sorted deposits carried forward by the meltwater from a continental glacier.

**overthrust** low angle fault with rocks on one side pushed a considerable distance over those of the opposite side by compressional forces; the wedge of rocks that have overridden others in this way.

**oxbow lake** a lake or pond found in a meander cut-off on a floodplain.

**oxidation** union of oxygen with other elements to form new chemical compounds.

**Oxisol** soil that develops over a long period of time in tropical regions with high temperatures and heavy annual rainfall.

**oxide** a mineral group composed of oxygen combining with other Earth elements, especially metals.

**oxygen-isotope analysis** a dating method used to reconstruct climate history; it is based on the varying evaporation rates of different oxygen isotopes and the changing ratio between the isotopes revealed in foraminifera fossils.

**ozone** gas molecule consisting of three atoms of oxygen ($O_3$); forms a layer in the upper atmosphere that serves to screen out ultraviolet radiation harmful at Earth's surface.

**ozone layer** part of the stratosphere that surrounds Earth, with concentrations of ozone that help to shield the surface from harmful short-wave energy from the sun, such as ultraviolet light.

**ozonosphere** also known as the ozone layer; this is a concentration of ozone gas in a layer between 20 and 50 km (13 to 50 miles) above Earth's surface.

**Pacific High** persistent cell of high atmospheric pressure located in the subtropics of the North Pacific Ocean, also called the Hawaiian High.

**pahoehoe** a smooth, ropy surface of a lava flow.

**paleogeography** study of the past geographical distribution of environments.

**paleomagnetism** the historic record of changes in Earth's magnetic field.

**palynology** method of determining past climatic conditions using pollen analysis.

**Pangaea** ancient continent that consisted of all of today's continental landmasses.

**parabolic dune** crescent-shaped sand dune with arms pointing upwind.

**parallel** circle on the globe connecting all points of equal latitude.

**parallelism of the axis** tendency of Earth's polar axis to remain parallel to itself at all positions in its orbit around the sun.

**parent material** residual (derived from bedrock directly beneath) or transported (by water, wind, or ice) mineral matter from which soil is formed.

**particulates** very small pieces of solid matter that can remain airborne for long time periods.

**passive-margin coast** coastal region that is far removed from the volcanism and tectonism associated with lithospheric plate boundaries.

**patch** a gap or area within a matrix (territory occupied by a dominant plant community) where the dominant vegetation is not supported due to natural causes.

**paternoster lake** chain of lakes connected by a postglacial stream occupying the trough of a glaciated mountain valley.

**patterned ground (frost polygons)** natural, repeating, often-polygonal designs of sorted sediment seen on the surface of periglacial environments.

**pediment** gently sloping surface of eroded bedrock, thinly covered with fluvial sediments, found at the base of an arid-region mountain.

**perched water table** a minor zone of saturation overlying an aquiclude that exists above the regional water table.

**percolation** subsurface water moving downward to lower zones by the pull of gravity.

**perennial flow** describes streams that conduct flow continuously all year.

**periglacial** pertaining to cold-region landscapes that are impacted by intense frost action but not covered by year-round snow or ice.

**perihelion** position of Earth at closest distance to sun during each Earth revolution.

**permafrost** permanently frozen subsoil and underlying rock found in climates where summer thaw penetrates only the surface soil layer.

**permeability** characteristic of soil or bedrock that determines the ease with which water moves through Earth material.

**pH scale** scale from 0 to 14 that describes the acidity or alkalinity of a substance and that is based on a measurement of hydrogen ions; pH values below 7 indicate acidic conditions; pH values above 7 indicate alkaline conditions.

**photosynthesis** the process by which carbohydrates (sugars and starches) are manufactured in plant cells; requires carbon dioxide, water, light, and chlorophyll (the green color in plants).

**physical geography** specialization in the systematic study of geography that focuses on the location, distribution, and spatial interaction of physical (environmental) phenomena.

**physical (mechanical) weathering** breakdown of rocks into smaller fragments (disintegration) by physical forces without chemical change.

**physical model** three-dimensional representation of all or a portion of Earth's surface.

**phytoplankton** tiny plants, algae and bacteria, that float and drift with currents in water bodies.

**pictorial/graphic model** representation of a portion of Earth's surface by means of maps, photographs, graphs, or diagrams.

**piedmont alluvial plain** a plain created by stream deposits at the base of an upland, such as a mountain, a hilly region, or a plateau.

**piedmont fault scarp** steep cliff due to movement along a fault that has offset unconsolidated sediment.

**piedmont glacier** an alpine glacier that extends beyond a mountain valley spreading out onto lower flatter terrain.

**pixel** the smallest area that can be resolved in a digital image. Pixels, short for "picture element," are much like pieces in a mosaic, fitted together in a grid to make an image.

**planar projection** map projection in which grid lines are transferred onto a plane, or flat surface; planar maps are most often used to show the polar regions, with the pole centrally located on the circular map, displaying one hemisphere.

**plane of the ecliptic** plane of Earth's orbit about the sun and the apparent annual path of the sun along the stars.

**planet** any of the eight largest bodies revolving about the sun, or any similar bodies that may orbit other stars.

**plankton** passively drifting or weakly swimming marine organisms, including both phytoplankton (plants) and zooplankton (animals).

**plant community** variety of individual plants living in harmony with each other and the surrounding physical environment.

**plastic solid** any solid material that changes its shape under stress, and retains that deformed shape after the stress is relieved.

**plate convergence** movement of lithospheric plates toward each other.

**plate divergence** movement of lithospheric plates away from each other.

**plate tectonics** theory that superseded continental drift and is based on the idea that the lithosphere is composed of a number of segments or plates that move independently of one another, at varying speeds, over Earth's surface.

**plateau** an extensive, flat-topped landform or region characterized by relatively high elevation, but low relief.

**playa** dry lake bed in a desert basin; typically fine-grained clastic (clay pan) or saline (salt crust).

**playa lake** a temporary lake that forms on a playa from runoff after a rainstorm or during a wet season.

**Pleistocene epoch** the interval of Earth history immediately before the present (Holocene) epoch that experienced times of glacial advance and lasted from about 2.6 million to 10,000 years ago.

**plucking** erosion process by which a glacier pulls rocks and sediment from the ground along its bed and into the flowing ice.

**plug dome** a steep-sided, explosive type of volcano with its central vent or vents plugged by the rapid congealing of its highly acidic lava.

**plunge pool** depression at the base of a waterfall formed by the impact of cascading water.

**pluton** see igneous intrusion.

**plutonic rock** see intrusive igneous rock.

**plutonism** the processes associated with the formation of rocks from magma cooling deep beneath Earth's surface.

**podzolization** soil-forming process of humid climates with long cold winter seasons. Spodosols, the typical end product of the process, are characterized by the surface accumulation of raw humus, strong acidity, and the leaching or eluviation of soluble bases and iron and aluminum compounds.

**point bar** deposit of alluvium found on the inside of a bend in a meandering stream channel.

**polar** referring to the North or South Polar regions.

**polar climate** climate region that does not have a warm season and is frozen much or all of the year.

**polar easterlies** easterly surface winds that move out from the polar highs toward the subpolar lows.

**polar front** the boundary, an often windy and stormy linear zone of low pressure, between cold air from the high latitudes and warm air from the tropics or subtropics.

**polar front jet stream** shifting boundary between cold polar air and warm subtropical air, located within the middle latitudes and strongly influenced by the polar jet stream.

**polar highs** high pressure systems located near the poles where air is settling and diverging.

**polar outbreak** strong flow of cold polar or arctic air that intrudes into the middle latitudes.

**polar zone** the latitudinal area poleward of 66½° (N or S).

**polarity reversal** a time in geologic history when the south magnetic pole became the north magnetic pole and vice versa.

**pollution** alteration of the physical, chemical, or biological balance of the environment that has adverse effects on the normal functioning of all life forms, including humans.

**porosity** characteristic of soil or bedrock that relates to the amount of pore space between individual peds or soil and rock particles and that determines the water storage capacity of Earth material.

**positive feedback** reaction to initial change in a system that reinforces the initial change and leads to imbalance in the system.

**potential evapotranspiration (potential ET)** hypothetical rate of evapotranspiration if at all times there is a more than adequate amount of soil water for growing plants.

**pothole** bedrock depression in a streambed drilled by the spinning of abrasive rocks in swirling flow.

**prairie** grassland regions of the middle latitudes. Tall-grass prairie varied from 0.5 to 3 meters (2 to 10 ft) in height and was native to areas of moderate rainfall; short-grass prairie of lesser height remains common in subhumid and semiarid (steppe) environments.

**precession cycle** changes in the time (date) of the year that perihelion occurs; the date is determined on the basis of a major period 21,000 years in length and a secondary period 19,000 years in length.

**precipitation** water in liquid or solid form that falls from the atmosphere and reaches Earth's surface.

**pressure belts** zones of high or low pressure that tend to circle Earth parallel to the equator in a theoretical model of world atmospheric pressure.

**pressure gradient** rate of change of atmospheric pressure horizontally with distance, measured along a line perpendicular to the isobars on a map of pressure distribution.

**prevailing wind** direction from which the wind for a particular location blows during the greatest proportion of the time.

**primary coastline** coast that has developed its present form primarily from land-based processes, especially fluvial and glacial processes.

**primary productivity** rate at which new organic material is created at a particular trophic level. Primary productivity through photosynthesis by autotrophs is at the first trophic level; secondary productivity is by heterotrophs at subsequent trophic levels.

**prime meridian** half of a great circle that connects the North and South Poles and marks zero degrees longitude. By international agreement the meridian passes through the Royal Observatory at Greenwich, England.

**principal meridian** one of many lines running N–S (meridians of longitude) designated by the U.S. Public Land Survey System for locating plots of land north or south of these meridians.

**prodelta** the portion of a delta that lies submerged in the standing body of water.

**profile** a graph of changes in height over a linear distance, such as a topographic profile.

**punctuated equilibrium** concept that periods of relative stability in many Earth systems are interrupted by short bursts of intense action causing major change.

**pyroclastic flow** airborne density current of hot gases and rock fragments unleashed by an explosive volcanic eruption.

**pyroclastic material (tephra)** pieces of volcanic rock, including cinders and ash, solidified from molten material erupted into the air.

**RADAR** *RAdio Detection And Ranging.*

**radiation** emission of electromagnetic wave energy through space; *see also* shortwave radiation and longwave radiation.

**radiation fog (ground or temperature-inversion fog)** fog produced by cooling of air in contact with a cold ground surface.

**rain** falling droplets of liquid water.

**rain shadow** dry, leeward side of a mountain range, resulting from the adiabatic warming of descending air.

**recessional moraine** end moraine deposited behind the terminal moraine marking a pause in the overall retreat of a glacier.

**recharge** replenishing the amount of stored water, particularly in the subsurface.

**recumbent fold** a fold in rock pushed over onto one side by asymmetric compressional forces; the axial plane of the fold is horizontal rather than vertical.

**recurrence interval** average length of time between events, such as floods, equal to or exceeding a given magnitude.

**region** area identified by certain characteristics it contains that makes it distinctive and separates it from surrounding areas.

**regional base level** the lowest level to which a stream system in a basin of interior drainage can flow.

**regional geography** specialization in the systematic study of geography that focuses on the location, distribution, and spatial interaction of phenomena organized within arbitrary areas of Earth space designated as regions.

**regolith** weathered rock material; usually covers bedrock.

**rejuvenated stream** a stream that has deepened its channel by erosion because of tectonic uplift in the drainage basin or lowering of its base level.

**relative humidity** ratio between the amount of water vapor in air of a given temperature and the maximum amount of water vapor that the air could hold at that temperature, if saturated; usually expressed as a percentage.

**relative location** location of an object in respect to its position relative to some other object or feature.

**relief** a measurement or expression of the difference between the highest and lowest location in a specified area.

**remote sensing** collection of information about the environment from a distance, usually from aircraft or spacecraft, for example, photography, radar, infrared.

**representative fraction (RF) scale** a map scale presented as a fraction or ratio between the size of a unit on the map to the size of the same unit on the ground, as in 1/24,000 or 1:24,000.

**reservoir** an artificial lake impounded behind a dam.

**residual parent material** rock fragments that form a soil and have accumulated in place through weathering.

**resolution (spatial resolution)** size of an area on Earth that is represented by a single pixel.

**reverse fault** high angle break with rocks on one side pushed up relative to those on the other side by compressional forces; hanging wall up, footwall down.

**revolution** motion of Earth along a path, or orbit, around the sun. One complete revolution requires approximately 365° days and determines an Earth year.

**rhumb line** line of true compass bearing (heading).

**ria** a narrow coastal embayment that occupies a submerged river valley.

**rift valley** major lowland consisting of one or more crustal blocks downfaulted as a result of tensional tectonic forces.

**rill** a narrow stream channel that even in a humid climate conducts flow only during precipitation events.

**rime** ice crystals formed along the windward side of tree branches, airplane wings, and the like, under conditions of supercooling.

**rip current** strong, narrow surface current flowing away from shore. It is produced by the return flow of water piled up near shore by incoming waves.

**ripples** small (centimeter-scale) wave forms in water or sediment.

**roche moutonnée** bedrock hill subjected to intense glacial abrasion on its up-ice side, with some plucking evident on the down-ice side.

**rock** a solid, natural aggregate of one or more minerals or particles of other rocks.

**rock cycle** a representation of the processes and pathways by which Earth material becomes different types of rocks.

**rock flour** rock fragments finely ground between the base of a glacier and the underlying bedrock surface.

**rock structure** the orientation, inclination, and arrangement of rock layers in Earth's crust.

**rockfall** nearly vertical drop of individual rocks or a rock mass through air pulled downward by the force of gravity.

**rockslide** rock unit moving rapidly downslope by gravity in continuous contact with the surface below.

**Rossby waves** horizontal undulations in the flow of the upper air winds of the middle and upper latitudes.

**rotation** turning of Earth on its polar axis; one complete rotation requires 24 hours and determines one Earth day.

**runoff** flow of water from the land surface, generally in the form of streams and rivers.

**Saffir-Simpson Hurricane Scale** A ranking (1–5) of the intensity of a hurricane, based on wind speeds and on the types of damage that occur, or may occur.

**salinas** *see* salt-crust playa.

**salinization** soil-forming process of low-lying areas in desert regions; the resulting soils are characterized by a high concentration of soluble salts as a result of the evaporation of surface water.

**salt crystal growth** weathering by the expansive force of salts growing in cracks in rocks; common in arid and coastal regions.

**saltation** the transportation by running water or wind of particles too large to be carried in suspension; the particles are bounced along on the surface or streambed by repeated lifting and deposition.

**salt-crust playa (salt flat)** an ephemeral lake bed composed mostly of salt minerals.

**sand (sandy)** sediment particles ranging in size from about 0.05 millimeter to 2.0 millimeters.

**sand dune** mound or hill of sand-sized sediment deposited and shaped by the wind.

**sand sea (erg)** an extensive area covered by sand dunes.

**sand sheet** an extensive cover of wind-deposited sand having little or no surface relief.

**sandstorm** strong winds blowing sand along the ground surface.

**Santa Ana wind** very dry downslope wind occurring in Southern California; *see also* foehn wind.

**satellite** any body that orbits a larger primary body, for example, the moon orbiting Earth.

**saturation** point at which sufficient cooling has occurred so that an air mass contains the maximum amount of water vapor it can hold. Further cooling produces condensation of excess water vapor.

**savanna** tropical vegetation consisting primarily of coarse grasses, often associated with scattered low-growing trees or patches of bare ground.

**scale** ratio between distance as measured on Earth and the same distance as measured on a map, globe, or other representation of Earth.

**Scattering** diffusion in all directions of incoming solar radiation by gas molecules, water droplets, or particles in the atmosphere.

**scientific method** answering questions (hypotheses) by collecting evidence, objectively analyzing the evidence, and proposing answers based on what the evidence supports.

**sclerophyllous** vegetation type commonly associated with the Mediterranean climate; characterized by tough surfaces, deep roots, and thick, shiny leaves that resist moisture loss.

**sea** steep, choppy, chaotic waves still forming under the influence of a storm.

**sea arch** span of rock extending from a coastal cliff under which the ocean or lake water freely moves.

**sea breeze** air flow by day from the sea toward the land; caused by the movement of air toward a zone of lower pressure associated with higher daytime temperatures over the land.

**sea cave** large wave-eroded opening formed near the water level in coastal cliffs.

**sea cliff (lake cliff)** steep slope of land eroded at its base by wave action.

**sea level** average position of the ocean shoreline.

**sea stack** resistant pillar of rock projecting above water close to shore along an erosion-dominated coast.

**seafloor spreading** movement of oceanic crust in opposite directions away from the midoceanic ridges, associated with the formation of new crust at the ridges and subduction of old crust at ocean margins.

**secondary coastline** coast that has developed its present form primarily through the action of coastal processes (waves, currents, and/or coral reefs).

**secondary productivity** the formation of new organic matter by heterotrophs, consumers of other life forms; *see* productivity.

**section** a square parcel of land with an area of 1 square mile as defined by the U.S. Public Lands Survey System.

**sedimentary rock** one of three major rock categories; formed by compaction and cementation of rock fragments, organic remains, or chemical precipitates.

**seif** a large, long, somewhat sinuous sand dune elongated parallel to the prevailing wind direction

**seismic wave** traveling wave of energy released during an earthquake or other shock.

**seismograph** instrument used to measure amplitude of passing seismic waves.

**selva** characteristic tropical rainforest comprising multistoried, broad-leaf evergreen trees with relatively little undergrowth.

**sextant** navigation instrument used to determine latitude by star and sun positions.

**shearing tectonic force** a force that works to move two objects past each other in opposite directions.

**sheet wash** thin sheet of unchannelized water flowing over land.

**shield volcano** dome-shaped accumulation of multiple successive lava flows extruded from one or more vents or fissures.

**shoreline** exact contact between the edge of a standing body of water and dry land.

**short-grass prairie** environment where the dominant vegetation type is short grasses.

**shortwave radiation** energy radiated in wavelengths of less than about 1 micrometer (one millionth of a meter); includes X-rays, gamma rays, ultraviolet rays, visible light and near infrared light.

**Siberian high** intensively developed center of high atmospheric pressure located in northern central Asia in winter.

**side-looking airborne radar (SLAR)** a radar system that is used for making maps of terrain features.

**silicate** the largest mineral group, composed of oxygen and silica and forming most of the Earth's crust.

**sill** a horizontal sheet of igneous rock intruded and solidified between other rock layers.

**silt (silty)** sediment particles with a grain size between 0.002 millimeter and 0.05 millimeter.

**sinkhole (doline)** roughly circular surface depression related to the solution of rock in karst areas.

**slash–and–burn (shifting cultivation)** typical subsistence agriculture of indigenous societies in the tropical rainforest. Trees are cut, the branches are burned, and crops are planted between the larger trees or stumps before rapid deterioration of the soil forces a move to a new area.

**sleet** form of precipitation produced when raindrops freeze as they fall through a layer of cold air; may also, locally, refer to a mixture of rain and snow.

**slide** fast mass wasting in which Earth material moves downslope in continuous contact with a discrete surface below.

**slip face** the steep, downwind side of a sand dune.

**slope aspect** direction a mountain slope faces in respect to the sun's rays.

**slow mass wasting** gravity-induced downslope movement of Earth material occurring so slowly that people cannot observe it directly.

**slump** thick unit of unconsolidated fine-grained material sliding downslope on a concave, curved slip plane.

**small circle** any circle that is not a full circumference of the globe. The plane of a small circle does not pass through the center of the globe.

**smog** combination of chemical pollutants and particulate matter in the lower atmosphere, typically over urban industrial areas.

**snow** precipitation in the form of ice crystals.

**snow line** elevation in mountain regions above which summer melting is insufficient to prevent the accumulation of permanent snow or ice.

**snowstorm** storm situation where precipitation falls in the form of snow.

**soil** a dynamic, natural layer on Earth's surface that is a complex mixture of inorganic minerals, organic materials, microorganisms, water, and air.

**soil (as a mass wasting material)** relatively thin unit of unconsolidated fine-grained slope material.

**soil fertilization** adding nutrients to the soil to meet the conditions that certain plants require.

**soil grade** classification of soil texture by particle size: clay (less than 0.002 mm), silt (0.002–0.05 mm), and sand (0.05–2.0 mm) are soil grades.

**soil horizon** distinct soil layer characteristic of vertical zonation in soils; horizons are distinguished by their general appearances and their specific chemical and physical properties.

**soil order** largest classification of soils based on development and composition of soil horizons.

**soil ped** naturally forming soil aggregate or clump with a distinctive shape that characterizes a soil's structure.

**soil profile** vertical cross section of a soil that displays the various horizons or soil layers that characterize it; used for classification.

**soil survey** a publication that includes maps showing the distribution of soils within a given area, in the U.S. usually a county.

**soil taxonomy** the classification and naming of soils.

**soil texture** the distribution of particle sizes in a soil that give it a distinctive "feel."

**soil water** water in the zone of aeration, the uppermost subsurface water layer.

**soil-forming regime** processes that create soils.

**solar constant** rate at which insolation is received just outside Earth's atmosphere on a surface at right angles to the incoming radiation.

**solar declination** latitude on Earth where the sun is directly overhead at solar noon (will always be within the tropics).

**solar energy** *see* insolation.

**solar noon** the time of day when the sun angle is at a maximum above the horizon (zenith).

**solar system** the system of the sun and the planets, their satellites, comets, meteoroids, and other objects revolving around the sun.

**solar wind** streams of hot ions (protons and electrons) traveling outward from the sun.

**solifluction** slow movement of saturated soil downslope by the pull of gravity; especially common in permafrost areas.

**solstice (summer and winter)** one of two times each year when the position of the noon sun is overhead at its farthest distance from the equator; this occurs when the sun is overhead at the Tropic of Cancer (about June 21) and the Tropic of Capricorn (about December 21).

**solution** dissolving material in a fluid, such as water, or the liquid containing dissolved material; water transports dissolved load in solution.

**solution sinkhole** topographic depression formed mainly by the solution and removal of soluble rock at the surface.

**sonar** a system that uses sound waves for location and mapping underwater.

**source** location high in the drainage basin, near the drainage divide, where a stream system's flow begins.

**source region** nearly homogeneous surface of land or ocean over which an air mass acquires its temperature and humidity characteristics.

**South Pole** maximum south latitude (90°S), at the point marking the axis of rotation.

**southeast trades** *see* trade winds.

**Southern Oscillation** the systematic variation in atmospheric pressure between the eastern and western Pacific Ocean.

**spatial distribution** location and extent of an area or areas where a feature exists.

**spatial interaction** process whereby different phenomena are linked or interconnected, and, as a result, impact one another through Earth space.

**spatial pattern** arrangement of a feature as it is distributed through Earth space.

**spatial science** term used when defining geography as the science that examines phenomena as it is located, distributed, and interacts with other phenomena throughout Earth space.

**specialist** an animal that can only survive on a single or very limited food type for its nutrition.

**specific heat** The amount of heat required to raise the temperature of 1 gram of material by 1° C.

**specific humidity** mass of water vapor present per unit mass of air, expressed as grams per kilogram of moist air.

**speleology** the scientific study of caverns.

**speleothem** general term for any cavern feature made by secondary (later) precipitation of minerals from subsurface water.

**spheroidal weathering** rounded shape of rocks often caused by preferential weathering along joints of cross-jointed rocks.

**spit** coastal landform of wave- and current-deposited sediment attached to dry land at one end.

**Spodosol** soil that develops in porous substrates such as glacial drift or beach sand.

**spring** natural outflow of groundwater to the surface.

**spring tide** the larger than average tidal range that occurs during new and full moon.

**squall line** narrow line of rapidly advancing storm clouds, strong winds, and heavy precipitation; usually develops in front of a fast-moving cold front.

**stability** condition of air when it is cooler than the surrounding atmosphere and resists the tendency to rise; the lapse rate of the surrounding atmosphere is less than that of stable air.

**stalactite** spire-shaped speleothem that hangs from the ceiling of a cavern.

**stalagmite** spire-shaped speleothem that rises up from a cavern floor.

**standard sea-level pressure** a reference standard for average atmospheric pressure at sea level, 1013.25 millibars, or 29.92 inches of mercury measured by a barometer.

**star dune** a large pyramid-shaped sand dune with multiple slip faces due to changes in wind direction.

**stationary front** frontal system between air masses of nearly equal strength; produces stagnation over one location for an extended period of time.

**steppe climate** characterized by middle-latitude semiarid vegetation, treeless and dominated by short bunch grasses.

**stock** an irregular mass of intrusive igneous rock (pluton) smaller than a batholith.

**storm** local atmospheric disturbance often associated with rain, hail, snow, sleet, lightning, or strong winds.

**storm surge** rise in sea level due to wind and reduced air pressure during a hurricane or other severe storm.

**storm track** path frequently traveled by a cyclonic storm as it moves in a generally eastward direction from its point of origin.

**stratification** distinct layers within sedimentary rocks, called strata.

**strato** signifies a low-level cloud (i.e., from the surface to 2000 m in elevation).

**stratopause** upper limit of stratosphere, separating it from the mesosphere.

**stratosphere** layer of atmosphere lying above the troposphere and below the mesosphere, characterized by fairly constant temperatures and ozone concentration.

**stratovolcano** *see* composite cone.

**stratus** uniform layer of low sheet-like clouds, frequently grayish in appearance.

**stream** general term for any natural, channelized flow of water regardless of size.

**stream capacity** the maximum amount of load that a stream can carry; varies with the stream's velocity.

**stream competence** the largest particle size that a stream can carry; varies with a stream's velocity.

**stream discharge** volume of water flowing past a point in a stream channel in a given unit of time.

**stream gradient** *see* gradient.

**stream hydrograph** plot showing changes in the amount of stream flow over time.

**stream load** material being transported by a stream at a given instant; includes bed load, suspended load, and dissolved load.

**stream order** numerical index expressing the position of a stream channel within the hierarchy of a stream system.

**stream terrace** former floor of a stream valley now abandoned and perched above the present valley floor and stream channel.

**stress (pressure)** force per unit area.

**striation** gouge, groove, or scratch carved in bedrock by abrading rock particles imbedded in a glacier.

**strike** compass direction of the line formed at the intersection of a tilted rock layer and a horizontal plane.

**strike-slip (lateral) fault** a fault with horizontal motion, where movement takes place along the strike of the fault.

**subarctic climate** climate type which produces a tundra landscape.

**subduction** process associated with plate tectonic theory whereby an oceanic crustal plate is forced downward into the mantle beneath a lighter continental plate when the two converge.

**sublimation** direct change of state of a material, such as water, from solid to gas.

**subpolar lows** east–west trending belts or cells of low atmospheric pressure located in the upper middle latitudes.

**subsurface horizon** a distinct horizontal soil layer that lies beneath the surface, characteristic of vertical zonation in soils.

**subsurface water** general term for all water that lies beneath Earth's surface, including soil water and groundwater.

**subsystem** separate system operating within the boundaries of a larger Earth system.

**subtropical highs** cells of high atmospheric pressure centered over the eastern portions of the oceans in the vicinity of 30°N and 30°S latitude; source of the westerlies poleward and the trades equatorward.

**subtropical jet stream** high-velocity air current flowing above the sinking air of the subtropical high pressure cells; most prominent in the winter season.

**succession** progression of natural vegetation from one plant community to the next until a final stage of equilibrium has been reached with the natural environment.

**sunspot** visible dark (cooler) spot on the surface of the sun.

**supercooled water** liquid water that exists below the freezing point of 0° C or 32° F.

**surf zone** the part of the nearshore area that consists of a turbulent bore of broken wave water.

**surface creep** wind-generated transportation consisting of pushing and rolling sediment downwind in continuous contact with the surface.

**surface of discontinuity** three-dimensional surface with length, width, and height separating two different air masses; also referred to as a front.

**surface inversion** a temperature inversion that forms close to the ground as a result of radiational cooling of the land surface during clear, calm nights.

**surface ocean current** an ocean current that flows on the surface, as opposed to a deepwater, undersea current.

**surface runoff** liquid water flowing over Earth's land surface.

**surge (glacial)** sudden shift downslope of glacial ice, possibly caused by a reduction of basal friction with underlying bedrock.

**suspended load** solid particles that are small enough to be transported considerable distances while remaining buoyed up in a moving air or water column.

**suspension** transportation process that moves small solids, often considerable distances, while buoyed up by turbulence in the moving air or water.

**sustainable development** development and natural resource use that meets present needs without compromising the ability of future generations to meet their needs.

**swallow hole** the site where a surface stream is diverted to the subsurface, such as into a cavern system.

**swash** thin sheet of broken wave water that rushes up the beach face in the swash zone.

**swash zone** the most landward part of the nearshore zone; where a thin sheet of water rushes up, then back down, the beach face.

**swell** orderly lake or ocean waves of rounded form that have traveled beyond the storm zone of wave generation.

**symbiotic relationship** relationship between two organisms which benefits both organisms.

**syncline** the downfolded element of folded rock structure.

**system** group of interacting and interdependent units that together form an organized whole.

**systems analysis** determining the parts of a system, the processes involved, and studying how changes in interactions among those parts and processes may affect the system and its operation.

**systems theory** a means of understanding how a system works by examining its important component parts, the processes involved, the interactions that take place, the interdependency of these factors, how they influence each other and ultimately their impact on the system as a whole.

**taiga** *see* boreal forest.

**tall-grass prairie** environment where the dominant vegetation type is tall grasses, with a few scattered trees.

**talus (talus slope, talus cone)** slope (sometimes cone-shaped) of angular, broken rocks at the base of a cliff deposited by rockfall.

**tarn** mountain lake in a glacial cirque.

**tectonic force** force originating within Earth that breaks and deforms Earth's crust.

**tectonic processes** processes that derive their energy from within Earth's interior and serve to create landforms by elevating, disrupting, and roughening Earth's surface.

**temperature** degree of heat or cold and its measurement.

**temperature gradient** rate of change of temperature with distance in any direction from a given point; refers to rate of change horizontally; a vertical temperature gradient is referred to as the lapse rate.

**temperature inversion** reverse of the normal pattern of vertical distribution of air temperature; in the case of inversion, temperature increases rather than decreases with increasing altitude.

**temperature lag** *see* daily temperature lag.

**tensional tectonic force** force originating within Earth that acts to pull two adjacent areas of rock away from each other (divergence).

**terminal moraine** end moraine that marks the farthest advance of an alpine or continental glacier.

**terra rossa** characteristic calcium-rich (developed over limestone bedrock) red-brown soils of the climate regions surrounding the Mediterranean Sea.

**terrestrial planets** the four closest planets to the sun—Mercury, Venus, Earth, and Mars.

**terrestrial radiation** the long wave thermal infrared radiation emitted from Earth as heat.

**thematic map** a map designed to present information or data about a specific theme, as in a population distribution map, a map of climate or vegetation.

**thermal expansion and contraction** notion that rocks can weather due to expansion and contraction effects of alternating heating and cooling.

**thermal infrared (TIR)** images made with scanning equipment that produces an image of heat differences.

**thermosphere** highest layer of atmosphere extending from the mesopause to outer space.

**Thornthwaite system** climate classification based on moisture availability and of greatest use at the local level; climate types are distinguished by examining and comparing potential and actual evapotranspiration.

**threshold** condition within a system that causes dramatic and often irreversible change for long periods of time to all variables in the system.

**thrust fault** low angle break with rocks on one side pushed over those of the other side by compressional forces.

**thunder** sound produced by the rapidly expanding, heated air along the channel of a lightning discharge.

**thunderstorm** intense convectional storm characterized by thunder and lightning, short in duration and often accompanied by heavy rain, hail, and strong winds.

**tidal interval** the time between successive high tides, or between successive low tides.

**tidal range** elevation difference between water levels at high tide and low tide.

**tide** periodic rise and fall of sea level in response to the gravitational interaction of the moon, sun, and Earth.

**till** sediment deposited directly by glacial ice.

**till plain** a broad area of low relief covered by continental glacier deposits.

**tilted fault block** crustal block between two parallel normal faults that has been uplifted along one fault and relatively downdropped along the other.

**time zone** Earth is divided into 24 time zones (24 hours) to coordinate time with Earth's rotation.

**tolerance** ability of a species to survive under specific environmental conditions.

**tombolo** strip of wave- and current-deposited sediment connecting the mainland to an island.

**topographic contour line** line on a map connecting points that are the same elevation above mean sea level.

**topography** the arrangement of high and low elevations in a landscape.

**tornado** small, intense, funnel-shaped cyclonic storm of very low pressure, violent updrafts, and converging winds of enormous velocity.

**tornado outbreak** when a thunderstorm(s) produce more than one tornado.

**tower karst** high, steep-sided hills formed by solution of limestone or other soluble rocks in karst areas.

**township** a square plot of land 6 miles on a side, further subdivided into 36 sections, laid out within the U.S. Public Lands Survey System.

**trace (of rain)** less than a measurable amount of rain or snow (i.e., less than 1 mm or 0.01 in.).

**traction** transportation process in moving water that drags, rolls, or slides heavy particles along in continuous contact with the bed.

**trade winds** consistent surface winds blowing in low latitudes from the subtropical highs toward the intertropical convergence zone; labeled northeast trades in the Northern Hemisphere and southeast trades in the Southern Hemisphere.

**transform movement** horizontal sliding of tectonic plates alongside and past each other.

**transpiration** transfer of moisture from living plants to the atmosphere by the emission of water vapor, primarily from leaf pores.

**transportation** movement of Earth materials from one site to another by gravity, water, wind, or glacial ice.

**transported parent material** rock fragments that form a soil and originated elsewhere and then were transported and deposited in the new location.

**transverse dune** a linear ridge-like sand dune that is oriented at right angles to the prevailing wind direction.

**transverse stream** a stream that flows across the general orientation or "grain" of the topography, such as mountains or ridges.

**travertine** calcium carbonate (limestone) deposits forming at springs and in caverns where groundwater is saturated with $CaCO_3$.

**tree line** elevation in mountain regions above which cold temperatures and wind stress prohibit tree growth.

**tributary** stream channel that delivers its water to another, larger channel.

**trophic level** number of feeding steps that a given organism is removed from the autotrophs (e.g., green plant—first level, herbivore—second level, carnivore—third level, etc.).

**trophic structure** organization of an ecosystem based on the feeding patterns of the organisms that comprise the ecosystem.

**Tropic of Cancer** parallel of latitude at 23½°N; the northern limit to the migration of the sun's vertical rays throughout the year.

**Tropic of Capricorn** parallel of latitude at 23½°S; the southern limit to the migration of the sun's vertical rays throughout the year.

**tropical climates** climate regions that are warm all year.

**tropical easterlies** winds that blow from the east in tropical regions.

**tropical monsoon climate** climate characterized by alternating rainy and dry seasons.

**tropical rainforest climate** hot wet climate which promotes the growth of rainforests.

**tropical savanna climate** warm, semi-dry climate which promotes tall grasslands.

**tropical zone** region on Earth lying between the Tropic of Cancer (23½°N latitude), and the Tropic of Capricorn (23½°S latitude).

**tropopause** boundary between the troposphere and stratosphere.

**troposphere** lowest layer of the atmosphere, exhibiting a steady decrease in temperature with increasing altitude and containing virtually all atmospheric dust and water vapor.

**trough** elongated area or "belt" of low atmospheric pressure; also glacial trough, a U-shaped valley carved by a glacier.

**trunk stream** the largest channel in a drainage system; receives inflow from tributaries.

**tsunami** wave caused when an earthquake, volcanic eruption, or other sudden event displaces ocean water; builds to dangerous heights in shallow coastal waters.

**tundra** high-latitude or high-altitude environments or climate regions that are not able to support tree growth because the growing season is too cold or too short.

**tundra climate** characterized by treeless vegetation of polar regions and very high mountains, consisting of mosses, lichens, and low-growing shrubs and flowering plants.

**turbulence** chaotic, mixing flow of fluids, often with an upward component.

**typhoon** a tropical cyclone found in the western Pacific, the same as a hurricane.

**U.S. Public Lands Survey System** a method for locating and dividing land, used in much of the Midwest and western United States. This system divides land into 6- by 6-mile-square townships consisting of 36 sections of land (each 1 sq mi). Sections can also be subdivided into halves, quarter sections, and quarter-quarter-sections.

**Ultisol** soil that has a subsurface clay horizon, is low in bases and is often red or yellow in color.

**unconformity** an interruption in the accumulation of different rock layers; often represents a period of erosion.

**uniformitarianism** widely accepted theory that Earth's geological processes operate today as they have in the past.

**unloading** physical weathering process whereby removal of overlying weight leads to rock expansion and breakage.

**uplift mechanisms** methods of lifting surface air aloft, they are orographic, frontal, convergence (cyclonic), and convectional.

**upper-level inversion** an atmospheric condition where air descending from aloft warms as it compresses during its descent, and overrides cooler air below.

**upslope fog** type of fog where upward flowing air, cools to form fog which hugs the slope of mountains.

**upwelling** upward movement of colder, nutrient-rich, subsurface ocean water, replacing surface water that is pushed away from shore by winds.

**urban heat island** *see* heat island.

**uvala (valley sink)** large surface depression resulting from coalescing of sinkholes in karst areas.

**valley breeze** air flow upslope from the valleys toward the mountains during the day. *See also* mountain breeze.

**valley glacier** an alpine glacier that extends beyond the zone of high mountain peaks into a confining mountain valley below.

**valley train** outwash deposit from glacial meltwater, resembling an alluvial fan confined by valley walls.

**variable** one of a set of objects and/or characteristics of objects, which are interrelated in such a way that they function together as a system.

**veering wind shift** the change in wind direction clockwise around the compass; for example, east to southeast to south, to southwest, to west, and northwest.

**vent** pipe-like conduit through which volcanic rock material is erupted.

**ventifact** rock displaying distinctive wind-abraded faces, pits, grooves, and polish.

**verbal scale** stating the scale of a map using words such as "one inch represents one mile."

**vernal equinox** the day when the subsolar point is at one of the lines of the tropics (23½° N and S) that marks the first days of summer and winter; summer begins on the June solstice for the Northern Hemisphere and winter begins on the December solstice. In the Southern Hemisphere the June solstice marks the beginning of winter and December the beginning of winter

**vertical exaggeration** a technique that stretches the height representation of terrain in order to emphasize topographic detail.

**vertical (direct) rays** sun's rays that strike Earth's surface at a 90° angle.

**Vertisol** soil that develops in regions with strong seasonality of precipitation.

**viscosity** a fluid's resistance to flowing.

**visualization models** a wide array of computer techniques used to vividly illustrate a place or concept, or the illustration produced by one of these techniques.

**volcanic ash** erupted fragments of volcanic rock of sand size or smaller (< 2.0 mm).

**volcanic neck** vertical igneous intrusion that solidified in the vent of a volcano.

**volcanism** the eruption of molten rock matter onto Earth's surface.

**volcano** mountain or hill created from the accumulation of erupted rock matter.

**V-shaped valley** the typical shape of a stream valley cross profile where the gradient is steep.

**warm current** a flow of sea water that moves like a river through the ocean and is relatively warmer than water in the ocean area that it flows through.

**warm front** leading edge of a relatively warmer, less dense air mass advancing upon a cooler, denser air mass.

**warping** broad and general uplift or settling of Earth's crust with little or no local distortion.

**wash (arroyo, barranca, wadi)** an ephemeral stream channel in an arid climate.

**water budget** relationship between evaporation, condensation, and storage of water within the Earth system.

**water mining** taking more groundwater out of an aquifer through pumping than is being replaced by natural processes in the same period of time.

**water table** upper limit of the zone of saturation below which all pore spaces are filled with water.

**water vapor** water in its gaseous form.

**wave base** water depth equal to half the length of a given wave; at smaller depths the wave interacts with the underwater substrate.

**wave crest** the highest part of a wave form.

**wave height** vertical distance between the trough and adjacent crest of a wave.

**wave period** time it takes for one wavelength to pass a given point.

**wave refraction** bending of waves in map view as they approach the shore, aligning themselves with the bottom contours in the breaker zone.

**wave steepness** ratio of wave height to wavelength for a given wave.

**wave trough** the lowest part of a wave form.

**wavelength** horizontal distance between two successive crests of a given wave.

**weather** atmospheric conditions, at a given time, in a specific location.

**weather radar** radar that is used to track thunderstorms, tornados, and hurricanes.

**weathering** physical (mechanical) fragmentation and chemical decomposition of rocks and minerals at and near Earth's surface.

**well** artificial opening that reaches the zone of saturation for the purpose of extracting groundwater.

**westerlies** surface winds flowing from the polar portions of the subtropical highs, carrying fronts, storms, and variable weather conditions from west to east through the middle latitudes.

**wet adiabatic lapse rate** rate at which a rising mass of air is cooled by expansion when condensation is taking place. The rate varies but averages 5°C/1000 meters (3.2°F/1000 ft).

**white frost** a heavy coating of white crystalline frost.

**wind** air in motion from areas of higher pressure to areas of lower pressure; movement is generally horizontal, relative to the ground surface.

**windward** location on the side that faces toward the wind and is therefore exposed or unprotected; usually refers to mountain and island locations.

**wind wave** wave on a water body created when air currents push the water surface along.

**xerophytic** vegetation type that has genetically evolved to withstand the extended periods of drought common to arid regions.

**yardang** aerodynamically shaped remnant ridge of wind-eroded bedrock or partly consolidated sediments.

**yazoo stream** a stream tributary that flows parallel to the main stream for a considerable distance before joining it.

**zenith** the highest angle above the horizon that the sun reaches at solar noon in a given place and time of year.

**zone of ablation** the lower elevation part of a glacier; where more frozen water is removed than added during the year.

**zone of accumulation** the higher elevation part of a glacier; where more frozen water is added than removed during the year.

**zone of aeration** uppermost layer of subsurface water where pore spaces typically contain both air and water.

**zone of depletion** top layer, or A horizon, of a soil, characterized by the removal of soluble and insoluble soil components through leaching and eluviation by gravitational water.

**zone of saturation** subsurface water zone in which all voids in rock and soil are always filled with water; the top of this zone is the water table.

**zone of transition** an area of gradual change from one region to another.

**zooplankton** tiny animals that float and drift with currents in water bodies.

# Index